T0172020

Electronic Circuit Design and Application

Stephan J. G. Gift • Brent Maundy

Electronic Circuit Design and Application

 Springer

Stephan J. G. Gift
Electrical and Computer Engineering
The University of the West Indies,
St. Augustine Campus
St. Augustine, Trinidad and Tobago

Brent Maundy
Electrical and Computer Engineering
University of Calgary
Calgary, AB, Canada

ISBN 978-3-030-46991-7   ISBN 978-3-030-46989-4   (eBook)
https://doi.org/10.1007/978-3-030-46989-4

© Springer Nature Switzerland AG 2021
All rights are reserved by the Publisher, whether the whole or part of the material is concerned, specifically the rights of translation, reprinting, reuse of illustrations, recitation, broadcasting, reproduction on microfilms or in any other physical way, and transmission or information storage and retrieval, electronic adaptation, computer software, or by similar or dissimilar methodology now known or hereafter developed.
The use of general descriptive names, registered names, trademarks, service marks, etc. in this publication does not imply, even in the absence of a specific statement, that such names are exempt from the relevant protective laws and regulations and therefore free for general use.
The publisher, the authors, and the editors are safe to assume that the advice and information in this book are believed to be true and accurate at the date of publication. Neither the publisher nor the authors or the editors give a warranty, expressed or implied, with respect to the material contained herein or for any errors or omissions that may have been made. The publisher remains neutral with regard to jurisdictional claims in published maps and institutional affiliations.

This Springer imprint is published by the registered company Springer Nature Switzerland AG
The registered company address is: Gewerbestrasse 11, 6330 Cham, Switzerland

S.J.G. Gift dedicates the book to his parents, his wife Sandra, and two sons Allister and Christopher.

Preface

Electronics is one of the most exciting areas of technology today. The semiconductor or microelectronics revolution that started with the invention of the transistor in 1947 and followed later by the introduction of the integrated circuit comprising many transistors on a single chip continues unabated. This most remarkable technology has resulted in the introduction of the microprocessor, the personal computer, cell phone, and MP3 player as well as the GPS, smart phone, the Internet, and the self-driving car. New and amazing electronic systems and sub-systems are introduced annually and the consumer is exposed to a dazzling array of truly amazing gadgets and devices.

This wonderful inventiveness has been fueled by unprecedented levels of transistor integration density with the number of transistors on a single chip growing from a few thousand in the early 1970s to over 10 billion in 2020. Moore's law, which holds that transistor IC density will double every 2 years, still has relevance some 50 years after it was first proposed. Thus, extremely powerful ICs are available that carry out a vast range of important functions.

The world in which we live is an analog one with signals generally being continuous over time. However, computers are digital devices that represent information in discrete forms. With the extensive use of computers and other digital systems, the need for analog systems that interface the analog world and the digital computer has become increasingly important. A smart phone for example has many analog sensors that interface with the advanced processor within. The internet of things that enables the wide collection and processing of large amounts of real-time data has spawned an array of sensors and ICs, many of which must be interconnected using analog circuits. There is an increasing demand for analog modules in these and other systems that carry out a multiplicity of functions. There is therefore a critical need for the development of skills in analog electronic circuit design and application.

Text Philosophy

This book treats with the design and application of a broad range of analog electronic circuits in a comprehensive and clear manner. The discussion of design and application in the text, we believe, brings out the inherently interesting nature of electronics. There are several books on the market that deal extensively with analog electronic circuits, but most tend to give theory, explanations, and analysis of circuit behavior and generally do not enable the reader to design complete real-world functioning circuits or systems. This text addresses this shortcoming while treating comprehensively with the subject. It first provides a foundation in the theory and operation of basic electronic devices including the diode, the bipolar junction transistor, the field effect transistor, and the operational amplifier. It then presents detailed instruction on the design of working real-world electronic circuits of varying levels of complexity including feedback amplifiers, power amplifiers, regulated power supplies, filters, oscillators, and waveform generators. Upon completion of the book, the reader will understand the operation of and be able to confidently design a broad range of functioning analog electronic circuits and systems.

With the proliferation of electronic systems today, the need for electronic design engineers is greater than before, and this book we believe fills an important niche in the world of analog electronics. It is intended primarily as an undergraduate text for University and College students enrolled in electrical, electronic, and computer engineering programs. It should also prove useful for graduate students and practicing engineers who need to design practical systems to meet research or real-world needs.

Text Features

This book enables the reader to analyze a variety of circuits, to develop a deep understanding of their operation, and to design and optimize a range of working circuits and systems. Many examples help the reader to quickly become familiar with key design parameters and design methodology for each class of circuits. Each chapter starts with fundamental ideas and develops them step-by-step into a broad range of applications of real-world circuits and systems. Each chapter ends with several circuit applications with a full discussion of design methods. Also, at the end of each chapter, there are research projects that are intended to stimulate the interests of the reader and encourage varying levels of investigation and experiment.

The attractive features of the book include the following:

- It comprehensively presents the design of working real-world analog electronic circuits for key systems and sub-systems.
- The material is clearly written and easily understood.
- Many worked examples of functioning circuits are presented.
- Design applications begin from the very first chapter and continue throughout the text.
- Research projects to stimulate and encourage further investigation are included at the end of each chapter.
- Ideas for further exploration are also included with these applications and research projects.
- Some simulations are used to demonstrate the functionality of the designed circuits

- Upon completion of the text, the reader will be able to confidently design important analog electronic circuits.

Text Overview

The book contains 14 chapters with content as follows:

1. *Semiconductor Diode* – Provides a sound introduction to semiconductor diodes and their use in several circuits. Applications and research projects are presented.

2. *Bipolar Junction Transistor* – Introduces the binary junction transistor, its theory of operation, and its various configurations. The use of the device in the design of single stage amplifier circuits is fully discussed. Applications and research projects are presented.

3. *Field Effect Transistor* – Introduces the field effect transistor, its theory of operation, and its various configurations. The use of the device in the design of single stage amplifier circuits is thoroughly discussed. Applications and research projects are presented.

4. *BJT and FET Models* – Explores electrical circuit models for representing the BJT and FET in accurately analyzing the behavior of circuits containing these devices. Applications and research projects are presented.

5. *Multiple-Transistor and Special Circuits* – Presents a comprehensive discussion of multiple transistor circuits including several special circuit configurations. Applications and research projects are presented.

6. *Frequency Response of Transistor Amplifiers* – Discusses the frequency behavior of single stage and multiple stage transistor circuits. Several important configurations are analyzed. Applications are presented.

7. *Feedback Amplifiers* – Introduces the concept of negative feedback, distortion, and frequency performance in amplifier systems and presents full discussion of its application in amplifier design. Applications and research projects are presented.

8. *Operational Amplifiers* – Presents the operational amplifier in its various configurations and discusses numerous applications of this versatile circuit element. Applications and research projects are presented.

9. *Power Amplifiers* – Thoroughly explores power amplifiers and the various classes and configurations with numerous design examples from low power to high power systems. Applications and research projects are presented.

10. *Power Supplies* – Provides a comprehensive discussion of unregulated and regulated power supplies, their operation and design using discrete and integrated components. Applications and research projects are presented.

11. *Active Filters* – Presents a full discussion on the theory and design of low-pass, high-pass, band-pass, band-stop, and all-pass filters using operational amplifiers. The Bessel, Butterworth, and Chebyshev filter

responses are considered. Applications and research projects are presented.

12. *Oscillators* – Presents a full discussion of positive feedback and the Barkhausen criterion in oscillator systems and explores the design of numerous oscillator types including Wien bridge, phase shift, LC, and crystal oscillators. Applications and research projects are presented.

13. *Waveform Generators and Non-Linear Circuits –* Presents the theory of operation and design of a full range of waveform generators and non-linear circuits including comparators, triangular wave generators, astable multivibrators, and precision rectifiers. Applications and research projects are presented.

14. *Special Devices* – This final chapter introduces the theory of operation and the circuit implementation of several special devices including photosensitive devices, opto-isolators, silicon controlled rectifier, triac, Shockley diode, diac, unijunction transistor, and the programable unijunction transistor. Several design examples are presented and applications and research projects are discussed.

Closing Remarks

We have prepared a text which we believe enables the reader to enjoy the wonderful world of electronics while learning and applying design rules that lead to practical working electronic systems. We have included many interesting ideas and research projects throughout the text that will hopefully excite the reader. We hope this book will satisfy most of the electronic circuit design and application needs of students and practicing engineers and invite feedback regarding its contents.

St. Augustine, Trinidad and Tobago   Stephan J. G. Gift
Calgary, AB, Canada   Brent Maundy

Acknowledgments

We would like to express our sincere thanks to our students who provided useful feedback over the years while using this material in classroom instruction and to our Universities for providing the support that enabled the preparation of this text. We wish to also thank Mr Itanie Gordon for drawing most of the diagrams in the text.

Contents

Author Biographies

Stephan J. G. Gift   graduated with BSc (First Class Honours) and PhD degrees in Electrical Engineering from the University of the West Indies in Trinidad and Tobago. He is Professor of Electrical Engineering and former Head of the Department of Electrical and Computer Engineering and Dean of the Faculty of Engineering at the University of the West Indies. He is currently Pro Vice Chancellor of Graduate Studies and Research at this University. He has published many papers in electronic circuit design and has performed reviews for many international journals. He is a senior member of the IEEE and a past president and Fellow of the Association of Professional Engineers of Trinidad and Tobago.

Brent Maundy   graduated with BSc (Upper Second Class Honours) in Electrical Engineering and MSc in Digital Electronics from the University of the West Indies. He has a PhD in Electronics form Dalhousie University, Nova Scotia. He is Professor of Electrical and Computer Engineering in the Department of Electrical and Computer Engineering at the University of Calgary, Alberta, Canada. He has published many papers in electronic circuit design and was an associate editor for the *IEEE Transactions on Circuits and Systems*.

Semiconductor Diode

The simplest electronic device is referred to as a diode. It consists essentially of two different materials in contact such that electric charge flows easily in one direction but is impeded in the other. Despite its simplicity, it performs an important role in electronic systems from the simple to the complex. In this chapter, we will discuss the nature and characteristics of the solid-state diode (i.e. one based on semiconductor material) as well as employ it in the design of modern electronic systems.

Firstly, the basic physical concepts of semiconductors are presented. Then the basic diode is analysed and its behaviour discussed. It is then applied in several designs in order to illustrate its implementation. At the end of the chapter, the reader will be able to:

- Understand the make-up and properties of intrinsic semiconductor material, n-type and p-type semiconductor material and the action of drift and diffusion currents
- Understand the general characteristics of various types of diodes including Zener diodes, light-emitting diodes, Schottky diodes, varactors, tunnel diodes and photodiodes
- Analyse diode circuits of moderate complexity
- Design diode circuits of moderate complexity
- Design circuits with various types of diodes

1.1 Theory of Semiconductors

Most electronic devices are constructed using solid-state material, i.e. conductors, semiconductors and insulators. It is important therefore that we understand the physics of these materials. All materials (matter) consist of atoms, which are the smallest unit of matter. Each atom comprises three basic particles, the proton, the neutron and the electron. The positively charged proton and uncharged neutron form the nucleus at the centre of the atom, while the negatively charged electron revolves around the nucleus in elliptical paths called shells. The outermost shell is called the valence shell, and the electrons that occupy this shell are called valence electrons. The Bohr models of hydrogen and helium, the two simplest atoms, are shown in Fig. 1.1.

The atom of hydrogen consists of a single proton and an electron in orbit around it. The helium atom consists of a nucleus containing two protons and two neutrons with two electrons moving around it. Each atom is built up from one other atom by adding electrons, protons and neutrons with the number of protons always being equal to the number of electrons so that the atom is electrically neutral. The number of protons is known as the atomic number, and the number of neutrons is called the neutron number.

© Springer Nature Switzerland AG 2021
S. J. G. Gift, B. Maundy, *Electronic Circuit Design and Application*,
https://doi.org/10.1007/978-3-030-46989-4_1

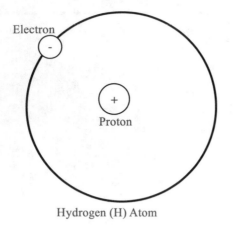 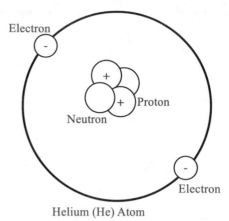

Fig. 1.1 Hydrogen and helium atoms

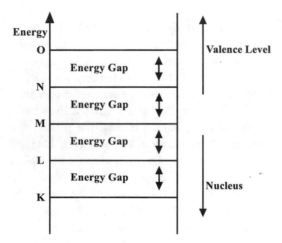

Fig. 1.2 Energy level of atom

Protons and neutrons are called nucleons, and the number of nucleons is called the mass number.

All elements are organized according to properties in a chart referred to as the periodic table. The difference between the atoms of the various elements making up this table lies in the number and arrangement of the constituent electrons, protons and neutrons. Each atom is made up of energy levels or shells lettered K, L, M, N, O... containing orbiting electrons as shown in Fig. 1.2. Each electron must possess a certain discrete quantity of energy in order to occupy a certain shell, the inner most shell requiring the electron to have the lowest energy. Hence, the shells are said to be at certain levels of energy.

The electron can move to a higher energy level by increasing its discrete amount of energy (such as by absorbing thermal energy) to the level associated with another energy level.

Once there, it will only return to the lower level if it loses the required amount of energy that will allow it to occupy a lower energy level. Thus, the electron can only absorb or lose energy in discrete amounts. These energy levels are numbered first, second, third... and higher. It follows therefore that the electron shells lettered K, L, M, N,... correspond to the energy levels numbered 1, 2, 3, 4,... respectively, each energy level represented by n. Each electronic shell comprises sublevels, the number of sublevels being equal to the number of the energy level. Thus, the K shell or first energy level has one sublevel referred to as the s sublevel, the L shell or second energy level has two sublevels referred to as the s and p sublevels, and the M shell or third energy level has three sublevels, namely, s, p and d sublevels, while the N shell or fourth energy level has four sublevels which are s, p, d and f. Each sublevel in a shell has a slightly different energy level. Finally, each sublevel is made up of orbitals each containing at most two electrons. The s sublevel has one orbital which is spherical, the p sublevel has three orbitals which are elliptical, the d sublevel has five orbitals, and the f sublevel has seven orbitals. The orbitals take the name of the sublevel in which they occur.

The electrons in the outermost shell, referred to as valence electrons, determine the chemical properties of the element. This number is equal to the number of the group in the periodic table of elements to which the atom belongs. Thus, lithium with atomic 3, sodium with atomic number 11 and potassium with atomic number 19 all have one valence electron and therefore belong to Group I in the periodic table.

1.1.1 Energy Levels

In a material, the valence electrons are bound in a crystalline structure. However, it may be possible in some materials if sufficient energy is applied for these electrons to break free of these inter-atomic or inter-molecular bonds and be capable of moving about in a conduction mode. Thus, these electrons can move from the valence band where they are bound to the conductions band where they are free. These two bands may or may not overlap.

When the bands overlap as in Fig. 1.3, there is no energy gap between the valence electrons and the conduction or free electrons, and hence in materials where this occurs, the valence electrons are loosely bound to the nucleus and can freely move about the structure. This movement of valence/conduction electrons, usually under the influence of an applied electric potential, is conduction, and such materials are referred to as conductors. The electric attraction between the electron cloud of conduction electrons and the atomic nuclei in the structure constitutes the metallic bond. Examples of conductors include the metals such as silver, gold, copper, zinc, aluminium and lead in which the valence electrons are freely available for conduction.

Where the conduction band and the valence band do not overlap as in Fig. 1.4, an energy gap or forbidden band exists between them. In this case, a discrete amount of energy is required to move the valence electron from its band in the structure into the conduction band. If the gap is narrow (requiring a relatively small amount of energy to overcome atomic attraction as provided, e.g. by ambient light and heat), then the material in which this occurs is called a semiconductor. Examples of semiconductors and the required energy for the generation of free carriers are germanium with gap energy $E_g = 0.67$ eV, silicon with $E_g = 1.11$ eV, gallium (III) arsenide (GaAs) with $E_g = 1.43$ eV and cadmium sulphide (CdS) with $E_g = 1.73$ eV. Some electrons are therefore available for conduction through excitation though not as many as in conductors.

Where the conduction band and the valence band are widely separated as in Fig. 1.5, the energy gap is large with $E_g \gg 1$. An extremely large amount of energy is required to move the

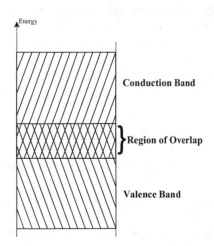

Fig. 1.3 Overlap of valence and conduction bands in conductors

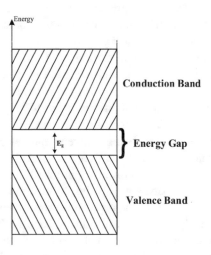

Fig. 1.4 Separation of valence and conduction bands in a semiconductor

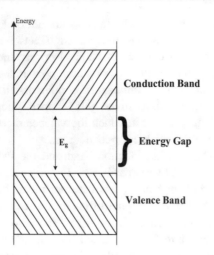

Fig. 1.5 Separation of valence and conduction bands in an insulator

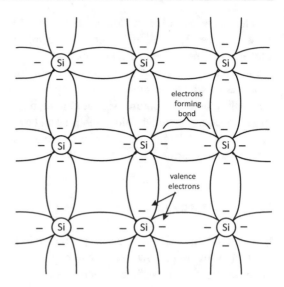

Fig. 1.6 Covalent bonding between atoms of silicon

valence electron from its band in the crystal structure into the conduction band. Such materials do not therefore normally conduct, as no conduction electrons are available, and are called insulators. Examples of such materials are mica, glass, porcelain and rubber.

If the temperature is raised in a conductor, the lattice vibrates and this results in more collisions with electrons. Such increased collisions manifest themselves as increasing resistance in the conductor. Conductors are therefore said to have a positive temperature coefficient of resistance. On the other hand, increased temperature in a semiconductor results in a greater number of electrons moving from the valence band to the conduction band, thereby increasing the conductivity of the semiconductor. Semiconductors therefore have a negative temperature coefficient of resistance.

Materials that are insulators at room temperature can become conductors at a suitably high temperature. The increased temperature would result in bonds being broken and electrons becoming free for conduction. The bond in semiconductors and insulators is called a covalent bond and involves the sharing of electrons from different atoms. In the germanium atom, there are 32 orbiting electrons, while silicon has 14 orbiting electrons. Both belong to Group IV of the periodic table, which means each has four electrons in the valence shell (and hence the ionization energy

required to remove any one of these electrons is lower than other electrons within the atoms). Hence in silicon (Fig. 1.6) or germanium material, the atoms are bound in a lattice structure in which each atoms shares its four valence electrons with four neighbouring atoms such that stable electron octets are formed in each atomic or valence shell. This sharing mechanism results in inter-atomic forces that bind the atoms in a crystalline structure. These bonds are called covalent bonds. While the valence electrons are bound tightly in the crystalline structure of the material, it is possible for them to absorb enough energy from ambient light or heat energy to break the covalent bond and become free for conduction. This corresponds to overcoming the energy gap E_g as shown in Fig. 1.2. The closer the atomic spacing in materials, the greater are the inter-atomic forces and hence the greater is the energy gap.

1.1.2   Intrinsic (Pure) Semiconductor

The two main semiconductor materials used in the manufacture of semiconductor devices are germanium and silicon. An atom of either substance may be represented by a central core of positive charge $4e^+$ surrounded by four orbiting electrons each having a negative charge e^-. In the solid-state structure, silicon and germanium form a lattice in which all atoms are equidistant from their immediate neighbours as shown in Fig. 1.6. Each atom having four neighbours meets this equidistant requirement. In the crystal lattice, each atom forms a covalent bond with its four neighbours, each bond involving two valence electrons, one from each atom. Thus, each bond pair traverses an orbit around both the parent atom and the neighbouring atom, thus enabling each atom to have effectively a stable state of eight electrons in its valence shell. This arrangement provides the lattice with a strong, stable, regular structure.

Despite the strength of the bonds, energy from ambient light and heat is sufficient to overcome the energy gap E_g and to cause some of the bonds to be broken and the electrons to be free. When a covalent bond is broken, a hole-electron pair is created, and both the hole and the free electron are known as charge carriers. They can thereafter move under the action of an applied electric field and contribute to electrical conduction. If the temperature of the material increases, thermal energy causes the breaking of additional covalent bonds and the consequent release of additional electrons. As a result the conductivity of the crystal increases. This means that as indicated earlier, a semiconductor material has a negative temperature coefficient of resistance.

When an electron escapes from a covalent bond, there is an absence of an electron, which, since this constitutes a missing negative charge e^-, is equivalent to a positive charge e^+. Such an area in the crystal lattice is called a **HOLE**. A hole exerts an attractive force on an electron, and it can be filled by an electron that has been thermally liberated form a covalent bond. This process is called **RECOMBINATION**, and it causes a continual loss of holes and free electrons. At a given temperature, a dynamic equilibrium is reached such that the rate of generation of electrons and holes is equal to the rate of recombination, this resulting in a constant number of holes and free electrons. Both the hole and the electron are known as charge carriers, though the electron conduction (electron current) is different from hole conduction (hole current). The motion of holes is in a direction opposite to the motion of

electrons, and the holes behave as positively charged particles. Holes and electrons move through a crystal by either diffusion or drift. Under normal circumstances, thermal energy causes random electron movement in a semiconductor with no net flow of charge. However, if an increased concentration of electron or holes occurs at one end of the semiconductor by some mechanism, this gives rise to a net flow referred to as a **DIFFUSION** current away from that area. **DRIFT** occurs when there is net charge movement under the action of an externally applied electric field. The velocity of the movement is called **DRIFT VELOCITY**, and this movement of holes and electrons through the crystal lattice is referred to as **DRIFT CURRENT**. The electrons travel faster than the holes since while the electrons undergo direct translation due to the field, the movement of the holes in the opposite direction of the electrons involves a series of discontinuous electron movements into and out of covalent bonds. Conduction of current in a pure semiconductor is known as intrinsic conduction. Since the charge carriers are thermally generated, the concentration of these carriers increases with temperature. As a result the conductivity of intrinsic semiconductors increases with temperature at the approximate rate of 5% per degree Celsius for germanium and 7% per degree Celsius for silicon.

1.1.3   Extrinsic (Impure) Semiconductor

The electrical characteristics of intrinsic semiconductor material can be significantly changed by the introduction of a small number (1 in 10^7) of certain atoms of impurity into the semiconductor material. This process of introducing impurity atoms into pure semiconductor crystal is called **DOPING**. A treated crystal is said to be **DOPED** and is called **EXTRINSIC** material. Since the number of impurity atoms is very much smaller than the number of atoms of the crystal, the crystal lattice is essentially undisturbed, and four crystal atoms surround each impurity atom. Two types of extrinsic materials of great importance in semiconductor device fabrication are n-type and p-type. Both types are formed by the addition of a specific amount of impurity atoms into a silicon or germanium base.

N-Type Material

N-type material shown in Fig. 1.7 is formed by the addition of impurity atoms of elements from Group 5 of the periodic table which have five valence electrons. Examples of these elements are arsenic, antimony and phosphorus. In a silicon crystal, each of the impurity atoms, for example, phosphorus, will take the place of one of the silicon atoms in the crystal lattice, and because the number of impurity atoms is small, the crystal lattice will be essentially undisturbed. As a result, four silicon atoms surround each impurity atom, and each phosphorus atom will establish covalent bonds with the four neighbouring silicon atoms. This means that one valence electron in the phosphorus atom is unused. This surplus electron is only loosely bound to its parent atom. It easily breaks free and is available for conduction. The impurity atom has therefore donated a conduction election to the crystal lattice. A free electron is created in the silicon crystal lattice without the creation of a corresponding hole. (A positive charge does remain in the parent atom, but since no bond is broken, the attraction of this charge for free electrons is not as a great as when a bond is broken.) Hole-electron pairs are however still produced by thermal energy. Because of the impurity,

the number of the free electrons in the lattice greatly exceeds the number of holes, and therefore the crystal is called n-type. The impurity atom is called a *DONOR* atom because it donates a free electron to the crystal lattice. In n-type material, electrons are the majority charge carriers, and holes are the minority charge carriers. The above discussion also holds for a germanium crystal lattice. Note that the crystal is still neutral overall. The effect of the doping process on the crystal conductivity can be described using the energy band diagram. As a result of the doping, a donor energy level appears in the forbidden band with gap energy E_g significantly less than that possessed by the intrinsic material. Thermal energy is sufficient at room temperature to move these electrons into the conduction bond, thereby increasing the conductivity of the semiconductor. At room temperature in intrinsic silicon, there is approximately one free electron for every 10^{12} atoms (1 to 10^7 for germanium). For a doping level of 1 donor atom in every 10^7 silicon atoms, the ratio $10^{12}:10^7$ indicates an increase in free electrons by a ratio 100,000:1.

P-Type Material

The p-type material shown in Fig. 1.8 is formed by the addition of impurity atoms of elements from Group 3 instead of Group 5 of the periodic table, these having three valence electrons. Examples of these elements are boron, gallium

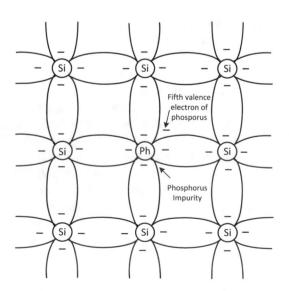

Fig. 1.7 Formation of n-type material

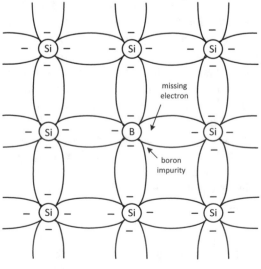

Fig. 1.8 Formation of p-type material

and indium. When boron impurity atoms are introduced into the silicon crystal lattice, each boron atom will again take the place of a silicon atom in the lattice with its three valence electrons. It will form covalent bonds with three of the four neighbouring silicon atoms. Since only three bonds are formed, one hole (position for bond formation) is introduced into the lattice for each impurity atom. This hole is free to conduct as thermally generated holes, which continue to be produced. Because of the impurity, the number of free holes is the crystal lattice is much greater than the electrons, and hence the material is called p-type. The impurity atoms are called *ACCEPTOR* atoms. Again the p-type material is electrically neutral. *HOLES* are the majority charge carriers, and electrons are the minority charge carriers in p-type semiconductor material.

1.2   Current Flow in Semiconductor Diodes

Consider p-type and n-type semiconductor materials that are brought into contact, forming a junction between them as shown in Fig. 1.9. Recall that both regions contain charge carriers of either sign though in the n-type region electrons are the majority carriers and in the p-type region holes are the majority carriers. Such an arrangement of p-type and n-type material in contact is referred to as a semiconductor diode. The symbol for the diode is shown in Fig. 1.10, with the arrow

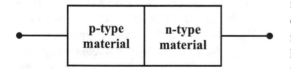

Fig. 1.9 P-type and n-type material in contact forming a semiconductor diode

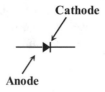

Fig. 1.10 Symbol of the diode

pointing in the direction p to n. The terminal on the p side is referred to as the anode, while that on the n side is referred to as the cathode. An external voltage applied to the two terminals is referred to as a bias. Below we consider the situations of zero bias (where the external voltage is zero), reverse bias (where the external voltage is applied such that the anode is negative relative to the cathode) and forward bias (where the external voltage is such that the anode is positive relative to the cathode).

1.2.1   Zero Bias

Consider the case where the externally applied voltage is zero as shown in Fig. 1.11. In both regions, there is a high probability that minority charge carriers will meet and recombine with majority charge carriers and as a result the lifetime of the minority charge carriers is short. While there is random movement of free electrons and holes in the lattice, the initial high concentration of electrons in the n-type and holes in the p-type results in a net movement of electrons from n-type to p-type and holes from p-type to n-type. This process is known as *DIFFUSION* and indicates the tendency for charge carriers to move away from areas of high concentration to lower concentration areas.

As the n-type region loses negative charge carriers and gains positive charge carriers and the p-type region loses positive charge carriers and gains negative charge carriers, the n-type region close to the junction becomes depleted of electrons and positively charged, and the p-type region close to the junction becomes depleted of holes and negatively charged. The region in the vicinity of the junction therefore becomes depleted of majority carriers and is called the *DEPLETION REGION*. It has a relatively high resistivity and is of the order of 0.001 mm wide. The migration also causes a build-up of positive charges in the n-type region close to the junction and negative charges in the p-type close to the junction. This results in a potential across the junction called a *BARRIER POTENTIAL*. It is directed such that further majority change diffusion across the junction is reduced though

Fig. 1.11 P-type and
n-type material in contact
with zero bias

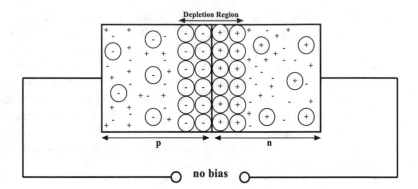

Fig. 1.12 Reverse-biased
diode

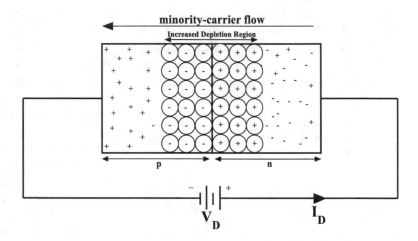

majority carriers with sufficient energy are able
to overcome the barrier potential and migrate to
the other side. Minority charge carriers
(electrons in the p-type material and holes in
the n-type material) that enter the depletion
region are swept across the junction to the
other side by the action of the barrier potential.
A dynamic equilibrium is reached such that the
majority charge carrier or *DIFFUSION* current
and minority charge currents are equal and hence
the net current across the junction is zero. It
should be noted that the crystal is electrically
neutral both before and after diffusion. The net
flow of charge in any direction for a semiconductor diode is zero, when the externally applied
voltage is zero.

1.2.2   Reverse-Biased Diode

The reverse bias condition exists when an external voltage V_D is applied across the p-n junction
with the positive potential being connected to the
n-type material and the negative potential being
connected to the p-type material. This is shown
in Fig. 1.12. Under the action of this potential,
majority charge carriers are attracted away from
the junction, thereby further depleting this
region. This increases both the magnitude of
the potential barrier and the width of the depletion region. Fewer majority carriers have sufficient energy to break through the barrier
potential, and hence the majority charge current
decreases   eventually   going   to   zero   with

increasing reverse bias. The minority charge current, which was aided by the barrier potential, increases quickly reaching its maximum value where all the minority charges which are thermally generated are swept across the junction. This current is called the *reverse saturation current* and is indicated as flowing from n to p in Fig. 1.12, according to conventional current flow. (Note that this current comprises holes flowing from n to p and electrons flowing from p to n.) With a high enough reverse bias, the current flowing across the junction rises to a maximum equal to the minority charge current or **REVERSE SATURATION CURRENT**. The reverse saturation current for germanium is typically 1–2 µA at room temperature, while that for silicon is much lower at 10–20 nA. This current approximately doubles in value for every 10 °C rise in temperature.

1.2.3 Forward-Biased Diode

The forward-biased condition exists when an external voltage V_D is applied across the p-n junction with the negative potential being connected to the n-type material and the positive potential being connected to the p-type material. This is shown in Fig. 1.13. Under these conditions, holes (majority carriers) are repelled from the p-type material towards the junction, and electrons (majority carriers) are repelled from the n-type material towards the junction. This movement of holes and electrons towards the junction results in a reduction of the width of the depletion layer and consequently the magnitude of the potential barrier.

Majority carriers from each material can now cross the junction under the action of the electric field. Even though the minority charge current flow remained constant, the increase in majority current flow results in a net increase in conventional current across the junction from p-type material to n-type material (holes from p to n and electrons from n to p). This current increases rapidly with increase in the forward bias V_D according to an exponential curve as shown in Fig. 1.14. The curve shows a sharp corner at about $V_D = 0.7$ V for silicon and $V_D = 0.3$ V for germanium. Below this voltage the current through the diode is quite small, while above this voltage, there is large current flow. This voltage can therefore be viewed as a threshold voltage above which the diode is on and V_D is constant and below which the diode is off.

1.3 General Characteristic of a Diode

From solid-state physics, the relationship between diode voltage V_D and diode current I_D is given by

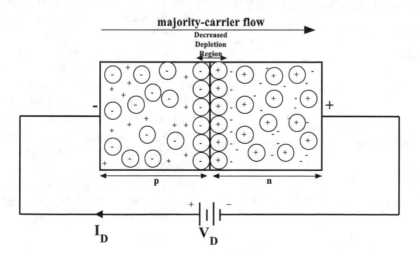

Fig. 1.13 Forward-biased diode

Fig. 1.14 Silicon diode
characteristic

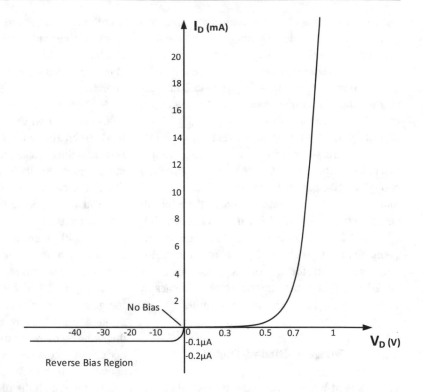

$$I_D = I_o \left(e^{\frac{qV_D}{mkT}} - 1 \right) \qquad (1.1)$$

where:

I_o = reverse saturation current
q = electron charge (1.6×10^{-19} Coulombs)
k = Boltzmann's constant (1.38×10^{-23} J/K)
T = absolute temperature in degrees Kelvin
m = an empirical constant between 1 and 2

If we let V_T be the thermal voltage given by
$V_T = \frac{kT}{q}$, then (1.1) becomes

$$I_D = I_o \left(e^{\frac{V_D}{mV_T}} - 1 \right) \qquad (1.2)$$

Equation (1.2) approximately represents the
diode characteristic Fig. 1.14. It can be written as

$$I_D = I_o e^{\frac{V_D}{mV_T}} - I_o \qquad (1.3)$$

Thus for positive valves of V_D (forward bias),
$I_D \gg I_o$ in which case

$$I_D = I_o e^{\frac{V_D}{mV_T}} \qquad (1.4)$$

The reverse saturation current I_o depends on
the level of doping and diode geometry. The
empirical constant m is a function of diode construction and can vary according to voltage and
current levels. For germanium diodes, $m = 1$,
while for silicon diodes, $m = 1$ for values of
current above that corresponding to the threshold voltage. (For $m = 1$, $mV_T = 26$ mV at 25 °C.)

The DC or static resistance of a diode
corresponds to the ratio V_d/I_d where V_d and I_d
are the operating DC voltage and current. This
parameter is of less significance than the AC or
dynamic resistance. At a particular operating
point, we can evaluate the AC or dynamic resistance $r_d = \Delta V_d/\Delta I_d$ of the diode where Δ
indicates a small change in the relevant quantity.
$\Delta V_d/\Delta I_d$ represents the inverse slope of the diode
characteristic at a particular operating point as
shown in Fig. 1.15. Thus, differentiating I_D in
(1.4) with respect to V_D gives

Fig. 1.15 Silicon diode characteristic showing dynamic resistance

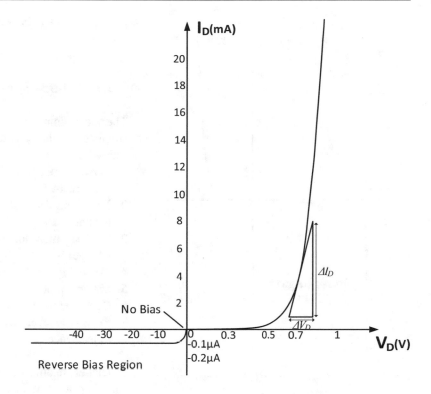

$$\frac{dI_D}{dV_D} = \frac{I_o e^{\frac{V_D}{mV_T}}}{mV_T} = \frac{1}{mV_T}(I_D + I_o)$$

$$\approx \frac{I_D}{mV_T}, I_D \gg I_o \qquad (1.5)$$

Hence,

$$r_d = \frac{dV_D}{dI_D} = \frac{mV_T}{I_D} = \frac{26\ \text{mV}}{I_D(\text{mA})} \qquad (1.6)$$

Example 1.1

Determine the dynamic resistance of a semiconductor diode for 1 mA current through the diode.

Solution Using Eq. (1.6), the dynamic resistance for a 1 mA current flow is given by $r_D = \frac{26\ \text{mV}}{1\ \text{mA}} = 26\ \Omega$.

In addition to the dynamic resistance, there are resistance of the semiconductor material of the diode and resistance between the diode material (body resistance) and the external metallic conductor (contact resistance). We use the designation r_F to denote the diode forward resistance comprising r_D, body resistance and contact resistance.

Breakdown Region

If the reverse bias voltage across a diode is increased beyond a certain value, the reverse current increases rapidly, and the junction is said to have broken down as shown in Fig. 1.16. This critical voltage is called the **BREAKDOWN VOLTAGE** of the junction. Two effects cause this breakdown:

(a) The **ZENER EFFECT** occurs when the electric field across the junction is strong enough to break covalent bonds in the lattice and thereby generate carriers. This effect occurs primarily at low levels of breakdown voltage or **ZENER VOLTAGE** V_Z corresponding to high doping levels in the p- and n-type materials.

(b) The **AVALANCH EFFECT** occurs when the charge carriers are accelerated by the electric field of the reverse bias to the extent where they create carriers by collision.

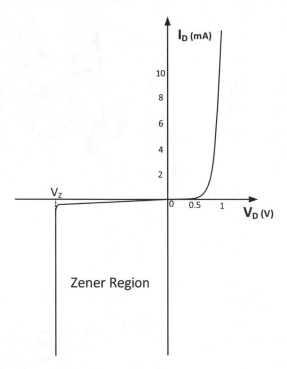

Fig. 1.16 Diode characteristic showing Zener region

For voltages less than about 5 volts, the Zener breakdown mechanism prevails, while for voltages above 5 volts, the avalanche breakdown mechanism is dominant. While the Zener breakdown mechanism is a significant contributor to the breakdown process only at low levels of V_Z, the breakdown region is called the **ZENER** region, and diodes designed to operate in this part of the p-n junction characteristic are called **ZENER DIODES** (see Sect. 1.4.1). Such diodes exploit the steep characteristic gradient to maintain a constant voltage even with changing reverse diode current. In general, the Zener region of the semiconductor diode must be avoided if the diode is not to be destroyed. For Zener diodes designed to operate in this region, current limiting must be introduced (usually by the inclusion of a resistor) in order to prevent diode failure. The maximum reverse bias potential that can be applied before breakdown is called the *peak*

inverse voltage (PIV) or the *peak reverse voltage* (PRV).

1.3.1 Diode Specifications

The data defining the characteristics of a diode are supplied by the manufacturer and include the following:
1. Forward voltage V_F (at a specified current and temperature)
2. Maximum forward current I_F (at a specified temperature)
3. Reverse saturation current I_R (at a specified voltage)
4. PIV rating (or PRV)
5. Maximum power dissipation
6. Operating temperature range
7. Reverse recovery time
8. Capacitance

Diodes are often rated according to their power handling capability. The physical construction of the diode determines its rating, and this information is included in the manufacturer's specification. Another diode rating is the current carrying capacity or forward current. The instantaneous power dissipated by a diode is given by the relation $P_D = I_D V_D$.

1.4 Diode Types

There is a variety of diodes available for electronic circuit design and application. These include Zener diodes, signal diodes, power diodes, varactor diodes, light-emitting diodes, photodiodes, PIN diodes and Schottky diodes. Each of these is briefly discussed in what follows.

1.4.1 Zener Diodes

The Zener diode is a diode designed to be operated in the Zener region as shown in Fig. 1.18. Its symbol is shown in Fig. 1.17, and

Fig. 1.17 Symbol for Zener diode

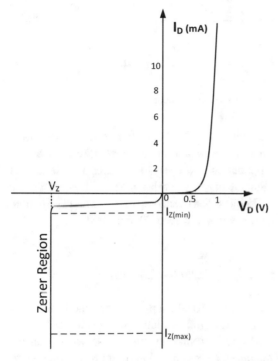

Fig. 1.18 Diode characteristics showing Zener operating region

it exploits the steep slope of the diode characteristic in the reverse bias region. As can be seen from the diode characteristic in Fig. 1.18, the voltage in this region is almost constant for varying current. The actual Zener voltage V_Z is determined by the doping levels in the Zener diode with increased doping resulting in a reduction of V_Z. Zener diode voltages V_Z vary from 1.8 volts to about 200 volts with a tolerance of about $\pm 5\%$ and power ratings from $\frac{1}{4}$ W to about 50 W. Silicon is usually preferred in the manufacture of Zener diodes due to its higher current and temperature capability.

In the on state, Zener diodes have a low dynamic resistance r_Z which is the reciprocal of the slope of the Zener curve given by $r_Z = \Delta V_Z / \Delta I_Z$, and the equivalent circuit may be shown as a battery of voltage V_Z in series with r_Z. In most applications, however, r_Z is quite small as compared with other circuit resistance and hence can

be ignored. This resistance is current-dependent and decreases with increased current. V_Z is itself temperature-dependent. Its value may increase or decrease with changes in temperature. The temperature coefficient T_C is an indicator of this and is given by

$$T_C = \frac{\Delta V_Z}{V_Z(T_1 - T_0)} \times 100\%/^{\circ}\mathrm{C} \qquad (1.7)$$

where ΔV_Z is the change in Zener voltage V_Z resulting from the temperature change from T_0 to T_1. T_C can be positive or negative reflecting, respectively, an increase of V_Z with temperature or a decrease of V_Z with temperature. It turns out that the temperature coefficient associated with the avalanche breakdown mechanism is positive, while that for the Zener breakdown diodes is negative. The result is that Zener diodes at the 5 volt mark tend to have very low temperature coefficients and are therefore best suited for use as stable reference voltages.

Example 1.2
A Zener diode has $V_Z = 8.2$ V at 25°C. If the diode has $T_C = 0.05\%/^{\circ}\mathrm{C}$, determine the Zener voltage at 75°C.

Solution Using equation (1.7), $\Delta V_Z = \frac{V_Z}{100} T_C$

$$(T_1 - T_0) = \frac{8.2}{100} \times 0.05 \times (75 - 25) = 0.205 \text{ V}.$$

Therefore V_Z at 75°C is $V_Z(75^{\circ}\mathrm{C}) = V_Z + 0.205 = 8.2 + 0.205 = 8.205$ V.

In operating the Zener diode, the reverse current through the device must not fall below a minimum value I_{Zmn} as shown in Fig. 1.16 in order that the Zener voltage does not fall below its operating value V_Z. Also, the reverse current must not exceed the maximum value I_{Zmx} set by the Zener power rating.

Example 1.3
If the 8.2 V Zener used in Example 1.2 has a power rating of 500 mW, determine the maximum reverse current I_{Zmx}.

Table 1.1 Standard Zener diode voltages

2.0	2.2	2.4	2.7	3.0	3.3
3.6	3.9	4.3	4.7	5.1	5.6
6.2	6.8	7.5	8.2	9.1	10
11	12	13	15	16	18

Solution The power P_D dissipated in a Zener diode is given by $P_D = V_Z I_Z$ where V_Z is the Zener voltage and I_Z is the Zener current. Therefore, the maximum Zener current I_{Zmx} is given by $I_{Zmx} = P_D/V_Z = 500$ mW/8.2 V $= 61$ mA.

The standard Zener voltages available are listed in Table 1.1.

Zener diodes whose voltages are a factor of ten times the values in the table are available up to 150 V.

1.4.2   Signal Diodes

Signal diodes are diodes designed to process voltages and currents of small value. Such devices are used in telecommunications, electronic signal processing and computer switching circuits. The characteristics of this class of diodes include low forward resistance and junction capacitance. The PIV rating is moderately high (25–250 volts), and the forward current rating is in the range 40–250 mA. These diodes are available singly or in arrays for convenience of implementation. One popular example is the 1N4148. This diode has a PIV of 100 V, a forward current of 300 mA, a power dissipation of 500 mW and a diode capacitance of 4 pF.

1.4.3   Power Diodes

This class of diodes is important in signal rectification, which is converting AC to DC for power applications. These include the half-wave rectifier and the full-wave rectifier discussed in Sects. 1.5.4, 1.5.5 and 1.5.6, circuits that find wide application in a range of electronic systems. The important parameters are the PIV, which is typically 50–1000 volts, the forward current that is of the order of 1–100 amperes and the low forward

Fig. 1.19 Varactor diode

resistance of less than 1 ohm. Power diodes come in single units and in packages of four for bridge applications. The 1N4001 series is a family of general purpose diodes rated at 1 A, while the 1N5400 series is rated up to 3 A. The PIV for both series varies between 50 and 1000 V.

1.4.4   Varactor Diodes

A varactor diode is a p-n junction diode designed to have a significant junction capacitance that can be varied by varying the reverse voltage on the junction. All p-n junctions comprise p-n semiconductor material bounding a region of high resistance called the depletion layer. This arrangement constitutes a capacitor whose value C given by $C = \frac{\varepsilon A}{W}$ where ε is he permittivity of the material constituting the capacitor, A is the junction area and W is the width of the depletion layer. The width increases with an increase in the applied reverse potential with the result that the junction capacitance of the diode decreases. The typical capacitance of these diodes is in the range 100–500 pF. These diodes are used for electronic tuning of a variety of circuits including tuners, oscillators and filters. The symbol for the varactor diode is shown in Fig. 1.19.

1.4.5   Light-Emitting Diodes

A light-emitting diode or LED is a diode that emits light as a result of the passage of current through the diode. The current flow excites electrons within the semiconductor material of

Table 1.2 Light-emitting diodes

Material	Colour	Forward voltage (V)
GaAsP	Red	1.8
GaP	Green	2.1
GaAsP	Yellow	2.0
GaAsP	Orange	2.0
SiC	Blue	3.0
GaN	White	4.1

Fig. 1.20 Symbol for light-emitting diode (LED)

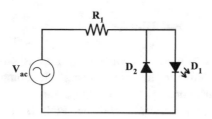

Fig. 1.21 Method of protection for LED

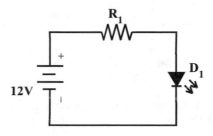

Fig. 1.22 Circuit for Example 1.4

the LED to higher energy levels after which they fall back to lower levels, thereby releasing energy in the form of light. The material of the LED determines the specific colour of the light as well as its turn-on voltage. LEDs are constructed from gallium phosphide (GaP), silicon carbide (SiC) and gallium arsenide phosphide (GaAsP).

The typical current rating of these devices is between 10 and 50 mA. The colour and associated forward voltage drops for several LED types are given Table 1.2. Its symbol is shown in Fig. 1.20.

The forward voltage drop during turn-on is higher than for ordinary silicon diodes. Moreover, the PIV rating is substantially lower and is typically between 3 and 5 volts. These devices must therefore not be subjected to large reverse voltages and can be protected by the inclusion of an ordinary diode D_2 in parallel with the LED as shown in Fig. 1.21. A series resistor R_1 is required to limit the forward current. The LED is a very efficient device and radiates very little heat during operation. It is unaffected by shock or vibration, and there is no surge current at turn-on. As a result the device has a long life.

Thus if the peak value of V_{ac} is 15 V, then if D_1 is a green LED with forward voltage 2.1 V, allowing a peak current of 10 mA, it follows that $R_1 = (15 - 2.1)/10$ mA $= 1.3$ k. Essentially the same calculation is used to determine the limiting resistor for a DC voltage source. In that case the protection diode D_2 would not be necessary.

Example 1.4

For the circuit shown in Fig. 1.22, determine the value of R_1 if the LED D_1 is red.

Solution Since a red LED has a forward voltage of 1.8 volts, for an LED current of 15 mA, the value of R_1 is given by $R_1 = (12 - 1.8)/15$ mA $= 680\ \Omega$.

1.4.6   Photodiodes

A photodiode is a diode that responds to light. The case of the diode is constructed with a transparent area such that incident light can fall on the p-n junction within. This energy generates additional holes and electrons from the crystal lattice,

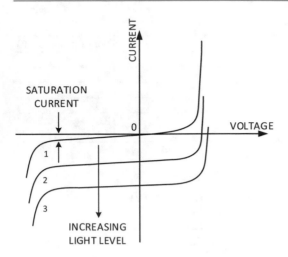

Fig. 1.23 Current (*I*) vs voltage (*V*) characteristics for a photodiode

Fig. 1.24 Symbol for photodiode

which are then swept away from the junction by an impressed reverse bias. The resulting current flow adds to the existing reverse saturation current, thereby increasing its value. Thus, light falling on a photodiode increases its reverse saturation current, and a typical family of curves corresponding to varying reverse bias voltages and reverse saturation current for various light intensities is shown in Fig. 1.23. As can be seen, for incident light at some fixed intensity, there is very little change in the current as the reverse bias is increased. In order to reduce the current to zero, the reverse bias must be reduced to zero and changed to a forward bias of typically 0.3 V at which point the current begins to drop. It falls to zero when the forward reaches about 0.5 V. The current that flows with no incident light is referred to as dark current. The change of current with light intensity corresponds to a change in diode

Fig. 1.25 Symbol for PIN diode

Fig. 1.26 Symbol for Schottky diode

resistance. The symbol for the photodiode is shown in Fig. 1.24.

1.4.7 PIN Diodes

The PIN diode shown in Fig. 1.25 is a diode designed for low capacitance. It comprises heavily doped p and n regions separated by a layer of intrinsic silicon (hence PIN). Because of its construction, the diode has a very low forward resistance, an extremely high reverse resistance, a high PIV rating and a low reverse capacitance. This low capacitance enables the diode to be employed in high-frequency switching applications.

1.4.8 Schottky Diodes

The Schottky or hot carrier diode is a semiconductor diode with a low forward voltage drop and fast switching action. It is constructed by fusing a metal region, e.g. platinum, to an n-type region. The resulting diode has virtually no charge storage and therefore has a short reverse recovery time. It is therefore useful in high-speed switching applications involving large forward currents. During forward bias, electrons move from the n-type region where they predominate to the metal region. The Schottky diode voltage-current characteristic is similar to that of a power diode. It however has a lower threshold voltage of about 0.3 V. As the reverse breakdown voltage increases for this diode, the forward voltage also

increases. The symbol for this diode is shown in
Fig. 1.26.

1.5 Diode Circuits

Having discussed the various types of diodes and
their characteristics, in this section we examine
the operation and design of several applications
of these devices.

1.5.1 DC Circuits

In general, a diode is in the "ON" state if the
current established by the applied source is such
that its direction matches that of the arrow in the
diode symbol. This generally corresponds to
$V_D \geq 0.7$ V for silicon and $V_D \geq 0.3$ V for
germanium. Analysing diode circuits with DC
inputs is quite simple and involves applying standard circuit laws.

Example 1.5

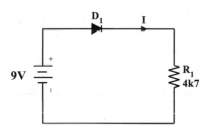

Fig. 1.27 Simple diode circuit

In the circuit shown in Fig. 1.27, find the current
I flowing through the resistor.

Solution In the circuit below, the voltage drop
across the diode is 0.7 volts, and hence the voltage across the resistor is $(9 - 0.7)$ volts. Therefore, the current I in the circuit is given by
$I = (9 - 0.7)/4k7 = 1.8$ mA. If D_1 is a red LED
with forward voltage 1.8 V, then $I = (9 - 1.8)/
4.7$ k $= 1.5$ mA.

1.5.2 Clippers

There is a class of diode networks called clippers
that have the ability to "clip" off a portion of a
signal without affecting the remaining part of the
alternating waveform. Such circuits are useful in
protecting electronic equipment against voltage
surges. The basic circuit of the clipper is shown
in Fig. 1.28.

As the input voltage goes more positive than
the reference voltage V_{ref}, the diode D_1 turns
on. Current then flows from the input signal
source through the diode and causes a voltage
drop across R_1 such that the excess input voltage
is reduced. The result is that the output voltage is
limited to a maximum value equal to $(V_{ref} + V_D)$.
For an input voltage that is less positive than V_{ref}
or of negative polarity, diode D_1 turns off, and the
full input signal waveform appears at the output
terminals of the circuit. Figure 1.29 shows the
clipped output waveform for an input sinusoidal
voltage waveform.

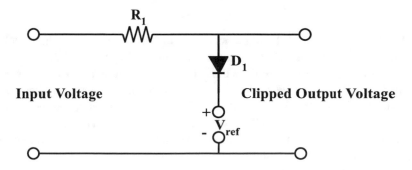

Fig. 1.28 Clipper circuit

Fig. 1.29 Waveform of
"clipped" voltage signal

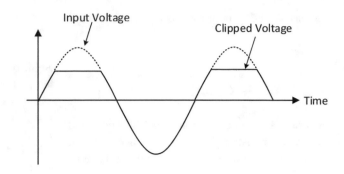

Fig. 1.30 Double polarity
clipping circuit

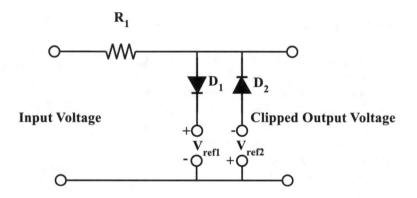

Fig. 1.31 Waveform of
voltage signal "clipped" for
negative and positive half-
cycles

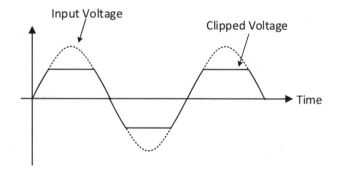

If the diode and the polarity of the reference voltage are reversed, the input voltage signal will be clipped on the negative rather than the positive half-cycle. If bi-directional diodes are placed in parallel as shown in Fig. 1.30, then both the positive and the negative half-cycles of the input waveform will be clipped as shown in Fig. 1.31. The result is that the amplitude of the input signal waveform can be restricted to a maximum value on both half-cycles. The two reference voltages can be of different magnitude. If they are both set to zero as shown in Fig. 1.32, the output amplitude is limited to ±0.7 volts. This circuit is used to protect some electronic circuits. The circuit can be further modified with the inclusion of Zener diodes to replace the reference voltages.

Fig. 1.32 Double polarity clipping circuit with zero reference voltages

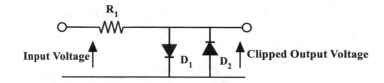

Fig. 1.33 Circuit and input waveform for Example 1.6

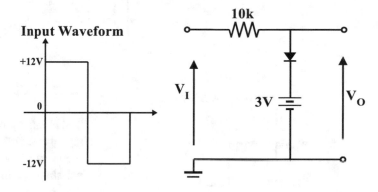

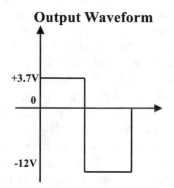

Fig. 1.34 Solution waveform for Example 1.6

Example 1.6

Sketch the output waveform of the circuit shown in Fig. 1.33. The input is a symmetrical square wave input signal of amplitude ± 12 volts.

Solution For the circuit in Fig. 1.33, as the input voltage goes to $+12$ V, the forward voltage of the diode and the battery voltage are both exceeded, and conduction through the diode occurs. The output voltage V_o is then $V_o = 3 + 0.7 = 3.7$ V. When the input voltage goes to -12 V, the diode turns off, and hence there is no conduction. The output voltage is then -12 volts. The output waveform is shown Fig. 1.34.

Example 1.7

Sketch the output waveform of the circuit shown in Fig. 1.35. The input is a symmetrical square wave input signal of amplitude ± 12 volts.

Solution For the circuit in Fig. 1.35, as the input voltage goes to $+12$ volts, the forward voltage of the diode and the reverse voltage of the Zener are both exceeded, and conduction through both devices occurs.

The output voltage V_o is then $V_o = 6.8 + 0.7 = 7.5$ volts. When the input voltage goes to -12 volts, the normal diode turns off, and hence so does the Zener. The output voltage is then -12 volts. The output waveform is shown Fig. 1.36.

Example 1.8

Sketch the output waveform of the circuit shown in Fig. 1.37. The input is a sine wave input signal of peak amplitude ± 10 volts.

Solution For the circuit in Fig. 1.37, on the positive half-cycle as the input voltage amplitude increases, the forward voltage of the Zener diode is exceeded, and conduction through the device occurs. The peak output voltage V_o is

Fig. 1.35 Circuit and input waveform for Example 1.7

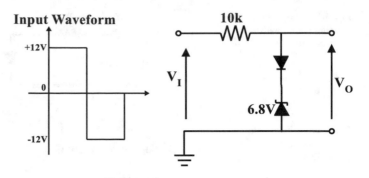

Input Waveform

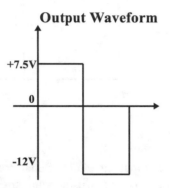

Output Waveform

Fig. 1.36 Solution waveform for Example 1.7

limited to 0.7 V. On the negative half-cycle of the input sine wave, as the amplitude increases negatively, the reverse voltage of the Zener is exceeded, and conduction in the reverse direction through the device occurs. The peak output voltage V_o is limited to −4.7 V. The resulting clipped waveform is shown Fig. 1.38.

There is another group of clippers in which the diode is in series with the output. The basic version to be considered here is shown in Fig. 1.39. For a positive going input waveform, the output signal assumes the value V_{ref} when the input signal amplitude is less than V_{ref} since then diode D_1 is off. However, D_1 conducts when the input amplitude exceeds V_{ref}. The output voltage amplitude is then the input amplitude less the diode voltage drop. The resulting waveform is shown in Fig. 1.40. If the diode and the reference voltage are reversed, then the waveform is also reversed. The most popular version of these is when $V_{ref} = 0$ which is called a rectifier. This will be dealt with in Sect. 1.5.4.

1.5.3 Clampers

The clamping network is one that will "clamp" a signal to a different DC level. It effectively shifts the level of the signal without changing the signal shape as in the clipping circuits. Two basic clamping circuits are shown in Figs. 1.41 and 1.42. The time constant CR of the circuit must be longer than the periodic time T of the input waveform, i.e. $CR \gg T$. This ensures that the voltage across the capacitor does not discharge significantly during the interval that the diode is non-conducting. In the circuit in Fig. 1.41, the diode is OFF when the input voltage is positive. When the input voltage is negative, the diode turns ON, and the output voltage is then equal to the small voltage drop V_D across the diode. The current flowing through the diode charges up the capacitor, and a voltage is developed across it, which is equal to the peak input voltage minus the diode voltage. When the next positive half-cycle arrives, the voltage across the output is equal to the sum of the applied input voltage and the voltage across the capacitor. (The capacitor attains its maximum voltage after a few cycles of the input waveform.) The result is that the output waveform has shifted in the positive direction by an amount equal to the capacitor voltage and the least positive value is clamped to just below 0 V (diode voltage). The most positive part of the output voltage is then equal to twice the peak input voltage minus the diode voltage drop, i.e. $(2V − V_D)$ volts. If the diode connection is reversed as in Fig. 1.42, the positive peak of the input voltage is clamped to just above 0 V. The output voltage waveforms

Fig. 1.37 Circuit and input waveform for Example 1.8

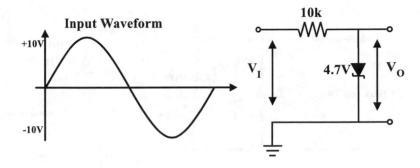

Fig. 1.38 Solution waveform for Example 1.8

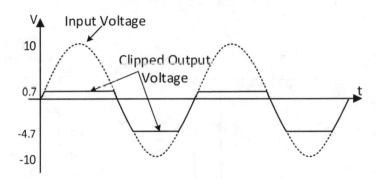

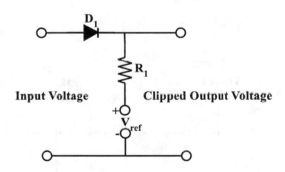

Fig. 1.39 Clipper with series diode

for a square wave input voltage are shown in Figs. 1.41 and 1.42.

If a reference voltage source V_{ref} is connected in series with the diode, an input voltage can be clamped to that voltage instead of to 0 V (approx.). The diagram in Fig. 1.43 shows a clamping circuit that includes a reference voltage V_{ref}. If the polarities of the diode and the reference voltage are reversed, then the input voltage will be clamped to $-V_{ref}$ instead of 0 V. It is interesting to note that the waveform is shifted in the

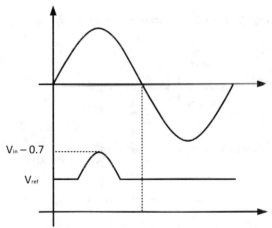

Fig. 1.40 Output waveform of clipper with series diode

direction of the diode and that amplitude above the zero reference is $(V_{ref} + V_D)$.

Example 1.9
Sketch the output waveform for the circuit shown in Fig. 1.44. The input is a symmetrical square wave signal of amplitude ± 12 V.

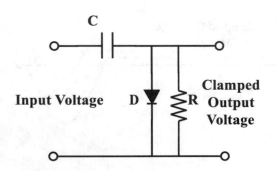

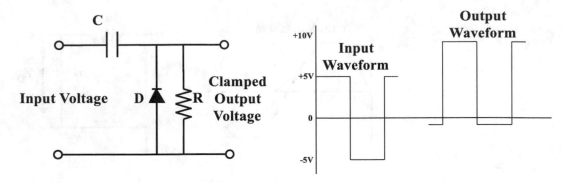

Fig. 1.41 Clamping circuit with positive shift

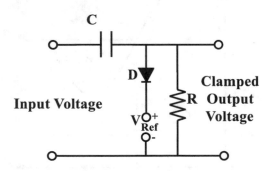

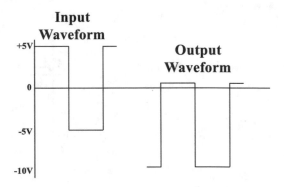

Fig. 1.42 Clamping circuit with negative shift

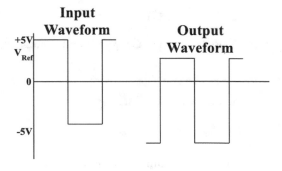

Fig. 1.43 Clamping circuit with reference voltage

Solution For the circuit in Fig. 1.44, as the input voltage goes to +12 volts, the capacitor charges to $V_C = 12 - 4 - 0.7 = 7.3$ volts through the forward-biased diode. The output voltage V_o is across the series battery-diode connection giving $V_o = 4.7$ volts. When the input voltage goes to -12 volts, the normal diode turns off, and the input voltage is in series with the voltage across the capacitor. This results in an output voltage of $V_o = -12 - 7.3 = -19.3$ volts. The effect is that the output waveform is shifted negative by 7.3 volts while still preserving the original waveform shape (Fig. 1.45).

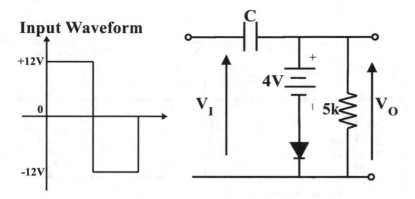

Fig. 1.44 Circuit and input waveform for Example 1.9

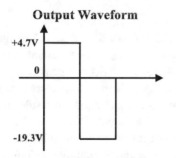

Fig. 1.45 Solution waveform for Example 1.9

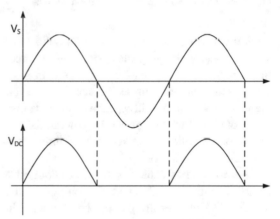

Fig. 1.47 Output waveform of half-wave rectifier

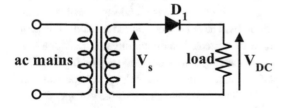

Fig. 1.46 Half-wave rectifier

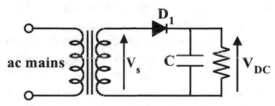

Fig. 1.48 Half-wave unregulated DC supply

1.5.4 Half-Wave Rectifier

The basic half-wave rectifier consists of a diode D_1 in series with an AC supply from a transformer and the load as shown in Fig. 1.46 where V_S is the rms voltage. On the half-cycle in which the AC signal V_S goes positive making the anode of the diode positive relative to its cathode, the diode conducts and current flows into the load in the direction of the diode. A DC voltage V_{DC} is developed across the load. On the negative half-cycle of the input voltage, diode D_1 is reversed-biased as its cathode is made more positive than

its anode. There is therefore no current flow through the diode into the load, and the output voltage across the load is zero. The voltage output consists of a series of half sine wave pulses as shown in Fig. 1.47. When the diode is off, the peak voltage across it is equal to $\sqrt{2}V_S$ the peak voltage of the transformer voltage, and hence the PIV rating of the diode must exceed this value.

The problem with this arrangement is that there is considerable variation in the output voltage.

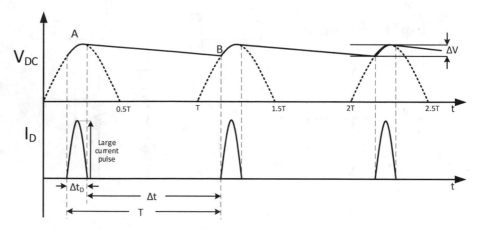

Fig. 1.49 Output voltage and diode charge current

Such a waveform is only suitable for applications such as battery charging where only the unidirectional nature of the voltage is important. In most equipment operated from such a supply however, the voltage variations will appear as hum and noise at the output of the equipment. In general, the DC output of a rectifier is required to have little variation and a minimum of ripple.

One method of achieving this is to connect a reservoir capacitor C across the output of the rectifier. The resulting supply shown in Fig. 1.48 is referred to as half-wave unregulated DC power supply. Each time V_S goes sufficiently positive, the diode conducts, and the voltage across the capacitor increases to $V_{pk} = \sqrt{2}V_S$. This corresponds to point A in Fig. 1.49. When V_S falls below the capacitor voltage, the diode turns off, and the capacitor discharges into the load at a rate determined by the time constant CR_L seconds where R_L is the load resistance. As the discharge occurs, the capacitor voltage V_{DC} is represented by the curve AB in Fig. 1.49. As V_S rises above the capacitor voltage at point B, the voltage at the anode of the diode exceeds the cathode voltage causing the diode to conduct. The capacitor is recharged and the cycle is repeated. The charging current I_D through diode D_1 takes the form of large pulses as shown in Fig. 1.49, which the diode must be able to handle.

For small load currents, the capacitor discharge between charging pulses is small, and hence the average output voltage is only slightly less than

$V_{DC} = \sqrt{2}V_S$. Since the capacitor voltage adds to the transformer voltage when the diode turns off, the PIV is increased to twice the peak secondary voltage, i.e. $2\sqrt{2}V_S$. A large load current causes C to discharge more rapidly, and this results in a greater variation in the output voltage. While increasing C will reduce this variation, the maximum value of capacitance that can be employed is limited since as the capacitance value is increased, an increasing current is required to recharge the capacitor. This means that the amplitude of the current pulses in Fig. 1.49 will increase and appropriately rated diodes will have to be employed to handle this larger forward current. The fluctuating unidirectional voltage appearing across the load may be regarded as a DC voltage with an AC voltage component. This AC voltage is known as the ripple voltage.

In order to calculate the peak-to-peak value ΔV of the output ripple and the average output voltage, certain approximations need to be made. Specifically, from Fig. 1.49 for the half-wave rectifier, the diode is off for most of the period T. During this time the capacitor delivers a load current I_{DC} and therefore discharges in the process. The decay is exponential but can be approximated as a linear decay. Also, the load current is assumed to be approximately constant during the discharge. For a capacitor, the relationship $I = CdV/dt$ holds. This can be approximated by $I \approx C\Delta V/\Delta t$ where ΔV is a small change in capacitor voltage and Δt is the associated small time interval over which this voltage change occurs. For the half-wave rectifier,

$$I_{DC} = C\frac{\Delta V}{\Delta t} = C\frac{\Delta V}{T_H} \qquad (1.8)$$

where ΔV is the change in voltage of the reservoir capacitor and T_H is the period of the half-wave rectified waveform. From this, the peak-peak output ripple is given by

$$\Delta V = \frac{I_{DC}T_H}{C} = \frac{I_{DC}}{Cf} \qquad (1.9)$$

where f is the frequency of the input voltage. From this, the average DC output voltage is

$$V_{DC} = V_{pk} - \Delta V/2 \qquad (1.10)$$

Example 1.10
Design a half-wave unregulated power supply delivering 12 volts DC at 1 A with less than 2.5 V peak-to-peak ripple.

Solution Since the peak transformer voltage is greater than the root mean square value by a factor of 1.414, it is reasonable to choose the transformer voltage V_{rms} to be equal to the required DC output voltage, i.e. $V_{rms} = V_{DC} = 12$ V. Also, to prevent transformer heating, the transformer secondary current rating should be larger than the maximum load current by at least

a factor of 1.5, i. e.; $I_{rms}(\text{sec}) = 1.5I_{DC}(\text{max}) = 1.5$ A. Choose a diode with $I_{AV} \geq I_{DC}(\text{max}) = 1$ A and $PIV > V_{pk} = 12\sqrt{2} = 16.8$ V. Capacitor C is given by $C = \frac{I_{DC}}{f\Delta V} = \frac{1}{60\times2.5} = 6666 \ \mu\text{F}$. A rough indication of the output voltage under full-load conditions is given by $V_{DC} = V_{pk} - \Delta V/2 = 16.8 - 2.5/2 \approx 16$.

1.5.5   Full-Wave Rectifier

In a full-wave rectifier, both half-cycles of the input waveform are utilized to produce a unidirectional load current. The basic circuit is shown in Fig. 1.50 and can be viewed as two half-wave rectifiers connected in parallel across the load. It utilizes a centre-tapped transformer having equal voltages V_S with diodes D_1 and D_2 connected to the load as shown. During the half-cycle of the input waveform where point A is made positive with respect to G and point B is made negative with respect to G, diode D_1 conducts, while D_2 is reversed-biased. As a result, current flows through A and D_1 from the transformer winding into the load resulting in the polarity across the load as shown (Fig. 1.50).

Similarly, when the input signal is such that point B is positive with respect to G and point A is negative relative to G, diode D_2 is now forward-biased, while D_1 is reversed-biased, and so current flows through B and D_2 in the load in the

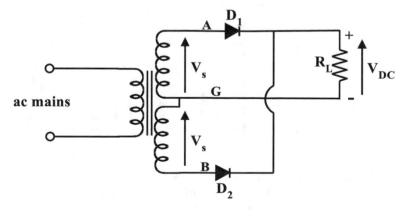

Fig. 1.50 Full-wave rectifier using a centre-tapped transformer

Fig. 1.51 Full-wave
rectifier with filter capacitor

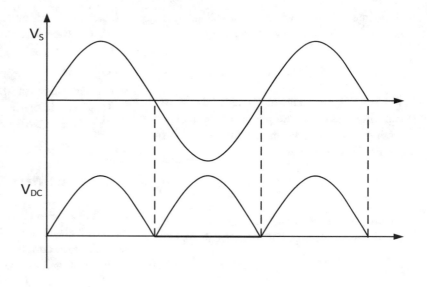

Fig. 1.52 Output
waveform of full-wave
rectifier

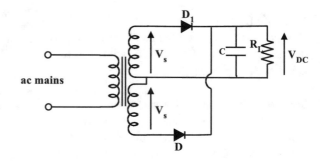

Fig. 1.53 Output voltage
and diode charge current

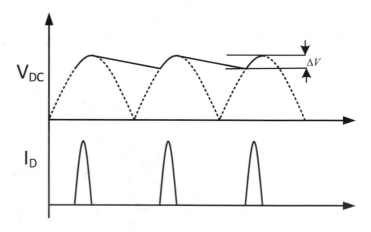

same direction as before. The waveform of the
output voltage is shown in Fig. 1.51. During the
conduction of either diode, the full transformer
voltage is applied across the other, and therefore
the PIV is again $2\sqrt{2}V_S$.

The peak load voltage is $\sqrt{2}V_S$. The inclusion
of a reservoir capacitor as shown in Fig. 1.52
reduces the ripple as shown in Fig. 1.53. It
functions in the same way as in the half-wave
rectifier. However, since two half-wave rectifiers

operating over alternate half-cycles are involved, the capacitor is recharged twice per input cycle instead of one. The result is that this circuit has fewer ripples than the half-wave rectifier. It however requires a centre-tapped transformer and two diodes for its implementation. For a full-wave rectifier, the period T_F of the rectified waveform is half that of the half-wave rectifier and therefore $T_F = 1/2f$. Hence for the full-wave rectifier, the peak-peak output ripple is given by

$$\Delta V = \frac{I_{DC}T_F}{C} = \frac{I_{DC}}{2f} \qquad (1.11)$$

Example 1.11

Design a full-wave unregulated power supply delivering 15 volts DC at 1 A with less than 1.5 V peak-to-peak ripple. Use a centre-tapped transformer.

Solution Since the peak transformer voltage is greater than the root mean square value by a factor of 1.414, it is reasonable to choose the transformer voltage V_{rms} to be equal to the required DC output voltage, i.e. $V_{rms}=V_{DC}=15$ V. Also, to prevent transformer heating, the transformer secondary current rating should be larger than the maximum load current by at least a factor of 1.5, i.e. $I_{rms}(\text{sec}) = 1.5 I_{DC}(\text{max}) = 1.5$ A. Choose a diode with $I_{AV} \geq I_{DC}(\text{max}) - 1$ A and $PIV > V_{pk} = 15\sqrt{2} = 21.2$ V . Capacitor C is given by $C = \frac{I_{DC}}{2f\Delta V} = \frac{1}{2\times 60 \times 1.5} = \mu F$. A rough indication of the output voltage under full-load conditions is given by $V_{DC} = V_{pk} - \Delta V/2 = 16.8 - 2.5/2 \approx 16$.

1.5.6   Bridge Rectifier

Full-wave rectification can be accomplished by an alternative circuit called a bridge rectifier shown in Fig. 1.54. The bridge rectifier circuit uses four diodes in a bridge arrangement that obviates the need for a centre-tapped transformer. During those half-cycles of the input that make point A positive with respect to point B, diodes D_2 and D_4 conduct, while D_1 and D_3 are non-conducting. Current therefore flows from point A to point B via D_2, the load and D_4. When A is negative relative to B, D_1 and D_3 conduct, and D_2 and D_4 do not conduct. Current then flows from point B to point A via D_3, the load and D_1. The resulting current flows in the same direction through the load resulting in a

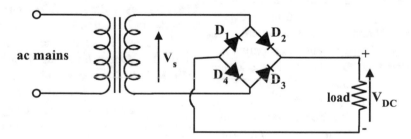

Fig. 1.54 Bridge rectifier

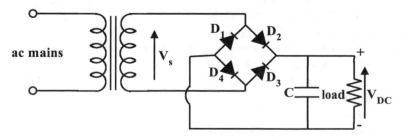

Fig. 1.55 Bridge rectifier with filter capacitor

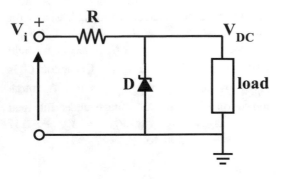

Fig. 1.56 Zener diode regulator

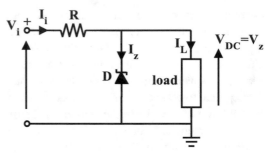

Fig. 1.57 Circuit for Example 1.12

fluctuating, unidirectional voltage across the load with a peak value $\sqrt{2}V_S$. This variation can again be reduced by the inclusion of a filter capacitor across the load. The output voltage waveform is essentially the same as in the previous full-wave rectifier. However, the PIV of each diode is only V_S, and no centre-tapped transformer is required. Figure 1.55 shows the bridge rectifier with a filter capacitor.

1.5.7   Zener Diode Regulators

Both the half-wave and full-wave rectifiers discussed above produce unregulated outputs when capacitive filtering is employed. An unregulated DC output suffers from several disadvantages including ripple and variation of output voltage with load current and supply voltage. The reverse bias characteristic of the Zener diode can be used to realize a regulated DC voltage output for use in low-power applications. The basic system consists of the Zener diode D being supplied current through a resistor R from a supply voltage V_i as shown in Fig. 1.56. The system develops a regulated voltage V_{DC} equal to the Zener voltage across the connected load. It functions as follows: The Zener diode D is operated in reverse bias mode with the resistor R providing the necessary bias current from the supply V_i. The breakdown voltage V_Z of the Zener appears across its terminals, the rest of V_i being dropped across R. In the breakdown region, the diode characteristic has a very steep slope such that large changes in current through the Zener

are accompanied by only small changes in the voltage across the Zener.

Thus, when a load is connected, the Zener voltage V_Z appears across the load, and current is diverted from the Zener into the load. Because of the steep diode characteristic, no change in V_Z accompanies this change in current through the Zener. Similarly, a reduction of current into the load forces an increased current into the Zener with no change in the Zener voltage. The maximum current that can be delivered by the supply is set by the minimum current I_{Zmn} that must be maintained through the Zener for it to maintain its terminal voltage. If the supply voltage changes, the current through the Zener again changes while still maintaining its terminal voltage.

Fixed Supply Voltage and Variable Load
Consider the situation in Fig. 1.56 in which the supply voltage V_i is fixed and the load current I_L varies. The current I_i flowing through R supplies both I_Z into the Zener and I_L into the load. Should the load current fall to zero, the current through the Zener will increase by I_L to I_{Zmx}, and therefore the Zener must be able to dissipate the associated energy. Hence the maximum Zener current is set by

$$P_D = V_Z I_{Zmx} \qquad (1.12)$$

where P_D is the power rating of the Zener. This yields a value of R given by

$$R = \frac{V_{in} - V_Z}{I_{Zmx}} \qquad (1.13)$$

The maximum load current I_{Lmx} is set by the minimum Zener current I_{Zmn} below which the Zener voltage begins to fall and is given by

$$I_{Lmx} = I_{Zmx} - I_{Zmn} \qquad (1.14)$$

Example 1.12

A voltage regulator is being designed using a 6.3 volt Zener diode that has a minimum current requirement of 5 mA and ½Watt power rating. The system shown in Fig. 1.57 will be powered from a 22 volt supply, and the load is variable. Calculate the required series resistance and the maximum current available from the regulator.

Solution Using (1.12), the maximum current that can be passed by the Zener is given by $P_D = V_Z I_{Zmx}$. This gives $I_{Zmx} = \frac{P_D}{V_Z} = \frac{0.5}{6.3} = 79.4$ mA. From (1.13), $R = \frac{22-6.3}{79.4 \text{ mA}} = 198 \ \Omega$. Therefore the maximum available load current is from (1.14) given by $I_{Lmx} = I_{Zmx} - I_{Zmn} = 79.4 - 5 = 74.4$ mA. In practice, the need for a safety margin will require R to be greater than 198 Ω such that the power dissipated in the Zener is substantially less than the rated value. This will of course result in the maximum load current being less than 74.4 mA.

Example 1.13

A 5 volt Zener diode with a minimum current requirement of 10 mA is to be used in a voltage regulator. The supply voltage is 18 volts, and the variable load current has a maximum value of 25 mA. Calculate the minimum power rating of the Zener diode, the required series resistance and the minimum load resistance that can be connected to the regulator.

Solution Since the load current is variable, the Zener must be able to accommodate the maximum load current (should the load resistor be removed) along with the minimum Zener current. This gives a maximum Zener current of $I_{Zmx} = I_{Lmx} + I_{Zmn} = 25 + 10 = 35$ mA. Hence the minimum power rating P_{Dmn} of the Zener is $P_{Dmn} = V_Z I_{Zmx} = 5 \text{ V} \times 35 \text{ mA} = 175$ mW. The series resistor is given by $R = \frac{18-5}{35 \text{ mA}} = 371 \ \Omega$.

Finally, the minimum load resistor that can be connected to the regulator is given by $R_{Lmn} = \frac{5}{I_{Lmx}} = \frac{5}{25} = 200 \ \Omega$.

Variable Supply Voltage and Fixed Load

We now consider the case involving a fixed load and a variable supply voltage. In this situation, the varying input voltage will cause a corresponding variation in the current through the associated resistor and hence the Zener diode. Note that since V_Z is constant, the load current does not experience any change. When the input voltage falls, the current through the Zener also falls but must not be allowed to fall below the minimum Zener current that enables it to function on the steep part of the slope of the breakdown characteristic and thereby maintain a constant voltage. This means that R must be selected such that the minimum (or greater) Zener current I_{Zmn} and the load current I_L flow with the input voltage at its minimum V_{imn}. Thus

$$R = \frac{V_{imn} - V_Z}{I_{Zmn} + I_L} \qquad (1.15)$$

When the supply voltage is at its maximum, the Zener current will also be at its maximum, and therefore the Zener power rating must exceed the maximum dissipation that will occur, i.e.

$$P_D \geq V_Z I_{Z\,max} \qquad (1.16)$$

Example 1.14

A 5 volt Zener diode has a minimum current requirement of 5 mA and is to be used in a voltage regulator. The supply voltage is 25 volts ±10%, and the fixed load current is 20 mA. Calculate the series resistance required, the minimum power rating of the Zener diode and the minimum instantaneous power dissipated in the Zener.

Solution The minimum value of the input voltage is $0.9 V_i = 22.5$ V. Therefore (1.15) gives $R = \frac{V_{imn} - V_Z}{I_{Zmn} + I_L} = \frac{22.5-5}{5+20} = \frac{17.5}{25 \text{ mA}} = 700 \ \Omega$. The maximum value of the input voltage is $1.1 V_i = 27.5$ V. This produces the maximum current I_{Rmx} through the series resistor given by $I_{R\,max} = \frac{27.5-5}{R} = \frac{22.5}{700} = 32$ mA. Since 20 mA flows into the fixed load, the maximum Zener current corresponding to the maximum input voltage is $I_{Zmx} = 32 - 20 = 12$ mA. The minimum power rating of the Zener is then $P_{Dmn} = V_Z I_{Zmx} = 5 \times 12 = 60$ mW. The minimum instantaneous power dissipated in the Zener occurs when the input voltage is at its minimum. This corresponds to the minimum Zener current flow $I_{Zmn} = 5$ mA. Hence, the minimum instantaneous power dissipated in the Zener is $V_Z I_{Zmn} = 5 \times 5 = 25$ mW.

Variable Supply Voltage and Load

For the situation in which there is varying load and varying supply voltage, the current I_i flowing through R supplies both $I_Z \geq I_{Zmn}$ into the Zener and I_L into the load. In the event the load current falls to zero, all of I_L will flow through the Zener whose current will increase. Since the input voltage is variable, this will cause a corresponding variation in the current through the associated resistor and hence the Zener diode. When the input voltage falls, the current through the Zener also falls but must not be allowed to fall below $I_{Zmn} + I_L$. This means that R must be selected such that the minimum Zener current I_{Zmn} and the load current I_L flow with the input voltage at its minimum V_{imn}. Thus

$$R = \frac{V_{imn} - V_Z}{I_{Zmn} + I_L} \qquad (1.17)$$

When the supply voltage rises to its maximum, the Zener current will also be at its maximum I_{Zmx}, and therefore the Zener power rating must exceed the maximum dissipation that will occur, i.e.

$$P_D \geq V_Z I_{Zmx} \qquad (1.18)$$

This is discussed in the example that follows.

Example 1.15

Design a Zener diode regulator to provide a maximum current of 10 mA at 9 volts using a half-wave unregulated supply. This system is to be used to replace the 9 volt battery used to power many small systems.

Solution The system configuration for the regulated supply is shown in Fig. 1.58. A transformer is used to drive a rectifying diode D_1 and a filter capacitor C_1. These components constitute an unregulated supply. The output then drives resistor R_1 which supplies the Zener diode D_2. In order to deliver 9 V, this Zener voltage must be 9 V. Capacitor C_2 provides extra filtering to remove high-frequency noise. In order to provide a sufficiently high supply voltage to ensure that the Zener diode is properly reverse-biased, a 12 V transformer is used. A current rating of 100 mA is more than adequate to deliver the load and Zener currents. The peak voltage on capacitor C_1 after rectification by D_1 is $12\sqrt{2} - 0.7 = 16.3$ V . Hence, the minimum PIV rating of diode D_1 is $2 \times 16.3 = 32.6$ V. A diode with a PIV rating of 100 V is a good choice. With a load current of 10 mA and allowing for a Zener current of 10 mA, the value of capacitor C_1 required to produce no more than a 1 V change in the voltage across this capacitor is $C = I/\Delta V f = 20 \times 10^{-3}/60 = 333$ µF. A value of 500 µF is chosen with a working voltage of 50 V to ensure safe operation. Here $f = 60$ Hz is the frequency of the mains

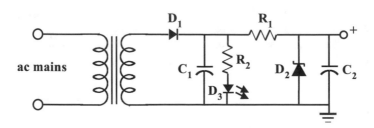

Fig. 1.58 Zener diode regulator

supply. The minimum voltage across C_1 is $16.3 - 1 = 15.3$ V. Allowing for transformer resistance, we use 15 V. Since resistor R_1 must allow a total of 20 mA to flow, its value is given by $R_1 = (15 - 9)/20$ mA $= 300$ Ω. The maximum input voltage is $12\sqrt{2} - 0.7 = 16.3$ V. Hence the maximum current through the Zener is $(16.3 - 9)/300$ Ω $= 24.3$ mA. Hence the maximum power dissipation in the Zener is 9 V $\times 24.3$ mA $= 218.7$ mW. A 500 mW Zener will provide good safety margin. This circuit is inherently short-circuit protected because of R_1. Thus, in the event of a short circuit at the output, the full voltage of capacitor C_1 will be applied across R_1 which will now be in parallel with C_1. Under this condition, the maximum current through this resistor will be 16.3 V/ 300 Ω $= 54.3$ mA. The maximum power that will be dissipated in this resistor is therefore 16.3 V $\times 54.3$ mA $= 886$ mW. A 2 W resistor for R_1 will provide an adequate safety margin. The LED D_3 will indicate when the system is on and together with R_2 will discharge capacitor C_1 when the system is off. Using a green LED for D_3 and a LED current of 10 mA, resistor $R_2 = (15 - 2.1)/10$ mA $= 1.3$ k. Finally, capacitor C_2 is used to remove high-frequency noise from the supply output, and a value of 0.1 μF will have a low reactance at frequencies of 10 kHz and above.

1.6 Applications

The circuits below are some simple applications of diodes in real-world systems that can be easily designed and implemented.

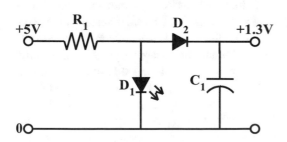

Fig. 1.59 1.3 volt supply

1.3 Volt Supply from 5 Volts

The circuit shown in Fig. 1.59 operates in the manner of a Zener diode regulator. It provides 1.3 volts from the widely available 5 volt supply as a replacement for small low-voltage cells. The resistor R_1 limits the current in the green LED D_1 to a few milliamps of say 10 mA which results in a diode voltage of about 2 volts across D_1. This gives $R_1 = (5 - 2)/10$ mA $= 300$ Ω. In order to reduce this 2 volts to the desired 1.3 volts, a signal diode D_2 such as the 1N4148 is included in the circuit as shown. A small capacitor $C_1 = 10$ μF is placed at the output in order to provide some basic filtering for noise signals above about 1 kHz. At this frequency, the reactance of the capacitor is given by $1/2\pi f C = 1/ 2\pi \times 10^3 \times 10 \times 10^{-6} = 16$ Ω which short-circuits signals above this frequency.

9 Volt Supply from 12 Volts

There are many portable devices that require 9 volts for their operation. The circuit in Fig. 1.60 converts the 12 volts available from the cigarette lighter in an automobile to 9 volts. Each of the four silicon diodes D_1 to D_4 drops 0.7 volts resulting in an output of just over 9 volts.

Ideas for Exploration (i) Modify the circuit in order to reduce the output of a 9 volt battery to 5 volts.

Battery Charger

A 12 volt battery charger circuit is shown in Fig. 1.61. The circuit involves a 24 volt centre-tapped 5 ampere transformer operating into a full-wave rectifier whose output is filtered by a 2000 μF capacitor C. The final circuit output is delivered to the battery under charge through a series 10 Ω variable power resistor VR_1 and an ammeter. At the start of charging, the variable

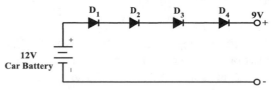

Fig. 1.60 9 volt supply

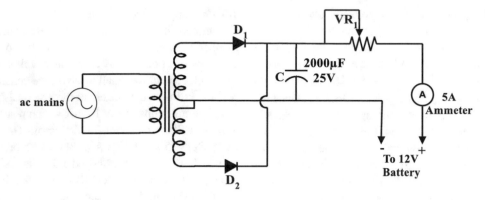

Fig. 1.61 Battery charger

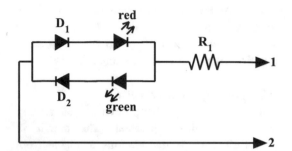

Fig. 1.62 Polarity indicator

Fig. 1.63 Simple light meter

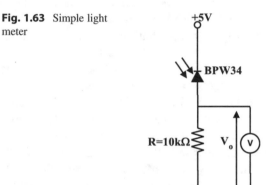

resistor is adjusted in order to limit the charge current to about 2 A. The peak capacitor voltage is $12\sqrt{2} = 17$ V, and the peak voltage change is $\Delta V = I/2Cf = 2/2 \times 2000 \times 10^{-6} \times 60 = 8.3$ V. Hence the average output voltage is $17 - (8.3/2) = 12.8$ V. Therefore, as the battery is charged, the current decreases and should drop to almost zero when the battery is fully charged.

Ideas for Exploration (i) Introduce reverse polarity protection to prevent incorrect battery connection by including a diode with suitable current rating in series with the positive output of the charger: positive output to anode and output at cathode to positive terminal of battery.

Polarity Indicator
The circuit in Fig. 1.62 is a polarity indicator. The resistor R_1 is chosen to limit the current in the diodes to a few milliamperes. The green LED lights when terminal 1 is at a positive potential

relative to terminal 2. The red LED lights when the reverse is true. Diodes D_1 and D_2 ensure that the leds are protected from reverse voltage. A resistor value of $R_1 = 1$ kΩ would allow the testing of voltages between 1 and approximately 15. The value would need to be increased in order to test the polarity of higher voltages.

Light Meter
An example of a photodiode is the BPW34 from Vishay Semiconductors. It is a high-speed device with high sensitivity in a miniature flat plastic package. Because of its water-clear epoxy construction, the diode is sensitive to radiation from the visible and the infrared regions of the electromagnetic spectrum. Using this device, a simple light meter is shown in Fig. 1.63. As light falls on the diode, the increase in current flows through the resistor, thereby resulting in a voltage drop across it. For the 10k resistor, the relationship between the light intensity (lumens per square

metre or lux) and voltage output is given by $lux = 1333V_o$. For example, a voltage from 0 to 0.75 volts will correspond to light intensity of 0–1000 lux. Hence, the voltmeter V connected across the resistor can be calibrated to read light intensity.

Ideas for Exploration (i) Use a simple analog multimeter as the voltmeter and calibrate the scale to read light intensity in units of lux.

Blown Fuse Indicator

The circuit in Fig. 1.64 indicates when a fuse is blown. During normal operation, the green LED is on, and the voltage across this device is insufficient to turn on the red LED in series with a normal signal diode D_1. Therefore, the red LED remains off. The current through the green LED is limited by resistor R_1 which can be of the order of kilo-ohms depending on the supply voltage.

For example, for a current of 10 mA, an input voltage of 12 volts will require $R_1 = (12 - 2)/$ 10 mA = 1 k. Here we have used an approximate LED voltage of 2 volts. A blown fuse disconnects the supply from the load and the green LED which goes off. This results in the full supply voltage being applied across the red LED in series with the signal diode, thereby causing this LED to turn on.

Diode Tester

This circuit, shown in Fig. 1.65, enables the polarity of unmarked semiconductor diodes to be determined. It is powered by a low-voltage transformer T_1. Resistor R_1 limits the current flowing through the diodes. When the diode under test is connected with the anode at A and the cathode at B, it will allow current to flow from A to B and through diodes D_1 and D_2 causing the green LED D_1 to light. If the test diode is connected in the reverse, then current will flow from terminal B to A and through

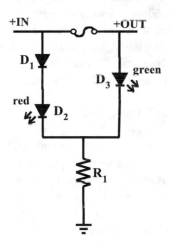

Fig. 1.64 Blown fuse indicator

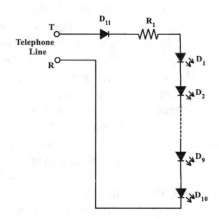

Fig. 1.66 Emergency telephone line light

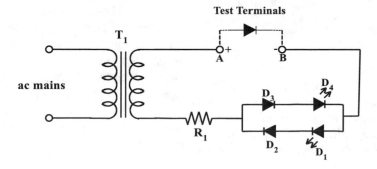

Fig. 1.65 Diode tester

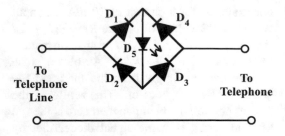

Fig. 1.67 Phone indicator

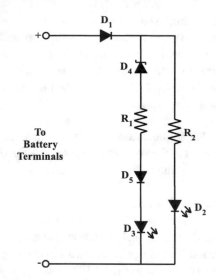

Fig. 1.68 Lead-acid battery monitor

diodes D_3 and D_4 causing the red LED D_4 to light. A short circuit exists if both leds light and an open circuit exists if neither LED lights. With a 12 volt transformer, allowing about 10 mA, then the resistor R_1 can be found using $R_1 = (12 - 0.7 - 0.7 - 2)/10\ mA = 860\ \Omega$.

Ideas for Exploration (i) Re-design the system for portable operation by replacing the transformer with a 9 volt battery and re-calculating R_1.

Emergency Phone-Line Light

This circuit, shown in Fig. 1.66, will provide light in a power outage situation. It utilizes the 48 volts DC available on a telephone line when on hook. It comprises 10 white leds D_1 to D_{10} connected in series with a current limiting resistor R_1 and a diode D_{11} to protect the leds from reverse voltage when the circuit is connected to the telephone tip (+ve) and ring (−ve). Since each LED has an on voltage of about 3.3 volts, then noting that D_{11} has a voltage drop of 0.7 volts, the voltage drop across R_1 is $(48 - 0.7 - (10 \times 3.3)) = 14.3$ V. The central office of the telephone system will detect currents on the telephone line in the range 6–25 mA as indicating a telephone off-hook condition. Therefore resistor R_1 must be chosen such that the current drawn from the telephone line is less than 5 mA. Using 4 mA, then $R_1 = 14.3/4\ mA = 3.6$ k.

Phone Alert

The circuit in Fig. 1.67 is designed to indicate when the telephone is in use. It is made up of a diode bridge D_1 to D_4 that ensures unidirectional current flow through the LED D_5. The circuit is connected between the telephone line terminals

and the telephone instrument. When the telephone is on hook, the phone line is open. Hence, no current flows and the LED is off. When the telephone goes off-hook, the telephone line is looped. As a result, current flows through the LED which lights, thereby indicating the off-hook condition. The telephone line resistance which is about 1k limits the current.

Lead-Acid Battery Monitor

The circuit in Fig. 1.68 enables the monitoring of the voltage of a lead-acid battery under charge. Diode D_1 ensures that the battery is connected with the correct polarity in which case the red LED D_2 will light. Resistor R_2 limits the current into this LED. A value of 1k will limit the current to about 12 mA. In the discharge state, the battery voltage will not be sufficient to enable conduction of the diode string comprising the green LED D_3 (1.8 V), the Zener diode D_4 (12 V) and diode D_5 (0.7 V), and hence, the green LED will be off. As the battery voltage rises above about 13.5 V, conduction in the diode string will begin, and the green LED will also start to glow. Resistor R_1 limits this current. Since most of the battery voltage will be dropped across the diodes in the diode string, a value of 100 for R_1 is sufficient as only the voltage above the diode voltages will be dropped across R_1.

Fig. 1.69 Zener diode regulator using voltage doubler

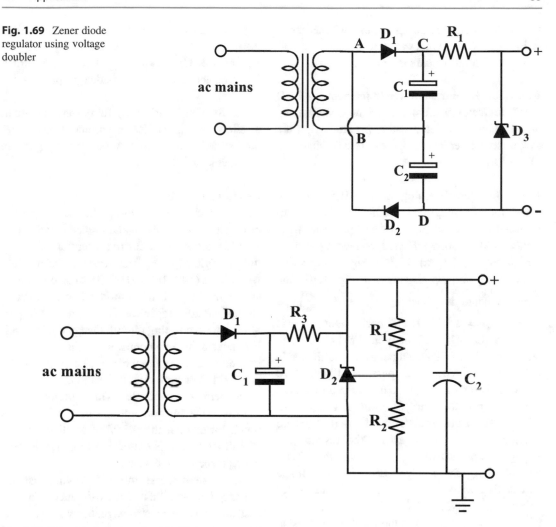

Fig. 1.70 Regulated supply using TL431A

Ideas for Exploration (i) Use this circuit to monitor the output voltage of the battery charger circuit discussed earlier.

Zener Diode Regulator Using a Voltage Doubler

The circuit in Fig. 1.69 is that of Zener diode voltage regulator that produces 12 volts DC at 25 mA from a 6 volt transformer. It operates using two half-wave rectifiers whose outputs are connected in series. Thus, during the half-cycle in which the transformer terminal A is positive with respect to transformer terminal B, diode D_1 turns on and charges capacitor C_1 to $6\sqrt{2} = 8.5$ V with the polarity shown. During this time diode D_2 is reversed-biased and therefore off. During the

half-cycle when terminal A is negative with respect to terminal B, diode D_2 turns on charging capacitor C_2 to $6\sqrt{2} = 8.5$ V. Diode D_1 turns off during this time. The result is that the voltage across terminals C and D is $2 \times 6\sqrt{2} = 17$ V. For a supply current of 25 mA, allowing a minimum Zener current of 5 mA, then the total Zener current must be 30 mA. For 30 mA into the 12 V Zener D_3, allowing a 1 V voltage drop on each capacitor, each capacitor must be $C = I/f\Delta V = 30$ mA/$60 \times 1 = 500$ μF. With this 1 V voltage drop on each capacitor, the value of R_1 that allows the flow of 30 mA into the 12 V Zener diode D_3 is given by $R_1 = (17 - 1 - 1 - 12)/30$ mA $= 100$ Ω. The maximum current into the Zener is $(17 - 12)/100 = 50$ mA, and therefore the maximum power dissipated in the Zener diode is $P_D = 12$ V $\times 50$ mA $= 600$ mW. Hence, a 1 W Zener is chosen for safe operation.

Ideas for Exploration (i) Implement an LED "ON" indicator at the output of the system using an LED and a series resistor. (ii) Re-design the system to deliver 9 volts at the output by changing D_3 and R_1.

Regulated Supply Using the TL431A

The TL431A integrated circuit (D_2) is a three-terminal device that functions as a programmable Zener diode whose effective voltage V_Z can be varied from 2.5 V to 36 V using two external resistors R_1 and R_2 as shown in Fig. 1.70. The Zener voltage is given by $V_Z = \left(1 + \frac{R_1}{R_2}\right) V_{ref}$ where $V_{ref} = 2.5$ V. The Zener current range is 1 mA to 100 mA, and the typical dynamic resistance is 0.22 Ω. The temperature coefficient is a low 0.4% at room temperature. Figure 1.70 shows the application of the device in a regulated power supply delivering 20 volts at a current of 50 mA. A 24 volt transformer is used to ensure adequate voltage drop across the resistor R_3. Using a current of 55 mA into the Zener and allowing for a 2 volt drop in the voltage across the reservoir capacitor C_1, the capacitor value is given by $C_1 = 55$ mA/$(2 \times 60) = 458$ μF. A value of 500 uF is used. The minimum value of the input voltage is $24\sqrt{2} - 2 = 32$ V. Hence $R_3 = (32 - 20)/55$

mA $= 218$ Ω. For a 20 volt output, $20 = (1 + R_1/R_2) \times 2.5$. Using $R_2 = 2$ k, then $R_1 = 14$ k. Capacitor $C_2 = 1$ μF removes any high-frequency noise and residual ripple.

Ideas for Exploration (i) Re-design the circuit to allow for a higher Zener current. This enables the system to supply more current to an external load.

Research Project

This research project involves the use of a germanium diode in the design and implementation of a simple radio receiver for the reception of amplitude-modulated (AM) broadcasts in the medium wave band (540–1600 kHz). The basic circuit is shown in Fig. 1.71. It comprises a coil (inductor) L_1 and variable capacitor VC_1 connected in parallel to form a tuning circuit. One end of the coil is connected to an antenna, while the other end is grounded to an external earth. The output of the parallel LC or tank circuit goes to diode D_1 that is a germanium diode. This semiconductor device has a turn-on voltage of 0.3 V that is much lower than the 0.7 V of a silicon diode. The diode output is connected to a high impedance piezoelectric earphone.

The system works in the following manner: An amplitude-modulated radio-frequency broadcast signal is an electromagnetic wave (carrier) whose frequency is in the radio-frequency band and whose amplitude is modulated by a superimposed audio frequency signal. The

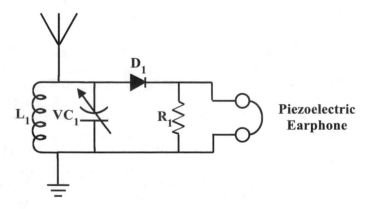

Fig. 1.71 Crystal radio

passage of such a wave induces an electrical signal in the antenna that flows into the inductor-capacitor tank circuit to ground. For a signal whose frequency is at the resonant frequency $f = \frac{1}{2\pi\sqrt{LC}}$ of the LC circuit, there is resonance such that a large potential difference develops across the tank circuit. This signal is rectified by the germanium diode, and the carrier component of the signal is filtered out by the capacitor associated with the piezoelectric earphone. The remaining signal is the original audio component, which drives the high impedance earphone.

The coil and the variable capacitor must enable changing the resonant frequency across the full medium wave band. A 365 pF variable capacitor and a 300 µH inductor are suitable. The coil can be constructed using a ferrite rod with 100 turns of no 26SWG (25AWG) enamelled copper wire. The germanium diode 1N34A is suitable for the detection stage as its turn-on voltage is only 0.3 V. The piezoelectric earphone converts the electrical signal to sound, and its high impedance ensures that frequency selectivity is not reduced. The resistor $R_1 \simeq 10$ k to 47 k provides a DC path for the proper operation of the diode. For effective operation, the antenna should be about 30 m long, about 7 m above ground and insulated from its supports.

The circuit must also have a good ground to the physical Earth using a solid copper rod driven into the ground. The circuit should be able to receive nearby AM stations, with difficulty being experienced with more distant ones.

Ideas for Exploration (i) Wind the antenna into a coil of wire and see how that affects reception. (ii) Replace the ferrite rod by a 1 inch former, thereby realizing an air-cored inductor. (iii) Experiment with coils with a reduced number of turns to attempt short wave reception (1.7–30 MHz).

Problems

1. Briefly explain the following:

   (i) The formation of n-type and p-type material
   (ii) Increased conductivity in n-type and p-type material
   (iii) Why there is conduction across a forward-biased p-n junction
   (iv) Why there is little or no conduction across a reverse-biased p-n junction
   (v) What happens when n-type and p-type material are brought into contact
   (vi) The effect of temperature on the conductivity of intrinsic semiconductor material
   (vii) Why a conductor conducts better than a semiconductor
   (viii) Why a semiconductor conducts better than an insulator
   (ix) The cause of the barrier potential in an np junction

2. Determine the dynamic resistance of a semiconductor diode for 0.5 mA current through the diode.

3. A Zener diode has $V_Z = 6.8$ V at 25°C. If the diode has $T_C = 0.03\%/°C$, determine the Zener voltage at 63°C.

4. An 10 V Zener has a power rating of 400 mW. Determine the maximum reverse current I_{Zmx}.

5. An LED is driven by a 9 volt battery in series with a resistor R_1. Determine the value of R_1 if the LED is green.

6. In the circuit of problem 5, if the LED is replaced by a silicon diode, what is the value of the current for the same value of R_1?

7. Sketch the output waveform for each of the circuits shown in Fig. 1.72. In each case V_i is a symmetrical square wave input signal of amplitude ± 10 volts.

8. For the circuits in problem 7, sketch the output waveform for a sine wave input signal of amplitude ± 15 volts.

9. Sketch the following circuits:

   (i) Full-wave rectifier using a centre-tapped transformer
   (ii) Full-wave bridge rectifier operating into a load
   (iii) Clipping circuit using one Zener diode

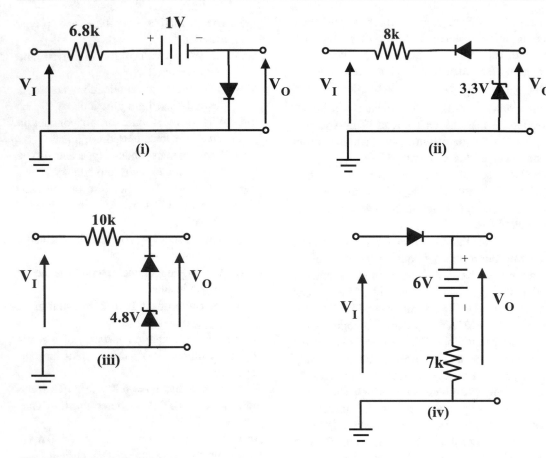

Fig. 1.72 Circuits for Question 7

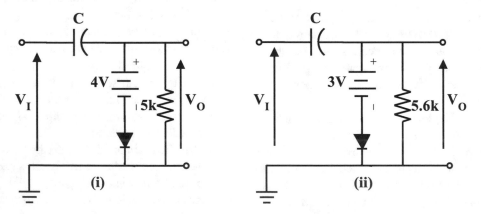

Fig. 1.73 Circuits for Question 10

(iv) Clamping circuit with a peak positive
     output of +0.7 volts

10. Sketch the output waveform for each of the
    circuits shown in Fig. 1.73. In each case the

input is a symmetrical square wave input
signal of amplitude ±12 volts.

11. With the aid of a circuit diagram, explain the
    operation of a half-wave rectifier operating

into a resistive load. Draw a voltage/time diagram of the output waveform. Explain the effect of a filter capacitor being placed across the load.

12. With the aid of a circuit diagram, explain the operation of a full-wave rectifier operating into a resistive load. Draw a voltage/time diagram of the output waveform. Explain the effect of a filter capacitor being placed across the load.

13. Compare the advantages and disadvantages of using a bridge rectifier with a normal transformer or two diodes and a centre-tapped transformer to realize a full-wave rectifier.

14. Design a half-wave unregulated power supply delivering 15 volts DC at 0.5 A with less than 2 V peak-to-peak ripple.

15. Using a centre-tapped transformer, design a full-wave unregulated power supply delivering 12 volts DC at 2 A with less than 1 V peak-to-peak ripple.

16. Re-design the circuit in problem 16 using a 15 volt transformer and a bridge rectifier.

17. A 6 volt Zener diode with a minimum current requirement of 5 mA is to be used in a voltage regulator. The supply voltage is 22 volts and the variable load current has a maximum value of 15 mA. Calculate the required series resistance and the minimum power rating of the Zener diode.

18. A 16 volt Zener diode with a minimum current requirement of 5 mA and a power rating of 800 mW is to be used in a voltage regulator.

The supply voltage is 24 volts and the load current is variable. Calculate the required series resistance and the minimum load resistor that can be connected to the supply.

19. A 5 volt Zener diode has a minimum current requirement of 10 mA and is to be used in a voltage regulator. The supply voltage is 20 volts ±10% and the fixed load current is 20 mA. Calculate the series resistance required, the minimum power rating of the Zener diode and the minimum instantaneous power dissipated in the Zener.

20. A 12 volt Zener diode has a minimum current requirement of 6 mA and is to be used in a voltage regulator. The supply voltage is 30 volts ±5% and the fixed load current is 15 mA. Calculate the series resistance required and the maximum instantaneous power dissipated in the Zener diode.

21. A 12 volt Zener diode with a minimum current requirement of 4 mA and a power rating of 600 mW is to be used in a voltage regulator. The supply voltage is 20 volts and the load current is variable. Calculate the required series resistance and explain what occurs if the supply is short-circuited.

22. A 9 volt Zener diode has a minimum current requirement of 3 mA and is to be used in a voltage regulator. The supply voltage is 24 volts ±5% and the variable load current has a maximum value of 18 mA. Calculate the series resistance required and the minimum power rating of the Zener diode.

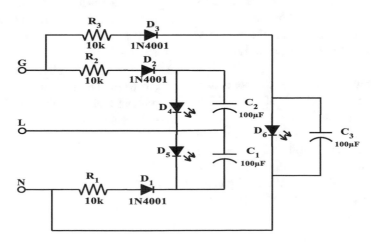

Fig. 1.74 Circuits for Question 28

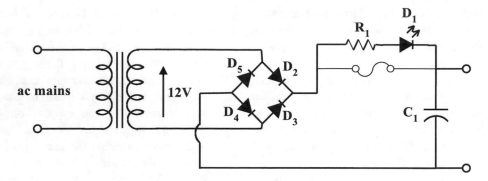

Fig. 1.75 Circuits for Question 29

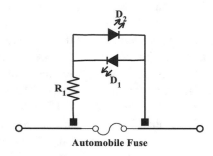

Fig. 1.76 Circuit for Question 30

23. A 6 volt Zener diode with a 1 W power rating has a minimum current requirement of 5 mA and is to be used in a voltage regulator. The supply voltage is 18 volts ±5% and the load current is variable. Calculate the series resistance required and the maximum current that can be delivered by the supply.

24. Design a Zener regulated power supply to deliver 12 volts at a current of 25 mA using an 18 volt transformer and a bridge rectifier.

25. Using the basic configuration of Fig. 1.60, design a circuit to deliver (i) 3 V from a 5 V DC supply and (ii) 6 V from a 12 V DC supply.

26. Determine how to arrange the components in the polarity indicator circuit of Fig. 1.62 if only one LED is available.

27. Using the basic Zener regulator configuration and a BZX55C2V0 2 volt Zener, design a circuit that supplies 1.3 volts from a 5 V supply that can replace 1.3 V mercury cells. Use an 1N4148 silicon diode to effect the voltage reduction from 2 V to 1.3 V.

28. The circuit shown in Fig. 1.74 is that of a mains wiring fault detector. Determine the condition of the leds (on/off) for the following fault conditions: (i) live disconnected; (ii) live and neutral interchanged; (iii) ground disconnected; (iv) live and ground interchanged; (v) neutral disconnected; and (vi) all connections good.

29. Resistor R_1 and LED D_1 in Fig. 1.75 constitute a blown fuse indicator for a power supply. Explain the operation of the arrangement and determine a suitable value for the resistor.

30. Discuss how the circuit shown in Fig. 1.76 can be used for testing fuses in an automobile without their removal and determine a suitable value of R_1.

Bipolar Junction Transistor

The bipolar junction transistor or BJT is a device capable of amplifying a voltage or current, something that diodes are not able to do. This amplifying characteristic makes the BJT suitable for a wide range of applications. The device was invented in 1947 by Walter H. Brattain, John Bardeen and William Shockley who were awarded the Nobel Prize in Physics in 1956 for this invention. It revolutionized the electronics industry by enabling miniaturization of electronic circuits and increased equipment portability. This chapter discusses the characteristics of the BJT and its use in elementary amplifier circuits. At the end of it, the student will be able to:

- Explain the basic operation of a BJT
- Describe the various configurations
- Describe the operation of a basic BJT amplifier
- Analyse single-stage transistor amplifiers
- Design single-stage transistor amplifiers

2.1 Transistor Construction and Operation

The bipolar junction transistor or BJT is basically a two-diode structure consisting of either p-n-p or n-p-n. The former is called a p-n-p transistor, while the latter is called an n-p-n transistor. The n-p-n structure involving n-type, p-type and n-type material in direct contact is shown in Fig. 2.1a, while the p-n-p structure consisting of p-type, n-type and p-type material in direct contact is shown in Fig. 2.1b. In the n-p-n structure, the n-type material referred to as the emitter region of the transistor is heavily doped, while the base and collector regions are lightly doped. In normal operation as an amplifier, the emitter-base junction is forward-biased by applying a potential V_{BE}, and the collector-base junction is reverse-biased by applying a potential V_{CB}. For the n-p-n transistor, electrons (I_E) are injected from the emitter into the base region because of the forward bias across the base-emitter junction. Most diffuse across the base to the collector where they (I_C) are collected by the reverse bias potential across the collector-base junction. A few electrons (I_B) flow out to the base connection because of the positive potential there. This process of carriers being generated in the emitter, traversing the base and being collected by the collector is referred to as transistor action. The p-n-p transistor behaves similarly with holes replacing electrons as the majority carriers. The symbols for n-p-n and p-n-p transistors are shown in Fig. 2.2.

In the arrangement in Fig. 2.3, V_{BE} forward-biases the base-emitter junction as before, but instead of a voltage V_{CB} directly across the collector-base junction, a voltage V_{CE} is applied across the collector and emitter, thereby referring both potentials to a common reference at the emitter. V_{CE} is chosen sufficiently larger than V_{BE} to ensure that the collector-base junction is reverse-biased. This is called the common emitter

© Springer Nature Switzerland AG 2021
S. J. G. Gift, B. Maundy, *Electronic Circuit Design and Application*,
https://doi.org/10.1007/978-3-030-46989-4_2

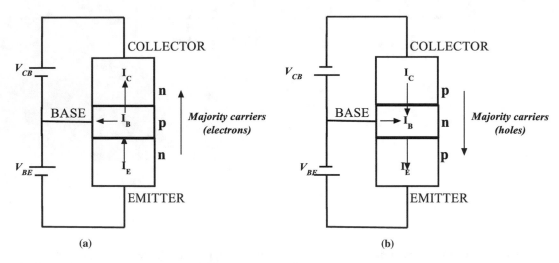

Fig. 2.1 NPN and PNP transistors. (**a**) NPN transistor. (**b**) PNP transistor

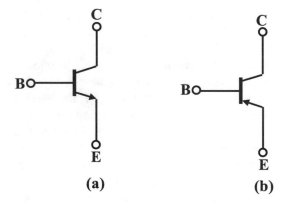

Fig. 2.2 NPN transistor (**a**) and PNP transistor (**b**)

configuration and is fundamental to transistor action. Using conventional current flow, according to Kirchhoff's current law for current flow,

$$I_E = I_C + I_B \qquad (2.1)$$

i.e. the emitter current is the sum of the collector current and the base current. The unique property of the transistor, which makes it useful, is the fact that the collector will collect carriers with an efficiency that is almost independent of the reverse bias collector-base voltage. That is, the value of the collector current I_C depends almost entirely on the emitter current from which it is derived which in turn depends on the base-emitter forward bias voltage V_{BE} or the base current I_B. This characteristic of the transistor can be used to produce amplifying action in the transistor.

Thus, the inclusion of a resistor R in the collector circuit as shown in Fig. 2.4 (where power supplies have been relabelled) will not affect the value of the collector current since the collector will still collect carriers with almost the same efficiency. Therefore, a changing base-emitter voltage corresponding to an input signal V_i as shown in Fig. 2.5 will produce a changing collector current. This results in a changing voltage drop across resistor R representing an output voltage signal V_o. For a sufficiently large resistor R, this output voltage V_o is larger than the input signal voltage V_i producing it, and hence signal amplification occurs. This important action will be discussed in detail in Transistor Amplifying Action on page 81. Note that the circuit of Fig. 2.5 is not useable in this form. Certain modifications are necessary to make it functional. These will be considered in Sect. 2.3.

Another important property of the transistor is that a large collector current I_C is associated with a small base current I_B according to

$$I_C = \beta I_B \qquad (2.2)$$

where β is the (common emitter) current gain, typically 100 for low-power transistors. Since most of the majority carriers from the emitter are

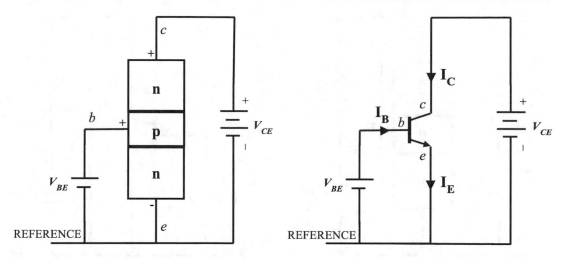

Fig. 2.3 Modified transistor arrangement with conventional current flow

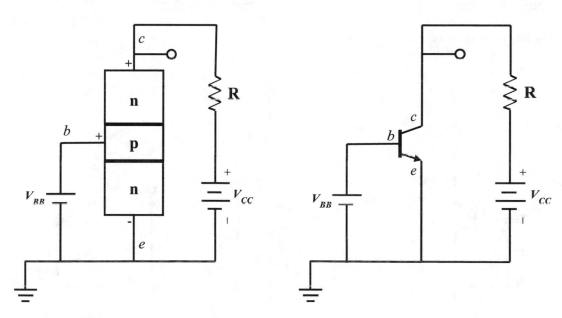

Fig. 2.4 Modified transistor arrangement with resistor R

collected by the collector, the collector current I_C is very nearly equal to the emitter current I_E,

$$I_C = \alpha I_E \qquad (2.3)$$

where $\alpha \leq 1$. α is referred to as the common base current gain. While the collector current I_C comprises mainly majority carriers from the emitter, there is a small component that is made up of minority carriers. This minority carrier component is called *leakage* current I_{CO}. Its value is

quite small relative to I_C and, like the reverse saturation current I_O for a diode, is temperature sensitive. Its effect can be quite critical in some designs. However, in modern devices, I_{CO} is often sufficiently small that it can be ignored.

2.2   Transistor Configurations

There are three transistor configurations, which provide useful amplification. These are the

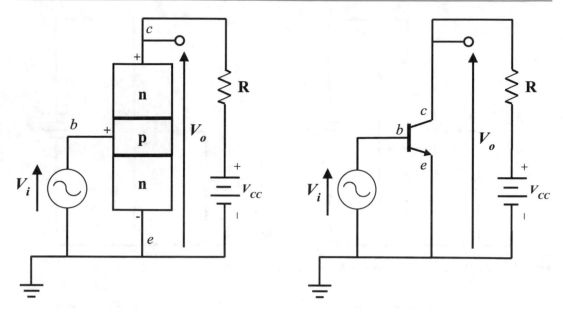

Fig. 2.5 Common emitter configuration

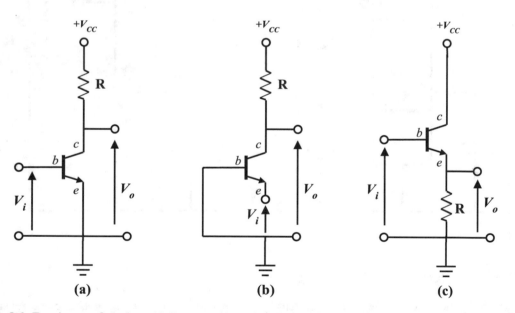

Fig. 2.6 Transistor configurations. (**a**) Common emitter (CE). (**b**) Common base (CB). (**c**) Common Collector (CC)

common emitter, common base and common collector configurations. These configurations using npn transistors are shown in Fig. 2.6. with input voltage V_i and output voltage V_o. The common emitter configuration shown in Fig. 2.6a is in fact that considered in Fig. 2.3 in discussing transistor action above. The name common emitter derives from the fact that the emitter terminal is common to both input and output circuits.

The common base configuration shown in Fig. 2.6b is so-called because the base terminal is common to both the input and the output. It corresponds to the circuit in Fig. 2.1 used to introduce transistor operation. The common collector configuration shown in Fig. 2.6c has the collector common to both the input and the output through the supply V_{CC} which is a short circuit for signals.

2.2.1   **Common Emitter Configuration**

Consider the common emitter configuration shown in Fig. 2.7. The input signal is applied at the base and the output signal taken at the collector as shown. This configuration can be viewed as fundamental to transistor operation since the other two configurations can be derived from this one using negative feedback. It is the most frequently used configuration.

In the analysis and design of transistor circuits using this configuration, static characteristic curves involving current-voltage plots are available that may be used. These curves give information on the value of current flowing into or out of one terminal in response to current into the base terminal or voltage across the base-emitter terminals. Three sets of characteristics can be plotted:

(i) Input characteristics
(ii) Mutual and transfer characteristics
(iii) Output characteristics

They can be experimentally determined using the circuit shown in Fig. 2.8. This circuit comprises a variable DC supply V_{CC} that supplies the collector-emitter circuit and a variable DC supply V_{BB} which supplies the base-emitter circuit. Voltmeter A measures the base-emitter voltage V_{BE}, while voltmeter B measures the collector-emitter voltage V_{CE}. The milliammeter measures the collector current I_C, while the microammeter measures the base current I_B. Each of three sets of characteristics will now be considered.

Common Emitter Input Characteristic
The common emitter input characteristic indicates the manner in which the base current I_B varies with changes in the base-emitter voltage V_{BE}, the collector-emitter voltage being held constant. Typical characteristics for various constant values of V_{CE} are shown in Fig. 2.9. The curve has a

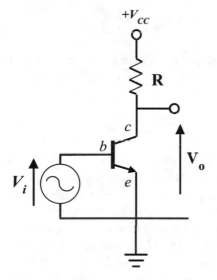

Fig. 2.7 Common emitter configuration

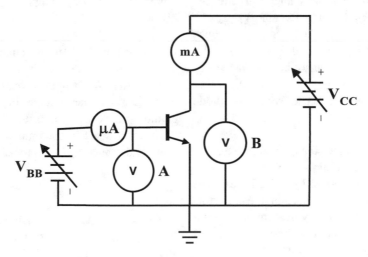

Fig. 2.8 Circuit for the determination of transistor characteristics

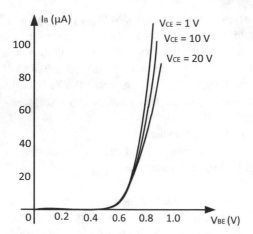

Fig. 2.9 Common emitter input characteristics

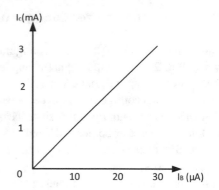

Fig. 2.10 Common emitter transfer characteristic

$$h_{fe} = \frac{\Delta I_C}{\Delta I_B}\bigg|_{V_{CE}\,\text{constant}} \qquad (2.5)$$

shape that is similar to that of the diode since like the diode, little or no base current flows below about 0.7 V, while above 0.7 V the base current increases sharply. Note also that varying V_{CE} causes only a small variation in the curves suggesting that the input characteristic is essentially independent of the collector-emitter voltage. The AC input resistance h_{ie} of the transistor for small variations in V_{BE} and I_B can be determined by taking the reciprocal of the curve at that point with V_{CE} held constant. It is given by

$$h_{ie} = \frac{\Delta V_{BE}}{\Delta I_B}\bigg|_{V_{CE}\,\text{constant}} \qquad (2.4)$$

where ΔV_{BE} and ΔI_B are small changes in V_{BE} and I_B. A parameter of less interest is the DC input resistance h_{IE} given by the ratio V_{BE}/I_B at a point. There may be considerable difference between h_{ie} and h_{IE}. V_{BE} and I_B are related by the diode equation and cannot be simultaneously controlled: One can be set, while the other assumes a value consistent with the diode equation.

Common Emitter Transfer Characteristic
The transfer characteristic shows the manner in which the collector current I_C varies in response to the base current I_B while keeping the collector-emitter voltage constant. A typical curve is shown in Fig. 2.10. The slope of this characteristic gives the AC current gain of the transistor

where ΔI_C and ΔI_B are small changes in I_C and I_B. The DC current gain β considered earlier is given by

$$\beta \equiv h_{FE} = \frac{I_C}{I_B} \qquad (2.6)$$

Again, the collector-emitter voltage has little or no effect on the common emitter transfer characteristic; the collector current is determined by the base current (or the base-emitter voltage) and not the collector-emitter voltage. Another interesting point is that this I_C/I_B characteristic is quite linear for a substantial part of the curve and therefore a (large) collector current flowing in response to a (small) base current can be quite linear. This represents (current) amplification with low distortion.

Common Emitter Mutual Characteristic
The mutual characteristic indicates how the collector current I_C changes in response to changes in the base-emitter voltage V_{BE}, the collector-emitter voltage being constant. The gradient of this characteristic is the mutual conductance or transconductance g_m of the transistor given by

$$g_m = \frac{\Delta I_C}{\Delta V_{BE}}\bigg|_{V_{CE}\,\text{constant}} \qquad (2.7)$$

where ΔI_C and ΔV_{BE} are small changes in I_C and V_{BE}. Typical characteristics for various values of V_{CE} are shown in Fig. 2.11. Once again, V_{CE} has

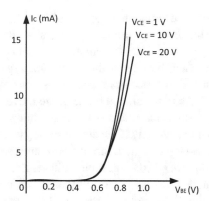

Fig. 2.11 Common emitter mutual characteristics

relatively little effect on the curves: I_C is set by V_{BE} (or I_B) and not V_{CE}. This curve is similar to the input characteristic and is non-linear. Therefore, the collector current I_C flowing in response to an input signal V_{BE} will experience increasing levels of distortion as the input amplitude increases. Note that g_m can be written as

$$g_m = \frac{\Delta I_C}{\Delta V_{BE}} = \frac{\Delta I_C}{\Delta I_B} \times \frac{\Delta I_B}{\Delta V_{BE}} = \frac{h_{fe}}{h_{ie}} \quad (2.8)$$

This is a useful relationship that will be referred to later.

Common Emitter Output Characteristics

The output characteristic illustrates the manner in which collector current I_C changes with collector-emitter voltage V_{CE} for (a) constant values of base current and (b) constant values of base-emitter voltage. The family of curves with base current as the varying parameter is shown in Fig. 2.12. The low value of the slope of these curves is yet another indication of the limited effect of V_{CE} on I_C. The slope at the approximately flat portion of these curves represents the output admittance h_{oe} of the transistor and is given by

$$h_{oe} = \frac{\Delta I_C}{\Delta V_{CE}}|_{I_B \text{constant}} \quad (2.9)$$

where ΔI_C and ΔV_{CE} are small changes in I_C and V_{CE}. The output impedance of the transistor in this configuration is the reciprocal of h_{oe}. These output characteristics can also be used to determine the (short circuit) current gain h_{fe} of the

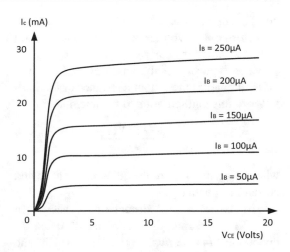

Fig. 2.12 Common emitter output characteristics for fixed values of I_B

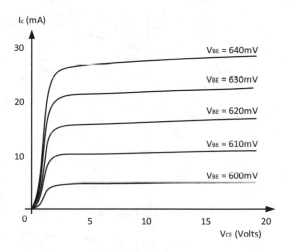

Fig. 2.13 Common emitter output characteristics for fixed values of V_{BE}

transistor, since, for a given value of collector-emitter voltage V_{CE}, the change in collector current ΔI_C and the corresponding change in base current ΔI_B can be obtained by projection on the curves. The family of curves with base-emitter voltage as the varying parameter is shown in Fig. 2.13.

Output impedance as well as transconductance can be determined using these curves. The flat portion of the output curves are all asymptotic to straight lines, which intercept the collector-emitter voltage axis at approximately the same point as shown in Fig. 2.14. The magnitude of the collector-emitter voltage where this interception occurs is called the early voltage V_A.

V_A is typically in the range $50 < V_A < 300$ volts. It enables the output resistance of the transistor in this configuration to be determined by

$$\frac{1}{h_{oe}} = \frac{V_A + V_{CEq}}{I_{Cq}} \tag{2.10}$$

where V_{CEq} and I_{Cq} are the quiescent voltage and current, respectively.

The basic output characteristics shown in Fig. 2.15 illustrate three regions of transistor operation: the *active region* in which the

emitter-base junction is forward-biased and the collector-base junction is reverse-biased, the *cut-off region* in which the emitter-base junction is not forward-biased and the *saturation region* in which the collector-base junction is not reverse-biased. As in the silicon diode, large movements of minority carriers across the base-emitter junction will occur only if the emitter-base junction is forward-biased. If therefore the base-emitter voltage $V_{BE} < 0.7$ V the threshold voltage, little or no conduction takes place (only leakage current made up of minority carriers), and the transistor is said to be *cut-off*. Thus for cut-off operation, $V_{BE} < 0.7$ V (Si), $I_B = 0$ and $I_C = I_{CEO}$. If $V_{BE} \geq 0.7$ V, majority carriers are injected from the emitter into the base region (where they become minority carriers), and their subsequent diffusion across the base into the collector depends on having an adequate reverse bias on the collector-base junction. If this junction is not reverse-biased, injection of carriers into the base will take place from the collector, and under these circumstances, the transistor is *saturated*. Thus, for saturated operation, $V_{BE} \geq 0.7$ V (Si) and the collector-base junction not reverse-biased. The

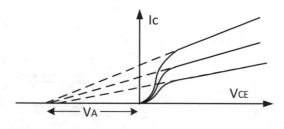

Fig. 2.14 Early voltage V_A

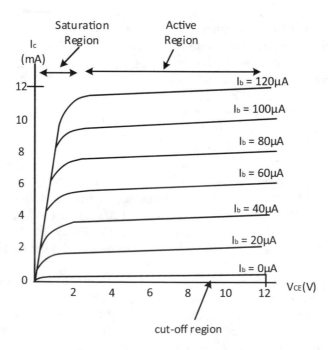

Fig. 2.15 Basic transistor output characteristics with operating regions

terminal currents in saturation are controlled by the external collector network and not the transistor.

The boundary between saturation and the active region depends on the magnitudes of the currents involved and by Kirchhoff's voltage law, $V_{CE} = V_{BE} + V_{CB}$. When the transistor is in conduction, V_{BE} is about 0.7 V, and to maintain collector action, V_{CB} must be of sufficient magnitude to collect the carriers injected into the base from the emitter. For signal level currents (units to tens of milliamps) collection requires $V_{CB} \approx 0.3$ V, so that active region operation requires $V_{CE} = 0.7 + 0.3 = 1$ V to avoid saturation, while for very small currents, V_{CE} can be a few tenths of a volt. Thus, for *active region operation*, $I_B > 0$ or $V_{BE} \approx 0.7$ V and $V_{CE} > V_{CEsat}$. On the curves, in the cut-off region where $I_B = 0$, the small collection current I_{CEO} that flows is leakage current. In the saturation region, V_{CE} is too small to maintain a reverse bias on the collector-base junction and I_C flows independently of I_B. In the active region, the curves are almost horizontal indicating very little dependence of I_C on V_{CE}.

2.2.2 Common Base Configuration

In the common base configuration shown in Fig. 2.16, the base terminal is common to both the input and the output circuits. The input signal is applied at the emitter and the output taken at the collector as shown. In this configuration, the output current is approximately equal to the input current, i.e. $I_C \approx I_E$. In the DC mode, as we had before

$$I_C = \alpha I_E \qquad (2.11)$$

note that $\alpha \approx 1$ for practical purposes. When $I_E = 0$, collector current continues to flow, and this is the leakage current I_{CBO} made up of minority carriers. It is usually quite small. In the common base configuration, an important parameter is the current gain of the transistor designated h_{fb}. It is defined as the ratio of a change in collector current ΔI_C to the change in

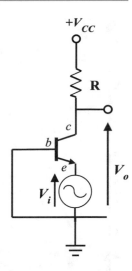

Fig. 2.16 Common base amplifier

emitter current ΔI_E producing it, the collector-base voltage being held constant

$$h_{fb} = \frac{\Delta I_C}{\Delta I_E}\Big|_{V_{CB}=\text{constant}} \qquad (2.12)$$

The DC current gain α considered earlier is given by

$$\alpha \equiv h_{FB} - \frac{I_C}{I_E} \qquad (2.13)$$

As in the case of the CE configuration, sets of characteristics can be developed in order to connect the currents and voltages into and out of the BJT in this configuration.

Common Base Input Characteristics
The input characteristic is shown in Fig. 2.17. It involves the input voltage V_{EB} and the input current which in this case is I_E. As is evident the curve is non-linear and displays a threshold value similar to the CE input characteristics and the diode curve. This is to be expected since the emitter-base junction is a diode.

Common Base Transfer Characteristic
This transfer characteristic demonstrates the manner in which collector current changes in response to the emitter current, the collector-base voltage being kept fixed. A typical curve is shown in Fig. 2.18, V_{CB} having negligible effect. The

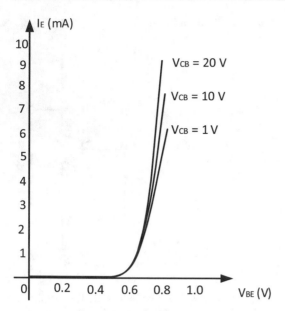

Fig. 2.17   Common base input characteristics

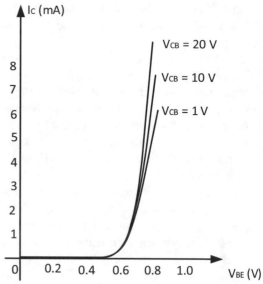

Fig. 2.19   Common base mutual characteristic

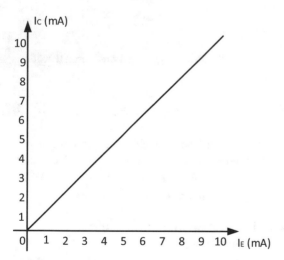

Fig. 2.18   Common base transfer characteristic

gradient of this characteristic gives the current gain h_{fb} of the transistor. The DC current gain α can also be determined. In this characteristic, the collector current is determined almost completely by I_E and is virtually independent of V_{CB}. Also, the I_C/I_E characteristic is extremely linear, even more so than the corresponding I_C/I_B characteristic of the CE characteristic. This improved linearity of the transfer characteristic is

attributable to negative feedback which shall be discussed in Chap. 7. $I_C/I_E \approx 1$, and therefore, this configuration produces no current amplification. This, of course, is consistent with the fact that $I_E = I_C + I_B \approx I_C$.

Common Base Mutual Characteristic

The mutual characteristic shows the manner in which the collector current changes in response to changes in the emitter-base voltage, the collector-base voltage held constant. Typical characteristics are shown in Fig. 2.19. The curve is non-linear as in the common emitter configuration though the effect of changes in V_{CB} is even less than the corresponding changes in the common emitter configuration. The slope is the transconductance g_m given by

$$g_m = \frac{\Delta I_C}{\Delta V_{EB}}|_{V_{CB}\text{constant}} \qquad (2.14)$$

and the value is approximately equal to that in the common emitter configuration.

Common Base Output Characteristics

The output characteristic for varying I_E is shown in Fig. 2.20. Note that these curves are flatter than the corresponding curves for the common emitter

Fig. 2.20 Common base output characteristics

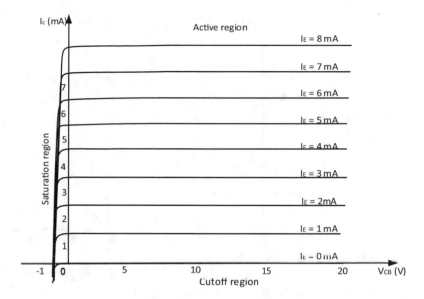

configuration. This indicates that the collector-base voltage in this configuration has an even lower effect on the collector current than the collector-emitter voltage in the common emitter configuration.

The output curves demonstrate the manner in which the collector current varies with collector-base voltage for constant values of (a) emitter current shown in Fig. 2.20 and (b) emitter-base voltage. The slope at a point on these curves is the output impedance h_{ob} of the transistor given by

$$h_{ob} = \frac{\Delta I_C}{\Delta V_{CB}}\Big|_{I_E=\text{constant}} \qquad (2.15)$$

The output impedance is $1/h_{ob}$ which is higher than that in the common emitter configuration (consistent with flatter curves). For this configuration, the regions of operation are defined as follows:

- *Active:* $I_E > 0$, base-emitter junction forward-biased and collector-base junction reverse-biased
- *Saturation:* $I_E > 0$, base-emitter junction forward-biased and collector-base junction forward-biased
- *Cut-off:* $I_E = 0$, $I_C = I_{CBO}$, base-emitter junction reverse-biased and collector-base junction forward-biased

Now

$$I_C = \alpha I_E + I_{CBO} = \alpha(I_C + I_B) + I_{CBO} \qquad (2.16)$$

Rearranging, we get

$$I_C = \frac{\alpha I_B}{1 - \alpha} + \frac{I_{CBO}}{1 - \alpha} \qquad (2.17)$$

In the common emitter configuration, when $I_B = 0$,

$$I_C|_{I_B=0} = \frac{I_{CBO}}{1 - \alpha} \qquad (2.18)$$

This corresponds to the collector current for zero base current, i.e. the leakage current I_{CEO}. Hence,

$$I_{CEO} = \frac{I_{CBO}}{1 - \alpha} \qquad (2.19)$$

This indicates that the leakage current in the common emitter configuration is much larger than the leakage current in the common base configuration.

2.2.3   Common Collector Configuration

In the common collector configuration shown in Fig. 2.21, the collector terminal is common to both the input and the output (since the voltage

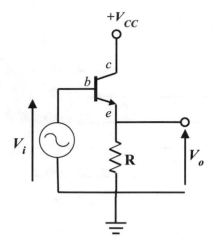

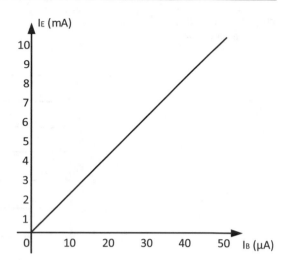

Fig. 2.21 Common collector amplifier

Fig. 2.22 Current transfer characteristic

source powering the circuit has no internal impedance and therefore short-circuits the collector to
ground). The input signal is applied at the base
and the output taken at the emitter as shown. In
this configuration, the input characteristic $V_B vs I_B$
depends on R and therefore is not plotted.

Common Collector Transfer Characteristic
There are two transfer characteristics: (a) the current transfer characteristic and (b) the voltage
transfer characteristic. The current transfer characteristic indicates the manner in which the emitter current changes in response to base current,
the collector-emitter voltage held constant, and
this is shown in Fig. 2.22. The slope of this
characteristic gives the current gain h_{fc} defined as

$$h_{fc} = \frac{\Delta I_E}{\Delta I_B}|_{V_{ce}=\text{constant}} \qquad (2.20)$$

Note that

$$h_{fc} = \frac{\Delta I_E}{\Delta I_B} = \frac{\Delta I_C + \Delta I_B}{\Delta I_B} = h_{fe} + 1 \qquad (2.21)$$

This characteristic is reasonably linear and is
independent of V_{CE}. The voltage transfer characteristic shown in Fig. 2.23 indicates the manner in
which the emitter voltage changes in response to
the base voltage, the collector-emitter voltage
kept constant. The slope of the characteristic
gives the short-circuit voltage gain A_V defined as

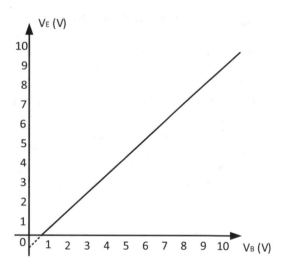

Fig. 2.23 Voltage transfer characteristic

$$A_V = \frac{\Delta V_E}{\Delta V_B}|_{V_{ce}}\text{ constant} \qquad (2.22)$$

This value is approximately one with no voltage amplification taking place. This characteristic
is even more linear than the current transfer characteristic and is attributable to negative feedback
to be discussed in Chap. 7. Also, the voltage
characteristic like the current characteristic is virtually independent of changes in V_{CE}.

Common Collector Output Characteristics

In the common collector configuration, the appropriate output signal is the voltage V_E. Hence, the output characteristic to be displayed is one in which the emitter voltage varies with the collector-emitter voltage. The collector-emitter voltage has essentially no effect on the emitter (output) voltage, and therefore, this family of curves is not particularly useful in practice. Note that $V_E = V_B - 0.7$. Hence $V_E \approx V_B$, that is, the output signal voltage at the emitter is approximately equal to the input signal voltage at the base. As a result, this configuration is referred to as the emitter follower. The regions of operation are:

- *Active:* $V_{EB} \approx 0.7$ V.
- *Saturation:* $V_{EB} \geq 0.7$ V, V_{CB} not reverse-biased.
- *Cut-off:* $V_{EB} < 0.7$ V, $I_E = 0$.

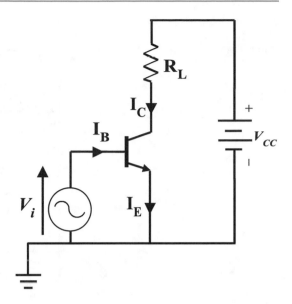

Fig. 2.24 Common emitter amplifier

2.3   Common Emitter Amplifier

Having discussed the characteristics of the binary junction transistor (BJT) in the various configurations, we now consider the application of each configuration in functioning circuits, starting with the common emitter configuration.

Transistor Amplifying Action

In order to demonstrate the amplifying action of the BJT, we utilize the common emitter configuration as shown in Fig. 2.24. Here a sinusoidal signal is applied at the base-emitter junction of the BJT, and in order to ensure that carriers generated in the emitter by the action of V_I are collected at the collector, a voltage V_{CC} is applied. Thus, as V_I goes positive (for the npn transistor), for a sufficiently large V_I (>0.7 V), the transistor is turned on (the base-emitter junction becomes forward-biased), and electrons flow from the emitter across the base region, most being collected by the collector. A small number flow out of the base terminal. These electron flows correspond to conventional currents I_E, I_C and I_B as

shown. Since a large collector current I_C flows in response to a small base current I_B, this corresponds to current amplification of I_B. Since most signals are in the form of a voltage, we are more interested in the amplification of the input voltage V_I which also influences I_C.

In order to utilize the output signal I_C conveniently, a resistor R_L is included in the collector circuit as shown. This will not affect the action of the transistor because of the excellent charge carrier collecting ability of the collector under conditions of varying collector voltage. The presence of R_L in the collector circuit means that I_C produces a voltage drop $V_C = I_C R_L$ across this resistor which can be readily measured. It turns out that in bipolar junction transistors, the ratio V_C/V_I can be quite large, typically of the order of a few hundred, thereby representing voltage amplification of V_I. This voltage amplification by the transistor is partly what has enabled it to revolutionize the electronics industry in a manner that continues today.

There are however two problems with the amplifying action just described. The first is that on the positive half-cycle of the input signal, V_I must exceed 0.7 V before the transistor is turned on resulting in a flow of (I_B and) collector current I_C. During the period before this turn-on occurs,

Fig. 2.25 Waveform distortion resulting from zero biasing

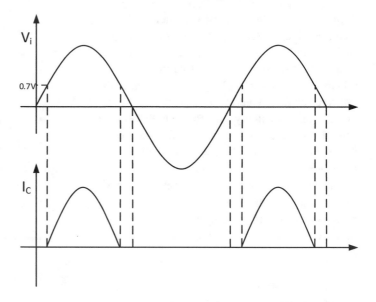

no collector current flows ($I_C = 0$), and hence there is no output voltage ($V_C = I_C R_L = 0$). This results in severe signal distortion. The second more serious problem occurs on the negative half-cycle for V_I. During the part of the signal cycle, the transistor remains off and hence there is no signal transmission and therefore no amplification occurs. The result as shown in Fig. 2.25 is that there is collector current flow for only one half-cycle of the input signal. The voltage developed across the collector resistor will therefore be severely distorted.

Both problems are easily solved by turning on the transistor with suitable DC voltages and establishing a standing current I_{Cq} in the device. This process is called BIASING, and the currents and voltages in the BJT are called bias or quiescent currents and voltages. This is shown in Fig. 2.26. The voltage source V_{BB} is used to turn on the BJT by forward biasing the base-emitter junction. For no input signal ($V_I = 0$), this results in DC currents I_{Cq} and I_{Bq}. When V_I is applied in series with V_{BB}, it modulates or changes the value of the transistor base voltage such that the base-emitter voltage of the BJT is varied about its quiescent value V_{BEq}. This results in a corresponding variation of the collector current about its quiescent value I_{Cq}. V_{BB} and V_I must be such that the transistor never turns off at any point during the signal cycle. As can be seen in

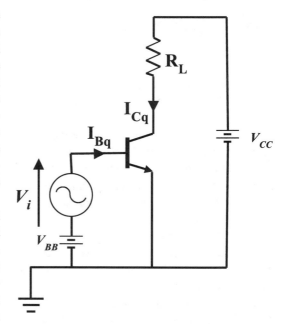

Fig. 2.26 Common emitter amplifier with bias

Fig. 2.27, I_C is fully reproduced eliminating both the dead bands on the positive half-cycle and the non-conduction on the negative half-cycle. This results in a full voltage waveform V_o at the collector of the transistor about the quiescent collector voltage V_{Cq} as shown in Fig. 2.28.

Note that the output voltage waveform is inverted relative to the input voltage waveform

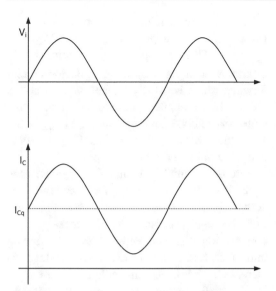

Fig. 2.27 No waveform distortion with biasing

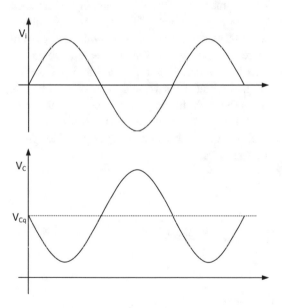

Fig. 2.28 Undistorted collector voltage

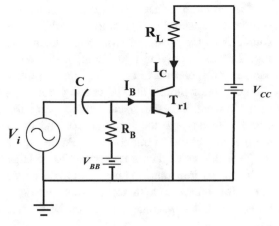

Fig. 2.29 Common emitter amplifier with modified biasing

since, as the collector current increases say, the voltage drop across the collector resistor R_L also increases and hence, by Kirchhoff's voltage law, the voltage at the collector of the transistor decreases ($V_o = V_{CC} - I_C R_L$). This biasing scheme has the disadvantage that DC flows through the signal source which is undesirable. A second problem is that setting the bias current I_{Cq} by adjusting V_{BB} is virtually impossible because of the steep slope of the $V_{EB} vs I_C$ curve

beyond the knee in Fig. 2.9. A practical biasing scheme that overcomes these problems and the selection of I_{Cq} and R_L in order to achieve optimum performance as part of the design process is discussed next.

Fixed Biasing

The basic scheme is shown in Fig. 2.29. It utilizes the fact that there is no knee on the common emitter transfer characteristic and control of I_C using I_B is quite practical. This approach is implemented in Fig. 2.29 where the bias voltage is applied to the base-emitter junction via resistor R_B. The effect of this is to fix I_B at some value. Thus, for the base circuit, applying Kirchhoff's voltage law

$$V_{BB} = I_B R_B + V_{BE} \qquad (2.23)$$

since the transistor is on, $V_{BE} \approx 0.7$ V and therefore

$$I_B = \frac{V_{BB} - 0.7}{R_B} \qquad (2.24)$$

Hence

$$I_C = \beta I_B$$
$$= \frac{\beta}{R_B}(V_{BB} - 0.7) \qquad (2.25)$$

Equation (2.25) is the bias-setting equation. Knowing β for the transistor, R_B can be selected to give the desired I_C.

The problem of DC bias through the signal source is addressed by the removal of the signal source from the position in which it is located in Fig. 2.26 and applying it to the input through a capacitor C, referred to as a coupling capacitor. This capacitor blocks the DC coming from V_{BB} from flowing into the signal source. This DC could both damage the signal source and upset the bias conditions in transistor Tr_1. At the signal frequencies, the capacitor is assumed to be large such that its impedance is negligible and hence does not interfere with signal flow into the transistor. Later we shall discuss choosing this value.

A second important equation is that defining the output circuit including V_{CE} and I_C for the BJT. Applying Kirchhoff's voltage law to the output circuit, we get

$$V_{CC} = I_C R_L + V_{CE} \qquad (2.26)$$

It is called the static load line for the circuit and represents the locus of all the possible operating V_{CE}/I_C points for the device in the particular circuit under consideration. This line is plotted on the common emitter output characteristics as shown in Fig. 2.30. From (2.26) when $I_C = 0$, $V_{CE} = V_{CC}$, and the transistor is cut-off. In such circumstances, there is no voltage drop across the collector resistor, and therefore the collector voltage is at V_{CC} the supply voltage. When $V_{CE} = 0$, $I_C = \frac{V_{CC}}{R_L}$. This corresponds to the transistor in saturation with the maximum collector current flowing. There is usually a small voltage across the collector-emitter terminals giving the collector-emitter saturation voltage $V_{CEsat} \approx 0.2$ V. Hence, biasing sets V_{CE} and I_C at some operating point Q on the load line.

In order to amplify a sinusoidal signal V_i, the signal is superimposed on the DC bias conditions at the base of the transistor as shown in Fig. 2.29. For the npn transistor, as V_i goes positive corresponding to conventional current flowing into the base, the base-emitter voltage is increased, and hence the collector current I_C increases. The voltage drop across R_L therefore increases, and since V_{CC} is fixed, V_{CE} is lowered. This is manifested as a movement up the load line. As V_i goes negative, the base-emitter voltage is reduced, and hence the collector current decreases with a resulting movement down the load line. When V_{CE} is reduced to $V_{CEsat}(\approx 200$ mV), the transistor is saturated. When V_{CE} is increased such that $V_{CE} = V_{CC}(I_C = 0)$, the transistor is cut-off. These values of V_{CE} represent the maximum and minimum voltages possible across the transistor.

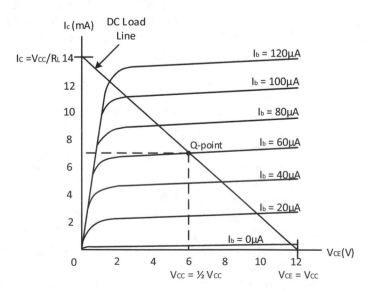

Fig. 2.30 Transistor DC load line

Fig. 2.31 Common
emitter amplifier

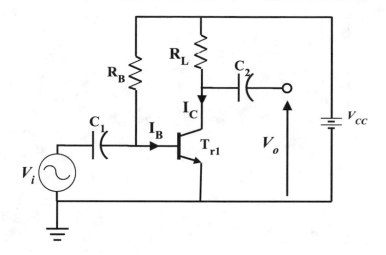

In the amplification of a signal, in order that the transistor faithfully reproduce the full waveform, V_{CE} must be able to increase and decrease by approximately the same value. This is referred to as maximum symmetrical swing. Thus, since $V_{CEmax} = V_{CC}$ and $V_{CEmin} \approx 0$ V, it follows that the maximum change in V_{CE} is V_{CC}. Therefore, for maximum symmetrical swing, we let $V_{CEq} = V_{CC}/2$, that is, set the quiescent collector voltage to be half of the supply voltage V_{CC}. In such a case, $I_{Cq} = I_{Cmax}/2$ where $I_{Cmax} = V_{CC}/R_L$. This will enable V_{CE} to increase by $V_{CC}/2$ to V_{CC} and decrease by the same amount $V_{CC}/2$ to zero. Also, instead of using two DC supplies V_{CC} and V_{BB}, a single supply or V_{CC} can be used as shown in Fig. 2.31 with R_B connected to V_{CC}. Then in the bias-setting Eq. (2.25), V_{BB} must be replaced by V_{CC}. Capacitor C_2, like C_1, is a coupling capacitor that couples the output signal to an external load while preventing the DC voltage at the collector of Tr_1 from interfering with or be interfered by that load.

Example 2.1

Using the circuit of Fig. 2.31, design a fixed-bias common emitter amplifier with a 20 volt supply and an npn silicon transistor having $\beta = 100$. Design for maximum symmetrical swing.

Solution   Choose $I_{Cq} = 1$ mA for effective transistor operation. For maximum symmetrical swing, $V_{CEq} = V_{CC}/2$. Hence $V_{CEq} = 20/$

$2 = 10$ V and $R_L = 10/1$ mA $= 10$ k. Since $\beta = 100$, and $I_B = I_C/\beta$, it follows that $I_{Bq} = \frac{1 \text{ mA}}{100} = 10 \mu A$. Kirchhoff's voltage law applied at the input yields $V_{CC} = I_{Bq}R_B + V_{BE}$. Therefore, $R_B = \frac{V_{CC} - V_{BE}}{I_{Bq}} = \frac{20 - 0.7}{10 \mu A} = 1.93 \text{ M}\Omega$, that is, $R_B = 1.93$ MΩ. The capacitors are chosen to be large say 50 µF. Hence with $R_B = 1.93$ MΩ, $I_{Cq} = 1$ mA, $V_{CEq} = 10$ V and the npn silicon transistor is biased for maximum symmetrical swing.

A signal voltage can now be applied at the input through coupling capacitor C_1; the output voltage is available at the collector through coupling capacitor C_2. Oscilloscope traces of the input and output are shown in Fig. 2.32. The input capacitor C_1 presents a reactance $X_{C_1} = 1/2\pi f C_1$ to the input signal of frequency f. If C_1 is sufficiently large, then this reactance is negligible for signal frequencies and therefore is practically a short circuit to such signals. It however blocks the flow of direct current, which would change the transistor bias current as well as possibly affect the signal source. This also applies to the output capacitor C_2. Methods of determining these values are discussed in Chap. 6.

The mutual characteristic shown in Fig. 2.33 illustrates the action of the transistor in the biased common emitter configuration. The Q point established by the bias is shown as I_{Cq} and V_{BEq}. The input signal V_i causes a variation ΔV_{BE} in V_{BE} which results in a corresponding change ΔI_C in I_C. Thus a change ΔV_{BE} in V_{BE} produces a change ΔI_C in the collector current I_C that results in a

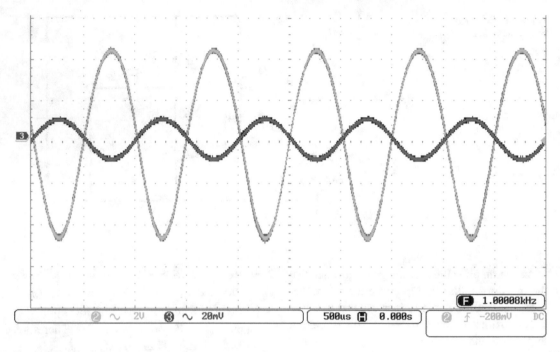

Fig. 2.32 Input and output signals of the common emitter amplifier

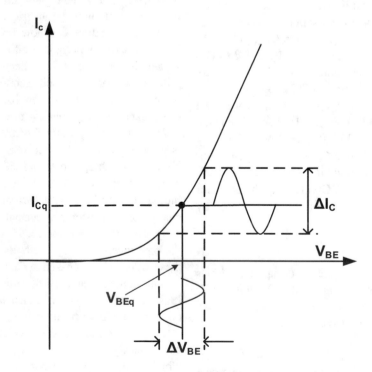

Fig. 2.33 Mutual characteristics of common emitter amplifier

change $\Delta V_C = -\Delta I_C R_L$ in the collector (output) voltage V_C. The negative sign indicates that as ΔV_{BE} increases, ΔI_C increases and hence the voltage drop across R_L increases. This corresponds to a lowering of the transistor collector voltage V_C by ΔV_C which is the output voltage V_o. It should be noted that the input signal can also be a current I_i into the transistor base resulting in a change ΔI_B in I_B. This results in a change ΔI_C in I_C and hence an output voltage $V_o = \Delta V_C$. However, most input signals are in the form of a voltage rather than a current, and hence, the focus is on input voltage signals.

Amplification

Now $\Delta I_C / \Delta V_{BE}$ is (approximately) the transconductance g_m of the transistor, and in order to evaluate the voltage gain $\Delta I_C R_L / \Delta V_{BE}$, we need to consider the diode equation

$$I_D = I_o \left(e^{\frac{qV_D}{mkT}} - 1 \right) \qquad (2.27)$$

Differentiating I_D with respect to V_D gives

$$\frac{dI_D}{dV_D} = \frac{q}{mkT} I_o e^{\frac{qV_D}{mkT}}$$

$$= \frac{q}{mkT} \left\{ I_o \left(e^{\frac{qV_D}{mkT}} - 1 \right) + I_o \right\} \qquad (2.28)$$

$$= \frac{q}{mkT} (I_D + I_o)$$

Noting that I_o is small, for values of I_D such that $I_D \gg I_o$, then

$$\frac{dI_D}{dV_D} = \frac{q}{mkT} I_D \qquad (2.29)$$

For the transistor, $I_D \equiv I_E \approx I_C$ and $V_D \equiv V_{BE}$. Also, for $m \approx 1$ and $T = 300$ K (room temperature) $\frac{q}{mkT} = 40\,\text{V}^{-1}$. Therefore using (2.29), we get

$$\frac{dI_D}{dV_D} = \frac{\Delta I_C}{\Delta V_{BE}} = 40\,I_C \qquad (2.30)$$

Hence $g_m = 40\ I_C$ A/V or $g_m = 40\ I_C$ mA/V where I_C is the current value in milliamperes. Note that this is approximately the same for all transistors. The voltage gain A_V of the simple common emitter voltage amplifier can now be found. It is given by

$$A_V = \frac{-\Delta I_C R_L}{\Delta V_{BE}} = -40\,I_C R_L = -g_m R_L \quad (2.31)$$

which is approximately the same for all transistors. Again, the negative sign indicates inversion of the output voltage relative to the input signal, i.e. these two signals are 180° out of phase as shown in Fig. 2.32. The input impedance Z_i at the base of the common emitter amplifier is the ratio of the input voltage change ΔV_{BE} to the corresponding input current change ΔI_{BE} flowing into the base circuit. Thus, noting that $I_C = \beta I_B$ and hence $\Delta I_C = \beta \Delta I_B$,

$$Z_i = \frac{\Delta V_{BE}}{\Delta I_B} = \frac{\Delta V_{BE}}{\Delta I_C / \beta} = \frac{\beta}{g_m} \qquad (2.32)$$

That is,

$$Z_i = \frac{\beta}{40 I_C} \qquad (2.33)$$

Example 2.2
For the circuit developed in Example 2.1, determine the voltage gain and input impedance.

Solution For the circuit of Fig. 2.31 $R_L = 10$ k and $I_C = 1$ mA. Hence using (2.31) $A_V = -40$ $I_C R_L = -40 \times 1 \times 10^{-3} \times 10 \times 10^3 = -400$. Using the peak values of the output and input voltages from Fig. 2.32, we find that the measured voltage gain is -350. Thus, the formula for gain is only an approximate one. Using $\beta = 100$, the input impedance Z_i is from (2.33) given by $Z_i = \frac{100}{40 \times 1 \times 10^{-3}} = 2.5\,\text{k}\Omega$. Thus, in the common emitter amplifying mode, the input impedance of the transistor is of moderate value only.

The amplified signal produced by this circuit while having reasonable fidelity for small-signal inputs does possess some level of distortion. Consider Fig. 2.33. It can be seen that equal changes $\pm \Delta V_{BE}$ of V_{BE} do not produce equal changes $\pm \Delta I_C$ of I_C. The positive-going change $+\Delta V_{BE}$ in V_{BE} produces a larger collector current change than the negative-going change $-\Delta V_{BE}$. This arises because of the non-linearity of the I_C/V_{BE} characteristic and results in the output voltage having a larger negative-going amplitude than the positive-going amplitude for equal amplitude bipolar inputs. This constitutes signal distortion. The effect can be reduced by (i) increasing I_{Cq} such that the Q point is at an increasingly linear part of the characteristic or (ii) reducing the variation of I_C and thereby restricting operation to an approximately linear part of the characteristic. Operating in the latter manner is called small-signal operation.

Circuit Analysis
It is sometimes necessary to analyse a given amplifier circuit in order to determine the voltages and currents in that circuit.

Example 2.3
Determine the collector current and voltage in the fixed-bias amplifier of Fig. 2.34, assuming transistor current gain of 100.

Solution In analysing this circuit, maximum symmetrical swing capability cannot be assumed. Therefore, the analysis is started by determining the base current I_B. Thus, applying Kirchhoff's voltage law to the base circuit, we have

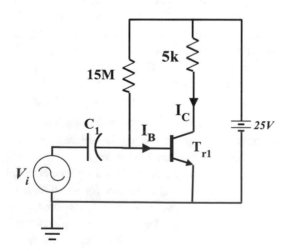

Fig. 2.34 Circuit for Example 2.3

$25 = I_B \times 1.5 \times 10^6 + 0.7$. This gives $I_B = 16.6$ μA. Hence the collector current is $I_C = 100 \times 16.6 \times 10^{-6} = 1.7$ mA. From this, $V(5 \text{ k}) = 1.7 \text{ mA} \times 5 \text{ k} = 8.5 \text{ V}$ and therefore the collector-emitter voltage $V_{CE} = 25 - V$ (5 k) $= 25 - 8.5 = 16.5$ V.

2.4   Alternative Biasing Methods

As we have discussed, in order that a BJT be able to amplify a signal, it must be turned on so that a suitable operating point is established. The simplest method and the one already discussed is called *fixed biasing* (or *constant current biasing*). The name is derived from the fact that the base current I_B is the parameter that is fixed (by fixing R_B and V_{CC}). Since $I_C = \beta I_B$, fixing I_B results in a particular value of I_C being established in the transistor. There are however several problems with this technique. Firstly, it is evident that since $I_C = \beta I_B$ and I_B is fixed, then I_C is dependent upon the value of β. For a specific transistor such as the 2N3904, the value of β varies widely from one device to another. Therefore, if the transistor is changed, even though it is the same number, β changes and hence so does I_C.

β also varies from one transistor type to another. For example, the 2N3904 has β typically at 150, while the 2N2222 has β at about 50. β also changes as a result of collector current changes as shown in Fig. 2.35. Finally, β is also temperature-dependent as can be seen in Fig. 2.35 and in response to a change of temperature may more than double its value. The effect of all β changes is a change in collector current. This results in a movement of the Q point, which reduces the maximum symmetrical swing and in extreme circumstances may push the transistor into saturation or close to cut-off.

Example 2.4
Given a DC supply of 10 V, a load resistor of 1 k and a silicon pnp transistor with $\beta = 100$ at room temperature, (a) bias the transistor for maximum

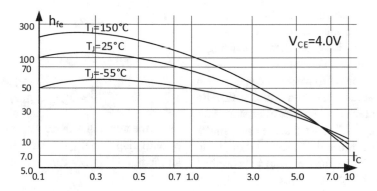

Fig. 2.35 Current gain vs collector current

symmetrical swing and (b) calculate the value of V_{CEq} at 50 °C if β goes to 75.

Solution

(a) By Kirchhoff's voltage law, $V_{CC} = I_C R_L + V_{CEq}$. For maximum symmetrical swing, $V_{CEq} = V_{CC}/2 = 10/2 = 5$ V. Hence since $R_L = 1$ k, then $I_C = (V_{CC} - V_{CEq})/R_L = (10 - 5)/1$ k $= 5$ mA. Now $I_B = I_C/\beta = 5$ mA/$100 = 50$ μA and $R_B = \frac{V_{CC} - V_{BE}}{I_B} = \frac{10 - 0.7}{50\,\mu A} = 186$ kΩ

(b) Since I_B is fixed and β has changed to 75, the collector current at 50 °C can be found as follows: $I_C = \beta I_B = 75 \times 50$ μA $= 3.75$ mA at 50 °C. The resulting quiescent collector voltage V_{CE} is given by $V_{CEq} = V_{CC} - I_{Cq} R_L = 10 - 3.75 = 6.25$ volts. When the BJT is biased for maximum symmetrical swing at $V_{CEq} = 5$ volts, the peak output voltage is 5Vpk or 10Vpk-pk. At 50 °C, when β increased, the increased collector current reduces the symmetrical swing to $3.75 \times 2 = 7.5$ volts.

Fixed biasing is therefore not a very good biasing method as it allows wide variation in I_C which affects the circuit performance. Its main advantages however are its simplicity and resulting high voltage gain.

Voltage Divider Biasing

Because of the wide variations in β, a different biasing scheme that is approximately independent

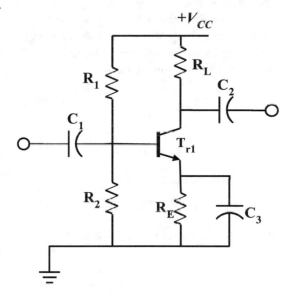

Fig. 2.36 Voltage divider-biased common emitter amplifier

of β is required. The voltage divider biasing method shown in Fig. 2.36 is one such technique. It exhibits excellent Q point stability and is little affected by β. As a result, it is one of the most commonly used in common emitter amplifiers. It comprises resistors R_1 and R_2 which form a voltage divider and thereby attempt to fix the base voltage rather than the base current. In addition, the resistor R_3 in the emitter circuit of the transistor, working in conjunction with the fixed base voltage created by R_1 and R_2, effectively stabilizes the collector current. Thus, with a fixed base voltage, if I_C tries to increase as a result of temperature changes say, then the voltage drop across R_E would increase, thereby increasing the emitter voltage of the transistor. Since the base voltage is fixed, then the base-emitter voltage of the transistor would decrease, thereby causing a decrease in the collector current close to its original value. Conversely if I_C tries to decrease, then the voltage drop across R_E would decrease, thereby reducing the emitter voltage of the transistor. With a fixed base voltage, it follows that the base-emitter voltage would increase and this would result in an increase in the collector current close to its original value. Capacitor C_3 short-circuits resistor R_E for signals, thereby nullifying the current stabilizing action of this resistor for signals. It is referred to as a bypass or decoupling capacitor.

Consider the circuit with the potential divider network comprising V_{CC}, R_1 and R_2 replaced by the Thevenin equivalent as shown in Fig. 2.37 (with capacitors removed). Here

$$V_{BB} = \frac{R_2}{R_1 + R_2} V_{CC} \qquad (2.34)$$

$$R_{BB} = \frac{R_1 R_2}{R_1 + R_2} \qquad (2.35)$$

Thus for the base-emitter loop,

$$V_{BB} = I_B R_B + V_{BE} + I_C R_E \qquad (2.36)$$

But $I_B = I_C/\beta$. Therefore, (2.36) becomes

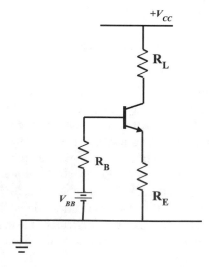

Fig. 2.37 Common emitter amplifier with Thevenin equivalent bias

$$V_{BB} = \frac{I_C}{\beta} R_B + V_{BE} + I_C R_E \qquad (2.37)$$

which when rearranged yields

$$I_C = \frac{V_{BB} - V_{BE}}{R_E + R_B/\beta} \qquad (2.38)$$

From the numerator in Eq. (2.38), the effect on I_C of variations of V_{BE} around 0.7 V caused by temperature changes can be reduced by making

$$V_{BB} \gg V_{BE} \qquad (2.39)$$

This ensures that $V_{BB} - V_{BE} \approx V_{BB}$ in (2.38), i.e. that V_{BE} is small compared with V_{BB}, thereby making I_C virtually immune to changes in V_{BE}. From the denominator in Eq. (2.38), the effect on I_C of variations of β can be reduced by making

$$R_E \gg R_B/\beta \qquad (2.40)$$

This results in $R_E + R_B/\beta \approx R_E$ in (2.38), thereby rendering I_C independent of β.

Now (2.39) will be satisfied if $V_{RE} = I_C R_E$ is made very much larger than V_{BE} say $V_{RE} \geq 5V_{BE} = 3.5$ V in which case $R_E = 3.5/I_{Cq}$. This value of V_{RE} may be too large if for a given V_{CC}, a particular peak-to-peak output swing is desired. For example, if $V_{CC} = 10$ V, then for $V_{RE} = 3.5$ V, the maximum peak-to-peak swing available is $10 - 3.5 = 6.5$ volts which is 3.25 volts. If a higher peak output swing is required, it may be necessary to lower V_{RE}. In general therefore, a good rule of thumb is to choose V_{RE} such that

$$V_{RE} = V_{CC}/10 \qquad (2.41)$$

Consider inequality (2.40), this can be rearranged to give

$$\frac{R_1 + R_2}{R_1 R_2} \gg \frac{1}{R_E}/\beta \qquad (2.42)$$

Multiplying both sides by the emitter voltage V_{RE} gives

$$\frac{V_{RE}}{R_2}\left(1 + \frac{R_2}{R_1}\right) \gg \frac{V_{RE}}{R_E}/\beta \qquad (2.43)$$

Since V_{RE} is made very much larger than V_{BE} and $V_{Bq} = V_{BE} + V_{RE}$, it follows that

$$V_{Bq} \approx V_{RE} \qquad (2.44)$$

Inequality (2.43) therefore becomes

$$\frac{V_{Bq}}{R_2}\left(1 + \frac{R_2}{R_1}\right) \gg \frac{V_{RE}}{R_E}/\beta \qquad (2.45)$$

Now V_{Bq}/R_2 is the current I_R through resistor R_2 of the potential divider, while V_{RE}/R_E is the collector current I_C. Thus (2.45) can be written

$$I_R\left(1 + \frac{R_2}{R_1}\right) \gg I_C/\beta = I_B \qquad (2.46)$$

This is satisfied if

$$I_R \gg I_B \qquad (2.47)$$

Inequality (2.47) requires that the current through the potential divider is much larger than the base current of the transistor which the divider supplies. This ensures that the potential divider voltage V_{Bq} is approximately fixed. Inequality (2.47) is approximately satisfied if

$$I_R \geq 10 I_B \qquad (2.48)$$

Since $\beta \approx 100$ for most (small-signal) transistors and $I_C = \beta I_B$, then

$$I_R \geq I_C/10 \qquad (2.49)$$

where I_R is the current flowing through the resistors R_1 and R_2 of the potential divider. Note however that I_R cannot be too large since these resistors affect the input impedance of the amplifier and may also represent a drain on the power supply. A simple design procedure for a common emitter amplifier can now be outlined:

1. Choose V_{CC} and I_{Cq}. V_{CC} may depend on the available supply. I_{Cq} is usually in the range 0.1 mA to 10 mA depending on noise, transistor current gain and frequency response characteristics.
2. To minimize effects of V_{BE} changes, $V_{RE} \geq 5V_{BE}$ which for silicon is $V_{RE} \geq 3.5$ volts. The required output voltage swing and available supply voltage may preclude the use

of this design rule. In general therefore, a good rule of thumb that is a reasonable compromise between large voltage swing and high quiescent current stability is $V_{RE} = \frac{1}{10}V_{CC}$.

3. $R_E \geq R_B/\beta$ where $R_B = R_1 R_2/(R_1 + R_2)$ in order to minimize changes in I_C due to changes in β. This criterion can be generally realized by choosing $I_R = \frac{1}{10}I_C$ where I_R is the current along the potential divider R_1, R_2.
4. $V_{Bq} = V_{RE} + V_{BE} = I_R R_2$ therefore, $R_2 = V_{Bq}/I_R$ $R_1 = (V_{CC} - V_{Bq})/I_R$
5. For maximum symmetrical swing, $V_{R_L} = \frac{V_{CC}-V_{RE}}{2}$, and therefore $R_L = \frac{V_{CC}-V_{RE}}{2I_{Cq}}$. V_{RE} is held constant by the capacity C_E.
6. Select large (of order of tens of μF) capacitor values for good low-frequency response.
7. Calculate the resulting voltage gain using $A_V = -g_m R_L$.

Example 2.5
Design a voltage divider-biased common emitter amplifier using an 18 V supply and an npn silicon transistor having $\beta = 100$.

Solution Choose $I_{Cq} = 1$ mA for good β and frequency response. Using rule-of-thumb (2.44) $V_{RE} = 18/10 = 1.8$ V. Use $V_{RE} = 2$ V. Hence, $R_E = 2/1$ mA $= 2$ k and $R_L = \frac{18-2}{1\,mA \times 2} = 8$ k . $V_{Bq} = V_{RE} + V_{BE} = 2.7$ volts, and from (2.52), $I_R = I_{Cq}/10 = 0.1$ mA. Therefore, $R_2 = \frac{2.7\,volts}{0.1\,mA} = 27$ k and $R_1 = \frac{18-2.7}{0.1\,mA} = 153$ k . Choose $C_E = 47$ µF. The resulting voltage gain is $A_V = -40 I_C R_L = -40 \times 1$ mA $\times 8$ k $= -320$. To complete the design, large coupling capacitors, say 47 µF, are used at the input and output. Note that the DC voltage at the emitter is kept constant by the capacitor C_3 and hence maximum peak-to-peak symmetrical swing is reduced from V_{CC} to $V_{CC} - V_{RE}$. A further point is that the resistors R_1 and R_1 are effectively in parallel with the input signal. Thus, while their values must be sufficiently low to ensure that $I_R \geq 10I_B$ for bias stability, they cannot be too low since they will load down the input signal.

Consider now the analysis of a voltage divider-biased common emitter amplifier circuit.

Example 2.6

In the common emitter voltage divider-biased-biased amplifier of Fig. 2.38, determine the collector current and the collector voltage.

Solution In analysing this circuit, once again maximum symmetrical swing cannot be assumed, and therefore the voltages and currents in the collector circuit are at this point unknown. Analysis must hence start at the base circuit where

conditions can be determined. Specifically, making the reasonable assumption that the base current is much less than the current through the voltage divider, then $V_B = \left(\frac{20\,\text{k}}{20\,\text{k}+180\,\text{k}}\right) \times 26\,\text{V} = 2.6\,\text{V}$. This gives $V_E = V_B - 0.7 = 1.9V$. Therefore, the collector current is $I_C = 1.9/1\,\text{k} = 1.9\,\text{mA}$ which gives $V(6\,\text{k}) = 1.9\,\text{mA} \times 6\,\text{k} = 11.4\,\text{V}$. Hence, collector voltage $V_C = 26 - V(6\,\text{k}) = 26 - 11.4 = 14.6\,\text{V}$. Finally, note that the collector voltage V_{CE} is given by $V_{CE} = V_C - V_E = 14.6 - 1.9 = 12.7\,\text{V}$.

Current Feedback Biasing

Another effective biasing scheme for bias stabilization is current feedback biasing (also called collector-base feedback biasing) shown in Fig. 2.39a. This technique results in almost the same level of stability as the voltage divider biasing method but in this form has the disadvantage of a reduced gain because of the feedback resistor R_B. The circuit functions as follows: If the collector current tries to increase, the increased voltage drop across R_L will result in a reduced collector voltage. This in turn reduces the base current I_B flowing through R_B into the transistor, thereby reducing I_C. The converse occurs for a decrease in I_C. The problem of reduced gain in this circuit can be overcome by splitting R_B and connecting a capacitor C_G to ground as shown in Fig. 2.39b.

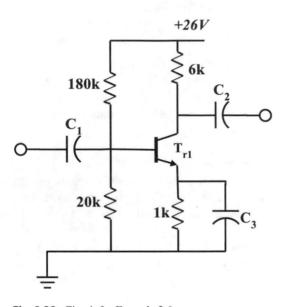

Fig. 2.38 Circuit for Example 2.6

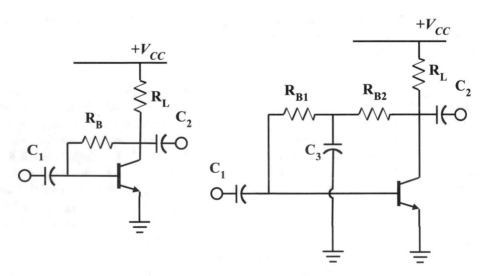

Fig. 2.39 Common emitter amplifier with collector-base feedback biasing

The effect of this capacitor is to remove the feedback action for signal frequencies while retaining it for DC. A simple design procedure is the following:

1. Choose a collector current I_{Cq}. Since β is large, the base current is small, and hence the current through R_L is approximately equal to I_{Cq}.
2. For maximum symmetrical swing, $V_{CEq} = V_{CC}/2$ and hence $R_L = V_{CC}/2I_{Cq}$.
3. For R_B, $V_{R_B} = V_{CEq} - V_{BE}$, and therefore $R_B = (V_{CEq} - V_{BE})/I_B$ where $I_B = I_{Cq}/\beta$. As in fixed biasing, β must be known.

Example 2.7

Using the configuration in Fig. 2.39, design a common emitter amplifier utilizing current feedback biasing. Use an npn silicon transistor having $\beta = 150$ and a 15 volt supply.

Solution Choose I_{Cq}1 mA. For maximum symmetrical swing, $V_{CEq} = V_{CC}/2 = 7.5$ volts. Hence, $R_L = 7.5/1$ mA $= 7.5$ k and $R_B = (7.5 - 0.7)/(1$ mA/150) $= 1.02$ MΩ. The capacitors are chosen to be large say 50 μF. The problem of reduced gain in this circuit can be overcome by splitting R_B and connecting a capacitor C_G of about 50 μF to ground as shown in Fig. 2.39. The effect of this capacitor is to remove the feedback action for signal frequencies.

Analysis of the collector-base feedback biased circuit is almost as straightforward as the other cases.

Example 2.8

Determine the collector current and voltage for the circuit shown in Fig. 2.40 assuming $\beta = 100$.

Solution Using Kirchhoff's voltage law, two equations for V_C can be written. They are $V_C = I_B \times 800$ k $+ 0.7$ and $V_C = 20 - I_C \times 2$ k $= 20 - \beta \times I_B \times 2$ k. Setting the right-hand side of these two equations equal and solving for I_B yield $I_B = 19.3$ μA, and therefore, $I_C = \beta I_B = 1.9$ mA. From this, $V_C = 20 - I_C \times 2$ k $= 20 - 1.9$ mA $\times 2$ k $= 16.2$ V.

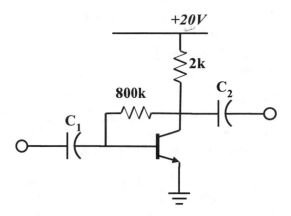

Fig. 2.40 Circuit for Example 2.8

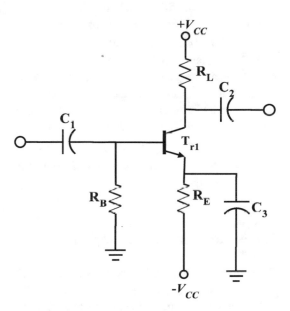

Fig. 2.41 Common emitter amplifier with bipolar supply

Biasing Using a Bipolar Supply It is possible to bias the common emitter amplifier using a bipolar power supply as shown in Fig. 2.41. Here the collector of the transistor goes to the positive supply $+V_{CC}$ through R_L as before. However, the emitter isn't returned to ground but instead goes to a negative supply $-V_{CC}$ through a resistor R_E which together set the collector current value. The resistor R_B which supplies base current to the transistor is connected to ground, thereby setting the base at an approximately zero potential.

Capacitors C_1 and C_2 are the usual input and output coupling capacitors, while C_3 is the decoupling capacitor that grounds the emitter for signals.

Fig. 2.42 Common base amplifier

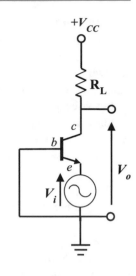

Example 2.9

Design a common emitter amplifier using the configuration in Fig. 2.41 with a $\pm 15V$ bipolar supply and a transistor with $\beta = 110$.

Solution Choose $I_{Cq} = 1$ mA for effective transistor operation. Noting that capacitor C_3 holds the emitter at a fixed voltage close to ground, for maximum symmetrical swing, $V_{CEq} = V_{CC}/2$. Hence, $V_{CEq} = 15/2 = 7.5$ V and $R_L = 7.5/1$ mA $= 7.5$ k. Since $\beta = 100$, and $I_B = I_C/\beta$, it follows that $I_{Bq} = \frac{1\,\text{mA}}{100} = 10\,\mu\text{A}$. Resistor R_B allows this bias current to flow into the base of the transistor. Its value should be of the order of 10 k. With the transistor base at approximately zero potential, Kirchhoff's voltage law applied to the emitter circuit yields $-0.7 - (-V_{CC}) = I_{Cq}R_E$. Therefore, $R_E = \frac{-0.7+15}{I_{Cq}} = \frac{15-0.7}{1\,\text{mA}} = 14.3$ k , that is, $R_E = 14.3$ k. The capacitors are chosen to be large say 50 μF.

2.5   Common Base Amplifier

The common base amplifier is shown in Fig. 2.42. As in the case of the common emitter amplifier, in order that the common base circuit operate properly, biasing is necessary to ensure that a complete signal cycle is reproduced. Therefore, a DC source V_{BB} is again used to turn on the transistor so that a quiescent collector current flows in the transistor before an input signal is applied. This arrangement is shown in Fig. 2.43. The signal input V_i then modulates this current such that a voltage change appears across resistor R_L. Similar to the common emitter circuit, this approach has the undesirable feature that DC current flows through the signal source. As it turns out, the various biasing schemes used in the common emitter configuration can all be utilized in the common base configuration with minimal adjustments. We will first consider voltage divider biasing.

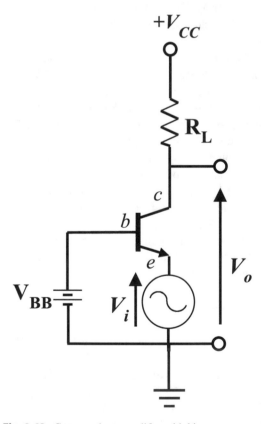

Fig. 2.43 Common base amplifier with bias

Voltage Divider Biasing
The voltage divider biasing technique applied to be common base configuration is shown in Fig. 2.44a. The components are the same as those used in the common emitter H-biased

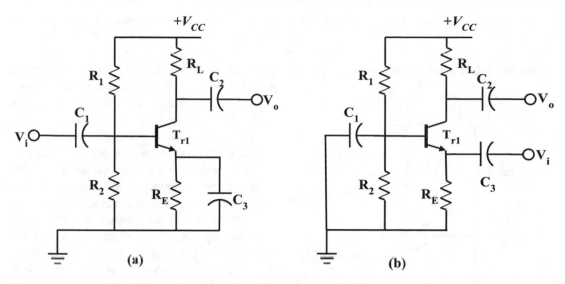

Fig. 2.44 Voltage divider-biased amplifiers. (**a**) Common emitter; (**b**) common base

amplifier. Note, however, that whereas in the common emitter amplifier C_E is grounded and the input signal applied via C_1 with the output at C_2, in the common base amplifier, C_1 is grounded and the input signal is applied at C_E. In fact, a common emitter voltage divider-biased amplifier may be converted to a common base voltage divider-biased amplifier by simply grounding the input capacitor C_1, ungrounding C_3 and applying the input signal via this capacitor as shown in Fig. 2.44b.

Example 2.10

Convert the common emitter amplifier of Example 2.5 shown in Fig. 2.45 to a common base amplifier.

Solution By grounding the input terminal at input capacitor C_1 and ungrounding decoupling capacitor C_2 and applying the input signal at this ungrounded terminal, the common emitter amplifier of Fig. 2.45 has been converted to a common base amplifier in Fig. 2.46.

It follows therefore that the design procedure developed for the voltage divider-biased common emitter amplifier is directly applicable to the voltage divider-biased common base amplifier.

Example 2.11

Design a voltage divider-biased common base amplifier using a 20 volt supply and an npn silicon transistor having $\beta = 125$.

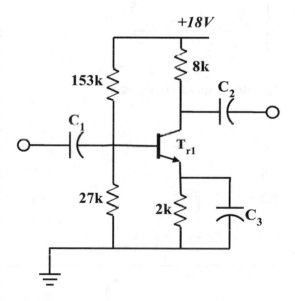

Fig. 2.45 Circuit for Example 2.10

Solution Using the circuit in Fig. 2.44b and following the procedure outlined for the common emitter case, choose $I_{Cq} = 1$ mA. Then $V_{R_E} = 20/10 = 2$ V. Hence $R_E = 2/1 = 2$ k. For maximum symmetrical swing, $V_{CEq} = V_{R_L} = (20 - 2)/2 = 9$ V. Therefore, $R_L = 9/1$ mA $= 9$ k. Now $V_{Bq} = V_{R_E} + V_{BEq} = 2.7$ V and $I_R = I_{Cq}/10 = 0.1$ mA. Hence, $R_2 = 2.7/0.1$ mA $= 27$ k and $R_1 = (20 - 2.7)/0.1$ mA $= 173$ k. Choose C_1, C_2 and C_3 large, say 100 μF.

The results of this circuit under test are shown in Fig. 2.47, where waveforms of the input and output voltage are shown. Before discussing the operation of the circuit, two points relating to the

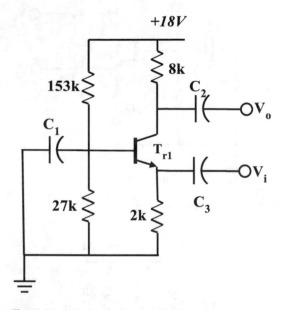

Fig. 2.46 Solution for Example 2.10

biasing technique must be made. Firstly, the emitter voltage, unlike the common emitter amplifier, is not fixed; it varies according to V_i. However, this change is generally small (of the order of hundreds of mV) as compared to the changes in the collector voltage, and therefore maximum symmetrical swing calculations, which assume fixed V_E, are still approximately correct. Secondly, in the common emitter design, in order to prevent loading down the input signal, R_1 and R_2 could not be too low. In this configuration, this restriction is removed since there is no signal input at the base of the transistor so that I_R may be made larger than $I_C/10$ in order to improve the bias stabilizing action of the circuit. However, note that an increased I_R represents an increased demand on the circuit power supply.

Example 2.12
Improve the bias stability in the common base amplifier in Example 2.11.

Solution For this circuit, $V_{Bq} = V_{R_E} + V_{BEq} = 2.7V$. Since there is no signal input at the

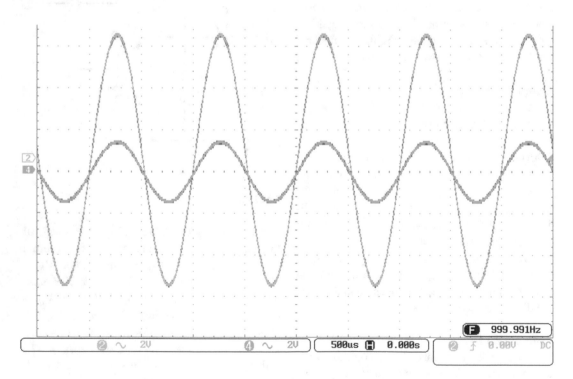

Fig. 2.47 Input and output signals for common base amplifier

transistor base, a higher current I_R through R_1 and R_2 is permissible since the associated lower values of R_1 and R_2 can be accommodated. Hence choose $I_R = I_{Cq}/2 = 0.5$ mA. This value is five times higher than the value used before but is not so high as to increase the current drain on the system power supply unduly. Hence $R_2 = 2.7/0.5$ mA $= 5.4$ k and $R_1 = (20 - 2.7)/0.5$ mA $= 34.6$ k.

Regarding the operation of the circuit, note that the capacitor C_1 grounds the base of the transistor for signals so that the input signal appears directly across the emitter-base junction. As V_i goes positive, the emitter-base voltage of the BJT is reduced, and this results in a reduction of the transistor collector current. The voltage drop across R_L is therefore reduced resulting in an increase in the collector voltage of the BJT, i.e. V_o goes positive. As V_i goes negative, the emitter-base voltage of the BJT is increased, and this results in an increase of I_C. The voltage across R_L increases, and hence, the collector voltage of the BJT goes down, i.e. V_o goes negative. The overall result is that V_i and V_o are in phase and there is no phase inversion of the voltage as occurs in the common emitter amplifier.

Let us now evaluate the voltage gain A_V, which using Eq. (2.30), is given by

$$A_V = \frac{V_o}{V_i} = \frac{\Delta I_C R_L}{\Delta V_{BE}} = 40 I_C R_L = g_m R_L \quad (2.50)$$

where $g_m = 40 I_C$. Thus the voltage gain of the common base circuit is the same as that of the common emitter circuit, given in Eq. (2.31) except there is no inversion. The input impedance Z_i at the emitter of the common base amplifier is the ratio of the input voltage change (ΔV_{EB}) to the corresponding input current change (ΔI_E) flowing into the emitter. It is given by

$$Z_i = \frac{\Delta V_{EB}}{\Delta I_E} \quad (2.51)$$

Noting that $\Delta I_E \approx \Delta I_C$ and from Eq. (2.30),

$$Z_i = \frac{\Delta V_{EB}}{\Delta I_E} = \frac{1}{g_m} = \frac{1}{40 I_C} \quad (2.52)$$

Thus, the input impedance of the common base circuit is a factor β lower than that of the common emitter configuration given in (2.32).

Example 2.13
For the circuit of Example 2.11, find the voltage gain and the input impedance.

Solution $A_V = +40 I_C R_L = 40 \times 1 \times 9$ k $= 360$ and $Z_i = 1/40 I_C = 1/40 \times 1 \times 10^{-3} = 25\ \Omega$.

As is evident, the input impedance of the common base configuration is quite low.

Constant Current or Fixed Bias
The common base amplifier using constant current or fixed bias is shown in Fig. 2.48. This circuit is slightly different from the common emitter fixed-bias circuit because of the presence of R_E which improves the bias stability of the common base circuit. An example illustrating the design procedure follows.

Example 2.14
Design a common base amplifier using fixed bias and a silicon npn transistor having $\beta = 100$. Use $V_{CC} = 22 volts$.

Solution Following the procedure outlined for the common emitter case, choose $I_{Cq} = 1$ mA.

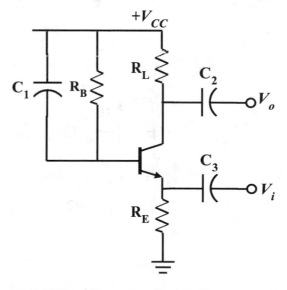

Fig. 2.48 Fixed-bias common base amplifier

Then $V_{RE} = 22/10 = 2.2$ volts. Use $V_{RE} = 2$ volts. Therefore $R_E = 2/1 = 2$ k. For maximum symmetrical swing, $V_{CEq} = V_{R_L} = (22 - 2)/2 = 10\,\text{V}$. Therefore $R_L = 10/1$ mA $= 10$ k. Now $V_{CC} = I_B R_B + V_{R_E} + V_{BEq}$ giving $I_C R_B / \beta = V_{CC} - V_{BEq} - V_{R_E} = 22 - 0.7 - 2 = 19.3$ volts. Hence $R_B = 19.3 \times 100/1$ mA $= 1.93$ M. Choose C_1, C_3 and C_2 large, say 100 µF.

Biasing Using a Bipolar Supply

Once again it is possible to convert the common emitter amplifier using bipolar supplies shown in Fig. 2.41 to a common base configuration by grounding the input capacitor C_1 and applying the input signal to an ungrounded capacitor C_3 as shown in Fig. 2.49. The output is taken from capacitor C_2. In this case both base resistor R_B and capacitor C_1 can be completely removed and the transistor base terminal directly grounded.

Example 2.15

Using the configuration shown in Fig. 2.49, design a common base amplifier to operate from a ±18 V supply. Determine (i) the voltage gain and (ii) input impedance of the circuit.

Solution Let the collector current be 2 mA. Since the transistor base is at ground potential, it follows that the voltage at the transistor emitter is -0.7 V. Hence $R_E = (-0.7 - (-18))/2$ mA $= 8.7$ k. The

voltage swing available at the collector of the transistor is approximately 18 volts since the emitter voltage experiences only small variations due to the signal. Therefore, for maximum symmetrical swing, $R_L = \frac{18/2}{2\,\text{mA}} = 4.5\,\text{k}$. Capacitors C_2 and C_3 are chosen to be large for good low-frequency response. Voltage gain is given by $A_V = +40 I_C R_L = 40 \times 2$ mA $\times 4.5$ k $= 360$. Input impedance is given by $Z_i = 1/40 I_C = 1/40 \times 2$ mA $= 12.5\ \Omega$.

2.6 Common Collector Amplifier

The common collector amplifier is shown in Fig. 2.50. This is the third configuration in which the BJT can be used, and this too needs to be biased in order that a complete signal cycle can be reproduced. Once again, a battery V_{BB} is used to establish a quiescent collector current in the transistor before an input signal is applied. This biasing arrangement is shown Fig. 2.51.

The signal input V_i then causes changes in this current such that a voltage change appears across resistor R_E. Similar to the common emitter and common base circuits, this approach allows DC current to flow through the signal source. Some of the biasing techniques employed in the common emitter and common base designs are applicable here but with some modifications.

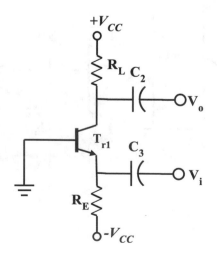

Fig. 2.49 Common base amplifier with bipolar supply

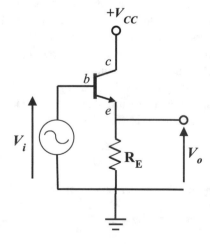

Fig. 2.50 Common collector amplifier

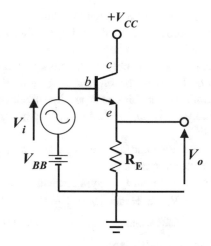

Fig. 2.51 Common collector amplifier with bias

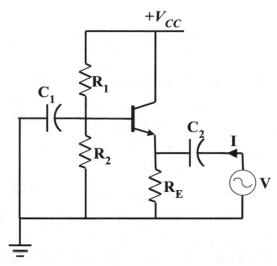

Fig. 2.53 Output impedance of common collector amplifier

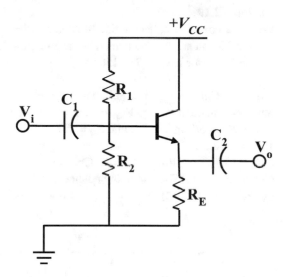

Fig. 2.52 Voltage divider-biased common collector amplifier

Voltage Divider Biasing

Voltage divider biasing applied to the common collector configuration is shown in Fig. 2.52. R_1 and R_2 fix the base voltage, and R_E provides current stabilizing action (feedback). Note that there is no collector resistor; the transistor collector is connected directly to the supply. It is therefore effectively grounded for signals through the DC supply. Signal input is applied at the base and output is taken at the emitter. A simply design procedure is outlined.

1. Choose V_{CC} and I_{Cq}. V_{CC} may depend on the available supply. I_{Cq} is influenced by the load

that this circuit must drive and is usually in the range 1–10 mA.
2. For maximum symmetrical swing, $V_{Eq} = V_{CC}/2$ giving $R_E = V_{CC}/2I_{Cq}$.
3. Choose $I_R = I_C/10$ in order to minimize changes in I_C due to changes in β. Then, $R_2 = V_{Bq}/I_R$ and $R_1 = (V_{CC} - V_{Bq})/I_R$ where $V_{Bq} = V_{CC}/2 + V_{BE}$
4. Choose C_1 and C_2 large.

Example 2.16

Design a common collector voltage divider-biased amplifier using an 18 volt power supply and an npn silicon transistor with $\beta = 100$.

Solution   Choose $I_{Cq} = 2$ mA and $V_{RE} = 18/2 = 9$ volts. Then $R_E = 9/2$ mA $= 4.5$ k. Choose $I_R = 2/10 = 0.2$ mA. Then $R_2 = 0.7/0.2$ mA $= 48.5$ k and $R_1 = (18 - 9.7)/0.2$ mA $= 41.5$ k. Choose C_1 and C_2 large, say 100 µF.

   In this circuit, V_o is in phase with V_i. In fact, since V_{BE} is approximately fixed, this results in $V_o \approx V_i$. As a result, this circuit is sometimes referred to as an emitter follower since the emitter follows the input. From circuit theory, the output impedance Z_o of this circuit can be found by short-circuiting the input and applying a voltage V and measuring the corresponding current I at the output as shown in Fig. 2.53.

Hence

$$Z_o = \frac{V}{I}\bigg|_{V_i=o} = \frac{\Delta V_{EB}}{\Delta I_E} \qquad (2.53)$$

Since $\Delta I_E \approx \Delta I_C$, this reduces to

$$Z_o = \frac{\Delta V_{EB}}{\Delta I_C} \qquad (2.54)$$

From Eq. (2.30), $\frac{\Delta I_C}{\Delta V_{BE}} = g_m$. Therefore,

$$Z_o = \frac{\Delta V_{EB}}{\Delta I_C} = \frac{1}{g_m} \qquad (2.55)$$

Example 2.17

Find the output impedance of a common collector amplifier in which the collector current is 1 mA.

Solution For $I_C = 1$ mA, $Z_o = 1/40I_C = 25\ \Omega$.

Thus, the output impedance of the common collector amplifier with low source impedance at the input is quite low. The input impedance is given by $Z_i = V_i/I_i$, where I_i is the input current. Now ignoring for the moment the two bias resistors R_1 and R_2,

$$V_i = \Delta V_{BE} + I_i(1+\beta)R_E \qquad (2.56)$$

$$Z_i = \frac{V_i}{I_i} = (1+\beta)R_E + \frac{\Delta V_{BE}}{I_i} \qquad (2.57)$$

But from (2.32), $\Delta V_{BE}/I_B = \beta/g_m = \beta/40I_C$. Hence

$$Z_i = \beta/40I_C + (1+\beta)R_E \qquad (2.58)$$

This input impedance is high because of the second term is quite high. The value is however reduced by resistors R_1 and R_2 which are effectively in parallel with Z_i.

Example 2.18

Determine the input impedance for the common collector circuit in Example 2.16.

Solution For that circuit, $Z_i = \beta/40I_C + (1+\beta)$ $R_E = 100/40 \times 2 \times 10^{-3} + (1+100) \times 4.5 \times 10^3$. This reduces to $Z_i = (1.25 + 454.5) \times 10^3 = 455.8$ k. This value is in parallel with resistors $R_1 = 41.5$ k and $R_2 = 48.5$ k and hence is reduced.

Constant Current or Fixed Biasing

Constant current or fixed biasing applied to the common collector configuration is shown in Fig. 2.54 where, as before, R_B supplies the base current. However, the presence of R_E improves the stability of the bias as it provides feedback to the emitter. A simple design procedure is the following:

1. Choose V_{CC} and I_{Cq}, the latter depending on load requirements.
2. For maximum symmetrical swing, $V_{CEq} = V_{CC}/2$. Hence $R_E = \frac{V_{CC}}{2I_{Cq}}$
3. $R_B = \frac{V_{CC}-V_{BEq}}{I_C/\beta}$ where $V_{BEq} = \frac{1}{2}V_{CC} + 0.7$.
4. Choose C_1 and C_2 large.

Example 2.19

Design a common collector fixed-bias amplifier using a 12 volt supply and a general purpose npn silicon transistor having $\beta = 150$.

Solution Choose $I_{Cq} = 2$ mA. Then for maximum symmetrical swing, $V_{Eq} = 12/2 = 6$ volts. Hence $R_E = 6/2$ mA $= 3$ k and $R_B = 6.7/(2$ mA$/150) = 502.5$ k.

$Z_o = 1/g_m = 1/40I_C = 12.5\ \Omega$. Choose C_1 and C_2 equal 47 μF. The input impedance is given by $Z_i = (1+150)3\text{k} + \frac{150}{40 \times 2 \times 10^{-3}} = 453\text{k}$. The

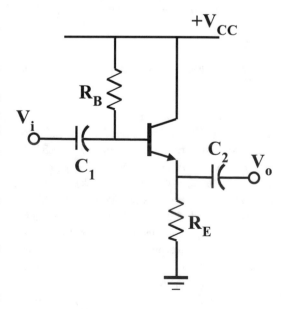

Fig. 2.54 Fixed-bias common collector amplifier

actual input impedance is obtained by placing Z_i in parallel with R_B giving 453 k//502 k.

The common collector amplifier with its high input impedance and low output impedance is excellent for serving as a buffer where a high-impedance source must drive a low-impedance load. When a load R_L is connected however, the peak-to-peak output voltage swing of the amplifier is reduced on the negative half-cycle for an npn (and the positive half-cycle for a pnp transistor). For the circuit shown in Fig. 2.55, on the positive half-cycle, the transistor supplies current to the load, and the peak output is limited only by the power supply rail. On the negative half-cycle however, the transistor must sink and not source current and the current flowing in from R_L and must be sunk by R_E. As this occurs, the transistor must progressively turn off since this current is now being supplied by R_L. Eventually when the transistor is completely off, the maximum current flows in from R_L. To determine the maximum value of the peak negative swing at the output at this time, we first note that the DC voltage V_C across the output coupling capacitor C_2 is $V_C = I_{Cq}R_E$ since no DC flows in R_L. Therefore when the transistor is cut-off on the negative half-cycle, the capacitor voltage is dropped across R_E and the load R_L, and this results in a load current I_L out of the load R_L and into R_E of

$$I_L = \frac{V_C}{R_E + R_L} = \frac{I_{Cq}R_E}{R_E + R_L} \quad (2.59)$$

The peak negative output voltage is hence

$$V_O^- = I_L R_L = I_{Cq}\frac{R_E R_L}{R_E + R_L} \quad (2.60)$$

which can be written as

$$V_O^- = I_{Cq}R_E\|R_L \quad (2.61)$$

If we use $V_{CEq} = V_{CC}/2$, then

$$I_{Cq} = \frac{V_{CC}/2}{R_E} \quad (2.62)$$

giving $V_o = \frac{V_{CC}}{2} \cdot \frac{R_L}{R_E + R_L}$. In order to maximize V_o, R_E should be made as small as is practicable so that $R_L/(R_E + R_L) \to 1$. Note however from (2.62) that this increases the quiescent current in the transistor. A better approach is to replace R_E by a constant current source. This will be discussed in Chap. 5.

Biasing Using a Bipolar Supply

Biasing of the common collector amplifier using a bipolar supply is shown in Fig. 2.56. Here the emitter is connected to the negative supply through resistor R_E, while the base bias is supplied via resistor R_B which is connected to ground. C_1 and C_2 are the usual input and output coupling capacitors. Note that the use of a bipolar supply in this configuration with the transistor base held at an approximately zero potential

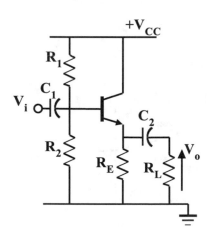

Fig. 2.55 Common collector amplifier with load

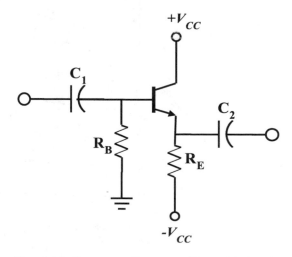

Fig. 2.56 Common collector amplifier with bipolar supply

ensures that the output at the emitter is biased for maximum symmetrical swing and permits a greater output voltage swing since the emitter can approach the positive and negative supply voltages.

Example 2.20
Design a common collector amplifier using a bipolar supply ± 12 V and a general purpose npn silicon transistor having $\beta = 150$.

Solution Choose $I_{Cq} = 5$ mA in order to drive external loads. Then $R_E = (-0.7 - (-12))/$ 5 mA $= 2.3$ k. Choose $R_B = 20$ k.

2.7   Transistor Operating Limits and Specifications

Each transistor has a region of operation in which it can function safely. This region is called the *safe operating area* (SOA) and is illustrated in the (CE) output characteristic of Fig. 2.57. The maximum collector current I_{Cmax} that a transistor can carry is set by the current carrying capacity of the fine wire used to connect the semiconductor regions to the terminal loads. Excessive current will melt these wires resulting in an open circuit at one or more of the device terminals. It is usually referred to as continuous collector current. The maximum collector-emitter voltage V_{CEO} is the maximum voltage that can be applied between the collector and emitter terminals of the transistor when the base terminal is open circuited. It is governed by the process of avalanche breakdown. In fact, beyond V_{CEO} of Fig. 2.57 is a region referred to as the *breakdown region*, which represents catastrophic failure of the transistor. The vertical line on the characteristic corresponds to V_{CEsat}, which is the minimum allowed collector-emitter voltage of the BJT if it is not to enter the non-linear *saturation region*. There is also the parameter V_{CBO} which is the maximum voltage that can be applied between the collector and base terminals when the emitter is open circuited.

The approximate horizontal line near the V_{CE} axis corresponds to I_{CEO}, which is the leakage current. The region beneath this line is the *cut-off* region that must also be avoided during (linear) amplifier operation. The maximum power dissipation P_{Dmax} in the transistor is given by

$$P_{D\,max} = V_{CE}I_C \qquad (2.63)$$

This is the equation of a hyperbola that completes the boundary of the SOA. It is especially

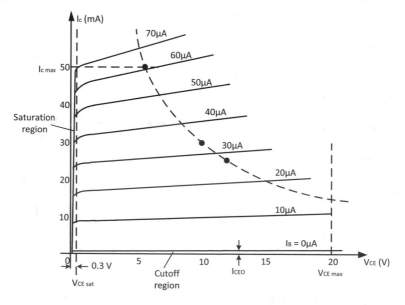

Fig. 2.57 Transistor safe operating area

important in power transistors where considerable power is dissipated during operation. Any rise in temperature of the device results in a movement of the hyperbola towards the origin. The effect is a reduction of the SOA with temperature increase. Appropriate strategies to mitigate these effects must be introduced in utilizing power transistors, the main strategy being the mounting of the power transistor on a heatsink. This is fully discussed in Chap. 10 on power amplifiers. Finally, the parameter f_T (sometimes referred to as the gain/bandwidth product) is the frequency at which the current gain h_{fe} of the transistor falls to unity. It gives an indication of the transistor performance at high frequencies.

2.8   Applications

In this section, several simple circuit implementations using the BJT are discussed. These are intended to enable the design of actual working circuits using BJTs.

Microphone Preamplifier 1

A preamplifier is an amplifier that increases the amplitude of a low-level signal such that it can drive a power amplifier. The circuit in Fig. 2.58 provides amplification for a dynamic microphone using a 9 V battery. Such microphones operate on the principle of electromagnetic induction and provide a nominal output of the order of 2 mV at a low output impedance. The transistor

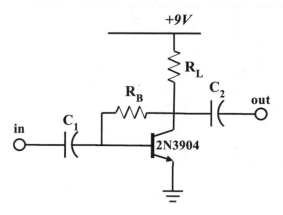

Fig. 2.58 Microphone preamplifier 1

employs collector-base feedback biasing and a collector current of 2 mA. Using the design procedure for collector-base feedback biasing, the value of R_L for maximum symmetrical swing is given by $R_L = (9/2)/2$ mA $= 2.25$ k. Using a current gain of 100 for the 2N3904 transistor, the feedback resistor R_B is given by $R_B = (4.5 - 0.7)/\frac{2\,\mathrm{mA}}{100} = 190$ k. The coupling capacitors C_1 and C_2 are about 10 μF. The circuit provides a gain of better than 100 which when converted to decibels is $20 \log 100 = 40$ dB. This means the output from the amplifier is greater than 200 mV, which is sufficient to drive a power amplifier. An interesting point here is that the low output impedance of the microphone reduces the effect of the feedback via R_B on the signal and hence allows for a higher gain without the need for splitting R_B and connecting a capacitor to ground.

Ideas for Exploration: (i) Use a loudspeaker with an impedance of 4–16 ohms to replace the dynamic microphone at the input. This device like the dynamic microphone operates on the principle of electromagnetic induction. While the microphone converts sound energy to an electrical signal, the loudspeaker converts an electrical signal to sound. However, its role is here reversed such that it is used as a microphone where it converts sound to an electrical signal which is connected to the input of the preamplifier. (ii) Use an electret microphone to replace the dynamic microphone. An electret or condenser microphone requires external power being fed through a resistor for its operation as shown in Fig. 2.59. The output signal is coupled to the input of the preamplifier. Because of the voltage at the output of this microphone, a non-polarized coupling capacitor should be used at the input of the preamplifier.

Microphone Preamplifier 2

Another dynamic microphone preamplifier is shown in Fig. 2.60. It comprises a common emitter amplifier using a small-signal transistor. Using the design technique developed in this chapter, resistors $R_1 = 73$ k and $R_2 = 17$ k set the base voltage of Tr1 to about 1.7 volts. This along with $R_3 = 1$ k establishes a quiescent current of 1 mA.

Biasing Tr1 for maximum symmetrical swing requires that resistor $R_4 = 4$ k. The capacitor values are large, $C_1 = C_2 = 10$ µF, $C_3 = 100$ µF, to allow adequate low-frequency response.

Ideas for Exploration: (i) Use a loudspeaker with an impedance of 4–16 ohms to replace the dynamic microphone. (ii) Use an electret microphone to replace the dynamic microphone. Again, because of the voltage at the output of this microphone, a non-polarized coupling capacitor should be used at the input of the preamplifier.

Speaker Protection Circuit

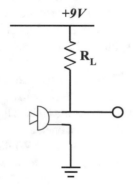

Fig. 2.59 Electret microphone

The circuit in Fig. 2.61 protects speakers at turn-on from the DC that may exist at the output of a power amplifier. It consists of a small-signal npn transistor such as the 2N3904 connected as an emitter follower. The input to the emitter follower is driven by the voltage across a capacitor C_2 whose potential increases at turn-on because of charge current flowing in through resistor VR_1. The output at the emitter of the transistor drives a relay coil the contacts of which connect the speaker to the output of the amplifier. The supply voltage for the circuit is derived from the supply voltage of the amplifier. Upon turn-on of the amplifier, capacitor C_2 charges up via potentiometer VR_1. Depending on the adjusted value of VR_1, the time for the capacitor to charge is sufficient to allow the output of the amplifier to stabilize. Values of $C_2 = 100$ µF and $VR_1 = 50$ kΩ will provide a charge time of the order of $VR_1 \times C_2 = 50 \times 10^3 \times 100 \times 10^{-6} = 5s$. When the capacitor voltage attains a value of about 9 volts, the relay is activated, and the contacts connect the speaker to the amplifier output. The extent of the delay can be adjusted by varying the potentiometer. Diode D_1 which can be an *IN*4001 diode protects the transistor from back emfs generated by the relay coil, while capacitor C_1 provides nominal filtering to the

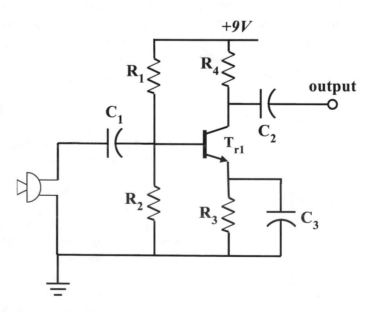

Fig. 2.60 Microphone preamplifier 2

Fig. 2.61 Speaker
protection circuit

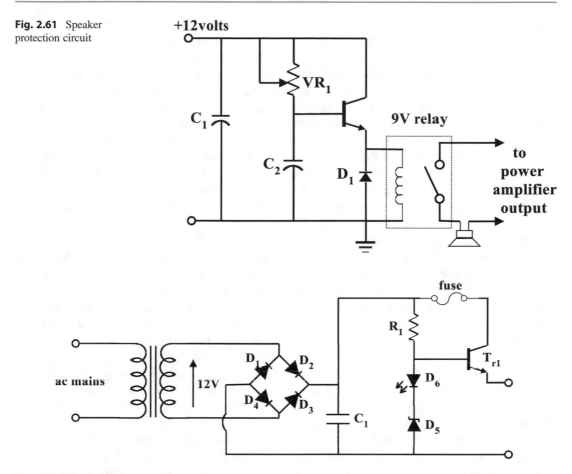

Fig. 2.62 Simple bench power supply

system supply voltage. A value of about 0.1 μF is suitable.

Ideas for Exploration: (i) Re-design the system using a 2N3906 pnp small-signal transistor. This would require reversal of the polarity of the power supply. (ii) Investigate the action of the diode in protecting the transistor.

Bench Power Supply

Figure 2.62 provides a fixed output of about 9 volts and a maximum current of about 250 mA. It uses a Zener diode regulator and boosts the current output with a transistor in emitter follower configuration. In order to ensure that sufficient voltage is available to deliver current to the associated Zener, a basic full-wave rectifier using a 12 volt transformer, a diode bridge comprising D_1 to D_4 and a filter

capacitor $C_1 = 1000$ μF provides the unregulated input. The peak voltage on the filter capacitor is $12\sqrt{2} - 1.4 = 15.6$ V. The maximum peak-to-peak ripple across the capacitor is given by $\Delta V = I/2fC = 100 \times 10^{-3}/ 2 \times 60 \times 1000 \times 10^{-6} = 0.8$ V giving a minimum unregulated voltage $15.6 - 0.8 = 14.8$ V. Using an 8.2 V Zener and a red LED as an on-indicator with a forward voltage of 1.8 V, the total voltage across the series arrangement is $8.2 + 1.8 = 10$ V. The voltage output at the emitter of the transistor would be 10 V less the 0.7 V of the transistor base-emitter junction giving about 9.3 V. For a Zener operating current of 10 mA, resistor R_1 is given by $(14.8 - 10)/ 10$ mA $= 480$ Ω. The 2N3055 power transistor boosts the current output capacity of the Zener diode by the factor of the current gain of the transistor which, for a 2N3055, is about 50. The maximum current available is calculated using the minimum allowable current for the Zener which is about 5 mA. Hence the available current from the Zener regulator is $10 - 5 = 5$ mA, and therefore the maximum output current is 5 mA $\times \beta = 5$ mA $\times 50 = 250$ mA where $\beta = 50$ is the current gain of the transistor. The fuse provides short-circuit protection.

Ideas for Exploration: (i) Modify the system to produce a variable bench supply by introducing a 10 k potentiometer across the series LED-Zener arrangement and connecting the wiper of the potentiometer to the transistor base. Identify any problems with this arrangement. (ii) Use a transistor with a higher gain and explain if this provides any performance improvement. (iii) Utilize the TL431A programmable Zener introduced in Chap. 1 to design the supply instead of the Zener diode.

Electronic Fuse

This circuit adds electronic over-current protection to the bench supply just discussed. It consists of a small-signal npn silicon transistor Tr_2 with a resistor R_2 between the base and emitter of the transistor. The arrangement is connected to the supply as shown in Fig. 2.63. As increased load current flows through R_2, the voltage drop across this resistor increases. When this value equals 0.7 V, transistor Tr_2 turns on drawing current away from the base of transistor Tr_1, thereby causing the supply current to remain

approximately constant. Any further small increase in the load current further turns on Tr_2, and this reduces the current to the Zener. The effect is that the Zener voltage falls and a constant voltage supply becomes a constant current supply. Then, $R_2 = 0.7/I_{Lmx}$ where I_{Lmx} is the maximum load current.

Ideas for Exploration: (i) Plot the voltage-current characteristic for the modified bench supply circuit to show the effect of the electronic fuse. (ii) Introduce a small capacitor across the output of the supply to improve filtering.

Blown Fuse Indicator

This circuit gives a visual indication when a protection fuse is blown. It is placed between the power supply and the circuit being protected and comprises a pnp transistor arranged as shown in Fig. 2.64. During normal operation with a continuous fuse, the voltage across the base-emitter junction of the transistor is approximately zero. The transistor is therefore off and the LED does not light. If the fuse is blown because of an over-current, then voltage drop across the base-emitter junction causes Tr1 to turn on. Current thereby flows through the LED which now turns on. When Tr1 is on, the full supply voltage is dropped across the LED and resistor. Thus, for a 24 volt supply and allowing an (red) LED voltage of 1.8 V and a current of 10 mA, $R_3 = (24 - 1.8)/10$ mA $= 2.2$ kΩ. Assuming a transistor current gain of 100, the base current of Tr1 is given by 10 mA/100 $= 0.1$ mA, and this flows through R_1

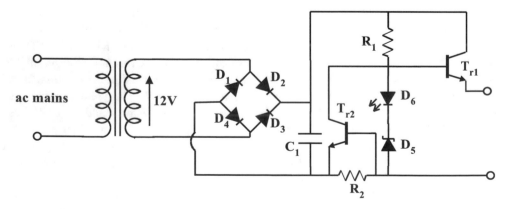

Fig. 2.63 Electronic fuse

Fig. 2.64 Blown fuse
indicator

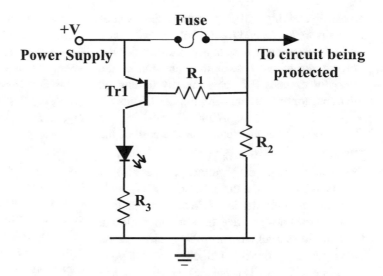

Fig. 2.65 Telephone
monitor

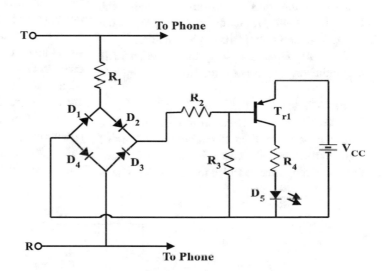

and R_2. Hence, $R_1 + R_2 = (24 - 0.7)/$
0.1 mA = 233 kΩ. To ensure transistor saturation, use 200 k. Then noting that R_2 is a load on
the supply, set $R_1 = R_2 = 100$ k.

Ideas for Exploration: (i) Introduce a green
LED in series with resistor R_2. During normal
operation there would be sufficient current
through this resistor to turn this LED on but it
goes off when the fuse is blown. This would
complement the action of the red LED.

Telephone Monitor
This circuit shown in Fig. 2.65 turns on an LED
when the telephone handset is lifted indicating

that the telephone is in use. The diode bridge
ensures that the DC signal into the transistor
from the telephone line is always of the same
polarity. When the telephone is on hook, the
telephone circuit is open, and 48 volts available
from the telephone exchange will be across tip
(T) and ring (R). At this time since the emitter of
the transistor is at V_{CC}, the resistor values R_1, R_2
and R_3 must be chosen such that the voltage at the
base of the transistor is greater than $V_{CC} - 0.7$
volts to ensure that the transistor is off. The LED
would therefore not be illuminated. When the
handset is lifted, the telephone circuit is closed,
and as a result of line resistance, the voltage

across tip and ring will drop. With suitable values for resistors R_1, R_2 and R_3, the voltage at the base of the transistor will drop to a value such that the voltage across the base-emitter junction of the transistor exceeds 0.7 volts, thereby turning on the transistor and illuminating the LED. Thus for $V_{CC} = 3$ volts, $R_1 = 470$ k, $R_2 = 1$ M and $R_3 = 100$ k, then the voltage at the base of the transistor is given by $\frac{R_3}{R_1+R_2+R_3} \times 48 = 3$ volts. At this voltage with the emitter voltage also at 3 volts, it follows that the base-emitter voltage is zero and the transistor is off. When the phone is lifted off the hook, the telephone circuit is closed. Current flows such that a voltage drop occurs along the line, thereby reducing the 48 volts across tip and ring. This reduces the base voltage and causes the transistor to turn on lighting the LED. Resistor R_4 limits the current through the LED. A value of about 100 Ω is suitable.

Ideas for Exploration: (i) Introduce an LED that indicates the on-hook condition.

Audio Mixer

This circuit for an audio mixer in Fig. 2.66 uses a common base amplifier to enable the mixing of signals from several sources. It consists of a common base amplifier in which the transistor base is held at a fixed voltage 1.4 V using two forward-biased diodes in series instead of a voltage divider. This arrangement provides a more stable voltage at the base. Current through these diodes is supplied from a 15 volt supply through a resistor $R_5 = 3.3$ k. Hence the current through diodes D_1 and D_2 is $(15 - 1.4)/3.3$ k $= 4$ mA. Capacitor C_5 ensures that the transistor base is grounded for signals. Choosing a collector current of 1 mA, then $R_6 = (1.4 - 0.7)/1$ mA $= 700$ Ω. For maximum symmetrical swing, the quiescent voltage across R_4 is made equal to the quiescent collector-emitter voltage giving $V_{CE} = V_{R_4} = (15 - 0.7)/2 = 7.2$ V . Hence $R_4 = 7.2/1$ mA $= 7.2$ k. Input and output coupling capacitors C_1 to C_4 are chosen to be large values for good low-frequency response. The input impedance at the emitter is $1/40I_C = 1/40 \times 1$ mA $= 25$ Ω which is quite low. By choosing signal input resistors R_1 to R_3 to be much greater than 25 Ω say 22 k, very good signal mixing occurs at the transistor emitter with very low channel interaction (cross-talk).

Ideas for Exploration: (i) Introduce 10 k potentiometers in each input circuit in order to adjust individual channel gain. (ii) Sacrifice maximum symmetrical swing for increased output

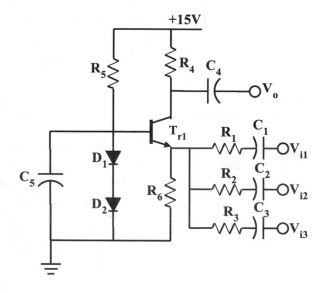

Fig. 2.66 Audio mixer

signal amplitude by increasing the value of resistor R_3. (iii) Re-design the circuit using a red or green LED instead of the two signal diodes.

Power Zener Diode

The power rating of a Zener diode can be increased by coupling the Zener with a power transistor as shown in Fig. 2.67. Resistor R_1 sets the current through the Zener at $0.7/R_1$. For 1 mA we get $R_1 = 700\ \Omega$. The effective Zener voltage V_Z' across the arrangement is $V_Z' = V_Z + 0.7$, which for a 4.3 V Zener gives $V_Z' = 4.3 + 0.7 = 5$ V.

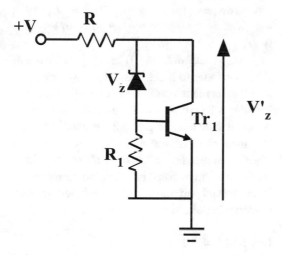

Fig. 2.67 Power Zener diode

While the current through the Zener is 1 mA, the power transistor can allow significantly higher currents, thereby increasing the power rating of the arrangement beyond the power rating of the Zener. For example, if the transistor current is set at 1A, then the power rating of the arrangement becomes approximately $P_D' = 5$ V $\times$ 1 A $= 5$ W. Note that the base current of the power transistor flows through the Zener in addition to the 1 mA supplied to the resistor, since this latter current is set by the base-emitter voltage of the transistor.

Ideas for Exploration: (i) Use this power Zener to design a high-current Zener diode regulator. (ii) Compare this regulator with that providing the bench power supply. Zener regulators are referred to as shunt regulators as the regulating device (Zener) is in parallel with the load, while the earlier regulator in the bench supply is a series regulator since the regulating element (transistor) is in series with the load. (iii) Implement this circuit using a pnp transistor.

Adjustable Zener Diode

The voltage specification of a Zener diode can be increased by coupling the Zener with a transistor as shown in Fig. 2.68. The voltage across resistor R_1 is $V_z + 0.7$, and therefore the current I through resistor R_1 is given by $I = (V_z + 0.7)/R_1$. Assuming a high-gain transistor, the base current will be

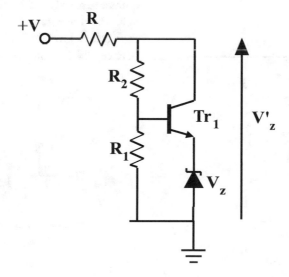

Fig. 2.68 Adjustable Zener diode

small and hence most of I flows through R_2. Therefore, the effective Zener voltage V_Z' across the arrangement is $V_Z' = \left(1 + \frac{R_2}{R_1}\right)(V_Z + 0.7)$. Thus using $R_1 = R_2 = 6.8$ k and a 6.8 V Zener gives $V_Z' = (1 + 1)(6.8 + 0.7) = 15$ V. The resulting current through R_1 and R_2 is $I = 7.5/6.8$ k $= 1.1$ mA. The power rating of the adjusted diode will then be a function of the power rating of the associated transistor. Note that if the Zener is replaced by a short circuit, the voltage output becomes $V_Z' = \left(1 + \frac{R_2}{R_1}\right)0.7$. This particular circuit arrangement is sometimes referred to as a V_{be} multiplier or *amplified diode* (see Chap. 5).

Ideas for Exploration: (i) Investigate the use of this adjustable Zener in the design of a variable voltage regulated supply.

Diode Tester

A transistor can be used in the testing of diodes as shown in Fig. 2.69. The diode to be tested is connected in the base biasing circuit. An LED along with a current limiting resistor is connected in the collector circuit. When a good diode is connected to the test terminals with the correct polarity, conduction through the diode and R_1 occurs, and a fraction of the supply voltage is dropped across resistor R_2. This turns on Tr_1 and hence the LED. If the connection is reversed, the diode becomes reverse-biased, and no conduction

occurs. As a result, no bias voltage is dropped across R_2, and hence the transistor remains off. The LED D_1 does not light. An open diode will cause no conduction with either connection and hence no lighting of the LED. A shorted diode will cause conduction and hence LED lighting with both connections. The values of R_1 and R_2 must be chosen such that the transistor is turned on when diode conduction takes place. $R_1 = 10$ k and $R_2 = 2.2$ k will give a voltage at the transistor base of $2.2 \times 9/(10 + 2.2) = 1.6$ V. This value is more than sufficient to turn on the transistor and will fall as base current is drawn from the junction of the two resistors. A value for resistor R_3 of 1 k will limit the current in the transistor to less than 9 V/1 k $= 9$ mA.

Ideas for Exploration: (i) Investigate the use of the diode tester to check transistors. Noting that a transistor is made up of two back-to-back diodes, it follows that both the base-emitter and base-collector junctions of a transistor should have a low forward resistance and a high reverse resistance, while the collector-emitter terminals should have high resistance in both connections. Both npn and pnp transistor can be tested but with appropriate polarities.

Dry Soil Indicator

The circuit in Fig. 2.70 functions as a dry soil detector. The two leads connected to the base and emitter terminals of the transistor are placed in the soil whose moisture level is to be monitored. If the water level is sufficient, the resistance between the two probes will be low. As a result,

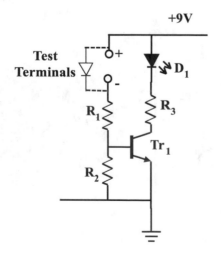

Fig. 2.69 Diode tester

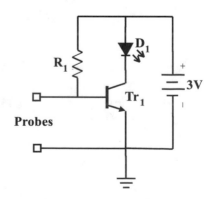

Fig. 2.70 Dry soil indicator

the voltage drop across the base-emitter junction is insufficient to turn on the transistor, most of the supply voltage being dropped across resistor R_1. As the soil becomes dry, the soil resistance will increase, and hence the voltage drop across the base-emitter junction will eventually be sufficient to turn on the transistor. The LED D_1 in the collector circuit will therefore turn on. Some experimentation may be necessary to determine a suitable value for R_1. A reasonable value to start with is $R_1 = 220$ k. Since the supply voltage is only 3 V, a current limiting resistor in series with the LED may not be necessary.

Ideas for Exploration: (i) Try the circuit with different types of soil.

Lie Detector

A lie detector is an instrument that attempts to detect when an individual is telling a lie by measuring changes in body characteristics such as skin resistance. It is based on the idea that skin resistance changes with emotional state, particularly as a result of perspiring. This resistance change can be easily detected and used as an indicator of lying. The lie detector circuit shown in Fig. 2.71 utilizes a transistor Tr1 with a milliammeter in the collector circuit. The probes connected to the collector and base terminals and held firmly in both hands by the candidate under test. Under normal conditions, the body resistance will allow a level of base current to flow. This is amplified by the transistor giving a collector current that is measured by the milliammeter. Upon questioning, if the candidate lies, the body resistance may fall resulting in an increased base current and hence a corresponding increase in collector current. This increase is measured by the meter and may be an indication that the individual may be lying. Almost any small-signal npn transistor such as the 2N3904 can be used. The resistor $R_1 = 1$ k in series with the milliammeter limits the current to less than 9 mA in the event that the probes are short-circuited. Note however that since R_1 is in the collector circuit, it has very little influence on the collector current. When in use, the range of the meter should be adjusted to accommodate the resulting current.

Ideas for Exploration: (i) Experiment with various individuals to evaluate the effectiveness of the circuit. (ii) Research other body characteristics used in lie detection including breathing rate, pulse rate and blood pressure.

Research Project 1

This project involves the use of the transistor to improve the performance of the simple radio receiver of Chap. 1. The circuit is shown in Fig. 2.72. Here the base-emitter junction of a germanium transistor Tr1 such as the NTE101 replaces the germanium diode in order to produce signal demodulation. The output of the tank circuit drives the base of the transistor, while the

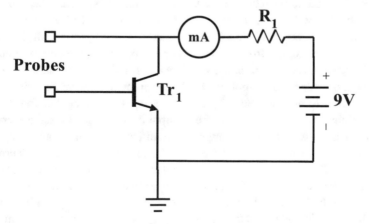

Fig. 2.71 Lie detector

Fig. 2.72 Improved
crystal radio

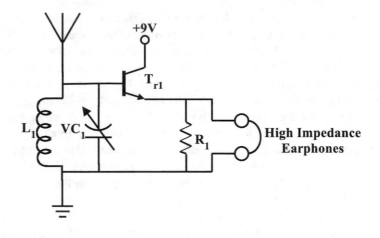

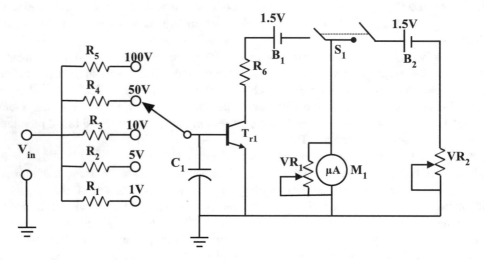

Fig. 2.73 High-impedance DC voltmeter

signal output at the emitter drives the piezo-electric earphone. The additional feature of this circuit compared to that of Chap. 1 is that the collector of the transistor is powered by a 9 volt battery. Thus, when a radio signal from the tank circuit turns on the base-emitter junction, the signal current into the transistor base is amplified by the factor β of the transistor. The effect is less loading of the tank circuit and hence greater frequency selectivity and increased signal output. Resistor $R_1 \simeq 10$ k is again required for proper circuit operation.

Ideas for Exploration: (i) Remove the power supply and connect the collector of the transistor to the base to realize a diode-connected transistor. In this way the circuit reverts to that of Chap. 1.

Research Project 2

In this research project, the transistor is used as a high-impedance DC voltmeter (Fig. 2.73). The circuit uses a germanium instead of a silicon transistor Tr1 in the common emitter mode and realizes a high input impedance of 200 kΩ/volt by employing high value resistors at the transistor base input. Upon application of a voltage at the input through one of these resistors, the transistor amplifies the small input base current, and the resulting larger collector current drives the microammeter M_1. The transistor is not biased on so that the threshold voltage of the transistor has to be overcome by the input signal for circuit operation. The use of a germanium transistor ensures that this turn-on voltage is low (0.3 V).

The need to overcome even this low voltage does result in some non-linearity at low input voltages. Resistors R_1 to R_5 are selected such that the same value of base current flows into the transistor with application of the full voltage on each range. That value of base current would result in the full-scale deflection of the meter. Thus, on the 1 volt range for $R_1 = 200$ k, ignoring the 0.3 V turn-on voltage, the maximum base current is given by 1 V/ 200 k = 5 μA. This current is amplified by the transistor current gain of about 60, giving a collector current of 300 μA. The sensitivity of the 100 μA meter M_1 is adjusted by $VR_1 = 10$ k to realize full-scale deflection for this current. Power supply B_1 provides power to the circuit, while supply B_2 along with $VR_2 = 100$ k is used to nullify transistor leakage current into the meter. The remaining resistors at the input are selected such that range-voltage/ range-resistor=5 μA. This results in $R_2 = 1$ M, $R_3 = 2$ M, $R_4 = 10$ M and $R_5 = 20$ M.

Ideas for Exploration: (i) Select R_6 to protect the meter from over-voltage. (ii) Determine C_1 to remove high-frequency noise from the input signal. (iii) Use a silicon transistor for Tr1 and compare the performance of the circuit with that using a germanium transistor.

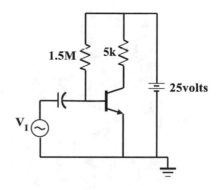

Fig. 2.74 Circuit for Question 3

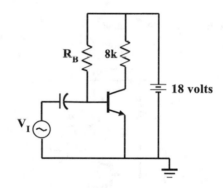

Fig. 2.75 Circuit for Question 5

Problems

1. Design a fixed-bias common emitter amplifier with a 24 volt supply and an npn silicon transistor having $\beta = 100$. Design for maximum symmetrical swing. Determine the voltage gain and input impedance.

2. Draw the load line for the circuit of problem 1, indicating the q point. What does the slope of the line represent?

3. Determine the collector current and voltage in the fixed-bias amplifier of Fig. 2.74, assuming transistor current gain of 100. If the transistor is replaced by one with current gain of 75, determine the new collector current and voltage.

4. Using a DC supply of 15 V, a load resistor of 2 k and a silicon pnp transistor with $\beta = 150$ at

room temperature, (a) bias the transistor for maximum symmetrical swing using fixed bias and (b) calculate the value of V_{CEq} at 50 °C if β goes to 110.

5. In the circuit of Fig. 2.75, determine the value of R_B in order to achieve a collector voltage of 10 V, assuming $\beta = 130$. Evaluate the input impedance for the circuit.

6. What are the two main sources of distortion in a common emitter amplifier? Discuss ways by which this distortion can be minimized.

7. Design a voltage divider-biased common emitter amplifier for maximum symmetrical swing using a small-signal npn silicon transistor with a current gain of 120 and a 25 volt supply. Justify all steps in your design. Evaluate the voltage gain of your circuit.

8. In the common emitter voltage divider-biased-biased amplifier of Fig. 2.76 determine the collector current and the collector voltage.

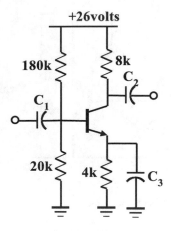

Fig. 2.76 Circuit for Question 8

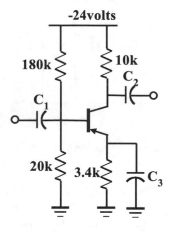

Fig. 2.77 Circuit for Question 9

9. In the common emitter voltage divider-biased-biased amplifier of Fig. 2.77 using a pnp transistor, determine the collector current, the collector voltage and the emitter voltage.

10. Design a common emitter amplifier utilizing current feedback biasing using an npn silicon transistor having $\beta = 200$ and a 12 volt supply.

11. Determine the collector current and voltage for the collector-base feedback circuit shown in Fig. 2.78 assuming $\beta = 100$.

12. Design a common emitter amplifier using a ± 20 V bipolar supply and a transistor with $\beta = 150$.

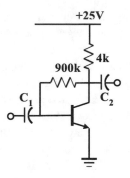

Fig. 2.78 Circuit for Question 11

13. Design a voltage divider-biased common base amplifier using a 28 volt supply and an npn silicon transistor having $\beta = 110$. Find the voltage gain and input impedance of the circuit.

14. Explain the phase relationship between the input and output voltages in a common base amplifier.

15. Design a common base amplifier using fixed bias and a silicon npn transistor having $\beta = 100$. Use $V_{CC} = 18$ volts.

16. Design a common base amplifier using a bipolar supply ± 20 V. Determine the voltage gain and input impedance.

17. Design a common base amplifier for maximum symmetrical swing using a small-signal pnp silicon transistor with a current gain of 125 and a 20 volt supply. Use a 3.3 V Zener to fix the base voltage. Justify all your design steps. Evaluate the input impedance and voltage gain of your circuit.

18. Design a voltage divider-biased-biased common collector amplifier for maximum symmetrical swing using a small-signal pnp silicon transistor with a current gain of 110 and a 22 volt supply. Justify all your design steps. Evaluate the input impedance and voltage gain of your circuit.

19. Design a voltage divider-biased-biased common collector amplifier for maximum symmetrical swing using a small-signal npn silicon transistor with a current gain of 125 and a 32 volt supply. Justify all your design steps. Determine the input impedance of your circuit.

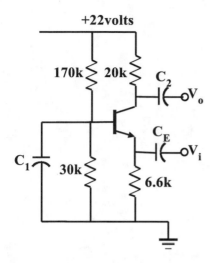

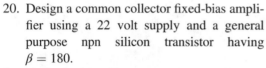

Fig. 2.79 Circuit for Question 21

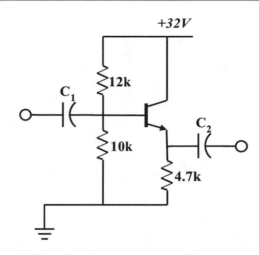

Fig. 2.80 Circuit for Question 24

20. Design a common collector fixed-bias amplifier using a 22 volt supply and a general purpose npn silicon transistor having $\beta = 180$.

21. Evaluate the collector current and voltage in the common base amplifier of Fig. 2.79.

22. Using the configuration of bench supply, design a regulated power supply giving 5 volts at 150 mA with electronic short-circuit protection.

23. Design a common collector amplifier using a pnp silicon transistor with a ± 15 V supply.

24. Determine the collector current and voltage in the common collector circuit of Fig. 2.80.

Bibliography

R.L. Boylestad, L. Nashelsky, *Electronic Devices and Circuit Theory*, 11th edn. (Pearson Education, New Jersey, 2013)

J. Millman, C.C. Halkias, *Integrated Electronics: Analog and Digital Circuits and Systems* (Mc Graw Hill Book Company, New York, 1972)

F.H. Mitchell, F.H. Mitchell, *Introduction to Electronics Design* (Prentice Hall of India, New Delhi, 1995)

D.A. Neeman, *Electronic Circuit Analysis and Design*, 2nd edn. (McGraw Hill, New York, 2001)

M. Yunik, *The Design of Modern Transistor Circuits* (Prentice Hall, New Jersey, 1973)

Field-Effect Transistor

3

The field-effect transistor or FET is a three-terminal semiconductor device that controls an electric current by an electric field. The FET actually pre-dates the BJT as the first patent was granted for such a device in 1928. Its impact on industry however was felt only about a decade after the development of the transistor in 1948. The FET is a unipolar device having only one p-n junction, and it differs from the BJT in several important respects, the main one being the FET's inherently high input impedance. There are two types of FETS: the junction gate FET (JFET or JUGFET) and the metal oxide semiconductor FET or MOSFET. The MOSFET, sometimes called the insulated gate FET or IGFET, itself comes in two versions: the depletion MOSFET and the enhancement MOSFET. Because of a difference in construction, the MOSFET has a higher input impedance than the JFET. The FET like the BJT can provide amplification of a signal and operate as a switch. It is important in many applications and forms the subject of this chapter. At the end of the chapter, the student will be able to:

- Describe and explain the characteristics and operation of a JFET
- Describe and explain the characteristics and operation of a MOSFET
- State the various FET amplifier configurations
- Design single stage amplifiers using a JFET
- Design single stage amplifiers using a MOSFET
- Explain the use of Shockley's equation

3.1 Operation of JFET

The basic structure of a JFET is shown in Fig. 3.1a. The device is a three-terminal device comprising a bar of n-type silicon material forming a channel with a ring of heavily doped p-type material at the centre forming a gate. The source (S) and drain (D) are located at each end of the n-type bar, and the gate (G) is connected to the p-type material. This type of JFET is referred to as an n-channel JFET. A p-channel JFET has a channel consisting of p-type material with an n-type gate, and this is shown in Fig. 3.1b. The symbols for n-channel and p-channel JFETS are shown in Fig. 3.2.

Considering the n-channel JFET, for normal operation the drain-source voltage V_{DS} is positive as shown in Fig. 3.3. Majority carriers enter the bar via the source terminal and leave the bar by way of the drain terminal. The magnitude of this drain current is inversely proportional to the resistance of the bar or wafer (which in turn depends on bar length and cross-sectional area). If, therefore, the cross-sectional area can be varied, then channel resistance and hence drain current can be influenced. From previous considerations, it is known that the region on either side of a p-n junction, known as the depletion region, is a region of high resistivity where width depends on the reverse-biased voltage applied to the junction. Through the action of this depletion layer, the reverse bias can exert control over the channel width and therefore channel current. This reverse

© Springer Nature Switzerland AG 2021
S. J. G. Gift, B. Maundy, *Electronic Circuit Design and Application*,
https://doi.org/10.1007/978-3-030-46989-4_3

Fig. 3.1 Basic structure of a JFET

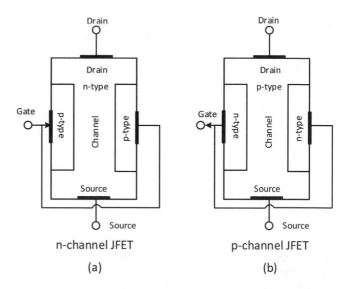

n-channel JFET

(a)

p-channel JFET

(b)

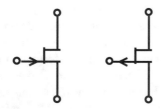

Fig. 3.2 Symbols for n-channel (left) and p-channel (right) JFETS

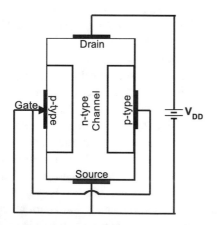

Fig. 3.3 N-channel JFET in operation

bias is applied via the gate-source terminals. The heavy doping of the p-type region ensures that the depletion region remains within the n-channel and does not extend into the p-type region.

When the gate is connected to the source corresponding to zero gate-source voltage and the drain-source voltage V_{DS} is increased slowly from zero, the channel behaves like a semiconductor resistor giving an approximately linear increase of current with applied voltage. The p-n junction becomes increasingly reverse-biased, causing the depletion layer to extend further into the channel. The reverse bias arises since the p-type region is a zero potential (relative to source), while the n-channel is progressively positive with increasing distance from the source to the drain. This potential gradient, which exists along the channel, produces an increased reverse bias of the p-n junction such that the width of the depletion region is increased. The region closer to the drain terminal where the potential is higher will be wider than the region closer to the source terminal where the potential is lower. The effect of this on the channel is shown in Fig. 3.4. Hence, increasing the drain-source voltage increases the depletion region and therefore increases channel resistance. The drain current is therefore reduced. Further increase in the drain-source voltage makes the depletion region extend further into the channel.

As a result of the depth of penetration of the depletion region, the width of the depletion region increases as we approach the drain, causing a constriction of the effective channel width. As the drain-source voltage V_{DS} is increased, a point is reached where the drain current remains constant at a value referred to as the saturation drain current with no further increase resulting from an increasing drain-source voltage.

The channel is now said to be "pinched off", and the FET is operating in the pinch-off or constant current region. Note that in this region the channel is not closed off completely since this would cause the drain current I_D to go to zero. The drain-source voltage at which pinch-off occurs corresponds in magnitude to what is called the pinch-off voltage V_P. If V_{DS} is increased still further, avalanche breakdown will eventually occur as in all p-n junctions. The onset of avalanche breakdown is very rapid and may cause irreparable damage to the junction. The resulting I_D/V_{DS} characteristic is shown in Fig. 3.5.

The reverse bias at the gate-channel p-n junction can be increased by the application of a negative potential to the gate terminal relative to the source. Thus, instead of $V_{GS} = 0$ as before, the gate is disconnected from the source and a negative bias applied at the gate with respect to the source as shown in Fig. 3.6. This reverse bias widens the existing depletion layer and as a result decreases channel width as well as increases channel resistance, even before V_{DS} is applied. Then as V_{DS} is increased from zero, the decreased channel width means that pinch-off will occur at a lower value of V_{DS} and a lower value of saturation drain current. Further negative increases in V_{GS}

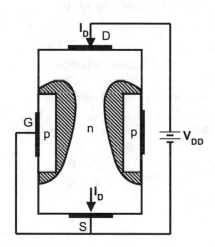

Fig. 3.4 Pinch-off in n-channel JFET

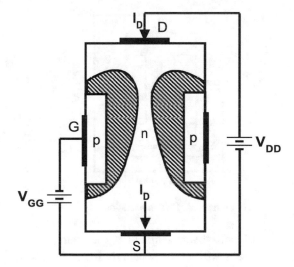

Fig. 3.6 JFET with negative gate-source voltage

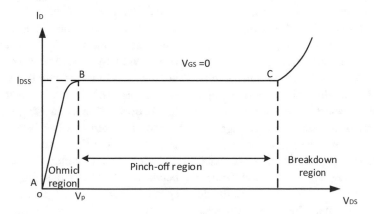

Fig. 3.5 Drain current vs drain-source voltage characteristic for JFET

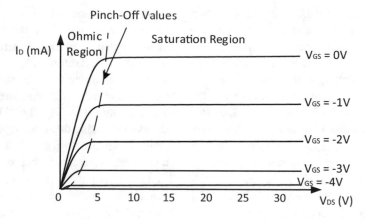

Fig. 3.7 JFET $I_D vs V_{DS}$ with V_{GS} as a varying parameter

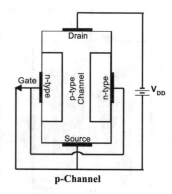

Fig. 3.8 P-channel JFET in operation

result in still lower values of V_{DS} at which pinch-off occurs with correspondingly lower values of saturation drain current. Eventually, at a sufficiently negative gate-source voltage $V_{GS} = -V_P$, pinch-off occurs at zero and all higher values of V_{DS} such that the drain current is zero. The result is a set of characteristics $I_D vs V_{DS}$ with V_{GS} as a varying parameter as shown in Fig. 3.7. They are the n-channel JFET output characteristics.

A p-channel junction FET operates in a similar manner except that the drain is driven by a negative potential with respect to the source and it is necessary to increase the gate-source voltage positively in order to reduce the drain current. This device is shown in Fig. 3.8 with the associated pinch-off action in Fig. 3.9 and output characteristics shown in Fig. 3.10. Both the p-channel and n-channel JFETs are operated with their gate-channel p-n junctions reverse-

biased. They therefore have very high input impedances of the order of $10^9 \, \Omega$.

3.2   JFET Characteristics

The JFET output characteristics consist of three distinct regions for any particular value of V_{GS} (Fig. 3.8):

- An approximately linear or Ohmic region
- An approximately pinch-off or saturation region
- An avalanche or breakdown region

The Ohmic region where $V_{DS} < V_P$ is one in which the JFET operates as a voltage-controlled resistor since varying V_{GS} changes the channel resistivity. The pinch-off or saturation region is the region where the drain current is almost completely controlled by the gate-source voltage and is virtually independent of the drain-source voltage. It is in this region that amplifying action is possible. The device is never operated in the breakdown region as catastrophic failure results. These characteristics can be obtained using the circuit shown in Fig. 3.12.

Using power supply P_1 and the voltmeter between the gate and source, V_{GS} is set at a specific negative voltage. Using power supply P_2 and the voltmeter between the drain and source, V_{DS} is varied. As V_{DS} is varied, the

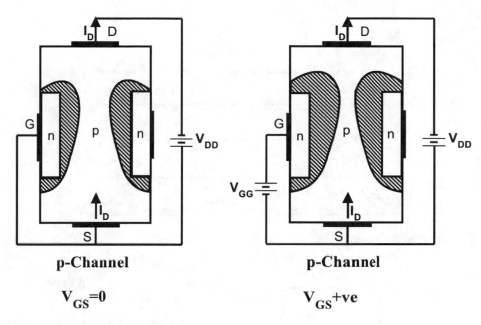

Fig. 3.9 Pinch-off action in p-channel JFET

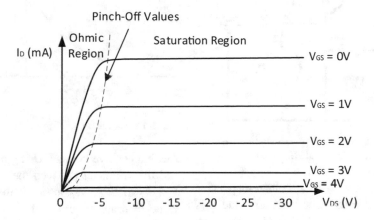

Fig. 3.10 JFET $I_D$vsV_{DS} with V_{GS} as a varying parameter

corresponding value of drain current I_D is recorded using the milliammeter in the drain circuit for various values of V_{DS}. This is repeated for various fixed values of V_{GS}. The result is a set of drain characteristics for the device.

As indicated earlier, the saturation drain current is the drain current that flows when the JFET is operated in the constant current region beyond pinch-off. The equation that defines the behaviour of the JFET is Shockley's equation which is given by

$$I_D = I_{DSS}\left(1 - \frac{V_{GS}}{V_P}\right)^2 \qquad (3.1)$$

where I_{DSS} is the saturation drain current which flows when the gate-source voltage is zero and V_P is the pinch-off voltage. This equation defines the relationship between the drain current of the FET and the gate-source voltage. In typical drain characteristics for the JFET shown in Fig. 3.11, each curve has a region of small values of V_{DS} in which I_D is proportional to V_{DS}. In this region the

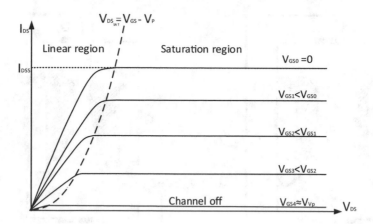

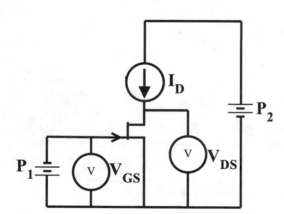

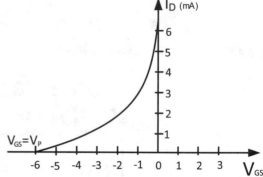

Fig. 3.11 JFET output characteristics

Fig. 3.12 Circuit for the determination of the characteristics of the n-channel JFET

Fig. 3.13 N-channel

device can be operated as a voltage-dependent resistance, i.e. as a resistance whose value V_{DS}/I_D depends on the value of V_{GS}. Here the gate-source voltage on the FET determines the channel resistance according to

$$r_d = \frac{r_o}{(1 - V_{GS}/V_P)^2} \qquad (3.2)$$

where r_o is the channel resistance r_d with $V_{GS} = 0$ V.

The mutual or transfer characteristic of a FET is a graph of drain current against gate-source voltage for constant values of drain-source voltage above pinch-off obtained using the circuit of Fig. 3.12. Using power supply P_2 and the voltmeter between the drain and source, V_{DS} is set at a

specific positive voltage. Using power supply P_1 and the voltmeter between the gate and source, V_{GS} is varied. As V_{GS} is varied, the corresponding value of drain current I_D is recorded using the milliammeter in the drain circuit for various values of V_{GS}. This is repeated for various fixed values of V_{DS}. The result is the mutual characteristic for the device.

This curve shows directly the effect of gate-source voltage on saturation drain current and indicates that the JFET is a voltage-controlled device. Mutual characteristics for n-channel and p-channel JFETS are shown in Figs. 3.13 and 3.14, respectively. From these curves, it can be seen that for $|V_{DS}| > |V_P|$ where I_D assumes its saturation value, as V_{GS} is changed from zero, the saturation drain current decreases from I_{DSS}. At some value $V_{GS} = V_P$, the drain current goes

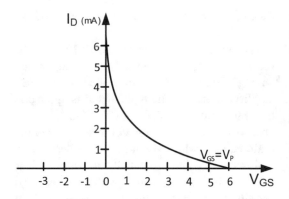

Fig. 3.14 P-channel

to zero, and this voltage is referred to as the pinch-off voltage for the FET. It is negative for n-channel JFETS and positive for p-channel JFETS. Therefore, the pinch-off voltage V_P is defined as that value of p-n junction reverse bias, which removes all the free charges from the channel; it is the reverse bias that will cause the depletion regions to pinch off. Now, V_{DS} is set at a value above pinch-off and V_{GS} and the corresponding value of drain current are recorded. There is very little variation in the curve for different values of V_{DS} above pinch-off. On examination of the saturation drain characteristic in Fig. 3.11, it can be seen that as the gate reverse bias is increased, pinch-off occurs at lower values of drain voltage, and at pinch-off, V_P can be approximated by the relation $|V_p| = |V_{DSp}| + |V_{GS}|$ where V_{DSp} is the drain-source voltage at the point of pinch-off. The values of transconductance and the drain-source resistance can be obtained from the drain characteristics, while the transconductance conductance can be determined from the mutual characteristic.

3.3   JFET Parameters

The important parameters of a JFET are its mutual conductance g_m, its input resistance R_i and its drain-source resistance r_{DS}. The mutual conductance or transconductance is defined as the ratio of a change in the drain current ΔI_D to the change

in the gate-source voltage ΔV_{GS} producing it, with the drain-source voltage V_{DS} held constant, i.e.

$$g_m = \frac{\Delta I_D}{\Delta V_{GS}}, V_{DS} \text{ constant} \qquad (3.3)$$

As in the BJT, the FET transconductance directly determines the gain of the FET when the device is connected as an amplifier. Typically, g_m has a value in the range 1 mA/V to 10 mA/V and is generally much smaller than that for a BJT. Its value can be found graphically using the slope of the mutual characteristic at particular operating conditions or analytically using Shockley's equation (3.4). Thus from Shockley's equation

$$I_D = I_{DSS}\left(1 - \frac{V_{GS}}{V_P}\right)^2 \qquad (3.4)$$

Differentiating I_D with respect to V_{GS} gives

$$g_m = \frac{dI_D}{dV_{GS}} = \frac{-2I_{DSS}}{V_P}\left(1 - \frac{V_{GS}}{V_P}\right) \qquad (3.5)$$

Using (3.4), this can be written as

$$g_m = \frac{-2I_{DSS}}{V_P}\sqrt{\frac{I_D}{I_{DSS}}} \qquad (3.6)$$

For $V_{GS} = 0$, (3.5) reduces to

$$g_{mo} = \frac{-2I_{DSS}}{V_P} \qquad (3.7)$$

The input impedance R_i of a JFET is the high value presented by the reverse-biased gate-channel p-n junction. Typically JFET input impedance is greater than 10^8 MΩ. Finally the drain-source resistance r_{DS} is the ratio of a change in the drain-source voltage ΔV_{DS} to the corresponding change in drain current ΔI_D, with the gate-source voltage V_{GS} held constant, i.e.

$$r_{DS} = \frac{\Delta V_{DS}}{\Delta I_D}, V_{GS} \text{ constant} \qquad (3.8)$$

r_{DS} may be in the range 40 kΩ to 1 MΩ and can be found from the inverse of the slope of the drain characteristic.

Example 3.1

A JFET has a signal voltage of 1.5Vpk value applied to its gate relative to its source. The drain current varies by ± 2 mA about its quiescent value. Calculate g_m.

Solution

$$g_m = \frac{2 \times 10^{-3}}{1.5} = 1.33 \text{ mA/V}$$

Example 3.2

An n-channel JFET has $I_{DSS} = 10$ mA and $V_P = -8$ V. For a drain current of 5 mA, determine g_m.

Solution Using (3.6), $g_m = \frac{-2 \times 10}{-8} \sqrt{\frac{5}{10}} = 1.77$ mA/V.

3.4   Using the JFET as an Amplifier

Like the BJT, the JFET can be operated in three configurations, the common source, common gate and common drain configurations, as shown in Fig. 3.15.

3.4.1   Common Source Configuration

This configuration is shown in Fig. 3.15a. It is the most frequently used amplifying configuration among FETS and is fundamental to FET amplifying action. The input signal is applied at the gate terminal of the FET relative to the grounded source. The effect of this is to vary the drain current, which similar to the collector current in the BJT must be initially non-zero. This is achieved by biasing which will be considered shortly. In order to convert the varying drain current I_D into a varying voltage, a resistor R_L is included in the drain circuit. This means that a change ΔI_D in I_D produces a drain voltage change $\Delta V_D = -\Delta I_D R_L$ in the drain circuit which can be easily monitored. The negative sign indicates that as I_D increases, the drain voltage V_D decreases since the voltage across R_L increases with the increase in drain current. The result is that a signal input $V_i = \Delta V_{GS}$ results in a signal output $V_O = \Delta V_{DS} = -\Delta I_D R_L$ and a full signal cycle can be reproduced. Now $\Delta I_D/\Delta V_{GS}$ is the transconductance g_m of the FET. Hence, the voltage $A_V = V_o/V_i$ for the JFET is given by

$$A_V = \frac{V_o}{V_i} = -\frac{\Delta I_D.R_L}{\Delta V_{GS}} = -g_m R_L \qquad (3.9)$$

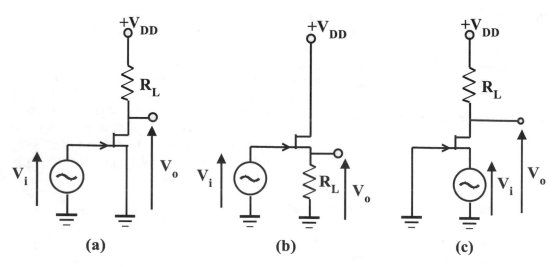

Fig. 3.15 JFET configurations: (**a**) common source, (**b**) common drain and (**c**) common gate

Example 3.3
If the FET above is used in a common source amplifier, determine the voltage gain if $R = 10$ k.

Solution

$$A_v = -g_m R = -2.5 \text{ mA} \times 10 \text{ k} = -25$$

It is instructive to note that the transconductance for a JFET is given by

$$g_m = \frac{-2I_{DSS}}{V_P}\sqrt{\frac{I_D}{I_{DSS}}} \qquad (3.10)$$

while that for the BJT is given by

$$g_m = -40I_C \qquad (3.11)$$

Fixed-Bias Configuration
Like the BJT, the FET needs to be operated with quiescent current so that changing the input voltage (V_{GS}) results in a changing drain current such that a full signal cycle can be reproduced. A fixed-bias arrangement for the common source JFET analogous to that for the common emitter BJT is shown in Fig. 3.16(a). The potential V_{GG} establishes the reverse bias V_{GS} on the gate-source junction, while resistor R_G connects the

gate to ground without short-circuiting V_i. This resistor does not inhibit the action of V_{GG} because of the high gate impedance of the FET. Resistor R_L converts drain current changes into drain voltage changes. In order to determine the quiescent current I_{Dq} use is made of Shockley's equation $I_D = I_{DSS}\left(1 - \frac{V_{GS}}{V_P}\right)^2$. Thus, knowing V_P and I_{DSS} for the FET, V_{GS} is substituted and I_D calculated or vice versa.

Example 3.4
Bias the n-channel JFET for a drain current of 5 mA using fixed-bias. For the FET, $I_{DSS} = 10$ mA and $V_P = -8$ V. Use $V_{DD} = 12$ V and choose R_L for maximum symmetrical swing.

Solution Using $I_{DSS} = 10$ mA, $V_P = -8$ V and $I_D = 5$ mA in Shockley's equation gives $V_{GS} = \left(1 - \sqrt{\frac{I_D}{I_{DSS}}}\right)V_P = \left(1 - \sqrt{\frac{5 \text{ mA}}{10 \text{ mA}}}(-8)\right) = 1 - 0.707)$ $(-8) = -2.34$ V.

Hence $V_{GG} = -2.34$ V. For maximum symmetrical swing with $V_{DD} = 12$ volts, $R_L = \frac{12/2}{5 \text{ mA}} - 1.2$ k. The input impedance of the JFET is greater than 10^8 Ω, and therefore R_G is chosen to be 1 MΩ to ensure that there is no voltage drop across this resistor arising from small currents

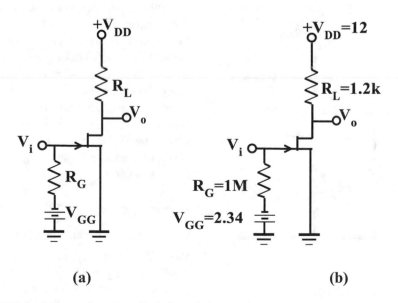

(a)   (b)

Fig. 3.16 (a) JFET fixed-bias configuration; (b) design solution

into the gate. The solution is shown in Fig. 3.16 (b). Input and output coupling capacitors must be used because of the voltages present on the gate and drain. Using (3.10), g_m is given by

$$g_m = \frac{2I_{DSS}}{V_P}\left(1 - \frac{V_{GS}}{V_P}\right) = 2 \cdot \frac{10}{8}\left(1 - \frac{2.34}{8}\right)$$

$$= 1.77 \text{ mA/V}.$$

Hence the voltage gain A_V is given by $A_v = -g_m R_L = -1.72 \times 1.2 \text{ k} = -2$. This of course is quite low as compared with the BJT and results from the comparatively low value of FET transconductance.

Self-Bias

One disadvantage of the fixed-bias method is the need for two voltage supplies. The voltage source V_{GG} can be eliminated by the inclusion of R_S in the source circuit of the FET as shown in Fig. 3.17.

This method of biasing is referred to as self-bias and operates in the following manner: Since the gate is connected to ground through R_G, the

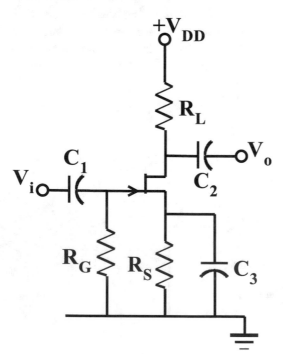

Fig. 3.17 Self-biased JFET amplifier

positive voltage developed across R_S as I_D flows results in the gate (0 V) being at a lower potential than the source ($+I_D R_S$) as in the fixed-bias configuration. V_{GG} is effectively replaced by $V_{GS} = -I_D R_S$. In order to determine the value of R_S to produce a desired I_D, Shockley's equation is again used. Thus, from (3.4),

$$I_D = I_{DSS}\left(1 - \frac{V_{GS}}{V_P}\right)^2$$

$$= I_{DSS}\left(1 + \frac{I_D R_S}{V_P}\right)^2 \qquad (3.12)$$

Re-arranging, this yields

$$R_S = \left(\sqrt{\frac{I_D}{I_{DSS}}} - 1\right)\frac{V_P}{I_D} \qquad (3.13)$$

Example 3.5

Using the FET of Example 3.4, find R_S to replace V_{GG}.

Solution From (3.13), $R_S = \left(\sqrt{\dfrac{5 \text{ mA}}{10 \text{ mA}}} - 1\right)$

$$\frac{-8}{5 \text{ mA}} = -.2929 \times \frac{-8}{5 \times 10^{-3}} = 469 \ \Omega.$$

With the presence of R_S, a bypass capacitor C_3 is now necessary across R_S in order to ground the FET source for signals and therefore not reduce the voltage gain. Coupling capacitors C_1 and C_2 are also required for DC blocking. In the presence of R_S, the load resistor R_L has to be re-calculated for maximum symmetrical swing. This is done by subtracting the source voltage of 2.34 V preserved by the capacitor from the full voltage swing 12 V before computing R_L: $R_L = (12 - 2.34)/2/5 \text{ mA} = 966 \ \Omega$

Example 3.6

Using Shockley's equation $I_D = I_{DSS}\left(1 - \frac{V_{GS}}{V_P}\right)^2$ and a 20 volt supply, design a common source amplifier using an n-channel JFET having a pinch-off voltage of -3 volts and $I_{DSS} = 8$ mA.

Use self-bias and a 2 mA quiescent current. Show that $g_m = -2\frac{\sqrt{I_D I_{DSS}}}{V_P}$ and use this to calculate the voltage gain of your circuit.

Solution Using $I_{DSS} = 8$ mA, $V_P = -3$ V and $I_D = 2$ mA in Shockley's equation gives

$$V_{GS} = \left(1 - \sqrt{\frac{I_D}{I_{DSS}}}\right)V_P = \left(1 - \sqrt{\frac{2 \text{ mA}}{8 \text{ mA}}}\right)(-3)$$
$$= (1 - 0.5)(-3) = -1.5 \text{ V}.$$

Hence $I_D R_S = 1.5$ V giving $R_S = 1.5$ V$/$ 2 mA $= 750$ Ω. For maximum symmetrical swing with $V_{DD} = 20$ volts, $R_L = \frac{(20-1.5)/2}{2 \text{ mA}} = 4.6$ k. The input impedance of the JFET is greater than 10^8 Ω, and therefore R_G is chosen to be 1 MΩ. The solution is shown in Fig. 3.18. Input and output coupling capacitors must be used because of the voltages present on the gate and drain. From (3.8), g_m is given by

$$g_m = \frac{-2I_{DSS}}{V_P}\sqrt{\frac{I_D}{I_{DSS}}} = -2\frac{\sqrt{I_D I_{DSS}}}{V_P} = -2\frac{\sqrt{2 \text{ mA} \times 8 \text{ mA}}}{-3}$$
$$= -2 \times \frac{4 \text{ mA}}{-3} = 2.67 \text{ mA}/\text{V}.$$

Hence the voltage gain A_V is given by $A_v = -g_m R_L = -2.67 \times 4.6$ k $= -12.3$.

Voltage Divider Biasing

A third FET biasing scheme is the voltage divider bias as shown in Fig. 3.19. Since there is no gate current,

$$V_G = \frac{R_2}{R_1 + R_2}V_{DD} \qquad (3.14)$$

and

$$V_{GS} = V_G - I_D R_S \qquad (3.15)$$

In order to determine the value of R_S to produce a desired I_D using Shockley's equation, it is easy to show that

$$R_S = \left(\sqrt{\frac{I_D}{I_{DSS}}} + \frac{V_G}{V_P} - 1\right)\frac{V_P}{I_D} \qquad (3.16)$$

This scheme, for the same R_S, permits higher values of I_D. Conversely, for a given value of I_D, this scheme permits higher values of R_S, which improves quiescent current stability.

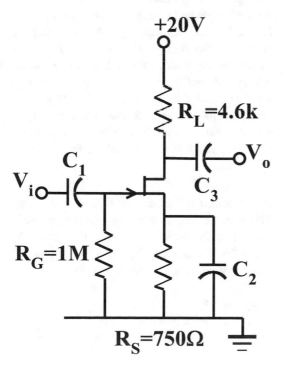

Fig. 3.18 Solution to Example 3.6

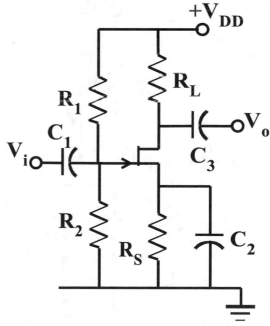

Fig. 3.19 FET voltage divider bias

Example 3.7

Using an n-channel JFET having $I_{DSS} = 8$ mA and $V_P = -4$ volts, bias the device using voltage divider biasing for a drain current of 1.2 mA. Design for a maximum symmetrical swing with $V_{DD} = 12$ volts.

Solution Using $V_G = V_{DD}/10 = 1.2$ V say, for $I_{DSS} = 8$ mA, $V_P = -4$ V and $I_D = 1.2$ mA, Shockley's equation gives $R_S = \left(\sqrt{\frac{1.2}{8}} + \frac{1.2}{-4} - 1\right)\frac{-4}{1.2} = 3.04$ k. Note $I_D R_S = 1.2 \times 3.04 = 3.65$ V. Hence for maximum symmetrical swing $R_L = \frac{(V_{DD}-3.65)/2}{1.2} = \frac{(12-3.65)/2}{1.2} = 3.48$ k. Since $V_G = 1.2$ volts, then for $R_2 = 200$ k and $R_1 = 10.8 \times 200$ k/1.2 = 1.8 M.

Biasing Using Bipolar Supply

The self-biased system of Fig. 3.17 can be improved by returning R_S to a negative supply voltage as shown in Fig. 3.20. This results in a larger value for R_S and hence greater bias stability as small changes in drain current result in larger corrective changes in gate-source voltage.

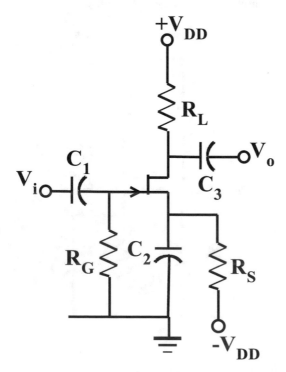

Fig. 3.20 Self-biasing using bipolar supply

Example 3.8

For the fixed-bias common source amplifier in Example 3.6, re-design the system to use a bipolar supply of ±20 V and return R_S to the negative supply voltage instead of ground.

Solution The modified system is shown in Fig. 3.20. Since R_S is being connected to −20 V instead of ground, its value has to be re-calculated. From the design in Example 3.6 where R_S was connected to ground, for $I_D = 2$ mA, $V_{GS} = -1.5$ V, and hence R_S was calculated from $I_D R_S = 1.5$. Now for R_S connected to −20 V, the full voltage drop across R_S is $1.5 - (-20) = 21.5$, and therefore R_S is calculated from $I_D R_S = 21.5$. This gives $R_S = 21.5/2$ mA = 10.8 k. All other component values remain the same.

3.4.2   Common Drain Configuration

This FET configuration is shown in Fig. 3.15(b) and is also referred to as a source follower configuration. It is analogous to the common collector or emitter follower configuration in the BJT. Here the input signal is at the gate and the output signal is at the source. The drain is grounded through the supply. In this configuration, the FET retains its high input impedance but now has reduced output impedance and near unity voltage gain. Figure 3.21 shows the configuration biased for signal amplification. Biasing is based on a voltage divider arrangement which sets the FET source voltage at half of the supply voltage for maximum symmetrical swing.

This gives $R_S = \frac{V_{DD}/2}{I_D}$. The gate-source voltage is determined for the desired drain current using Shockley's equation as before, and the gate voltage set by the voltage divider is then given by $V_G = \frac{V_{DD}}{2} + V_{GS}$. Coupling capacitors C_1 and C_2 take the signal into and out of the circuit.

Example 3.9

Design a source follower circuit using a JFET with $I_{DSS} = 8$ mA, $V_P = -4$ V and a 12 V supply.

Solution Since $V_{DD} = 12$ V, then the quiescent source voltage is 6 V. Hence if we choose

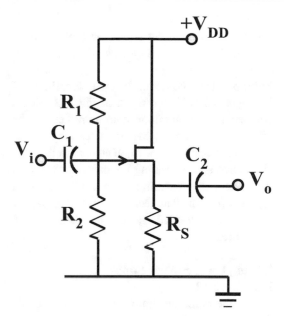

Fig. 3.21   JFET common drain amplifier

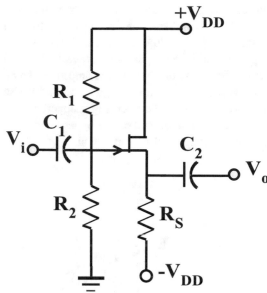

Fig. 3.22   JFET common drain biased with bipolar supply

$I_D = 2$ mA, then $R_S = 6$ V/2 mA $= 3$ k. For $I_D = 2$ mA, $V_{GS} = \left(1 - \sqrt{\frac{2\ mA}{8\ mA}}\right)(-4\ V) = -2$ V. Hence $V_G = 6 - 2 = 4$ V. From this it follows that $R_1 = 2$ M and $R_2 = 1$ M will produce this gate voltage.

Biasing Using Bipolar Supply

The common drain amplifier of Fig. 3.21 using voltage divider biasing can be improved by returning R_S to a negative supply voltage $-V_{DD}$ as shown in Fig. 3.22. This results in a larger value for R_S and hence greater bias stability as well as a larger output voltage swing since the available peak-to-peak voltage swing increases from V_{DD} to $2V_{DD}$.

Example 3.10

For the common drain amplifier in Example 3.9, re-design the system to use a bipolar supply of ± 12 V and return R_S to the negative supply voltage instead of ground.

Solution The modified system is shown in Fig. 3.22. Since R_S is being connected to -12 V instead of ground, its value has to be re-calculated. From the design in Example 3.9 where R_S was connected to ground, for maximum

symmetrical swing, the source voltage was set at half the supply voltage giving $R_S = 6V/I_D$. Now for R_S connected to -12 V, for maximum symmetrical swing, the source voltage is set at 0 V giving a voltage drop across R_S of $0 - (-12) = 12$ V. Therefore R_S is calculated from $R_S = 12/2$ mA $= 6$ k. All other component values remain the same.

3.4.3   Common Gate Configuration

The last configuration of the FET is the common gate configuration shown in Fig. 3.15(c). Here the input signal is applied to the source and the output signal taken from the drain of the FET. The gate is grounded and hence common to both input and output. This configuration is analogous to the common base BJT amplifier and is frequently used in high-frequency amplifiers and where low input impedance is required. A practical implementation of this configuration in an amplifier circuit is shown in Fig. 3.23. The design of this circuit follows the same procedure as the common source circuit discussed in Sect. 3.4.1. In the self-bias case, the resistor R_G is now not necessary, and the gate can be connected directly to ground. In the voltage divider bias, the gate is grounded using a capacitor. In both cases the signal is injected at the source using a capacitor, and the output signal is available at the drain. Hence, a common source amplifier can be converted to a common gate amplifier by grounding the input coupling capacitor and lifting the grounded end of the bypass capacitor and there applying the input signal.

Example 3.11
Show how a common source amplifier can be converted to a common gate amplifier.

Solution The common source amplifier of Example 3.6 is shown again in Fig. 3.24(a). Its conversion to a common gate amplifier can be effected by lifting the bypass capacitor C_3 from ground and applying the input signal through it to the source and grounding the capacitor going to the gate. The resulting circuit is shown in Fig. 3.24(b). The gate resistor and input capacitor are no longer needed and can be replaced by a short circuit.

Biasing Using Bipolar Supply
Example 3.12
   For the common drain amplifier in Example 3.11, re-design the system to use a bipolar supply of ± 20 V and return R_S to the negative supply voltage instead of ground.

Solution The modified system is shown in Fig. 3.25. Since R_S is being connected to -20 V instead of ground, its value has to be re-calculated. Thus from the design in Example 3.11 where R_S was connected to ground, for $I_D = 2$ mA, $V_{GS} = -1.5$ V and hence R_S was

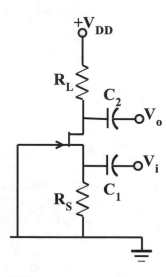

Fig. 3.23 JFET common gate amplifier

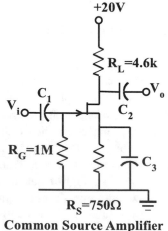

Common Source Amplifier

(a)

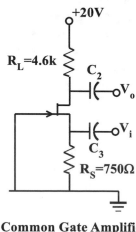

Common Gate Amplifier

(b)

Fig. 3.24 Common source amplifier converted to a common gate amplifier

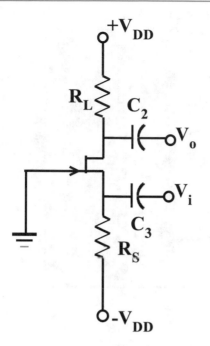

Fig. 3.25 Common gate amplifier biased with bipolar supply

calculated from $I_D R_S = 1.5$. Now with R_S connected to -20 V, the full voltage drop across R_S is $1.5 - (-20) = 21.5$, and therefore R_S is calculated from $I_D R_S = 21.5$. This gives $R_S = 21.5/2$ mA $= 10.8$ k. All other component values remain the same. As is evident, this procedure is exactly that for the bipolar biasing of a self-biased common source amplifier.

3.5   The MOSFET

In addition to the JFET, there is the MOSFET which is of two types – depletion and enhancement – and both of these are available in n-channel and p-channel types. This device achieves extremely high input impedance because of an insulating layer of silicon dioxide between the gate and the conduction channel instead of the reverse bias method of the JFET.

3.5.1   Depletion-Type MOSFET

In the depletion-type MOSFET shown in Fig. 3.26, a channel is diffused between source and drain, with the same type of impurity as used

for the source and drain. Thus, the n-channel type consists of a lightly doped p-type substrate into which two highly doped n regions along with relatively lightly doped n-channel are diffused. A thin silicon dioxide (SiO_2) layer is also placed over the surface and aluminium contacts introduced for gate, source and drain terminals. In order ensure that the p-n junction formed between the channel and substrate is reverse-biased, the substrate is held at a negative potential relative to the channel. This is accomplished either by connecting the substrate to the source internally as occurs in most devices or by way of an external connection. The effect is that as the drain source voltage V_{DS} is increased from zero, an increase in a voltage drop along the channel results with a consequent flow in drain current through the channel. Since the p-type substrate is connected to the source, the voltage drop along the channel results in a reverse bias between the n-type channel and the p-type substrate. This creates a depletion region in the n-type channel that extends into the channel to a depth that depends on the magnitude of V_{DS}. Since the voltage drop along then channel is greater at the drain as compared with the source, the width of penetration of the depletion layer is greater near the drain than near the source. With the gate-source voltage held at zero, as the drain-source voltage V_{DS} is increased, the voltage drop along the channel causes the depletion layer to extend further into that part of the channel region nearest to the drain than across the part nearest the source to an extent that depends on the magnitude of the drain-source voltage as shown in Fig. 3.27(a). This penetration determines the resistance of the channel to current flow.

The overall effect is that an increase of drain-source voltage initially causes an increase of drain current. Further increase results in an increase in the width of the depletion region, and this leads to saturation of the drain current as the width of the depletion region extends almost completely across the drain. Such a condition is referred to as pinch-off and prevents further increase of the drain current with increase of the drain-source voltage. The resulting drain current vs drain-source voltage is shown in Fig. 3.28.

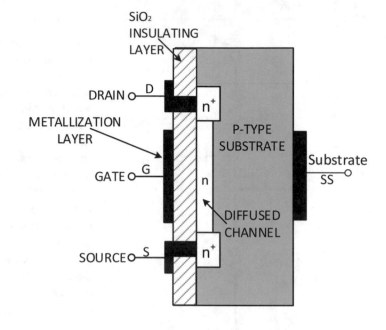

Fig. 3.26 N-channel depletion MOSFET structure

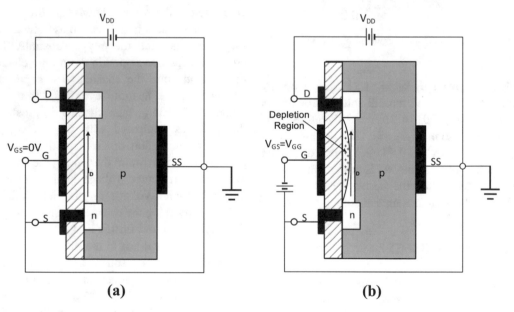

(a)   **(b)**

Fig. 3.27 N-channel depletion MOSFET action

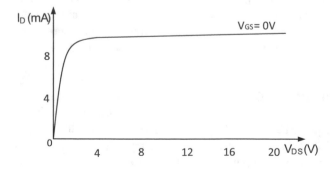

Fig. 3.28 Drain characteristics of MOSFET

If the gate-source voltage is made negative relative to the source as shown in Fig. 3.27(b), electrons are repelled out of the channel into the n⁺ region. This reduces the number of free electrons which are available for conduction in the channel region and so the channel resistance is increased. As a result, pinch-off occurs at a lower drain-source voltage and saturation current.

If the gate-source voltage is changed from zero and made positive, electrons will be attracted into the channel from the heavily doped n⁺ regions. The number of conduction electrons is increased and so the channel resistance is reduced. This will cause a greater drain current to flow for a given drain-source voltage and means that a larger value of drain-source voltage is required to cause pinch-off. Therefore, the overall effect of the gate-source voltage increase is an increase in pinch-off voltage and saturation current. This mode is referred to as the enhancement mode. Thus, the gate-source voltage controls the saturation drain current of the MOSFET. Moreover, because of the silicon dioxide layer, an n-channel depletion MOSFET operates in both an enhancement and a depletion mode, and it is possible to change the polarity of the gate-source voltage. The transfer and output characteristics are shown in Figs. 3.29 and 3.30, respectively, and are similar to those of the JFET except that the JFET has only the depletion mode. Shockley's equation given by $I_D = I_{DSS}\left(1 - \frac{V_{GS}}{V_P}\right)^2$ also defines the behaviour of the depletion MOSFET.

The p-channel depletion MOSFET is shown in Fig. 3.31. It operates in a manner similar to the n-channel depletion MOSFET except that all polarities are reversed. The associated drain characteristics are shown in Fig. 3.32 with reversed voltage polarities. The mutual characteristic is shown in Fig. 3.33, and here too the polarity change in the gate-source voltage can be seen. As previously observed, the JFET is also a depletion mode or normally on device. It

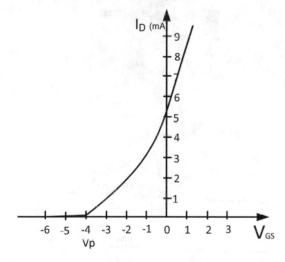

Fig. 3.29 N-channel mutual characteristics

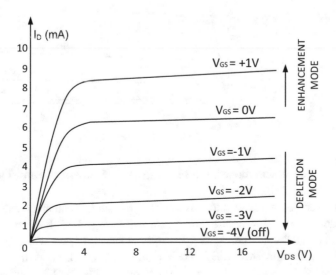

Fig. 3.30 Drain characteristics of n-channel depletion MOSFET

Fig. 3.31 p-channel
depletion MOSFET
structure

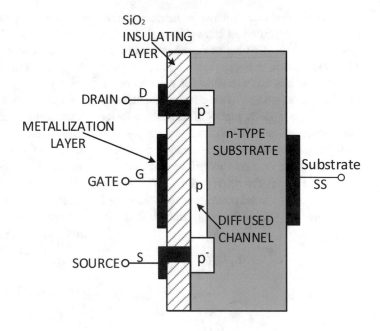

Fig. 3.32 Drain
characteristics of p-channel
depletion MOSFET

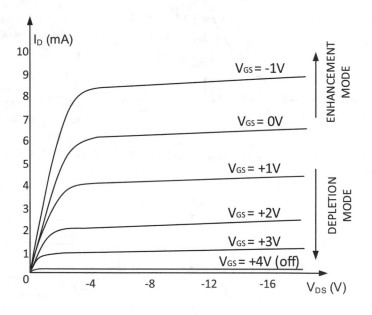

cannot be used in the enhancement mode because
this would involve forward biasing the gate-
source junction. This could result in excessive
gate-current flow and would lower the input
impedance or even damage the device.

Symbols for the depletion MOSFET are
shown in Fig. 3.34.

3.5.2   Enhancement-Type MOSFET

The n-channel enhancement-type MOSFET is
shown in Fig. 3.35. It consists of a lightly doped
p-type substrate into which two highly doped n^+
regions are diffused. A thin layer of insulating
silicon dioxide (SiO_2) is grown over the surface

and aluminium contacts made as shown in the diagram. Like the enhancement-type MOSFET, the substrate and source are connected together in order to ensure that the channel substrate junction is reverse-biased. Unlike the depletion-type MOSFET, no channel exists between the two highly doped drain and source regions. Because of this if no potential is applied between gate and source, there is essentially no conduction through the channel with the application of a drain-source voltage. In fact for $V_{GS} \leq 0$, the channel current is of the order of a few nano-amperes (since the channel consists of two back-to-back diodes). With a positive voltage applied between gate and source as shown in Fig. 3.36, a virtual channel is created between the drain and the source. The positive voltage attracts negative carriers into the p-type substrate from the n-type regions, and this results in significantly increased conductivity between drain and source.

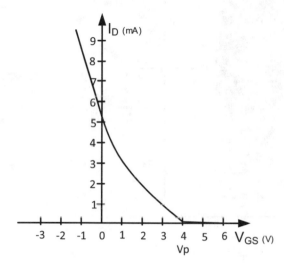

Fig. 3.33 P-channel mutual characteristics

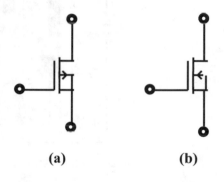

(a)   **(b)**

Fig. 3.34 Symbol for the depletion MOSFET (**a**) n-channel and (**b**) p-channel

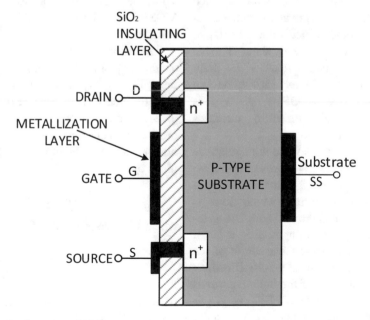

Fig. 3.35 n-channel enhancement MOSFET structure

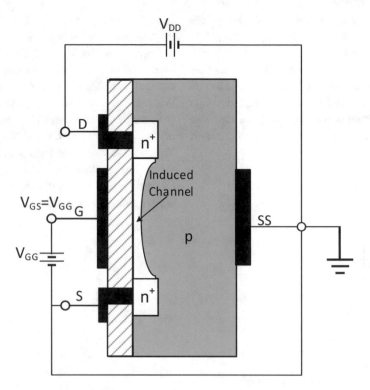

Fig. 3.36 N-channel enhancement MOSFET action

The minimum positive gate-source voltage that is necessary for the creation of the virtual channel so that conduction is possible is called the threshold voltage. Its value at which the drain current I_D reaches some defined value, say 10 μA, is often quoted and is typically about 2 volts. Thus, the positive voltage enhances the drain current, and such a device is called an ENHANCEMENT-type MOSFET. As V_{GS} is made positive, the drain current I_D increases slowly then rapidly.

An increase in the gate-source voltage above the threshold value will attract more electrons into the channel region. Once the virtual channel has been formed, the drain current which flows is influenced by the magnitude of both the gate-source and drain-source voltages. For a given $V_{GS} > V_T$ as V_{DS} is increased, the width of the depletion region increases until pinch-off occurs, thereby causing saturation of the drain current as occurred in the depletion MOSFET. As V_{GS} is increased above V_T, channel conductivity increases, and therefore, pinch-off occurs at the higher values of V_{DS} and I_D as shown in Fig. 3.37. Thus, the drain current of the enhancement-type MOSFET can be controlled by the gate-source voltage.

The enhancement-type MOSFET has a mutual characteristic Fig. 3.38 that is very different from the JFET and the depletion-type MOSFET. The main difference arises because of the need for the gate-source voltage to exceed a threshold value in order that channel conduction occurs. The defining equation is different from the Shockley and is given by

$$I_D = k(V_{GS} - V_T)^2 \qquad (3.17)$$

where k is a constant. It is found using

$$k = \frac{I_{D^{(on)}}}{\left(V_{GS^{(on)}} - V_T\right)^2} \qquad (3.18)$$

where $I_{D^{(on)}}$ and $V_{GS^{(on)}}$ are particular drain current and associated gate-source voltage values for the

Fig. 3.37 Drain characteristics of n-channel enhancement MOSFET

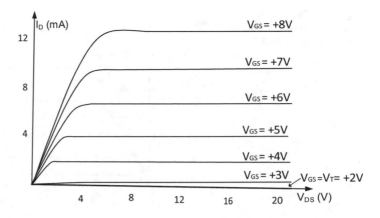

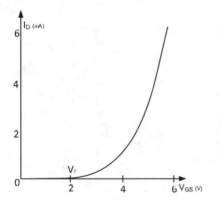

Fig. 3.38 Mutual characteristic of n-channel enhancement MOSFET

device. Specification sheets usually provide V_T, $V_{GS^{(on)}}$ and $I_{D^{(on)}}$. Having found k then Eq. (3.16) can be used to find V_{GS} for a given I_D.

The p-channel enhancement-type MOSFET operates in a similar manner except that the polarities are reversed. The corresponding drain and mutual characteristics are shown in Figs. 3.39 and 3.40. It is important to note that the enhancement-type MOSFET has no conducting channel and one has to be created by enhancement process. Therefore, like the JFET, the gate-source voltage can only be of one polarity. Also, the existence of a turn-on threshold gate-source voltage in the enhancement-type MOSFET is similar to the existence of a turn-on threshold base-emitter voltage in the BJT. Symbols for the enhancement-type MOSFET are shown in Fig. 3.41.

3.5.3 MOSFET Parameters

The MOSFET has a set of parameters similar to the JFET. Specifically these are its transconductance g_m which is typically in the range 1–10 mA/V, its input resistance R_i which is usually higher than about 10^{10} Ω and its drain source resistance r_{DS} which is of the order 5–50 kΩ. The flat nature of the output characteristics is an indication of the independence of I_D of V_{DS} for a given V_{GS}. In practice however increasing V_{DS} after pinch-off has occurred causes the channel pinch-off point to migrate away from the drain towards the source. This causes an effective channel length reduction, a phenomenon referred to as channel length modulation. The overall effect is to cause an increase in drain current with further increase in drain-source voltage in the saturation region. This manifests itself as output characteristics that have a slightly positive slope instead of being flat. As occurred with the BJT, these curves when extrapolated intercept the V_{DS} axis at a common point $V_{DS} = -V_A$. The curves enable the determination of the drain output resistance.

3.6 MOSFET Amplifiers

The MOSFET can also be used in three configurations. These are discussed in this section with several examples given of practical applications.

Fig. 3.39 Drain characteristics of p-channel enhancement MOSFET

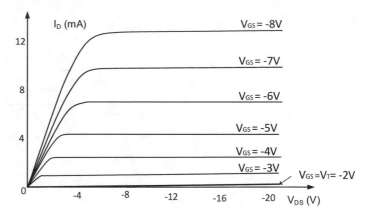

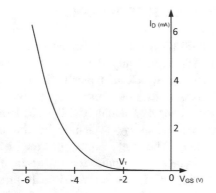

Fig. 3.40 Mutual characteristic of p-channel enhancement MOSFET

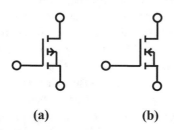

Fig. 3.41 Symbol for the enhancement MOSFET (**a**) n-channel and (**b**) p-channel

3.6.1   Depletion-Type MOSFET Common Source Amplifier

The depletion-type MOSFET which also obeys Shockley's equation can be biased in a similar manner to the JFET. This is shown Fig. 3.42 where self-bias is used.

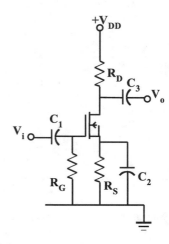

Fig. 3.42 Depletion MOSFET common source amplifier using self-bias

Example 3.13

Using self-bias, design a biasing network for a common source n-channel depletion-type MOSFET having $I_{DSS} = 8$ mA and $V_P = -8$ V.

Solution The circuit is shown in Fig. 3.42. Choosing a quiescent drain current of 2 mA, Shockley's equation gives $V_{GS} = \left(1 - \sqrt{\frac{I_D}{I_{DSS}}}\right) V_P = \left(1 - \sqrt{\frac{2\,\text{mA}}{8\,\text{mA}}}\right)(-8\,\text{V}) = -4\,\text{V}.$

Hence $R_S = 4$ V/2 mA $= 2$ k. We determine R_D in order to achieve maximum symmetrical swing. Therefore, $R_D = (V_{DD} - I_D R_S)/I_D$. This yields $R_D = (20 - 4)/2$ mA $= 4$ k. Finally, resistor R_G can be set at 1 M because of the high input impedance of the device.

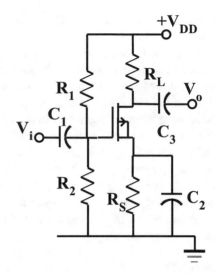

Fig. 3.43 MOSFET voltage divider bias

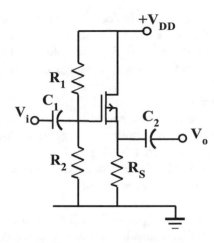

Fig. 3.44 MOSFET common drain amplifier

A depletion-type common source amplifier using voltage divider bias is shown in Fig. 3.43. The design approach is similar to that used in the JFET amplifier in Fig. 3.17. In Fig. 3.43 since there is no gate current,

$$V_G = \frac{R_2}{R_1 + R_2} V_{DD} \quad (3.19)$$

and

$$V_{GS} = V_G - I_D R_S \quad (3.20)$$

The value of R_S required to produce a desired I_D is determined by using Shockley's equation. This gives

$$R_S = \left(\sqrt{\frac{I_D}{I_{DSS}}} + \frac{V_G}{V_P} - 1 \right) \frac{V_P}{I_D} \quad (3.21)$$

This scheme, for the same R_S, permits higher values of I_D than is possible with self-bias. Conversely, for a given value of I_D, this scheme permits higher values of R_S, which generally improves quiescent current stability.

Example 3.14

Using an n-channel depletion MOSFET having $I_{DSS} = 6$ mA and $V_P = -3$ V and the circuit shown in Fig. 3.43, bias the device using voltage

divider biasing for a drain current of 5 mA. Design for a maximum symmetrical swing with $V_{DD} = 18$ volts.

Solution Using $V_G = V_{DD}/10 = 1.8$ V, for $I_{DSS} = 6$ mA, $V_P = -3$ V and $I_D = 5$ mA, Shockley's equation gives $R_S = \left(\sqrt{\frac{1.8}{6}} + \frac{1.8}{-3} - 1 \right) \frac{-3}{5} = 631$ Ω. Note $I_D R_S = 5 \times 0.631 = 3.2$ V. Hence for maximum symmetrical swing $R_L = \frac{(V_{DD} - 3.2)/2}{5} = \frac{(18 - 3.2)/2}{5} = 1.48$ k . Since $V_G = 1.8$ volts, then for $R_2 = 200$ k and $R_1 = 18 \times 200$ k/1.8 − 200 k = 1.8 M.

3.6.2 Depletion-Type MOSFET Common Drain Amplifier

This MOSFET configuration is shown in Fig. 3.44. It is similar to the common drain or source follower configuration using the JFET in Fig. 3.21. As in the JFET common drain configuration, the MOSFET retains its high input impedance but now has reduced output impedance and approximately unity voltage gain. Figure 3.44 shows the configuration biased for signal processing. Biasing employs a voltage divider network which sets the MOSFET source voltage at half of the supply voltage for maximum symmetrical swing. This gives $R_S = \frac{V_{DD}/2}{I_D}$. The gate-source voltage is again determined for the desired drain current using Shockley's equation, and the gate voltage set by the voltage divider is then given by $V_G = \frac{V_{DD}}{2} + V_{GS}$. Capacitors C_1 and C_2 are the input and output coupling capacitors.

Example 3.15

Using the circuit shown in Fig. 3.44, design a common drain amplifier using an n-channel depletion MOSFET having $I_{DSS} = 7$ mA and $V_P = -5$ volts. Design for maximum symmetrical swing with $V_{DD} = 20$ volts.

Solution Since $V_{DD} = 20$ V, then the quiescent source voltage must be 10 V for maximum symmetrical swing. Hence if we choose $I_D = 4$ mA, then $R_S = 10$ V/4 mA $= 2.5$ k. For $I_D = 4$ mA,
$$V_{GS} = \left(1 - \sqrt{\frac{4\text{ mA}}{8\text{ mA}}}\right)(-4\text{ V}) = -1.2\text{ V} .$$ Hence $V_G = 10 - 1.2 = 8.8$ V. Resistors $R_1 = 880$ k and $R_2 = 1.12$ M will produce this gate voltage.

3.6.3   Depletion-Type MOSFET Common Gate Amplifier

The third configuration in which the MOSFET can be used is the common gate configuration shown in Fig. 3.45. This configuration is similar to the common gate JFET amplifier of Fig. 3.23.

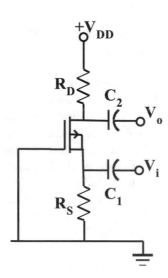

Fig. 3.45   MOSFET common gate amplifier

The design of this circuit follows the same procedure as the common source circuits discussed earlier. In the self-bias case, the resistor R_G is no longer necessary, and the gate can be connected directly to ground as shown in Fig. 3.45. In the voltage divider bias, the gate is grounded using the input capacitor. In both cases the signal is injected at the source using a capacitor, and the output signal is taken at the drain. Hence, the common source amplifier can be converted to a common gate amplifier by grounding the input coupling capacitor and lifting the grounded end of the bypass capacitor and there applying the input signal.

Example 3.16

Design a common gate amplifier using an n-channel depletion MOSFET with $I_{DSS} = 2$ mA and $V_P = -3$ volts. Design for maximum symmetrical swing using a 24 V supply.

Solution The circuit is shown in Fig. 3.45. Choosing a quiescent drain current of 1 mA, Shockley's equation gives $V_{GS} = \left(1 - \sqrt{\frac{I_D}{I_{DSS}}}\right)V_P = \left(1 - \sqrt{\frac{1\text{ mA}}{2\text{ mA}}}\right)(-3\text{ V}) = -0.88\text{ V}.$

Hence, $R_S = 0.88$ V/1 mA $= 880$ Ω. We determine R_D in order to achieve maximum symmetrical swing.

Therefore, $R_D = (V_{DD} - I_D R_S)/I_D$. This yields $R_D = (24 - 0.88)/1$ mA $= 23$ k. Note that no gate resistor R_G is needed as the gate can be directly connected to ground.

3.6.4   Enhancement-Type MOSFET Common Source Amplifier

As indicated earlier, the equation defining the characteristics of the enhancement-type MOSFET is different from the depletion-type device. Recall that the drain current is zero for zero gate-source voltage since there is initially no conducting channel. The gate-source voltage V_{GS} must exceed a threshold voltage V_T for channel conduction to occur. Then, the drain current is given by

$$I_D = k(V_{GS} - V_T)^2 \qquad (3.22)$$

The specification sheet for the device typically provides the threshold voltage V_T along with a drain current value and the corresponding gate-source voltage. The constant k can be found using these values after which the equation can be used to determine gate-source voltage for achieving a particular value of drain current. One method of biasing the enhancement-type MOSFET is feedback biasing as shown in Fig. 3.46. Here since there is no gate current, there is no voltage drop across R_G and therefore $V_{GS} = V_{DS}$.

Example 3.17

Using the circuit shown in Fig. 3.46, design a feedback biasing network for an enhancement-type MOSFET having $V_T = 3$ V, $I_{Don} = 6$ mA and $V_{GSon} = 8$ V. Use $V_{DD} = 15$ V.

Solution   The circuit is shown in Fig. 3.46. Using the data provided, k is given by $k = 6$ mA/ $(8\ V - 3\ V)^2 = 0.24$ mA/V^2. For a quiescent drain current of 3 mA, $V_{GS} = V_T + \sqrt{\frac{I_D}{k}} =$

$3 + \sqrt{\frac{3\ mA}{0.24\ mA/V^2}} = 6.54$ V. Hence, $V_{GS} = V_{DS} =$

6.54 V, and therefore $R_D = (V_{DD} - V_{DS})/ I_D = (15 - 6.54)/3$ mA $= 2.8$ k. Note that $V_{RD} = 15 - 6.54 = 8.46$ V while $V_{DS} = 6.54$ V. This is not biased for maximum symmetrical swing, though it will deliver a reasonable output. The extremely high input impedance of the FET means that R_G can be a high-value resistor, say 10 M.

The bias arrangement in Fig. 3.47 is a modified version of that in Fig. 3.46. The two resistors R_{G1} and R_{G2} result in a lower voltage at the gate of the FET as compared with the drain and hence allow maximum symmetrical swing design.

Example 3.18

Using the configuration in Fig. 3.47, design a modified feedback biasing network for an enhancement-type MOSFET having $V_T = 3$ V, $I_{Don} = 6$ mA and $V_{GSon} = 8$ V. Use $V_{DD} = 20$ V.

Solution   The circuit is shown in Fig. 3.47. Using the data provided, k is given by $k = 6$ mA/ $(8\ V - 3\ V)^2 = 0.24$ mA/V^2. For a quiescent

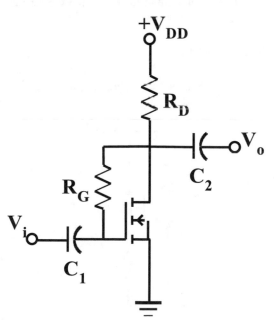

Fig. 3.46   Feedback biasing in the enhancement-type MOSFET

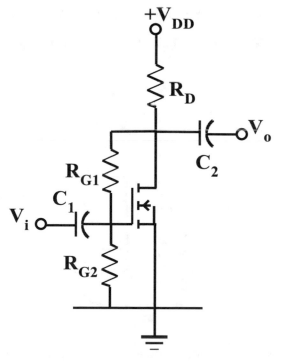

Fig. 3.47   Modified enhancement-type MOSFET bias

drain current of 3 mA, $V_{GS} = V_T +$ $\sqrt{\frac{3\,mA}{0.24\,mA/V^2}} = 6.54$ V. For maximum symmetrical swing, $V_D = V_{DD}/2 = 20/2 = 10$ V and therefore $R_D = V_D/I_D = 10/3$ mA $= 3.3$ k. Since the drain is at 10 V and the gate is at 6.54 V, the resistors R_{G1} and R_{G2} can now be calculated. The extremely high input impedance of the FET means that no current flows into the gate. Therefore $6.54/R_{G2} = 10/(R_{G1} + R_{G2})$ which for $R_{G2} = 10$ M gives $R_{G1} = 5.3$ M.

Another popular biasing method for the enhancement MOSFET is voltage divider biasing shown in Fig. 3.48. As in the case of the BJT, the resistor R_S provides improved stability of the quiescent drain current.

Example 3.19
Using the configuration in Fig. 3.48, design a voltage divider-biased common source amplifier using an enhancement-type MOSFET having $V_T = 2$ V, $k = 0.05$ mA/V^2 and use $V_{DD} = 15$ V.

Solution The circuit is shown in Fig. 3.48. For a quiescent drain current of 10 mA, $V_{GS} = V_T + \sqrt{\frac{.01}{0.05}} = 2.45$ V. Using the rule-of-thumb

$V_S = V_{DD}/10$ in order to allow large output voltage swing, $V_S = 15/10 = 1.5$ V. Then $R_S = 1.5/10$ mA $= 150\ \Omega$. For maximum symmetrical swing $V_{R_D} = (V_{DD} - V_S)/2 = (15 - 1.5)/2 = 6.75$ V and therefore $R_D = 6.75/I_D = 6.75/10\,mA = 675\,\Omega$. The resistors R_{G1} and R_{G2} can now be calculated. $V_G = V_{GS} + V_S = 2.45 + 1.5 = 3.95$ V. The extremely high input impedance of the FET means that no current flows into the gate. Therefore $3.95/R_{G2} = 15/(R_{G1} + R_{G2})$ which for $R_{G2} = 1$ M gives $R_{G1} = 2.8$ M.

3.6.5   Enhancement-Type MOSFET Common Drain Amplifier

The configuration shown in Fig. 3.49 is the common drain amplifier using the enhancement-type MOSFET. It is similar to that for the other FET types. Here it is biased using voltage divider biasing. The standard approach of setting the source voltage to half the supply voltage in order to make possible maximum symmetrical swing can be employed. Another approach is to set the gate to half of the supply voltage by selecting $R_{G1} = R_{G2}$ and then using Eq. (3.22) to determine the value of R_S to give the desired value of drain current.

Example 3.20
Using the configuration in Fig. 3.49, design a voltage divider-biased common drain amplifier

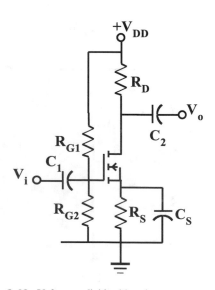

Fig. 3.48 Voltage divider-biased common source enhancement-type MOSFET

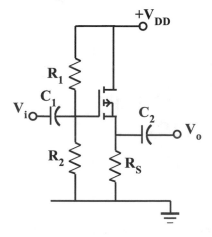

Fig. 3.49 Common drain amplifier

using an enhancement-type MOSFET having $V_T = 2$ V, $k = 0.25$ mA/V² and use $V_{DD} = 20$ V.

Solution The circuit is shown in Fig. 3.49. Setting $R_{G1} = R_{G2} = 1$ M this gives $V_G = 10$ V. For a quiescent drain current of 2 mA, $V_{GS} = V_T + \sqrt{\frac{I_D}{k}} = 2 + \sqrt{\frac{2\ mA}{0.25\ mA/V^2}} = 4.8$ V . Hence, $V_S = V_G - V_{GS} = 10 - 4.8 = 5.2$ V and therefore $R_S = 5.2/2$ mA $= 2.6$ k.

3.6.6 Enhancement-Type MOSFET Common Gate Amplifier

The common gate configuration shown in Fig. 3.50 can be derived from the voltage divider-biased common source in Fig. 3.48. Once again by grounding the input capacitor C_1, ungrounding the bypass capacitor C_S and injecting the input signal through this latter capacitor, the circuit now functions as a common gate amplifier. The signal output is taken at the drain.

Example 3.21
Using the configuration in Fig. 3.50, design a voltage divider-biased common gate amplifier using an enhancement-type MOSFET having $V_T = 2$ V, $k = 0.05$ mA/V², and use $V_{DD} = 15$ V as used in the voltage divider-biased common source amplifier in Example 3.19.

Solution The design procedure used in Example 3.19 for the voltage divider-biased common source amplifier is employed to produce the design shown in Fig. 3.48. However, now capacitor C_1 is grounded and the input signal applied at capacitor C_S which is ungrounded. The resulting circuit is shown in Fig. 3.51.

3.7 Applications

In this section, we discuss several circuits in which field-effect transistors are applied. Before doing so, we summarize and compare the characteristics of the BJT, JFET and MOSFET in Table 3.1. These varying characteristics enable these devices to be used in a wide range of applications, some of which will be discussed below.

Headphone Amplifier
The circuit shown in Fig. 3.52 is a common source amplifier using a JFET that provides an amplified signal to a pair of high-impedance headphones such as the widely available piezo-electric earphones. It uses the design principles outlined in Sect. 3.4.1 and incorporates a volume control potentiometer in the drain circuit instead of a fixed resistor. Capacitor C_2 ensures that none of the DC interferes with the headset. As a rule of thumb, the coupling capacitors are chosen such

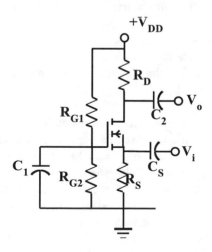

Fig. 3.50 Common gate amplifier

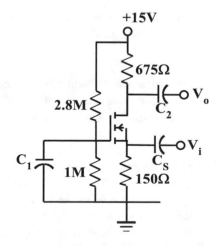

Fig. 3.51 Solution for Example 3.21

Table 3.1 Comparison of characteristics

Characteristic	BJT	JFET	MOSFET
Control type	Current (and voltage) controlled	Voltage controlled	Voltage controlled
Controlling terminal	Base	Gate	Gate
Current flow	Bipolar (majority and minority carriers)	Unipolar (majority carriers)	Unipolar (majority carriers)
Junction	One forward- and one reverse-biased junctions	One reverse-biased junction	One reverse-biased junction
Input resistance	Moderate	High	Very high
Gain	High	Low	Low
Noise	Moderate (low at low currents)	Low	Low
Mutual characteristic	Non-linear	Non-linear	Non-linear

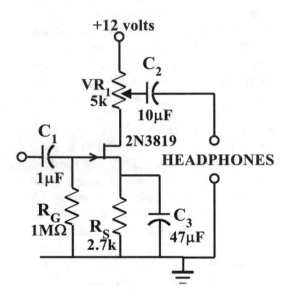

Fig. 3.52 Headphone amplifier

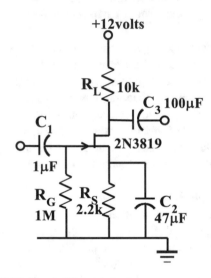

Fig. 3.53 Preamplifier circuit

that their reactance value is equal to the imped-ance of the load at the lowest desired frequency of operation.

Ideas for Exploration Explain why the head-phone volume goes down as the wiper of the potentiometer moves towards the positive supply voltage.

JFET Preamplifier

The circuit shown in Fig. 3.53 is another common source amplifier that amplifies the signal from a crystal microphone or other signal source. The design principles outlined in Sect. 3.4.1 again apply. The output is used to drive a power amplifier.

Ideas for Exploration

(i) Re-design the circuit using a p-channel JFET such as the 2N5460.

(ii) Simulate the circuit in Multisim and deter-mine the gain.

Crystal Microphone Lead Extender

Piezoelectric or crystal microphones have a high output impedance of about 1 M. As a result, they cannot be used with long leads into the amplifier. The circuit shown in Fig. 3.54 has a low output impedance and allows the crystal microphone to be used with long leads. It comprises a common drain or source follower amplifier using a 2N4393 JFET with $I_{DSS} = 10$ mA and $V_T = -2$ V. In order to make it portable, a 9 volt battery is used to

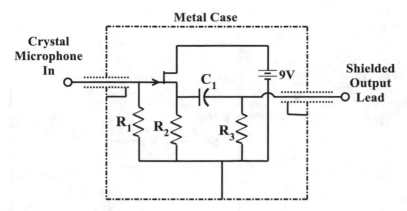

Fig. 3.54 Crystal microphone lead extender

provide power. Choosing a drain current of 2 mA, Shockley's equation gives $V_{GS} = \left(1 - \sqrt{\frac{2}{10}}\right) \times (-2) = -1.1$ V. Then $R_S = 1.1$ V/2 mA $= 550$ Ω. Let $R_1 = 2.2$ M giving the circuit a high input impedance. The low output impedance of the JFET requires that a large electrolytic capacitor $C_1 = 100$ µF value be employed at the output. Resistor $R_3 = 10$ k connects the negative plate of the capacitor to ground, thereby ensuring that it is properly polarized.

Ideas for Exploration

(i) Modify the circuit by using a bipolar supply of ±9 V. Such voltages can be supplied by two 9 volt batteries for portable use. In this design, R_S connects the source terminal of the JFET to the negative supply and not ground as before. For $I_D = 2$ mA the required gate-source voltage is $V_{GS} = -1.1$ V. Since the gate is at zero potential and must be 1.1 V lower than the source, it follows that the source is at 1.1 V. Hence, the voltage drop across R_S is 1.1 V $-$ (-9 V) $= 10.1$ V. The value of R_S is therefore given by $R_S = 10.1$ V/2 mA $= 5$ k. The increased supply voltage provides increased symmetrical swing.

(ii) Test both circuits using several meters of shielded cable feeding into a power amplifier.

JFET DC Voltmeter

The circuit shown in Fig. 3.55 employs a JFET in a DC voltmeter application that is able to measure voltages from 0 to 100 V with an input impedance of 1M on all of five ranges. It comprises a JFET in the source follower mode driving a microammeter in a Wheatstone bridge arrangement. Resistors R_8 and R_9 and potentiometer VR_2 form a potential divider across the supply such that a zero-reference voltage at point A means that point B is at -4 V and point C is at $+8$ V. Selecting a drain current of 1 mA, Shockley's equation for the 2N3819 JFET gives $V_{GS} = -2.8$ V. Hence with the gate connected to the reference voltage point A through the resistors R_1 to R_6, the voltage drop across R_7 is 2.8 $-$ (-4) $= 6.8$ V which gives $R_7 = 6.8/1$ mA $= 6.8$ k. Thus Tr1, R_7 along with R_8, VR_2 and R_9 form a Wheatstone bridge network and VR_2 is adjusted such that for zero signal into the JFET, the bridge is balanced and no current flows through the meter. A positive voltage signal at the input is delivered at the source of the JFET, and this causes the bridge to go out of balance such that current proportional to the input signal flows in the meter. Using $R_9 = 1.5$ k, then the current through the R_8, VR_2, R_9 chain is 4/1.5 k $= 2.67$ mA. For a 2 k potentiometer for VR_2, then the series arrangement of R_8 and VR_2 has a voltage drop given by 2.67 mA $\times$ (R_8 + 2 k) $= 8$ V giving $R_8 = 1$ k. This allows a voltage drop across VR_2 of 5.3 V which enables

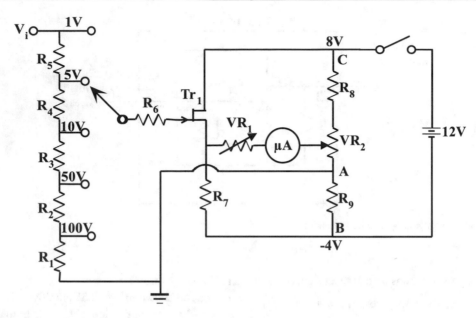

Fig. 3.55 JFET DC voltmeter

balancing of the bridge since the source voltage is 2.8 V. VR_1 is chosen to be a 2.5 k potentiometer that enables calibration of the meter. The potential divider network R_1 to R_5 provides ranges of 1 V, 5 V, 10 V, 50 V and 100 V. These values are $R_1 = 10$ k, $R_2 = 10$ k, $R_3 = 80$ k which is made up with standard resistor values 68 k + 12 k, $R_4 = 100$ k and $R_5 = 800$ k which is made up with standard resistor values 680 k + 120 k. Therefore, when on the 5 V range, 5 V at the input would result in 1 V at the junction of R_3 and R_2. Similarly, when on the 50 V range, 50 V at the input would result in 1 V at the junction of R_2 and R_1. Thus, with no signal applied, VR_2 is adjusted for zero current through the meter. With the meter switched to the 1 V range, a 1 V DC is applied and VR_1 adjusted for full-scale deflection. In order to avoid excessive drift, this circuit will benefit from the use of a Zener-regulated power supply. Resistor $R_6 = 1$ M protects the JFET from over-voltage at the input.

Ideas for Exploration (i) Expand the voltage ranges of the voltmeter from five to seven to cover the range 1 V to 1000 V, using the attenuator network shown in Fig. 3.56.

JFET AC Voltmeter
The circuit of an AC voltmeter using a JFET is shown in Fig. 3.56. The circuit utilizes a 2N4868 JFET in the common source configuration such that voltage signals at the gate are amplified and delivered at the drain. The pinch-off voltage for his FET is $V_P = -2$ V with $I_{DSS} = 2$ mA. Hence for a drain current of 0.4 mA, Shockley's equation gives $V_{GS} = \left(1 - \sqrt{\frac{0.4}{2}}\right)(-2) = -1.1$ V . Thus $R_{10} = 1.1/0.4 = 2.75$ k. In order to allow sufficient voltage swing at the output, resistor R_9 is set at 10 k. The voltage drop across this resistor is therefore 4 V, and with 1.1 across R_{10} that leaves about 4 V across the transistor drain-source junction. The output of the drain drives diodes D_1 and D_2 through coupling capacitor $C_2 = 1$ μF. On the positive half-cycle at the output, diode D_2 turns off, while diode D_1 turns on. Current through R_9 flows through diode D_1. On the negative half-cycle, diode D_1 turns off and diode D_2 turns on. Current then flows through the 100 μA meter-potentiometer-D_2 connection into the drain of the FET. $VR_1 = 20$ k is used to calibrate the meter. Resistors R_1-R_7 form an input attenuator which reduces the voltage to be measured before

Fig. 3.56 JFET AC voltmeter

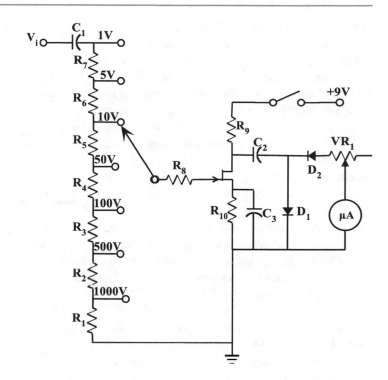

it is applied to the gate of the JFET. This divider network provides seven ranges of 1 V, 5 V, 10 V, 50 V, 100 V, 500 V and 1000 V such that the JFET gate voltage is 1 V when the maximum voltage on any range is applied at the measurement input. For this, $R_7 = 800$ k, $R_6 = 100$ k, $R_5 = 80$ k, $R_4 = 10$ k, $R_3 = 8$ k, $R_2 = 1$ k, $R_1 = 1$ k. Therefore, on the 5 V range, 5 V at the input would result in 1 V at the junction of R_3 and R_2. Similarly, on the 100 V range, 100 V at the input would result in 1 V at the junction of R_2 and R_1. The input attenuator presents an input impedance of 1M on all ranges. Resistor $R_8 = 1$ M protects the JFET from over-voltage at the input. Coupling capacitor $C_1 = 1$ μF prevents DC from entering the system. Coupling capacitor $C_2 = 10$ μF and bypass capacitor $C_3 = 100$ μF.

Ideas for Exploration (i) Test the frequency response of the system by applying a fixed input voltage, varying the frequency from about 20 Hz to about 100 kHz and noting the response on the meter.

MOSFET Amplifier

This amplifier circuit shown in Fig. 3.57 employs an enhancement MOSFET in a simple

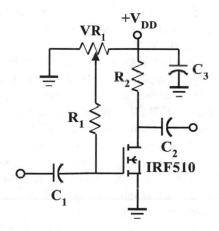

Fig. 3.57 MOSFET amplifier

configuration that is quite versatile. The potentiometer VR_1 is adjusted in order to overcome the threshold voltage and turn on the device. Its value is not critical and can be between 10 k and 100 k. Resistor R_1 ensures that the input impedance to the amplifier is sufficiently high. Its value can be quite high, say 100 k to 1 M since a very tiny current flows through into the gate of the MOSFET because of its inherently high input resistance. Through this adjustment, the drain voltage is set at half the supply voltage for maximum symmetrical swing. If R_2 is sufficiently low, the circuit can be used to drive a speaker for audio output. In that case the quiescent current will be high, and therefore a heatsink will be required to keep the MOSFET cool. For example, using a 9 volt power supply and $R_2 = 50\ \Omega$, adjustment of VR_1 for maximum symmetrical swing gives a drain voltage of $9/2 = 4.5$ volts. Hence, the quiescent current through the MOSFET is $4.5/50 = 90$ mA. In practice V_{DD} can be in the range of about 9 volts to 24 volts. Because of the high input resistance of the MOSFET, capacitor C_1 can be a relatively low value of 1 μF. Since capacitor C_2 couples the output signal to a possibly low-impedance load such as headphones or a speaker, its value must be higher. $C_2 = 220$ μF has a reactance at 100 Hz of $10^6/2\pi \times 100 \times 220 = 7\ \Omega$ which is comparable to an 8Ω speaker load. A higher value capacitor will improve the low-frequency response of the circuit. Capacitor C_3 provides power supply filtering if necessary. The IRF510 MOSFET used in the circuit has a maximum drain-source voltage rating of 100volts and can handle drain currents of up to 5.6 A. It can dissipate 125 W of power. The threshold turn-on voltage is typically 3volts. As the quiescent current is increased, the increased power dissipation in the MOSFET may necessitate a heatsink (see Chap. 9). Finally, since the MOSFET can sink large current values on the negative half-cycle of the output signal while on the positive half-cycle output current must be sourced through resistor R_2, the output voltage will experience some asymmetry with the negative half-cycle being greater in amplitude than the positive half-cycle. This asymmetry will increase as the load resistor value decreases. A circuit example is shown in Fig. 3.58.

Ideas for Exploration (i) Replace R_2 with an 8ohm speaker and operate the circuit. Note that current will always be flowing through the coil of the speaker. Investigate the issues associated with this, particularly speaker size.

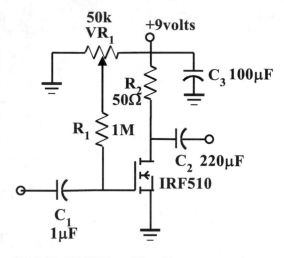

Fig. 3.58 MOSFET amplifier with component values

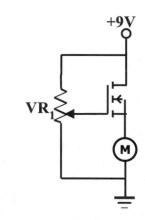

Fig. 3.59 DC motor speed controller

DC Motor Speed Controller
This simple project is a DC motor speed controller based on an enhanced MOSFET such as the STP55NF06 in the source follower configuration (Fig. 3.59). The potentiometer $VR_1 = 100$ k connected across the power supply applies a variable positive gate-source voltage that turns on the MOSFET. The motor is connected between the source of the MOSFET and ground, while the drain is connected to the 9 V supply. Thus, as the voltage on the wiper of the potentiometer is increased, the gate voltage is increased until the threshold voltage of the MOSFET is exceeded

and the device turns on. This voltage is transferred to the low-voltage DC motor which then operates. The speed of the motor increases as the voltage is increased. The value of the potentiometer is not critical, and the high input impedance of the FET allows the use of a high-value potentiometer such as 100 k or higher.

Ideas for Exploration

(i) Replace the motor by a 9 volt lamp in order to realize one form of a lamp dimmer.
(ii) Remove the motor and use the system as a variable bench supply by taking the voltage output from the source of the MOSFET. How does the threshold voltage of the MOSFET affect the operation of the circuit?

Light Dimmer
This is another simple using the IRF540N enhanced MOSFET with a minimum threshold voltage of 2 V but now in the common drain mode (Fig. 3.60). A potentiometer VR_1 in series with a fixed resistor R_1 is connected across the power supply. The wiper of the potentiometer applies a variable positive gate-source voltage that turns on the MOSFET. The lamp is connected between the drain of the MOSFET and the 12 V supply. Thus, as the voltage on the wiper of the potentiometer is increased, the

gate-source voltage is increased until the threshold voltage of the MOSFET is exceeded and the device turns on lighting the lamp. The brightness of the lamp is controlled by the amount of current flowing through the MOSFET which increases as the voltage is increased. The value of the potentiometer is not critical, and the high input impedance of the MOSFET allows the use of a high-value potentiometer such as 100 k or higher. If $VR_1 = 500$ k, then for $R_1 = 47$ k, the minimum voltage on the gate will be 1 V which is just below the threshold voltage for the MOSFET.

Ideas for Exploration (i) Replace the lamp by a 12 volt DC motor and experiment with the resulting motor speed controller.

Adjustable Timer
The circuit in Fig. 3.61 is an adjustable timer. It used the IRF510N enhanced MOSFET in the common drain configuration. When the switch is depressed, capacitor C_1 charges up to the supply voltage and turns on the MOSFET, thereby lighting the LED. Resistor $R_2 = 1$ k limits the LED current to about 10 mA. When the switch is released, the capacitor C_1 discharges through R_1. When the capacitor voltage falls below the threshold voltage of the MOSFET, it turns off and turns off the LED. For $C_1 = 500$ μF and $R_1 = 10$ k, the discharge time is about 10 seconds.

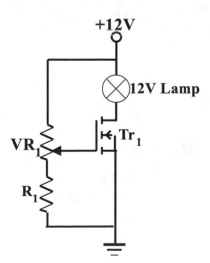

Fig. 3.60 Light dimmer

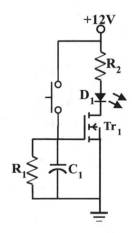

Fig. 3.61 Adjustable timer

Ideas for Exploration

(i) Derive the equation $t = R_1 C_1 \ln\left(\frac{V_I}{V_T}\right)$ for the discharge time of the capacitor through the resistor, given the initial voltage V_I and the threshold voltage of the MOSFET V_T.

(ii) Vary the resistor and capacitor values to achieve different delay times.

(iii) Experiment with different loads between the supply and the drain of the MOSFET.

Research Project 1

This project involves the design of a circuit for checking inductance values. It utilizes the principle of series resonance which occurs in a series connection of an inductor and a capacitor when, at a particular frequency, the magnitude of the reactance of the capacitor is equal to the magnitude of the reactance of the capacitor. In such a case, the voltages across these two components cancel and the circuit impedance is at a minimum. Current flow through the circuit is therefore at a maximum. The circuit in Fig. 3.62 operates on this basis. The JFET is connected in the source follower mode with a signal of variable frequency applied at the input. The output drives a capacitor $C_3 = 0.01\ \mu F$ and the inductor L to be tested with a microammeter inserted to measure current. In operation, a sinusoidal signal from an oscillator is passed into the system. Potentiometer VR_1 enables the adjustment of the amplitude of the signal. The frequency of the signal is adjusted

over a wide range, and at the resonant frequency of the LC circuit, the current in the microammeter will increase sharply. The signal frequency and amplitude are adjusted until the maximum current increase occurs corresponding to almost full-scale deflection on the meter. At resonance, $2\pi fL = 1/2\pi fC$ giving $L = 10^9/395f^2$ in Henrys where f is the frequency reading on the signal oscillator.

Ideas for Exploration

(i) Test the system with some known inductances.

(ii) Does the use of the JFET confer any particular advantage to the circuit?

Research Project 2

This second research project investigates the internal structure of the MOSFET and in particular identifies and utilizes the internal capacitance between the gate and the source. The data sheet of the IRF1405 provides information on the gate-source capacitance for this device. The project involves utilizing this device in a touch-sensitive switch that exploits this junction capacitance as shown in Fig. 3.63. It employs the IRF1405 enhancement MOSFET in the common source configuration. The relay is a 12 V device whose contacts connect a lamp or other appliance to the mains supply. Diode D_1 protects the MOSFET from back emf from the relay coil. When the contacts A are bridged by touch, charge is

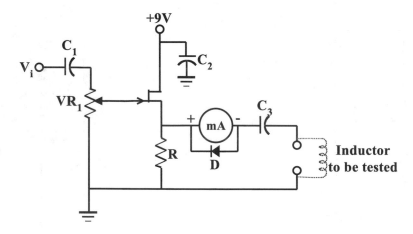

Fig. 3.62 Inductance tester

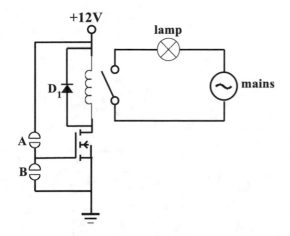

Fig. 3.63 Touch-sensitive switch

transferred from the supply to the internal gate-source capacitance of the MOSFET, thereby turning on the device. When the contacts B are touched, the capacitor is discharged and the MOSFET turns off. This system can be used in a variety of applications.

Ideas for Exploration

(i) Investigate the internal structure of the enhancement MOSFET and the junction capacitances that are used in the switching.

(ii) Try using the system to switch on other devices such as a motor or radio.

Problems

1. Briefly explain the following:
   (a) The high input impedance of a JFET
   (b) The high input impedance of a MOSFET
   (c) The main difference between enhancement and depletion mode MOSFETS
   (d) Pinch-off in a JFET
   (e) The difference between n-channel and p-channel JFETS
   (f) The main similarity between a BJT and an enhancement-type MOSFET
   (g) The main similarity between a JFET and a depletion-type MOSFET

2. A JFET has a signal voltage of 0.75Vpk value applied to its gate relative to its source.

The drain current varies by ± 1.5 mA about its quiescent value. Calculate g_m.

3. An n-channel JFET has $I_{DSS} = 6$ mA and $V_P = -5$ V. For a drain current of 4 mA, determine g_m using Shockley's equation. If this JFET is used in a common source amplifier with its source grounded for signals and drain resistance $R_L = 10$ k, determine the voltage gain from gate to drain.

4. Using fixed-bias, bias a n-channel JFET for a drain current of 3 mA using a device having $I_{DSS} = 7$ mA and $V_P = -4$ V. Use $V_{DD} = 14$ V and choose R_L for maximum symmetrical swing.

5. Bias a p-channel JFET for a drain current of 2 mA using fixed-bias. For the FET, $I_{DSS} = 12$ mA and $V_P = +5$ V. Use $V_{DD} = -18$ V and select R_L for maximum symmetrical swing.

6. Using the FET of problem 4 above, find R_S to replace the fixed-bias voltage source V_{GG}.

7. Using Shockley's equation $I_D = I_{DSS}\left(1 - \frac{V_{GS}}{V_P}\right)^2$ and a 26 volt supply, design a common source amplifier based on self-bias, using an n-channel JFET having a pinch-off voltage of -2.5 volts and $I_{DSS} = 5$ mA. Use a 3 mA quiescent current. Use $g_m = -2\frac{\sqrt{I_D I_{DSS}}}{V_P}$ to calculate the voltage gain of your circuit.

8. Using an n-channel JFET having $I_{DSS} = 3$ mA and $V_P = -5.5$ volts, bias the device using voltage divider biasing for a drain current of 2.5 mA. Design for a maximum symmetrical swing with $V_{DD} = 18$ volts.

9. State the advantages and disadvantages of self-bias as compared with voltage divider bias in JFETS.

10. For the self-biased common source amplifier in problem 7, re-design the system to use a bipolar supply of ± 15 V, returning R_S to the negative supply voltage.

11. Using Shockley's equation $I_D = I_{DSS}\left(1 - \frac{V_{GS}}{V_P}\right)^2$ and a 27 volt supply, design a common source amplifier using an p-channel JFET having a pinch-off voltage

of +4 volts and I_{DSS} = 6 mA. Use a 1 mA quiescent current and fixed-bias. Noting $g_m = -2\frac{\sqrt{I_D I_{DSS}}}{V_P}$, determine the voltage gain of your circuit.

12. Using Shockley's equation $I_D = I_{DSS}\left(1 - \frac{V_{GS}}{V_P}\right)^2$ and an 18 volt supply, design a common source amplifier using an n-channel JFET having a pinch-off voltage of −3 volts and I_{DSS} = 4 mA. Use voltage divider bias and a 2 mA quiescent current. Explain the functions of the input gate resistor and the source resistor.

13. Using Shockley's equation and a 24 volt supply, design a common source amplifier using an n-channel depletion MOSFET having a pinch-off voltage of −4 volts and I_{DSS} = 5 mA. Use a 5 mA quiescent current and self-bias. Calculate the voltage gain of your circuit using $g_m = -2\frac{\sqrt{I_D I_{DSS}}}{V_P}$.

14. For the fixed-bias common source amplifier in problem 13, re-design the system to use a bipolar supply of ±22 V and return R_S to the negative supply voltage instead of ground.

15. Determine the drain current in the circuit shown in Fig. 3.64. if the quiescent gate-source voltage for the JFET is −2.8 V.

16. Design a common drain amplifier using an n-channel JFET having I_{DSS} = 4.1 mA and $V_P = -5.3$ V and voltage divider biasing. Design for a maximum symmetrical swing with V_{DD} = 27 V.

17. Design a source follower circuit using a p-channel JFET with I_{DSS} = 8 mA, $V_P = +2.3$ V and a −12 V supply.

18. For the common drain amplifier in problem 16, re-design the system to use a bipolar supply of ±18 V and return R_S to the negative supply voltage instead of ground.

19. Design a common gate amplifier using an n-channel JFET having I_{DSS} = 3.2 mA and $V_P = -1.9$ V. Design for a maximum symmetrical swing with V_{DD} = 18 V.

20. Using self-bias, design a biasing network for a common source n-channel depletion-type

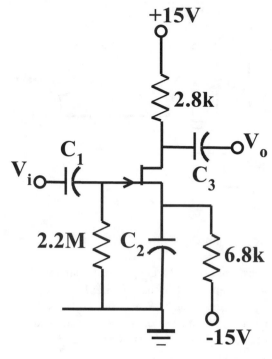

Fig. 3.64 Circuit for Question 15

MOSFET having I_{DSS} = 2.9 mA and $V_P = -4.7$ V. Design for maximum symmetrical swing.

21. Using an n-channel depletion MOSFET having I_{DSS} = 3.7 mA and $V_P = -3.9$ V, bias the device for common source operation using voltage divider biasing and a drain current of 3 mA. Design for a maximum symmetrical swing with V_{DD} = 16 V.

22. Repeat the design in problem 21 using a p-channel depletion MOSFET having I_{DSS} = 4.1 mA and $V_P = +4.7$ V.

23. Design a common drain amplifier using an n-channel depletion MOSFET having I_{DSS} = 5.9 mA and $V_P = -6.1$ V. Design for maximum symmetrical swing with V_{DD} = 32 V.

24. Design a common gate amplifier using an n-channel depletion MOSFET with I_{DSS} = 1.3 mA and $V_P = -2.9$ V. Design for maximum symmetrical swing using a 14 V supply.

25. Using the circuit shown in Fig. 3.65, design a feedback biasing network for an

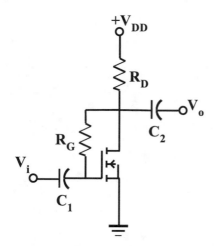

Fig. 3.65 Circuit for Question 25

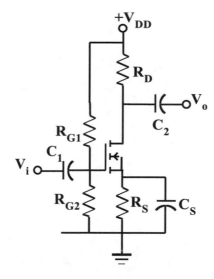

Fig. 3.67 Circuit for Question 28

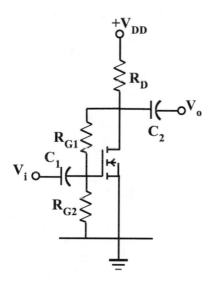

Fig. 3.66 Circuit for Question 27

enhancement-type MOSFET having $V_T = 2.7$ V, $I_{Don} = 5$ mA and $V_{GSon} = 7$ V. Use $V_{DD} = 15$ V.

26. Design a feedback biasing network for a common source enhancement-type p-channel MOSFET having $V_{GST} = 2.5$ V, $I_{Don} = 4$ mA and $V_{GSon} = 7$ V. Use $V_{DD} = 25$ V.

27. Using the configuration in Fig. 3.66, design a modified feedback biasing network for a common source enhancement-type MOSFET having $V_T = 3.3$ V, $I_{Don} = 6.4$ mA and

$V_{GSon} = 5.3$ V. The supply voltage is $V_{DD} = 12$ V.

28. Using the configuration in Fig. 3.67, design a voltage divider-biased common source amplifier using an enhancement-type MOSFET having $V_T = 2.4$ V and $k = 0.07$ mA/V^2, and use $V_{DD} = 22$ V.

29. Using the configuration in Fig. 3.49, design a voltage divider-biased common drain amplifier using an enhancement-type n-channel MOSFET having $V_T = 2.4$ V and $k = 0.29$ mA/V^2, and use $V_{DD} = 25$ V.

30. Using the configuration in Fig. 3.50, design a voltage divider-biased common gate amplifier using an enhancement-type MOSFET having $V_T = 2.1$ V and $k = 0.05$ mA/V^2, and use $V_{DD} = 30$ V.

Bibliography

R.L. Boylestad, L. Nashelsky, *Electronic Devices and Circuit Theory*, 11th edn. (Pearson Education, New Jersey, 2013)

J. Eimbinder, *FET Applications Handbook*, 2nd edn. (Tab Books, Pasadena, 1970)

F.G. Rayer, *50 (FET) Field Effect Transistor Projects* (Babani Press, London, 1977)

BJT and FET Models

In Chaps. 2 and 3, the BJT and FET were introduced and their operation discussed. The need for biasing was established and various biasing schemes presented. Also, the operation of the BJT and FET as amplifiers each in three configurations was discussed. While the methods utilized gave reasonably good answers, more precise design requires the use of BJT and FET models or equivalent circuits. The models we adopt are the hybrid parameter or h-parameter model and the y-parameter model. These transistor parameters can in general be obtained from manufacturer's data sheets. At the end of this chapter, the student will be able to:

- Explain h- and y-parameters
- Use h- and y-parameters in the analysis of transistor circuits

4.1   H-Parameters and the BJT

Small-signal amplifiers are amplifiers that are assumed to operate over the linear range of the active device (BJT or FET) by using only small-signal voltages and currents. Therefore the AC operation may be deduced by representing the transistor by an equivalent circuit consisting of linear signal generators and circuit elements. In deducing this equivalent circuit, the transistor is regarded as an active two-port network as shown in Fig. 4.1, i.e. it has two input terminals

and two output terminals. The BJT is considered first.

There are four external variables, the input voltage and current V_i and I_i and output voltage and current V_o and I_o. Since operation takes place over the linear range of transistor characteristics (small-signal operation), V_i, I_i, V_o and I_o can be related to each other by constant parameters. Thus V_i and I_o can be related to V_o and I_i by four constants, called hybrid or h-parameters giving

$$V_i = h_i I_i + h_o V_o \tag{4.1}$$

$$I_O = h_f I_i + h_o V_o \tag{4.2}$$

which can be represented as shown in Fig. 4.2. (Note that there are other ways of relating the four variables. See Sect. 4.5.)

Under small-signal conditions,

$$h_i = \frac{V_i}{I_i}\big|_{V_o=0} \text{ input impedance with short -circuited output} \tag{4.3}$$

$$h_f = \frac{I_o}{I_i}\big|_{V_o=0} \text{ forward current gain with short -circuited output} \tag{4.4}$$

$$h_r = \frac{V_i}{V_o}\big|_{I_i=0} \text{ voltage feedback ratio with open -circuited input} \tag{4.5}$$

$$h_o = \frac{I_o}{V_o}\big|_{I_i=0} \text{ output admitance with open -circuited input} \tag{4.6}$$

The values of these parameters depend on the transistor configuration. In order to distinguish

© Springer Nature Switzerland AG 2021
S. J. G. Gift, B. Maundy, *Electronic Circuit Design and Application*,
https://doi.org/10.1007/978-3-030-46989-4_4

among the three configurations, i.e. common emitter (CE), common base (CB) and common collector (CC), a second subscript is added to each h-parameter. Thus for the CE configuration, h_i becomes h_{ie}; for the CB, it is h_{ib}; for the CC, it is h_{ic}; and so on for the other parameters.

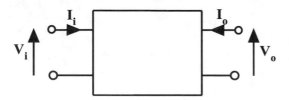

Fig. 4.1 Transistor as an active two-port network

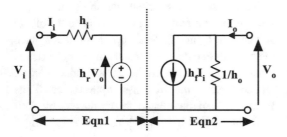

Fig. 4.2 Transistor h-parameter equivalent circuit

Now for the common emitter configuration, h_{ie} is the ratio of a small change in input voltage (ΔV_i) to the resulting small change in input current (ΔI_i) which for this configuration is the input impedance. Hence from Eq. (2.33) in Chap. 2,

$$h_{ie} = \frac{h_{fe}}{40 I_C} \qquad (4.7)$$

Here h_{fe} is the ratio of the change in output current to the change in input current producing it and hence

$$h_{fe} = \beta \qquad (4.8)$$

The actual value of h_{fe} is a function of the transistor collector current I_C, temperature and signal frequency. Figure 4.3 shows a typical plot of h_{fe} against collector current $I_C(mA)$.

It can be seen that h_{fe} initially increases with collector current but reaches a maximum at around 20 mA. Manufactures generally state the maximum value of h_{fe} and the associated collector current at which it attains this value. A plot of current gain against temperature is shown in Fig. 4.4 where h_{fe} increases with temperature by as much as twice its value over a temperature

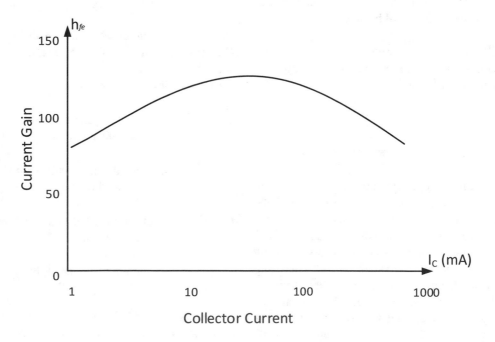

Fig. 4.3 Transistor current gain against collector current

change from 20 to 80 °C. Finally, the current gain h_{fe} of a transistor varies with signal frequency, its value falling off at higher frequencies. This is illustrated in Fig. 4.5. f_β is the frequency at which the current gain falls by 3 dB of its low-frequency value and is referred to as the cut-off frequency of the transistor. The frequency at which the current gain falls to unity is called the transition frequency f_t.

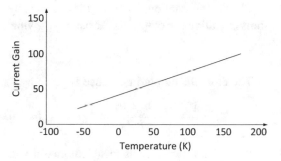

Fig. 4.4 Transistor current gain against temperature

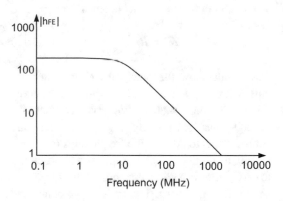

Fig. 4.5 Transistor current gain against frequency

4.2   Analysis of a Common Emitter Amplifier Using H-Parameters

We are now ready to use the h-parameter equivalent circuit to analyse an amplifier circuit. Specifically, we wish to study the CE amplifier configuration. Now the h-parameter model is valid only for linear systems and since the transistor characteristics are nonlinear, the signals must be restricted to small voltage and current variations about the Q point such that the transistor operates approximately linearly. In such circumstances, the h-parameters can be used to analyse transistor networks for the small time-varying components of signals and not the DC component.

Consider the basic CE circuit shown in Fig. 4.6 in which the biasing components have been omitted for simplicity. In an AC equivalent circuit, the battery is a short circuit to AC. Thus replacing the transistor by the h-parameter equivalent circuit yields Fig. 4.6(a) which becomes Fig. 4.6(b) after rearranging. This is called the AC equivalent circuit for the CE amplifier. Since $h_{re} \approx 0$ and assuming $1/h_{oe} \gg R_L$, then this circuit can be reduced to that shown in Fig. 4.7.

All of the linear circuit laws including those of Ohm, Thevenin, Norton, Kirchhoff and the Superposition theorem can now be applied since the non-linear transistor has been linearized about the operating point. We can now derive an expression for the voltage gain A_v of the amplifier, where $A_v = V_o/V_i$.

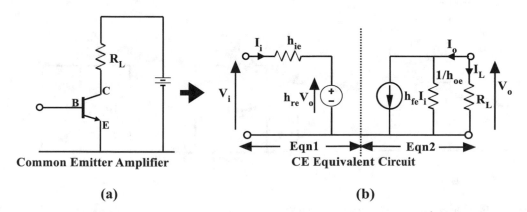

Common Emitter Amplifier

(a)

CE Equivalent Circuit

(b)

Fig. 4.6 Common emitter amplifier and its equivalent circuit

From Fig. 4.7, noting that $I_i = I_b$ where I_b is the signal current flowing into the base of the transistor, and $I_o = I_c$ where I_c is the signal current flowing into the collector,

$$V_o = -I_o R_L = -h_{fe} I_b R_L \qquad (4.9)$$

$$V_i = h_{ie} I_b \qquad (4.10)$$

Therefore the voltage gain $A_v = V_o/V_i$ is given by

$$A_V = \frac{V_o}{V_i} = \frac{-h_{fe} R_L}{h_{ie}} \qquad (4.11)$$

The negative sign is indicative of an inversion between the output voltage and the input voltage. Similarly, the current gain $A_i = I_o/I_i$ is given by

$$A_i = \frac{I_o}{I_i} = \frac{h_{fe} I_b}{I_b} = h_{fe} \qquad (4.12)$$

The input impedance Z_i can be obtained from (4.10). Z_i is the ratio of the applied input voltage V_i to the input current I_b, i.e.

$$Z_i = \frac{V_i}{I_b} = h_{ie} \qquad (4.13)$$

In order to determine the output impedance Z_o, we first find the Thevenin equivalent of the amplifier output. Then $Z_o = V_{os}/I_{sc}$ where V_{oc} is the open-circuit voltage and I_{sc} is the short-circuit current. The open-circuit output voltage is

$$V_{oc} = -h_{fe} I_b R_L \qquad (4.14)$$

The short-circuit output current is obtained by short-circuiting the output, i.e. resistor R_L giving

$$I_{sc} = -h_{fe} I_b \qquad (4.15)$$

Therefore the output impedance is given by

$$Z_o = \frac{V_{oc}}{I_{sc}} = \frac{-h_{fe} I_b R_L}{-h_{fi} I_b} = R_L \qquad (4.16)$$

If the parameter $1/h_{oe}$ is included, then it will appear in parallel with R_L, and the open-circuit output voltage V_{oc} becomes $V_{oc} = -h_{fe} I_b R_L // \frac{1}{h_{oe}}$. The output impedance then becomes

$$Z_o = R_L // \frac{1}{h_{oe}} \qquad (4.17)$$

Consider now the practical CE amplifier circuit with fixed bias and coupling capacitors shown in Fig. 4.8(a). We assume that the capacitors are sufficiently large so that they are short circuits to AC for the frequencies involved. Noting again that the DC supply is a short circuit to AC, the h-parameter equivalent circuit is shown in Fig. 4.8(b). The resistor R_B connected

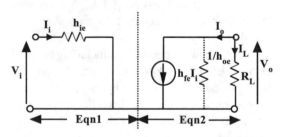

Fig. 4.7 Modified equivalent circuit

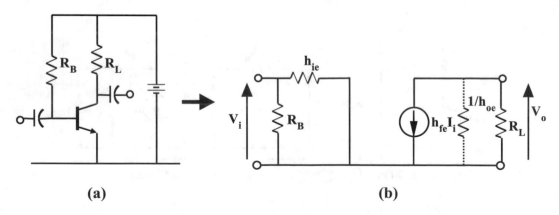

(a)   **(b)**

Fig. 4.8 Fixed-bias common emitter amplifier with equivalent circuit

to the transistor base is grounded through the DC supply. The expressions for the voltage gain and the output impedance are unaffected, but since R_B appears in parallel with h_{ie}, the input impedance is reduced. Hence

$$Z_i = h_{ie} \| R_B \qquad (4.18)$$

The current gain I_o/I_i is slightly reduced since some of the input signal current flow into R_B. However since $R_B \gg h_{ie}$, this reduction is very small.

Example 4.1

For the common emitter amplifier shown in Fig. 4.9, draw the equivalent circuit using h-parameters. Then for a collector current of 1 mA, determine the voltage gain and input impedance.

Solution The equivalent circuit is shown in Fig. 4.10. For this circuit, $h_{ie} = \frac{h_{fe}}{40 I_C} = \frac{100}{40 \times 1 \text{ mA}} = 2.5 \text{ k}$. Hence from (4.18), the input impedance is given by $Z_i = 2.5 \text{ k} \| 1.93 \text{ M} \approx 2.5 \text{ k}$. From (4.11) the voltage gain is given by $A_V = \frac{-100 \times 10 \text{ k}}{2.5 \text{ k}} = -400$.

Example 4.2

The circuit of Example 4.1 is used to drive a 20 k load as shown in Fig. 4.11. Draw the equivalent circuit using h-parameters and determine the voltage gain and input impedance.

Solution The only change in the equivalent circuit is the inclusion of the 20 k load resistor in parallel with the 10 k collector resistor as shown in Fig. 4.12. Thus, the input impedance remains the same but the voltage gain is reduced to

$$A_V = \frac{-100 \times 10 \text{ k} // 20 \text{ k}}{2.5 \text{ k}} = -267$$

For the common emitter amplifier with voltage-divider bias shown in Fig. 4.13, we approach the drawing of the AC equivalent as before.

We short-circuit all capacitors and the DC supply. The resulting AC equivalent circuit is shown in Fig. 4.14. Bias resistors R_1 and R_2 appear in parallel with h_{ie} with no other changes. Hence, the voltage gain remains the same as in the

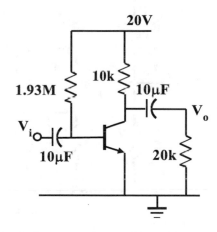

Fig. 4.11 Common emitter amplifier driving a load

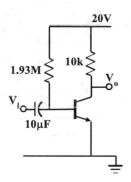

Fig. 4.9 Common emitter amplifier

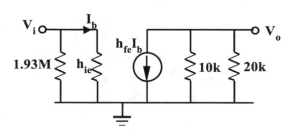

Fig. 4.12 Equivalent circuit of common emitter amplifier driving a load

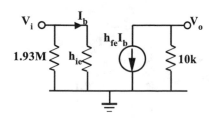

Fig. 4.10 Equivalent circuit of common emitter amplifier

fixed bias case, but the input impedance becomes $Z_i = R_1 \| R_2 \| h_{ie}$. The current gain changes since some of the signal current is lost in the bias resistors R_1 and R_2. The fraction that flows into the base I_b/I_i is using the current divider theorem given by

$$\frac{I_b}{I_i} = \frac{R_1 \| R_2}{h_{ie} + R_1 \| R_2} \qquad (4.19)$$

Hence

$$A_i = \frac{I_o}{I_i} = \frac{I_o}{I_b} \times \frac{I_b}{I_i} = h_{fe} \times \frac{R_1 \| R_2}{h_{ie} + R_1 \| R_2} \quad (4.20)$$

Thus the current gain suffers attenuation when bias components are introduced. For the H-biased common emitter circuit involving a source

resistor R_S connected in series with the input as shown in Fig. 4.15, the equivalent circuit is shown in Fig. 4.16. R_S in conjunction with the resistors R_1, R_2 and h_{ie} form a potential divider and hence introduce voltage signal alternation. The voltage gain A_V is given by

$$A_V = \frac{V_o}{V_i} = \frac{V_o}{V_b} \times \frac{V_b}{V_i} \qquad (4.21)$$

where V_b is the signal voltage at the base of the transistor. Now

$$\frac{V_o}{V_b} = \frac{-h_{fe} R_L}{h_{ie}} \qquad (4.22)$$

and

$$\frac{V_b}{V_i} = \frac{Z_i'}{R_S + Z_i'} \qquad (4.23)$$

where

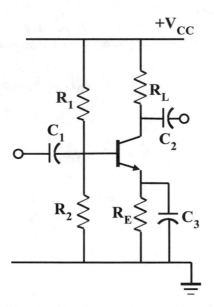

Fig. 4.13 Voltage divider-biased common emitter amplifier

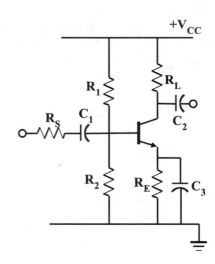

Fig. 4.15 Common emitter amplifier with input resistor

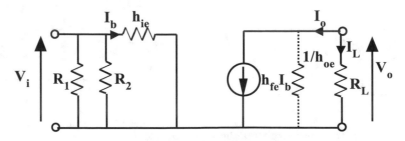

Fig. 4.14 Equivalent circuit of voltage divider-biased common emitter amplifier

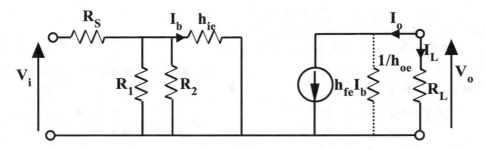

Fig. 4.16 Equivalent circuit of common emitter amplifier with input resistor

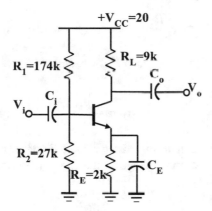

Fig. 4.17 Voltage divider-biased common emitter amplifier

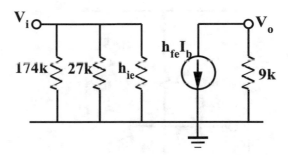

Fig. 4.18 Equivalent circuit of amplifier in Fig. 4.17

$$Z_i' = R_1//R_2//h_{ie} \qquad (4.24)$$

Hence

$$A_V = \frac{-Z_i'}{R_S + Z_i'} \cdot \frac{h_{fe}R_L}{h_{ie}} \qquad (4.25)$$

The input impedance Z_i would now be $Z_i = R_S + Z_i'$

Example 4.3
For the circuit shown in Fig. 4.17, draw the h-parameter equivalent circuit and determine voltage gain, circuit gain, input impedance and output impedance.

Solution The equivalent circuit is shown in Fig. 4.18. The collector current is easily shown to be 1 mA and hence $h_{ie} = \frac{h_{fe}}{40I_C} = \frac{100}{40 \times 1 \text{ mA}} = 2.5 \text{ k}$. Hence the voltage gain is given by $A_V = \frac{-100 \times 9 \text{ k}}{2.5 \text{ k}} = -360$. Using (4.20) current gain $A_i = h_{fe} \times \frac{R_1 \| R_2}{h_{ie} + R_1 \| R_2} = 100 \times$

$\frac{27 \text{ k}//173 \text{ k}}{2.5 \text{ k} + 27 \text{ k}//173 \text{ k}} = 90$. From (4.24) the input impedance is given by $Z_i = R_1//R_2//h_{ie} = 27 \text{ k}// 173 \text{ k}//2.5 \text{ k} = 2.3 \text{ k}$. The output impedance is simply the $9k$ load resistor.

If a 15 k external load resistor is added to the circuit of Fig. 4.17, then the equivalent circuit in Fig. 4.18 is modified by the inclusion of a 15 k load resistor in parallel with the 9 k collector resistor. The voltage gain becomes $A_V = \frac{-100 \times 9 \text{ k}//15 \text{ k}}{2.5 \text{ k}} = -225$. The current gain defined by the ratio of the current into the 15 k resistor to the input current is given by $A_i = h_{fe} \times \frac{R_1 \| R_2}{h_{ie} + R_1 \| R_2} \times \frac{9 \text{ k}}{9 \text{ k} + 15 \text{ k}} = 90 \times .375 = 33.8$. The input impedance remains the same.

4.2.1 Common Emitter Amplifier with Partial Decoupling

The next amplifier to be discussed is the CE amplifier shown in Fig. 4.19 in which the emitter resistor is only partially bypassed. The equivalent circuit is shown in Fig. 4.20. Note that while R_E is short-circuited by the capacitor C_3, R_e is not and therefore appears in the equivalent circuit. Note that both I_b and $h_{fe}I_b$ flow into R_e. Hence, at the input we have

$$V_i = I_b h_{ie} + (I_b + h_{fe}I_b)R_e \\ = I_b(h_{ie} + (1 + h_{fe})R_e) \tag{4.26}$$

At the output

$$V_o = -h_{fe}I_b R_L \tag{4.27}$$

Hence the voltage gain of the circuit is given by

$$A_V = \frac{V_o}{V_i} = \frac{-h_{fe}R_L}{h_{ie} + (1 + h_{fe})R_e} \tag{4.28}$$

Since h_{fe} is large $1 + h_{fe} \approx h_{fe}$. Therefore, dividing through by h_{fe} we get

$$A_V = \frac{-R_L}{R_e + r_e}, \text{where } r_e = h_{ie}/h_{fe} \tag{4.29}$$

If R_e is large, then $R_e \gg r_e$ and (4.29) reduces to

$$A_V \approx \frac{-R_L}{R_e} \tag{4.30}$$

The output impedance is the ratio of the open-circuit output voltage to the short-circuit output current, i.e.

$$Z_o = \frac{-h_{fe}I_b R_L}{-h_{fe}I_b} = R_L \tag{4.31}$$

The circuit input impedance is found by first finding the ratio V_i/I_b which is from (4.26)

$$Z_i' = \frac{V_i}{I_b} = h_{ie} + (1 + h_{fe})R_e \tag{4.32}$$

From Fig. 4.20, this impedance is clearly in parallel with R_1 and R_2 and hence

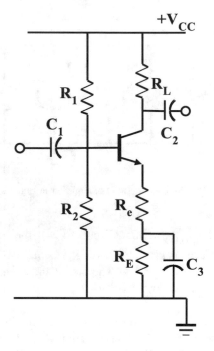

Fig. 4.19 Common emitter amplifier with partially decoupled emitter resistor

Fig. 4.20 Equivalent circuit to Fig. 4.19

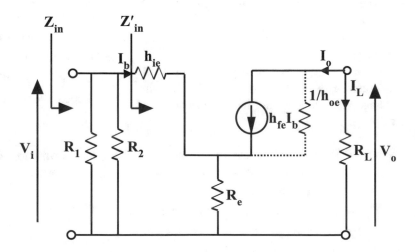

$$Z_i = R_1 \| R_2 \| \{h_{ie} + (1 + h_{fe})R_e\} \qquad (4.33)$$

Comparing this circuit with the CE amplifier with completely bypassed emitter resistor, firstly, the voltage gain magnitude goes from $\frac{h_{fe}R_L}{h_{ie}} = \frac{R_L}{r_e}$ to $\frac{R_L}{r_e + R_e}$, i.e. the presence of R_e significantly reduces the voltage gain. Secondly the input impedance changes from h_{ie} to $h_{ie} + (1 + h_{fe})R_e$ (ignoring the effect of R_1 and R_2), i.e. R_e causes a significant increase in the input impedance, looking into the transistor base. In practice, however, R_1 and R_2 limit the input impedance increase.

Example 4.4

In the circuit of Fig. 4.17, consider the case where the emitter resistor is split such that only 1.6 k is bypassed. Determine the voltage gain and input impedance.

Solution Using the circuit values of Fig. 4.17 and the nomenclature of Fig. 4.19, $R_e = 400$ and $R_E = 1.6$ k. Then from (4.29) the voltage gain is given by $A_V = \frac{-9}{0.4 + 0.025} = -21$. This is significantly reduced from the original value of $A_V = \frac{-9}{0.025} = -360$ as a result of the presence of the unbypassed 400 Ω resistor. This resistor also affects the input resistance which becomes $Z_i = 27$ k$\|173$ k$\|\{2.5$ k $+ (1 + 100)$ 0.4 k$\} = 27$ k$//173$ k$//42.9$ k $= 11$ k.

The reduction of gain resulting from the inclusion of R_e in the emitter circuit of a common emitter amplifier allows the design of such amplifiers for both maximum symmetrical swing and a specific (reduced) gain.

Example 4.5

Using the partially bypassed configuration in Fig. 4.19, design a common emitter amplifier having a gain of 50, using a 2N3904 npn transistor and a 24 V power supply. Determine the input impedance.

Solution Choosing a quiescent current of 1 mA, the design procedure for maximum symmetrical swing in a voltage-divider common emitter

amplifier in Chap. 2 yields $R_E = 2.4$ k, $R_L = 10.8$ k, $R_1 = 209$ k and $R_2 = 31$ k. Using Eq. (4.29) for the voltage gain of the partially bypassed configuration and noting that $r_e = 1/40I_C = 25\ \Omega$, $50 = \frac{-10800}{R_e + 25}$ giving $R_e = 191\ \Omega$. Thus the emitter resistor $R_E = 2.4$ k must be split into $R_E = 2.2$ k and $R_e = 191\ \Omega$. The input impedance is given by $Z_i = 31$ k$\|209$ k$\|\{2.5$ k $+ (1 + 100)$ 0.191 k$\} = 31$ k$//209$ k$//21.8$ k $= 12.1$ k

Bootstrapping

In order to reduce the effect of R_1 and R_2 and thereby realize higher input impedance, bootstrapping can be employed in the partially bypassed circuit of Fig. 4.19. The technique involves (i) the introduction of resistor R_B between the transistor base and the junction of R_1 and R_2 and (ii) the introduction of a capacitor C_B between the transistor emitter and the junction of R_1, R_2 and R_B as shown in Fig. 4.21. As usual, all capacitors are assumed to be large enough to be short circuit to AC. Using the simplified model of the BJT, the equivalent circuit representing the bootstrapped circuit is shown in Fig. 4.22. Let $Z_i = R_1 \| R_2 \| R_e = R'_e$. From KVL,

$$I_b h_{ie} = I_x R_B \qquad (4.34)$$

giving

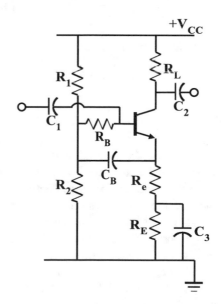

Fig. 4.21 Bootstrapping in a common emitter amplifier

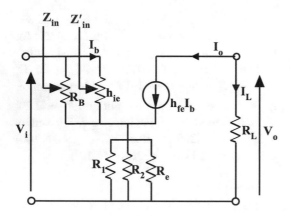

Fig. 4.22 Equivalent circuit of bootstrapped common emitter amplifier

$$I_x = h_{ie}I_b/R_B \qquad (4.35)$$

Using KVL at the input of the circuit,

$$V_i = I_b h_{ie} + I_b\left(1 + h_{fe} + \frac{h_{ie}}{R_B}\right)R'_e \qquad (4.36)$$

Hence the input impedance to the circuit is given by

$$Z_i = \frac{V_i}{I_b + I_x} = \frac{V_i}{I_b(1 + h_{ie}/R_B)}$$

$$= \frac{h_{ie} + \left(1 + h_{fe} + \frac{h_{ie}}{R_B}\right)}{1 + \frac{h_{ie}}{R_B}} \qquad (4.37)$$

If $R_B \gg h_{ie}$, then (4.37) reduces to

$$Z_i = h_{ie} + \left(1 + h_{fe}\right)R'_e \qquad (4.38)$$

Thus, the original circuit input impedance $Z_i = R_1\|R_2\|(h_{ie} + (1 + h_{fe})R_e)$ becomes $Z_i = h_{ie} + (1 + h_{fe})(R_1\|R_2\|R_e)$, and the effect of R_1 and R_2 is reduced. Note however that bias stability is also reduced since the effective source resistance of the voltage divider bias changes from $R_1\|R_2$ to $R_1\|R_2 + R_B$. Because of this, the DC voltage across R_B should be small compared with the DC voltage across $R_E + R_e$.

The voltage gain of the circuit can also be determined. From Fig. 4.22, the output voltage is given by

$$V_o = -h_{fe}I_b R_L \qquad (4.39)$$

Hence, using Eq. (4.36) for V_i, the voltage gain A_V is given by

$$A_V = \frac{V_o}{V_i} = -\frac{h_{fe}R_L}{h_{ie} + \left(1 + h_{fe} + h_{ie}/R_B\right)R'_e} \qquad (4.40)$$

This reduces to

$$A_V = -\frac{h_{fe}R_L}{h_{ie} + \left(1 + h_{fe}\right)R'_e}, R_B \gg h_{ie} \qquad (4.41)$$

Thus, the gain of the original circuit is virtually unchanged. If a resistor R'_L is connected to C_2 and grounded, this represents a load on the CE amplifier. This resistor appears in parallel with R_L in the equivalent circuit. Hence $R_L \rightarrow R_L\|R'_L$. The effect of this is to reduce the voltage gain A_V. Thus, loading the CE amplifier reduces the voltage gain.

Example 4.6
Improve the input impedance of the partially bypassed configuration in Example 4.5 using bootstrapping.

Solution Bootstrapping requires the introduction of resistor R_B into the circuit. In order to maintain quiescent current stability, the voltage across R_B must be significantly less than the voltage at the emitter which is 2.4 V. Using $V_{R_B} = 0.24$ V , then with a base current of 1 mA/$h_{fe} = 0.01$ mA, the value of R_B is $R_B = 0.24$ V/0.01 mA = 24 k. Hence the new input impedance is given by $Z_i = 2.5$ k + (1 + 100) (31 k$\|$209 k$\|$0.19 k) = 21.5 k which is significantly larger than the original value.

Alternative Amplifier Configuration
The amplifier discussed in Fig. 4.19 in which the emitter resistor is only partially bypassed had a lower but less variable gain. We now consider an alternative version shown in Fig. 4.23 in which the emitter resistor R_E is bypassed by a resistor R_e in series with capacitor C_3. The form of equivalent circuit is shown in Fig. 4.20. Note that R_E is not short-circuited by the capacitor C_3 which is in series with resistor R_e. Therefore, in the

equivalent circuit, R_E and R_e appear in parallel, C_3 being a short circuit for signals. Hence the equivalent circuit is essentially that of Fig. 4.24, and the expressions for voltage gain and input impedance are the same for both circuits with $R_e \to R_E \| R_e$ which reduces to R_e for $R_E \gg R_e$.

4.2.2   Common Emitter Amplifier with Collector-Base Feedback Bias

The h-parameter equivalent circuit will be used to analyse the common emitter amplifier with collector-base feedback biasing. The configuration is shown in Fig. 4.25, and the equivalent circuit is shown in Fig. 4.26. Here the resistor R_F connects the collector to the base. From Fig. 4.26, the input and output voltages are given by

$$V_i = I_b h_{ie} \tag{4.42}$$

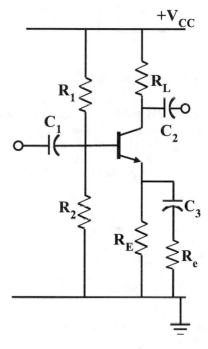

Fig. 4.23 Alternative common emitter amplifier

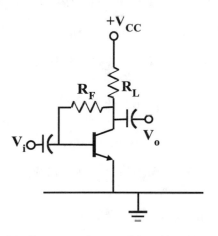

Fig. 4.25 Common emitter amplifier with collector-base feedback bias

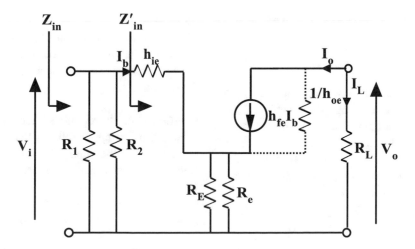

Fig. 4.24 Equivalent circuit of alternative common emitter amplifier

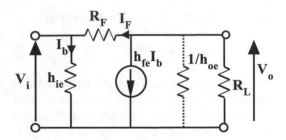

Fig. 4.26 Equivalent circuit of amplifier with collector-base feedback bias

$$V_o = (-h_{fe}I_b - I_F)R_L \qquad (4.43)$$

where

$$I_F = (V_o - V_i)/R_F \qquad (4.44)$$

Substituting (4.44) in (4.43), we get

$$V_o = I_b \frac{R_L R_F}{R_L + R_F} \frac{(h_{ie} - h_{fe}R_F)}{R_F}$$

$$= I_b R_L//R_F \frac{(h_{ie} - h_{fe}R_F)}{R_F} \qquad (4.45)$$

Hence the voltage gain is given by

$$A_V = \frac{V_o}{V_i} = \frac{R_L//R_F(h_{ie} - h_{fe}R_F)/R_F}{h_{ie}}$$

$$= R_L//R_F\left(1 - \frac{h_{fe}R_F}{h_{ie}}\right)/R_F \qquad (4.46)$$

Since $h_{fe}R_F/h_{ie} \gg 1$, (4.46) reduces to

$$A_V = -h_{fe}R_L//R_F/h_{ie} \qquad (4.47)$$

Note that if $R_F \rightarrow \infty$ (i.e. removed from the circuit), then Eq. (4.47) reduces to

$$A_V = -h_{fe}R_L/h_{ie} \qquad (4.48)$$

which is the standard equation for the voltage gain of the fixed or voltage divider-biased common emitter amplifier. The input impedance to this circuit can be found from $Z_i = V_i/I_i$ where I_i is the total current flowing into the input of the circuit. Noting that $I_i = I_b - I_F$ then the input admittance $Y_i = 1/Z_i$ is given by

$$Y_i = \frac{I_i}{V_i} = \frac{I_b - I_F}{V_i} = \frac{I_b}{V_i} + \frac{-I_F}{V_i}$$

$$= Y_b + Y_F \qquad (4.49)$$

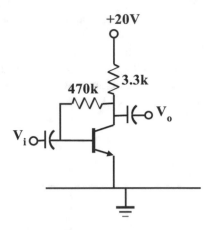

Fig. 4.27 Circuit for Example 4.7

where $Y_b = I_b/V_i$ and $Y_F = -I_F/V_i$ are the input admittances making up Y_i. From (4.42), $Y_b = I_b/V_i = 1/h_{ie}$. From (4.44),

$$I_F = (V_o - V_i)/R_F = (A_V - 1)V_i/R_F \qquad (4.50)$$

Hence,

$$Y_F = -I_F/V_i = (1 - A_V)/R_F \qquad (4.51)$$

Therefore

$$Y_i = \frac{I_i}{V_i} = Y_b + Y_F = \frac{1}{h_{ie}} + \frac{1 - A_V}{R_F} \qquad (4.52)$$

This corresponds to two impedances in parallel given by

$$Z_i = 1/Y_i = h_{ie}//R_F/(1 - A_V) \qquad (4.53)$$

Example 4.7
The common emitter amplifier in Fig. 4.27 is biased for maximum symmetrical swing. Determine the voltage gain and the input impedance for the circuit.

Solution Since the circuit is biased for maximum symmetrical swing, it follows that the collector voltage is half the supply voltage, i.e. 10 volts. Hence the transistor collector current is 10/3.3 k = 3 mA. The transistor base current is (10 − 0.7)/470 k = 0.02 mA. Hence the current gain of the transistor is h_{fe} = 3 mA/ 0.02 mA = 150. Therefore, h_{ie} for Tr1 is $h_{ie} = h_{fe}/40I_C$ = 150/40 × 3 mA = 1.25 k.

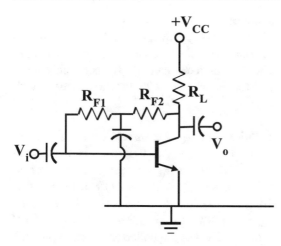

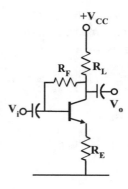

Fig. 4.29 Collector-base feedback bias amplifier with emitter resistor

Fig. 4.28 Common emitter amplifier with modified collector-base feedback bias

From the expression (4.46), the voltage gain for the circuit of Fig. 4.27 is given by $A_V = -h_{fe}R_L/h_{ie} = -150 \times 3.3/1.25 = -396$

Thus, the input impedance of the common emitter amplifier with collector-base feedback bias is reduced from h_{ie} to $h_{ie}\|R_F/(1 - A_V)$ where the feedback network causes the reduction. This effect can be virtually eliminated by splitting R_F into two resistors and connecting the junction to ground with a capacitor as shown in Fig. 4.28. With this arrangement, the changed input impedance becomes $h_{ie}\|R_{F_1}$ which is approximately h_{ie} since $R_{F_1} \gg h_{ie}$.

The gain of the common emitter amplifier with collector-base feedback biasing can be stabilized by the inclusion of an emitter resistor as shown in Fig. 4.29. The equivalent circuit is shown in Fig. 4.30. Here the resistor R_E is introduced in the emitter circuit. From Fig. 4.30, the input and output voltages are given by

$$V_i = I_b \left[h_{ie} + (1 + h_{fe})R_E \right] \qquad (4.54)$$

$$V_o = \left(-h_{fe}I_b - I_F \right) R_L \qquad (4.55)$$

where

$$I_F = (V_o - V_i)/R_F \qquad (4.56)$$

Substituting (4.56) and (4.54) in (4.55), we get

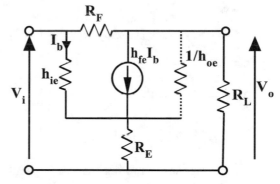

Fig. 4.30 Equivalent circuit of collector-base feedback bias amplifier with emitter resistor

$$V_o = -h_{fe}I_bR_L - \frac{R_L}{R_F}(V_o - V_i) = I_b \frac{R_L R_F}{R_L + R_F}$$

$$\left(-h_{fe} + \frac{h_{ie} + (1 + h_{fe})R_E}{R_F} \right) \qquad (4.57)$$

Hence the voltage gain is given by

$$A_V = \frac{V_o}{V_i}$$

$$= \frac{R_L//R_F \left(-h_{fe} + \frac{h_{ie} + (1+h_{fe})R_E}{R_F} \right)}{h_{ie} + (1 + h_{fe})R_E} \qquad (4.58)$$

For $R_F \gg R_E, h_{ie}, R_L$, (4.58) reduces to

$$A_V = \frac{V_o}{V_i} = \frac{-h_{fe}R_L}{h_{ie} + (1 + h_{fe})R_E}$$

$$= \frac{-R_L}{R_E + r_e} \approx -\frac{R_L}{R_E}, R_E \gg r_e \qquad (4.59)$$

Determination of the input impedance follows the steps of the circuit without R_E with the input admittance given by

$$Y_i = \frac{I_i}{V_i} = \frac{I_b - I_F}{V_i} = \frac{I_b}{V_i} + \frac{-I_F}{V_i} = Y_h + Y_F \quad (4.60)$$

From (4.42),

$$Y_b = I_b/V_i = 1/\big(h_{ie} + (1 + h_{fe})R_E\big) \quad (4.61)$$

From (4.44),

$$I_F = (V_o - V_i)/R_F = (A_V - 1)V_i/R_F \quad (4.62)$$

where $A_V = -R_L/(R_E + r_e)$

Hence,

$$Y_F = -I_F/V_i = (1 - A_V)/R_F \quad (4.63)$$

Therefore

$$Y_i = \frac{I_i}{V_i} = Y_b + Y_F$$

$$= \frac{1}{h_{ie} + (1 + h_{fe})R_E} + \frac{1 - A_V}{R_F} \quad (4.64)$$

This corresponds to two impedances in parallel given by

$$Z_i = 1/Y_i = \big\{h_{ie} + (1 + h_{fe})R_E\big\}//R_F/(1 - A_V) \quad (4.65)$$

If the R_F-splitting arrangement used above is employed here, then Z_i becomes $Z_i = 1/Y_i = \big\{h_{ie} + (1 + h_{fe})R_E\big\}\|R_{F_1}$ which is higher.

Example 4.8
Using a supply voltage of 24 volts, design a collector-base feedback-biased common emitter circuit with fixed gain of 10 using an emitter resistor as in Fig. 4.29 and a transistor with $h_{fe} = 125$.

Solution Let $I_{Cq} = 2$ mA. For maximum symmetrical swing $V_{R_L} \simeq V_{CE} = a$. Further, for a gain of 10, $R_L = 10R_E$ and therefore the quiescent voltage across R_E is $V_{R_E} = a/10$. It follows that $24 = a + a + a/10$ which yields $a = 11.4$ V. Hence $R_L = 11.4/2$ mA $= 5.7$ k and $R_E = 1.14/2$ mA $= 570\ \Omega$. The collector voltage V_C of the transistor is $V_C = V_{CE} + V_{R_E} = 11.4 + 1.14 = 12.54$ V, the base voltage $V_B = 1.14 + 0.7 = 1.84$ V, and the base current is $I_{Bq} = I_{Cq}/h_{fe} = 2$ mA$/125 = 0.016$ mA.

Therefore $R_F = (V_C - V_B)/I_{Bq} = (12.54 - 1.84)/0.016$ mA $= 669$ k.

4.3   Analysis of the Common Collector Amplifier Using H-Parameters

In this section we will utilize the common emitter h-parameter BJT equivalent circuit to analyse the common collector amplifier. Consider the basic common collector circuit shown in Fig. 4.31a in which bias components have been excluded. The equivalent circuit using common emitter

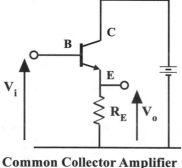

Common Collector Amplifier

(a)

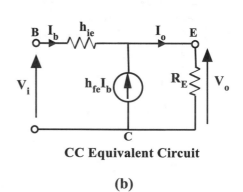

CC Equivalent Circuit

(b)

Fig. 4.31 Common collector amplifier circuit (**a**) and equivalent circuit (**b**)

h-parameters is shown in Fig. 4.31b. R_S and V_S represent the Thevenin equivalent of the driving stage. Using previous approximations, since $h_{re} \approx 0$, $h_{re}V_{ce}$ can be neglected, and if $1/h_{oe} \gg R_E$, it can also be neglected. Using Kirchhoff's voltage law,

$$V_i = I_b(h_{ie}) + (1 + h_{fe})i_b R_E \qquad (4.66)$$

$$V_o = (1 + h_{fe})I_b R_E \qquad (4.67)$$

Hence the voltage gain A_V from base to emitter is given by

$$A_v = \frac{V_o}{V_i} = \frac{(1 + h_{fe})I_b R_E}{\{(1 + h_{fe})R_E + h_{ie}\}I_b}$$

$$= \frac{R_E}{R_E + r_e} \qquad (4.68)$$

From (4.68), it follows that $A_V < 1$. Since $r_e \ll R_E$ in most instances, $A_V \approx 1$ for the common collector amplifier. Note that output and input voltages are in phase. To calculate the input impedance, we need to find the ratio V_i/I_b. From Eq. (4.66),

$$Z_i = \frac{V_i}{I_b} = h_{ie} + (1 + h_{fe})R_E \qquad (4.69)$$

Thus the input impedance is quite large since $h_{fe} \gg 1$. For voltage divider biasing, two bias resistors R_1 and R_2 would now appear in parallel with Z_i such that the reduced input impedance is given by

$$Z_i = R_1 // R_2 // (h_{ie} + (1 + h_{fe})R_E) \qquad (4.70)$$

In order to calculate the output impedance, we short-circuit the input voltage source V_S and apply a voltage V at the output and find the corresponding current I flowing into the emitter (ignoring R_E) as shown in Fig. 4.32. Here R_S is the source resistance.

Then,

$$V = I_b(h_{ie} + R_S) \qquad (4.71)$$

$$I = I_b(1 + h_{fe}) \qquad (4.72)$$

Hence

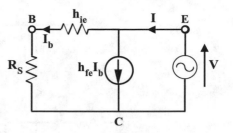

Fig. 4.32 Determining output impedance of common collector amplifier

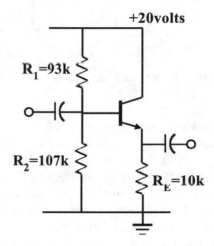

Fig. 4.33 Common collector amplifier for Example 4.9

$$Z_o = \frac{V}{I} = \frac{R_S + h_{ie}}{1 + h_{fe}} = \frac{R_S}{h_{fe}} + r_e \qquad (4.73)$$

Since this is in parallel with R_E, then $Z_o = R_E \| Z_o$. Usually, $Z_o \ll R_E$ giving $Z_o = \frac{R_S}{h_{fe}} + r_e$. The output impedance of the common collector amplifier is quite low.

Example 4.9

For the amplifier circuit shown in Fig. 4.33, determine the voltage gain, input impedance and output impedance.

Solution Using (4.68) the voltage gain is given by $A_v = \frac{R_E}{R_E + r_e} = \frac{10\,k}{10\,k + .025\,k} \approx 1$. The input impedance is given by $Z_i = 93\ k//107\ k// (2.5\,k + (1 + 100)10\,k) = 47.4\,k$ where $h_{ie} = 2.5\,k$. The output impedance is $Z_o = r_e = 25\,\Omega$.

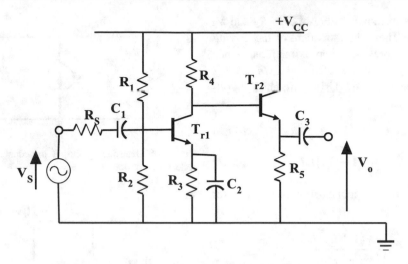

Fig. 4.34 Two-stage amplifier using the common collector amplifier

The high input impedance and low output impedance of the common collector amplifier makes it ideal for coupling the output of the common emitter amplifier to a load as shown in Fig. 4.34. The high input impedance prevents the loading of the collector of the common emitter amplifier which would lower the voltage gain, while the low output impedance means that most of the output voltage is dropped across the external load. A further advantage of this two-transistor arrangement is that the common collector amplifier can be directly coupled to the collector of the common emitter amplifier thereby eliminating the need for a coupling capacitor between these stages. The voltage gain from the input of the first transistor to the output of the second remains approximately the same. This arrangement will be explored in Chap. 5.

Bootstrapping

It was seen in Eq. (4.70) that the input impedance of the common collector amplifier is reduced by the voltage-divider biasing resistors. The bootstrapping technique used in the partially bypassed common emitter amplifier to overcome this problem can also be used in the common collector amplifier as shown in Fig. 4.35. The analysis is essentially the same with the emitter resistor R_E in the common collector amplifier replacing the un-bypassed emitter resistor R_e.

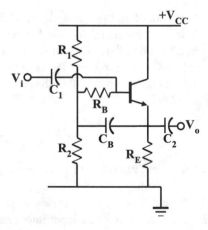

Fig. 4.35 Bootstrapped common collector amplifier

Example 4.10

Improve the input impedance of the common collector amplifier in Example 4.9 using bootstrapping.

Solution Bootstrapping requires the introduction of resistor R_B into the circuit. In order to maintain quiescent current stability, the voltage across R_B must be significantly less than the voltage at the emitter which is 2.4 V. Using $V_{R_B} = 0.24$ V, then with a base current of 1 mA/$h_{fe} = 0.01$ mA, the value of R_B is $R_B = 0.24$ V/0.01 mA $= 24$ k $\gg$ $h_{ie} = 2.5$ k. Hence the new input impedance is given by $Z_i = 2.5$ k $+ (1 + 100)(93$ k$\|107$

k∥10 k) = 844 k which is significantly larger than the original value of 47 k.

4.4 Analysis of the CB Amplifier Using H-Parameters

Here we analyse the third configuration, namely, the common base amplifier, using common emitter h-parameters. A biased common base amplifier is shown in Fig. 4.36. Assuming $h_{re}V_{ce} \approx 0$ and $1/h_{oe}$ is large, the equivalent circuit reduces to that shown in Fig. 4.37. The input voltage V_i is given by

$$V_i = -I_b h_{ie} \qquad (4.74)$$

while

$$V_o = -I_b h_{fe} R_L \qquad (4.75)$$

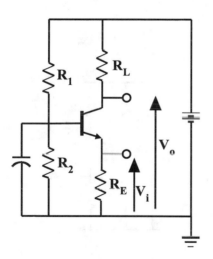

Fig. 4.36 Common base amplifier

Hence the voltage gain is given by

$$A_v = \frac{V_o}{V_i} = \frac{-I_b h_{fe} R_L}{-I_b h_{ie}} = \frac{h_{fe} R_L}{h_{ie}} = \frac{R_L}{r_e} \qquad (4.76)$$

Note that the voltage gain for the common base amplifier has the same magnitude as that for the common emitter amplifier, but unlike the common emitter amplifier, the output and input voltages are in phase. In the common emitter amplifier, the output voltage is 180° out of phase with the input voltage (represented by a negative sign), while in the common base amplifier the output voltage is in phase with the input voltage.

To calculate the input impedance, Z_i, we evaluate the input current I flowing into the emitter of the BJT:

$$I_i = -I_b(1 + h_{fe}) \qquad (4.77)$$

$$V_i = -I_b h_{ie} \qquad (4.78)$$

hence

$$Z_i = \frac{V_i}{I_i} = \frac{h_{ie}}{1 + h_{fe}} = r_e \qquad (4.79)$$

Hence, the input impedance of the common base amplifier is quite low and is comparable in value to the output impedance of the common collector amplifier. This value is lowered even further by R_E which is effectively in parallel with Z_i, i.e. $Z_i = r_e \| R_E$. By short-circuiting the input and applying a voltage at the output, the output impedance Z_o is easily shown to be $Z_o = R_L$.

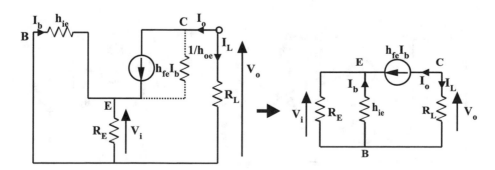

Fig. 4.37 H-parameter equivalent circuit of common base amplifier

We wish to determine the output impedance Z_o of a common base amplifier looking in at the transistor collector (and ignoring the collector resistor R_L), with an input signal that is associated with a high source impedance, i.e. a current source input signal as shown in Fig. 4.38. The equivalent circuit is shown in Fig. 4.39 with $R_E \gg h_{ie}$. With an open-circuited input, a voltage V is applied at the collector and the resulting current I determined. Then

$$I = h_{fe}I_b + I_x = -I_b \qquad (4.80)$$

which yields

$$I_x = -(1 + h_{fe})I_b \qquad (4.81)$$

$$V = I_x(1/h_{oe}) - I_b h_{ie}$$
$$= -(1 + h_{fe})I_b(1/h_{oe}) - I_b h_{ie} \qquad (4.82)$$

Hence

$$Z_o = \frac{V}{I} = \frac{-I_b[(1 + h_{fe})(1/h_{oe}) + h_{ie}]}{-I_b}$$
$$= (1 + h_{fe})(1/h_{oe}) + h_{ie} \qquad (4.83)$$

Hence the output impedance of a current-driven common base amplifier is extremely high.

Example 4.11

For the common base amplifier shown in Fig. 4.40, determine the voltage gain, input impedance and output impedance.

Solution The voltage gain is $A_v = \frac{9\,k}{.025} = 360$, while the input impedance is given by $Z_i = r_e \| R_E$. This yields $Z_i = 0.025\,k \| 2\,k \approx 25$. The output impedance $Z_o = 9\,k$.

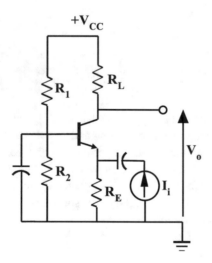

Fig. 4.38 Current-driven common base amplifier

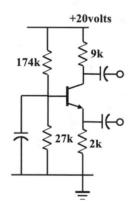

Fig. 4.40 Common base amplifier for Example 4.11

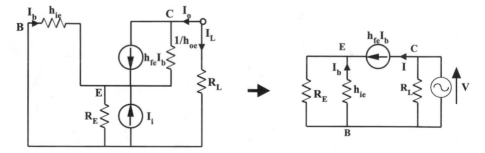

Fig. 4.39 Determination of output impedance of common base amplifier

4.5   Y-Parameters and the FET

As in the case of the BJT, small-signal operation of the FET assumes device operation over the linear range. Hence the AC operation may be realized by representing the FET by an equivalent circuit comprising linear signal generators and other linear circuit elements. The FET is regarded as an active two-port network with the input and output currents and voltages related by constant parameters. If the voltages are represented as the independent variables and the currents the dependent variables, then the constants are referred to as y-parameters with

$$I_o = y_f V_i + y_o V_o \qquad (4.84)$$

$$I_i = y_i V_i + y_r V_o \qquad (4.85)$$

which can be represented as shown in Fig. 4.41.

We use these parameters to derive the equivalent circuit for the FET shown in Fig. 4.42. From Chap. 3, the output and input currents were given by

$$I_o = g_m V_i + \frac{1}{r_d} V_o \qquad (4.86)$$

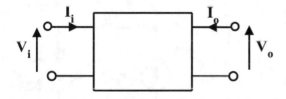

Fig. 4.41 Linear circuit representation of FET

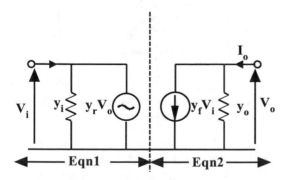

Fig. 4.42 Equivalent circuit representation of FET

$$I_i = \frac{1}{r_i} V_i \approx 0 \qquad (4.87)$$

where g_m is the transconductance of the FET, r_d is the output resistance and r_i is the input resistance. Comparing Eqs. (4.84)–(4.85) and (4.86)–(4.87) gives $y_f \equiv g_m$, $y_o \equiv 1/r_d$, $y_i \equiv 1/r_i$ and $y_r = 0$. The BJT h-parameter equivalent circuit can also be represented using y-parameters. Thus, using $h_{re} \approx 0$, then from Sect. 4.1,

$$I_o = \frac{h_f}{h_i} V_i + h_o V_o \qquad (4.88)$$

$$I_i = \frac{1}{h_i} V_i \qquad (4.89)$$

and the BJT y parameters can be determined by comparing Eqs. (4.88) and (4.89) with the defining Eqs. (4.86) and (4.87). Here $y_f \equiv \frac{h_f}{h_i}$ and this is the trans-conductance g_m of the BJT. This result was stated in Chap. 4. $y_o \equiv h_o$ is the output admittance and $y_i = 1/h_i$ is the input admittance. The resulting equivalent circuit is shown in Fig. 4.43.

4.6   Analysis of the Common Source JFET Amplifier

The basic common source amplifier with a fixed bias is shown in Fig. 4.44a. The corresponding y-parameter model is shown in Fig. 4.44b.

The analysis of this circuit is quite straightforward. The output voltage is given by

$$V_o = -g_m V_{GS} R_L // r_d \approx -g_m V_{GS} R_L, R_L \ll r_d \qquad (4.90)$$

Hence the voltage gain A_V is given by

$$A_V = \frac{V_o}{V_i} = -\frac{g_m V_{GS} R_L}{V_{GS}} = -g_m R_L \qquad (4.91)$$

The transconductance g_m is usually small, and hence the voltage gain of a common source FET is usually lower than the voltage gain of a common emitter BJT. Note that the input impedance is R_G since the internal input impedance of the FET is of the order of 10^{10} Ω which is normally

Fig. 4.43 Y-parameter equivalent circuit of BJT

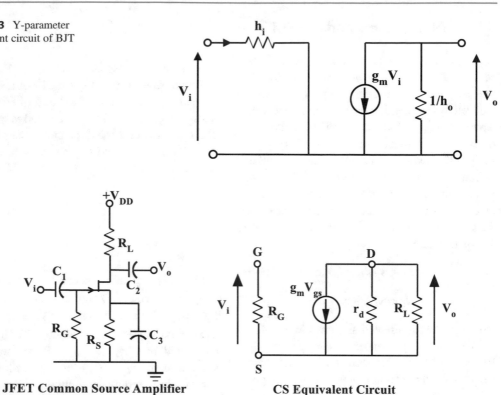

JFET Common Source Amplifier

(a)

CS Equivalent Circuit

(b)

Fig. 4.44 JFET common source amplifier (**a**) and equivalent circuit (**b**)

much greater than R_G. r_d corresponds to the output impedance seen at the drain of the FET and is of the order of 100 kΩ. It can be found using the channel resistance equation $r_d = \frac{r_o}{(1 - V_{GS}/V_P)^2}$ from Chap. 3.

If part of the resistor R_S of the FET is un-bypassed or if capacitor C_3 is removed, then the equivalent circuit becomes that shown in Fig. 4.45. If we assume $r_d \gg R_L$, then

$$V_o = -g_m V_{GS} R_L \qquad (4.92)$$

$$V_i = V_{GS} + g_m V_{GS} R_S \qquad (4.93)$$

Hence

$$\frac{V_o}{V_i} = -\frac{g_m V_{GS} R_L}{(1 + g_m R_S) V_{GS}} = -\frac{g_m R_L}{1 + g_m R_S} \qquad (4.94)$$

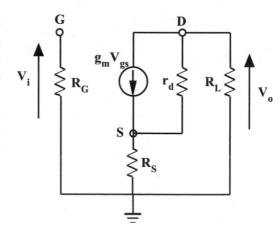

Fig. 4.45 Equivalent circuit of common source amplifier with un-bypassed source resistor

Thus, as in the BJT, inclusion of an un-bypassed source resistor R_S reduces the voltage gain of the common source FET amplifier.

Example 4.12

For the circuit shown in Fig. 4.44 with $R_L = 966\,\Omega$, $R_S = 468\,\Omega$, $R_G = 1\,M\Omega$ and $g_m = 5\,mA/V$ find the voltage gain and the input impedance.

Solution Using Eq. (4.91) the voltage gain is given by $A_V = -g_m R_L = 5 \times 0.966 = -4.83$. The input impedance is set by resistor $R_G = 1\,M\Omega$.

Example 4.13

If the bypass capacitor in Example 4.12 is removed, determine the changed value of the voltage gain.

Solution Using Eq. (4.94), the changed voltage gain is given by

$$A_V = -\frac{g_m R_L}{1 + g_m R_S} = -\frac{5 \times 0.966}{1 + 5 \times 0.468} = -1.45.$$

4.7   Analysis of the Common Drain JFET Amplifier

The basic circuit of a common drain or source follower JFET amplifier is shown in Fig. 4.46. The equivalent circuit is shown in Fig. 4.47. The output voltage is given by

$$V_o = g_m V_{GS} R_S \qquad (4.95)$$

and the input voltage V is given by

$$V_i = V_{GS} + V_o = V_{GS} + g_m V_{GS} R_S \qquad (4.96)$$

Hence the voltage gain A_V becomes

$$A_V = \frac{V_o}{V_i} = \frac{g_m V_{GS} R_S}{(1 + g_m R_S)V_{GS}}$$

$$= \frac{g_m R_S}{1 + g_m R_S} \qquad (4.97)$$

If $g_m R_S \gg 1$, then $A_V \approx 1$. That is, the gain of the common drain amplifier is approximately unity which means that the source follows the gate.

In order to evaluate the output impedance Z_o looking in at the source, we find the ratio of the open-circuit output voltage and short-circuit current, i.e. $Z_o = V_{os}/I_{sc}$. The open-circuit voltage is given by

$$V_{oc} = A_V V_i \qquad (4.98)$$

and the short-circuit current is given by

$$I_{sc} = g_m V_{GS} = g_m V_i \qquad (4.99)$$

since, when the output is short-circuited, $V_{GS} = V_i$. Hence

$$Z_o = \frac{V_{os}}{I_{sc}} = \frac{g_m R_S}{1 + g_m R_S} \frac{V_i}{g_m V_i}$$

$$= \frac{R_S}{1 + g_m R_S} \qquad (4.100)$$

Note that the output impedance of the JFET is independent of any source resistance and dependent only on the transconductance and the value of R_S. Note also that the input impedance of this circuit is very high because of the inherently high input impedance at the gate of the FET. The source follower is therefore analogous to the emitter follower. It possesses very high input impedance, approximately unity gain and relatively low output impedance.

Example 4.14

For the circuit of Fig. 4.46, $R_1 = 2\,M\Omega$, $R_2 = 1\,M\Omega$, $R_S = 3\,k\Omega$ and $g_m = 5\,mA/V$ find

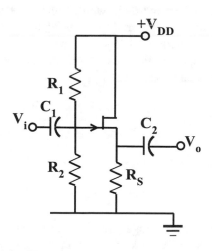

Fig. 4.46 JFET common drain amplifier

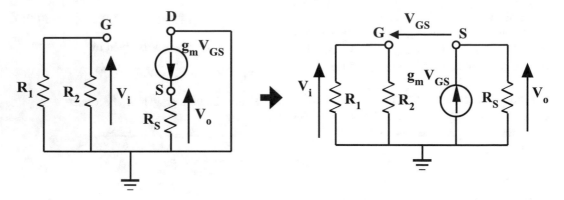

Fig. 4.47 Equivalent circuit of common drain amplifier

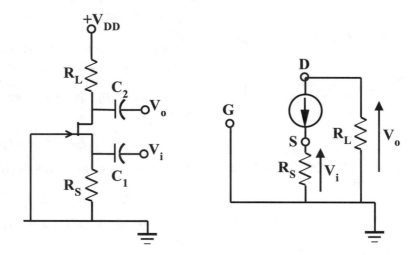

Fig. 4.48 Common gate amplifier

the voltage gain, input impedance and output impedance.

Solution Using (4.97) the voltage gain is given by $A_V = \frac{g_m R_S}{1 + g_m R_S} = \frac{5 \times 3}{1 + 5 \times 3} = 0.94$. The input impedance is $R_1 // R_2 = 1 // 2 \ \text{M}\Omega = 667 \ \text{k}$. From (4.100) the output impedance is given by $Z_o = \frac{R_S}{1 + g_m R_S} = R_S // 1 / g_m = 3 \ \text{k} // 1 / 5 \ \text{k} = 188 \ \Omega$.

4.8   Analysis of the Common Gate JFET Amplifier

The common gate amplifier configuration and its equivalent circuit are shown in Fig. 4.48. The input signal is at the source, and the output signal

is taken from the drain with the gate grounded. This circuit is generally used in high-frequency applications. It is analogous to the common base BJT amplifier. For this circuit, the output and input voltages are given by

$$V_o = -g_m V_{GS} R_L \qquad (4.101)$$

$$V_i = -V_{GS} \qquad (4.102)$$

Therefore

$$A_V = \frac{V_o}{V_i} = g_m R_L \qquad (4.103)$$

Thus the voltage gain of the common gate JFET amplifier has the same magnitude as the common source amplifier but opposite sign; there is no signal inversion in the common gate amplifier. The input impedance Z_i at the source is given by

$$Z_i = \frac{V_i}{I_i} = \frac{-V_{GS}}{-g_m V_{GS}} = \frac{1}{g_m} \qquad (4.104)$$

Actually, the resistor R_S appears in parallel with $1/g_m$ giving

$$Z_i = \frac{1}{g_m} // R_S \qquad (4.105)$$

Example 4.15
For the circuit shown in Fig. 4.48 with $R_L = 966\,\Omega$, $R_S = 468\,\Omega$ and $g_m = 5$ mA/V find the voltage gain and the input impedance.

Solution Using Eq. (4.103), the voltage gain is given by $A_V = g_m R_L = 5 \times 0.966 = 4.83$. From (4.105) the input impedance given by $Z_i = 1/g_m //$ $R_S = 1/5//0.468 = 140\,\Omega$.

4.9   Depletion MOSFET Common Source Amplifier

The depletion MOSFET operates in a manner similar to the JFET. In the circuit shown in Fig. 4.49, a common source MOSFET is considered. The circuit uses no gate-source bias and therefore the drain current is equal to I_{DSS}. This is allowable since the gate voltage can be both

positive and negative for this kind of MOSFET. The equivalent circuit is similar to that for the JFET and is also shown in Fig. 4.49.
The voltage gain is easily shown to be

$$A_V = \frac{V_o}{V_i} = -g_m r_d // R_L \approx -g_m R_L, r_d \gg R_L \qquad (4.106)$$

If an un-bypassed bias resistor R_S is included in the source circuit, then the voltage gain becomes

$$A_V = \frac{g_m r_d // R_L}{1 + g_m R_S} \qquad (4.107)$$

Example 4.16
For the circuit shown in Fig. 4.49 with $R_L = 5$ k, $R_G = 10$ MΩ and $g_m = 5$ mA/V find the voltage gain and the input impedance.

Solution Using Eq. (4.106), the voltage gain is given by $A_V = -g_m R_L = 5 \times 5 = -25$. The input impedance is set by resistor $R_G = 10$ MΩ.

This device may be operated in the same manner as a JFET as shown in Fig. 4.50. Here R_S is included in order to reduce the drain current to a value less than I_{DSS}. Because of the bypass capacitor C_3, the equivalent circuit is the same as that in Fig. 4.49, and therefore the expressions for voltage gain and input impedance are also the same.

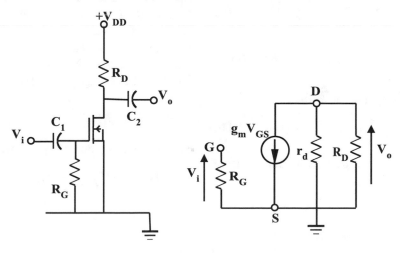

Fig. 4.49 Depletion MOSFET common source amplifier

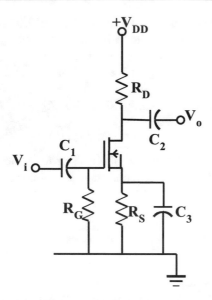

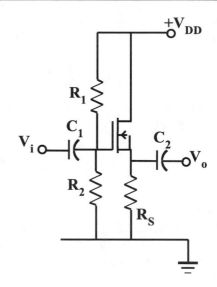

Fig. 4.50 Depletion MOSFET common source amplifier with bias resistor

Fig. 4.51 Depletion MOSFET common drain amplifier

$$Z_o = \frac{R_S}{1 + g_m R_S} \qquad (4.111)$$

Input impedance Z_i is given by

$$Z_i = R_G \qquad (4.112)$$

4.10 Depletion MOSFET Common Drain Amplifier

Because of the similarity between the equivalent circuits of the JFET and the depletion MOSFET, the equivalent circuit for the depletion MOSFET common drain configuration shown in Fig. 4.51 follows exactly that for the JFET in Fig. 4.46 and is given in Fig. 4.52. For this circuit,

$$V_o = -g_m V_{GS} R_S \qquad (4.108)$$

$$V_i = V_{GS} + V_o = V_{GS}(1 + g_m R_S) \qquad (4.109)$$

Therefore the voltage gain is therefore given by

$$A_V = \frac{V_o}{V_i} = \frac{g_m V_{GS} R_S}{(1 + g_m R_S) V_{GS}}$$
$$= \frac{g_m R_S}{1 + g_m R_S} \qquad (4.110)$$

which goes to unity if $g_m R_S \gg 1$. Output impedance Z_o is given by

4.11 Depletion MOSFET Common Gate Amplifier

Again, since the equivalent circuits of the JFET is similar to that for the Depletion MOSFET, the equivalent circuits for the Depletion MOSFET common gate configuration is exactly that for the JFET in Fig. 4.48. The MOSFET common gate amplifier and its equivalent circuit are shown in Fig. 4.53. The voltage gain is given by

$$A_V = \frac{V_o}{V_i} = g_m R_L \qquad (4.113)$$

and the input impedance is given by

$$Z_i = \frac{1}{g_m} // R_S \qquad (4.114)$$

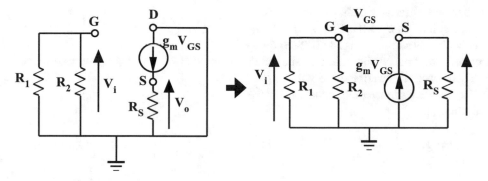

Fig. 4.52 Equivalent circuit of depletion MOSFET common drain amplifier

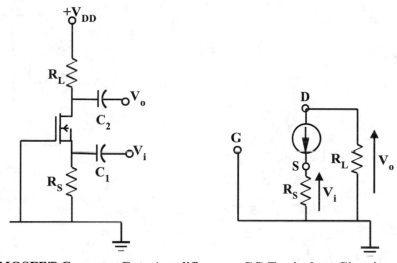

Depletion MOSFET Common Gate Amplifier   CG Equivalent Circuit

Fig. 4.53 Depletion MOSFET Common Gate Amplifier and Equivalent Circuit

4.12   Enhancement MOSFET Common Source Amplifier

The enhancement MOSFET can also be used in amplifier design. Since the conduction channel has to be created, the device requires a biasing network to overcome the threshold voltage and turn the device on. The circuit of a common source amplifier using an enhancement MOSFET is shown in Fig. 4.54. The voltage divider arrangement is used. The equivalent circuit for this arrangement is shown in Fig. 4.55.

The analysis of this configuration is straightforward and follows that given for the other field effect transistor circuits. The value of g_m is obtained using Eq. (3.16) for the drain current given by $I_D = k(V_{GS} - V_{GST})^2$. Since g_m is defined by $g_m = \frac{\Delta I_D}{\Delta V_{GS}}$, its value can be derived by differentiating I_D with respect to V_{GS}:

$$g_m = \frac{dI_D}{dV_{GS}} = \frac{d}{dV_{GS}} k(V_{GS} - V_{GST})^2$$

$$= 2k(V_{GS} - V_{GST}) \tag{4.115}$$

This reduces to

$$g_m = 2k(V_{GS} - V_{GST}) \tag{4.116}$$

The value of k must be determined for a specific operating point using the device specification sheet.

Another biasing arrangement that can be used to bias the enhancement MOSFET is the drain-gate feedback bias shown in Fig. 4.56. The equivalent circuit for this configuration is shown in Fig. 4.57. In order to determine the voltage gain of the circuit, we write the node equation for the drain terminal:

$$I_i = g_m V_{gs} + \frac{V_o}{r_d // R_D} \qquad (4.117)$$

Also $V_{gs} = V_i$ and $I_i = \frac{V_i - V_o}{R_G}$. Hence

$$\frac{V_i - V_o}{R_G} = g_m V_i + \frac{V_o}{r_d // R_D} \qquad (4.118)$$

which yields

$$V_o \left(\frac{1}{r_d // R_D} + \frac{1}{R_G} \right) = V_i \left(\frac{1}{R_G} - g_m \right) \quad (4.119)$$

from which the voltage gain A_V is given by

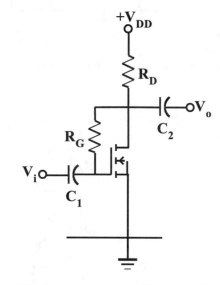

Fig. 4.56 Common source amplifier using drain-gate feedback bias

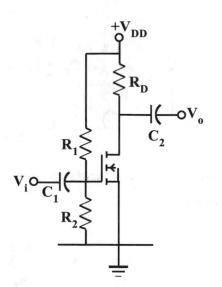

Fig. 4.54 Enhancement MOSFET Common Source Amplifier

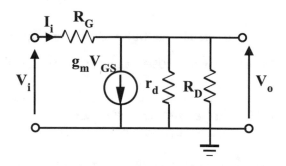

Fig. 4.57 Equivalent circuit of circuit in Fig. 4.56

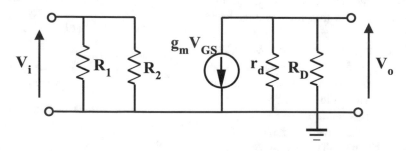

Fig. 4.55 Equivalent circuit of common source amplifier

$$A_V = \frac{V_o}{V_i} = \frac{\frac{1}{R_G} - g_m}{\frac{1}{R_G} + \frac{1}{r_d//R_D}} = \frac{\frac{1}{R_G} - g_m}{\frac{1}{R_G//r_d//R_D}}$$

$$\approx -g_m R_G // r_d // R_D, g_m \gg \frac{1}{R_G} \qquad (4.120)$$

In order to find the input impedance to this circuit, we use $I_i = \frac{V_i - V_o}{R_G}$ to get

$$V_o = V_i - I_i R_G \qquad (4.121)$$

Substituting (4.121) in (4.117) gives

$$I_i = g_m V_i + \frac{V_i}{r_d//R_D} - I_i \frac{R_G}{r_d//R_D} \qquad (4.122)$$

$$I_i\left(1 + \frac{R_G}{r_d//R_D}\right) = V_i\left(g_m + \frac{1}{r_d//R_D}\right) \qquad (4.123)$$

Therefore, the input impedance $Z_i = V_i/I_i$ is given by

$$Z_i - V_i/I_i - \frac{R_G + r_d//R_D}{1 + g_m r_d//R_D} \qquad (4.124)$$

Example 4.17
The MOSFET shown in Fig. 4.58. has $k = 0.31 \times 10^{-3}$ A/V^2, $V_{GST} = 3.5$ V and $V_{GSq} = 10$ V. Determine the voltage gain and input impedance of the circuit.

Solution From (4.116) the transconductance g_m is given by $g_m = 2k(V_{GS} - V_{GST}) = 2 \times 0.31 \times 10^{-3}(10 - 3.5) = 4.03$ mA/V. From (4.120), the expression for the voltage gain of the common source amplifier in Fig. 4.58 is given by $A_V \simeq - g_m R_D, R_D \ll R_G, r_d$. Therefore $A_V = - 4.03 \times 5$ k $= 20.2$. From (4.124), the input impedance is $Z_i = \frac{R_G + R_D}{1 + g_m R_D}, r_d \gg R_D$. This gives $Z_i = \frac{8 M}{1 + 20.2} = 377$ k.

4.13   Enhancement MOSFET Common Drain Amplifier

The circuit for an enhancement MOSFET common drain amplifier is shown in Fig. 4.59, while the equivalent circuit is shown in Fig. 4.60. Apart from the need to exceed the threshold voltage in order to turn on the MOSFET, the arrangement is similar to the common drain circuit using the depletion MOSFET and therefore has similar characteristics.
For this circuit,

$$V_o = -g_m V_{GS} R_S \qquad (4.125)$$

$$V_i = V_{GS} + V_o = V_{GS}(1 + g_m R_S) \qquad (4.126)$$

Therefore the voltage gain is therefore given by

$$A_V = \frac{V_o}{V_i} = \frac{g_m V_{GS} R_S}{(1 + g_m R_S)V_{GS}}$$

$$= \frac{g_m R_S}{1 + g_m R_S} \qquad (4.127)$$

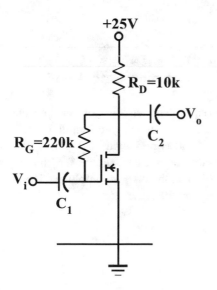

Fig. 4.58 Circuit for Example 4.17

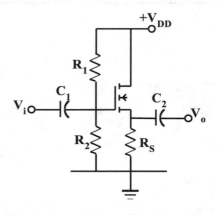

Fig. 4.59 Common drain amplifier sing enhancement MOSFET

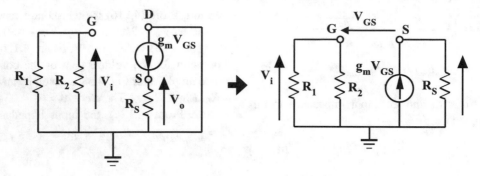

Fig. 4.60 Equivalent circuit of common drain amplifier using enhancement MOSFET

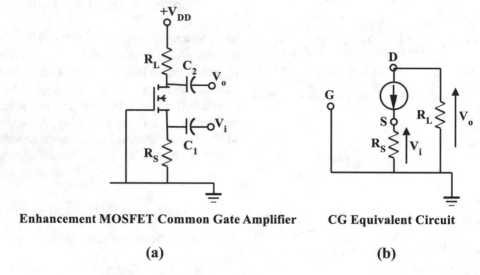

Enhancement MOSFET Common Gate Amplifier   **CG Equivalent Circuit**

(a)   **(b)**

Fig. 4.61 Enhancement MOSFET Common Gate Amplifier and Equivalent Circuit

which goes to unity if $g_m R_S \gg 1$. Output impedance Z_o is given by

$$Z_o = \frac{R_S}{1 + g_m R_S} \qquad (4.128)$$

Input impedance Z_i is given by

$$Z_i = R_G \qquad (4.129)$$

4.14   Enhancement MOSFET Common Gate Amplifier

The equivalent circuit for the enhancement
MOSFET common gate amplifier shown in
Fig. 4.61(a) is presented in Fig. 4.61(b). This is
the same equivalent circuit for the depletion
MOSFET common gate amplifier. The analysis
is therefore the same.

The voltage gain is given by

$$A_V = \frac{V_o}{V_i} = g_m R_L \qquad (4.130)$$

and the input impedance is given by

$$Z_i = \frac{1}{g_m} // R_S \qquad (4.131)$$

4.15 Applications

Several simple applications involving use of the transistor models are presented in this section.

Linear AC Voltmeter

A simple AC voltmeter can be designed using a BJT in a configuration employing negative feedback as shown in Fig. 4.62. The circuit consists of a common emitter amplifier with collector-base feedback biasing. The output of the circuit taken at the transistor collector drives a rectifier bridge through a coupling capacitor C_2 with a moving coil meter connected to the bridge. This ensures that unidirectional current flows through the meter. The output from the bridge is applied at the base of the transistor. This constitutes a feedback loop that helps to overcome the nonlinearity caused by the diode bridge and thereby linearize the meter operation, particularly at low values of input voltage. In order to make the system portable, the power supply is a 9 V battery. A collector current of 0.3 mA is used in order to reduce the

drain on the battery. For maximum symmetrical swing, $V_{R_1} = 9/2 = 4.5$ V. Hence $R_1 = 4.5$ V/ 0.3 mA = 15 k. The associated base current is 0.3/ 100 = 3 uA and therefore $R_2 = (4.5 - 0.7)/$ 3 μA = 1.3 M. The input resistor R_m sets the sensitivity of the circuit. The relationship between V_i and the meter current I_m can be found using the equivalent circuit for the system shown in Fig. 4.63. After manipulation, the input resistor R_m is given by $R_m = 0.9 \frac{V_{fsd}}{I_{fsd}}$ where V_{fsd} is the maximum input voltage and I_{fsd} is the maximum value of meter current. Thus, for a maximum voltage of 1 V and a 100uA microammeter, R_m is given by $R_m = 0.9 \frac{1\,V}{100\,\mu A} = 9$ k. The coupling capacitors are chosen to be large.

Ideas for Exploration (i) Determine a set of values for R_m corresponding to voltage ranges from 1volt to 1000 V.

Audio Mixer

The circuit of Fig. 4.64 is an audio system that mixes the signals from several sources. The

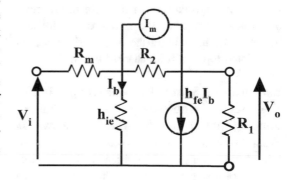

Fig. 4.63 Equivalent circuit of Fig. 4.62

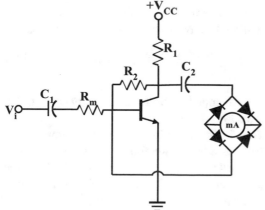

Fig. 4.62 Linear AC voltmeter

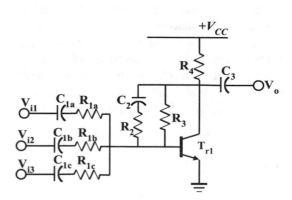

Fig. 4.64 Audio mixer

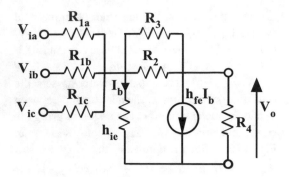

Fig. 4.65 Equivalent circuit of Fig. 4.64

transistor is in the common emitter configuration and biased using collector-base feedback via R_3. There is a resistor R_2 that is connected in series with capacitor C_2 that places R_2 in parallel with R_3 for signals only. Using a supply voltage of 15 volts and 0.5 mA collector current, the collector resistor R_4 is $R_4 = 7.5$ V/0.5 mA $= 15$ k. Resistor $R_3 = (7.5 - 0.7)/5$ µA $= 1.5$ M. With $R_{1a} = R_{1b} = R_{1c} = 100$ k and $R_2 = 100$ k, the equivalent circuit is shown in Fig. 4.65 and yields an output voltage given by $V_o = V_{i1} + V_{i2} + V_{i3}$.

Ideas for Exploration (i) Increase the number of inputs to five and derive the relationship between the inputs and the output; (ii) research the related topic of inter-channel crosstalk.

Bootstrapped Source Follower Using a JFET
The circuit of a bootstrapped common drain or source follower using a JFET is shown in Fig. 4.66. It comprises the voltage divider common drain circuit of Fig. 4.46 with (i) the introduction of resistor R_B between the transistor gate and the junction of R_1 and R_2 and (ii) the introduction of capacitor C_B between the transistor source and the junction of R_1, R_2 and R_B. Using the simplified model of the FET, the equivalent circuit representing the bootstrapped circuit is shown in Fig. 4.67. Let the current flowing into the parallel combination of resistors $R' = R_1 \| R_2 \| R_S$ be I_x. Then from KVL,

$$V_i = V_{GS} + I_x R' \qquad (4.132)$$

with

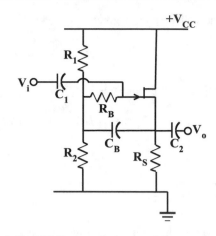

Fig. 4.66 JFET bootstrapped source follower

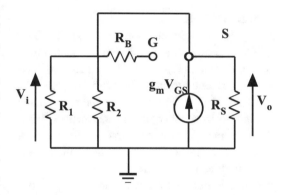

Fig. 4.67 Equivalent circuit of Fig. 4.66

$$V_{GS} = I_i R_B \qquad (4.133)$$

and from KCL

$$I_x = I_i + g_m V_{GS} \qquad (4.134)$$

Substituting for I_x and V_{GS} in (4.132) and finding the input impedance $Z_i = V_i/I_i$ yields

$$Z_i = V_i/I_i = R_B + (1 + g_m R_B)R' \qquad (4.135)$$

Bootstrapped Source Follower Using a MOSFET
The circuit shown in Fig. 4.68 is a bootstrapped version of the source follower using an enhancement MOSFET from Fig. 4.59. It is similar to the circuit using the JFET and has the same equivalent circuit. The input impedance is therefore

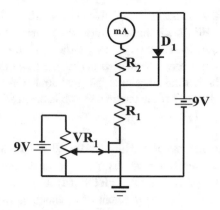

Fig. 4.70 JFET tester

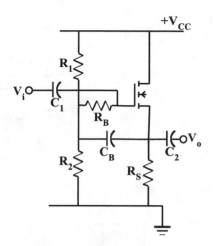

Fig. 4.68 EMOSFET source follower

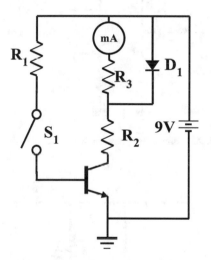

Fig. 4.69 Transistor gain tester

$$Z_i = R_B + (1 + g_m R_B)R' \qquad (4.136)$$

Research Project 1

This research project is the design of a transistor gain tester that enables the current gain of binary junction transistors to be measured. The basic circuit for npn transistors is shown in Fig. 4.69. Using a 9-volt battery, when switch S_1 is closed, resistor $R_1 = 820$ k passes a current of $(9 - 0.7)/820$ k $= 10$ μA into the base of the transistor under test. This causes a collector current $I_C = 10$ μA $\times \beta$ which is measured in milliamperes by

the milliammeter. Hence the current gain β is given by $\beta = I_C$ mA $\times 10^{-3}/10$ μA $\times 10^{-6} = 100 I_C$ where I_C is the actual reading on the milliammeter in milliamperes. Therefore, the current gain of the transistor being tested is obtained by multiplying the reading on the milliammeter by 100. Resistor $R_3 = 120$ Ω and diode D_1 protect the meter in the event that a shorted transistor is introduced. $R_2 = 330$ Ω limits the battery current under such a condition.

Ideas for Exploration (i) Introduce a switch that changes the battery polarity to enable the testing of pnp transistors. (ii) Design a 9-volt Zener-regulated mains-powered supply to replace the battery in order to prevent inaccuracies arising from a falling battery voltage.

Research Project 2

This second research project is the design of an instrument for the determination of I_{DSS} and V_P for JFETs. The basic circuit for testing n-channel JFETs is shown in Fig. 4.70. and comprises the JFET under test with its gate and source connected such that a variable gate-source voltage can be applied via potentiometer $VR_1 = 100$ k. In testing, potentiometer VR_1 is adjusted so that the gate-source voltage to the FET is zero and the current in the milliammeter measured. Providing the drain-source voltage exceeds pinch-off, this drain current with zero gate-source voltage corresponds to I_{DSS}. The potentiometer is then adjusted to apply a negative gate-source voltage to the FET such that an almost zero current flows through the device as indicated on the milliammeter. A voltmeter across the gate-source terminals then measures the gate-source voltage that pinches off the channel which is the pinch-off voltage V_P.

Ideas for Exploration (i) Introduce a switch that changes the battery polarity to enable the testing of p-channel JFETS. (ii) Explain why this circuit is very tolerant of a changing supply voltage. (iii) Explore the development of a circuit using similar ideas for the testing of depletion MOSFETS.

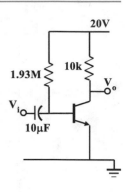

Fig. 4.71 Circuit for Question 1

Problems

1. Draw the h-parameter equivalent circuit for the common emitter amplifier shown in Fig. 4.71, and determine the voltage gain and input impedance. If a load resistor of 10 k is connected at the output, what will the new voltage gain?
2. Draw the h-parameter equivalent circuit of a common emitter H-biased amplifier circuit, and derive the expression for the voltage gain, input impedance and output impedance.
3. Draw the equivalent circuit for the amplifier shown in Fig. 4.72 and determine the voltage gain. If a load resistor of 5 k is connected at the output, what will be the new voltage gain?
4. For the circuit in question 3, determine the voltage gain if a 1 k resistor is connected in series with the input.
5. Draw the equivalent circuit for the partially decoupled common emitter amplifier shown in Fig. 4.73, and determine the voltage gain and the input impedance.
6. Using the partially bypassed configuration in Fig. 4.73 of question 5, design a common emitter amplifier having a gain of 40, using a 2N3904 npn transistor and a 18 V power supply. Determine the input impedance.
7. Show that bootstrapping can increase the input impedance of a common emitter amplifier in

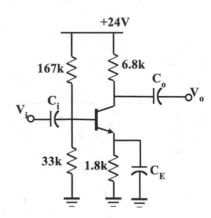

Fig. 4.72 Circuit for Question 3

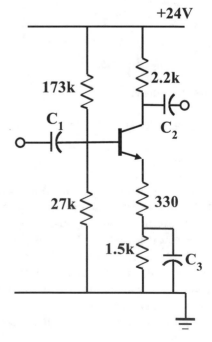

Fig. 4.73 Circuit for Question 5

the partially bypassed configuration. Use this technique to increase the input impedance of the design in question 6.

8. Using the configuration in Fig. 4.74, design a common emitter amplifier having a gain of 25, using an npn transistor and a 12 V power supply.

9. The common emitter amplifier in Fig. 4.75 is biased for maximum symmetrical swing. Determine the voltage gain for the circuit and the current gain of the transistor.

10. Using a supply voltage of 16 volts, design a collector-base feedback biased common emitter circuit with fixed gain of 12 using an emitter resistor and a transistor with $h_{fe} = 150$.

11. For the common collector amplifier shown in Fig. 4.76 determine the voltage gain, input impedance and output impedance.

12. Improve the input impedance of the common collector amplifier in question 11 using bootstrapping.

13. For the common base amplifier shown in Fig. 4.77, determine the voltage gain, input impedance and output impedance.

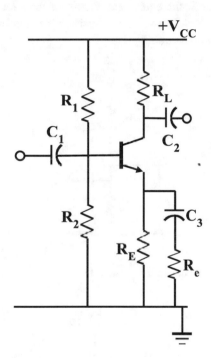

Fig. 4.74 Circuit for Question 8

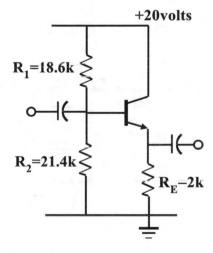

Fig. 4.76 Circuit for Question 11

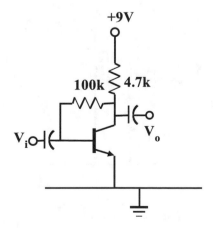

Fig. 4.75 Circuit for Question 9

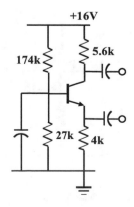

Fig. 4.77 Circuit for Question 13

14. For the circuit shown in Fig. 4.78, draw the equivalent circuit. For $R_L = 2$ k, $R_S = 1$ k, $R_G = 2$ M and $g_m = 8$ mA/V, calculate the voltage gain and input impedance of the circuit.

15. If the bypass capacitor is removed from the circuit, determine the new value of the voltage gain.

16. Draw the equivalent circuit of the common drain JFET amplifier in Fig. 4.79 and derive its voltage gain.

17. For the circuit of Fig. 4.79, $R_1 = 1.8$ MΩ, $R_2 = 1.5$ MΩ, $R_S = 2.2$ kΩ and $g_m = 4$ mA/ V, find the voltage gain, input impedance and output impedance.

18. For the common gate amplifier shown in Fig. 4.80 if $R_L = 3.2$ k, $R_S = 2.5$ k and $g_m = 5$ mA/V, find the voltage gain and the input impedance.

19. The common source depletion mode MOSFET in Fig. 4.81 has $R_L = 10$ k, $R_G = 10$ M and $g_m = 9$ mA/V. Draw the equivalent circuit and determine the voltage gain.

20. Draw the equivalent circuit for the circuit in Fig. 4.81 with a bias resistor inserted in the source circuit.

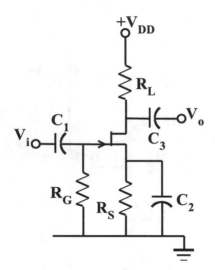

Fig. 4.78 Circuit for Question 14

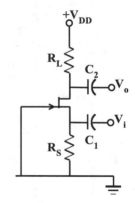

Fig. 4.80 Circuit for Question 18

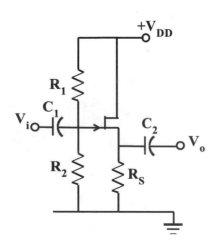

Fig. 4.79 Circuit for Question 16

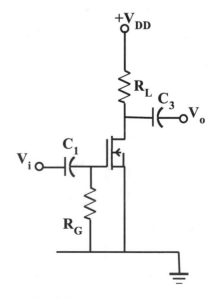

Fig. 4.81 Circuit for Question 19

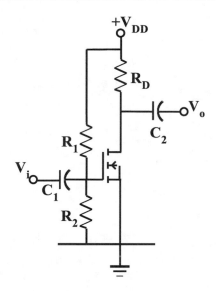

Fig. 4.82 Circuit for Question 23

21. Draw the equivalent circuit of the depletion MOSFET common drain amplifier.
22. Draw the equivalent circuit and derive the voltage gain and input impedance of the depletion MOSFET common gate amplifier.
23. Draw the equivalent circuit for the enhancement mode MOSFET common source amplifier shown in Fig. 4.82. Hence derive the expression for the voltage gain.
24. Using the equation for the drain current, derive an expression for g_m.
25. Draw the equivalent circuit for the enhancement MOSFET common drain amplifier and derive the voltage gain.

26. Draw the equivalent circuit for the common gate enhancement MOSFET amplifier and determine the voltage gain and input impedance.
27. If the gate of an n-channel JFET is connected to the source, the gate-source voltage will be held at zero, and therefore if the drain-source voltage exceeds the pinch-off voltage, the drain current will be constant at IDSS. Explain how this arrangement can be used as a constant current source.
28. Is this arrangement possible with a p-channel JFET and if so how?
29. Discuss the use of an n-channel enhancement mode MOSFET in the source follower mode in place of an emitter follower BJT in enhancing a Zener diode regulator.
30. Is this arrangement possible with a p-channel enhancement mode MOSFET and if so how?

Bibliography

R.L. Boylestad, L. Nashelsky, *Electronic Devices and Circuit Theory*, 11th edn. (Pearson Education, New Jersey, 2013)

M. Yunik, *The Design of Modern Transistor Circuits* (Prentice Hall, New Jersey, 1973)

Multiple Transistor and Special Circuits

5

In this chapter, several transistor circuit configurations are considered. These circuits generally involve more than one transistor and can be implemented in a wide range of applications as well as perform some special functions. In this chapter, the design and application of many of these circuits will be discussed. After completing the chapter, the reader will be able to:

- Explain the operation of cascaded amplifiers, in particular two common emitter amplifiers, and determine overall voltage gain and other characteristics
- Design several multitransistor amplifier circuits using BJTs and FETs
- Explain the operation of and utilize a range of special transistor circuits in practical applications
- Design simple electronic switches using BJTs and FETs
- Utilize the JFET as a voltage-controlled resistor

5.1 Cascaded Amplifiers

The single transistor amplifier has been considered in Chaps. 2, 3 and 4. The BJT single transistor amplifier produces a higher voltage gain than its FET counterpart, because of the higher transconductance of the BJT as compared with the FET. In either case, if a higher voltage gain is

required than is achievable using a single device, then a cascade connection of two single stage amplifiers can be used.

Consider the block diagram of a cascade amplifier shown in Fig. 5.1. Here the output of amplifier A_1 feeds directly into the input of amplifier A_2. The overall gain G of the cascaded system is

$$G = \frac{V_o}{V_i} = \frac{V_o}{V_1} \times \frac{V_1}{V_i} = G_1 \times G_2 \quad (5.1)$$

where G_1 is the gain of amplifier A_1 and G_2 is the gain amplifier A_2. Equation (5.1) indicates that the gain G of the cascaded amplifier is the product of the gains of the individual amplifiers. This holds true for any number of amplifiers. However, Eq. (5.1) assumes no interaction between the cascaded amplifiers, that is, the gain amplifier A_1 is unaffected by amplifier A_2 and vice versa. This assumption holds true for operational amplifiers since the output impedance of these devices is quite low. However, it does not necessarily hold for voltage amplifiers based on discrete BJTs where the output impedance may be higher. In multi-stage amplifiers using BJTs, there is considerable interaction between these stages. In particular, the input impedance of the succeeding stage changes the effective value of the load of the preceding stage thereby changing its gain.

Consider the two-stage amplifier in Fig. 5.2 comprising two cascaded common emitter

© Springer Nature Switzerland AG 2021
S. J. G. Gift, B. Maundy, *Electronic Circuit Design and Application*,
https://doi.org/10.1007/978-3-030-46989-4_5

amplifiers. Here capacitor C_2 couples the output of the first amplifier built around transistor Tr_1 to the input of the second amplifier built around transistor Tr_2. Because this capacitor and other resistors are involved, this type of coupling is referred to as RC coupling. The voltage gain G_1 of the first stage for a signal going from the base to the emitter of Tr_1 is given by

$$G_1 = -g_{m1}\overline{R}_{L1} \qquad (5.2)$$

where $\overline{R}_{L1}$ is the effective load of Tr_1. Because of the presence of Tr_2, this load is made up of R_{L1} along with the load presented by the input to Tr_2 which involves R_3, R_4 and h_{ie2} associated with Tr_2. [It will be shown shortly that $\overline{R}_{L1} = R_{L1}//R_3//R_4//h_{ie2}$.] Thus the voltage gain of Tr_1 is reduced because of the loading effect resulting from the input to Tr_2. However, the gain G_2 of Tr_2 is given by

$$G_2 = -g_{m2}R_{L2} \qquad (5.3)$$

and the overall gain G is given by

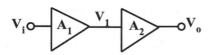

Fig. 5.1 Cascaded amplifiers

$$G = G_1 \times G_2- = g_{m1}g_{m2}\overline{R}_{L1}R_{L2} \qquad (5.4)$$

This gain can be as large as 1000 and higher. Thus, despite the loading effect of Tr_2, the cascading of two BJT amplifier stages can produce high gains.

Example 5.1
Draw the equivalent circuit of the two-stage amplifier in Fig. 5.3. Determine the voltage gain V_o/V_i and hence the current gain I_o/I_i. Assume $h_{fe} = 100$ for both transistors.

Solution The equivalent circuit is shown in Fig. 5.4. The voltage gain is given by $\frac{V_o}{V_i} = A_{V1} \times A_{V2}$ where A_{V1} is the voltage gain of Tr_1 and A_{V2} is the voltage gain of Tr_2. In order to proceed, the values of h_{ie1} for Tr_1 and h_{ie2} for Tr_2 must be determined. Let V_{B1} and V_{B2} be the DC voltages at the bases of the two transistors. Then, because of the voltage divider, $V_{B1} = \frac{47}{153+47} \times 20 = 4.7$ V . Hence $V_{E1} = 4$ and $I_{C1} = 4/4 = 1$ mA. Thus $h_{ie1} = \frac{100}{40\times1} = 2.5$ k. Similarly, $V_{B2} = \frac{37}{163+37} \times 20 = 3.7$ V giving $V_{E2} = 3$, $I_{C2} = 3/6 = 0.5$ mA and $h_{ie2} = \frac{100}{40\times0.5} = 5$ k. The load R_{L1} of Tr_1 is $R_{L1} = 8$ k$//$ 37 k$//$163 k$//h_{ie2} = 8$ k$//37$ k$//163$ k$//5$ k $= 2.8$ k. Hence the gain A_{V1} of Tr_1 is

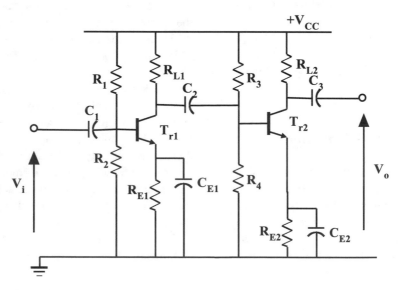

Fig. 5.2 Two cascaded common emitter amplifiers

Fig. 5.3 Two-stage amplifier

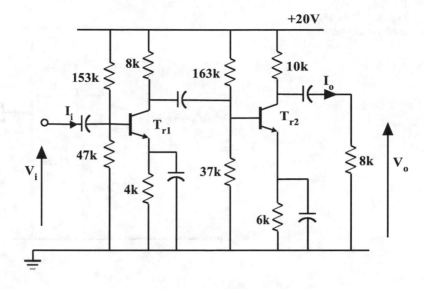

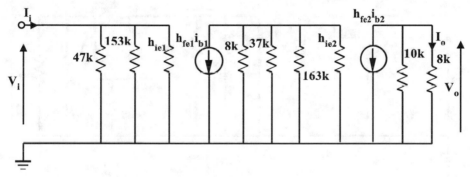

Fig. 5.4 Equivalent circuit of two-stage amplifier

$A_{V1} = -40I_{C1}R_{L1} = -40 \times 1 \times 2.8 = -112$. The load R_{L2} of the second transistor is $R_{L2} = 10 \text{ k}//8 \text{ k} = 4.4 \text{ k}$. Therefore, the voltage gain A_{V2} of Tr_2 is given by $A_{V2} = -40I_{C2}R_{L2} = -40 \times 0.5 \times 4.4 = -88$. Hence the overall voltage gain A_V is $A_V = -A_{V1} \times A_{V2} = -112 \times -88 = 9856$. In order to determine the current gain I_o/I_i, we note that $\frac{V_o}{V_i} = \frac{I_o R_L}{I_i Z_i}$ where R_L is the 8 k resistor into which I_o flows and Z_i is the input impedance at the base of Tr_1. Z_i at the base of Tr_1 is $Z_i = 47 \text{ k}//$ $153 \text{ k}//h_{ie1} = 47 \text{ k}//153 \text{ k}//2.5 \text{ k} = 2.3 \text{ k}$. Rearranging gives $\frac{I_o}{I_i} = \frac{V_o}{V_i} \times \frac{Z_i}{R_L} = 9856 \times \frac{2.3 \text{ k}}{8 \text{ k}} = 2834$.

Example 5.2

Draw the equivalent circuit of the two-stage amplifier in Fig. 5.5. Determine the voltage gain

V_o/V_s and hence the current gain I_o/I_s. Assume $h_{fe} = 120$ for both transistors.

Solution The equivalent circuit is shown in Fig. 5.6. The voltage gain is given by $\frac{V_o}{V_s} = \rho \times A_{V1} \times A_{V2}$ where ρ is the factor by which the input signal is attenuated before it arrives at the base of Tr_1, A_{V1} is the voltage gain of Tr_1 and A_{V2} is the voltage gain of Tr_2. In order to proceed, the values of h_{ie1} and h_{ie2} must be determined. Let V_{B1} and V_{B2} be the DC voltages at the bases of the two transistors. Then, because of the voltage divider, $V_{B1} = \frac{37}{300} \times 30 = 3.7$ V. Hence $V_{E1} = 3$ and $I_{C1} = 3/6 = 0.5$ mA. Thus $h_{ie1} = \frac{120}{40 \times 0.5} = 6k$. Similarly, $V_{B2} = \frac{23.5}{150} \times 30 = 4.7V$ giving $V_{E2} = 4$, $I_{C2} = 4/(3.6 + 0.4) = 1mA$ and $h_{ie2} = \frac{120}{40 \times 1} = 3 \text{ k}$. The load R_{L1} of Tr_1 is $R_{L1} = 6 \text{ k}//$

Fig. 5.5 Two-stage
common emitter amplifier

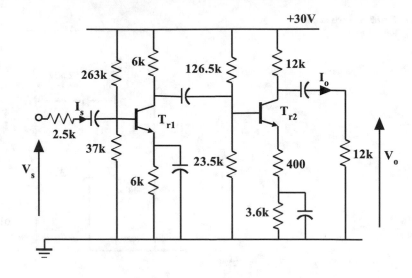

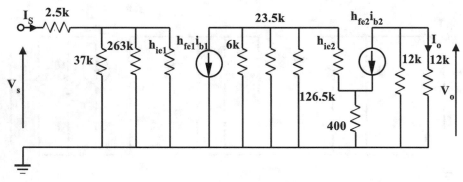

Fig. 5.6 Equivalent circuit of two-stage common emitter amplifier

23.5 k//126.5 k//($h_{ie2} + h_{fe2}.400$) = 6 k//23.5 k//
126.5 k//(3 k + 120 × 400) = 4.2 k. Hence the
gain A_{V1} of Tr$_1$ is A_{V1} = $-40I_{C1}R_{L1}$ =
$-40 × 0.5 × 4.2 = -84$. The load R_{L2} of the
second transistor is R_{L2} = 12 k//12 k = 6 k.
Therefore, the voltage gain A_{V2} of Tr$_2$ is given
by $A_{V2} = \frac{-R_{L2}}{400+h_{ie2}/h_{fe2}} = \frac{6 \text{ k}}{400+3 \text{ k}/120} = -14.1$. The
input impedance Z_i at the base of Tr$_1$ is Z_i = 37 k//
263 k//h_{ie1} = 5 k. Hence the attenuation factor ρ
is given by $\rho = \frac{Z_i}{2.5 \text{ k}+Z_i} = \frac{5 \text{ k}}{7.5 \text{ k}} = 0.67$. Therefore,
the overall voltage gain A_V is A_V = $-\rho$
× A_{V1} × A_{V2} = 0.67 × 84 × 14.1 = 794. In
order to determine the current gain I_O/I_S, we note
that $\frac{V_O}{V_S} = \frac{I_O R_L}{I_S(R_S+Z_i)}$ where R_L is the 12 k resistor
into which I_O flows and R_S is the 2.5 k series
resistor at the input to Tr$_1$. Rearranging this
gives $\frac{I_O}{I_S} = \frac{V_O}{V_S} × \frac{R_S+Z_i}{R_L} = 794 × \frac{(2.5 \text{ k}+5 \text{ k})}{12 \text{ k}} = 496$.

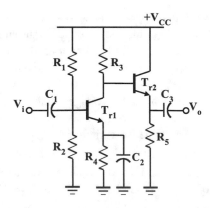

Fig. 5.7 Common emitter amplifier with emitter follower
output

Another interesting two-transistor circuit
involves the use of a common collector amplifier
to reduce the loading on the output of a common
emitter amplifier. The circuit is shown in Fig. 5.7.

Here the common emitter amplifier Tr_1 is followed by a common collector amplifier Tr_2. A feature of this circuit is the use of the collector of Tr_1 to bias transistor Tr_2 by direct connection, thereby eliminating the need for biasing components and capacitive coupling between the two transistors. Such an arrangement is referred to as **direct coupling** and, when used throughout an amplifier, confers the advantage of enabling the amplifier to respond down to DC. Since the input impedance $Z_{i2} = [h_{ie2} + (1 + h_{fe2})R_5]$ to Tr_2 is high, the load on Tr_1 is small so that the effective collector resistance for Tr_1 is $R_3//Z_{i2} \approx R_3$. Hence the gain of the first stage is preserved at $G_1 - -g_{m1}R_3$. Since the second stage is an emitter follower, the gain of this stage is approximately unity and therefore the overall gain G is $G = -g_m R_3$. This circuit has the additional advantage of low output impedance because of the common collector output amplifier. Its value Z_o is given by

$$Z_o = \left(\frac{h_{ie2} + R_3}{1 + h_{fe}}\right)//R_5 \qquad (5.5)$$

Example 5.3

Using the two-stage design in Fig. 5.7, design a two-stage amplifier having the ability to drive low-impedance loads.

Solution The design of the first stage follows the steps for the standard common emitter stage. For a 20-volt supply, the design approach of section (Sect. 2.4 in Chap. 2) yields $R_1 = 174$ k, $R_2 = 26$ k, $R_3 = 9$ k and $R_4 = 2$ k. For this circuit, the collector voltage of Tr_1 is 11 volts. This is therefore the base voltage of Tr_2 resulting in an emitter voltage for Tr_2 of 10.3 volts. Therefore if we choose a current of 10 mA for Tr_2 in order to ensure a large current drive capability, then $R_5 = 10.3/10$ mA $= 1$ k.

Example 5.4

Modify the design in Example 5.3 in order to set the voltage gain at 50.

Solution In order to lower the voltage gain to 50, an un-bypassed resistor R_6 is introduced in the emitter of Tr_1 as shown in Fig. 5.8. The voltage gain now becomes $A_V = \frac{R_3}{R_6 + r_e}$ where $r_e = 1/40I_C = 1/40 = 25$ Ω. Hence $50 = \frac{9 \text{ k}}{R_6 + 25}$ giving $R_6 = 155$ Ω.

An emitter follower can also be used to preserve the gain of a single FET amplifier while providing a reduced output impedance. Such a circuit is shown in Fig. 5.9. The FET Tr_1 is a common source amplifier with a high input impedance, the exact value for which is established by bias resistor R_1.

Example 5.5

Design a two-stage amplifier using the topology of Fig. 5.9.

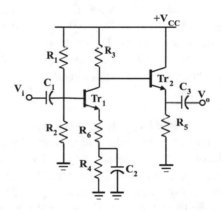

Fig. 5.8 Common emitter amplifier with reduced gain

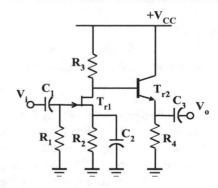

Fig. 5.9 Common source JFET amplifier with emitter follower output

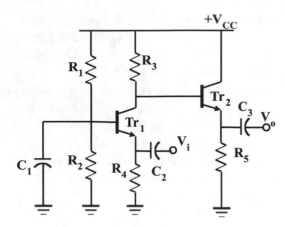

Fig. 5.10 BJT design

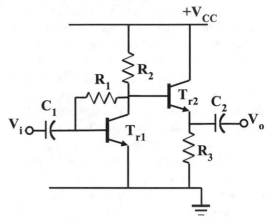

Fig. 5.11 Direct coupled amplifier

Solution The design of the JFET stage follows the steps used in the design of the common source JFET amplifier in Chap. 3. The values obtained for a 18-volt supply and a 1 mA quiescent current are $R_1 = 1$ M, $R_2 = 2$ k and $R_3 = 8$ k. The DC voltage at the drain of Tr_1 is 10 volts. Therefore the base of the BJT is 10 volts giving 9.3 at the emitter. Choosing a collector current of 5 mA for large current drive capability, $R_4 = 9.3/5$ mA $= 2$ k.

The circuit of Fig. 5.10 utilizes a common base amplifier in the first stage and a common emitter amplifier in the second stage. This circuit produces the same voltage gain as that in Fig. 5.7 using a common emitter amplifier in the first stage but does not produce signal inversion. Also, its input impedance is significantly lower than that of Fig. 5.7.

Example 5.6
Using the circuit configuration in Fig. 5.10, design a two-stage amplifier using an 18-volt supply.

Solution The design follows exactly the design of the circuit of Fig. 5.7 except that capacitor C_1 is grounded and the input signal is applied through an ungrounded capacitor C_2. In other words, the design of Fig. 5.7 can be converted to that in Fig. 5.10 by simply adjusting capacitors C_1 and C_2 as done previously.

The configuration in Fig. 5.11 involves the collector-base feedback biasing common emitter amplifier with an emitter follower output that prevents loading on the amplifying stage. The voltage at the collector of Tr_1 is usually $V_{CC}/2$ for maximum symmetrical swing and hence the determination of R_3 easily follows as R_3 equals $(V_{CC}/2 - 0.7)$ divided by the quiescent collector current of Tr_2.

5.1.1   Direct Coupled High-Gain Configurations

The direct coupled configurations discussed thus far all give gains in the range 40 dB–50 dB. Even though the configurations comprise two transistors, this gain is provided by a single common emitter stage. In order to achieve higher gains as discussed previously, at least two stages must provide gain. The circuit shown in Fig. 5.12 comprises two common emitter amplifiers that have direct coupling between them with the collector of the first stage Tr_1 being connected to the base of the second stage Tr_2. Resistor R_3 provides bias stability in Tr_2 while resistor R_4 provides bias to Tr_1 in a manner that is essentially collector-base feedback bias for this stage. However, by connecting resistor R_4 at the emitter of Tr_2 instead of the collector of Tr_1, transistor Tr_2 is also included in the bias-stabilizing feedback loop.

Fig. 5.12 Direct coupled
high-gain amplifier

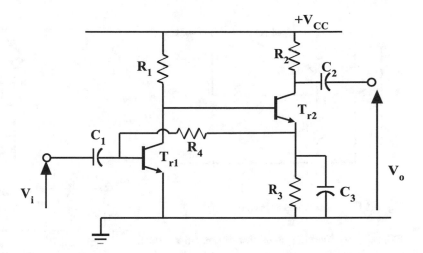

Fig. 5.13 Equivalent
circuit of direct coupled
high-gain amplifier

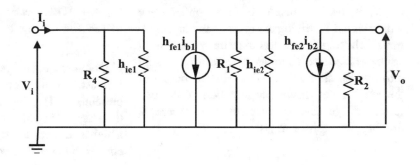

As a result, bias stability for the two transistors is achieved since any quiescent current change in either transistor is corrected by the loop. Note that for signals, capacitor C_3 grounds the emitter of Tr_2 thereby disconnecting the feedback loop. The equivalent circuit is shown in Fig. 5.13. The voltage gain of the first stage is $A_{V1} = -40I_{C_1}R_1 \| h_{ie2}$, while the gain of the second stage is $A_{V2} = -40I_{C_2}R_2$. Hence the overall gain is the product of these two gains, and this can be quite high.

Example 5.7
Using the configuration in Fig. 5.12, design a direct coupled amplifier with high gain.

Solution Using a 20-volt supply, since the two stages have voltage gain, it follows that the voltage swings at the collector of Tr_2 are greater than those at the collector of Tr_1. Therefore, maximum symmetrical swing consideration are more

important for Tr_2 than Tr_1. Applying the design techniques of Chap. 2 for a common emitter amplifier, let the emitter voltage be $V_{CC}/10 = 20/10 = 2$ V. Choosing 2 mA quiescent current for this stage, then $R_3 = 2$ V/2 mA $= 1$ k. For maximum symmetrical swing, the voltage drop across R_2 is $(20 - 2)/2 = 9$ V. Hence $R_2 = 9$ V/2 mA $= 4.5$ k. The voltage at the base of Tr_2 is the same voltage at the collector of Tr_1 which is 2.7 V. Choosing a quiescent current in Tr_1 of 1 mA, then $R_1 = (20 - 2.7)/1$ mA $= 17.3$ k. Finally, R_4 is calculated from $(2 - 0.7)$ $R_4 = 1$ mA/100 where the current gain of Tr_1 is taken as 100. This gives $R_4 = 130$ k. The gain for this circuit is about 76 dB.

Bootstrapping
A method of achieving high gain in a single common emitter stage is to increase the value of the load resistance. This can be accomplished for signals by the *bootstrapping* method used in

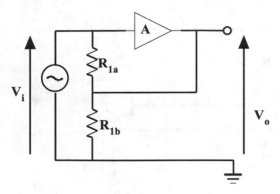

Fig. 5.14 Bootstrapping technique

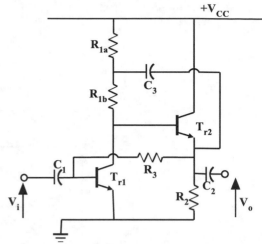

Fig. 5.15 Direct coupled amplifier with bootstrapping to increase gain

Chap. 4 to increase amplifier input impedance. The method is shown in Fig. 5.14. Here the resistor R_1 is the load resistor in an amplifier stage with a signal V_i driving this resistor and an amplifier A whose gain ρ is less than unity such as an emitter or source follower. The resistor is split into two resistors R_{1a} and R_{1b} with the output of the amplifier A driving the junction of the two resistors. For this arrangement, the current I_R flowing into resistor R_{1a} is

$$I_R = (V_i - V_o)/R_{1a} \qquad (5.6)$$

where $V_o = \rho V_i$.

Hence the effective resistance seen by the input signal V_i is

$$R_1' = \frac{V_i}{I_R} = R_{1a}/(1 - \rho) \qquad (5.7)$$

Since $\rho < 1$, it follows that $R_1' > R_{1a}$. For example, for $\rho = 0.95$, then $R_1' = R_{1a}/(1 - 0.95) = 20R_{1a}$. Therefore for $R_{1a} = 2k$ say, then $R_1' = 20 \times 2\,\text{k} = 40\,\text{k}$. Thus the effective resistance of R_1 as seen by the signal can be made quite large by this method.

The application of this technique to a common emitter amplifier is shown in Fig. 5.15. This configuration consists of a common emitter driving a common collector. The collector resistor R_1 of the first stage is split into R_{1a} and R_{1b}, and capacitor C_3 connects the emitter of Tr_2 to the junction of the two resistors. Its reactance at the lowest frequency of interest must be less than $R_{1a}//(R_{1b} + h_{fe2}R_2)$. Since the other end of resistor R_{1b} is connected to the base of Tr_2, this resistor

is therefore bootstrapped. The effect is that an increased value of collector resistance is presented to Tr_1, this resulting in a high gain for this stage. This gain is preserved by Tr_2 which has a high input impedance since it is configured as an emitter follower. The voltage gain is about 66 dB.

Example 5.8

Using the configuration in Fig. 5.15, design a direct coupled amplifier with high gain.

Solution Let $V_{CC} = 20$ V. Since only the first stage has gain, then maximum symmetrical swing is applied to this stage. Using a collector current of 1 mA in Tr_1, then for maximum symmetrical swing, $R_1 = 10$ V/1 mA $= 10$ k. Splitting this into two resistors then $R_{1a} = R_{1b} = 5$ k. The emitter voltage of Tr_2 is $10 - 0.7 = 9.3$ V. Using a current of 2 mA in Tr_2, then $R_2 = 9.3/2 = 4.7$ k. For R_3, $(9.3 - 0.7)/R_3 = 1$ mA/100 giving $R_3 = 860$ k. In evaluating the gain of the circuit, we first note that R_{1a} is in parallel with R_2 and part of the emitter load of Tr_2. Hence the effective resistance seen by Tr_2 emitter is 4.7 k// 5 k $= 2.4$ k. Therefore, the voltage gain of Tr_2 is $g_m R_E/(1 + g_m R_E) = 40 \times 2 \times 2.4/(1 + 40 \times 2 \times 2.4) = 0.995$. Therefore $R_{1b}' = 5$ k$/(1 - 0.995) = 1$ M. This very high resistance is presented to the transistor collector. Unfortunately, it is not the only resistance seen

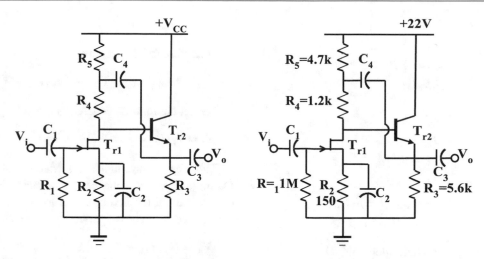

Fig. 5.16 Increasing the gain of a JFET amplifier with bootstrapping

by Tr$_1$ collector; there are two others. The first of these is the input impedance of the emitter follower Tr$_2$ which is $h_{fe} \times 2.4$ k $= 240$ k. The second is the output impedance $1/h_{oe}$ of Tr$_1$ which we have generally omitted because it is large compared with other resistors. However, this is no longer the case here. From the specifications for the 2N3904, this value is of the order of 100 k. Hence the effective collector load of Tr$_1$ is 1 M//240 k//100 k $= 66$ k. Therefore, the voltage gain of Tr$_1$ is $40 \times 1 \times 66 = 2640 = 68$ dB.

This bootstrapping arrangement can also be used to increase the gain of a JFET common source amplifier as shown in Fig. 5.16. This configuration is similar to Fig. 5.15, except that the JFET with fixed biasing replaces the BJT in the first stage.

Exercise 5.1

Analyse the design shown in Fig. 5.16 involving a JFET having $V_P = -3$ V and $I_{DSS} = 3$ mA.

5.2   Darlington Pair

An important two-transistor circuit is the Darlington pair shown in Fig. 5.17. It consists of two BJTs in which the two collectors are

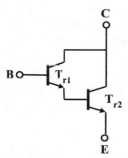

Fig. 5.17 Darlington pair

connected while the emitter of the first feeds the base of the second. The arrangement acts as a single transistor with base B (the base of the first transistor), collector C (the two connected collectors) and emitter E (the emitter of the second transistor). This unit has an enhanced current gain that is approximately the product of the current gain of the individual transistors. In order to see this, consider the Darlington pair in Fig. 5.17. For this,

$$I_{E1} = I_{C1} + I_{B1}$$
$$I_{C1} = \beta_1 I_{B1} \tag{5.8}$$

$$I_{E2} = I_{C2} + I_{B2}$$
$$I_{C2} = \beta_2 I_{B2} \tag{5.9}$$

But $I_{B2} = I_{E1}$ and $I_C = I_{C1} + I_{C2}$. Therefore

$$I_C = \beta_1 I_{B1} + \beta_2 I_{B2}$$
$$= \beta_1 I_{B1} + \beta_2 I_{E1}$$
$$= \beta_1 I_{B1} + \beta_2 (1 + \beta_1) I_{B1} \qquad (5.10)$$
$$\approx \beta_1 \beta_2 I_{B1}$$

and

$$I_E = I_{E2} = \beta_2 I_{B2} + I_{B2}$$
$$= (1 + \beta_2) I_{B2}$$
$$= (1 + \beta_2) I_{E1} \qquad (5.11)$$
$$= (1 + \beta_2)(1 + \beta_1) I_{B1}$$
$$\approx (1 + \beta_1 \beta_2) I_{B1}$$

Thus for the Darlington pair,

$$I_C = \beta_D I_B \qquad (5.12)$$

where $\beta_D = \beta_1 \beta_2$ and $I_E = (1 + \beta_D) I_B$. Therefore, the Darlington pair functions as though it is a single transistor with very high current gain β_D. It is possible to obtain two BJTs forming a Darlington pair in a single package. The MPSA29 from Motorola is an example of such a device. This transistor has a typical current gain of 10,000.

In order to determine the voltage gain and the input impedance of the Darlington Pair in the *common emitter* configuration, consider the equivalent circuit of the arrangement as shown in Fig. 5.18. The input voltage is given by

$$V_i = I_{b1} h_{ie1} + I_{b2} h_{ie2} \qquad (5.13)$$

From the equivalent circuit,

$$I_{b2} = (1 + h_{fe1}) I_{b1} \qquad (5.14)$$

and therefore (5.13) becomes

$$V_i = I_{b1} (h_{ie1} + (1 + h_{fe1}) h_{ie2}) \qquad (5.15)$$

Hence, the input impedance is given by

$$Z_i = V_i / I_{b1} = h_{ie1} + (1 + h_{fe1}) h_{ie2} \qquad (5.16)$$

Therefore, the input impedance of the Darlington pair is enhanced by a factor of the current gain of the first transistor as compared to a single device. The output voltage is given by

$$V_o = (h_{fe1} I_{b1} + h_{fe2} I_{b2}) R_L$$
$$= \left[(h_{fe1} + h_{fe2}(1 + h_{fe1})) \right] I_{b1} R_L \qquad (5.17)$$

Therefore, the voltage gain is given by

$$A_V = V_o / V_i$$
$$= \frac{(h_{fe1} + h_{fe2}(1 + h_{fe1})) R_L}{h_{ie1} + (1 + h_{fe1}) h_{ie2}} \qquad (5.18)$$

For $h_{fe1} h_{ie2} >> h_{ie1}$, this reduces to

$$A_V \simeq \frac{h_{fe2} R_L}{h_{ie2}} \qquad (5.19)$$

Thus, the voltage gain of the Darlington pair is similar to that of a single transistor. In the presence of an un-bypassed emitter resistor R_e, Eq. (5.15) for the input voltage becomes

$$V_i = I_{b1} \left[h_{ie1} + (1 + h_{fe1}) h_{ie2} + (1 + h_{fe1})(1 + h_{fe2}) R_e \right]$$
$$(5.20)$$

From this the input impedance is given by

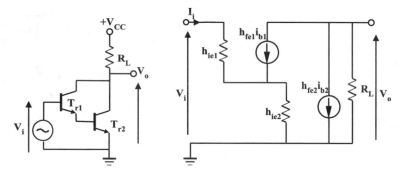

Fig. 5.18 Darlington pair in a common emitter amplifier

$$Z_i = V_i/I_{b1}$$
$$= h_{ie1} + (1 + h_{fe1})h_{ie2} + (1 + h_{fe1})$$
$$\times (1 + h_{fe2})R_e$$
$$\simeq h_{fe1}h_{fe2}R_e \qquad (5.21)$$

and the voltage gain is given by

$$A_V = V_o/V_i = \frac{(h_{fe1} + h_{fe2}(1 + h_{fe1}))R_L}{h_{ie1} + (1 + h_{fe1})h_{ie2} + (1 + h_{fe1})(1 + h_{fe2})R_e}$$
$$(5.22)$$

This reduces to

$$A_V \simeq \frac{R_L}{R_e}, \quad h_{fe2}R_e \gg h_{ie2} \qquad (5.23)$$

Example 5.10
Design a common emitter amplifier using a Darlington package having $\beta_D = 1000$ and a 24-volt supply. Calculate the voltage gain of your circuit.

Solution For the circuit in Fig. 5.19, choose $I_{Cq} = 1$ mA for good β_D and frequency response. Using rule-of-thumb (2.44), $V_{RE} = 24/10 = 2.4$ V. Hence $R_E = 2.4/1$ mA $= 2.4$ k, and for maximum symmetrical swing, $R_L = \frac{24-2.4}{1\,mA \times 2} = 10.8$ k. $V_{Bq} = V_{RE} + 2V_{BE} = 3.8$ volts and $I_R = 10I_{Rq} = 10 \times 1$ mA$/1000 = 0.01$ mA. Therefore, $R_2 = \frac{3.8\,volts}{0.01\,mA} = 380$ k and $R_1 = \frac{24-3.8}{0.01\,mA} = 2.02$ M . The resulting voltage gain is $A_V = -40I_CR_L = -40 \times 1$ mA $\times 10.8$ k $= -432$. To complete the design, large coupling

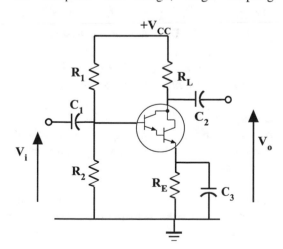

$+V_{CC}$

V_i

V_o

Fig. 5.19 Circuit for Example 5.10

capacitors, say 47 μF, are used at the input and output. Note that the resistors R_1 and R_1 are effectively in parallel with the input signal. Thus while their values must be sufficiently low to ensure that $I_R \geq 10I_B$ for bias stability, they normally cannot be too low since they will load down the input signal. However, because of the Darlington pairs, it is possible to use $I_R = 50I_{Bq} = 50 \times 1$ mA$/1000 = 0.05$ mA. Then, $R_2 = \frac{3.8\,volts}{0.05\,mA} = 76$ k and $R_1 = \frac{24-3.8}{0.05\,mA} = 404$ k .These values of resistors are still reasonably high.

Example 5.11
Modify the circuit in Example 5.10 to achieve a gain of 20.

Solution Using (5.23), the voltage gain of 20 is achieved by including an un-bypassed resistor R_e such that $A_V \simeq \frac{R_L}{R_e} = \frac{10.8\,k}{R_e} = 20$ giving $R_e = 540\,\Omega$. In order to accommodate this resistor in series with R_E, this resistor is reduced to $R_E = 2.4$ k $- .54$ k $= 1.9$ k.

Darlington Pair in Common Collector Configuration
We wish now to determine the voltage gain and the input impedance of the Darlington pair in the *common collector* configuration. Consider the equivalent circuit of the arrangement as shown in Fig. 5.20.

From the equivalent circuit, the input voltage is given by

$$V_i = I_{b1}h_{ie1} + I_{b2}h_{ie2} + (1 + h_{fe})I_{b2}R_E \quad (5.24)$$

But

$$I_{b2} = (1 + h_{fe1})I_{b1} \qquad (5.25)$$

Hence

$$V_i = I_{b1}[h_{ie1} + (1 + h_{fe1})h_{ie2} + (1 + h_{fe1})(1 + h_{fe2})R_E]$$
$$(5.26)$$

The output voltage is given by

$$V_o = (1 + h_{fe2})I_{b2}R_E$$
$$= (1 + h_{fe1})(1 + h_{fe2})I_{b1}R_E \qquad (5.27)$$

Hence the voltage gain is given by

Fig. 5.20 Darlington pair in common collector configuration

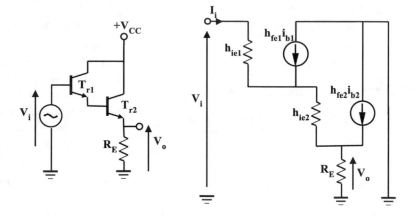

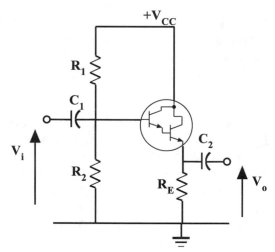

$$A_V = V_o/V_i = \frac{(1+h_{fe1})(1+h_{fe2})I_{b1}R_E}{[h_{ie1} + (1+h_{fe1})h_{ie2} + (1+h_{fe1})(1+h_{fe2})R_E]I_{b1}}$$

$$(5.28)$$

This reduces to

$$A_V = V_o/V_i = \frac{(1+h_{fe1})(1+h_{fe2})R_E}{[h_{ie1} + (1+h_{fe1})h_{ie2} + (1+h_{fe1})(1+h_{fe2})R_E]}$$

$$(5.29)$$

which can be about 0.99. This gain is slightly higher than that of a single-transistor emitter follower which is $(1 + h_{fe})R_E/(h_{ie} + (1 + h_{fe})R_E)$. The input impedance is given by

$$\begin{aligned} Z_i &= V_i/I_{b1} \\ &= h_{ie1} + (1+h_{fe1})h_{ie2} + (1+h_{fe1}) \\ &\quad \times (1+h_{fe2})R_E \\ &\simeq h_{fe1}h_{fe2}R_E \end{aligned} \qquad (5.30)$$

which is quite high.

Example 5.12
Design a common collector amplifier using a Darlington package having $\beta_D = 1000$ and a 24-volt supply. Calculate the input impedance of your circuit.

Solution For the circuit in Fig. 5.21, choose $I_{Cq} = 5$ mA in order to supply an external load. For maximum symmetrical swing, $V_{RE} = 24/2 = 12$ V. Hence $R_E = 12/5$ mA $= 2.4$ k. $V_{Bq} = V_{RE} + 2V_{BE} = 13.4$ volts and $I_R = 10I_{Bq} = 10 \times 5$ mA/1000 $= 0.05$ mA.

Fig. 5.21 Common collector amplifier using Darlington pair

Therefore, $R_2 = \frac{13.4 \text{ volts}}{0.05 \text{ mA}} = 268$ k and $R_1 = \frac{24 - 13.4}{0.05 \text{ mA}} = 212$ k. To complete the design, large coupling capacitors, say 47 µF, are used at the input and output. The input impedance is $R_1//R_2//1000R_E = 268$ k//212 k//1000 $\times$ 2.4 k $= 113$ k.

Example 5.13
Apply bootstrapping to the input of the circuit of Example 5.12 and estimate the final input impedance.

Solution The bootstrapped case is shown in Fig. 5.22. The voltage gain of the Darlington pair is about 0.99. Hence a value of R_B of 100 k

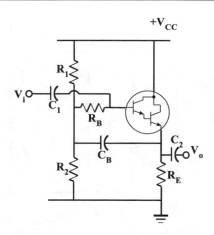

Fig. 5.22 Bootstrapped common collector amplifier using Darlington pair

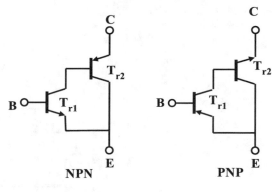

Fig. 5.23 Feedback pair

will present an impedance to the input signal of $R_B/(1 - 0.99) = 10$ M. This value is in parallel with the input impedance seen at the transistor base which is about $\beta_D R_E = 5000 \times 2.4\,\text{k} = 12\,\text{M}$. The input impedance is therefore 10 M// 12 M = 5.5 M

5.3   Feedback Pair

The feedback pair or compound Darlington is a two-transistor configuration that behaves in a manner similar to the Darlington pair. It comprises an npn transistor and a pnp transistor, the emitter of the npn connected to the collector of the pnp and the collector output of the npn connected to the base of the pnp. The arrangement shown in Fig. 5.23 functions effectively as an npn transistor. The PNP version is also shown in Fig. 5.23.

Consider the circuit shown in Fig. 5.23.

$$I_{E1} = I_{C1} + I_{B1}$$
$$I_{C1} = \beta_1 I_{B1} \tag{5.31}$$

$$I_{E2} = I_{C2} + I_{B2}$$
$$I_{C2} = \beta_2 I_{B2} \tag{5.32}$$

But $I_{C1} = I_{B2}$ and $I_E = I_{E1} + I_{C2}$. Hence

$$I_C = I_{E2}$$
$$= (1 + \beta_2)I_{B2}$$
$$= (1 + \beta_2)\beta_1 I_{B1} \tag{5.33}$$
$$\approx \beta_1\beta_2 I_{B1}$$

Also,

$$I_E = \beta_2 I_{B2} + (1 + \beta_1)I_{B1}$$
$$= \beta_1\beta_2 I_{B1} + (1 + \beta_1)I_{B1} \tag{5.34}$$
$$= (1 + \beta_1\beta_2)I_{B1}$$

Thus for the feedback pair,

$$I_C = \beta_F I_B \tag{5.35}$$

where $\beta_F = \beta_1\beta_2$ and $I_E = (1 + \beta_F)I_B$.

The voltage gain and input impedance of the feedback pair in the *common emitter* configuration are now determined. Consider the equivalent circuit of the arrangement as shown in Fig. 5.24. The input voltage is given by

$$V_i = I_{b1}h_{ie1} \tag{5.36}$$

Hence, the input impedance is given by

$$Z_i = V_i/I_{b1} = h_{ie1} \tag{5.37}$$

This is the same as the single transistor. The output voltage is given by

$$V_o = (1 + h_{fe2})I_{b2}R_L \tag{5.38}$$

From the equivalent circuit,

Fig. 5.24 Feedback pair in a common emitter amplifier

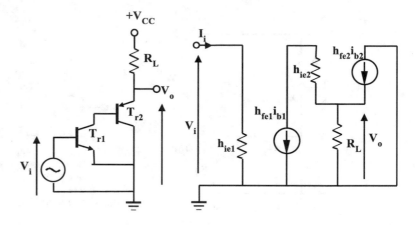

$$I_{b2} = -h_{fe1}I_{b1} \tag{5.39}$$

and therefore (5.38) becomes

$$V_o = -h_{fe1}(1 + h_{fe2})I_{b1}R_L \tag{5.40}$$

Therefore, the voltage gain is given by

$$A_V = V_o/V_i = \frac{-h_{fe1}(1 + h_{fe2})R_L}{h_{ie1}}$$

$$= -h_{fe2}g_{m1}R_L \tag{5.41}$$

This result indicates that the voltage gain of the feedback pair in the common emitter configuration is that of the first transistor ($-g_{m1}R_L$), multiplied by the current gain of the second transistor. This suggests gains of the order of 20,000! In practice however, the output resistance $1/h_{oe1}$ of the first transistor which was omitted from the analysis limits the achievable gain. In order to evaluate the effect of this parameter, we re-introduce it into the equivalent circuit for Tr_1. The effect is that all of the collector current of Tr_1 does not flow into the base of Tr_2; some of this current is lost through $1/h_{oe1}$. Using the current divider theorem, Eq. (5.39) becomes

$$I_{b2} = -h_{fe1}I_{b1}\frac{1/h_{oe1}}{1/h_{oe1} + [h_{ie2} + (1 + h_{fe2})R_L]} \tag{5.42}$$

where the term in square brackets in the denominator is the input impedance at the base of Tr_2. Using the modified Eq. (5.33) for I_{b2}, the expression for the voltage gain becomes

$$A_V = -h_{fe2}g_{m1}R_L\frac{1/h_{oe1}}{1/h_{oe1} + [h_{ie2} + (1 + h_{fe2})R_L]} \tag{5.43}$$

For $h_{fe2} >> 1$, this reduces to

$$A_V = -h_{fe2}g_{m1}R_L\frac{1}{1 + h_{oe1}(h_{ie2} + h_{fe2}R_L)} \tag{5.44}$$

If $h_{fe2}R_L >> h_{ie2}$, then (5.44) further reduces to

$$A_V = -h_{fe2}g_{m1}R_L\frac{1}{1 + h_{oe1}(h_{ie2} + h_{fe2}R_L)}$$

$$= -\frac{h_{fe2}g_{m1}R_L}{1 + h_{oe1}h_{fe2}R_L} \simeq -g_{m1}1/h_{oe1} \tag{5.45}$$

Since $1/h_{oe1}$ is usually larger than R_L, the gain of the feedback pair in the common emitter configuration is higher than the gain of the Darlington pair (or a single transistor) in the same configuration. This very interesting result seems rarely to appear in textbooks.

Example 5.14
Determine the voltage gain for the feedback pair in the common emitter configuration in Fig. 5.25. where the first transistor has a current of 0.7 mA.

Solution The 2N3904 transistor in the first stage of the feedback pair has h_{oe1} of 1 to 40 µmhos. Using $h_{oe1} = 20$ µmhos, then $1/h_{oe1} = 1/20 \times 10^{-6} = 50$ k which is greater than the load

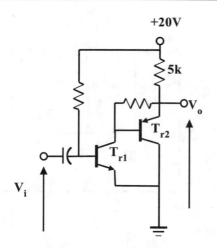

Fig. 5.25 Circuit for Example 5.13

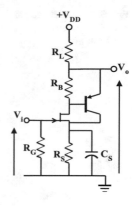

Fig. 5.26 Feedback pair using JFET and BJT in common source amplifier

resistance of 5 k. Therefore, the voltage gain of the circuit is given by $A_V = -g_{m1}l/h_{oe1} = -40 \times 2 \times 50 = -4000$.

Note that the gain of a single transistor with the same collector current 2 mA operating into the same load resistor $R_L = 5$ k is $A_V = -g_{m1}R_L = -40 \times 2 \times 5 = -400$ which is an order of magnitude lower than 4000. In the presence of an un-bypassed emitter resistor R_e, Eq. (5.36) for the input voltage becomes

$$V_i = I_{b1}\left[h_{ie1} + \left(1 + h_{fe1}\right) + h_{fe1}h_{fe2}\right)]R_e \tag{5.46}$$

From this the input impedance is given by

$$\begin{aligned} Z_i &= V_i/I_{b1} \\ &= h_{ie1} + \left(1 + h_{fe1} + h_{fe1}h_{fe2}\right)R_e \\ &\simeq h_{fe1}h_{fe2}R_e \end{aligned} \tag{5.47}$$

Thus, both the Darlington pair and the feedback pair have high current gains $\beta_F = \beta_1\beta_2$. The feedback pair has the advantage of having only one V_{BE} voltage drop between the base and emitter, while the Darlington pair has two. Also, in the common emitter configuration, the voltage gain of the feedback pair is significantly higher than that of the Darlington pair. However, the Darlington pair tends to have a better frequency response.

The configuration of the feedback pair may be used to enhance the effective transconductance of a JFET. Thus, a JFET can replace the first BJT in the circuit while retaining the second BJT as shown in Fig. 5.26. The result is that the overall gain of the circuit in a common source configuration increases from $-g_{m1}R_L$ to $-h_{fe2}g_{m1}R_L$ where g_{m1} is the transconductance of the JFET. However, as occurred in the feedback pair involving two BJTs, the drain resistance of the JFET will limit the gain increase. Resistor R_B is used to provide adequate bias current to the JFET.

Example 5.15

Design a common source using a JFET-BJT feedback pair in the topology shown in Fig. 5.26. Use a 20-volt supply and an n-channel JFET having $V_P = -2$ V and $I_{DSS} = 6$ mA. Determine the voltage gain of the circuit.

Solution We follow the design procedure for a self-biased common source amplifier in Chap. 3. Using $I_{DSS} = 6$ mA, $V_P = -3$ V and $I_D = 1$ mA in Shockley's equation gives $V_{GS} = \left(1 - \sqrt{\frac{I_D}{I_{DSS}}}\right)V_P$

$$= \left(1 - \sqrt{\frac{2 \text{ mA}}{8 \text{ mA}}}\right)(-3) = (1 - 0.4)(-2)$$

$$= -1.2 \text{ V}$$

. Hence $I_D R_S = 1.2$ V giving $R_S = 1.2$ V/ 1 mA $= 1.2$ k. For maximum symmetrical swing with $V_{DD} = 20$ volts, we allow for the

Fig. 5.27 Feedback pair in common collector configuration

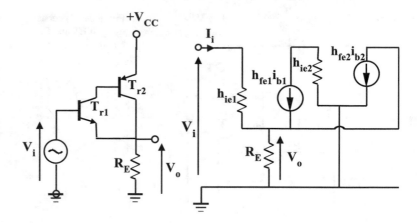

0.7 V across R_B. Hence $R_L = \frac{(20-1.2-0.7)/2}{1\text{ mA}} = 9$ k. Resistor $R_B = 0.7$ V/1 mA $= 700\ \Omega$. The input impedance of the JFET is greater than $10^8\ \Omega$, and therefore R_G is chosen to be 1 MΩ. From (3.8), g_m is given by $g_m = \frac{-2I_{DSS}}{V_P}\sqrt{\frac{I_D}{I_{DSS}}} = -2\frac{\sqrt{I_D I_{DSS}}}{V_P} = -2\frac{\sqrt{1\text{ mA}\times 6\text{ mA}}}{-2} = -2\times\frac{2.45}{-2} = 2.45$ mA/V . Hence the voltage gain A_V is given by $A_v \simeq -g_m r_D = -2.45\times 50$ k $= -122.5$.

Feedback Pair in Emitter Follower Configuration

We now determine the voltage gain and the input impedance of the feedback pair in the *common collector* configuration. Consider the equivalent circuit of the arrangement as shown in Fig. 5.27.

From the equivalent circuit, the input voltage is given by

$$V_i = I_{b1}h_{ie1} + \left[(1 + h_{fe1})I_{b1} - h_{fe2}I_{b2}\right]R_E \quad (5.48)$$

But

$$I_{b2} = -h_{fe1}I_{b1} \quad (5.49)$$

Hence

$$V_i = I_{b1}\left[h_{ie1} + (1 + h_{fe1} + h_{fe1}h_{fe2})R_E\right] \quad (5.50)$$

The output voltage is given by

$$V_o = I_{b1}(1 + h_{fe1} + h_{fe1}h_{fe2})R_E] \quad (5.51)$$

Hence the voltage gain is given by

$$A_V = V_o/V_i$$
$$= \frac{(1 + h_{fe1} + h_{fe1}h_{fe2})R_E}{\left[h_{ie1} + (1 + h_{fe1} + h_{fe1}h_{fe2})R_E\right]} \quad (5.52)$$

which is almost exactly unity: a near-perfect emitter follower! The input impedance is given by

$$Z_i = V_i/I_{b1}$$
$$= h_{ie1} + (1 + h_{fe1} + h_{fe1}h_{fe2})R_E$$
$$\simeq h_{fe1}h_{fe2}R_E \quad (5.53)$$

which also is quite high and can become as high as 10 M or higher with bootstrappping.

The feedback pair may be adjusted to provide gain in a manner that is different from the common emitter configuration discussed so far. This is done by the introduction of resistors as shown in Fig. 5.28. It is sometimes called the series feedback pair because it involves the application of a kind of negative feedback referred to as voltage-series feedback. It comprises an npn transistor Tr$_1$ operating in the common emitter mode with the emitter resistor un-bypassed. The output of Tr$_1$ is coupled to another common emitter transistor Tr$_2$ this employing a pnp transistor.

The unique arrangement in this circuit is the connection of the collector resistor R_5 to the emitter of transistor Tr$_1$ rather than to ground. The gain can be found by considering the equivalent circuit shown in Fig. 5.29. Here the bias components R_1, R_2 and R_3 have been omitted for simplicity. From this,

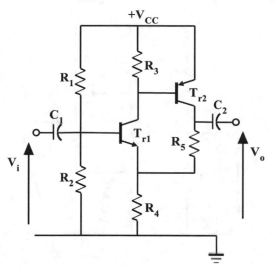

Fig. 5.28 Series feedback pair

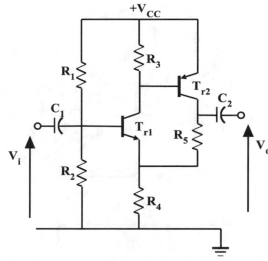

Fig. 5.30 Practical feedback pair with gain

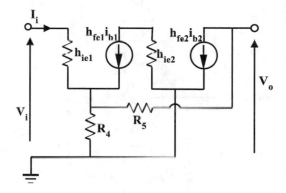

Fig. 5.29 Equivalent circuit of series feedback pair (simplified)

$$V_o = -h_{fe2}I_{b2}R_5 \\ + \left[(1 + h_{fe1})I_{b1} - h_{fe2}I_{b2}\right]R_4 \quad (5.54)$$

Noting that $I_{b2} = -h_{fe1}I_{b1}$, (5.39) reduces to

$$V_o = h_{fe1}h_{fe2}I_{b1}R_5 \\ + \left[(1 + h_{fe1})I_{b1} + h_{fe1}h_{fe2}I_{b1}\right]R_4 \quad (5.55)$$

At the input,

$$V_i = I_{b1}h_{ie1} \\ + \left[(1 + h_{fe1})I_{b1} + h_{fe1}h_{fe2}I_{b1}\right]R_4 \quad (5.56)$$

Therefore, the voltage gain $A_V = V_o/V_i$ is given by

$$A_V = \frac{h_{fe1}h_{fe2}R_5 + (1 + h_{fe1} + h_{fe1}h_{fe2})R_4}{h_{ie1} + (1 + h_{fe1} + h_{fe2})R_4}$$

$$(5.57)$$

Since $h_{fe1}h_{fe2} >> 1 + h_{fe1}$ and assuming $h_{fe1}h_{fe2}R_4 >> h_{ie1}$, then (5.57) reduces to

$$A_V = 1 + R_5/R_4 \quad (5.58)$$

A practical implementation of this basic circuit is shown in Fig. 5.30. Resistors R_1, R_2 and R_4 provide voltage-divider bias to Tr_1 while resistor R_3 provides bias current for Tr_1 as the base current of Tr_2 is generally not sufficient.

Example 5.16

Design a feedback pair using Fig. 5.30 to provide a fixed gain of 5 using a 20-volt supply.

Solution Let the collector current of each transistor be 1 mA. Then $R_3 = 0.7$ V/1 mA $= 700$ Ω. For bias stability of Tr_1, let the emitter voltage be $V_{CC}/10 = 20/10 = 2$ V. Then, noting that the collector currents of both transistors flow into R_4, the value of this resistor is $R_4 = 2$ V/ 2 mA $= 1$ k. For maximum symmetrical swing, the voltage at the collector of Tr_2 which is the output must be $2 + (20-2)/2 = 11$. From this, $R_5 = 9$ V/1 mA $= 9$ k. This gives a gain of

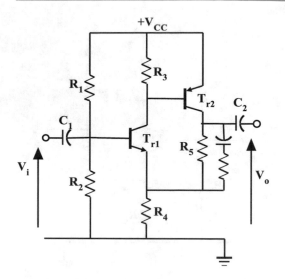

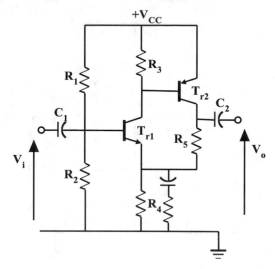

Fig. 5.31 Feedback pair with variable gain

$A_V = 1 + R_5/R_4 = 10$. In order to achieve a gain of 5, a 5 k portion of $R_5 = 9$ k can be bypassed with a large capacitor leaving 4 k in the signal path. This results in $A_V = 1 + R_5/R_4 = 1 + 4$ k/1 k $= 5$. Also, using $I = 1$ mA/10 $= 0.1$ mA in resistors R_1 and R_2 for bias stability gives $R_1 = 173$ k and $R_2 = 27$ k.

The signal gain of this basic circuit may be changed by connecting series-connected resistor and capacitor in parallel with either R_4 for increased gain or R_5 for decreased gain as shown in Fig. 5.31. The feedback pair utilizing a JFET for the first transistor can also produce gain by the introduction of resistors R_4 and R_5 as shown in Fig. 5.32. The effective gain is $A_V = 1 + R_5/R_4$.

Example 5.17
Determine the gain of the JFET-BJT feedback arrangement in Fig. 5.33.

Solution Using $A_V = 1 + R_5/R_4$, the gain of this amplifier is $A_V = 1 + 10^4/100 = 101$.

5.4 Current Sources

A current source is a current supply that maintains its current value, regardless of the load. This implies infinite output impedance. However, a

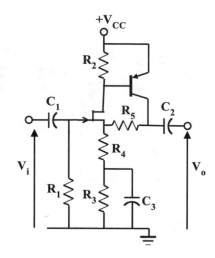

Fig. 5.32 BJT-JFET feedback pair with variable gain

practical current source has a finite output resistance R as shown in Fig. 5.34. The value of R is generally high: $R \to \infty$ corresponds to an ideal current source. While the terminals of a voltage source should in general not be short-circuited lest an infinite current flows, the terminals AB of the current source should in general not be open-circuited lest an infinite voltage develops across its terminals. Current sources are widely used in electronic circuits design and are particularly important in integrated circuit design. They can be realized using BJTs and FETS.

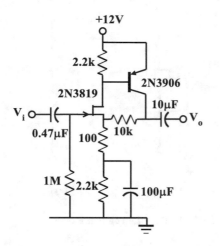

Fig. 5.33   Circuit for Example 5.16

Fig. 5.34   Practical current
source

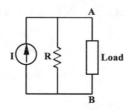

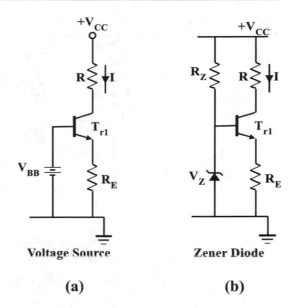

Voltage Source   **Zener Diode**

(a)   **(b)**

Fig. 5.35   BJT current source with **(a)** voltage source and
(b) Zener diode

BJT Current Source 1

A constant current source using a BJT is shown in
Fig. 5.35. It consists in Fig. 5.35 (a) of a transistor
Tr$_1$ with a resistor R_E connecting the emitter to
ground and its base fixed at a potential V_{BB} by a
suitable voltage source as shown in Fig. 5.35 (a).
It follows therefore that

$$V_{RE} = (V_{BB} - V_{BE}) = I_E R_E \qquad (5.59)$$

where I_E is the emitter current in the transistor.
Therefore

$$I_E = (V_{BB} - 0.7)/R_E \approx I_C \qquad (5.60)$$

The voltage V_{BB} can be supplied by a Zener
diode as shown in Fig. 5.35 (b). Here $V_{BB} = V_Z$
and R_Z is chosen such that the Zener diode is
properly reverse-biased.

Example 5.18

Calculate the constant current I_C in the circuit of
Fig. 5.35 (b) for $V_Z = 6.2, R_Z = 1$ k and $R_E = 5.5$ k,
and determine the current through the Zener
diode.

Solution   The voltage V_{RE} across R_E is given by
$V_{RE} = 6.2 - 0.7 = 5.5$ V. Hence $I_C = 5.5$ V/
5.5 k = 1 mA. The current I_Z in the Zener is given
by $I_Z = (15 - 6.2)/1$ k = 8.8 mA.

BJT Current Source 2

Another BJT constant current source utilizes two
BJTs. The configuration is shown in Fig. 5.36.
The base emitter function of Tr$_2$ maintains a
voltage of 0.7 V across the resistor R_1 thereby
yielding a current

$$I_R = 0.7/R_1 \qquad (5.61)$$

Ignoring the base currents of the two
transistors, the current I_{R1} is equal to the emitter
current of Tr$_1$ and hence

Fig. 5.36 Two-transistor current source

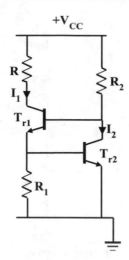

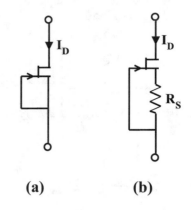

(a)   **(b)**

Fig. 5.37 JFET constant current source

$$I_1 = I_{R_1} = 0.7/R_1 \qquad (5.62)$$

R_2 supplies current to the base of Tr_1 and the collect of Tr_2.

Example 5.19
Using the configuration in Fig. 5.36, design a constant current source that produces a current of 1 mA from a 12-volt supply.

Solution For a current of 1 mA, $R_1 = 0.7$ V/ 1 mA $= 700$ Ω. For proper operation of Tr_2, choose $I_2 = 1$ mA. Then $R_2 = (12 - 1.4)/ 1 = 10.6$ kΩ. Resistor R represents the load through which the constant current flows.

JFET Current Source
The JFET functions as an extremely effective constant source when operated above pinch-off. Thus in an n-channel JFET for $V_{GS} = 0$, for example, and $V_{DS} > V_P$, the drain current is constant at I_{DSS}. This constant current value is that achieved in the circuit shown in Fig. 5.37 (a). Here $V_{GS} = 0$ and $I_D = I_{DSS}$ for $V_{DS} > V_P$. Note that this arrangement constitutes a two-terminal device. If the gate-source voltage is made negative, then the saturation drain current is reduced, and hence a new constant current results. This arrangement is shown in Fig. 5.37 (b) where the resistor R_S included in the source circuit produces $V_{GS} = - I_D R_S$. Hence a variable constant current two-terminal source can be achieved by making

R_S variable. The exact value of I_D corresponding to a particular value of R_S can be found using Shockley's equation which is given by

$$I_D = I_{DSS}\left(1 - \frac{V_{GS}}{V_P}\right)^2 \qquad (5.63)$$

Example 5.20
Using a JFET having $V_P = -5$ volts and $I_{DSS} = 8$ mA, design a constant current source with current value 2 mA.

Solution Using Shockley's equation for $I_D = 2$ mA, we have $2 = 8\left(1 - \frac{V_{GS}}{-5}\right)^2$. Solving for V_{GS} gives $V_{GS} = - 2.5$ volts. Since $I_D R_S = 2.5$, then $R_S = 1.25$ k.

The JFET connected as a constant current source is available commercially and is referred to as a constant current diode or current-regulating diode. It can operate for voltages ranging from 2 volts to 300 volts. When operated with reverse polarity, the constant current diode conducts as a junction diode, and therefore two such devices can be connected in series (one with normal polarity and the other with reverse polarity) to regulate alternating current. Siliconix manufactures two-terminal FET constant current diodes in plastic packages with currents ranging from 0.24 mA (J500) to about 4.7 mA (J511) and in metal TO-18 packages with currents ranging from 1.6 mA (CR160) to 4.7 mA (CR470).

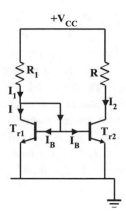

Fig. 5.38 Widlar current mirror

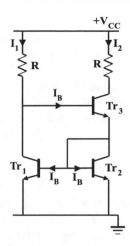

Fig. 5.39 Wilson current mirror

Central Semiconductor offers the CCL0035-CCL5750 series with currents from 35 μA to 6 mA with peak voltage ratings of 100 volts.

5.5   Current Mirror

The circuit shown in Fig. 5.38 is called a current mirror (based on a circuit developed by Bob Widlar) since the current established in Tr_1 via resistor R_1 is mirrored in transistor Tr_2 and hence flows through R regardless of its value. The circuit can therefore function as an effective constant current source. The current through R_1 is given by

$$I_1 = \frac{V_{cc} - V_{BE}}{R_1} \qquad (5.64)$$

Transistors Tr_1 and Tr_2 have the same base-emitter voltage since their base-emitter junctions are connected in parallel. If we assume the two transistor are matched (i.e. have approximately the same characteristics), it follows that they both have the same emitter/collector currents, i.e. $I = I_2$. Since $I = \beta I_B$

$$I = I_1 - 2I_B \approx I_1 \qquad (5.65)$$

Hence

$$I_2 = I \approx I_1 \qquad (5.66)$$

A better current mirror is the Wilson current mirror shown in Fig. 5.39. This is a three-

transistor arrangement that is more accurate than the current mirror just considered. For this circuit,

$$I_1 = \frac{V_{cc} - 2V_{BE}}{R_1} \qquad (5.67)$$

Now Tr_1 and Tr_2 again have the same collector currents I_C and base currents I_B. For Tr_3 the emitter current is given by

$$I_{E3} = \beta I_B + 2I_B = (2 + \beta)I_B \qquad (5.68)$$

Hence,

$$I_{B3} = \frac{I_{E3}}{(1 + \beta)} = \frac{2 + \beta}{1 + \beta} I_B \approx I_B \qquad (5.69)$$

Therefore

$$I_1 = I_{C1} + I_B = I_C + I_B \qquad (5.70)$$

Now

$$I_2 = I_{E3} - I_B \qquad (5.71)$$

But

$$I_{E3} = I_{C2} + 2I_B \qquad (5.72)$$

Therefore

$$I_2 = I_{C2} + 2I_B - I_B = I_C + I_B \qquad (5.73)$$

Hence

$$I_1 = I_2 \qquad (5.74)$$

Example 5.21

Design a Wilson current mirror to deliver a constant current of 0.5 mA using a 24-volt supply.

Solution Using Eq. (5.67), $R = (V_{CC} - 2V_{be})/I_1$. Since $I_2 = 0.5$ mA, then $I_1 = 0.5$ mA, and therefore, $R = (24 - 1.4)/0.5$ mA $= 4.2$ k.

Example 5.22

In Fig. 5.40 determine I_1, I_2, V_1 and V_2.

Solution The current through the 19.3 k resistor in the collector of Tr_1 is $I(19.3$ k$) = (20 - 0.7)/19.3$ k $= 1$ mA. Since Tr_1 and Tr_2 form a current mirror, $I_1 = I = 1$ mA. Also, Tr_3 and Tr_4 form a current mirror, and therefore the current through the 11.1 k resistor is equal to I_1. Therefore, the voltage across this resistor is V (11.1 k) $= 1$ mA $\times 11.1$ k $= 11.1$ V. Hence $V_1 = 11.1 - 0.7 = 10.4$ V. The current through Tr_5 is $I(Tr_5) = V_1/20.8$ k $= 10.4/20.8 = 0.5$ mA. This current develops a voltage drop across the 6 k resistor in the collector of Tr_5 given by V (6 k) $= 0.5$ mA $\times 6$ k $= 3$ V. As a result, the voltage at the base of Tr_6 is $V_B(Tr_6) = 20 - 3 = 17$ V giving $V_2 = 17 - 0.7 = 16.3$ V. Finally, $I_2 = 16.3$ V/32.6 k $= 0.5$ mA.

5.6   V$_{BE}$ Multiplier

The V_{BE} multiplier is single transistor circuit that effectively multiplies the base-emitter voltage of a transistor. It is sometimes referred to as a variable Zener. The circuit is shown in Fig. 5.41.

Here the base-emitter voltage of Tr_1 sets a fixed voltage 0.7 V across R_2 thereby producing a current $I_{R_2} = 0.7/R_2$. If the base current of Tr_1 is small compared with I_{R2}, then $I_{R_1} \approx I_{R_2}$. Hence

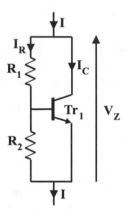

Fig. 5.41   V$_{BE}$ multiplier

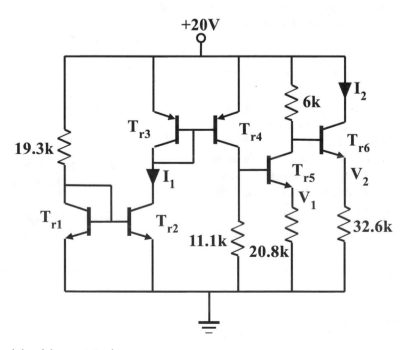

Fig. 5.40   Circuit involving current mirrors

$$V_{R_1} = I_{R_1}R_1 = 0.7\frac{R_1}{R_2} \qquad (5.75)$$

Therefore

$$V_Z = V_{R_1} + V_{R_2}$$
$$= 0.7 + 0.7\frac{R_1}{R_2}$$
$$= 0.7\left(1 + \frac{R_1}{R_2}\right) \qquad (5.76)$$
$$= \left(1 + \frac{R_1}{R_2}\right)V_{BE}$$

Thus the V_{BE} of Tr_1 is multiplied by a factor $(1 + R_1/R_2)$, setting a voltage V_Z across the transistor. The circuit therefore functions as a Zener diode. Note that any increase in I_R will increase the voltage across R_2 which in turn will tend to increase V_{BE}. This will then turn the transistor further on thereby increasing I_C and hence diverting the additional current through the transistor.

Example 5.23
Determine values for R_1 and R_2 such that the voltage across the transistor in the V_{BE} multiplier of Fig. 5.40 is 2.8 volts.

Solution From Eq. (5.61), for $R_2 = 1$ k then 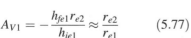 $2.8 = \left(1 + \frac{R_1}{1}\right)0.7$. This gives $R_1 = 3$ k.

5.7   Cascode Amplifier

An important amplifier circuit configuration utilizing multiple transistors is the cascode amplifier shown in Fig. 5.42. In this circuit, a common emitter amplifier Tr_1 drives a common base amplifier Tr_2. The base of Tr_2 is held at a potential V_B which can be provided by a Zener diode or other voltage source arrangement. Tr_1 is shown here utilizing fixed biasing, but other biasing schemes can be used. The input impedance of Tr_2 is r_{e2}, and this is the load of Tr_1.

Therefore, the gain of common emitter amplifier Tr_1 is

$$A_{V1} = -\frac{h_{fe1}r_{e2}}{h_{ie1}} \approx \frac{r_{e2}}{r_{e1}} \qquad (5.77)$$

where

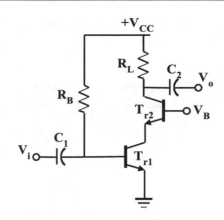

Fig. 5.42 Cascode amplifier

$$r_{e1} \approx \frac{h_{ie1}}{h_{fe1}} \qquad (5.78)$$

The gain of the second-stage Tr_2 is given by

$$A_{V2} = +\frac{h_{fe2}R_L}{h_{ie2}} \approx \frac{R_L}{r_{e2}} \qquad (5.79)$$

The overall gain A_V of the circuit is therefore

$$A_V = A_{V1} \times A_{V2} = \frac{r_{e2}}{r_{e1}} \times \frac{R_L}{r_{e2}} = \frac{R_L}{r_{e1}}$$
$$= g_{m1}R_L \qquad (5.80)$$

where $g_{m1} = 1/r_{e1}$. The input impedance of the circuit is that of Tr_1 in the common emitter configuration which in this case is $h_{ie1}//R_B$. Because of the presence of Tr_2, the configuration has the transfer characteristic of a common emitter amplifier but the output characteristic (high output impedance) of a common base amplifier. Note that voltage divider biasing may be applied to Tr_1 and the voltage V_B may be provided by a Zener or a potential divider with appropriate grounding capacitor.

One major advantage of the configuration is that since the collector of Tr_1 is held at an almost constant voltage, the Miller effect (multiplication of the collector-base capacitance C_{cb} of Tr_1) is reduced (this is discussed in Chap. 7). The result is that the effect of C_{cb} is minimized and results in an improved frequency response for the circuit. Another major advantage of this circuit is the reduction of early effect (modulation of the collector-base capacitance) distortion arising from the reduction of the voltage change at the collector of Tr_1.

Example 5.24

Using the configuration shown in Fig. 5.43, design a cascode amplifier circuit. Use $V_{CC} = 24$ V.

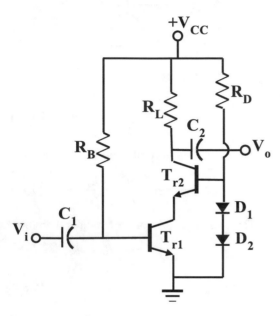

Fig. 5.43 Practical cascode amplifier

Solution The practical circuit is shown in Fig. 5.43. Here the diodes D_1 and D_2 fix the voltage at the base of Tr_2 at 1.4 volts, and R_D supplies current to the diodes. Choose this current to be about 5 mA, and then $R_D = \frac{24 - 1.4}{5 \text{ mA}} = 4.5$ k . Choose $I_{Cq}(Tr_2) = 1$ mA . The maximum peak-to-peak voltage swing for this transistor is approximately $24 - 1.4 = 22.6$ volts . Therefore, for maximum symmetrical swing, $R_L = \frac{22.6}{2}/1$ mA $= 11.3$ k . Using $\beta(Tr_1) = 100$, $R_B = \frac{24 - 0.7}{(1 \text{ mA}/100)} = 2.3$ M . Voltage gain for this circuit is $A_V = -40 I_C R_L = -40 \times 1 \times 11.3 = -452$.

A modified cascode amplifier utilizing a BJT and an FET which does not require a bias voltage V_B is shown in Fig. 5.44. It functions essentially as a three-terminal device. It is self-biasing, provided the BJT collector current is such that the resulting FET gate-source voltage is sufficient to keep the BJT out of saturation. A third configuration utilizes two JFETS in the configuration in Fig. 5.44. It is also self-biasing provided the gate-source voltage of the second FET exceeds the pinch-off voltage of the first, thereby keeping it operating in the saturation region. All of these cascode circuits can be used to produce very high output impedance current sources.

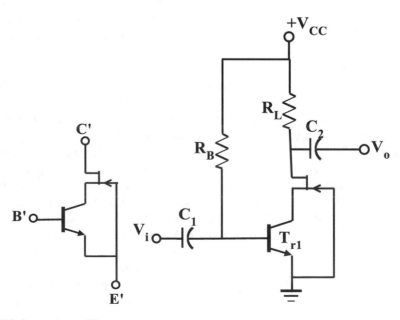

Fig. 5.44 Modified cascode amplifier

5.8   Improved Emitter Follower

The emitter follower has a low output impedance which enables it to drive low-impedance loads without a reduction of the output voltage. The configuration also has high current drive capability and is therefore widely used in the output stage of amplifiers. One drawback of the circuit discussed in Fig. 2.55 in Chap. 2 is its inability to deliver full output voltage swing on both half-cycles of a sinusoidal waveform. Specifically, an emitter follower using an npn transistor can deliver an output voltage into a load R_L limited only by the power supply voltage. However, on the negative half-cycle, the output voltage $V_O{}^-$ is limited to

$$V_O{}^- = I_{Cq}R_E\|R_L \qquad (5.81)$$

where I_{Cq} is the quiescent current in the transistor. This situation can be improved by replacing the emitter resistor R_E by a constant current source as shown in Fig. 5.45. In this circuit, all of the current I_{Cq} is available for sinking load current so that the peak negative output voltage increases to

$$V_O{}^- = I_{Cq}R_L \qquad (5.82)$$

The constant current source can be any of those discussed earlier.

Example 5.25
Design an improved emitter follower using the circuit of Fig. 5.45.

Solution A possible configuration is shown in Fig. 5.46. The constant current source is

implemented around Tr_2 with diodes D_1 and D_2 setting the fixed voltage across resistor R_3 and the transistor base-emitter junction. We choose $I_{Cq} = 2$ mA so that the circuit can drive low-impedance loads. Then $R_3 = 0.7$ volts/ 2 mA = 350 Ω. Resistor R_2 supplies current to the diodes, and if we allow about 5 mA, then $R_2 = (15 - 1.4)/5$ mA = 2.7 k. Finally, resistor R_1 supplies bias current into Tr_1 and a value of about 10 k ensures that the resulting offset voltage is small.

An even better arrangement is the complimentary emitter follower circuit shown in Fig. 5.47

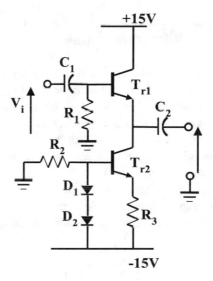

Fig. 5.46 Circuit of improved emitter follower

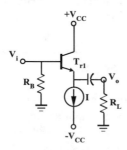

Fig. 5.45 Improved emitter follower

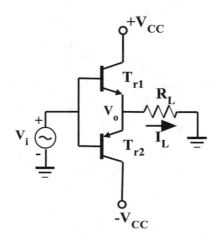

Fig. 5.47 Complimentary symmetry emitter follower

utilizing complimentary transistors. Here the emitter resistor or constant current source is replaced by another transistor in emitter follower configuration but having opposite polarity. On the positive half-cycle, the npn transistor as before delivers a positive-going output voltage and sources a high current, while on the negative half-cycle, the pnp transistor delivers a negative-going output voltage and sinks a high current. Proper operation requires that a bias voltage be applied between the two transistor base terminals so that the two transistors are turned on during the quiescent state. This circuit is discussed more fully in Chap. 10.

5.9   Differential Amplifier

One of the most important and widely used circuits is the differential amplifier or "long tail" pair shown in Fig. 5.48. It comprises two transistors of the same polarity connected at their emitters where a constant current source is connected to supply bias current to each. Each transistor is essentially in a common emitter mode with an input signal supplied to its base and an output signal taken at its collector. The system operates in the following manner: Assuming the

input to Tr_2 is set to zero, as the input signal V_{i1} to Tr_1 increases positively, the base-emitter voltage drop V_{be1} of Tr_1 increases and V_{be2} of Tr_2 decreases. The result of this is an increase in the collector current through Tr_1 which, because of the constant current source, causes a corresponding decrease in the collector current of Tr_2. Therefore, the collector voltage of Tr_1 goes down while that of Tr_2 goes up, by the same magnitude. If V_{i1} increases negatively, the reverse occurs resulting in a decrease in the collector current of Tr_1 and an increase in the collector current of Tr_2. This in turn results in an increase in the voltage at the collector of Tr_1 and a decrease in the voltage at the collector of Tr_2.

The effect of the input signal V_{i1} then is to produce an inverted signal voltage at the collector of Tr_1 and an in-phase signal voltage at the collector of Tr_2. For signals V_{i1} and V_{i2} at both inputs, $(V_{i1} - V_{i2})$ is the differential input voltage, and $(V_{o1} - V_{o2})$ is the differential output voltage. Since there are two inputs and two outputs, the amplifier is said to have a double-ended input and a double-ended output. It is important to note that the system does not respond to a common mode signal, i.e. a signal common to both inputs. To see this, we tie both inputs together and apply a signal, the constant current source prevents any change in collector current and hence there is no output signal in response. The circuit may be simplified by replacing the constant current source with a resistor from the connected emitters to the negative supply. Also, if an output is needed from one collector only, then the resistor in the collector of the unused output may be removed and replaced by a short circuit. This modified circuit is shown in Fig. 5.49. A more complete discussion of the differential amplifier is presented in Chap. 8.

5.10   BJT Switch

The binary junction transistor can function as a switch which is an important application in which a voltage supply is connected across a load. In this mode the transistor is operated either in saturation (fully on) in which case the voltage drop

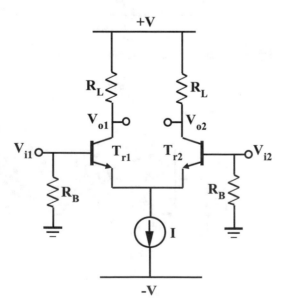

Fig. 5.48 Differential amplifier

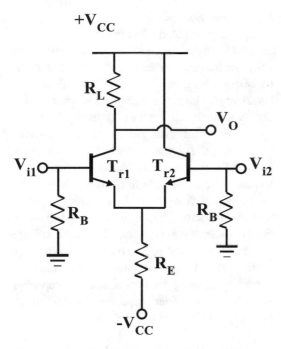

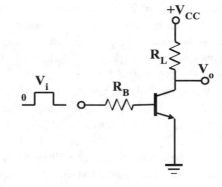

Fig. 5.50 Transistor switch

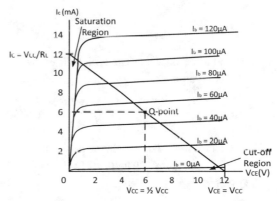

Fig. 5.49 Modified differential amplifier

Fig. 5.51 Transistor output characteristics showing saturation and cutoff regions

across it is quite small or in cut-off (fully off) in which case the current through the device is virtually zero. The basic circuit is shown in Fig. 5.50.

As we have seen before in the common emitter configuration, with the load resistor R_L collector current and collector-emitter voltage operate along a load line such that if the base current (or base-emitter voltage) is at zero, the transistor is off and the collector voltage is at its maximum value the supply voltage. The collector current is then the very small collector leakage current I_{CEO}. If the base current is increased, then the collector current increases correspondingly thereby causing the collector-emitter voltage to fall. If the base current is increased sufficiently, the collector-emitter voltage falls to a very small value referred to as the saturation voltage V_{CEsat} which is typically $V_{CEsat} \approx 0.2$ V with $V_{BEsat} \approx 0.7$ V. The collector current is then at its maximum value $I_C \approx V_{CC}/R_L$ limited only by the resistor R_L, and the transistor is said to be in saturation. These ON-OFF regions are shown in the output characteristics in Fig. 5.51.

Thus when functioning as an amplifier, the transistor operates in the active region where it

is never in saturation or in cut-off so that full reproduction of the amplified signal is assured. However when operated as a switch, the transistor is driven into either the fully ON state by supplying sufficient base current or the fully OFF state by reducing the base current to zero. This switching can be conveniently done by the application of a positive-going voltage pulse to the transistor base terminal through the base resistor R_B. When the voltage pulse is at zero, no base current flows into the transistor, and therefore it is off. Since there is no collector current, no voltage therefore appears across R_L. When the voltage pulse goes positive, base current is injected into the transistor via the resistor R_B, and the transistor is switched on. As a result almost the full supply voltage appears across the load resistor. The value of resistor R_B determines the actual value of base current that flows and must therefore be carefully chosen. The input voltage that switches the transistor on and off is often supplied by a digital circuit such as a microcomputer or other logic circuit where the voltage switches from zero corresponding to logical o and +5 volts corresponding to logical 1.

Let the magnitude of the positive switching voltage be V_L. We now determine the value of the resistor R_B for the transistor to be fully turned on when the input voltage is at V_L. Under full saturation, the collector-emitter voltage is at its saturation value, and most of the supply voltage is across R_L. Thus the collector current at saturation is given by

$$I_{Csat} = \frac{V_{CC} - V_{CEsat}}{R_L} \approx \frac{V_{CC}}{R_L} \quad (5.83)$$

The value of base current which flows when $V_i = V_L$ is given by

$$I_{Bsat} = \frac{V_L - V_{BEsat}}{R_B} = \frac{V_L - 0.7}{R_B} \quad (5.84)$$

To ensure that the transistor is driven into saturation, $I_{Bsat} \geq \frac{I_{Csat}}{\beta}$ which from (5.83) and (5.84) yields

$$\frac{V_L - 0.7}{R_B} \geq \frac{V_{CC}}{\beta R_L} \quad (5.85)$$

This reduces to

$$R_B \leq \frac{V_L - 0.7}{I_{Csat}/\beta} = \frac{V_L - 0.7}{V_{CC}/R_L\beta} \quad (5.86)$$

Example 5.26

A bipolar junction transistor is being used to switch 16 volts across a 5 k load. Using an npn transistor with $\beta = 100$, design a simple switching circuit based on the circuit in Fig. 5.51 that is fully on when driven by a 5-volt input.

Solution The value of the collector current when the transistor is saturated is given by $I_{Csat} = 16/5 = 3.2$ mA. Hence the required input resistor R_B is given by $R_B \leq \frac{16-0.7}{3.2 \text{ mA}/100} = 478 \text{ k}\Omega$. Use a 400 k$\Omega$ to ensure that the transistor is saturated. This lower value of R_B will result in a somewhat higher value of base current than that calculated for saturation.

BJT Device Switching

The power dissipated when the transistor is off is approximately zero because of the very small collector leakage current that flows through the transistor even though the collector-emitter voltage is large. The power dissipated when it is on is also very small because the collector-emitter voltage is very small even though the collector current is large. There can however be significant power dissipation in the transistor during the transition from one state to the other. In order to minimize this dissipation, the switching needs to take place rapidly. The actual switching time – the time between the application of the base voltage and the changed state of the transistor – is heavily influenced by junction capacitors within the transistor, specifically the collector-base capacitor and the base-emitter capacitor as shown in Fig. 5.52.

Consider the application of a voltage pulse through resistor R_B to the transistor base as shown in Fig. 5.52. Because of the presence of the base-emitter junction capacitor C_{BE}, the input voltage does not cause an instantaneous rise in the base-emitter voltage, and, as a result, the collector-emitter voltage does not change instantaneously. This capacitor must be charged through R_B such that the base-emitter voltage rises to approximately 0.7 volts and this causes a delay after which the collector-emitter voltage starts to fall. The time duration between the application of the voltage pulse and the fall of the

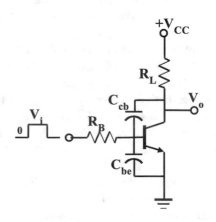

Fig. 5.52 Transistor switch with junction capacitors

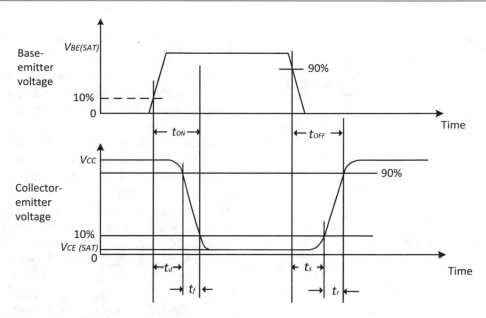

Fig. 5.53 Transistor switching characteristics

collector-emitter voltage from V_{CC} to $0.9V_{CC}$ is referred to as delay time t_d. During the time the transistor was off with the base at zero volts, the collector-base junction capacitor C_{CB} was charged to V_{CC} through the load resistor R_L, and this capacitor voltage appears across the transistor collector. Therefore, as the collector voltage falls in response to the base-emitter voltage, this junction capacitor must be discharged through the transistor in order that the collector voltage continues to fall. This further delays the decrease of the collector-emitter voltage which eventually reaches its saturation value V_{CEsat}. The time taken for the collector voltage to fall from $0.9V_{CC}$ to $0.1V_{CC}$ is referred to as the fall-time t_f, and the sum of the delay time and the fall time is the turn-on time t_{on}.

After the collector voltage attains its saturation value, the collector current cannot be further increased, and therefore any additional increase in the base current will result in excess charge storage in the base region of the transistor. When the input voltage pulse is reduced to zero, the need to remove this excess charge results in a delay in the turning off of the transistor. The base-emitter capacitor C_{BE} must be discharged through R_B so that the base-emitter voltage can fall and turn-off can begin. The time duration

between the removal of the voltage pulse and the rise of the collector-emitter voltage to $0.1V_{CC}$ is the storage delay t_S. After this time the transistor begins to turn off and the collector voltage rises. The rate of rise is limited by the recharging of the collector-base junction capacitor which takes place through the load resistor R_L. The time taken for the collector voltage to rise from $0.1V_{CC}$ to $0.9V_{CC}$ is the rise time t_R, and the sum of the storage time and the rise time is called the turn-off time t_{off}. A plot of transistor switching voltages is shown in Fig. 5.53. Typical values for the turn-on times is 65 nanoseconds and turn-off time is 240 nanoseconds.

BJT Switching Applications

Several applications of the transistor switch are possible. Two examples are shown in Fig. 5.54 and Fig. 5.55. In the first, the transistor is used to switch on a small incandescent lamp. Here the lamp replaces the load resistor in the basic circuit. When the input voltage goes high, the transistor is turned on, and the full supply voltage is placed across the lamp causing it to light. The current through the lamp on turn-on needs to be known so that the base resistor can be determined. The transistor must be able to handle the current load, and many small to medium power

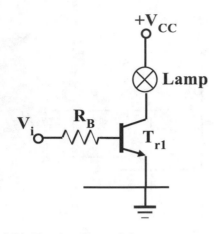

Fig. 5.54 Transistor lamp switch

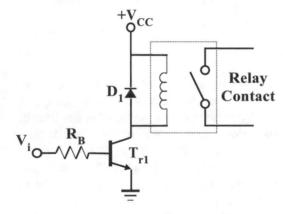

Fig. 5.55 Transistor relay switch

transistors will suffice. Because the value of R_B depends on the transistor current gain and this parameter is variable, the minimum specified current gain value should be used in the calculation.

Example 5.27
A microcomputer operating on 5 volts is being used to control a low-power incandescent lamp in an industrial application involving a supply of 12 volts and a 12 volt, 10 mA lamp. Calculate resistor R_B such that this circuit is functional.

Solution When the transistor is saturated, the value of the collector current is 10 mA. Hence using a 2N3904 small signal transistor with $\beta = 100$, the required input resistor R_B is given by $R_B \leq \frac{12-0.7}{10 \text{ mA}/100} = 113 \text{ k}\Omega$. Use a 100 k$\Omega$ to

ensure that the transistor is saturated. If a Darlington pair with $\beta = 10000$ is used, then $R_B \leq \frac{12-0.7}{10 \text{ mA}/100000} = 1.1 \text{ M}\Omega$. This higher value of resistor would mean that the current being drawn from the output port of the microcomputer is significantly reduced and this is very desirable.

In the second application in Fig. 5.55, the transistor load is a relay coil which when activated closes a set of contacts. This arrangement enables complete isolation between the transistor circuit and the circuit in which the contacts are connected. Again the current through the coil when the full supply is applied needs to be known in order to calculate R_B. This circuit needs a diode across the coil as shown. This is because high voltages are generated in a coil when the current through the coil is suddenly interrupted. These voltages can result in the destruction of the transistor. The diode provides a current path such that these voltages do not damage the transistor.

5.11 FET Switch

Like the BJT, the FET can be operated as a switch. This can be done by controlling the gate-source voltage of the FET. In the case of an n-channel JFET, for example, the device is switched ON when the gate-source voltage is zero and the drain-source voltage is less than the pinch-off voltage. Under these conditions the channel is conducting and no pinch-off has occurred. The device is OFF when the gate-source voltage is negative with a magnitude equal to the pinch-off voltage. In such a case the channel is pinched off by the negative gate-source voltage, and therefore no drain current flows for any value of drain-source voltage. The basic switching circuit is shown in Fig. 5.56 with these conditions illustrated in the output characteristics shown in Fig. 5.57. In order that the ON condition be operational, the value of load resistor R_L must be such that the resulting drain current is less than I_{DSS} the maximum possible drain current, i.e. $V_{DD}/R_L \leq I_{DSS}$ or $R_L \geq V_{DD}/I_{DSS}$. This ensures that the device stays in the

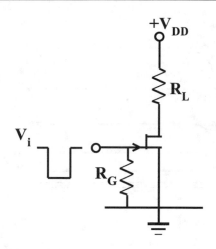

Fig. 5.56 Basic JFET switch

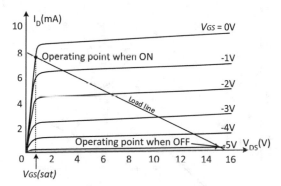

Fig. 5.57 Output switching characteristics of JFET

linear or ohmic region where the channel resistance is of the order of hundreds of ohms and does not enter the saturation region. Resistor R_G provides the gate-source connection necessary for application of the gate-source voltage while having a high input impedance to prevent loading the input voltage source.

JFET Switch

As in the case of the BJT, the junction capacitors of the JFET, namely, the gate-source capacitor and the drain-gate capacitor, heavily influence the switching times of the device. In Fig. 5.56, consider the application of a negative-going voltage pulse. Initially the voltage is zero and the FET is ON. As the voltage pulse goes negative, in order that the device starts to turn off the gate-source junction capacitor must be charged to

Fig. 5.58 Enhancement-type MOSFET switch

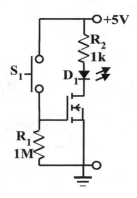

beyond pinch-off. The absence of any resistor in the gate lead means that the delay caused by this charging action is small. Once the device starts to turn off, the rise of the drain voltage will be limited by the increase of the voltage across the drain-gate capacitor which must be charged through R_L which largely determines the rise time. The resulting delay is more significant than that caused by the charging of the gate-source capacitor. When the input voltage returns to zero, the reverse occurs. In order that the device starts to turn ON, the gate-source capacitor must be discharged, and this introduces a turn-on delay. After this occurs, the drain current will increase, and the drain voltage will fall. The drain-gate capacitor will discharge through the FET channel, and since the channel resistance r_D in the linear region is low, the fall time is low. Thus since $R_L > r_D$ the turn-off time is much shorter than the turn-on time.

MOSFET Switch

In the case of the n-channel enhancement-type MOSFET shown in Fig. 5.58, a zero gate-source voltage corresponding to the switch off means that there is no channel conduction and hence the device is off. A positive gate-source voltage greater than the threshold voltage (corresponding to switch on) turns on the device and conduction results. For the depletion-type MOSFET, a zero gate-source voltage is insufficient to turn it off; a negative gate-source voltage is necessary for turn-off. The need for a bipolar signal for turn-on and turn-off in the case of the depletion-type MOSFET makes this MOSFET less useful as a switch than the enhancement type where only one

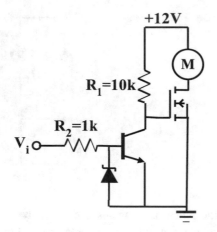

Fig. 5.59 Cascaded transistor and MOSFET switch

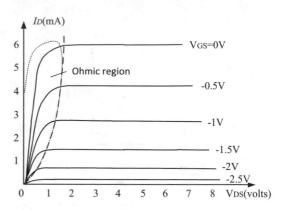

Fig. 5.60 I_D vs V_{DS} of JFET

signal polarity is necessary for turn-on and turn-off.

The enhancement MOSFET switch has large power capability and very high input impedance which make it very suitable for connection to microcontroller or other logic circuit output in order to control high-power systems. However, the +5 volts may be insufficient to overcome the threshold voltage of the MOSFET to turn it on. This problem may be solved by cascading a BJT switch and a MOSFET switch as shown in Fig. 5.59. Turning off the BJT applies +12 volts to the gate of the MOSFET and turns it on. The input of the MOSFET can be damaged by electrostatic discharge. In order to protect the gate from over-voltage, a Zener diode clipping circuit D_1 can be used as shown in Fig. 5.59. An over voltage at the input will force the Zener diode into conduction thereby limiting the gate voltage. The excessive voltage is then dropped across R_2.

5.12   Voltage-Controlled Resistor

As has been already observed, the FET operates as a linear resistor in the ohmic region to the left of the pinch-off locus in Fig. 5.60. By varying the gate-source voltage, the value of this resistance can be varied thereby making the FET a voltage-controlled resistor. This action is evident in Fig. 5.61 for an n-channel JFET where for

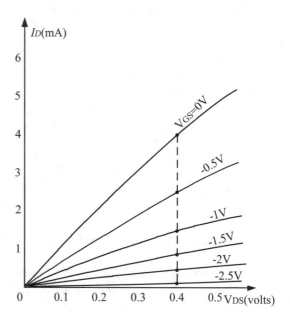

Fig. 5.61 Ohmic or linear region of JFET

drain-source voltage less than the pinch-off voltage represented on the locus, the slope of the curves and hence the channel resistance vary with the gate-source voltage.

As V_{GS} is made more negative, the curves become flatter corresponding to a higher resistance. An approximate value of this resistance can be obtained from Shockley's equation

$$I_D = I_{DSS}\left(1 - \frac{V_{GS}}{V_P}\right)^2 \quad (5.87)$$

This resistance r_d is given by

$$r_d = \frac{V_{DS}}{I_D} = \frac{V_{DS}}{I_{DSS}} \frac{1}{\left(1 - \frac{v}{V_P}\right)^2} \qquad (5.88)$$

Using $r_o = V_{DS}/I_{DSS}$, this can be written as

$$r_d = \frac{r_o}{(1 - V_{GS}/V_P)^2} \qquad (5.89)$$

Thus minimum r_d corresponds to $V_{GS} = 0$ and is determined by the geometry of the FET. A device having a channel with small cross-sectional area will exhibit high r_o and low I_{DSS}. A family of n-channel FETs designed for use as voltage-controlled resistors is available from Siliconix. These devices have r_o of between 20 ohms and 4000 ohms with the VCR2N in the range 20–60, the VCR4N in the range 200–600 and the VCR7N in the range 4 k–8 k. One application of the VCR is a voltage-controlled attenuator shown in Fig. 5.62. By changing the negative gate-source voltage, the resistance r_d varies resulting in an output signal V_o given by $V_o = V_i \frac{r_d}{R+r_d}$. A requirement of this circuit is that the voltage across the FET does not exceed the pinch-off voltage so that the device remains in the linear region.

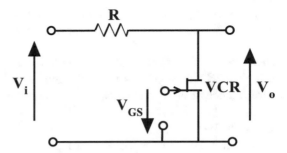

Fig. 5.62   Voltage-controlled attenuator using VCR

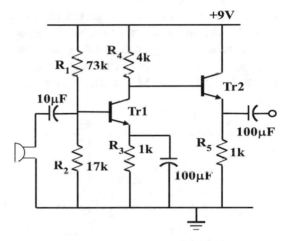

Fig. 5.63   Microphone preamplifier 1

5.13   Applications

Several circuits implementing multiple transistors and FETS are discussed in this section.

Microphone Preamplifier 1

A simple microphone preamplifier is shown in Fig. 5.63. It comprises a common emitter amplifier using a small-signal transistor followed by an emitter follower amplifier to provide a low output impedance. Using the design technique developed in Chap. 2, resistors R_1 and R_2 set the base voltage of Tr_1 to about 1.7 volts. This along with R_3 establishes a quiescent current of 1 mA in Tr_1. Biasing Tr_1 for maximum symmetrical swing requires that resistor R_4 be 4 k. Selecting R_5 as 1 k sets a current in Tr_2 of 5 mA. The capacitor values are large to allow adequate low-frequency response.

Ideas for Exploration (i) Re-design the circuit using the 2N3906 pnp transistor, and compare the performance of the two circuits.

Microphone Preamplifier 2

Figure 5.64 shows the circuit of another microphone preamplifier. It employs a collector-base feedback-biased common emitter amplifier around Tr_1 and an emitter follower to reduce the loading on the collector of Tr_1 and reduce the output impedance of the circuit. For maximum symmetrical swing, the voltage across R_2 must be $9/2 = 4.5$ V. Using a collector current of 1 mA for Tr_1, then $R_2 = 4.5/1$ mA $= 4.5$ k. Using $\beta = 100$ for Tr_1, then the associated base current is 1 mA/$100 = 10$ μA. Hence $R_1 = (4.5 - 0.7)/10$ μA $= 380$ k. Since the base of Tr_2 is directly connected to the collector of Tr_1, then the emitter of Tr_2 is at $4.5 - 0.7 = 3.8$ V. Choosing a current of 2 mA for Tr_2 to enable sufficient drive current for external loads, then $R_3 = 3.8$ V/2 mA $= 1.9$ k.

Ideas for Exploration (i) Re-design the circuit using the 2N3906 pnp transistor, and compare the performance of the two circuits.

Logic Probe
This circuit shown in Fig. 5.65 is a logic probe that enables the determination of the state of a logic signal which can attain either a high state (+5 volts) or a low state (0 volts). Probe P2 is connected to the earth of the logic circuit, while probe P1 is connected to the logic signal. When there is no signal at P1 (floating probe), providing V_{CC} is less than the sum of the voltage drops across the LEDs and the transistor base-emitter

voltages, both transistors will be off and hence so will both LEDs. When P1 goes to a high state that exceeds the turn-on voltage of Tr_1 (0.7 V) and the red LED (1.7 V) which is approximately 2.4 V, then the red LED will turn on. Tr_2 will remain off. When P1 goes to a low voltage that is less than $V_{CC} - 2.8$, then there will be sufficient voltage across the GREEN LED (2.1 V) and the base-emitter junction of Tr_2 to turn them on. Resistors R_3 and R_4 serve to limit the current, and resistors R_1 and R_2 limit the base current into the transistors. Using a supply voltage of $V_{CC} = 3$ volts, maximum input voltage of 5 volts and maximum LED current of 10 mA, then $R_3 = (5 - 0.7 - 1.7)/10$ mA $= 260 \, \Omega$ where the LED voltage is 1.7 V. For low input voltage of zero and allowing an LED current of 2 mA, then $R_4 = (3 - 0.7 - 2.1)/2$ mA $= 100 \, \Omega$. Resistors R_1 and R_2 can be about 1 k each in order to protect the transistor inputs from over-voltage.

Ideas for Exploration (i) Re-design the circuit to operate with a 9-volt battery.

Telephone Use Indicator
This circuit shown in Fig. 5.66 indicates telephone usage and is an improvement on the single-transistor telephone monitor of Fig. 2.65 in Chap. 2. When the telephone is in use diode D_5 (red LED) is on and diode D_6 (green LED) is off. When the telephone is not being used, diode D_5 (red LED) is off and diode D_6 (green LED) is on. The diode bridge ensures that the DC signal

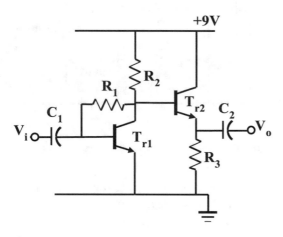

Fig. 5.64 Microphone preamplifier 2

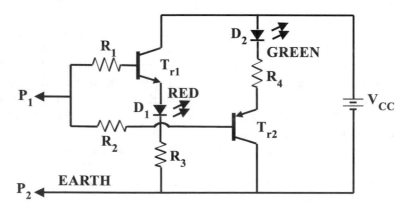

Fig. 5.65 Logic probe

Fig. 5.66 Telephone use
indicator

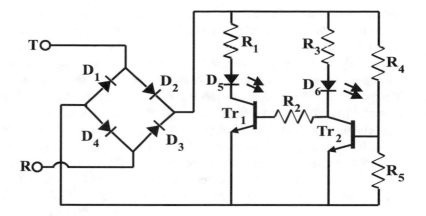

into the transistor from the telephone line is
always of the same polarity, regardless of the
polarity orientation of the connection to the telephone line. With the telephone on hook, the telephone circuit is open and 48 volts from the
telephone exchange will appear across tip
(T) and ring (R). This voltage is applied to potential divider R_4 and R_5 producing a voltage drop
across R_5 that turns on transistor Tr_2 and hence
also LED D_6(green). For $R_5 = 2.2$ k and
$R_4 = 47$ k, the voltage across R_5 is $\{2.2/
(47 + 2.2)\} \times 48 = 2.1$ V which will ensure that
Tr_2 turns on. During operation, this voltage will
be limited to less than this value as Tr_2 turns on
and current flows through R_4. Resistor R_3 limits
the current through LED D_6 and a value of
$R_3 = 27$ k restricts the transistor collector current
to about 2 mA. With Tr_2 saturated, its collector
voltage is low, and hence transistor Tr_1, whose
base is connected to Tr_2 collector via R_2, is off. As
a result, LED D_5 is off. When the handset is lifted,
the telephone circuit is closed and as a result of
line resistance, the voltage across tip and ring will
drop to about 9 V. This causes the voltage across
resistor R_5 to fall to about 0.4 V. This causes Tr_2
to turn off. A current will now flow through the
path R_3, D_6, R_2 into the base of Tr_1. For
$R_2 = 27$ k, the value of this current is
$(9 - 2.1 - 0.7)/54$ k $= 115$ µA. This current is
not sufficient to light D_6 which therefore goes off.
Transistor Tr_1 will now turn on with a collector
current given by $115 µA \times \beta(Tr_1)$ which for
$\beta = 100$ is 11.5 mA. However, resistor $R_1 = 1$ k
in the collector circuit of Tr_1 will limit this current

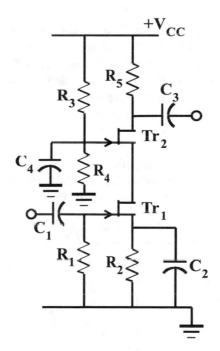

Fig. 5.67 FET cascode amplifier

to less than 9 V/1 k $= 9$ mA. This will cause diode
D_5 to turn on.

Ideas for Exploration (i) Compare this system
with the telephone monitor circuit in Chap. 2.

FET Cascode Amplifier
This amplifier shown in Fig. 5.67 utilizes two
junction field effect transistors in a cascode configuration similar to that of Sect. 5.7 using BJTs.
Field effect transistor Tr_1 operates in the normal
common source mode with resistors R_1 and R_2

selected according to the design rules in Chap. 3. Note that the gate of Tr_1 is held at zero volts by R_1 and current through R_2 causes the voltage at the source of Tr_1 to go positive thereby reverse-biasing the gate-source junction of Tr_1 and limiting the drain current to a selected value. The voltage $V_G(Tr_2)$ at the gate of Tr_2 set by R_3 and R_4 is equal to the drain-source voltage across Tr_1. Thus, if the voltage at the source of Tr_1 is ΔV, then the voltage at the source of Tr_2 is $V_G(Tr_2) + \Delta V$. Finally, the load resistor R_5 is determined to ensure maximum symmetrical swing.

Ideas for Exploration (i) Modify the circuit by using a JFET for Tr_2 that has a sufficient gate-source voltage to enable self-biasing of Tr_2 and the operation of Tr_1 in the saturation region. This would mean that components R_3, R_4 and C_2 would not be needed and the gate of Tr_2 would be connected to the source of Tr_1.

BJT Cascode Amplifier
The amplifier shown in Fig. 5.68 utilizes two binary junction transistors in a cascode

configuration similar to that of Sect. 5.7 using BJTs but biased in a more conventional manner. Transistor Tr_1 operates in the normal common emitter mode. Choosing a quiescent current of 2 mA and using one tenth of the supply voltage as the emitter voltage of Tr_1 (2 V), resistor $R_5 = 1$ k. Allowing 1 V across Tr_1 for proper operation, for maximum symmetrical swing, the voltage across R_4 and Tr_2 is $(20 - 3)/2 = 8.5$ V. Therefore $R_4 = 8.5/2 = 4.3$ k. Since the resistor chain R_1-R_2-R_3 is supplying base current to two transistor, we let the current I down this chain be one fifth (instead of one tenth) of the collector current giving $I = 0.4$ mA. Since the emitter of Tr_1 is at 2 V, the base of Tr_1 is at 2.7 V and hence $R_3 = 2.7$ V/0.4 mA $= 6.8$ k. Since the voltage at the emitter of Tr_2 is 3 V, the base of Tr_2 is 3.7 V and therefore the voltage across R_2 is 1 V. Hence $R_2 = 1$ V/0.4 mA $= 2.5$ k. The voltage across R_1 is $20 - 3.7 = 16.3$ V and hence $R_1 = 16.3/0.4$ mA $= 40.8$ k. The gain of this circuit is given by -40×2 mA $\times 4.3$ k $= 344 = 51$ dB, and its upper cut-off frequency was measured at 19 MHz.

Ideas for Exploration (i) Modify the circuit by using a JFET for Tr_2 that has a sufficient gate-source voltage to enable self-biasing of Tr_2 and the operation of Tr_2 in the active region. This would mean that capacitor C_2 would not be needed though it can be retained to provide supply filtering to the base of Tr_1. The gate of Tr_2 would then be connected to the emitter of Tr_1.

Touch-Sensitive Indicator
The next system to be discussed is a circuit that switches on when touched. The circuit is shown in Fig. 5.69. It consists of two transistors arranged such that the emitter of the first is connected to the base of the second (Darlington pair). There are current limiting resistors in the collector of each transistor with the second transistor including an LED in its collector. Touching the base and supply voltage simultaneously will cause a small current to flow in the base of Tr_1 where it is amplified by a factor of the transistor current gain. This amplified current now flows into the

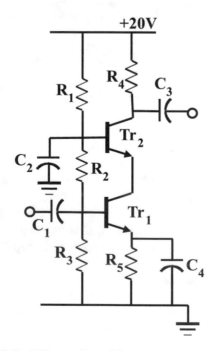

Fig. 5.68 BJT cascode amplifier

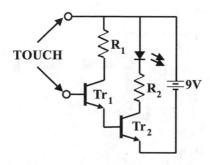

Fig. 5.69 Touch-sensitive indicator

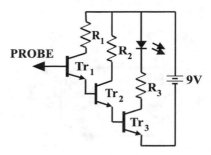

Fig. 5.70 Electromagnetic field detector

base of Tr_2 and is amplified further by a factor equal to the transistor current gain. The result is that a tiny current into the circuit at the base of Tr_1 is amplified by a factor of about $100 \times 100 = 10^4$ and results in a significant current flowing in the collector of Tr_2. The LED in the collector of Tr_2 is therefore illuminated. R_2 is determined by the current level required to turn on the LED. Thus, for a 9-volt supply and allowing about 15 mA to activate the LED, $R_2 = (9 - 2.1)/15$ mA $= 460\ \Omega$. Using the minimum value of the transistor current gain, the maximum current in Tr_1 is given by 15 mA/100 $= 0.15$ mA. Hence $R_1 = 9/0.15$ mA $= 60$ k. A value of 47 k is suitable.

Ideas for Exploration (i) Replace resistor R_2 and the LED by a relay coil (with protection diode), and use the circuit to switch on a mains connected device.

Electromagnetic Field Detector

This circuit is an enhancement of that in Fig. 5.69 and detects electromagnetic fields associated with house wiring, static electricity and other sources. The circuit is shown in Fig. 5.70 and bears some similarity to a Darlington pair. It consists of three transistors arranged such that the emitter of the first is connected to the base of the second and the emitter of the second connected to the base of the third. There are current-limiting resistors in the collector of each transistor with the final transistor including an LED in its collector. Any current induced in the base of Tr_1 is amplified by a factor of the transistor current gain. This amplified current now flows into the base of Tr_2 and is amplified further by a factor equal to the transistor

current gain. This amplification is continued in Tr_3 with the result that a tiny current into the circuit at the base of Tr_1 is amplified by a factor of about $100 \times 100 \times 100 = 10^6$ results in a significant current flowing in the collector of Tr_3. The LED in the collector of Tr_3 is therefore illuminated. R_3 is determined by the current level required to turn on the LED. Thus, for a 9-volt supply and allowing about 15 mA to activate the LED, $R_3 = (9 - 2.1)/15$ mA $= 460\ \Omega$. Using the minimum value of the transistor current gain, the maximum current in Tr_2 is given by 15 mA/100 $= 0.15$ mA. Hence $R_2 = 9/0.15$ mA $= 60$ k. A value of 47 k is suitable. Similarly, $R_1 = 9/0.0015$ mA $= 6$ M, and a value of 1 M is suitable.

Nickel-Cadmium Battery Charger

The circuit on Fig. 5.71 is that of a nickel-cadmium battery charger. These batteries require a constant current for a fixed period for proper recharging. The circuit comprises an unregulated supply from a 40 V centre-tapped transformer that provides about $20\sqrt{2} = 28$ V across the filter capacitor C_1. A value of 1000 uF ensures very little fall in voltage as currents up to 100 mA are drawn from the supply. Diodes D_3 and D_4 provide a fixed voltage at the base of Tr_1 such that there is a 0.7 V across resistor R_2. This results in a constant current of $0.7/R_2$ in the collector of Tr_1. The value of R_2 is determined by the constant current to be supplied. For 50 mA, $R_2 = 0.7/50$ mA $= 14\ \Omega$. Resistor R_1 supplies current to diodes D_3 and D_4 and to the base of Tr_1. For 5 mA through these diodes, $R_1 = (28 - 1.4)/5$ mA $= 5.3$ k. This constant current source can accommodate

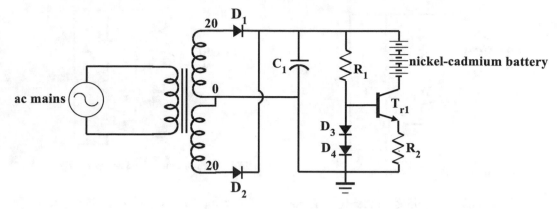

Fig. 5.71 Nickel-cadmium battery charger

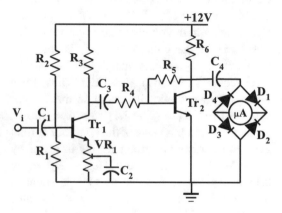

Fig. 5.72 Audio level meter

increase in battery voltage up to about 20 V. Significant heat is dissipated in the transistor, so it should be a power transistor such as the 2N3055 and should be mounted on a heatsink in order to dissipate the heat. Note that the system does not have an end-of-charge feature to turn off the system. This has to be done manually.

Ideas for Exploration (i) Recalculate the value of R_2 in order to charge batteries requiring different charging currents and introduce a single-pole multi-throw switch to enable easy changing of the charging current.

Audio Level Meter

This circuit shown in Fig. 5.72 is an audio level meter that measures the amplitude of an audio signal coming from a preamplifier which is in the range 20 Hz to 20 kHz. It utilizes two common emitter stages that amplify the audio signal. For the first stage, choosing a collector current of 0.5 mA and allowing 1 V at the emitter of Tr_1, the resistance of VR_1 is 1 V/0.5 mA = 2 k. A 2 k potentiometer is used and connected in the manner shown to enable variation of the gain of this stage. The voltage at the base of Tr_1 is 1.7 V. Let the current down the R_1-R_2 chain be .05 mA which is one tenth the collector current. Then $R_1 = 1.7/50\ \mu A = 34$ k and $R_2 = (12 - 1.7)/.05$ mA = 206 k. The available voltage swing at the collector of Tr_1 is $12 - 1 = 11$ V. Hence for maximum swing, $R_3 = 5.5/0.5$ mA = 11 k. In the second transistor, choosing a collector current of 0.5 mA, then for maximum symmetrical swing $R_6 = 6/0.5 = 12$ k. The associated base current is 0.5 mA/100 = 5 μA. Therefore $R_5 = (6 - 0.7)/5\ \mu A = 1$ M. Resistor R_4 is selected to allow calibration for full-scale deflection using VR_1. A value $R_4 = 10$ k is used. At the input, $C_1 = 10\ \mu F$ couples signals into the system. At the output of Tr_2, capacitor $C_4 = 10\ \mu F$ couples the signal to the diode bridge which allows unidirectional current flow through the microammeter. The high output impedance (12 k) of the second stage ensures that there is a controlled current flow through the microammeter. During operation with C_2 large (100 μF), VR_1 is adjusted such that a 100 mV input signal results in full-scale deflection of the meter (100 μA).

Ideas for Exploration (i) Introduce a resistor in series with capacitor C_1 that allows the system to be driven by the output of a power amplifier where the output voltage goes up to 50 V corresponding to power amplifiers with outputs just over 100 Watts.

Speaker to Microphone Converter

The circuit in Fig. 5.73 converts a loudspeaker into a very sensitive microphone. Transistor Tr_1 is connected in the common base mode with collector-base feedback biasing while Tr_2 is connected as an emitter follower to provide a low output impedance. A current of 1 mA in Tr_1 means that $R_3 = 4$ k results in 5 V at the collector of the transistor. Resistor $R_1 = 500\ \Omega$ ensures that the signal input is not shorted to ground. This means that the transistor emitter is at 0.5 V. Hence $R_2 = (9 - 4 - 0.7 - 0.5)/5\ \mu A = 760$ k. A transistor current gain of 200 is assumed. Resistor R_2 may need to be adjusted for different

gains. Tr_1 collector voltage of 5 V results in the emitter voltage of Tr_2 as 4.3 V. For 2 mA in Tr_2, $R_4 = 4.3$ V/2 mA = 2.2 k. Capacitors C_1 and C_2 are coupling capacitors, while capacitor C_3 grounds the base of Tr_1 for operation as a common base amplifier.

Ideas for Exploration (i) Compare this circuit with the speaker-to-microphone circuit of Chap. 2; (ii) introduce bootstrapping of a fraction of R_3 in order to increase the gain of the circuit thereby making the system an even more sensitive microphone.

Sound to Light Converter

This circuit in Fig. 5.74 converts sound to light signals. An electret microphone at the input converts sound into an electrical signal that is fed into transistor Tr_1 via capacitor C_1. Tr_1 is connected as a simple common emitter amplifier with collector-base feedback. The collector current is set at 2 mA to ensure that sufficient drive current is available to drive the four connected transistors. For maximum symmetrical swing, the voltage across R_3 is 4.5 V. Hence $R_3 = 4.5$ V/ 2 mA = 2.2 k. Resistor $R_2 = (4.5 - 0.7)/ 0.02$ mA = 190 k where a current gain of 100 is used. The collector of Tr_1 is direct coupled to Tr_2-Tr_5, each of which has an LED in the collector and a 1 k resistor in the emitter. This ensures that the input impedance to these transistor amplifiers is high (better than 100 k) thereby allowing Tr_1 to drive several such transistors. This resistor also serves to define the current in

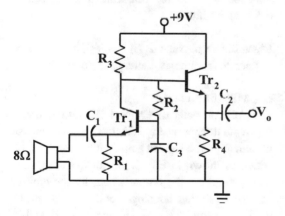

Fig. 5.73 Speaker to microphone converter

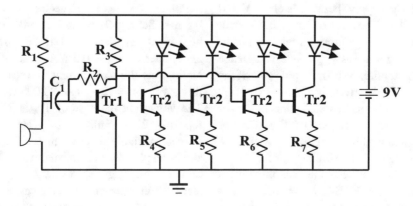

Fig. 5.74 Sound to light converter

Fig. 5.75 Low drift DC voltmeter

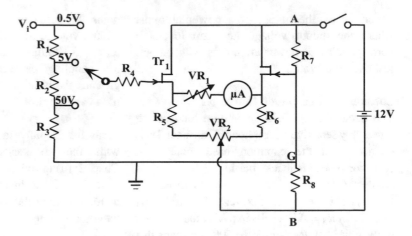

each of these transistors in response to the signal from Tr_1. These leds vary in brightness in response to the input from the microphone.

Ideas for Exploration (i) Replace the microphone by the output from the auxiliary output from a cell phone, and observe the effect of playing music through the system. The lights should respond to the varying music signal amplitude.

Low Drift DC Voltmeter

The circuit in Fig. 5.75 is that of a low drift voltmeter. It is an improved version of the JFET voltmeter introduced in Chap. 3 as it is less subject to drift. It comprises two 2N3919 JFETs connected as a differential amplifier so that any drift occurs in both devices with little effect on the system. Resistors $R_7 = 68$ k and $R_8 = 33$ k provide a reference point G as zero potential such that, with a 12 V supply, point A is at +8 V and point B is at -4 V. The gates of both JFETS are at zero potential and for a drain current of 1 mA, Shockley's equation gives $V_{GS} = -2.8$ V. Hence with a -4 V potential, resistor R_5 plus part of the resistance of VR_2 is given by $(2.8 - (-4))$V/ 1 mA $= 6.8$ k. The same calculation holds for R_6 plus part of the resistance of VR_2. Choose $VR_2 = 10$ k. Then half of this plus $R_5 = 1.8$ k gives the required 6.8 k. Similarly, $R_6 = 1.8$ k. Resistor $R_4 = 100$ k protects the gate of the input device, while $R_1 = 1.8$ M, $R_2 = 180$ k, and $R_3 = 20$ k form an input attenuator. Potentiometer

VR_2 zeros the meter for zero input voltage. For a 0.5 V signal at the gate of the FET, most of this is dropped across the meter circuit. Hence the resistance VR_1 in series with the meter is given by $VR_1 = 0.5$ V/100 μA $= 5$ k. This potentiometer must be adjusted for full-scale deflection with 0.5 V at the input.

Ideas for Exploration (i) Re-design the system to operate from a 9 V battery for portability.

AC Millivoltmeter

The AC voltmeter of Chap. 4 (Fig. 4.62) had a full-scale deflection for 1 V. In order to increase the sensitivity, a second amplifying stage must be added as shown in Fig. 5.76. For quiescent current stability in the output stage, let the emitter voltage be 2 V. Choosing a collector current of 0.5 mA for low-current operation, then $R_6 = 4$ k, and with maximum symmetrical swing, $R_5 = 3.5/ 0.5$ mA $= 7$ k. Let the collector current of Tr_1 be 0.5 mA. Setting $R_4 = 1.5$ k in order to achieve a high input impedance to Tr_1, then the voltage at the emitter or Tr_1 is 0.5 mA $\times$ 1.5 k $= 0.75$ V. Noting that the base current of Tr_1 is 0.005 mA, this gives $R_2 = (2 - 1.45)/0.005$ mA $= 110$ k. The voltage at the base of Tr_2 is 2.7 V. Hence $R_3 = (9 - 2.7)/0.5$ mA $= 12.6$ k. Resistor $R_A = 1$ k is made variable in order to set the f.s.d. input signal amplitude. For signals, it is effectively in parallel with R_4 giving $R = R_4//R_A$. Using equivalent circuit representation, the value of this resistor R for full-scale deflection with an input signal

Fig. 5.76 AC millivoltmeter

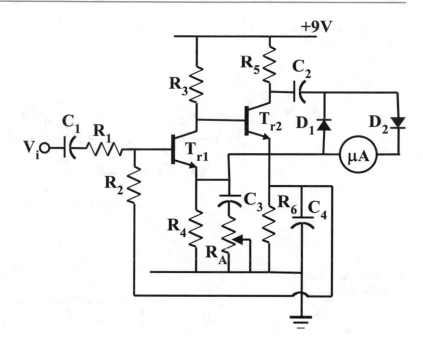

V_i is given by $R \simeq 0.75 \frac{V_i}{I_m}$. Thus for $I_m = 100$ μA and $V_i = 100$ mV, $R \simeq 0.75 \frac{0.1 \text{ V}}{1 \times 10^{-6}} = 750$ Ω. The correct value is likely to be lower than this and is attained by adjusting R_A. Resistor $R_1 = 1$ k provides some level of protection to the input transistor from overvoltage.

Ideas for Exploration (i) Introduce an attenuator to extend the range of the meter to 100 V.

Research Project 1

The circuits considered up to this point generally utilize small signal transistors. This project involves the design of an amplifier using the design principals presented thus far that will accept signal from a source such as the auxiliary output of a cell phone and drive a loudspeaker to give audible output. This is a prelude to the particularly interesting subject of power amplifier design which will be fully introduced in Chap. 9. The circuit is shown in Fig. 5.77. It comprises a common emitter amplifier using a small signal transistor Tr_1, driving a power transistor Tr_2 also connected in the common emitter mode. Tr_2 drives an 8 ohm loudspeaker connected in its collector circuit. Resistor R_3 should result in about 120 mA in the collector of

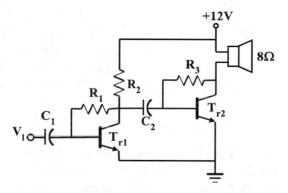

Fig. 5.77 Low-power amplifier

Tr_2. This will produce about 50 mW into an 8 ohm loudspeaker. Some experimentation here with the value of R_3 may be necessary. The power transistor can be a 2N3055. It should be mounted on a small heatsink in order to dissipate any heat produced. Note also that the collector current of the power transistor flows through the speaker resulting in a permanent displacement of the speaker cone. It should be noted that speakers with small diameters do not operate well under this condition.

Ideas for Exploration (i) Replace the power transistor by a power Darlington such as the

Fig. 5.78 Low-power amplifier with volume control

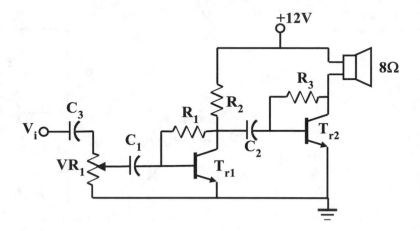

Fig. 5.79 Advanced crystal radio

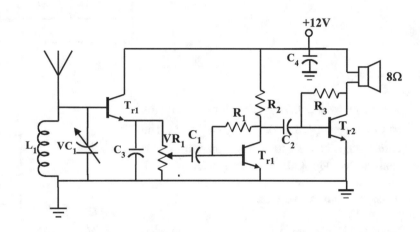

MJ3000. Resistor R_3 would have to be increased for the same value of collector current. The Darlington will provide a higher input impedance which reduces loading on Tr_1 and thereby increase the sensitivity of the system, i.e. a lower signal voltage will drive the system; (ii) introduce a volume control by including a potentiometer $VR_1 = 10k$ at the input. The signal source then drives the potentiometer and the potentiometer wiper is connected to the amplifier input as shown in Fig. 5.78; (iii) using the MOSFET amplifier of Fig. 3.57 in Chap. 3, replace the power BJT and implement the power amplifier.

Research Project 2

This project is a further development of the crystal radio project. The system from Chap. 2 where a germanium transistor was used instead of a germanium diode is employed here as the input signal source to the low-power amplifier of research project 1 above. The combined system is shown in Fig. 5.79. Capacitor C_3 must be about 470 pF and filters the carrier from the modulated radio frequency signal. Capacitor C_4 provides power supply decoupling such that the inductance associated with the power leads does not affect the system performance. Its value can be about 100uF. Resistor R_1 in the previous system is here replaced by potentiometer $VR_1 = 10$ k.

Ideas for Exploration (i) Experiment by reducing the length of the antenna; (ii) introduce a radio frequency signal booster as shown in Fig. 5.80. This simple circuit uses an MPF102 or other suitable JFET in the common source mode. Resistor $R_1 = 1$ M ensures that the gate is grounded for JFET biasing. $C_1 = C_2 = 1000$ pF are coupling capacitors which can be small since the circuit is operating at high frequencies (540 kHz to 1600 kHz).

The inductor $L_1 = 500\ \mu\text{H}$ at 1 MHz say has a reactance $X_L = 2\pi fL = 3.1\ \text{k}$. Therefore with $g_m \simeq 5$ mA/V for the MPF102, the circuit gain for signals at 1 MHz is $g_m X_L = 5\ \text{mA} \times 3.1\ \text{k} = 15.5$. Capacitor $C_3 = 0.1\ \mu\text{F}$ provides decoupling of the power supply to ensure reduced noise. The antenna to this circuit can now be a portable telescopic type. The output of this RF signal booster goes to the tank circuit of the crystal radio and replaces the long-wire antenna previously used; (iii) use the RF signal booster to

replace the long-wire antenna in the earlier forms of the crystal radio in Chaps. 1 and 2, and check if reception is improved.

Problems

1. Draw the equivalent circuit of the two-stage amplifier in Fig. 5.81. Determine the voltage gain V_o/V_s and hence the current gain I_o/I_s. Assume $h_{fe} = 100$ for both transistors.
2. Draw the equivalent circuit of the two-stage amplifier in Fig. 5.82. Determine the voltage gain V_o/V_s and hence the current gain I_o/I_s. Assume $h_{fe} = 200$ for both transistors.
3. Draw the equivalent circuit of the two-stage amplifier in Fig. 5.83. Determine the voltage gain V_o/V_s and hence the current gain I_o/I_s. Assume $h_{fe} = 100$ for both transistors.
4. Draw the equivalent circuit of the two-stage amplifier in Fig. 5.84. Determine the voltage gain V_o/V_s and hence the current gain I_o/I_s. Assume $h_{fe} = 100$ for both transistors.
5. Using the two-stage design in Fig. 5.7 and a 40-volt supply, design a two-stage amplifier that can drive a low-impedance load. Determine the voltage gain of the amplifier. Determine the value of the resistor that can be introduced in the emitter of the first stage as

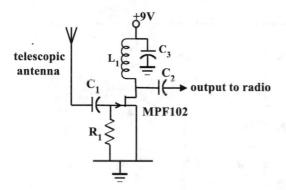

Fig. 5.80 Radio frequency signal booster

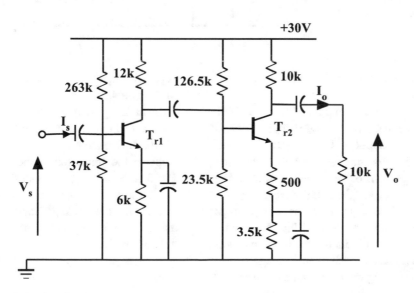

Fig. 5.81 Circuit for Question 1

Fig. 5.82 Circuit for Question 2

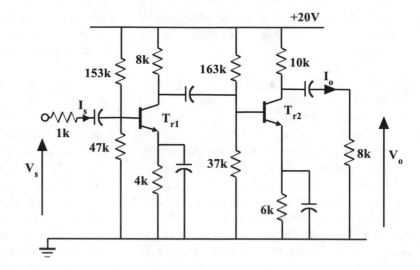

Fig. 5.83 Circuit for Question 3

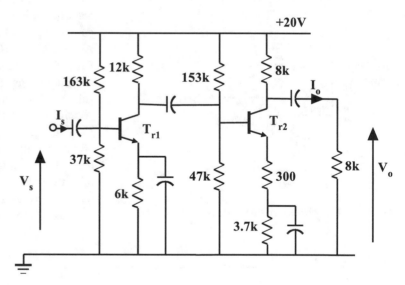

shown in Fig. 5.8 in order to lower the voltage gain to -25.

6. Design a two-stage amplifier having a JFET input stage using the topology of Fig. 5.9 and a 25-volt supply. Use a JFET with $V_P = -5$ volts and $I_{DSS} = 5$ mA. Determine the voltage gain of the circuit, and state the main advantage of the JFET stage.

7. Design a low input impedance two-stage amplifier using the circuit of Fig. 5.10 and a 20-volt supply. Determine the input and output impedances as well as the voltage gain.

8. Using the configuration in Fig. 5.85, design a direct coupled amplifier with high gain. Use a

supply voltage of 26 V, and calculate the gain of your circuit.

9. Explain the principle of bootstrapping, and use the technique to increase the gain of the circuit in Question 8.

10. Replace the transistor in the first stage of Fig. 5.85 by a JFET, and re-design the circuit.

11. Design a common emitter amplifier using a pnp Darlington package having $\beta_D = 2000$ and a 30-volt supply. Calculate the voltage gain of your circuit.

12. Introduce partial bypassing of the emitter resistor in Question 11 in order to achieve a gain of 50.

Fig. 5.84 Circuit for
Question 4

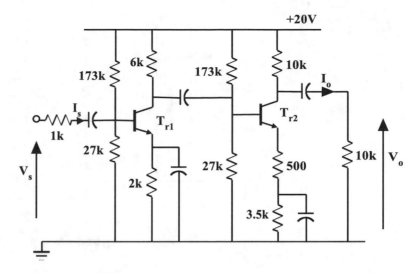

Fig. 5.85 Circuit for
Question 8

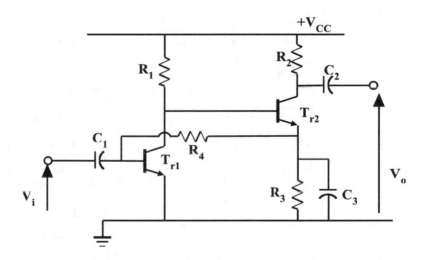

13. Show that the current gain of a Darlington pair is approximately equal to the product of the current gains of the two transistors making up the pair.

14. Design a common collector amplifier using a Darlington package having $\beta_D = 1000$ and a 32-volt supply. Calculate the input impedance of your circuit.

15. Apply bootstrapping in order to increase the input impedance of the circuit and estimate the value of this impedance.

16. Show that current gain of a feedback pair is approximately equal to the product of the current gains of the two transistors making up the pair.

17. Derive the voltage gain of a feedback pair in a common emitter configuration, and show that it is significantly higher than either the Darlington pair or a single BJT.

18. Design a circuit using BJTs in a feedback pair to have a gain of 8. Use a supply voltage of 15 volts.

19. Repeat the design in Question 18 with a JFET as the input transistor.

20. Using the configuration in Fig. 5.86, design a constant current source that produces a current of $I = 5$ mA from a 15-volt supply.

21. Using the basic circuit of Question 20, show how a pnp transistor can be used to supply a constant current to a grounded load.

Fig. 5.86 Circuit for
Question 20

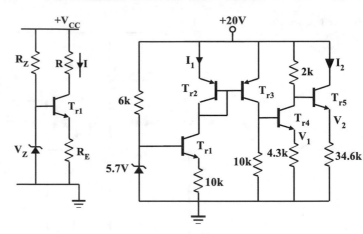

Fig. 5.88   Circuit for Question 27

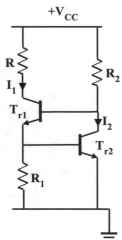

Fig. 5.87   Circuit for Question 22

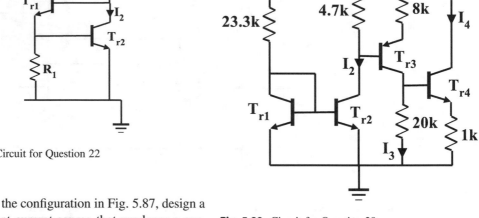

Fig. 5.89   Circuit for Question 28

22. Using the configuration in Fig. 5.87, design a
    constant current source that produces a cur-
    rent of 0.5 mA from a 15-volt supply.

23. Repeat the design in Question 22 using pnp
    transistors.

24. Using a JFET having $V_P = -4$ volts and
    $I_{DSS} = 4$ mA, design a constant current
    source with current value 1 mA. Determine
    the current if $R_S = 0$.

25. Design a two-transistor current mirror giving
    a constant current of 0.5 mA using a 12-volt
    supply.

26. Design a Wilson current mirror to deliver a
    constant current of 1.5 mA using a 16-volt
    supply.

27. In Fig. 5.88 determine I_1, I_2, V_1 and V_2.

28. In Fig. 5.89 determine I_1, I_2, I_3 and I_4.

29. In Fig. 5.90 determine I_1, I_2, V_3 and V_4.

30. In Fig. 5.91 determine I_1, I_2, I_3 and I_4.

31. Using the circuit of Fig. 5.92, design a
    cascode amplifier to operate from a 32-volt
    supply, and determine the voltage gain of the
    circuit.

32. Using a bipolar 15-volt supply, set up a
    two-transistor differential amplifier with
    each transistor having 1 mA quiescent cur-
    rent. Use matched transistors, equal collector

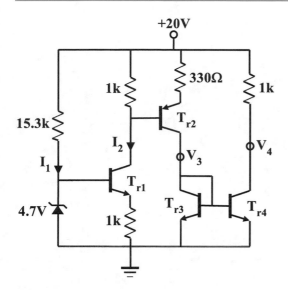

Fig. 5.90 Circuit for Question 29

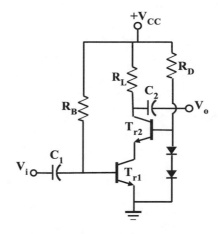

Fig. 5.92 Circuit for Question 31

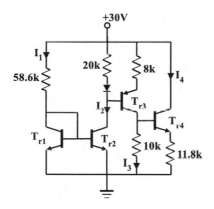

Fig. 5.91 Circuit for Question 30

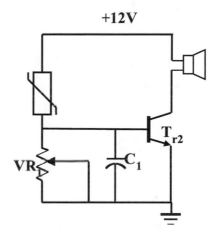

Fig. 5.93 Circuit for Question 35

resistors and design for maximum symmetrical swing.

33. A bipolar junction transistor is being used to switch 12 volts across a 10 k load. Using an npn transistor with $\beta = 150$, design a simple switching circuit that is fully on when driven by a 5-volt input.

34. Using an enhancement MOSFET design a switch to drive a 12-volt 5 mA relay from a 12-volt supply. The device has a threshold voltage of 3 volts.

35. Investigate the properties and characteristics of a negative temperature coefficient thermistor. Using such a thermistor and the circuit shown in Fig. 5.93, design a heat sensing system that sounds an alert when the ambient temperature reaches a particular value.

Bibliography

R.L. Boylestad, L. Nashelsky, *Electronic Devices and Circuit Theory*, 11th edn. (Pearson Education, New Jersey, 2013)

Vishay Siliconix, FETS as Voltage-Controlled Resistors (1997, March)

Frequency Response of Transistor Amplifiers

The discussions in the previous chapters concerned the mid-frequency performance of an amplifier. At these frequencies, the coupling and bypass capacitors pass the signals virtually unimpeded, while the transistor junction capacitors are considered to be open circuits. The BJT and FET models provided useful tools with which to analyse these circuits. The performance at low and high frequencies however requires further consideration. In the case of the low frequencies, the effect of the coupling and bypass capacitors needs to be determined while at high frequencies the response which is largely determined by transistor junction capacitances needs to be ascertained. In this chapter therefore, more complex equivalent circuits are introduced in order to examine the full frequency response characteristics of BJTs and FETs. While the analysis is done using the JFET, it applies in general to the MOSFET also. After completing the chapter, the reader will be able to

- Determine the low-frequency response of transistor amplifiers
- Determine the high-frequency response of transistor amplifiers

6.1 BJT Low-Frequency Response

In the previous chapters on the BJT and the FET, all coupling and bypass capacitors were chosen to be large such that these capacitors represented short-circuits to the signal currents passing through them. If however the signal frequency is sufficiently low, the reactance of these capacitors becomes significant and can affect signal levels in the circuit. In this section, the circuits are analysed in order to determine the effect of the capacitors so that appropriate values may be chosen for optimum circuit response. In preparation for doing so, we consider the RC circuit in Fig. 6.1 with an input sinusoidal signal V_i and an output signal V_o. The reactance of the capacitor in the complex frequency domain s is $1/sC$ where $s = j\omega$ and ω is the angular frequency that is related to frequency f by $\omega = 2\pi f$. The transfer function V_o/V_i for the circuit is given by

$$A_V = \frac{V_o}{V_i} = \frac{R}{R + 1/sC} = \frac{sCR}{1 + sCR} \quad (6.1)$$

The magnitude of this transfer function is given by

$$|A_V| = \left| \frac{sCR}{1 + sCR} \right| = \frac{|sCR|}{|1 + sCR|}$$

$$= \frac{\omega CR}{\sqrt{1 + \omega^2 C^2 R^2}} \quad (6.2)$$

From (6.2), when $\omega CR >> 1$, then $|A_V| = 1$ and when $\omega CR << 1$, then $|A_V| = \omega CR$. Thus for $\omega CR << 1$, converting $|A_V|$ to decibels using $|A_V|$ dB $= 20 \log |A_V|$, we have

© Springer Nature Switzerland AG 2021
S. J. G. Gift, B. Maundy, *Electronic Circuit Design and Application*,
https://doi.org/10.1007/978-3-030-46989-4_6

Fig. 6.1 CR circuit

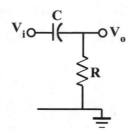

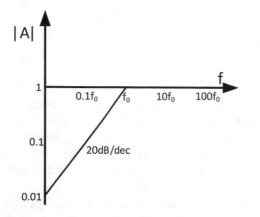

Fig. 6.2 Frequency response plot of CR circuit

$$|A_V|\mathrm{dB} = 20 \log \omega CR$$
$$= 20 \log \omega + 20 \log CR \qquad (6.3)$$

This can be written as

$$|A_V|\mathrm{dB} = 20 \log f + 20 \log 2\pi CR$$
$$= 20 \log f - 20 \log f_o \qquad (6.4)$$

where $f_o = 1/2\pi RC$ corresponding to $\omega_o = 1/RC$. Equation (6.4) is that of a straight line on a $|A_V|$ dB $-\log f$ graph with slope +20 dB/ decade and $|A_V|$dB $= 0$ at $f = f_o$ as shown in Fig. 6.2. On the same graph, we have $|A_V| = 1$ for $\omega CR > 1$ which is equivalent to $f > f_o$. At $f = f_o$,

$$|A_V| = \frac{\omega CR}{\sqrt{1 + \omega^2 C^2 R^2}} = \frac{1}{\sqrt{1 + j}}$$
$$= 1/\sqrt{2} = 0.707 = -3\,\mathrm{dB} \qquad (6.5)$$

This response is that of a high-pass filter and is essentially the response of input and output coupling capacitors at low frequencies. We now consider these responses.

6.1.1 Input Coupling Capacitor

Consider the fixed-bias common emitter amplifier circuit shown in Fig. 6.3 where a coupling capacitor C_i passes the input signal V_i into the base of the transistor. The h-parameter equivalent circuit is shown in Fig. 6.4 with the coupling capacitor C_i included. The reactance of the capacitor in the complex frequency domain s is $1/sC_i$ where $s = j\omega$ and ω is the angular frequency that is related to frequency f by $\omega = 2\pi f$. From the equivalent circuit,

$$V_i = I_b(h_{ie} + 1/sC_i) \qquad (6.6)$$

$$V_o = -I_b h_{fe} R_L \qquad (6.7)$$

Therefore, the voltage gain A_V is given by

$$A_V = \frac{V_o}{V_i} = -\frac{h_{fe} R_L I_b}{(h_{ie} + 1/sC) I_b}$$
$$= -\frac{h_{fe} R_L s C_i}{1 + s C_i h_{ie}} \qquad (6.8)$$

In order to plot the gain magnitude $|A_V|$ against frequency f, consider the following:

$$|A_V| = \left| \frac{-h_{fe} R_L s C_i}{1 + s C_i h_{ie}} \right| = \frac{|-h_{fe} R_L s C_i|}{|1 + s C_i h_{ie}|}$$
$$= \frac{h_{fe} R_L \omega C_i}{\sqrt{1 + \omega^2 C_i^2 h_{ie}^2}} \qquad (6.9)$$

When $\omega C_i h_{ie} >> 1$, then

$$|A_V| \to \frac{h_{fe} R_L \omega C_i}{\omega C_i h_{ie}} = \frac{h_{fe} R_L}{h_{ie}} \qquad (6.10)$$

which is the common emitter mid-frequency gain when C_i is treated as a short-circuit derived in Chap. 4. The magnitude of this gain converted to decibels is given by

$$|A_V|_{\mathrm{dB}} = 20 \log \left(\frac{h_{fe} R_L}{h_{ie}} \right) \qquad (6.11)$$

Now let $\omega_L = 1/C_i h_{ie}$ which means $f_L = 1/2\pi C_i h_{ie}$. Then $\omega C_i h_{ie} >> 1$ corresponds to $\omega >> \omega_L$ or $f >> f_L$. For $\omega C_i h_{ie} << 1$ which corresponds to $\omega << \omega_L$ or $f << f_L$, then

Fig. 6.3 Common emitter amplifier

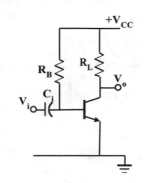

Fig. 6.4 H-parameter equivalent circuit of common emitter amplifier

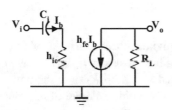

$$A_V(s) = -\frac{h_{fe}R_L s C_i}{1 + s C_i h_{ie}}$$

$$= -\frac{h_{fe}R_L}{h_{ie}}\frac{s C_i h_{ie}}{1 + s C_i h_{ie}} \quad (6.15)$$

At $f = f_L$ corresponding to $\omega = \omega_L$, (6.15) becomes

$$A_V(f_L) = -\frac{h_{fe}R_L}{h_{ie}}\frac{j\omega C_i h_{ie}}{1 + j\omega C_i h_{ie}}$$

$$= -\frac{h_{fe}R_L}{h_{ie}}\frac{j}{1 + j} \quad (6.16)$$

$$|A_V(f_L)|_{dB} = 20\log\left(\frac{h_{fe}R_L}{h_{ie}}\frac{1}{\sqrt{2}}\right)$$

$$= 20\log\left(\frac{h_{fe}R_L}{h_{ie}}\right)$$

$$- 20\log\sqrt{2} \quad (6.17)$$

This gives

$$|A_V(f_L)|_{dB} = \left[20\log\left(\frac{h_{fe}R_L}{h_{ie}}\right) - 3\right]_{dB} \quad (6.18)$$

$$|A_V| = \frac{h_{fe}R_L}{h_{ie}}\omega C_i h_{ie} = \frac{h_{fe}R_L}{h_{ie}}\frac{\omega}{\omega_L} \quad (6.12)$$

which in decibels becomes

$$|A_V|_{dB} = 20\log\left(\frac{h_{fe}R_L}{h_{ie}}\frac{\omega}{\omega_L}\right)$$

$$= 20\log\left(\frac{h_{fe}R_L}{h_{ie}}\frac{f}{f_L}\right) \quad (6.13)$$

This can be written as

$$|A_V|_{dB} = 20\log f + 20\log\frac{h_{fe}R_L}{h_{ie}}$$

$$- 20\log f_L \quad (6.14)$$

On a $|A_V|_{dB}$ vs $\log f$ plot, Eq. (6.14) represents a straight line of slope 20 which at $f = f_L$ gives $|A_V|_{dB} = 20\log\frac{h_{fe}R_L}{h_{ie}}$ where it intersects the line (6.11). The slope has units dB/decade and is 20 dB/decade since for every decade or factor of 10 increase in the frequency, the gain increases by 20 dB which is itself a factor of 10. Finally, from (6.8),

Therefore, at the frequency $f = f_L$, the magnitude of the gain falls by 3 dB from its mid-band value. This frequency is referred to as the low-frequency cut-off, and the reduction in gain as the frequency falls below this frequency arises because of the effect of the coupling capacitor. The frequency response plot for this circuit is shown in Fig. 6.5.

Thus, the value of C_i sets the lower cut-off frequency of the common emitter amplifier of Fig. 6.3 and is given by

$$f_L = \frac{1}{2\pi h_{ie} C_i} \quad (6.19)$$

In general, the lower cut-off frequency f_L should be lower than some value $f_L{}^*$ hence

$$C_i \geq \frac{1}{2\pi h_{ie} f_L{}^*} \quad (6.20)$$

Fig. 6.5 Frequency response plot for amplifier circuit

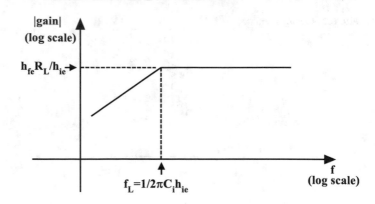

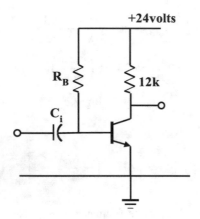

Fig. 6.6 Amplifier circuit biased for maximum symmetrical swing

Example 6.1

For the amplifier circuit shown in Fig. 6.6 which is biased for maximum symmetrical swing, determine a suitable value of coupling capacitor to realize a maximum lower cut-off frequency of 50 Hz.

Solution Since the circuit is biased for maximum symmetrical swing, it follows that $V_{CEq} = 24/2 = 12$ volts, and hence $I_{Cq} = 12$ volts$/12$ k $= 1$ mA. This gives $h_{ie} = h_{fe}/40 I_C = 100/40 \times 1$ mA $= 2.5$ k. Noting that $h_{ie} < < R_B$, it follows from (6.20) that $C_i \geq \frac{1}{2\pi h_{ie} f_L^*} = \frac{1}{2 \times \pi \times 2.5 \times 10^3 \times 50} = 1.27 \,\mu\text{F}$. We therefore choose $C_i = 2 \,\mu\text{F}$.

We wish to note two additional points. The first is that if the common emitter amplifier uses

voltage divider biasing instead of fixed biasing involving base resistors R_1 and R_2, then h_{ie} must be changed to $h_{ie}//R_1//R_2$ since these two resistors are generally within an order of magnitude of h_{ie}. More generally, if Z_i is the input impedance of the amplifier, then it is easily sown that the formula for the lower cut-off frequency becomes

$$f_L = \frac{1}{2\pi Z_i C_i} \tag{6.21}$$

The second point is that if the signal source is driving the amplifier through a resistor R_S, then the voltage gain becomes

$$A_V(s) = -\frac{h_{fe} R_L s C_i}{1 + s C_i (R_S + h_{ie})}$$

$$= -\frac{h_{fe} R_L}{R_S + h_{ie}} \frac{s C_i (R_S + h_{ie})}{1 + s C_i (R_S + h_{ie})} \tag{6.22}$$

from which f_L is given by

$$f_L = \frac{1}{2\pi (R_S + h_{ie}) C_i} \tag{6.23}$$

and the mid-band gain becomes $A_V = -\frac{h_{fe} R_L}{R_S + h_{ie}}$.

6.1.2   Output Coupling Capacitor

Consider now the common emitter amplifier in Fig. 6.7 with output coupling capacitor C_o connected to an external load R_X. In order to determine the effect of capacitor C_o on the low-frequency response of the amplifier, C_i is assumed to be very large.

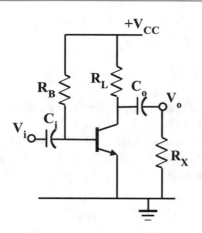

Fig. 6.7 Common emitter amplifier with output coupling capacitor

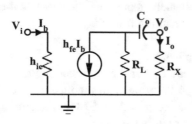

Fig. 6.8 H-Parameter equivalent circuit including output capacitor

From the h-parameter equivalent circuit shown in Fig. 6.8,

$$V_i(s) = I_b h_{ie} \tag{6.24}$$

$$V_o(s) = I_o R_X \tag{6.25}$$

where

$$I_o = -I_b h_{fe} \frac{R_L}{R_L + 1/sC_o + R_X}$$

$$= -\frac{h_{fe} I_b s C_o R_L}{1 + sC_o(R_L + R_X)} \tag{6.26}$$

Hence the gain is given by

$$A_V = \frac{I_o R_X}{I_b h_{ie}}$$

$$= -\frac{h_{fe} I_b s C_o R_L}{1 + sC_o(R_L + R_X)} \frac{R_X}{I_b h_{ie}}$$

$$= -\frac{\frac{h_{fe} R_X}{h_{ie}} . s C_o R_L}{1 + sC_o(R_L + R_X)} \tag{6.27}$$

After manipulation (6.27) becomes

$$A_V = -\frac{\frac{h_{fe} \overline{R_L}}{h_{ie}} . s C_o(R_L + R_X)}{1 + sC_o(R_L + R_X)} \tag{6.28}$$

where $\overline{R_L} = R_L // R_X$. Thus the mid-frequency gain is $-\frac{h_{fe} R_L // R_X}{h_{ie}}$ and the lower cut-off frequency f_L is given by

$$f_L = \frac{1}{2\pi(R_L + R_X)C_o} \tag{6.29}$$

If the output of the amplifier drives the input of a second amplifier with input impedance Z_i instead of an external load, then (6.29) becomes

$$f_L = \frac{1}{2\pi(R_L + Z_i)C_o} \tag{6.30}$$

6.1.3   Emitter Bypass Capacitor

The most effective biasing scheme used with the common emitter amplifier was voltage divider biasing shown in Fig. 6.9. This circuit includes an input coupling capacitor C_i, an output coupling capacitor C_o and a bypass capacitor C_E. The low-frequency effects of C_i and C_o have already been determined. In order to determine the effect of C_E, C_i and C_o are assumed to be very large and therefore can be replaced by short-circuits. Capacitor C_E ensures that the emitter of the transistor is grounded for signal voltages. However, as the signal frequency falls, the reactance of C_E increases, and this capacitor no longer represents a signal short-circuit.

Consider the equivalent circuit shown in Fig. 6.10. Here

$$V_i(s) = I_b h_{ie} + (1 + h_{fe}) I_b R_E // 1/sC_E$$

$$= h_{ie} I_b + (1 + h_{fe}) I_b \frac{R_E}{1 + sC_E R_E} \tag{6.31}$$

$$V_o(s) = -I_b h_{fe} R_L \tag{6.32}$$

Therefore, the voltage gain is given by

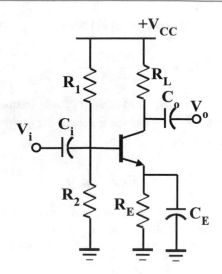

Fig. 6.9 H-biased common emitter amplifier

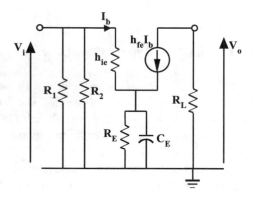

Fig. 6.10 Equivalent circuit including bypass capacitor

$$A_V(s) = -\frac{h_{fe}R_L}{\left[h_{ie} + (1 + h_{fe})\frac{R}{1 + sC_ER_E}\right]} \quad (6.33)$$

After manipulation this becomes

$$A_V(s) = -\frac{\frac{h_{fe}R_L}{h_{ie}+(1+h_{fe})R_E}(1 + sC_ER_E)}{1 + \frac{sC_ER_Eh_{ie}}{h_{ie}+(1+h_{fe})R_E}} \quad (6.34)$$

Noting that $1 + h_{fe} \approx h_{fe}$ and $h_{ie}/h_{fe} = r_e$, then

$$A_V(s) = -\frac{\frac{R}{r_e+R_E}(1 + sC_ER_E)}{1 + \frac{sC_ER_Er_e}{r_e+R_E}}$$

$$= -\frac{R_L}{r_e + R_E}\frac{1 + sC_ER_E}{1 + sC_ER_E//r_e} \quad (6.35)$$

In general, $R_E >> r_e$ and therefore $R_E//r_e \approx r_e$. Hence (6.35) becomes

$$A_V(s) = -\frac{R_L}{r_e + R_E}\frac{1 + sC_ER_E}{1 + sC_Er_e} \quad (6.36)$$

A frequency response plot of (6.36) is shown in Fig. 6.11.

Note that the magnitude of the mid-frequency gain is R_L/r_e which starts to fall as frequency decreases at

$$f_1 = \frac{1}{2\pi C_E r_e} \quad (6.37)$$

However, at $f_2 = \frac{1}{2\pi C_E R_E}$ the gain tends to $\frac{R_L}{r_e+R_E}$. Since $R_E >> r_e$ then $f_1 > f_2$ and frequency $f_L = f_1$ approximately represents the lower cut-off frequency resulting from C_E.

Example 6.2
Determine C_E for the common emitter amplifier in Fig. 6.12 in order to produce a lower cut-off frequency of less than 50 Hz. Assume all coupling capacitors are large.

Solution $f_L = \frac{1}{2\pi C_E r_e} \leq 50\,\text{Hz}$. Therefore $C_E \geq \frac{1}{2\pi r_e 50}$. For this circuit, the voltage V_B at the base of the transistor is $V_B = \frac{26\,\text{k}}{174\,\text{k}+26\,\text{k}} \times 20 = 2.6\,\text{volts}$. Hence the emitter voltage $V_E = 2$ volts giving $I_C = 2/2\,\text{k} = 1$ mA and $r_e = 1/40I_C = 1000/40 = 25\,\Omega$. This gives $C_E \geq \frac{1}{2\pi r_e 50} = \frac{1}{2\times\pi\times25\times50} = 127\,\mu\text{F}$. Use $C_E = 150\,\mu\text{F}$.

Exercise 6.1
For the case where the common emitter amplifier is driven through a source resistor R_S, show that the frequency f_L becomes $f_L = \frac{1}{2\pi C_E\left(r_e+\frac{R}{h_{fe}}\right)}$.

In practice, in the voltage divider-biased common emitter amplifier, the coupling and bypass capacitors all influence the low frequency response of the amplifier. In the determination of these capacitor values, one approach is to use C_E to set f_L, C_E being the largest of the capacitors and select C_i and C_o sufficiently large such that the associated cut-off frequencies are well below that created by C_E.

Fig. 6.11 Frequency response plot showing the effect of C_E

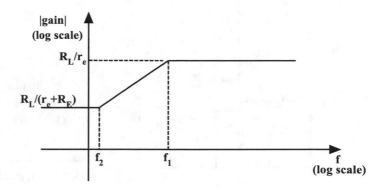

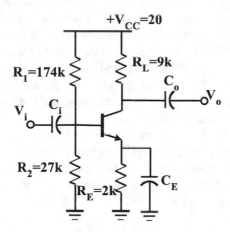

Fig. 6.12 Common emitter amplifier

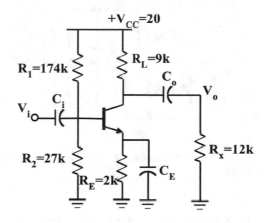

Fig. 6.13 Common emitter amplifier with load

Example 6.3

Determine the capacitors C_i, C_E and C_o in the circuit of Fig. 6.13. To give $f_L = 50$ Hz. Assume $h_{fe} = 100$.

Solution For this circuit, $I_C = 1$ mA and hence $r_e = 25\ \Omega$ and $h_{ie} = 2.5$ k. Using (6.21), $C_i = \frac{1}{2\pi Z_i 50}$ where $Z_i = R_1//R_2//h_{ie} = 26$ k//174 k// 2.5 k $= 2.25$ k. Hence $C_i = \frac{1}{2\times\pi\times2.25\,k\times50} = 1.4\,\mu F$. Using (6.29), $C_o = \frac{1}{2\pi(9\,k+12\,k)50} = 0.15\,\mu F$. Finally using (6.37), $C_E = \frac{1}{2\times\pi\times25\times50} = 127\,\mu F$. Since C_E is the largest capacitor, we use this to determine the lower cut-off frequency and choose it to be $C_E = 150\ \mu F$. We then select the other two capacitors to be about ten times the calculated values so that the resulting cut-off frequencies are well below $f_L = 50$ Hz. Hence $C_i = 15\ \mu F$ and $C_o = 2\ \mu F$.

6.2   FET Low-Frequency Response

The determination of the coupling and bypass capacitors in a common source FET amplifier is similar to the common emitter BJT amplifier. Consider the common source JFET amplifier shown in Fig. 6.14. Similar to the BJT, the lower cur-off frequency set by C_i is given by

$$f_L = \frac{1}{2\pi R_G C_i} \qquad (6.38)$$

where R_G replaces h_{ie} and again C_i and R_G together form a high-pass filter. Similarly, the lower cut-off frequency created by the output coupling capacitor C_o is given by

$$f_L = \frac{1}{2\pi(R_L + R_X)C_o} \qquad (6.39)$$

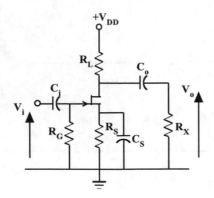

Fig. 6.14 Common source JFET amplifier

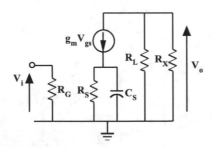

Fig. 6.15 Equivalent circuit of common source JFET amplifier

In order to determine the bypass capacitor C_S, the equivalent circuit shown in Fig. 6.15 needs to be analysed. Thus

$$V_i = V_{GS} + g_m V_{GS} R_S // 1/sC_S$$

$$= V_{GS} + g_m V_{GS} \frac{R_S}{1 + sC_S R_S} \qquad (6.40)$$

$$V_o = -g_m V_{GS} R_L // R_X \qquad (6.41)$$

Therefore

$$A_V(s) = -\frac{g_m R_L // R_X}{1 + g_m \frac{R}{1 + sC_S R_S}} \qquad (6.42)$$

After manipulation this becomes

$$A_V(s) = -\frac{g_m R_L // R_X}{1 + g_m R_S} \cdot \frac{1 + sC_S R_S}{1 + sC_S \frac{R}{1 + g_m R_S}} \qquad (6.43)$$

Thus at low frequencies $A_V \rightarrow -\frac{g_m R_L // R_X}{1 + g_m R_S}$ and at high frequencies $A_V \rightarrow -g_m R_L // R_X$.

From (6.43), there is a pole at

$$f_L = \frac{1 + g_m R_S}{2\pi R_S C_S} \qquad (6.44)$$

and a zero at $f_2 = \frac{1}{2\pi R_S C_S}$. Since $f_L > f_2$, the lower cut-off frequency is determined by f_L. The Bode plot is shown in Fig. 6.16.

6.3 Hybrid-Pi Equivalent Circuit

The most widely used model for transistor high-frequency behaviour is the hybrid-pi equivalent circuit shown in Fig. 6.17. This model was first published by L.J. Giaceletto in the RCA Review of 1954. The resistor $r_{bb'}$ is the ohmic base resistance, $r_{cb'}$ is the collector-base resistor which is quite large (several megohms) while capacitors $C_{b'e}$ and $C_{b'c}$ are the emitter-base and collector-base junction capacitors.

At low frequencies, since $r_{bb'}$ is small and $r_{cb'}$ is large, the hybrid-pi equivalent circuit becomes that shown in Fig. 6.18. This is very similar to the h-parameter low-frequency equivalent circuit of Chap. 4 where

$$h_{ie} = r_{bb'} + r_{b'e} \qquad (6.45)$$

$$1/h_{oe} = r_{ce} \qquad (6.46)$$

and

$$g_m v_{b'e} = h_{fe} i_b \qquad (6.47)$$

The effect of eliminating $r_{cb'}$ in the hybrid-pi equivalent circuit is essentially the same as neglecting h_{re} in the h-parameter equivalent circuit.

In the hybrid-pi equivalent circuit, the resistance $r_{cb'}$ is usually neglected giving the equivalent circuit in Fig. 6.19 where resistor R_L is the usual collector resistor. The capacitance $C_{cb'}$ is the capacitance between the collector-base junction of the transistor. The reverse bias on this junction creates a depletion region at the junction with charges at the boundaries of the region. The effect is that of a capacitor whose value varies with the magnitude of the reverse bias V_{cb} at the

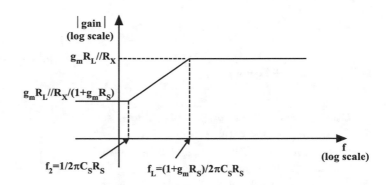

Fig. 6.16 Frequency response plot for common source JFET amplifier

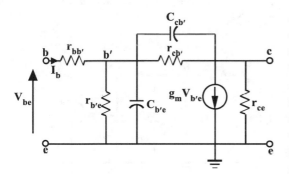

Fig. 6.17 BJT hybrid-pi equivalent circuit

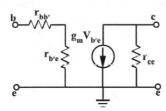

Fig. 6.18 BJT low-frequency equivalent circuit

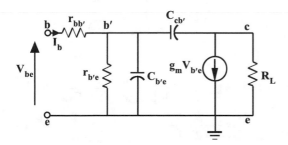

Fig. 6.19 Simplified BJT hybrid-pi equivalent circuit

junction. $C_{cb'}$ varies between 0.8 pF to about 50 pF. The capacitance $C_{b'e}$ is the capacitance between the base-emitter junction. This capacitance arises as a result of a mechanism referred to

as DIFFUSION CAPACITANCE. This phenomenon occurs because of time delay during the diffusion of majority carriers form the emitter of the transistor to the base region. This delay mechanism limits the rate at which the carriers can move between the emitter and base regions and constitutes a capacitance. Its value depends on the emitter DC current I_E, increasing with this current as charge increases. $C_{b'e}$ ranges between approximately 20 pF to 0.01 µF. Its value is typically 100 times greater than $C_{cb'}$. The simplified hybrid-pi circuit of Fig. 6.19 is used to examine the high-frequency behaviour of the transistor.

Single Pole Transfer Function

In discussing the high-frequency response of single stage transistor amplifiers, the voltage gain expressions $A_V = V_o/V_S$ will often reduce to a transfer function of the form

$$A_V(j\omega) = \frac{k}{1 + jf/f_o} \qquad (6.48)$$

where k is the low-frequency gain of the amplifier stage and f_o is the frequency of the pole. In order to represent this transfer function on a frequency response plot, we express the magnitude in decibels given by

$$|A|_{dB} = 20 \log |A|$$
$$= 20 \log k - 20 \log |1 + jf/f_o| \qquad (6.49)$$

This reduces to

$$|A|_{dB} = 20 \log k, f \ll f_o \qquad (6.50)$$

$$|A|_{dB} = 20 \log k - 20 \log \frac{f}{f_o}$$

$$= -20 \log f + 20 \log k$$
$$+ 20 \log f_o, \ f$$
$$>> f_o \qquad (6.51)$$

Using these equations, $|A|_{dB}$ is plotted against $\log f$. For frequencies f well below the pole frequency such that $f << f_o$, the response approaches the horizontal line $|A|_{dB} = 20 \log k$. For frequencies well above the pole frequency such that $f >> f_o$, the response approaches the line whose slope is -20 dB/decade. These lines are shown in Fig. 6.20. Note that at $f = f_o$, both lines intersect since they both have the same magnitude value $|A|_{dB} = 20 \log k$. For the transfer function, at $f = f_o$,

$$|A| = \frac{k}{\sqrt{1 + (f/f_o)^2}} = \frac{k}{\sqrt{1+1}}$$
$$= k/\sqrt{2} = 0.707 \, k \qquad (6.52)$$

Hence

$$|A|_{dB} = 20 \log |A|$$

$$= 20 \log k - 20 \log \sqrt{2}$$
$$= 20 \log k - 3 \qquad (6.53)$$

Thus at the pole frequency, the magnitude of the transfer function drops to 0.707 of its low-frequency value k, i.e. by 3 dB as shown in Fig. 6.20.

Consider now the short-circuit current gain β, which is defined as

$$\beta(j\omega) = I_c/I_b \qquad (6.54)$$

This parameter is actually frequency-dependent and its value can be found using the hybrid-pi equivalent circuit. Shorting the collector and emitter yields the equivalent circuit of Fig. 6.21 where $C_{cb'}$ is now placed in parallel with $C_{b'e}$:

$$I_c = g_m V_{b'e} \qquad (6.55)$$

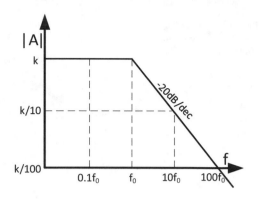

Fig. 6.20 Frequency response plot of single stage amplifier

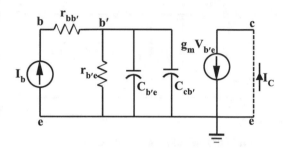

Fig. 6.21 Equivalent circuit with short-circuited collector and emitter

$$I_b = \frac{V_{b'e}}{Z_{b'e}} \qquad (6.56)$$

where

$$Z_{b'e} = r_{b'e} // \frac{1}{s(C_{b'e} + C_{b'c})}$$

$$= \frac{r_{b'e}}{1 + s r_{b'e}(C_{b'e} + C_{b'c})} \qquad (6.57)$$

Hence

$$\beta(j\omega) = g_m Z_{b'e} = \frac{g_m r_{b'e}}{1 + s r_{b'e}(C_{b'e} + C_{b'c})}$$

$$= \frac{\beta_o}{1 + jf/f_\beta} \qquad (6.58)$$

where

$$\beta_o = g_m r_{b'e} \approx h_{fe} \qquad (6.59)$$

and

$$f_\beta = \frac{1}{2\pi r_{b'e}(C_{b'e} + C_{b'c})} \quad (6.60)$$

Equation (6.58) is of the single pole form (6.48). The quantity β_o is referred to as the low-frequency beta or h_{fe} that we have used in our BJT h-parameter analysis. Also, f_β is called the beta cut-off frequency. It is that frequency at which the magnitude of β falls to 3 dB down from the low-frequency value β_o. From (6.58), the frequency f_T at which β goes to unity is given by

$$1 = \frac{\beta_o f_\beta}{jf_T}, \ f \gg f_\beta \quad (6.61)$$

Hence

$$|1| = \left|\frac{\beta_o f_\beta}{jf_T}\right| = \beta_o \frac{f_\beta}{f_T} \quad (6.62)$$

giving

$$f_T = \beta_o f_\beta \quad (6.63)$$

or

$$f_T = \frac{\beta_o}{2\pi r_{b'e}(C_{b'e} + C_{b'c})} \quad (6.64)$$

The frequency f_T is called the transition frequency or gain-bandwidth product of the transistor. Figure 6.22 illustrates f_β and f_T.

Since $C_{b'e} \gg C_{b'c}$, eq. (6.64) becomes

$$f_T = \frac{\beta_o}{2\pi r_{b'e} C_{b'e}} \quad (6.65)$$

But

$$r_{b'e} \approx h_{ie} = \frac{h_{fe}}{40I_c} = \beta_o r_e \quad (6.66)$$

where

$$r_e = \frac{1}{40I_c} \quad (6.67)$$

giving

$$f_T = \frac{1}{2\pi r_e C_{b'e}} \quad (6.68)$$

f_T varies with collector current I_C as shown in Fig. 6.23.

Example 6.4

A silicon npn transistor has $f_T = 400$ MHz. Calculate $C_{b'e}$ for $I_C = 1$ mA.

Solution $r_e - \frac{1}{40I_c} - \frac{1}{40} - 25\,\Omega$. Hence $f_T - \frac{1}{2\pi 25 C_{b'e}}$ giving $C_{b'e} = \frac{1}{2\times\pi\times25\times400\times10^6} = 15.9\,\text{pF}$.

6.4   Miller Effect

In order to analyse the common emitter amplifier using the hybrid-pi equivalent circuit, a new theorem is introduced that helps to further simplify the equivalent circuit. Consider the basic amplifier shown in Fig. 6.24 in which feedback admittance Y_f is connected from output to input.

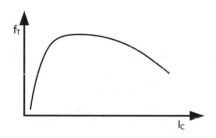

Fig. 6.23 Variations of f_T with collector current I_C

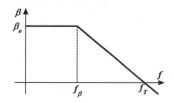

Fig. 6.22 Frequency response plot of the magnitude of β

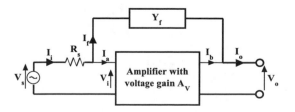

Fig. 6.24 Amplifier with feedback admittance Y_f

The amplifier is driven by a voltage source V_S in series with a source R_S and has a voltage gain $A_V = V_o/V_i$. The input admittance of the basic amplifier given by

$$Y_a = I_a/V_i \qquad (6.69)$$

while the output admittance is

$$Y_b = -I_b/V_o \qquad (6.70)$$

With Y_f connected, the resulting input admittance Y_{in} becomes

$$Y_{in} = I_i/V_i \qquad (6.71)$$

and the resulting output admittance Y_{out} is given by

$$Y_{out} = -I_o/V_o \qquad (6.72)$$

From (6.71),

$$Y_{in} = \frac{I_i}{V_i} = \frac{I_f + I_a}{V_i} \qquad (6.73)$$

But

$$I_f = (V_i - V_o)Y_f \qquad (6.74)$$

Therefore,

$$Y_{in} = \frac{(V_i - V_o)Y_f + I_a}{V_i}$$

$$= Y_a + (1 - A_V)Y_f \qquad (6.75)$$

Based on (6.75), the input admittance Y_{in} seen at the amplifier input is the input admittance Y_a of the basic amplifier in parallel with an admittance whose value is that of the feedback admittance multiplied by the factor $(1 - A_V)$. If the amplifier is inverting, then $A_V \rightarrow -A_V$ and the multiplying factor $(1 - A_V)$ becomes $(1 + A_V)$.

At the output of the amplifier, we have

$$Y_{out} = -\frac{I_o}{V_o} = -\frac{(I_f + I_b)}{V_o} \qquad (6.76)$$

Substituting for I_f using (6.74) gives

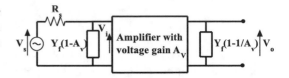

Fig. 6.25 Equivalent amplifier circuit

$$Y_{out} = -\left[\frac{(V_i - V_o)Y_f + I_b}{V_o}\right]$$

$$= Y_b + Y_f\left(1 - \frac{1}{A_V}\right)$$

$$\approx Y_b + Y_f, |A_V| \gg 1 \qquad (6.77)$$

Thus, the output admittance Y_{out} of the amplifier with feedback Y_f is the sum of the basic admittance Y_b and the admittance Y_f of the feedback admittance providing the magnitude of the gain of the amplifier is significantly greater than unity. Results (6.75) and (6.77) constitute Miller's theorem. Based on this theorem, the amplifier circuit of Fig. 6.24 can be re-drawn in the equivalent form of Fig. 6.25 in which the feedback admittance Y_f is replaced by an admittance $Y_f(1 - A_V)$ at the input and an admittance $Y_f\left(1 - \frac{1}{A_V}\right)$ at the output. This result enables a more simplified analysis of the original amplifier.

6.5   Common Emitter Amplifier

The results that have been derived will now be used to analyse the circuit of the common emitter amplifier shown in Fig. 6.26.

The equivalent circuit is shown in Fig. 6.27 in which the collector-emitter resistor $r_{ce} = 1/h_{oe}$ is assumed to be very large compared with R_L and therefore has been omitted. Since $C_{b'e}$ is connected between b and c in order to apply Miller's Theorem, we determine the voltage gain A_V between b and c given by

$$A_V = \frac{V_o}{V_{b'e}} \qquad (6.78)$$

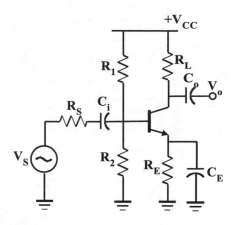

Fig. 6.26 Common emitter amplifier

Fig. 6.27 Equivalent
circuit of common emitter
amplifier

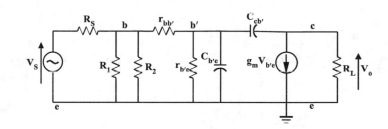

Now

$$V_o = -g_m V_{b'e} R'_L \qquad (6.79)$$

where R'_L is R_L including the effect of $C_{cb'}$.
Therefore

$$A_V = -\frac{g_m V_{b'e} R'_L}{V_{b'e}} = -g_m R'_L \qquad (6.80)$$

Since the feedback admittance Y_f is

$$Y_f = j\omega C_{cb} \qquad (6.81)$$

using Miller's Theorem, Y_f across the collector-
base junction can be replaced by $Y_f(1 - A_V)$ at the
input and $Y_f(1 - 1/A_V)$ at the output. At the input,

$$Y_f(1 - A_V) = j\omega C_{cb'}(1 + g_m R'_L) \qquad (6.82)$$

which is equivalent to a capacitance $C_{M'}$ given by

$$C_{M'} = C_{cb'}(1 + g_m R'_L) \qquad (6.83)$$

At the output,

$$Y_f(1 - 1/A_V) = j\omega C_{cb'}(1 + 1/g_m R'_L)$$
$$\approx j\omega C_{cb'}, g_m R'_L \gg 1 \qquad (6.84)$$

which is equivalent to a capacitance $C_{cb'}$ at the
output. The capacitance $C_{M'}$ at the input is called
the Miller capacitance and the resulting equivalent circuit is shown in Fig. 6.28.

Because of the multiplying effect of the
$(1 + g_m R'_L)$ factor, its value is comparable to $C_{b'e}$
between the base emitter junction.

Evaluating the transfer function V_o/V_b for the
amplifier,

$$V_o = -g_m V_{b'e} R'_L \qquad (6.85)$$

where

$$R'_L = R_L // 1/sC_{b'e} \qquad (6.86)$$

$$V_{b'e} = \frac{r_{b'e} // 1/sC_M}{r_{b'b} + r_{b'e} // 1/sC_M} V_b \qquad (6.87)$$

where

Fig. 6.28 Equivalent
circuit after application of
Miller's theorem

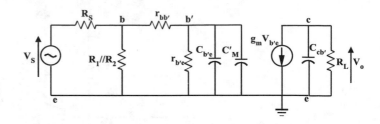

Fig. 6.29 Equivalent
circuit with simplified
source

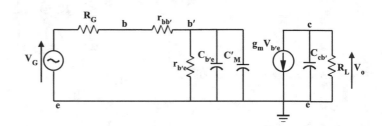

$$C_M = C_{M'} + C_{b'e} \qquad (6.88)$$

Hence

$$\frac{V_o}{V_b} = \frac{V_o}{V_{b'e}}\frac{V_{b'e}}{V_b} = -g_m R'_L \frac{r_{b'e}//1/sC_M}{r_{bb'} + r_{b'e}//1/sC_M}$$

$$= -g_m R'_L \frac{\dfrac{r_{b'e}1/sC_M}{r_{b'e} + 1/sC_M}}{r_{bb'} + \dfrac{r_{b'e}1/sC_M}{r_{b'e} + 1/sC_M}} = \frac{-g_m R'_L r_{b'e}}{r_{bb'} + r_{b'e} + r_{bb'}r_{b'e}sC_M}$$

$$= \frac{-g_m R'_L \dfrac{r}{r_{bb'} + r_{b'e}}}{1 + sC_M r_{bb'}//r_{b'e}}$$

$$(6.89)$$

The circuit to the left of b involving the source
voltage V_S as well as R_S, R_A and R_B can be
replaced by the Thevenin equivalent V_G in series
with R_G as shown in Fig. 6.29 where

$$R_G = R_1//R_2//R_S \qquad (6.90)$$

and

$$V_G = \frac{R_1//R_2}{R_S + R_1//R_2} V_S \qquad (6.91)$$

Thus, R_G can be treated as part of $r_{bb'}$ giving

$$\frac{V_o}{V_G} = -g_m R'_L \frac{\dfrac{r_{b'e}}{r_{bb'} + R_G + r_{b'e}}}{1 + sC_M(r_{bb'} + R_G//r_{b'e})} \qquad (6.92)$$

Substituting for V_G in (6.92) gives

$$\frac{V_o}{V_S} = \frac{V_o}{V_G}\frac{V_G}{V_S}$$

$$= -g_m R'_L \frac{\dfrac{r_{b'e}}{r_{b'e} + R_E + r_{b'e}}}{1 + sC_M(r_{bb'} + R_G)//r_{b'e}}\rho$$

$$(6.93)$$

where

$$\rho = \frac{R_A//R_B}{R_S + R_A//R_B} \qquad (6.94)$$

From (6.42),

$$R'_L = R_L//1/sC_{cb'} = \frac{R_L}{1 + sR_L C_{cb'}} \qquad (6.95)$$

If $C_{cb'}$ is small, then the pole created by $R_L C_{cb'}$
is at a high frequency and can be ignored, in
which case $R'_L \to R_L$ in (6.95).

Thus from (6.93), transfer function V_o/V_S for
the common emitter amplifier can be written as

$$\frac{V_o}{V_S} = \frac{k}{1 + jf/f_o}$$

where the low-frequency gain k of the amplifier is
given by

$$k = g_m R_L \rho \frac{r_{b'e}}{r_{bb'} + R_G + r_{b'e}} \qquad (6.96)$$

and the cut-off frequency f_o is given by

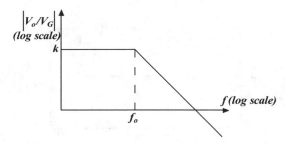

Fig. 6.30 Frequency response plot of the common emitter amplifier

$$f_o = \frac{1}{2\pi C_M (r_{bb'} + R_G)//r_{b'e}} \qquad (6.97)$$

A frequency response plot for the amplifier is shown in Fig. 6.30. f_o is the upper cut-off frequency of the amplifier. It is the frequency at which the gain of the amplifier falls by 3 dB below its low-frequency value. If the attenuation factor rho is ignored and R_G set to zero, the low-frequency gain k in (6.96) becomes

$$k = g_m R_L \rho \frac{r_{b'e}}{r_{bb'} + R_G + r_{b'e}} = g_m R_L \qquad (6.98)$$

the value obtained in Chap. 2.

From eq. (6.97), the frequency response of the amplifier can be increased by reducing the source resistance R_S. Thus if $R_S = 0$ corresponding to a voltage source, then f_o becomes

$$f_o' = \frac{1}{2\pi C_M r_{bb'}//r_{b'e}} \approx \frac{1}{2\pi C_M r_{bb'}}$$
$$> f_o, r_{bb'} \ll r_{b'e} \qquad (6.99)$$

If however R_S is very large corresponding to a current source, then for $R_S > R_1//R_2$, f_o becomes

$$f_o'' = \frac{1}{2\pi C_M (r_{bb'} + R_1//R_2)//r_{b'e}}$$
$$\approx \frac{1}{2\pi C_M R_1//R_2//r_{b'e}} < f_o, r_{bb'}$$
$$\ll R_1//R_2 \qquad (6.100)$$

The cut-off frequency is also influenced by the gain $g_m R_L$ of the amplifier since $C_M' \approx C_{cb'} g_m R_L$.

If the gain is reduced by reducing R_L, the cut-off frequency is increased. A large transition frequency f_T reduces C_{be} in (6.68) and therefore also improves the frequency response of the amplifier. Finally, a small C_{cb} also improves the frequency response by decreasing the Miller capacitance C_M.

Thus, in order to realize a high cut-off frequency f_o in the amplifier design, the following steps can be taken:

- Use a low source impedance, i.e. voltage-drive the amplifier
- Use a low load resistance R_L
- Use a transistor with a large f_T
- Use a transistor with a low C_{cb}

Example 6.5
Determine the upper cut-off frequency for the common emitter amplifier shown in Fig. 6.31 where a 2 N3904 transistor is used.

Solution The 2 N3904 has $f_T = 270$ MHz, $C_{cb'} = 4$ pF and $\beta_o = 100$. The quiescent current for this circuit is easily determined to be 1 mA. Then $r_{b'e}$ is given by $r_{b'e} = \beta_o r_e = \frac{100}{40 \times 1\,\text{mA}} = 2.5$ k. The capacitance $C_{b'e}$ can be found from $C_{b'e} = 1/2\pi f_T r_e$ giving $C_{b'e} = 1/2\pi f_T r_e = \frac{1}{2\pi \times 270 \times 10^6 \times 25} = 23.6$ pF. $r_{bb'}$ is assumed to be about 20 Ω. From (6.88), $C_M = C_{b'e}$

$$+C_{cb'}(1 + g_m R_L) = 23.6\,\text{pF}$$
$$+ (1 + 40 \times 1\,\text{mA} \times 9\,\text{k})4\,\text{pF}$$
$$= 1468\,\text{pF}$$

. From (6.90) $R_G = R_A//R_B//R_S = 27$ k//173 k//600 ≈ 600 Ω. Hence from (6.97), $f_o = \frac{1}{2\pi C_M}$

$$(r_{bb'} + R_G)//r_{b'e} =$$

$$\frac{1}{2\pi \times 1468\,\text{pF} \times (20 + 600)//2500}$$
$$= 218140\,\text{Hz}$$

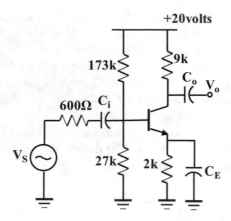

Fig. 6.31 Common emitter amplifier

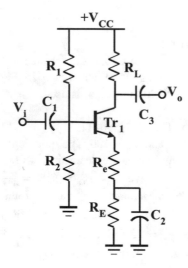

Fig. 6.32 Common emitter amplifier with partially bypassed emitter resistor

6.6   Common Emitter Amplifier with Local Series Feedback

The circuit of Fig. 6.32 shows a common emitter amplifier with only a partially decoupled emitter resistor. The resistor R_e appears in series with the emitter terminal. In order to evaluate the effect of the resistor on the frequency response of the stage, the transistor is again replaced by the simplified hybrid-pi equivalent circuit shown in Fig. 6.33. The circuit can be further simplified by

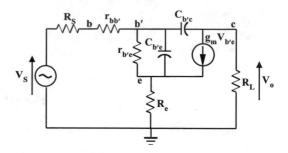

Fig. 6.33   Equivalent circuit of amplifier in Fig. 6.32

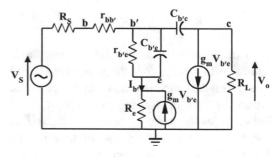

Fig. 6.34   Simplified equivalent circuit

replacing the dependent current source $g_m V_{b'e}$ by two such sources as shown in Fig. 6.34. The voltage $V_{b'}$ at base terminal b' with respect to ground is given by

$$V_{b'} = V_{b'e} + (I_{b'} + g_m V_{b'e})R_e \qquad (6.101)$$

Now

$$I_{b'} = V_{b'e}/Z_{b'e} \qquad (6.102)$$

where

$$Z_{b'e} = r_{b'e}//1/sC_{b'e} \qquad (6.103)$$

Therefore

$$V_{b'} = V_{b'e} + (V_{b'e}/Z_{b'e} + g_m V_{b'e})R_e$$
$$= V_{b'e}\{1 + R_e(g_m + y_{b'e})\} \qquad (6.104)$$

where

$$y_{b'e} = 1/Z_{b'e} = 1/r_{b'e} + sC_{b'e} \qquad (6.105)$$

Using (6.104) and (6.102), we get

$$\frac{V_{b'}}{I_{b'}} = \{1 + R_e(g_m + y_{b'e})\}Z_{b'e}$$

$$= \frac{1 + R_e(g_m + y_{b'e})}{y_{b'e}} \quad (6.106)$$

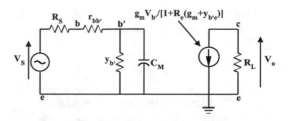

Equation (6.106) represents the effective impedance between b' and ground. Hence the components $r_{b'e}$, $C_{b'e}$, R_e and $g_m V_{b'c}$ can all be replaced by an admittance $y_{b'}$ of value

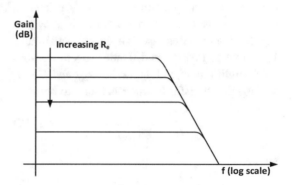

Fig. 6.35 Equivalent circuit of amplifier in Fig. 6.32

$$y_{b'} = \frac{y_{b'e}}{1 + R_e(g_m + y_{b'e})} \quad (6.107)$$

The voltage gain A_V between b' and c is given by

$$A_V = \frac{V_o}{V_{b'}} = \frac{-g_m V_{b'e} R_L}{V_{b'e}[1 + R_e(g_m + y_{b'e})]}$$

$$= \frac{-g_m R_L}{1 + R_e(g_m + y_{b'e})} \quad (6.108)$$

Therefore using Miller's theorem, the capacitor $C_{b'c}$ between b' and c can be replaced by C_M across b' and ground where

$$C_M = \frac{g_m R_L}{1 + R_e(g_m + y_{b'e})} C_{b'c} \quad (6.109)$$

and a capacitor $C_{b'c}$ across c and ground. Also, from (6.104)

$$V_{b'e} = \frac{V_{b'}}{1 + R_e(g_m + y_{b'e})} \quad (6.110)$$

The resulting equivalent circuit is shown in Fig. 6.35.

The transfer function V_O/V_S is given by

$$\frac{V_o}{V_S} = \frac{-g_m R_L}{1 + R_e(g_m + y_{b'e}) + (R_S + r_{bb'})(y_{b'e} + sC_{b'e}g_m R_L)} \quad (6.111)$$

Substituting for $y_{b'e}$ using (6.105), (6.111) becomes

$$\frac{V_o}{V_S} = \frac{-k}{1 + jf/f_o} \quad (6.112)$$

where

The caption on the right:

Fig. 6.36 Effect of local series feedback on bandwidth

$$k - \frac{g_m R_L r_{b'e}}{R_S + r_{bb'} + r_{b'e} + R_e(1 + g_m r_{b'e})} \quad (6.113)$$

and

$$f_o = \frac{1}{2\pi C_{b'e} r_{b'e}(R_e + \alpha)/[R_S + r_{b'b} + r_{b'e}(1 + g_m R_e)]} \quad (6.114)$$

with

$$\alpha = (R_S + r_{bb'})(1 + g_m R_L C_{b'c}/C_{b'e}) \quad (6.115)$$

From (6.114), for $R_e < < \alpha$, as R_e increases, f_o increases as shown in Fig. 6.36. Also, the gain-bandwidth product kf_o is

$$kf_o = \frac{g_m R_L r_{b'e}}{2\pi C_{b'e} r_{b'e}(R_e + \alpha)}$$

$$= \frac{g_m R_L}{2\pi[R_e C_{b'e} + (R_S + r_{bb'})(C_{b'e} + C_{b'c}g_m R_L)]} \quad (6.116)$$

This product is maximum for $R_S = 0$ and therefore this stage is best driven from a voltage source.

6.7   Common Emitter Amplifier with Local Shunt Feedback

The circuit of Fig. 6.37 shows a common emitter amplifier with a resistor connecting collector and base and driven by a current source I_S of internal resistance R_S. The equivalent circuit is shown in Fig. 6.38, and in this case, it is the trans-resistance V_o/I_S that is investigated. Under several reasonable conditions including $R_F \gg r_{bb'}$ and $C_{b'e} \gg C_{b'c}$, this transfer function is found to be

$$\frac{V_o}{I_S} = \frac{R_m}{1 + jf/f_o} \quad (6.117)$$

where

$$R_m = \frac{g_m r_{b'e} R_S R_F R_L}{(R_S + r_{bb'} + r_{b'e})(R_F + R_L) + g_m r_{b'e} R_S R_L} \quad (6.118)$$

and

$$f_o = \frac{1}{2\pi r_{b'e} \frac{(r_{bb'}+R_S)}{(R_S+r_{bb'}+r_{b'e})(R_F+R_L)+g_m R_L r_{b'e} R_S} [C_{b'e}(R_F+}$$

$$R_L) + C_{b'c} g_m R_L R_F] \quad (6.119)$$

This reduces to

$$f_o = \frac{1}{2\pi r_{b'e} \frac{(r_{bb'}+R_S)}{(R_S+r_{bb'}+r_{b'e})+g_m(R_L//R_F)r_{b'e} R_S/R_F} [C_{b'e}+}$$

$$C_{b'c} g_m R_L//R_F] \quad (6.120)$$

Reducing R_F corresponds to increasing feedback. Eventually $R_F < < R_L$ and f_o becomes

$$f_o = \frac{1}{2\pi r_{b'e} \frac{(r_{bb'}+R_S)}{(R_S+r_{bb'}+r_{b'e})+g_m r_{b'e} R_S} [C_{b'e} + C_{b'c} g_m R_F]}$$

$$(6.121)$$

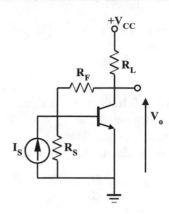

Fig. 6.37 Common emitter amplifier with local shunt feedback

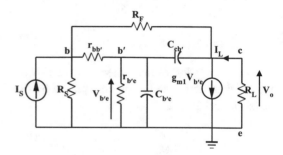

Fig. 6.38 Equivalent Circuit of Common Emitter Amplifier

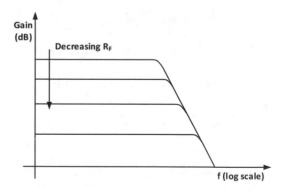

Fig. 6.39 Effect of local shunt feedback on bandwidth

From (6.121), f_o increases as R_F is reduced, i.e. the bandwidth increases with increased feedback as shown in Fig. 6.39. It is convenient to identify the current gain $A_I = I_L/I_S$ for the circuit which is easily obtained by dividing V_o/I_S by the load resistor R_L. Thus, the current gain-bandwidth product $A_I f_o$ for this circuit is given by

$$A_I f_o = \frac{g_m R_S R_F}{(r_{bb'} + R_S)[C_{b'e}(R_F + R_L) + C_{b'e} g_m R_F R_L]}$$
(6.122)

which is a maximum when $R_S \to \infty$. That is, the common emitter amplifier with local shunt feedback is best driven from a current source in order to improve the frequency response characteristics.

6.8   High-Frequency Response of the Cascode Amplifier

Another method of improving the frequency response of the common emitter amplifier is to operate the configuration as a Cascode amplifier shown in simplified form in Fig. 6.40 with equivalent circuits in Figs. 6.41 and 6.42. The configuration consists essentially of a common emitter amplifier Tr_1 operation into a common base amplifier Tr_2. The input of Tr_2 is the load of Tr_1. The Miller capacitance C_M' arising from $C_{cb'1}$ is given by

$$C_M' = (1 + g_m R_L) C_{cb'1}$$
(6.123)

Since R_{L1} is the input impedance r_{e2} of Tr_2, its value is given by $R_{L1} = r_{e2}$. Because of the high

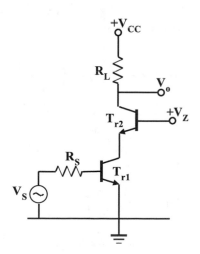

Fig. 6.40 Cascode amplifier

frequency response of the common-base amplifier compared with the common emitter amplifier, the frequency effect associated with r_{e2} can be ignored. The DC collector current of Tr_1 is approximately that of Tr_2. Therefore

$$g_{m1} \approx 1/r_{e2}$$
(6.124)

and hence the gain of the common emitter stage is unity. Hence

$$C_M' = (1 + 1) C_{cb1} = 2 C_{cb'1}$$
(6.125)

and the total effective capacitance C_M' between b_1 and e_1 is

$$C_M' = C_{b'e1} + C_{cb'1}$$
(6.126)

The upper cut-off frequency therefore becomes

$$\bar{f}_o = \frac{1}{2\pi C_M'(R_S + r_{b'b})//r_{b'e}} > f_o$$

$$- \frac{1}{2\pi C_M(R_S + r_{b'b})//r_{b'e}}$$
(6.127)

since $C_M' < C_M$. i.e., the cut-off frequency of the common-emitter stage of the Cascode amplifier is higher than that of the normal common emitter amplifier because the voltage gain of the common-emitter stage in the Cascode circuit is unity, and therefore there is no multiplication of the collector-base junction capacitor; the value of C_M' is significantly reduced.

Now

$$V_o = -g_{m2} V_{b'e2} R_L'$$
(6.128)

Therefore

$$\frac{V_o}{V_S} = \frac{V_o}{V_{e2b2}} \frac{V_{c1e1}}{V_S}$$

$$= \frac{\frac{-r_{b'e}}{r_{bb'} + R_S + r_{b'e}}}{1 + s C_M(r_{bb'} + R_S)//r_{b'e}}$$

$$\times \frac{g_m R_L}{(1 + s C_{cb2} R_L)}$$
(6.129)

which has the same low-frequency voltage gain expression as the common emitter amplifier but a higher cut-off frequency.

Fig. 6.41 Equivalent
circuit of Cascode amplifier

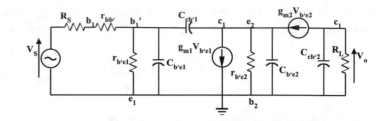

Fig. 6.42 Simplified
equivalent circuit of
Cascode amplifier

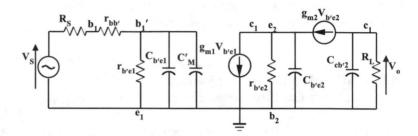

6.9   High-Frequency Response of the Common Base Amplifier

In the common emitter amplifier, the base emitter input capacitance along with the Miller capacitance serve to limit the high frequency response of the amplifier. In the common base amplifier of Fig. 6.43 now under consideration, the absence of any significant capacitance between the collector and the emitter of the transistor means that the Miller effect in this configuration is negligible. A higher frequency response is therefore expected.

The hybrid-pi equivalent circuit for this configuration is shown in Fig. 6.44. Since $r_{bb'}$ is small compared with the adjacent circuit components, a further simplification of the equivalent circuit is its removal such that b is connected directly to ground. Thus the capacitance $C_{b'c}$ only appears across the output load R_L and there is no Miller effect. To simplify the analysis, the voltage sources V_S is converted to a current source V_S/R_S. This simplified equivalent circuit is shown in Fig. 6.45.

Thus,

$$V_o = -g_m R_L//1/sC_{b'e} V_{b'e}$$

$$= \frac{-g_m R_L}{1 + sC_{b'e} R_L} V_{b'e} \qquad (6.130)$$

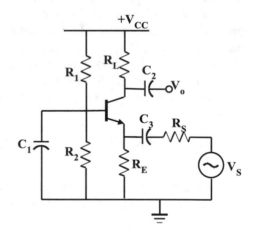

Fig. 6.43   Common base amplifier

$$V_{b'e} = -(g_m V_{b'e} + V_S/R_S)1/\left(\frac{1}{R_S} + \frac{1}{r_{b'e}} + sC_{b'e}\right)$$

$$= -(g_m V_{b'e} + V_S/R_S)\frac{R_S r_{b'e}}{r_{b'e} + R_S + sR_S r_{b'e} C_{b'e}} \qquad (6.131)$$

From this,

$$V_{b'e}\left(1 + \frac{g_m R_S r_{b'e}}{r_{b'e} + R_S + sR_S r_{b'e} C_{b'e}}\right)$$

$$= -V_S \frac{r_{b'e}}{r_{b'e} + R_S + sR_S r_{b'e} C_{b'e}} \qquad (6.132)$$

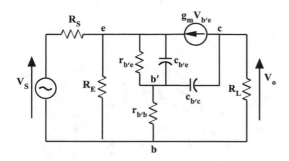

Fig. 6.44 Equivalent circuit of common base amplifier

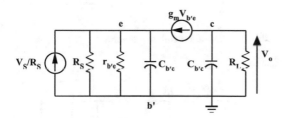

Fig. 6.45 Simplified equivalent circuit of common base amplifier

$$\frac{V_o}{V_S} = \frac{V_o}{V_{b'e}} \frac{V_{b'e}}{V_S}$$

$$= \frac{g_m Z_L r_{b'e}}{r_{b'e} + R_S + g_m r_{b'e} R_S + s R_S r_{b'e} C_{b'e}}$$

$$= \frac{g_m Z_L \frac{r}{r_{b'e} + (1+h_{fe}) R_S}}{1 + s C_{b'e} \frac{R_S r_{b'e}}{r_{b'e} + (1+h_{fe}) R_S}}$$

$$\tag{6.133}$$

where

$$g_m r_{b'e} = h_{fe} \tag{6.134}$$

and

$$Z_L = R_L // 1/s C_{b'e} \tag{6.135}$$

Dividing the numerator and denominator in (6.133) by $(1 + h_{fe})$ yields

$$\frac{V_o}{V_S} = \frac{\frac{Z_L}{r_e + R_S}}{1 + s C_{b'e} \left(\frac{R_S r_e}{R_S + r_e}\right)}$$

$$= \frac{\frac{R_L}{r_e + R_S}}{\left[1 + s C_{b'e} \left(\frac{R_S r_e}{R_S + r_e}\right)\right](1 + s C_{b'c} R_L)}$$

$$\tag{6.136}$$

For a transistor with small $C_{b'c}$ and R_L, the upper cut-off frequency of the common base amplifier is

$$f_H = \frac{1}{2\pi C_{b'e} R_S // r_e} \tag{6.137}$$

The transition frequency f_T of a transistor is given by

$$f_T = \frac{1}{2\pi C_{b'e} r_e} \tag{6.138}$$

and therefore $f_H > f_T$. However, in practice the second pole of (6.136) corresponding to the cut-off frequency

$$f'_H = \frac{1}{2\pi C_{b'c} R_L} \tag{6.139}$$

sets the upper cut-off frequency with $f'_H < f_H, f_T$. Thus, the frequency response of the common base amplifier can be of the order of f_T and therefore much higher than the frequency response of the common emitter amplifier. However, the low input impedance of the common base amplifier limits its applicability.

Example 6.6
Determine the upper cut-off frequency for the common base amplifier shown in Fig. 6.46 where a 2 N3904 transistor is used.

Solution The transistor is again the 2 N3904 and it is biased at 1 mA. Therefore the characteristics are the same as in Example 6.5 i.e. $I_{cq} = 1$ mA, $f_T = 270$ MHz, $C_{cb'} = 4\,\text{pF}$, $C_{b'e} = 23.6\,\text{pF}$, $r_{b'e} = 2.5\,\text{k}$ and $r_e = 1/40 I_c = 25\,\Omega$. Hence

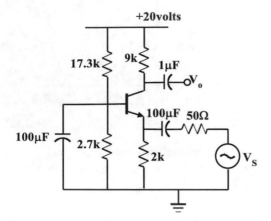

Fig. 6.46 Common base amplifier

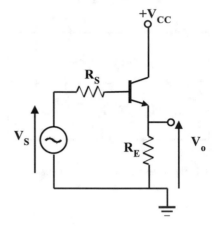

Fig. 6.47 Common collector amplifier

using (6.139), the upper cut-off frequency is given by $f'_H = \frac{1}{2\pi C_{b'c}R_L} = \frac{1}{2\pi \times 4 \times 10^{-12} \times 9000} = 4.4\,\text{MHz}$. Note that the pole frequency given by (6.138) is $f_H = \frac{1}{2\pi C_{b'e}r_e} = \frac{1}{2\pi \times 23.6 \times 10^{-12} \times 25} = 262\,\text{MHz}$ is much higher than $f'_H = 4.4\,\text{MHz}$.

6.10   High-Frequency Response of the Common Collector Amplifier

The common collector amplifier or emitter follower also has a better frequency response than the common emitter amplifier. In this section we analyse the performance of this configuration. The basic circuit is shown in Fig. 6.47 where biasing components are omitted for simplicity. Using the hybrid-pi equivalent circuit, the emitter follower can be represented as shown in Fig. 6.48. From this,

$$V_o = \left(g_m V_{b'e} + \frac{V_{b'e}}{r_{b'e}//1/sC_{b'e}}\right)R_E$$

$$= V_{b'e}\left(g_m + \frac{1 + sr_{b'e}C_{b'e}}{r_{b'e}}\right)R_E$$

$$= V_{b'e}\left(\frac{g_m r_{b'e} + 1 + sr_{b'e}C_{b'e}}{r_{b'e}}\right)R_E \quad (6.140)$$

Therefore

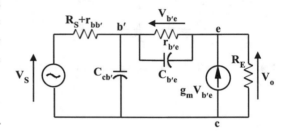

Fig. 6.48 Equivalent circuit

$$V_{b'e} = \frac{r_{b'e}}{R_E}\left(\frac{1}{1 + g_m r_{b'e} + sr_{b'e}C_{b'e}}\right)V_o \quad (6.141)$$

Now

$$V_{b'c} = V_o + V_{b'e}$$

$$= V_o + V_o\frac{r_{b'e}}{R_E}$$

$$\times \frac{1}{1 + g_m r_{b'e} + sr_{b'e}C_{b'e}} \quad (6.142)$$

Hence

$$\frac{V_o}{V_{b'c}} = \cfrac{1}{1 + \cfrac{r}{R_E}\cfrac{1}{1 + g_m r_{b'e} + sr_{b'e}C_{b'e}}}$$

$$= \frac{R_E(1 + g_m r_{b'e} + sr_{b'e}C_{b'e})}{R_E + r_{b'e} + R_E g_m r_{b'e} + sr_{b'e}R_E C_{b'e}}$$

$$= \cfrac{R_E(1 + g_m r_{b'e})\left(1 + \cfrac{sC_{b'e}}{1 + g_m r_{b'e}}\right)}{(R_E + r_{b'e} + R_E g_m r_{b'e})\left(1 + \cfrac{sC_{b'e}r_{b'e}R_E}{R_E + r_{b'e} + R_E r_{b'e}g_m}\right)}$$

$$(6.143)$$

This reduces to

$$\frac{V_o}{V_{b'c}} = \frac{R_E}{R_E + r_{b'e}} \frac{1 + g_m r_{b'e}}{1 + g_m r_{b'e}//R_E}$$
$$\times \frac{1 + sC_{b'e}\frac{r}{1+g_m r_{b'e}}}{1 + s\frac{C_{b'e}r_{b'e}//R_E}{1+g_m r_{b'e}//R_E}} \qquad (6.144)$$

But

$$\frac{R_E}{R_E + r_{b'e}} \frac{1 + g_m r_{b'e}}{1 + g_m r_{b'e}//R_E}$$
$$= \frac{R_E(1 + g_m r_{b'e})}{R_E + r_{b'e} + g_m r_{b'e}R_E}$$
$$= \frac{R_E(1 + g_m r_{b'e})}{R_E\left(1 + \frac{r}{R_E} + g_m r_{b'e}\right)} \approx 1, R_F >$$
$$> \frac{1}{g_m} \qquad (6.145)$$

Hence

$$\frac{V_o}{V_{b'c}} \approx \frac{1 + sC_{b'e}\frac{r}{1+g_m r_{b'e}}}{1 + s\frac{C_{b'e}r_{b'e}//R_E}{1+g_m r_{b'e}//R_E}}$$
$$= \frac{1 + sC_{b'e}\frac{r}{1+g_m r_{b'e}}}{1 + s\frac{C_{b'e}r_{b'e}}{1+g_m r_{b'e}+\frac{r}{R_E}}} \qquad (6.146)$$

Finally, since the input impedance seen at the transistor base is high, the ratio $V_{b'c}/V_i \approx \frac{1/sC_{cb'}}{R_S + r_{bb'} + 1/sC_{cb'}} = \frac{1}{1 + sC_{cb'}(R_S + r_{bb'})}$. Then the voltage gain is given by

$$\frac{V_o}{V_i} = \frac{1}{1 + sC_{cb'}(R_S + r_{bb'})} \cdot \frac{1 + sC_{b'e}\frac{r}{1+g_m r_{b'e}}}{1 + s\frac{C_{b'e}r_{b'e}}{1+g_m r_{b'e}+\frac{r}{R_E}}}$$
$$\qquad (6.147)$$

Thus if $R_E >> r_{b'e}$ then $V_o/V_{b'c} \approx 1$. This suggests a very high frequency response for the common collector amplifier if driven from a low source impedance and has a high value emitter resistor in which case the time constant $C_{b'c}r_{bb'}$ is small compared with $C_{b'e}r_{b'e}$. Then $V_{b'c} \approx V_i$. Under such circumstances, the effect of $r_{bb'}$ and $C_{b'c}$ come into play and limit the overall frequency response. Then the circuit bandwidth is primarily set by $r_{bb'}$ and $C_{b'c}$ at a frequency

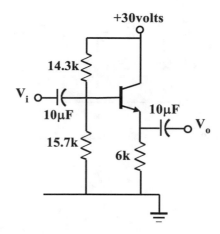

Fig. 6.49 Common collector amplifier

$$f_U = \frac{1}{2\pi(R_S + r_{bb'})C_{cb'}} \qquad (6.148)$$

Example 6.7
Determine the upper cut-off frequency for the common collector amplifier shown in Fig. 6.49 where a 2 N3904 transistor is used.

Solution For the emitter follower Fig. 6.49 driven from a low source impedance of 50 Ω, the upper cut-off frequency f_U is given by

$$f_U = \frac{1}{2\pi(R_S + r_{bb'})C_{cb'}}$$
$$= \frac{1}{2\pi \times (50 + 20) \times 4 \times 10^{-12}} = 568\,\text{MHz}$$

6.11   High-Frequency Response of a Common Source FET Amplifier

The equivalent circuit for the JFET at high frequency is shown in Fig. 6.50. There are two main junction capacitances, namely, the gate-source capacitance C_{gs} and the gate-drain capacitance C_{gd}. Consider the common source configuration shown in Fig. 6.51. In this configuration, the equivalent circuit becomes that shown in Fig. 6.52. Based on Miller's theorem, the capacitor C_{gd} may be replaced by a capacitor $C_{gd}(1 + g_m R_L)$ across C_{gs} where $R'_L = R_L//r_{ds}$.

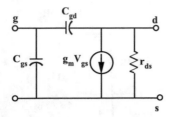

Fig. 6.50 High-frequency equivalent circuit of JFET

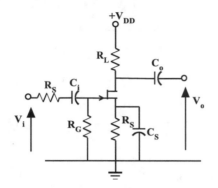

Fig. 6.51 Common source JFET amplifier

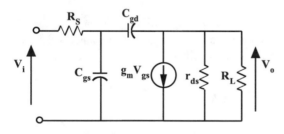

Fig. 6.52 Equivalent circuit of JFET amplifier

In such a case the source resistance R_S and the capacitance $C = C_{gs} + C_{gd}\left(1 + g_m R'_L\right)$ form a low-pass filter with cut-off frequency

$$f_{H1} = \frac{1}{2\pi R_S\left[C_{gs} + \left(1 + g_m R'_L\right)C_{gd}\right]} \quad (6.149)$$

This frequency can be increased by reducing R_S which corresponds to driving the circuit from a low-impedance source. Note also that by Miller's theorem a capacitor of approximate value C_{gd} appears across the output between the drain and

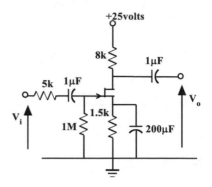

Fig. 6.53 Common source JFET amplifier

ground. This capacitor introduces another pole at a frequency given by

$$f_{H2} = \frac{1}{2\pi R_L//r_{ds}C_{gs}} \quad (6.150)$$

This frequency would typically be higher than that associated with R_S in (6.149).

Example 6.8
Find the cut-off frequency of the common source amplifier shown in Fig. 6.53. The JFET has the following characteristics:

$$C_{gs} = 50\,\text{pF}, C_{gd} = 6\,\text{pF}, r_d = 100\,\text{k}\Omega, g_m$$
$$= 5\,\text{mA/V}, R_S = 5\,\text{k}$$

Solution Using (6.149), $f_{H1} \approx$ $\frac{1}{2\pi R_S\left[C_{gs}+\left(1+g_m R'_L\right)C_{gd}\right]}$, the upper cut-off frequency is given by $f_{H1} \approx \frac{1}{2\pi R_S\left[C_{gs}+\left(1+g_m R'_L\right)C_{gd}\right]} =$ $\frac{1}{2\pi 5000[50+(1+5\times7.4)6]\times10^{-12}} = 114\,\text{kHz}$. The other pole frequency is given by $f_{H2} = \frac{1}{2\pi R'_L//r_{ds}C_{gs}} =$ $\frac{1}{2\pi\times8\,\text{k}//100\,\text{k}\times50\,\text{pF}} = 430\,\text{kHz}$.

6.12   High-Frequency Response of a Common Gate FET Amplifier

In the common source amplifier, the gate-source capacitance and the Miller capacitance both limited the high frequency response of the circuit. In

Fig. 6.54 Common gate FET amplifier

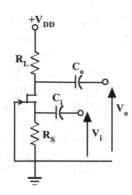

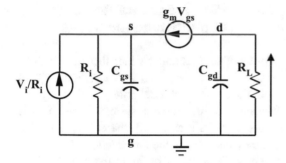

Fig. 6.55 Equivalent circuit of common gate amplifier

the common gate amplifier of Fig. 6.54 the capacitance between the drain and the source is small and therefore the Miller effect is insignificant. As in the case of the common base configuration, a good high-frequency response is expected. The equivalent circuit is shown in Fig. 6.55. To simplify the analysis, the voltage source V_i is converted to a current source V_i/R_i and the bias-setting resistor R_S is omitted since in generally, $R_S > > 1/g_m$ where $1/g_m$ is the input impedance of the configuration. Thus,

$$V_o = -g_m\left(R_L//1/sC_{gd}\right)V_{gs}$$

$$= -\frac{g_m R_L}{1 + sC_{gd}R_L}V_{gs} \qquad (6.151)$$

$$V_{gs} = -\left(g_m V_{gs} + V_i/R_i\right)\frac{1}{1/R_i + sC_{gs}} \qquad (6.152)$$

This becomes

$$V_{gs}\left(1 + \frac{g_m R_i}{1 + sR_i C_{gs}}\right)$$

$$= V_{gs}\left(\frac{1 + g_m R_i + sR_i C_{gs}}{1 + sR_i C_{gs}}\right)$$

$$= -V_i\frac{1}{1 + sR_i C_{gs}} \qquad (6.153)$$

Hence

$$\frac{V_o}{V_i} = \frac{V_o}{V_{gs}}\cdot\frac{V_{gs}}{V_i}$$

$$= -\frac{g_m R_L}{1 + sC_{gd}R_L}\cdot\frac{1}{1 + g_m R_i + sR_i C_{gs}} \qquad (6.154)$$

This reduces to

$$\frac{V_o}{V_i} = -\frac{g_m R_L}{1 + g_m R_i}\cdot\frac{1}{\left(1 + sC_{gd}R_L\right)\left(1 + sC_{gs}R_i//1/g_m\right)} \qquad (6.155)$$

From (6.155), the upper cut-off frequency is given by

$$f_L = \frac{1}{2\pi R_L C_{gd}} \qquad (6.156)$$

since the pole introduced by C_{gs} and $R_i//1/g_m$ is in general at a higher frequency.

Example 6.9
For a common gate JFET amplifier with $R_S = 1$ k and $R_L = 9$ k, using the JFET of Example 6.8, find the upper cut-off frequency of the circuit.

Solution For the JFET of Example 6.8, $C_{gd} = 6$ pF. From Eq. (6.156), the upper cut-off frequency is given by $f_L = \frac{1}{2\pi R_L C_{gd}} = \frac{1}{2\pi\times 9\,\text{k}\times 6\,\text{pF}} = 2.9\,\text{MHz}$.

6.13 High-Frequency Response of a Common Drain FET Amplifier

We now consider the common drain or source follower configuration shown in Fig. 6.56. The associated equivalent circuit is presented in Fig. 6.57. Here, R_G and r_{ds} have been omitted since both are generally quite large.

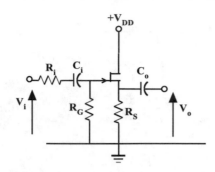

Fig. 6.56 Common drain FET amplifier

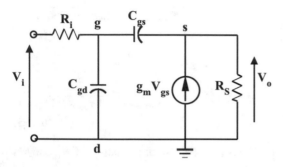

Fig. 6.57 Equivalent circuit of common drain amplifier

From Fig. 6.57,

$$V_o = V_{gs}(g_m + sC_{gs})R_S \qquad (6.157)$$

From which

$$V_{gs} = \frac{V_o}{R_S(g_m + sC_{gs})} \qquad (6.158)$$

Now

$$V_{gd} = V_o + V_{gs}$$

$$= V_o\left(\frac{1 + (g_m + sC_{gs})R_S}{(g_m + sC_{gs})R_S}\right) \qquad (6.159)$$

Therefore

$$\frac{V_o}{V_{gd}} = \frac{(g_m + sC_{gs})R_S}{1 + (g_m + sC_{gs})R_S}$$

$$= \frac{g_m R_S}{1 + g_m R_S} \cdot \frac{1 + sC_{gs}/g_m}{1 + s\frac{g_m R_S}{1+g_m R_S}C_{gs}/g_m} \qquad (6.160)$$

If $g_m R_S >> 1$, then $\frac{g_m R_S}{1+g_m R_S} \approx 1$ and in (6.160) $V_o/V_{gd} \to 1$. This suggests a very high-frequency response for the common drain amplifier when driven by a low source impedance R_i (in which case $V_{gd} \approx V_i$) and has a high value of FET source resistor R_S. Note however that a non-zero R_i in conjunction with C_{gs} and the small Miller capacitance resulting from C_{gd} form a low-pass filter which limits the eventual frequency response of this circuit.

6.14 High-Frequency Response of Multistage Amplifiers

For amplifier systems in which there are cascaded stages such as discussed in Chap. 5, each stage introduces at least one pole in the frequency response of the overall system. When two or more stages are cascaded, the effect on the frequency response is the introduction of additional poles at the same or different frequencies. Thus, an amplifier comprising two common emitter stages will have at least two poles given by

$$A_V(j\omega) = \frac{k}{(1 + jf/f_{o1})(1 + jf/f_{o2})} \qquad (6.161)$$

where k is the low-frequency gain of the cascaded amplifier and f_{o1} and f_{o2} are the two poles of the system. A frequency response plot illustrating the position of the poles is shown in Fig. 6.58. It can be shown that for $f >> f_{o1}, f_{o2}$, the slope of the curve approaches -40 dB/decade which is twice the value of a single pole system. Similarly, if a third stage is added, the resulting system will have at least three poles resulting in a high-frequency roll-off rate of -60 dB/decade as shown in Fig. 6.58. These poles may be close or separated in frequency depending on the configuration of the particular cascaded amplifier. Each pole introduces phase shift (lag) between the output signal and the input signal and this must be considered when designing feedback amplifiers if instability is to be avoided in such systems. This is fully discussed in Chap. 7 on feedback amplifiers.

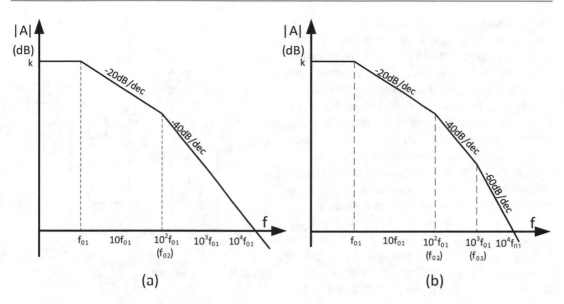

Fig. 6.58 Frequency plots of two (2)- and three (3)-pole system

Example 6.10
The transfer function of a three-pole amplifier system is given in eq. (6.162).

$$A_V(j\omega) = \frac{10^4}{(1 + jf/10)(1 + jf/10^3)(1 + jf/10^4)}$$
(6.162)

Plot the asymptotic curve for this system, indicating the low-frequency gain and the position of each pole.

Solution From the transfer function (6.162), there are three poles. The break frequency resulting from each pole is given by $f = f_o$ where f_o is the defined by the pole function $(1 + jf/f_o)$. Examination of each of these poles in (6.162) reveals the pole frequencies as $f_{o1} = 10$ Hz, $f_{o2} = 1$ kHz and $f_{o3} = 10$ kHz. The low-frequency gain $A_V(DC)$ is found when $f \to 0$ which corresponds to $A_V(DC) = 10^4$. Hence the asymptotic curve is easily found by drawing a horizontal line starting at 10^4 at some frequency below $f_{o1} = 10$ Hz and after passing through the frequency $f_{o1} = 10$ Hz changes slope from zero to -20 db/decade. As the frequency increases, it continues at this slope and changes to -40 dB/decade after passing through the frequency $f_{o2} = 1$ kHz. With further increase in

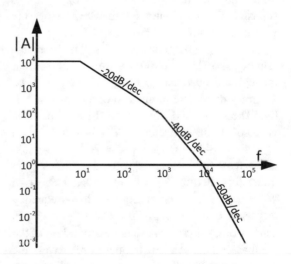

Fig. 6.59 Solution for Example 6.10

frequency the slope changes to -60 dB/decade as the line passes through $f_{o3} = 10$ kHz. This is the final slope change. This is shown in Fig. 6.59.

6.15   Applications

In this section, several circuits discussed in previous chapters are simulated in order to illustrate high frequency response characteristics.

Direct Coupled Ring-of-Three

The circuit in Fig. 6.60 is a three-stage direct coupled amplifier that has a very high gain. It comprises three common emitter gain stages that are directly coupled and operating from a 9 V supply as shown in Fig. 6.60. The output of Tr_1 drives Tr_2, and the output of Tr_2 drives Tr_3. The resistor chain $R_6 - R_5 - R_1$ provides bias current to the base of Tr_1. It represents a DC feedback loop that ensures that there is bias stability at the output. Any change in the quiescent voltage at the collector of Tr_3 is corrected by changes in Tr_1 and Tr_2. Capacitor C_2 short-circuits the feedback signal to ground for AC. Selecting a current of 0.5 mA for Tr_1, it follows that $R_2 = 0.7$ V/ 0.5 mA $= 1.4$ k. Similarly, with a current of 0.5 mA for Tr_2, then $R_3 = 0.7$ V/0.5 mA $= 1.4$ k. In the final stage, since this is the output that drives an external load, a current of 1 mA is selected. For maximum symmetrical swing, the voltage across R_4 must be $9/2 = 4.5$ V.

Hence $R_4 = 4.5/1$ mA $= 4.5$ k. The base current of Tr_1 is 0.5 mA/100 $= 5$ µA. Let the current through the bias resistor string $R_6 - R_5 - R_1$ be 10 times this value in order to ensure bias stability. Then $R_1 = 0.7/50$ µA $= 14$ k. Since the voltage at the collector of Tr_3 is 4.5 V, then $R_5 + R_6 = (4.5 - 0.7)/50$ µA $= 76$ k. This resistance may have to be adjusted for higher gain transistors. Setting $R_5 = R_6$ gives $R_5 = R_6 = 76$ k/2 $= 38$ k. For adequate low-frequency response, coupling capacitors $C_1 = C_2 = 10$ µF. Capacitor C_4 is chosen to ensure grounding of the feedback signal at a low frequency relative to the signals being amplified. A value of 47 uF is used. This circuit has a gain of about 100 dB and a bandwidth of 300 kHz.

Ideas for Exploration (i) Simulate the system and determine the gain by injecting a low-amplitude sine wave at 1 kHz. It will be observed that the high gain means that the amplifier output will be driven into clipping for all but the lowest signal amplitudes. (ii) Measure the gain for one or two lower-gain amplifiers discussed previously for comparison.

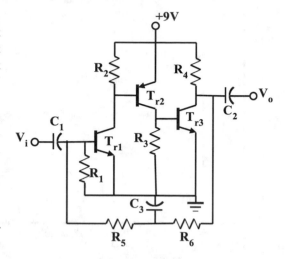

Fig. 6.60 Direct coupled ring-of-three

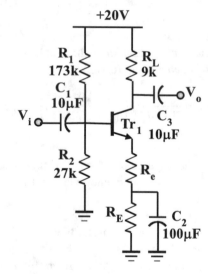

Fig. 6.61 Common emitter amplifier with partially bypassed emitter resistor

Common Emitter Amplifier with Local Series Feedback

The circuit in Fig. 6.61 is a voltage divider-biased common emitter amplifier designed using the principles of Chap. 2. Its frequency response for varying values of R_e was simulated using multisim while keeping the sum $R_e + R_E = 2$ k constant so that the bias conditions remain unchanged. The results are shown in Table 6.1.

Table 6.1 Results of frequency response tests of CE amplifier with local series feedback

Resistor R_e	100 Ω	200 Ω	300 Ω	400 Ω	500 Ω
Bandwidth	1.9 MHz	2.9 MHz	3.7 MHz	4.4 MHz	4.9 MHz
Voltage gain	36.2 dB	31.4 dB	28.3 dB	26 dB	24.3 dB

As R_e increased from 100 ohms to 500 ohms, the bandwidth of the circuit increased from 1.9 MHz to 4.9 MHz, while the gain decreased from 36.2 dB to 2.6 dB as illustrated in Fig. 6.36. This confirms the theory discussed in Sect. 6.6 and verifies the use of local series feedback to extend the bandwidth while stabilizing the gain of a common emitter amplifier.

Ideas for Exploration (i) Simulate the system and measure frequency response of the circuit for various values of R_e. (ii) Measure the gain for the system and observe that the gain becomes more stable as R_e increases.

Common Emitter Amplifier with Local Shunt Feedback

The circuit in Fig. 6.62 is a collector-base feedback biased common emitter amplifier designed using the principles of Chap. 2. The feedback component R_f in series with the capacitor C_f allows the changing of the effective feedback resistor R_F for signals while maintaining bias conditions thereby applying local shunt feedback. Its frequency response for varying values of R_f was tested with a source resistance of 10 k. The results are shown in Table 6.2. As R_f decreased from 100 k to 1 k, the bandwidth of the circuit increased from 1 MHz to 87.2 MHz while gain decreased from 18.6 dB to -20 dB as illustrated in Fig. 6.39. This confirms the theory discussed in Sect. 6.7 and verifies the use of local shunt feedback to extend the bandwidth while stabilizing the gain of a common emitter amplifier.

Ideas for Exploration (i) Simulate the system and measure frequency response of the circuit for various values of R_f. (ii) Measure the gain for the system and observe that the gain becomes more stable as R_f decreases.

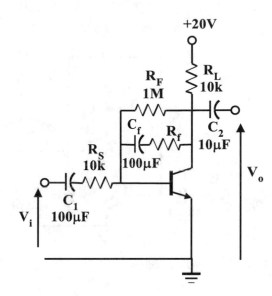

Fig. 6.62 Common emitter amplifier with local shunt feedback

Cascode Amplifier Using BJTs

The circuit in Fig. 6.63 is a Cascode amplifier comprising a BJT in a common emitter configuration driving another transistor in a common base mode. The first transistor is biased with a collector current of 1 mA using R_B, while the base of the second transistor is held at a fixed voltage using two 1 N4148 silicon diodes. A source resistance of $R_S = 1$ k was used. The frequency response of the circuit was tested using a 10 mV sinusoidal input. The bandwidth of the system was measured at 5.6 MHz. A common emitter amplifier with 1 mA current and the same load resistor and source resistor has a bandwidth of 343 kHz which is significantly lower. Thus, the Cascode arrangement prevents the multiplication of the collector-base capacitance of the first transistor and subsequent formation of a low-pass filter with the 1 k resistor at the input. This confirms the theory discussed in Sect. 6.8 and

Table 6.2 Results of frequency response tests of CE amplifier with local shunt feedback

Resistor R_f	100 k	50 k	20 k	10 k	1 k
Bandwidth	1 MHz	2 MHz	5 MHz	9.6 MHz	87.2 MHz
Voltage gain	18.6 dB	13.2 dB	5.6 dB	0 dB	−20 dB

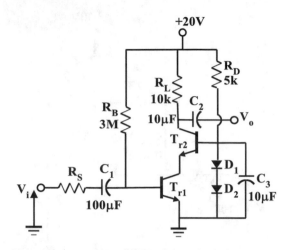

Fig. 6.63　Cascode amplifier

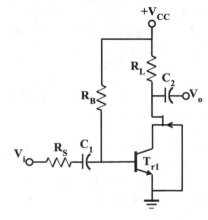

Fig. 6.64　BJT-JFET Cascode amplifier

verifies the use of the Cascode for wide-band amplification.

Ideas for Exploration (i) Simulate the system and measure frequency response of the circuit for various values of R_S. (ii) Replace the diode bias arrangement with a potential divider with filter capacitor and examine the system performance.

Cascode Amplifier Using BJT and JFET
The circuit in Fig. 6.64 is a Cascode amplifier comprising a 2 N3904 BJT in a common emitter configuration driving a JFET in a common gate mode. The first transistor is biased with a collector current of 1 mA using $R_B = 2.7$ M, while the gate of the JFET is connected to the emitter of the BJT as shown in Fig. 6.64. The 2 N3458 JFET with a 1 mA drain current has a gate-source voltage of −1.91 V which is sufficient for the proper operation of the BJT. A source resistance $R_S = 1$ k is used. The frequency response of the circuit was tested using a 10 mV sinusoidal input. The bandwidth of the system was measured at 2.2 MHz. This again is better than a common emitter amplifier with 1 mA current and the

same load resistor and source resistor which has a bandwidth of 343 kHz. Because the common emitter BJT is operating into the low input impedance source of the JFET, the significant multiplication of the collector-base capacitance of the BJT is prevented. The advantage of this form of the Cascode amplifier is that no additional biasing arrangements for the JFET are necessary.

Ideas for Exploration (i) Simulate the system and measure frequency response of the circuit for various values of R_S. (ii) Try different JFETS while ensuring that the gate-source voltage at the particular quiescent current is sufficient to properly operate the BJT.

Problems

1. Determine the input coupling capacitor for the circuit shown in Fig. 6.65 in order that the amplifier have a low frequency cutoff frequency of 150 Hz. Assume transistor $\beta = 125$.

2. For the amplifier circuit shown in Fig. 6.66 which is biased for maximum symmetrical

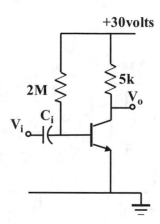

Fig. 6.65 Circuit for Question 1

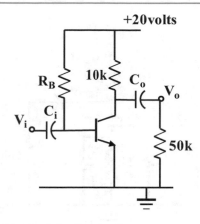

Fig. 6.67 Circuit for Question 3

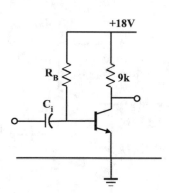

Fig. 6.66 Circuit for Question 2

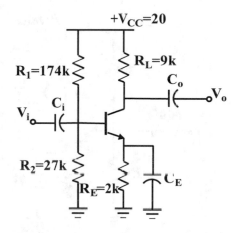

Fig. 6.68 Circuit for Question 4

swing, determine a suitable value of coupling capacitor to realize a maximum lower cut-off frequency of 70 Hz.

3. Determine the output coupling capacitor C_o for the circuit shown in Fig. 6.67 in order that the amplifier has a low-frequency cut-off frequency of 50 Hz. Assume transistor $\beta = 100$.

4. Determine C_E for the common emitter amplifier in Fig. 6.68 in order to produce a lower cut-off frequency of less than 25 Hz. Assume all coupling capacitors are large.

5. Determine the capacitors C_i, C_S and C_o in the circuit of Fig. 6.69 to give $f_L = 80$ Hz. Assume $h_{fe} = 200$.

6. Evaluate the lower cut-off frequency resulting from each of the capacitors shown in Fig. 6.70.

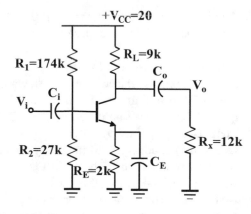

Fig. 6.69 Circuit for Question 5

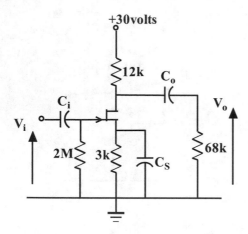

Fig. 6.70 Circuit for Question 6

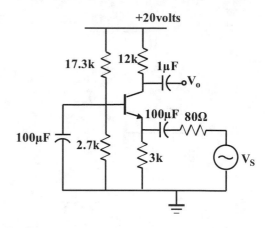

Fig. 6.72 Circuit for Question 10

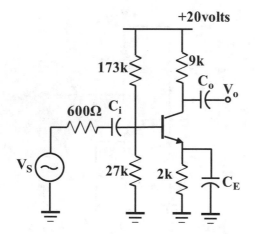

Fig. 6.71 Circuit for Question 9

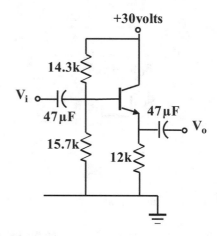

Fig. 6.73 Circuit for Question 11

7. A silicon NPN transistor has $f_T = 200$ MHz. Determine $C_{b'e}$ for a collector current of 2 mA.

8. State and verify Miller's theorem.

9. For the common emitter amplifier shown in Fig. 6.71, determine the upper cut-off frequency. The transistor used has $f_T = 250$ MHz, $C_{cb'} = 7$ pF and $\beta = 125$.

10. Determine the upper cut-off frequency for the common base amplifier shown in Fig. 6.72 where a 2 N3904 transistor is used.

11. Determine the upper cut-off frequency for the common collector amplifier shown in Fig. 6.73 where a 2 N3904 transistor is used.

12. Find the cut-off frequency of the common source amplifier shown in Fig. 6.74. The JFET has the following characteristics: $C_{gs} = 70$ pF, $C_{gd} = 8$ pF, $r_d = 150$ kΩ, $g_m = 10$ mA/V.

13. Draw the equivalent circuit of the circuit in Fig. 6.74.

14. For a common gate JFET amplifier with $R_S = 5$ k and $R_L = 6$ k, using a JFET having $C_{gd} = 4$ pF, find the upper cut-off frequency of the circuit.

15. Explain the excellent wideband characteristics of the Cascode amplifier

16. Sketch a common emitter amplifier in which local series feedback is used to improve the

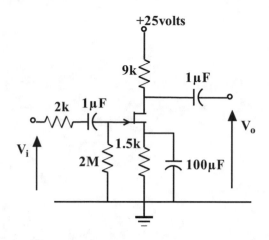

Fig. 6.74 Circuit for Question 12

amplifier bandwidth. For the maximum bandwidth, what should be the nature of the source resistance?

17. Sketch a common emitter amplifier in which local shunt feedback is used to improve the amplifier bandwidth. For the maximum bandwidth, what should be the nature of the source resistance? For the realization of a Cascode amplifier using a JFET as the second stage, what is the necessary characteristic of the JFET for this arrangement.

Bibliography

S.S. Hakim, *Feedback Circuit Analysis* (Iliffe Books Ltd, London, 1966)

J. Millman, C.C. Halkias, *Integrated Electronics: Analog and Digital Circuits and Systems* (McGraw Hill, New York, 1972)

A.S. Sedra, K.C. Smith, *Microelectronic Circuits*, 6th edn. (Oxford University Press, Oxford, 2011)

Feedback Amplifiers

7

A feedback amplifier is one in which a portion of the output is fed back to the system input where it is combined with the input signal. The basic concept is shown in Fig. 7.1. At the amplifier input, the feedback signal which may be a voltage or current is combined with the input signal which is also a voltage or current through a summing or mixing network, and the resulting signal is passed into the amplifier system. The summing network is either a series circuit which mixes feedback voltage with source voltage or a shunt circuit which mixes feedback current with source current while the feedback network is most often a passive network, usually resistive.

The basic feedback principle in feedback systems is the sampling of a portion of the output signal such that when mixed with the input improves the overall characteristics of the amplifier. When any increase in the output signal results in a feedback signal into the input in such a way as to cause a decrease in the output signal, the amplifier is said to have negative feedback. Negative feedback is generally useful because it can improve the characteristics of the various amplifier topologies presented earlier. For example, feedback can produce a significant improvement in the linearity and frequency response of the amplifier and lower distortion and noise. Another advantage of negative feedback is that it stabilizes the gain of the amplifier against variations in then device characteristics of the transistors or other active devices used in the amplifier. The price paid for these improvements is the lowered gain

of the amplifier with feedback (closed loop gain) compared with the gain of the amplifier without feedback (open-loop gain). Feedback can however result in amplifier instability, and therefore appropriate steps need to be taken to prevent this.

In this chapter we introduce the concept of feedback and apply it to amplifier systems. The resulting circuits display a range of improved characteristics, which are discussed. At the end of the chapter, the student will be able to:

- Explain the feedback concept
- Discuss the different types of amplifiers and feedback
- Design feedback amplifiers

7.1 Classification of Amplifiers

Amplifier systems may be conveniently classified into four types: voltage amplifiers, current amplifiers, transconductance amplifiers and trans-resistance amplifiers. This classification is based on the magnitudes of the input and output impedances of an amplifier relative to the source and load impedances, respectively.

7.1.1 Voltage Amplifier

The voltage amplifier accepts a voltage at its input and delivers a voltage at its output. It is

© Springer Nature Switzerland AG 2021
S. J. G. Gift, B. Maundy, *Electronic Circuit Design and Application*,
https://doi.org/10.1007/978-3-030-46989-4_7

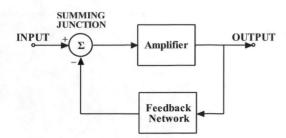

Fig. 7.1 Basic feedback system

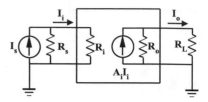

Fig. 7.3 Norton equivalent circuit of a current amplifier

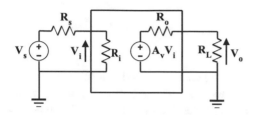

Fig. 7.2 Thevenin equivalent circuit of a voltage amplifier

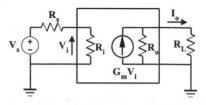

Fig. 7.4 Thevenin equivalent input and Norton equivalent output of a transconductance amplifier

convenient to represent the amplifier as a Thevenin equivalent circuit as shown in Fig. 7.2.

In order to amplify effectively, the input resistance R_i of the voltage amplifier must be high so that most of the input voltage is dropped across the input of the amplifier and little is lost across the source resistance R_s, i.e. $R_i \gg R_s$. In such a case, $V_i \approx V_s$. Ideally, the input resistance must be infinite. In order to effectively deliver the voltage to a load, the output impedance R_o of the system must be low so that most of the output voltage is delivered to the load R_L and little is dropped across the output impedance, i.e. $R_o \ll R_L$. Under these conditions, the output voltage is $V_o \simeq A_v V_i$. Then, the voltage at the amplifier output is proportional to the voltage at the input, the proportionality factor being independent of the magnitudes of the source and load resistances. Ideally, the output impedance of a voltage amplifier is zero.

7.1.2   Current Amplifier

The current amplifier accepts a current at its input and delivers a current at its output. It is convenient to represent this amplifier as a Norton equivalent circuit as shown in Fig. 7.3. In order to

effectively amplify the current signal, the input impedance R_i of the current amplifier must be low so that most of the input current flows into the input of the amplifier and little is lost through the source resistance R_s, i.e. $R_i \ll R_s$. In such circumstances, $I_i \simeq I_s$. In order to effectively deliver the amplified current to a load, the output impedance R_o of the system must be high so that most of the output current flows into the load R_L and little is lost through the output impedance. Under these conditions, the output current $I_o \simeq A_i I_i$, and the current at the output is proportional to the current input, the proportionality factor being independent of the magnitudes of the source and load resistances. Ideally, the input resistance of a current amplifier is zero, and the output resistance is infinite.

7.1.3   Transconductance Amplifier

We turn now to the transconductance amplifier which accepts a voltage at its input and delivers a current at its output. For this system, it is convenient to represent the amplifier as a Thevenin equivalent input and a Norton equivalent output as shown in Fig. 7.4. In order to operate effectively, the input resistance R_i must be high so that most of the input voltage is dropped across the input of the amplifier and little is lost across the source resistance R_s, i.e. $R_i \gg R_s$. For these conditions, the input voltage $V_i \simeq V_s$. In order to effectively deliver the output current to a load, the output resistance R_o of the system must be high so that most of the current flows into the load R_L and little flows into the shunt output R_o resulting in an output current $I_o \simeq G_M V_i$. Under these conditions, the output current is proportional to the input voltage, and the proportional factor is independent of the magnitudes of the source and load resistances. Ideally, the input and output resistances of a transconductance amplifier are both infinite.

7.1.4   Trans-resistance Amplifier

Finally, the trans-resistance amplifier accepts a current at its input and delivers a voltage at its output. It is convenient to represent this system as a Norton equivalent input and Thevenin equivalent output as shown in Fig. 7.5. In order to amplify effectively, the input resistance R_i of the trans-resistance amplifier must be low so that most of the input current flows into the input of the amplifier and little is lost through the source resistance R_s. In order to effectively deliver the output voltage to a load, the output resistance R_o of the system must be low so that most of this voltage is delivered to the load R_L and little is

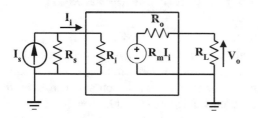

Fig. 7.5 Norton equivalent input and Thevenin equivalent output of a trans-resistance amplifier

dropped across the output impedance. Then, the input current $I_i \simeq I_s$ and the output voltage $V_o \simeq R_M I_i$. As a result the voltage at the output is proportional to the input current, and the factor of proportionality is independent of the magnitudes of the source and load resistances. Ideally, the input and output resistances of a trans-resistance amplifier are both zero. Table 7.1 summarizes the characteristics of the four amplifier types.

7.2   Feedback Amplifier Topologies

Feedback may be applied to anyone of the amplifier topologies presented in the previous section in order to improve their performance. Thus feedback around a voltage amplifier will increase its normally high input resistance and lower its normally low output resistance. The method of application depends on the topology, and in this section we explore such methods. In each of the four configurations considered, the output signal may be sampled using an appropriate sampling network and this signal applied at the input by way of a suitable feedback network. At the amplifier input, the feedback signal is mixed with the external signal being amplified via a summing network and the resulting signal passed into the amplifier. The summing network is either a series mixer in which feedback voltage is compared with source voltage or a shunt mixer in which feedback current is compared with source current.

Feedback around the four basic amplifier topologies is shown in Figs. 7.6, 7.8, 7.10 and 7.12. The voltage amplifier, trans-resistance amplifier, transconductance amplifier and current amplifier have, respectively, voltage-series feedback, voltage-shunt feedback, current-series feedback and current-shunt feedback.

Table 7.1 Amplifier characteristics

Parameter	Voltage	Current	Transconductance	Trans-resistance
Transfer function	$V_o = A_v V_i$	$I_o = A_i I_i$	$I_o = G_M V_i$	$V_o = R_M I_i$
Input resistance	∞	0	∞	0
Output resistance	0	∞	∞	0

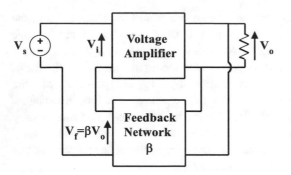

Fig. 7.6 Feedback voltage amplifier

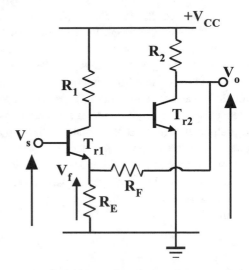

7.2.1   Voltage-Series Feedback

For the voltage amplifier in Fig. 7.6, the output signal being sampled is a voltage, and this sampling is done by connecting across the output.

The resulting feedback voltage V_f must be subtracted from the input voltage V_s by placing the signal in series with the input voltage. Note that the input and output signal voltages must be in phase for voltage subtraction to take place. The result is voltage-series feedback. From Fig. 7.6,

$$V_i = V_s - V_f \qquad (7.1)$$

where $V_f = \beta V_o$. The output voltage is given by

$$\begin{aligned} V_o &= A_v V_i = A_v\left(V_s - V_f\right) \\ &= A_v V_s - A_v \beta V_o \end{aligned} \qquad (7.2)$$

This gives

$$V_o(1 + \beta A_v) = A_v V_s \qquad (7.3)$$

Therefore, system voltage gain with feedback $A_f = \frac{V_o}{V_s}$ is given by

$$A_f = \frac{A_v}{1 + \beta A_v} \qquad (7.4)$$

The result (7.4) shows that voltage-series feedback around a voltage amplifier results in the amplifier gain reduced by the factor $(1 + \beta A_v)$. An example of the application of voltage-series feedback is shown in the simplified two-transistor circuit of Fig. 7.7 consisting of a common emitter amplifier Tr_1 driving another common emitter amplifier Tr_2. A portion of the output voltage at the collector of Tr_2 is returned through the feedback network $R_F R_E$ to the emitter of Tr_1 where it is subtracted from the input voltage.

Fig. 7.7 Voltage-series feedback

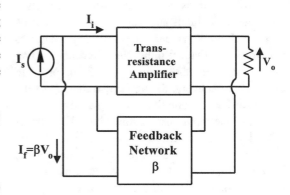

Fig. 7.8 Feedback trans-resistance amplifier

7.2.2   Voltage-Shunt Feedback

In the case of the trans-resistance amplifier of Fig. 7.8, the output signal being sampled is a voltage and again the sampling is accomplished by connecting across the output. However, the sampled voltage must first be converted to a current since it must be subtracted from the input signal current. The subtraction is effected by connecting the feedback current in shunt with the input current. Note that the output voltage must be inverted relative to the voltage associated with the input current for current subtraction to take place. The result is voltage-shunt feedback. From Fig. 7.8

$$I_i = I_s - I_f \qquad (7.5)$$

where $I_f = \beta V_o$. The output voltage is given by

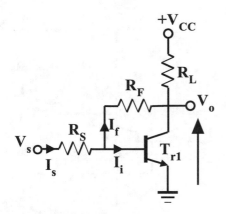

Fig. 7.9 Voltage-shunt feedback

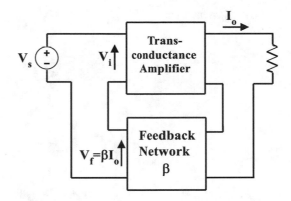

Fig. 7.10 Feedback transconductance amplifier

$$V_o = R_M I_i = R_M(I_s - I_f)$$
$$= R_M I_s - R_M \beta V_o \qquad (7.6)$$

This gives

$$V_o(1 + \beta R_M) = R_M I_s \qquad (7.7)$$

Therefore, system trans-resistance with feedback $R_f = \frac{V_o}{I_s}$ is given by

$$R_f = \frac{R_M}{1 + \beta R_M} \qquad (7.8)$$

The result (7.8) shows that voltage-shunt feedback around a trans-resistance amplifier results in the amplifier trans-resistance reduced by the factor $(1 + \beta R_M)$. An example of the application of voltage-shunt feedback is the single transistor common emitter amplifier shown in Fig. 7.9, where the output voltage at the collector of the transistor develops a feedback current in resistor R_F which is connected between the collector and base of the transistor.

7.2.3 Current-Series Feedback

For the transconductance amplifier in Fig. 7.10, the output signal is a current rather than a voltage, and sampling this signal requires connecting directly in the current output path, i.e. in series. The sampled current must then be converted to a voltage in order to subtract it from the input signal

voltage by connecting in series. The input voltage must be in phase with the associated output voltage for the voltage subtraction process to take place. The result is current-series feedback. From Fig. 7.10,

$$V_i = V_s - V_f \qquad (7.9)$$

where $V_f = \beta I_o$. The output current is given by

$$I_o = G_M V_i = G_M(V_s - V_f)$$
$$= G_M V_s - G_M \beta I_o \qquad (7.10)$$

This gives

$$I_o(1 + \beta G_M) = G_M V_s \qquad (7.11)$$

Therefore, system transconductance with feedback $G_f = \frac{I_o}{V_s}$ is given by

$$G_f = \frac{G_M}{1 + \beta G_M} \qquad (7.12)$$

The result (7.12) shows that voltage-shunt feedback around a transconductance amplifier results in the amplifier transconductance reduced by the factor $(1 + \beta G_M)$. An example of the application of this type of feedback is the single transistor common emitter circuit of Fig. 7.11. Here the output current I_o is sampled by the emitter resistor R_E where a feedback voltage V_f is developed. This voltage appears in series with the input voltage V_s from which it is subtracted. It should be noted that the output current develops a voltage across the collector resistor R_L which is available at the collector of the transistor.

Fig. 7.11 Current-shunt
feedback

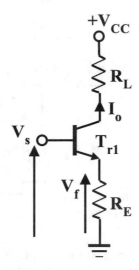

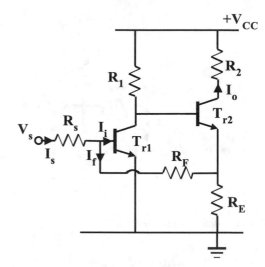

Fig. 7.13 Current-shunt feedback

$$I_o = A_i I_i = A_i (I_s - I_f) = A_i I_s - A_i \beta I_o \quad (7.14)$$

This gives

$$I_o(1 + \beta A_i) = A_i I_s \quad (7.15)$$

Therefore, system current gain with feedback
$A_f = \frac{I_o}{I_s}$ is given by

$$A_f = \frac{A_i}{1 + \beta A_i} \quad (7.16)$$

The result (7.16) shows that current-shunt feedback around a current amplifier results in the
amplifier current gain reduced by the factor
$(1 + \beta A_i)$. An example of current-shunt feedback
is demonstrated in the simplified two-transistor
circuit of Fig. 7.13. The output current I_o is sampled by the feedback network $R_F R_E$ and a feedback current I_f flows through R_F. This current is
subtracted from the input current.

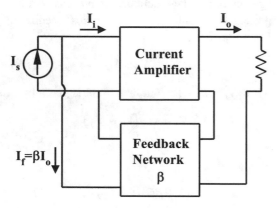

Fig. 7.12 Feedback current amplifier

7.2.4   Current-Shunt Feedback

Finally, for the current amplifier in Fig. 7.12, the
output signal is again a current, and therefore the
sampling must be effected by connecting in the
signal path, i.e. in series. The sampled current is
then subtracted from the input current by
connecting the feedback current in shunt with
the input current. For the current subtraction to
take place, the associated input and output
voltages must be out of phase. The resulting connection is referred to as current-shunt feedback.
From Fig. 7.12

7.3   Transfer Gain with Feedback

In this section, we discuss the transfer function of
the various amplifiers with negative feedback. We
do so using a generalized amplifier system shown
in Fig. 7.14 which represents the four amplifier
types considered, namely, voltage, current,
transconductance and trans-resistance. In the

$$I_i = I_s - I_f \quad (7.13)$$

where $I_f = \beta I_o$. The output current is given by

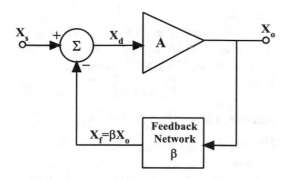

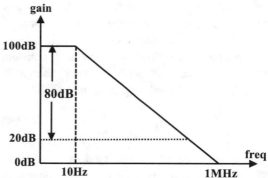

Fig. 7.14 Basic feedback amplifier system

Fig. 7.15 Open-loop response of 741 op amp

diagram, the (open-loop) transfer gain A is for the four amplifier types Λ_V, Λ_I, G_M, R_M, while the feedback network has a reverse transmission factor β. For the amplifier, we have

$$X_o = AX_d \quad (7.17)$$

$$X_f = \beta X_o \quad (7.18)$$

and

$$X_d = X_S - X_f \quad (7.19)$$

where X_S is the input signal, X_o is the output signal, X_f is the feedback signal and X_d is the error signal. Using (7.19) and (7.18) in (7.17) gives the closed-loop transfer gain $A_f \equiv \frac{X_o}{X_S}$ as

$$A_f = \frac{A}{1 + \beta A} \quad (7.20)$$

Note that for negative feedback, $|1 + \beta A| > 1$ and therefore $|A_f| < |A|$, i.e. the closed-loop gain of the amplifier, is less than the open-loop gain. The opposite is true for positive feedback as it occurs in oscillators. For the feedback system, the quantity $A\beta$ is called the loop gain and represents the gain experienced by a signal that travels through the amplifier A and back to the input via the feedback network β. Thus from (7.20),

$$20 \log A_f = 20 \log A - 20 \log (1 + A\beta)$$
$$\approx 20 \log A - 20 \log A\beta \quad (7.21)$$

$$20 \log A\beta = 20 \log A - 20 \log A_f \quad (7.22)$$

i.e. the loop gain in decibels is equal to the difference between the open-loop gain in decibels and the closed-loop in decibels. The loop gain is therefore an indication of the amount of feedback around the amplifier.

A simple application of this involves operational amplifiers. In the 741, for example, the open-loop response is shown in Fig. 7.15 with a low-frequency gain of 10^5 or 100 dB. For a closed-loop gain of 10 corresponding to 20 dB, it follows that the amount of feedback applied to the op-amp is $100dB - 20dB = 80dB$. If the closed-loop gain is reduced to 1 or 0 dB, then the feedback level increases to 100 dB.

Three conditions must be satisfied for the above feedback analysis to hold true:

(i) The input signal is transmitted to the output through the amplifier A and not through the network β.

(ii) The feedback signal is transmitted from the output to the input through the β block and not through the amplifier A.

(iii) The reverse transmission factor β of the feedback network is independent of the load and the source resistances.

7.4   Gain Stabilization Using Negative Feedback

The components in an amplifier system are likely to experience changes in values arising from environmental changes in temperature, ageing, component replacement and other factors. These changes in component values will generally result in changes in the gain of the amplifier. In many applications, amplifier gain changes are unacceptable. Feedback stabilizes amplifier gain against component value change. To see this, consider the basics amplifier with feedback shown in Fig. 7.14. The closed-loop gain from (7.20) is

$$A_f = \frac{A}{1 + \beta A} \qquad (7.23)$$

In the presence of a large amount of negative feedback, the loop gain $A\beta$ is much greater than one and therefore (7.23) reduces to

$$A_f \approx \frac{A}{\beta A} = \frac{1}{\beta} \qquad (7.24)$$

i.e. for $A\beta \gg 1$, the closed-loop gain is $1/\beta$ and is independent of the open-loop gain A. Since β can be made independent of A, it follows from (7.24) that for $A\beta \gg 1$, changes in A do not affect the closed-loop gain A_f. The feedback renders the closed-loop amplifier virtually immune to changes in open-loop gain.

In order to quantify this gain-stabilizing effect, consider again Eq. (7.23). For a constant feedback factor β, differentiating A_f with respect to A gives

$$\frac{dA_f}{dA} = \frac{(1 + \beta A) - \beta A}{(1 + \beta A)^2} = \frac{1}{(1 + \beta A)^2} \qquad (7.25)$$

Equation (7.25) leads to

$$dA_f = \frac{dA}{(1 + \beta A)^2} \qquad (7.26)$$

Using Eq. (7.23), we have

$$\frac{dA_f}{A_f} = \frac{dA/(1 + \beta A)^2}{A/(1 + \beta A)} = \frac{1}{(1 + \beta A)} \frac{dA}{A} \qquad (7.27)$$

The absolute value of Eq. (7.27) is

$$\left| \frac{dA_f}{A_f} \right| = \frac{1}{(1 + \beta A)} \left| \frac{dA}{A} \right| \qquad (7.28)$$

Equation (7.28) shows that the magnitude of the fractional change in gain with feedback $|dA_f/A_f|$ is reduced by the factor $1 + A\beta \approx A\beta$ or the loop gain, compared to the fractional change without feedback $|dA/A|$.

Example 7.1

An amplifier with gain 1000 has a gain change of 20% due to temperature. Calculate the change in gain of the feedback amplifier if the feedback factor $\beta = 0.1$.

Solution Using Eq. (7.28), $\left| \frac{dA_f}{A_f} \right| = \frac{1}{(1 + \beta A)} \times \left| \frac{dA}{A} \right| = \frac{1}{(1 + 0.1 \times 1000)} \times 20 = 0.198 \approx 0.2\%$.

Thus, while the amplifier open-loop gain changes from 1000 by 20%, the closed-loop amplifier changes by only 0.2%. This represents a 100-fold improvement. The factor $D = 1 + A\beta$ is sometimes called the de-sensitivity.

7.5   Increase in Bandwidth Using Negative Feedback

Consider an amplifier with low-frequency gain k and bandwidth f_o given by

$$A = \frac{k}{1 + jf/f_o} \qquad (7.29)$$

Then substituting (7.29) in (7.23) gives

$$A_f = \frac{A}{1 + \beta A} = \frac{\frac{k}{1 + jf/f_o}}{1 + \beta \frac{k}{1 + jf/f_o}} \qquad (7.30)$$

This becomes

$$A_f = \frac{k}{1 + k\beta + jf/f_o}$$
$$= \frac{k/(1 + \beta k)}{1 + jf/f_o(1 + k\beta)} \qquad (7.31)$$

This finally reduces to

$$A_f = \frac{k'}{1 + jf/f_o'} \qquad (7.32)$$

where

$$k' = k/(1 + \beta k) \qquad (7.33)$$

and

$$f_o' = f_o(1 + \beta k) \qquad (7.34)$$

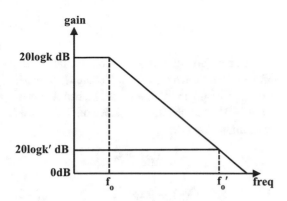

Fig. 7.16 Amplifier with negative feedback

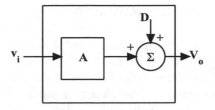

Fig. 7.17 Amplifier without feedback

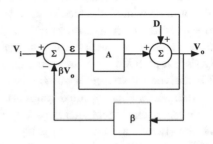

Fig. 7.18 Amplifier with feedback

From (7.33), the closed-loop low-frequency gain is $k' = \frac{k}{1+\beta k} \approx \frac{1}{\beta}$ for $k\beta \gg 1$, and the closed-loop bandwidth is $f_o' = f_o(1 + \beta k) \approx f_o k\beta$ again for $k\beta \gg 1$. This means that the bandwidth of the amplifier is increased by the factor $k\beta$ as a result of the negative feedback. This occurs at the expense of low-frequency gain which is reduced by the same factor shown in Fig. 7.16.

Example 7.2

An amplifier with a transfer function (7.29) has a low-frequency open-loop gain of 100,000 and an open-loop bandwidth of 10 Hz. For a closed-loop gain of 10, calculate the closed-loop bandwidth of the amplifier.

Solution For this amplifier $f_o = 10$ Hz and $k = 100, 000$. Since $k \gg 1$, to a very good approximation the closed-loop gain is $1/\beta$ giving $\beta = 0.1$. Therefore, the closed-loop bandwidth $f_o' = f_o(1 + \beta k) \approx f_o k\beta = 10 \times 100, 000 \times 0.1 = 100, 000$ Hz.

7.6   Feedback and Harmonic Distortion

Distortion in an amplifier results from nonlinearity associated with the active elements in the system. This nonlinearity increases as the amplitude of the signal increases and is therefore greatest at the output of the amplifier. The effect is the generation of harmonics of the input signal that were not present originally. This is called harmonic distortion which can also be reduced by the use of negative feedback. Consider the open-loop amplifier shown in Fig. 7.17.

The distortion produced in this amplifier is represented as the signal D added at the output. This assumes that the amplifier is approximately linear such that the principle of superposition applies. Thus,

$$V_o = Av_i + D \qquad (7.35)$$

If feedback is now applied as shown in Fig. 7.18, then

$$V_o = \varepsilon A + D \qquad (7.36)$$

and

$$\varepsilon = V_i - \beta V_o \qquad (7.37)$$

Substituting (7.37) in (7.36) yields

$$V_o = \frac{AV_i}{1 + A\beta} + \frac{D}{1 + A\beta} \qquad (7.38)$$

From (7.38) and (7.35), it can be seen that for the same output V_o, the distortion D produced at the output has been reduced by a factor of the loop gain. Negative feedback reduces output-generated noise by the same factor as it does output stage distortion.

7.7   Input Resistance

Negative feedback directly influences the input resistance of an amplifier. The nature of that influence depends on the method of application of the feedback. Specifically, if the feedback is series applied (voltage), then since the feedback voltage opposes the input signal voltage, it follows that the input current is reduced from what it would be without the feedback, and therefore the overall effect is an increase in input resistance. This result, which we will quantify shortly, is not affected by whether the feedback signal is derived by voltage or current sampling. If the feedback is shunt applied (current), then since the feedback current increases the signal current, the effect is a decrease in the input resistance to this type of amplifier. Again, the effect is independent of the method of sampling of the feedback signal. In the sections to follow, the effect of negative feedback on amplifier input resistance is examined and quantified.

7.7.1   Voltage-Series Feedback

The Thevenin equivalent of the voltage-series feedback amplifier is shown in Fig. 7.19. Since the amplifier input resistance is being considered, the source resistance is set to zero. The amplifier input resistance R_{if} is then given by $R_{if} = V_s/I_i$. In order to determine this, we have

$$V_s = V_f + V_i = \beta V_o + I_i R_i \qquad (7.39)$$

and

$$V_o = \frac{A_v V_i R_L}{R_o + R_L} = A_V I_i R_i \qquad (7.40)$$

where

$$A_V = \frac{V_o}{V_i} = \frac{A_v R_L}{R_o + R_L} \qquad (7.41)$$

is the open-loop gain of the amplifier, taking the effect of the external load into account, and A_v is the open-loop gain with no load. Hence $A_V \rightarrow A_v$ as $R_L \rightarrow \infty$. Substituting (7.40) in (7.39) and rearranging we get

$$\begin{aligned} R_{if} &= \frac{V_s}{I_i} = \frac{V_i(1 + \beta A_V)}{I_i} \\ &= R_i(1 + \beta A_V) \qquad (7.42) \end{aligned}$$

From (7.42) it can be seen that the input resistance of the amplifier is increased by the loop gain value as a result of the negative feedback.

7.7.2   Voltage-Shunt Feedback

The Norton equivalent input-Thevenin equivalent output representation of the voltage-shunt feedback amplifier is shown in Fig. 7.20. Since the amplifier input resistance is being considered, the source resistance is set to infinity (removed). In order to determine the amplifier input resistance R_{if} from Fig. 7.20, we have

$$I_s = I_f + I_i = \beta V_o + I_i \qquad (7.43)$$

and

$$V_o = \frac{R_m I_i R_L}{R_o + R_L} = R_M I_i \qquad (7.44)$$

where

$$R_M = \frac{V_o}{I_i} = \frac{R_m R_L}{R_o + R_L} \qquad (7.45)$$

is the open-loop trans-resistance of the amplifier, taking the effect of the external load into account

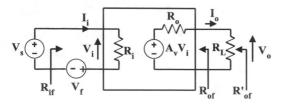

Fig. 7.19 Thevenin Equivalent of Voltage-Series Feedback Amplifier

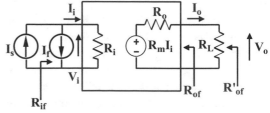

Fig. 7.20 Voltage-Shunt Feedback Amplifier

and R_m is the open-loop gain with no load. Therefore $R_M \rightarrow R_m$ as $R_L \rightarrow \infty$. Substituting (7.44) in (7.43) and rearranging, we get

$$R_{if} = \frac{V_i}{I_s} = \frac{V_i}{I_i(1 + \beta R_M)} = \frac{R_i}{1 + \beta R_M} \quad (7.46)$$

From (7.46) it can be seen that the input resistance of the amplifier is lowered by the factor of the loop gain as a result of the negative feedback.

7.7.3   Current-Series Feedback

The Thevenin equivalent input-Norton equivalent output representation of the current-series feedback amplifier is shown in Fig. 7.21. Since the amplifier input resistance is being considered, the source resistance is again set to zero. In order to determine the amplifier input resistance R_{if} from Fig. 7.21, we have

$$V_s - V_f + V_i - \beta I_o + V_i \quad (7.47)$$

and

$$I_o = \frac{G_m V_i R_o}{R_o + R_L} = G_M V_i \quad (7.48)$$

where

$$G_M = \frac{I_o}{V_i} = \frac{G_m R_o}{R_o + R_L} \quad (7.49)$$

is the open-loop transconductance of the amplifier, taking the effect of the external load into account, and G_m is the open-loop gain with no load. Therefore $G_M \rightarrow G_m$ as $R_L \rightarrow 0$. Substituting (7.48) in (7.47) and rearranging, we get

$$R_{if} = \frac{V_s}{I_i} = \frac{V_i(1 + \beta G_M)}{I_i}$$
$$= R_i(1 + \beta G_M) \quad (7.50)$$

From (7.50) it can be seen that the input resistance of the amplifier is increased by the factor of the loop gain as a result of the negative feedback.

7.7.4   Current-Shunt Feedback

The Norton equivalent of the current-shunt feedback amplifier is shown in Fig. 7.22. Since the amplifier input resistance is being considered, the source resistance is once again set to infinity (removed). In order to determine the amplifier input resistance R_{if} from Fig. 7.22, we have

$$I_s = I_f + I_i = \beta I_o + I_i \quad (7.51)$$

and

$$I_o = \frac{A_i I_i R_o}{R_o + R_L} = A_I I_i \quad (7.52)$$

where

$$A_I = \frac{I_o}{I_i} = \frac{A_i R_o}{R_o + R_L} \quad (7.53)$$

is the open-loop current gain of the amplifier, taking the effect of the external load into account, and I_i is the open-loop gain with no load. Therefore $A_I \rightarrow A_i$ as $R_L \rightarrow 0$. Substituting (7.52) in (7.51) and rearranging, we get

$$R_{if} = \frac{V_i}{I_s} = \frac{V_i}{I_i(1 + \beta A_I)} = \frac{R_i}{1 + \beta A_I} \quad (7.54)$$

From (7.54) it can be seen that the input resistance of the amplifier is decreased by the factor of the loop gain as a result of the negative feedback.

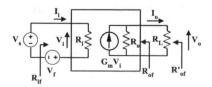

Fig. 7.21 Current-Series Feedback Amplifier

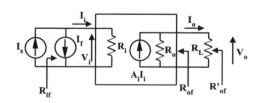

Fig. 7.22 Current-shunt feedback amplifier

7.8   Output Resistance

Negative feedback also directly influences the output resistance of an amplifier. The nature of that influence depends on the method of deriving the feedback. In particular, if the feedback is acquired by sampling an output voltage, then regardless of the method of application of the feedback, the output resistance is reduced in an attempt to stabilize that output voltage. If on the other hand the feedback is acquired by sampling an output current, then the effect is an increase in output resistance in an attempt to maintain the output current. Again, the effect is independent of the method of application of the feedback signal. In the following sections, the effect of negative feedback on amplifier output resistance is examined and quantified.

7.8.1   Voltage-Series Feedback

Using the Thevenin equivalent of the voltage-series feedback amplifier Fig. 7.19, we determine the output resistance R_{of} of this topology. In order to do so, the load R_L must be disconnected and the input signal removed by setting $V_s = 0$. A voltage V is then applied to the output terminals as in Fig. 7.23 and the resulting current I flowing into the output terminal determined. Then the ratio V/I gives the output resistance R_{of}. From Fig. 7.19, we have

$$I = \frac{V - A_v V_i}{R_o} = \frac{V + A_v \beta V}{R_o} \quad (7.55)$$

Since with $V_s = 0$, $V_i = -\beta V$. It follows from (7.55) that R_{of} is given by

$$R_{of} \equiv \frac{V}{I} = \frac{R_o}{1 + \beta A_v} \quad (7.56)$$

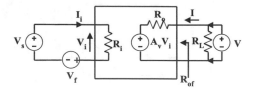

Fig. 7.23 Determining output resistance

This result confirms that the output resistance R_{of} of the amplifier with feedback is reduced. It is reduced by the factor $1 + \beta A_v$. R_{of} is the output resistance seen by the load R_L. If R_L is treated as part of the amplifier, then the output resistance with feedback becomes R_{of} in parallel with R_L, that is

$$R'_{of} = \frac{R_{of} R_L}{R_{of} + R_L} \quad (7.57)$$

Substituting for R_{of} from (7.56) yields

$$R'_{of} = \frac{\frac{R}{1+\beta A_v} R_L}{\frac{R_o}{1+\beta A_v} + R_L} = \frac{R_o R_L}{R_o + R_L + \beta A_v R_L} \quad (7.58)$$

This reduces to

$$R'_{of} = \frac{R_o R_L / (R_o + R_L)}{1 + \beta A_v R_L / (R_o + R_L)}$$

$$= \frac{R'_o}{1 + \beta A_V} \quad (7.59)$$

Here $R'_o = \frac{R_o R_L}{R_o + R_L}$ is the parallel resistance of R_o and R_L representing the amplifier output resistance with feedback with R_L treated as part of the amplifier, and $A_V = A_v R_L / (R_o + R_L)$ is the amplifier gain with feedback, taking the effect of R_L into account. Note that (7.59) takes the same form as (7.56).

7.8.2   Voltage-Shunt Feedback

Using the Norton equivalent input-Thevenin equivalent output representation of the voltage-shunt feedback amplifier in Fig. 7.20, we determine the output resistance R_{of} of this topology using a similar approach. We again disconnect the load R_L from the amplifier output and set the input signal $I_s = 0$. With a voltage V applied to the output the resulting current I flowing into the output terminal determined is given by

$$I = \frac{V - R_m I_i}{R_o} = \frac{V + R_m \beta V}{R_o} \quad (7.60)$$

since with $I_s = 0$, $I_i = -\beta V$. It follows from (8.44) that R_{of} is given by

$$R_{of} \equiv \frac{V}{I} = \frac{R_o}{1 + \beta R_m} \tag{7.61}$$

This result shows that the output resistance R_{of} of the amplifier without feedback is reduced by the factor $1 + \beta R_m$. If R_L is treated as part of the amplifier, then the output resistance with feedback becomes R_{of} in parallel with R_L, which reduces to

$$R'_{of} = \frac{R'_o}{1 + \beta R_M} \tag{7.62}$$

Here R'_o is again the parallel resistance of R_o and R_L representing the amplifier output resistance without feedback with R_L treated as part of the amplifier, and $R_M = R_m R_L/(R_o + R_L)$ is the amplifier trans-resistance without feedback, taking the effect of R_L into account.

7.8.3   Current-Series Feedback

We here use the Thevenin equivalent input-Norton equivalent output representation of the current-series feedback amplifier shown in Fig. 7.21 in order to determine the output resistance. We disconnect the load R_L from the amplifier output and set the input signal $V_s = 0$. A voltage V applied to the output produces a current I flowing into the output terminal. With $V_s = 0$, $V_f = -\beta I$ and therefore $V_i = -V_f = \beta I$. Hence the current I is given by

$$I = \frac{V}{R_o} - G_m V_i = \frac{V}{R_o} - G_m \beta I \tag{7.63}$$

It follows from (7.63) that R_{of} is given by

$$R_{of} \equiv \frac{V}{I} = R_o(1 + \beta R_m) \tag{7.64}$$

Result (7.64) indicates that the output resistance of the amplifier is increased by a factor of the loop gain of the amplifier.

7.8.4   Current-Shunt Feedback

In analysing the current-shunt feedback configuration, we use the Norton equivalent representation of the feedback amplifier shown in Fig. 7.22.

We disconnect the load R_L from the amplifier output and set the input signal $I_s = 0$. A voltage V applied to the output produces a current I flowing into the output terminal. Since for $I_s = 0$ $I_f = -\beta I$ and therefore $I_i = -I_f = \beta I$, then the current I is given by

$$I = \frac{V}{R_o} - A_i I_i = \frac{V}{R_o} - A_i \beta I_i \tag{7.65}$$

It follows from (7.65) that R_{of} is given by

$$R_{of} \equiv \frac{V}{I} = R_o(1 + \beta A_i) \tag{7.66}$$

Again, the current sampling in this amplifier configuration results in an increased amplifier output resistance.

7.9   Analysis of Feedback Amplifiers

In this section we apply negative feedback to actual transistor amplifier circuits and discuss methods of analysing the various configurations. Specifically, we consider the voltage-series, voltage-shunt, current-shunt and current-series modes of feedback systems.

In analysing a feedback amplifier, the open-loop gain A and the feedback factor β need to be determined. These two parameters allow the determination of the closed-loop gain and other characteristics of the closed-loop system. The open-loop transfer function A represents the gain of the amplifier with no feedback but considering the loading of the β network. In order to determine A, the following rules must be applied:

Input Circuit
1. For voltage sampling, set the output voltage to zero. This corresponds to short-circuiting the output terminal to ground.
2. For current sampling, set the output current to zero. This corresponds to open-circuiting the output terminal.

Output Circuit
1. For series mixing, set the input current to zero. This corresponds to opening the input terminal.

2. For shunt mixing, set the input voltage to zero. This corresponds to short-circuiting the input terminal.

These steps ensure that the negative feedback is reduced to zero while still taking into account the loading of the feedback circuitry on the basic amplifier.

7.9.1   Voltage-Series Feedback

Consider the voltage amplifier comprising two common emitter amplifiers that are cascaded to produce a higher overall voltage gain as shown in Fig. 7.24. The biasing circuitry is omitted for simplicity. We first note that because of the signal inversion produced by each stage, the output voltage at the collector of the second stage is in phase with the input voltage at the base of the first stage. We note also that a signal injected at the emitter of the first stage will produce an inverted output signal and therefore the feedback signal must be applied at the emitter of Tr_1. Voltage sampling is thus effected by sampling the output voltage at the collector of the second stage and applying series feedback mixing with the input voltage at the emitter of the first stage. The β network therefore comprises resistors R_F and R_E. One of the assumptions necessary for the

feedback analysis to hold is that there must be no signal feed-forward through the feedback network. The emitter current of $Tr1$ is approximately equal to the input current and flows forward through the feedback network $R_E//R_F$. However, this current is generally small as compared with the current I_F from the output. In order to determine the basic amplifier without feedback, we first determine the input circuit of the amplifier by setting the output voltage to zero. The effect of this is that R_F appears in parallel with R_E. Following this, the output circuit is determined by setting the input current to zero. This corresponds to opening the connection of the junction of R_E and R_F to the emitter of $Tr1$ with the result that R_F appears in series with R_E. The resulting open-loop voltage amplifier is shown in Fig. 7.25. The feedback factor β is given by

$$\beta = \frac{V_f}{V_o} = \frac{R_E}{R_F + R_E} \qquad (7.67)$$

The amplifier open-loop gain can now be found in the usual way following which feedback analysis can proceed. It should be noted that there is (local) current-series feedback in the first stage of the amplifier because of the presence of the un-bypassed emitter resistance. This kind of feedback is analysed shortly. An example follows.

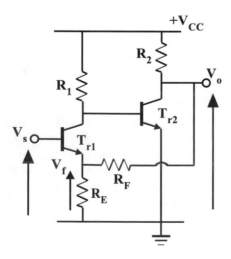

Fig. 7.24 Voltage-series feedback amplifier

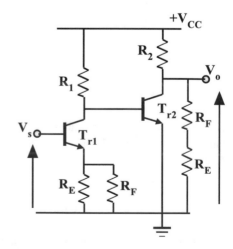

Fig. 7.25 Open-loop amplifier circuit with feedback network loading

Fig. 7.26 Voltage-series
feedback amplifier

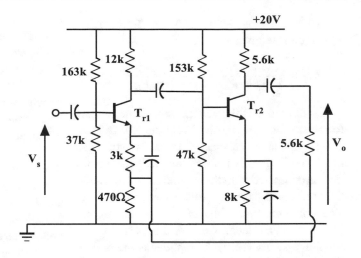

Example 7.3

Evaluate the voltage gain, input resistance and
output resistance for the feedback voltage amplifier in Fig. 7.26. For the transistors $h_{fe} = 100$,
$h_{ie} = 2.5k$, $h_{re} = h_{oe} = 0$.

Solution

(i) The first step is the evaluation of the open-
loop gain of the amplifier, taking into
account the loading effect of the feedback
network. The effective load R_{L1} of Tr_1 is
$R_{L1} = 12$ k//153 k//47 k//2.5 k $= 1.96$ k.
The effective emitter resistance R_e of Tr_1 is
$R_{e1} = 470//5.6$ k $= 434$ Ω. Hence the voltage gain A_{V1} of the first stage is $A_{V1} =$
$-\frac{h_{fe}R_{L1}}{h_{ie} + (1 + h_{fe})R_{e1}} = -\frac{100 \times 1.96\,k}{2.5\,k + 101 \times 0.434\,k} = -4.27$.
The effective load R_{L2} of Tr_2 is $R_{L2} = 5.6$ k//
$(5.6 + 0.47)$ k $= 2.91$ k. Hence the voltage
gain A_{V2} of the second stage is $A_{V2} =$
$-\frac{h_{fe}R_{L2}}{h_{ie}} = -\frac{100 \times 2.91\,k}{2.5\,k} = -116.4$. Therefore,
the overall open-loop voltage gain A_V is
$A_V = A_{V1} \times A_{V2} = 4.27 \times 116.4 = 497$.
Now $\beta = \frac{R_E}{R_F + R_E} = \frac{470}{5600 + 470} = 0.077$. Also
$1 + \beta A_V = 1 + 0.077 \times 497 = 39.3$. There-
fore the closed-loop voltage gain A_{Vf} is
found to be $A_{Vf} = \frac{A_V}{1 + \beta A_V} = \frac{497}{39.3} = 12.6$.
This is quite close to the value for very
large open-loop gain given by
$A_{Vf}(A_V \to \infty) = 1/\beta = 12.9$.

(ii) Looking in at the base of the first transistor
(i.e. omitting for the moment the two base
biasing resistors), the open-loop input resistance R_i is $R_i = h_{ie} + (1 + h_{fe})R_{e1} = 2.5$
k $+ 101 \times 0.47$ k $= 49.97$ k. There the
input resistance R_{if} of the closed-loop amplifier is $R_{if} = (1 + \beta A_V)$
$R_i = 39.3 \times 49.97k = 1.96M$. The effective
input resistance to the closed-loop amplifier
would now be $R_{if}//163$ k//37 k, i.e. R_{if} in
parallel with the two base biasing resistors
which are outside the feedback loop of the
amplifier.

(iii) The open-loop input resistance R_o is
$R_o = R_{L2} = 2.91k$. Therefore, the closed-
loop output resistance R_{of} is $R_{of} = \frac{R_o}{1 + \beta A_V} = \frac{2.91\,k}{39.3} = 74\,Ω$.

Another voltage amplifier in which voltage-
series feedback is employed is the emitter follower. In this circuit the full output voltage
constitutes the feedback voltage and is subtracted
from the input voltage. The error voltage is that
dropped across the base emitter junction of the
transistor. It is possible to apply the feedback
analysis to this circuit to arrive at the gain expression and other characteristics. This analysis
shows that the emitter follower can be viewed as
a common emitter amplifier with collector resistor
R_E and open-loop gain $A_V = \frac{h_{fe}R_E}{h_{ie}} = g_m R_E$ around
which 100% voltage-series feedback
corresponding to $\beta = 1$ is applied. The resulting

closed-loop voltage gain is given by $A_{Vf} = \frac{A_V}{1+\beta A_V}$.
This therefore gives

$$A_{Vf} = \frac{h_{fe}R_E/h_{ie}}{1 + \beta h_{fe}R_E/h_{ie}} = \frac{g_m R_E}{1 + g_m R_E} \quad (7.68)$$

which of course is the voltage gain of the emitter
follower found previously. The circuit's increased
input impedance, decreased output impedance,
improved frequency response and reduced distortion as compared with the common emitter amplifier also result from this voltage-series feedback.
The common source amplifier is an analogous
case using FET technology. A third example of
a voltage-series feedback amplifier is the operational amplifier which will be discussed in
Chap. 8.

7.9.2   Current-Shunt Feedback

An example of current-shunt feedback is the current amplifier comprising two common emitter
amplifiers as shown in Fig. 7.27. The biasing
circuitry is again omitted for simplicity. We note
that because of the signal inversion produced by
the first stage, the signal output at the emitter of
the second stage is out of phase with the input
signal at the base of the first stage. Current sampling of the output current in Tr_2 is achieved by

inclusion of the resistor R_E in the emitter of Tr_2
through which the output current I_o flows. In
conjunction with resistor R_F, the two resistors
form a current divider (corresponding to the voltage divider in the voltage-series feedback case)
which results in a feedback current I_f flowing
away from the base of Tr_1 and subtracting from
the input current I_S. Because the output current is
sampled and a feedback current is mixed at the
input, this arrangement is current-shunt feedback.
The β network therefore comprises resistors R_F
and R_E. Note however that in this particular configuration, the output current is converted to a
voltage in R_2 giving $V_o = I_o R_2$. Resistor R_2 is
however outside the feedback loop.

To determine the basic current amplifier without feedback, we determine the input circuit of the
amplifier by reducing the output current to zero.
This is done by opening the output current loop at
the emitter of Tr_2. The effect of this is that R_E
appears in series withR_F from the base of Tr_1 to
ground. The output circuit is determined by
setting the input voltage to zero. This corresponds
to short-circuiting the input terminal of Tr_1 to
ground. The result is that R_F appears in parallel
with R_E at the emitter of Tr_2. The resulting open-
loop amplifier is shown in Fig. 7.28. The signal
source is represented as a Norton equivalent circuit since the feedback signal is a current. The

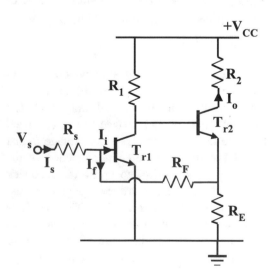

Fig. 7.27 Current-shunt feedback amplifier

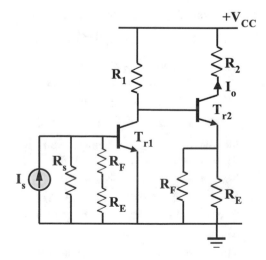

Fig. 7.28 Open-loop amplifier circuit with feedback network loading

open-loop current gain can now be found. Noting that the collector current is much larger than the base current of Tr_2, it follows that

$$I_f = \frac{R_E}{R_F + R_E} I_o \qquad (7.69)$$

From this the feedback factor beta is given by

$$\beta = \frac{I_f}{I_o} = \frac{R_E}{R_F + R_E} \qquad (7.70)$$

Because the input resistance of this feedback amplifier is low, the circuit is easily converted into a voltage amplifier by driving the input with a voltage signal V_S operating into a resistor $R_S \gg R_{if}$ which gives $I_S = V_S/R_S$.

Example 7.4

For the current amplifier shown in Fig. 7.29, determine the closed-loop current gain, input resistance and output resistance. For the transistors $h_{fe} = 100$, $h_{ie} = 2.5k$, $h_{re} = h_{oe} = 0$. Determine R_S to give a voltage gain of 15.

Solution

(i) The open-loop current gain must first be determined. This is given by $A_I = \frac{I_o}{I_S} = \frac{I_o}{I_{b2}} \frac{I_{b2}}{I_{c1}} \frac{I_{c1}}{I_{b1}} \frac{I_{b1}}{I_S}$. From the circuit $\frac{I_o}{I_{b2}} = h_{fe} = -100$; $\frac{I_{b2}}{I_{c1}} = -\frac{R_1}{R_1 + R_{i2}}$ where (noting that $R_F \ll R_B$) we have $R_{i2} = h_{ie} + (1 + h_{fe})R_E //$

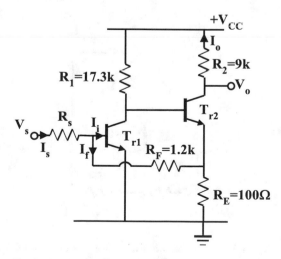

Fig. 7.29 Current-shunt feedback amplifier example

$R_F = 2.5 + 101 \times 0.1 // 1.2 = 11.8$ k. Hence $\frac{I_{b2}}{I_{c1}} = -\frac{17.3}{17.3 + 11.8} = -0.59$; $\frac{I_{c1}}{I_{b1}} = h_{fe} = 100$ and $\frac{I_{b1}}{I_S} = \frac{R_F + R_E}{R_F + R_E + h_{ie}} = \frac{1.2 + 0.1}{1.2 + 0.1 + 2.5} = 0.34$. The result is $A_I = \frac{I_o}{I_S} = 100 \times 0.59 \times 100 \times 0.34 = 2006$. The feedback factor is $\beta = \frac{R_E}{R_F + R_E} = \frac{0.1}{1.2 + 0.2} = 0.08$ and $1 + \beta A_I = 1 + 0.08 \times 2006 = 161.5$. This gives the closed-loop current gain $A_{If} = \frac{A_I}{1 + \beta A_I} = \frac{2006}{161.5} = 12.4$. This can be compared with the value $A_{If}(A_I \to \infty) = 1/\beta = 12.5$.

(ii) The open-loop input resistance to the current amplifier is given by $R_i = h_{ie} // (R_F + R_E) = 2.5 // 1.3 = 0.855$ k. Therefore, the closed-loop feedback is given by $R_{if} = \frac{R_i}{1 + \beta A_I} = \frac{855}{161.5} = 5.3 \Omega$. As a current amplifier with feedback, the input resistance is small as expected. To enable the circuit to accept a voltage signal input, a resistor $R_S \gg R_{if}$ can be added in series with the input. For a voltage gain of 15, $V_o/V_s = I_o R_2 / I_S R_S = A_{IF} \times \frac{R_2}{R_S} = 12.4 \times \frac{9k}{R_S} = 15$. This gives $R_S = 7.4$ k.

(iii) The open-loop output resistance R_o looking in at the collector of the second transistor is $R_o = \infty$ since $h_{oe} = 0$. Hence the closed-loop output resistance R_{of} looking at the collector is $R_{of} = R_o(1 + \beta A_I) = \infty$. It follows therefore that the output resistance of the amplifier is $R_{of} // R_2 = R_2 = 9$ k.

Another example of current-shunt feedback is the common base amplifier configuration. Here the full output current constitutes the feedback current that is subtracted from the input current. The error current flows in/out of the base of the transistor. By viewing the configuration this way, it is possible to apply the feedback analysis to this circuit to arrive at the current gain expression and other characteristics. This analysis shows that the common base amplifier can be viewed as a common emitter amplifier with open-loop current gain $A_I = h_{fe}$ around which 100% current-shunt feedback corresponding to $\beta = 1$ is applied. The resulting closed-loop current gain is given by $A_{If} = \frac{A_I}{1 + \beta A_I}$. This therefore gives

$$A_{If} = \frac{h_{fe}}{1 + \beta h_{fe}} = \frac{h_{fe}}{1 + h_{fe}} \qquad (7.71)$$

which of course is the current gain of the common base amplifier. The circuit's reduced input impedance, increased output impedance, improved frequency response and reduced distortion as compared with the common emitter amplifier also result from the current-shunt feedback. The common gate amplifier is an analogous case using FET technology.

7.9.3   Current-Series Feedback

The example of current-series feedback that we wish to consider is the single transistor circuit shown in Fig. 7.30. The sampled current signal is the output current I_o flowing in resistor R_E where it develops a feedback signal voltage V_f across resistor R_E. Because of the arrangement in which feedback voltage V_f is in series with the input voltage, then V_f is subtracted from the input signal voltage, the error voltage driving the base-emitter junction of the transistor. Applying the rules of analysis, the input circuit of the open-loop amplifier is derived by opening the output loop. The result of this is that resistor R_E appears in the input circuit, and this affects the open-loop transconductance of the amplifier. The output

circuit is obtained by opening the input loop. The result is that R_E also appears in the output circuit, but this has practically no effect on the output current. The open-loop circuit is shown in Fig. 7.31.

The feedback factor is given by

$$\beta = \frac{V_f}{I_o} = \frac{-I_o R_E}{I_o} = -R_E \qquad (7.72)$$

The open-loop transconductance G_M is given by

$$G_M = \frac{I_o}{V_s} = \frac{-h_{fe} I_b}{I_b (h_{ie} + R_E)} = -\frac{h_{fe}}{h_{ie} + R_E} \qquad (7.73)$$

Hence the closed-loop transconductance becomes

$$G_{Mf} = \frac{I_o}{V_s} = \frac{G_M}{1 + \beta G_M}$$
$$= \frac{-h_{fe}}{h_{ie} + (1 + h_{fe})R_E} = -\frac{1}{r_e + R_E} \qquad (7.74)$$

It is possible to derive (7.74) using the h-parameter equivalent circuit for the closed-loop amplifier without using the equivalent circuit in Fig. 7.31. It is interesting to note that since $h_{fe} \gg 1$ then for sufficiently large R_E, $(1 + h_{fe}) R_E \gg h_{ie}$ and (7.74) becomes $G_{Mf} \approx -1/R_E$ which is $1/\beta$. This represents stabilization of the transconductance consistent with the effects of negative feedback. The output current develops a voltage given by $V_o = I_o R_L$. Hence the voltage gain of this circuit is

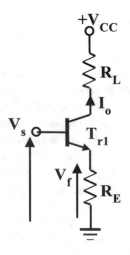

Fig. 7.30 Current-series feedback amplifier

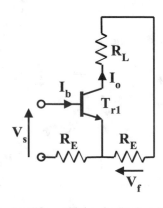

Fig. 7.31 Amplifier without feedback with network loading

$$A_{Vf} = \frac{V_o}{V_s} = \frac{I_o R_L}{V_s} = G_{Mf} R_L$$

$$= -\frac{h_{fe} R_L}{h_{ie} + (1 + h_{fe}) R_E} = -\frac{R_L}{r_e + R_E} \quad (7.75)$$

obtained in Chap. 4. The open-loop input imped-ance of the amplifier is $R_i = h_{ie} + R_E$, and therefore the closed-loop input impedance is given by

$$R_{if} = (1 + \beta G_M)(h_{ie} + R_E)$$
$$= h_{ie} + (1 + h_{fe}) R_E \quad (7.76)$$

This is exactly the expression obtained using h-parameter analysis. Looking into the collector of the transistor the output impedance without feedback (in the absence of the output admittance $1/h_{oe}$) is infinite, and therefore the addition of feedback cannot increase it further. However, the output impedance involving the collector resistor R_L is simply the infinite output impedance in parallel with R_L which is R_L.

Example 7.5
For the transconductance amplifier of Fig. 7.32 determine the closed-loop transconductance G_{Mf}, the closed-loop input impedance and the closed-loop output impedance.

Solution We proceed in the manner developed in the case of the other two amplifiers. Thus the open-loop transconductance is $G_M = \frac{I_o}{V_s} =$

$-\frac{h_{fe}}{h_{ie} + R_E} = -\frac{100}{2.5 + 2} = -22.22\,\mathrm{mA/V}$. $\beta = -R_E = -2\,\mathrm{k}$. And hence $1 + \beta G_M = 1 - 2 \times (-22.22) = 45.44$. Therefore $G_{Mf} = \frac{G_M}{1 + \beta G_M} = \frac{-22.22}{45.44} = -0.49\,\mathrm{mA/V}$. Looking in at the base of the transistor the open-loop input impedance is $R_i = h_{ie} + R_E = 2.5 + 2 = 4.5\,\mathrm{k}$. Therefore, the closed-loop input impedance is $R_{if} = (1 + \beta G_M) R_i = 45.44 \times 4.5\,\mathrm{k} = 204\,\mathrm{k}$. The effective input impedance is $R_{if}//173//27 = 21\,\mathrm{k}$. Thus, the high input impedance of this circuit is reduced by the biasing resistors which are outside the feedback loop. Looking into the collector of the amplifier, the open- and closed-loop output impedance is ∞ since $h_{oe} = 0$. Therefore, the output impedance of the amplifier is $R_{of} = \infty//R_L = 9\,\mathrm{k}$. Once again, the output current of this amplifier is converted to a voltage in R_L yielding $V_o = I_o R_L$. Therefore the voltage gain of the circuit is $V_o/V_S = I_o R_L/V_S = G_{Mf} R_L = -0.49 \times 9 = -4.4$, which is precisely the value obtained using the Eq. (7.75).

7.9.4   Voltage-Shunt Feedback

The single transistor amplifier circuit of Fig. 7.33 is an example of voltage-shunt feedback. The sampled output voltage at the collector of the transistor develops a feedback current in the feedback resistor R_F. Because of the inverting action of the configu-ration, this current is subtracted from the input signal at the base of the transistor. The rules of feedback analysis give the input circuit of the open-loop amplifier by shorting the output node

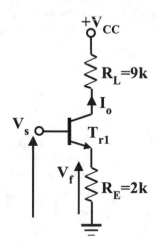

Fig. 7.32 Transconductance amplifier example

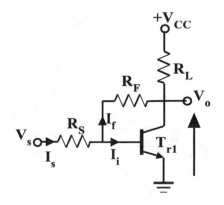

Fig. 7.33 Voltage-shunt feedback amplifier

corresponding to setting $V_o = 0$. This results in R_F being placed between the base and emitter of the transistor. The output circuit of the open-loop amplifier is determined by shorting the input node corresponding to setting $V_i = 0$. This places R_F across the collector and emitter of the transistor. The signal source is represented as a Norton equivalent circuit since the feedback signal is a current. The resulting circuit is shown in Fig. 7.34.

From Fig. 7.33, since the output voltage V_o is much greater than the input voltage V_i at the transistor base and out of phase with it, for a positive going input voltage the feedback current I_f flows such that

$$I_f = \frac{V_i - V_o}{R_F} \approx -\frac{V_o}{R_F} \qquad (7.77)$$

Therefore

$$\beta = \frac{I_f}{V_o} = -1/R_F \qquad (7.78)$$

Assuming $R_F \gg R_L$, the open-loop trans-resistance R_M is given by

$$R_M = \frac{V_o}{I_i} = \frac{-h_{fe}I_b R_L R_F}{I_b(h_{ie} + R_F)} = -\frac{h_{fe}R_L R_F}{h_{ie} + R_F} \quad (7.79)$$

Hence the closed-loop trans-resistance becomes

$$R_{Mf} = \frac{V_o}{I_i} = \frac{R_M}{1 + \beta R_M}$$
$$= \frac{-h_{fe}R_L R_F}{h_{ie} + R_F + h_{fe}R_L} \qquad (7.80)$$

If $R_F \gg h_{ie}$ then (7.80) becomes

$$R_{Mf} = \frac{V_o}{I_i} = -h_{fe}R_L//R_F \qquad (7.81)$$

The open-loop input impedance R_i seen by the current source from Fig. 7.34 is $R_i = R_F//h_{ie}$, and if $R_F \gg h_{ie}$, then $R_i = h_{ie}$. The closed-loop input impedance R_{if} is then

$$R_{if} = \frac{R_i}{1 + \beta R_M} = \frac{h_{ie}}{1 + h_{fe}R_L/R_F} \qquad (7.82)$$

Because the input resistance of this feedback amplifier is low, the circuit is easily converted into a voltage amplifier by driving the input with a voltage signal V_S operating into a resistor $R_S \gg R_{if}$ which yields $I_S = V_S/R_S$. The output impedance for the closed-loop amplifier is given by

$$R_{of} = \frac{R_o}{1 + \beta R_M} = \frac{R_L//R_F}{1 + \beta R_M} \qquad (7.83)$$

The current feedback amplifier is another example of a trans-resistance amplifier with voltage-shunt feedback. This circuit will be discussed in Chap. 8.

Example 7.6
For the trans-resistance amplifier of Fig. 7.35, determine R_{Mf}, R_{if}, R_{of} and the value of R_S to give the system a voltage gain of 15. Use $h_{fe} = 100$ and $h_{ie} = 250$.

Solution In order to evaluate the closed-loop trans-resistance R_{Mf}, the open-loop trans-resistance R_M must first be determined. Using (7.79)
$R_M = \frac{V_o}{I_i} = -\frac{h_{fe}R_L R_F}{h_{ie} + R_F} = -\frac{100 \times 1 \times 93}{.25 + 93} = -99.73\,\text{k}.$

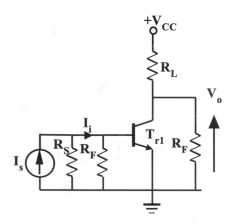

Fig. 7.34 Open-loop amplifier circuit with feedback network loading

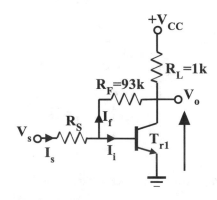

Fig. 7.35 Voltage-shunt feedback amplifier example

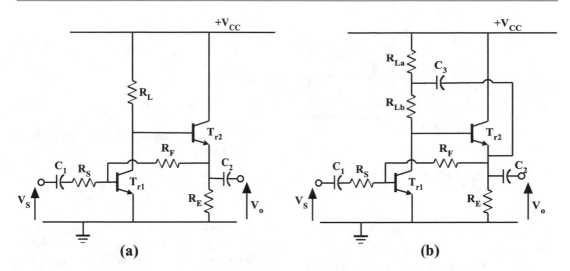

Fig. 7.36 Two transistor configuration with voltage shunt feedback

For this system $\beta = -\frac{1}{R_F} = -\frac{1}{93}$. Hence $1 + \beta R_M = 1 + \frac{99.73}{93} = 2.07$.   Therefore $R_{Mf} = \frac{R_M}{1+\beta R_M} = -\frac{99.73}{2.07} = -48.2\,\text{k}$. The open-loop input resistance $R_i = R_F//h_{ie} = 93//0.25 \approx 250\,\Omega$, and hence the closed-loop input resistance is $R_{if} = \frac{R_i}{1+\beta R_i} = \frac{250}{2.07} = 121\,\Omega$. Since R_{if} is low, if $R_S \gg R_{if}$ then $I_S = V_S/R_S$ where V_S is a voltage signal. Then the system voltage gain is $V_o/V_S = V_o/I_S R_S = \frac{R_{Mf}}{R_S}$. For a voltage gain of -10, $\frac{R_{Mf}}{R_S} = -10$ and therefore $R_S = \frac{-48.2}{-10} = 4.8\,\text{k}$. Note that 4.8 k $\gg$ 121 Ω. Finally, the output resistance is given by $R_{of} = \frac{R_o}{1+\beta R_M} = \frac{R_L//R_F}{1+\beta R_M} = \frac{1//93}{2.07} = 483\,\Omega$.

The configuration shown in Fig. 7.36a involves voltage-shunt feedback as in Fig. 7.35 except that the use of an emitter follower prevents loading of R_L by R_F. Hence the assumption $R_F \gg R_L$ is no longer necessary. Therefore, the transfer function V_o/I_S is given by

$$R_{Mf} = \frac{V_o}{I_i} = \frac{R_M}{1+\beta R_M}$$

$$= \frac{-h_{fe}R_L R_F}{h_{ie} + R_F + h_{fe}R_L} \qquad (7.84)$$

Since $R_S \gg R_{if}$ which yields $I_i = V_S/R_S$, then

$$V_o/V_S = V_o/I_i R_S$$

$$= \frac{-h_{fe}R_L R_F}{h_{ie} + R_F + h_{fe}R_L} / R_S. \qquad (7.85)$$

Hence

$$\frac{V_o}{V_S} = \frac{-h_{fe}R_L R_F/R_S}{h_{ie} + R_F + h_{fe}R_L}$$

$$= \frac{-R_F/R_S}{1 + (h_{ie} + R_F)/h_{fe}R_L}$$

$$= \frac{-R_F/R_S}{1 + R_F/h_{fe}R_L}, R_F \gg h_{ie} \qquad (7.86)$$

For the bootstrapped version shown in Fig. 7.36b, $h_{fe}R_L \to h_{fe}R_L'$ where $R_L' \gg R_L$ and therefore

$$\frac{V_o}{V_S} = \frac{-R_F/R_S}{1 + (h_{ie} + R_F)/h_{fe}R_L'} = -\frac{R_F}{R_S} \qquad (7.87)$$

7.10   Voltage Amplifiers

Following the general consideration of the four types of amplifiers and the application of negative feedback around them, we wish now to consider in greater detail the amplifier type of the set that is most widely used: the voltage amplifier. The voltage amplifier can be found in an amazing array of products and applications. The operational amplifier (see Chap. 8) is one example of a voltage amplifier that is manufactured in vast numbers for a wide range of applications including audio systems, industrial systems and domestic applications. These devices are available for almost any conceivable requirement including low noise, low distortion, high voltage-output,

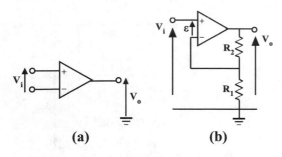

Fig. 7.37 Voltage amplifier (**a**) without feedback; (**b**) with voltage-series feedback

high current output or high-frequency applications.

As pointed out previously, a voltage amplifier has a high input impedance and low output impedance. The input signal is a voltage and the output signal is also a voltage. The feedback type is voltage-series as shown in Fig. 7.37. This is often referred to as the *non-inverting* configuration since the output signal is in phase with the input signal. The sample V_f of the output signal V_o given by $V_f = \beta V_o$ where $\beta = R_1/(R_2 + R_1)$ is subtracted from the input signal V_i. The resulting error signal $\varepsilon = V_i - V_f$ is amplified by the voltage amplifier. This type of feedback makes a good voltage amplifier better by increasing the bandwidth and the (already high) input impedance and decreasing the distortion and the (already low) output impedance.

The transfer function $A_v = V_o/V_i$ of the open-loop amplifier (i.e. amplifier without feedback) in Fig. 7.37a is represented here by a single pole system given by

$$A_v = \frac{A_o}{1 + jf/f_o} \tag{7.88}$$

where A_o is the low-frequency gain and f_o is the open-loop break frequency or bandwidth. The transfer function $A_f = V_o/V_i$ of the closed-loop amplifier in Fig. 7.37b is given by

$$A_f = \frac{A_v}{1 + A_v\beta} \tag{7.89}$$

After substituting for A_v using (7.88) and manipulating, we get

$$A_f = \frac{\left(1 + \frac{R}{R_1}\right) \cdot \frac{A_o\beta}{1+A_o\beta}}{1 + jf/f_o(1 + A_o\beta)} \tag{7.90}$$

where

$$\beta = R_1/(R_1 + R_2) \tag{7.91}$$

is the feedback factor. At low frequencies, Eq. (7.90) reduces to

$$A_f(DC) = \left(1 + \frac{R_2}{R_1}\right) \cdot \frac{A_o\beta}{1 + A_o\beta} \tag{7.92}$$

In operational amplifiers and many voltage amplifiers, A_o is usually very large and hence the loop gain $A_L = A_o\beta$ is usually such that $A_o\beta \gg 1$. Equation (7.90) therefore reduces to

$$A_f = \frac{1 + R_2/R_1}{1 + jf/\bar{f}_o}, A_o\beta \gg 1 \tag{7.93}$$

where

$$\bar{f}_o = f_o A_o \beta \tag{7.94}$$

is the closed-loop bandwidth. From (7.93), the low-frequency closed-loop gain $A_f(DC)$ for $A_o\beta \gg 1$ is given by

$$A_f(DC) = 1 + R_2/R_1 = 1/\beta \tag{7.95}$$

Note that as expected, Eq. (7.92) also reduces to

$$A_f(DC) = 1 + R_2/R_1, A_o\beta \gg 1 \tag{7.96}$$

Example 7.7

A voltage amplifier has a low-frequency open-loop gain of 10^4. If $\beta = 0.1$, determine the low-frequency closed-loop gain using (7.92) and compare with (7.95).

Solution From (7.92), $A_f(DC) = \left(1 + \frac{R_2}{R_1}\right) \cdot \frac{A_o\beta}{1+A_o\beta} = \frac{A_o}{1+A_o\beta} = \frac{10^4}{1+10^4 \times 0.1} = 9.99$. Using (7.95), the low-frequency closed-loop gain is given by $A_f(DC) = 1 + R_2/R_1 = 1/\beta = 1/0.1 = 10$. These two values are within 0.1%.

As can be seen from (7.94), the closed-loop bandwidth $\bar{f}_o$ varies with the loop gain $A_o\beta$.

Alternatively, since A_o is fixed, by expressing (7.94) in terms of $A_f(DC)$ as

$$\bar{f}_o = f_o A_o / A_f(DC) \qquad (7.97)$$

it is clear that the closed-loop bandwidth $\bar{f}_o$ is inversely proportional to the closed-loop low-frequency gain $A_f(DC)$. The loop gain $A_L = A_o \beta$ expressed in dB is given by

$$
\begin{aligned}
A_L(dB) &= 20\log(A_o\beta) \\
&= 20\log A_o + 20\log\beta \\
&= 20\log A_o - 20\log(1/\beta) \\
&= 20\log A_o - 20\log A_f(DC) \qquad (7.98) \\
&- \text{low-freq open-loop gain } (dB) \\
&- \text{low-freq closed-loop gain } (dB)
\end{aligned}
$$

As can be seen from Fig. 7.38 the loop gain A_L is an indication of the amount of feedback around the amplifier. Since from (7.75), the minimum value of $A_f(DC)$ is $A_f(DC) = 1$ corresponding to $\beta = 1$, the maximum. This represents the amount of feedback applied in the system. The maximum feedback occurs when $\beta = 1$ and is given by $A_{Lmax} = 20\log A_o$. This is the unity gain configuration and is obtained by setting $R_1 \rightarrow \infty$ and/or $R_2 = 0$ in Fig. 7.37. The corresponding bandwidth is given by

$$\bar{f}_{o(max)} = f_o A_o \qquad (7.99)$$

which is referred to as the gain bandwidth product (GBP) of the voltage amplifier.

The loop gain and closed-loop bandwidth are shown in Fig. 7.38. Finally, from (7.97),

$$A_f(DC)\bar{f}_o = f_o A_o = GBP = const \qquad (7.100)$$

This equation relates the closed gain with the closed-loop bandwidth, based on the GBP of the particular device.

Example 7.8
For example, for an amplifier with a GBP of 100 kHz, determine the closed-loop bandwidth if the closed-loop gain is 10.

Solution Using Eq. (7.98), the closed-loop bandwidth would be 100 kHz/10 = 10 kHz.

Example 7.9
A voltage amplifier has an open-loop characteristic shown in Fig. 7.39. For a closed-loop gain of 10 (20 dB), determine the following:
 (i) The open-loop bandwidth
 (ii) The open-loop low-frequency gain
 (iii) The GBP for the amplifier
 (iv) The closed-loop bandwidth
 (v) The amount of applied feedback in the system

Solution (i) From the characteristic shown in Fig. 7.39, (i) the open-loop bandwidth is 1 kHz; (ii) the open-loop low-frequency gain is 80 dB; (iii) the GBP is given by $GBP = f_o A_o = 10^3 \times 10^4 = 10^7$ Hz; (iv) for a closed-loop gain of 10, the closed-loop bandwidth $\bar{f}_o$ is found using (8.77a) which yields $\bar{f}_o = GBP/A_f(DC) = 10^7/10 = 1\,MHz$; (v) the amount of feedback is open - loop gain (dB) –

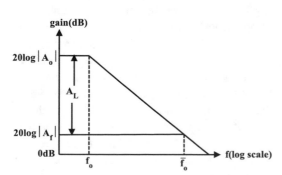

Fig. 7.38 Frequency response plot showing loop gain

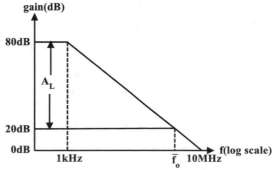

Fig. 7.39 Diagram for Example 7.9

closed - loop gain (dB) = 80 dB − 20 dB = 60 dB.

The open-loop characteristic shown in Fig. 7.39 corresponding to a voltage amplifier with a single pole characteristic will have a maximum phase shift between output and input of 90 degrees. It can be shown that such a system will not become unstable with feedback and can accommodate 100% feedback. As a result, it is sometimes referred to as being unity gain stable. However, steps have to be taken with (multistage) amplifiers in order to realize conditions that ensure system stability when feedback is applied. These methods are discussed in Sect. 7.12. Also, the analysis developed is directly applicable to practical operational amplifiers since the assumed high input impedance and low output impedance that characterize a good voltage amplifier are easily met by modern operational amplifiers. These will be more extensively discussed in the next chapter.

7.11   Transistor Feedback Amplifier

The feedback concept can be applied in the design of a voltage amplifier system using discrete devices. One basic topology is shown in Fig. 7.40. It comprises three amplifying stages: the

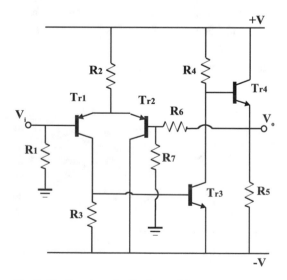

Fig. 7.40 Transistor feedback amplifier

input stage made up of the differential amplifier Tr_1 and Tr_2, the intermediate common emitter stage Tr_3 and the emitter follower output stage Tr_4. Signals applied at the input of the differential amplifier $Tr_1 Tr_2$ are amplified and passed via R_3 to the second stage Tr_3. This common emitter amplifier provides the bulk of the voltage gain. The output stage Tr_4 is an emitter follower that provides a low-impedance output drive. In this configuration, the loading effects of the feedback network are negligible.

Differential Amplifier

In the differential amplifier stage, two transistors Tr_1 and Tr_2 are connected at their emitters where an approximately constant current I through resistor R_2 is connected to supply bias current to each. Each transistor is basically in a common emitter mode with an input supplied to its base and an output taken from its collector. At the two base terminals of the differential amplifier are the input signal $V_i = v_{i1}$ at the base of Tr_1 and the feedback signal $V_f = v_{i2}$ at the base of Tr_2. The output v_{o1} is taken at the collector of Tr_1 where resistor R_3 is connected. It is possible to obtain an output v_{o2} at the collector of Tr_2 by the inclusion of another resistor R_3. Thus $v_{i1} - v_{i2}$ is the differential input voltage and v_{o1} is the single-ended output voltage. Since there are two inputs and one output, the amplifier is said to have a double-ended input and a single-ended output.

In examining the small-signal behaviour of this circuit, the output voltage v_{o1} at the collector of Tr_1 due to each input acting alone, that is, with the opposite input grounded, will be determined and then the superposition principle will be applied to determine the output due to both acting simultaneously. Consider the case where input 2 is grounded ($v_{i2} = 0$) and a small signal applied to input 1. We assume perfectly matched transistors Tr_1 and Tr_2 and therefore that $I_{C1} = I_{C2}$. Since Tr_1 is essentially a common emitter amplifier, v_{i1} is amplified and inverted. For amplification of v_{i1}, Tr_2 functions as a common base amplifier. Hence the input impedance r_e of Tr_2 is the emitter resistance of Tr_1. Therefore, the voltage gain of Tr_1 is

$$A_{V1} = \frac{v_{o1}}{v_{i1}} = -\frac{h_{fe}R_L}{h_{ie} + (1 + h_{fe})r_e} \qquad (7.101)$$

where

$$r_e = \frac{h_{ie}}{1 + h_{fe}} = \frac{0.025}{I} \qquad (7.102)$$

is the input impedance to the common base amplifier Tr_2 and R_L is the load at the collector of Tr_1. Therefore,

$$A_{V1} = \frac{v_{o1}}{v_{i1}} = -\frac{h_{fe}R_L}{2h_{ie}} \approx -\frac{R_L}{2r_e} \qquad (7.103)$$

Since the current $I = I_{E1} + I_{E2}$ in resistor R_2 is approximately constant, it follows that the change in I_{E1} in response to v_{i1} results in a corresponding change (of opposite sign) in I_{E2}. This can be used to produce an amplified voltage v_{o2} at the collector of Tr_2 by the inclusion of a resistor, that is in phase with v_{i1}. If v_{i1} is now set to zero and an input signal v_{i2} applied at the base of Tr_2, then Tr_2 acts like a common emitter amplifier with an unbypassed emitter resistor r_e of Tr_1 with Tr_1 as a common base amplifier. Since there is no resistor in the collector of Tr_2, the change in collector current I_{E2} does not produce an output voltage in Tr_2. Since the current I is constant, it follows that the change in I_{E2} in response to v_{i2} results in a corresponding change (of opposite sign) in I_{E1}. This therefore produces an amplified voltage v_{o1} at the collector of Tr_1 given by

$$A_{V2} = \frac{v_{o1}}{v_{i2}} = \frac{R_L}{2r_e} \qquad (7.104)$$

Thus for the output v_{o1} of Tr_1, from (7.103) the component due to v_{i1} is $-\frac{R_L}{2r_e}v_{i1}$ and from (7.104) that due to v_{i2} is $\frac{R_L}{2r_e}v_{i2}$. Using the superposition principle,

$$v_{o1} = -\frac{R_L}{2r_e}(v_{i1} - v_{i2}) \qquad (7.105)$$

Hence the gain [single-ended out/differential in] is given by

$$\frac{v_{o1}}{v_{i1} - v_{i2}} = -\frac{R_L}{2r_e} \qquad (7.106)$$

If a resistor R_3 is included in the collector of Tr_2, then by similar arguments the output v_{o2} from the

collector of Tr_2 is given by $v_{o2} = -v_{o1}$. Therefore from (7.105)

$$v_{o2} = \frac{R_L}{2r_e}(v_{i1} - v_{i2}) \qquad (7.107)$$

Hence the differential output $(v_{o1} - v_{o2})$ is given by

$$v_{o1} - v_{o2} = -\frac{R_L}{r_e}(v_{i1} - v_{i2}) \qquad (7.108)$$

i.e. the differential gain is given by

$$\frac{v_{o1} - v_{o2}}{v_{i1} - v_{i2}} = -\frac{R_L}{r_e} \qquad (7.109)$$

In the circuit of Fig. 7.40, the load is resistor R_3 in parallel with the input impedance h_{ie} to transistor $Tr3$. Thus, the voltage gain of the input stage of Fig. 7.40 is given by

$$A_{v1} = -\frac{R_3//h_{ie3}}{2r_e} \qquad (7.110)$$

and

$$h_{ie3} = \frac{h_{fe}}{40I_{C3}} \qquad (7.111)$$

Common Emitter Amplifier

The second stage of the amplifier consists of transistor Tr_3 in a common emitter amplifier configuration with load R_4. This stage amplifies the output from the differential amplifier stage and provides most of the voltage gain of the overall amplifier. The gain of the stage is given by

$$A_{v2} = -40I_{C3}R_L \qquad (7.112)$$

where I_{C3} is the current in transistor Tr_3 and R_L is the load made up of R_4 in parallel with the input impedance of Tr_4. Since Tr_4 is in the emitter follower configuration, its input impedance is high and therefore does not significantly affect R_4. Hence $R_L \approx R_4$.

Emitter Follower

The output stage of the amplifier consists of an emitter follower whose gain is approximately one. It presents low output impedance and provides the amplifier with the ability to deliver current without loading down the amplifier and reducing the open-loop gain.

The overall gain of the open-loop amplifier is given by the product of the open loop gain of the three stages. The closed-loop amplifier is shown in Fig. 7.40. The negative feedback being applied is voltage-series feedback as in the non-inverting operational amplifier. For a sufficiently large open-loop gain, the closed-loop gain is given by

$$A_V = 1 + \frac{R_6}{R_7} \qquad (7.113)$$

In order that the circuit be stable, a small capacitor must be placed from the collector of transistor Tr_3 to its base. This is called compensation and is discussed in Sect. 7.12.

Example 7.10

Design a feedback amplifier with a closed-loop gain of 10 using the configuration shown in Fig. 7.40. Use small signal transistors having $h_{fe} = 100$ and a ± 25-volt supply. Justify all steps in your design. Calculate the open-loop voltage gain of your amplifier.

Solution Let the current in Tr_1 and Tr_2 be 0.5 mA. This is a reasonable current that results in good transistor frequency response and current gain. This results in 1 mA flowing through resistor R_2. The voltage at the base of Tr_1 is approximately 0 volts and hence $R_2 = \frac{25-0.7}{1 \text{ mA}} = 24.3 \text{ k}$. The base current of Tr_1 flowing through R_1 is 0.5 mA/ $100 = 5$ μA. Allowing a maximum 50 mV drop across R_1 gives $R_1 = 50 \times 10^{-3}/5 \times 10^{-6} = 10$ k. Since the base-emitter voltage of Tr_3 is across R_3, $R_3 = 0.7/0.5 \text{ mA} = 1.4$ k. The voltage at the output of the amplifier must be zero under quiescent conditions. The voltage at the base of Tr_4 is therefore 0.7 volts. This gives the voltage drop across R_4 as $25 - 0.7 = 24.3$ volts. We choose a current 2 mA in Tr_3 which is somewhat higher than the currents in the differential stage. Then $R_4 = 24.3/2 \text{ mA} = 12$ k. In order to drive external loads, we choose an even larger current of 5 mA in Tr_4. Then $R_5 = 25/5 \text{ mA} = 5$ k. For a closed-loop gain of 10, we assume that the open-loop gain is sufficiently large such that the closed-loop gain is

given by $A_{Vf} = 10 = 1 + R_7/R_6$. For $R_6 = 1$ k then $R_7 = 9$ k. In this design, a compensation capacitor between the collector and base of Tr_3 is generally needed for stability. Its value, which is likely to be in the range 1 pF $-$ 100 pF, is adjusted for a flat frequency response (see Example 7.12).

The open-loop voltage gain is determined as follows: The voltage gain A_{V1} of the differential stage is $A_{V1} = \frac{R_3//h_{ie3}}{2r_e}$ where r_e is associated with Tr_1 and Tr_2 and h_{ie3} is associated with Tr_3. $r_e = 1/40I_C = 1/40 \times 0.5 = 50$ Ω and $h_{ie3} = 100/40 \times 2$ mA $= 1.25$ k. Therefore $A_{V1} = \frac{1.4 \text{ k}//1.25 \text{ k}}{2 \times 50} = 10$. The voltage gain A_{V2} of the second and main voltage gain stage is given by $A_{V2} = 40I_C R_L$ where $I_C = 2$ mA and R_L is the load resistance of Tr_3. This load is $R_4 = 12$ k in parallel with the input impedance of the emitter follower Tr_4. The input impedance of Tr_4 is higher than $h_{fe}R_5 = 100 \times 5 \text{ k} = 500$ k. Therefore $R_L = 12$ k//500 k ≈ 12 k and $A_{V2} = 40 \times 2 \times 12 = 960$. Therefore, the open-loop gain of the amplifier is $A_V = A_{V1}A_{V2} = 10 \times 960 = 9600$. Using the complete formula, $A_{Vf} = \frac{A_V}{1+\beta A_V} = \frac{9600}{1+0.1 \times 9600} = 9.99$. This is very close to 10 and therefore justifies our use of $A_{Vf} = 1 + R_7/R_6$.

The design of this amplifier is not complete. The circuit has at least three poles, each of which can introduce significant phase shift such that instability can occur upon application of feedback. Steps must therefore be taken to ensure that the system with feedback is stable. This is the subject of the next section.

7.12 Stability and Compensation

In a negative feedback amplifier with open-loop gain A and feedback factor β, the closed-loop gain is given by $A_f = \frac{A}{1+\beta A}$. If the magnitude of the loop gain $A\beta$ is very large, i.e. $|A\beta| \gg 1$, then $A_f \to 1/\beta$ which is the closed-loop low-frequency gain of an amplifier. This analysis is however generally inadequate since A and sometimes β are frequency dependent. This means that at certain frequencies $|A\beta|$ may not be very much greater than unity. Under such circumstances, the situation needs to be further examined. Thus, the closed-loop gain becomes

$$A_f(s) = \frac{A(s)}{1 + \beta(s)A(s)} \qquad (7.114)$$

For physical frequencies $s = j\omega$ and the closed-loop gain takes the form

$$A_f(j\omega) = \frac{A(j\omega)}{1 + \beta(j\omega)A(j\omega)} \qquad (7.115)$$

The loop gain now becomes $A(j\omega)\beta(j\omega)$ which is a complex function that can be represented by its magnitude and phase

$$A(j\omega)\beta(j\omega) = |A(j\omega)\beta(j\omega)|e^{j\varphi(\omega)} \qquad (7.116)$$

where $\varphi(\omega)$ is the phase shift of a signal exiting the feedback network relative to a signal entering the amplifier. The stability of the closed-loop system depends on the loop-gain. If the magnitude of the loop-gain is unity and the phase shift is $180°$, then

$$|A(j\omega)\beta(j\omega)|e^{j\varphi(\omega)} = 1.e^{-j180°} = -1 \qquad (7.117)$$

Hence the closed-loop gain becomes $\frac{A(j\omega)}{1-1} \to \infty$. This suggests that the system will produce an output without an input and is therefore unstable. If the magnitude of the loop gain is less than one when $\varphi = -180°$, the system will not sustain the output without an input and is stable. The difference in dB between unity loop gain corresponding to 0 dB and the loop gain value at $\phi = -180°$ is the *Gain Margin* which is the amount by which the gain can be increased before instability ensues. The difference between the phase at the frequency at which the loop gain magnitude is unity and $\phi = -180°$ is the *Phase Margin*. It is the amount by which the phase lag can be increased before exciting instability. Both of these parameters give an indication of the level of stability of the feedback system, the larger the values the greater the level of stability. They are illustrated in Fig. 7.41. Multi-stage feedback amplifiers with several poles are all subject to instability because of the phase shift contribution of the various stages.

A simple approach to investigating stability involves the separate construction of Bode plots for the amplifier transfer function $A(j\omega)$ and the factor $1/\beta(j\omega)$ which represents the closed-loop gain for high-gain systems. Noting that

$$20\log|A\beta| = 20\log|A(j\omega)|$$
$$- 20\log\left|\frac{1}{\beta(j\omega)}\right| \qquad (7.118)$$

we can see that the difference between the two graphs is actually the loop gain expressed in dB. Therefore, the stability of the feedback system can be assessed by determining the difference between the |A| plot and the |1/β| plot. For | $A\beta$| = 1, then 20 log |$A\beta$| = 0 and hence the point of intersection of the two curves corresponds to the point at which |$A\beta$| = 1. An example will illustrate the method:

Consider a three-pole amplifier whose open-loop transfer function is given by

$$A(j\omega) = \frac{10^5}{\left(1 + j\frac{f}{10^3}\right)\left(1 + j\frac{f}{10^4}\right)\left(1 + j\frac{f}{10^5}\right)}$$
$$(7.119)$$

The magnitude $|A(j\omega)|$ at a frequency $f(Hz)$ is found from

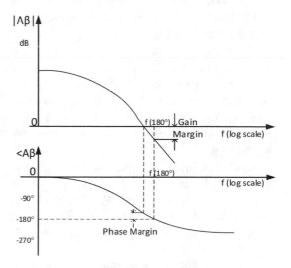

Fig. 7.41 Bode Plot of Amplifier Loop Gain

Fig. 7.42 Bode plot for three (3) pole amplifier

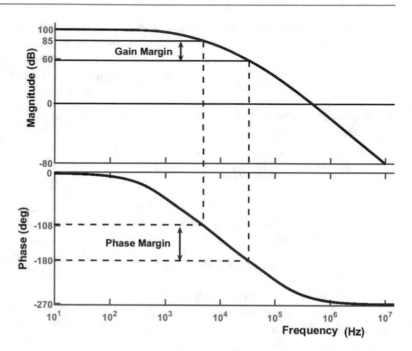

$$|A(f)| = \frac{10^5}{\left|1 + j\frac{f}{10^3}\right|\left|1 + j\frac{f}{10^4}\right|\left|1 + j\frac{f}{10^5}\right|} \quad (7.120)$$

while the phase φ is given

$$\varphi = -\left[\tan^{-1}(f/10^3) + \tan^{-1}(f/10^4) + \tan^{-1}(f/10^5)\right] \quad (7.121)$$

Consider a frequency-independent feedback factor β that yields a closed-loop gain of $20 \log (1/\beta) = 85$ dB. Plots of $|A(j\omega)|$ and $|1/\beta|$ are shown in Fig. 7.42. At the point of intersection of the curves, the frequency is 1.6 kHz and the phase angle is $-108°$. Therefore, at a gain of 85 dB, the phase margin is $180° - 108° = 72°$ and the amplifier is stable. The gain margin can be found by finding the difference between the $|1/\beta|$ plot (at the intersection point) and the value of the $|A|$ plot at $\phi = -180°$. Since the open-loop gain at which the phase is $-180°$ is approximately 60 dB, it follows that the gain margin is 85 dB $-$ 60 dB $=$ 25 dB. For a closed-loop gain of 50 dB, $\phi < -180°$ at the point of intersection and therefore the amplifier is unstable.

It is possible to determine the minimum closed-loop gain for a stable amplifier. It is

given by the intersection point at which $\phi = -180°$. This represents a theoretical stability limit and not a practical limit since phase angles of less than this limit for decreasing gains result in decreasing levels of stability manifested as humps in the closed-loop response as shown in Fig. 7.43.

It can be shown that a phase shift approaching $-180°$ will only occur on the -40 dB/dec segment of the $|A|$ plot, and hence instability can be avoided by the following simple rule: The feedback amplifier will be stable if the 20 log $|1/\beta|$ curve intersects the 20 log $|A|$ curve at a -20 dB/dec segment. This results in a phase margin of approximately $45°$, a value considered acceptable for a stable amplifier. In the case where β is also frequency dependent, the rule becomes the feedback amplifier will be stable providing the relative slopes of the 20 log $|1/\beta|$ and the 20 log $|A|$ curves at the point of intersection does not exceed -20 dB/dec.

7.12.1 Compensating Feedback Amplifiers

As we discussed, a multi-pole amplifier is increasingly likely to be unstable as the closed-

Fig. 7.43 Gain-frequency plot for varying closed-loop gains in a feedback amplifier

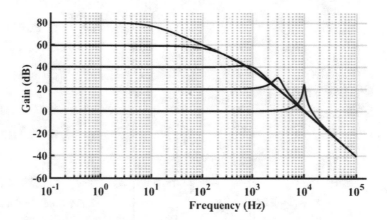

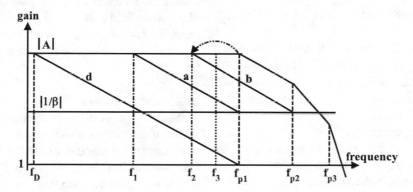

Fig. 7.44 Gain-frequency plot for three compensating methods

loop gain is reduced. In order to ensure stability of the closed-loop system with acceptable levels of frequency response peaking, the magnitude and phase responses of the open-loop amplifier and the feedback network can be adjusted such that $\phi < -180^\circ$ when $|A\beta| = 1$ or alternatively that $|A\beta| < 1$ when $\phi = -180^\circ$. This process is referred to as *Compensation*. The three methods to be discussed are (i) dominant pole compensation, (ii) miller compensation and (iii) lead compensation.

Dominant Pole Compensation

This is perhaps the simplest method and involves the introduction of a pole at a frequency that is lower than all the system poles such that the effects of these poles on the open-loop magnitude response is minimal and therefore the changed

open-loop gain response intersects the closed-loop plot at a relative slope of -20 dB/dec. In order to illustrate the method, consider the open-loop $20 \log |A|$ response of Fig. 7.44 having three poles at f_{p1}, f_{p2} and f_{p3} and a closed $20 \log |1/\beta|$ plot as shown. The $|1/\beta|$ plot intersects the $|A|$ plot at a slope greater than -20 dB/dec, and our rule-of-thumb for a stable feedback amplifier is not satisfied. In order to correct this, a new pole is introduced at the maximum frequency f_1 that causes the intersection of the changed $|A|$ and the $|1/\beta|$ plot at a relative slope of -20 dB/dec and at a frequency before the effect of the lowest pole at f_{p1} (curve a). The closed-loop amplifier is now stable. For general purpose operational amplifiers that must be stable for all values of feedback (unity-gain stable), the pole at f_1 would have to be moved to f_D so that the modified plot

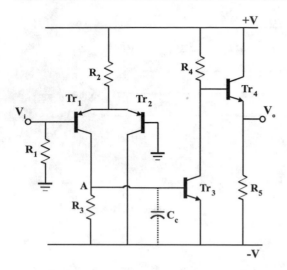

Fig. 7.45 Amplifier with dominant pole compensation

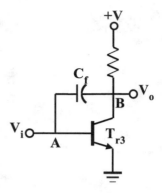

Fig. 7.46 Miller compensation around Tr_3

intersects the 0*dB* line at -20 dB/dec (curve d). The advantage of this method is its simplicity, but its disadvantage is that gain across the frequency spectrum is reduced and therefore less feedback is available at these frequencies.

A better approach to compensating a feedback amplifier is to create the dominant pole by shifting the pole at the lowest frequency f_{p1} to a lower frequency rather than introducing a new one. Thus we can move the first pole at f_{p1} to a frequency f_2 such that the effect of the next lowest pole at f_{p2} is avoided before intersection with the | $1/\beta$| plot (curve b). As can be seen in Fig. 7.44, the new |A| plot intersects the |1/β| plot at -20 dB/ dec, but the resulting dominant pole frequency f_2 is greater than the previous dominant pole frequency f_1 making more negative feedback available across the frequency range. For the circuit shown in Fig. 7.45, the lowest frequency pole is set at node A and is given by $f_{p1} = 1/2\pi R_A C_A$. Here R_A is the total resistance between node A and the emitter of Tr_3 which is approximately R_3//

h_{ie3} and C_A is primarily made up of the base-emitter capacitance of Tr_3. The pole-shifting technique can be implemented by introducing a capacitor C_C across the base-emitter junction of Tr_3 as shown. The adjusted pole is now at $f_2 = 1/2\pi R_A(C_A + C_C)$.

Miller Compensation

A third method of compensation that is superior to the other two is Miller compensation. It is implemented by placing a capacitor C_f between the collector and base of Tr_3 in Fig. 7.46. A simplified equivalent circuit of the input and output of transistor Tr_3 is shown in Fig. 7.46. Here R_A and C_A are the total resistance and capacitance between node A and ground, and R_B and C_B are the total resistance and capacitance between node B and ground. Without the compensation capacitorC_f the poles created by the transistor are

$$f_{p1} = 1/2\pi R_A C_A$$
$$f_{p2} = 1/2\pi R_B C_B \qquad (7.122)$$

With the introduction of C_f, the transfer function from the input of Tr_1 to the output of Tr_3 becomes

$$\frac{V_o}{V_i} = g_{m1} \frac{(g_{m2} - sC_f)R_A R_B}{1 + s[C_A R_A + C_B R_B + C_f(g_{m2}R_A R_B + R_A + R_B)] + s^2[C_A C_B + C_f(C_A + C_B)]R_A R_B}$$
$$(7.123)$$

where g_{m1} is the transconductance of the differential stage and g_{m2} is the transconductance of the second stage. The zero in this transfer function is

located at a frequency $f_z = g_{m2}/2\pi C_f$which for typical transistor parameters is generally at a high frequency and away from the main system poles. Considering the denominator polynomial and noting that it is second-order, it can be approximated by

$$D(s) = \left(1 + \frac{s}{\omega'_{p1}}\right)\left(1 + \frac{s}{\omega'_{p2}}\right)$$

$$= 1 + s\left(\frac{1}{\omega'_{p1}} + \frac{1}{\omega'_{p2}}\right) + \frac{s^2}{\omega'_{p1}\omega'_{p2}} \quad (7.124)$$

where ω'_{p1} and ω'_{p2} are the representative poles of the system. One pole is usually dominant, i.e. $\omega'_{p1} \ll \omega'_{p2}$. Therefore $D(s)$ becomes

$$D(s) \approx 1 + \frac{s}{\omega'_{p1}} + \frac{s^2}{\omega'_{p1}\omega'_{p2}} \quad (7.125)$$

Comparing the denominator of (7.123) and (7.125) and equating coefficients yields

$$f'_{p1} = \frac{1}{2\pi(C_A R_A + C_B R_B + C_f(g_{m2} R_A R_B + R_A + R_B))} \approx \frac{1}{2\pi g_{m2} R_B C_f R_A}$$

$$f'_{p2} \approx \frac{g_{m2} C_f}{C_A C_B + C_f(C_A + C_B)} \quad (7.126)$$

An increase in C_f reduces f'_{p1} but increases f'_{p2}. This effect is called pole-splitting and allows movement of f'_{p1} to a dominant frequency while also moving f'_{p2} to a higher frequency away from the first pole as shown in Fig. 7.47 (curve c).

The overall effect is that the available compensated open-loop gain is increased. It should also be noted in (7.126) that the capacitance C_f has effectively been multiplied by the Miller-effect factor $g_{m2} R_B$ which is the low-frequency gain of the stage. The effect overall is the increase of the value of the capacitance in parallel with R_A to $C'_f = g_{m2} R_B C_f$ such that f'_{p1} can be written as $f'_{p1} = 1/2\pi C'_f R_A$. The capacitance C_f therefore need not be very large

and is usually much smaller than the value C_c used in the pole-shifting method of compensation applied at node A in Fig. 7.45.

In order to determine the value of C_f needed to compensate a particular amplifier, we note that (7.123) can be written as

$$\frac{V_o}{V_i}(s) = \frac{g_{m1} R_A g_{m2} R_B \left(1 - \frac{sC_f}{g_{m2}}\right)}{\left(1 + \frac{s}{\omega_{p1}}\right)\left(1 + \frac{s}{\omega_{p2}}\right)} \quad (7.127)$$

At frequencies close to the dominant pole, frequency (7.127) becomes

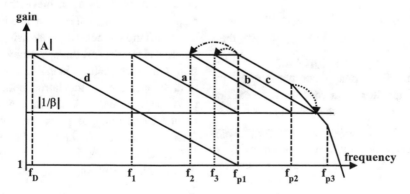

Fig. 7.47 Gain-frequency plot showing miller compensating method (curve c)

$$A(s) = \frac{V_o}{V_i}(s) \approx \frac{g_{m1}R_A g_{m2}R_B}{\left(1 + \frac{s}{\omega_{p1}}\right)} \quad (7.128)$$

The magnitude of this transfer function using (7.126) is

$$|A(j\omega)| = \frac{g_{m1}R_A g_{m2}R_B}{\frac{\omega}{\omega_{p1}}}$$

$$= \frac{g_{m1}g_{m2}R_A R_B}{\omega g_{m2}R_B C_f R_A} = \frac{g_{m1}}{\omega C_f} \quad (7.129)$$

At the frequency ω_T where the $|1/\beta|$ curve intersects the $|A|$ curve, we have $|A(\omega_T)\beta| = 1$ and therefore

$$\frac{\beta g_{m1}}{\omega_T C_f} = 1 \quad (7.130)$$

This gives

$$C_f = \frac{g_{m1}\beta}{2\pi f_T} \quad (7.131)$$

Example 7.11
Consider the amplifier circuit in Fig. 7.45 with an open-loop transfer function corresponding to Eq. (7.123). The objective is to compensate the amplifier so that the closed-loop amplifier with resistive feedback is unity-gain stable, i.e.it is stable for all values of β up to the maximum value of unity. The second stage T_{r3} has $C_A = 1000$ pF, $C_B = 10$ pF and $g_m = 40$ mA/V. It is assumed that the pole at f_{p1} is associated with the input of T_{r3} and the pole at f_{p2} is introduced by the output of T_{r3}.

Solution Using $f_{p1} = 10^4 = \frac{1}{2\pi C_A R_A}$ gives $R_A = \frac{1}{10^4 \times 2\pi \times 1000 \times 10^{-12}} = 15.9$k . For $f_{p2} = 10^5$ then $10^5 = \frac{1}{2\pi C_B R_B}$ gives $R_B = \frac{1}{10^5 \times 2\pi \times 10 \times 10^{-12}} = 159$k. Using dominant pole compensation by introducing capacitor C_C across the base-emitter junction of T_{r3} means that the pole at f_{p1} is moved to a new frequency f_D given by $f_D = \frac{1}{2\pi R_A(C_A + C_C)}$. In order to determine the actual frequency f_D, a line of slope -20 dB/dec is drawn backwards

from the 0 dB gain at frequency $f_{p2} = 10^5$ and the frequency at which it intersects the open-loop characteristic is noted. In this case the intersection occurs at 1 Hz. Thus $f_D = 1 = \frac{1}{2\pi R_A(C_A + C_C)}$ yields $C_C = .009$ µF. Instead of dominant pole compensation, Miller compensation can be used by adding capacitor C_f across the collector-base junction of T_{r3}. This results in new poles at $f'_{p1} \approx \frac{1}{2\pi g_{m2}R_B C_f R_A}$ and $f'_{p2} \approx \frac{g_{m2}C_f}{C_A C_B + C_f(C_A + C_B)}$. Let f'_{p2} coincide with the third system pole at $f_{p3} = 10^6$ Hz. Then in order to determine the frequency f'_{p1}, a line of slope -20 dB/dec is drawn backwards from the 0 dB gain at frequency $f_{p3} = 10^6$ and the frequency at which it intersects the open-loop characteristic is noted. This turns out to be $f'_{p1} = 10$ Hz . Hence $f'_{p1} \approx \frac{1}{2\pi g_{m2}R_B C_f R_A} = 10$ giving $C_f = 157$ pF. Then $f'_{p2} \approx \frac{g_{m2}C_f}{C_A C_B + C_f(C_A + C_B)} = 37$ MHz which exceeds $f_{p3} = 10^6$ Hz.

Example 7.12
Calculate the compensation capacitor C_f for the design in Example 7.10.

Solution For the circuit of Example 7, $r_e = 1/40 \times 0.5$ mA $= 50$ Ω and $A_V = 10$. Therefore, for $f_c = 20$ MHz, the capacitor C_f is given by $C_f = \frac{g_m}{2\pi A_V f_c} = \frac{1}{2r_e \times 2\pi \times 20\,\text{MHz}} = 79.5$ pF.

Lead Compensation
The two methods discussed thus far all involve modifying the shape of the forward transfer function A in order to achieve the desired $A\beta$ response. The shape of $A\beta$ can also be influenced by modifying the beta response using appropriate reactive elements and thereby achieve compensation. One such method is called lead compensation. It operates by the introduction of leading phase in the feedback network β designed to reduce the amount of phase lag caused by the forward network A. In the case of a voltage amplifier, lead compensation can be implemented by the placement of a capacitor C across the

feedback resistor as shown in Fig. 7.48. Experimental adjustment of the value to maximize bandwidth while minimizing frequency response peaking is a simple approach to utilizing the technique.

7.13 Three-Transistor Feedback Amplifier

The three-transistor circuit shown in Fig. 7.49 is another useful voltage amplifier configuration. It is a single-ended design that requires capacitor coupling. The first stage is a common emitter amplifier that provides modest gain. Emitter resistor R_1 is included to allow the application of voltage-series feedback around the amplifier. The second stage is another common emitter amplifier which provides the main voltage gain. Resistor R_6 is a feedback loop that provides DC bias to Tr_1. The loop is disabled at medium and

high frequencies by capacitor C_2. The final stage is an emitter follower that prevents loading of R_4 and gives the amplifier a low-impedance output. Resistor R_7 along with R_1 is the overall feedback loop. Capacitor C_3 is necessary for DC blocking in the feedback loop. C_1 and C_4 are coupling capacitors and C_5 is required for stability.

Example 7.13
Design a transistor amplifier with a gain of 10 using the configuration of Fig. 7.49. Use a 30 volt supply and transistors with current gain of 100.

Solution We choose currents of 1 mA for Tr_1, 2 mA for Tr_2 and 5 mA for Tr_3. We also select $R_1 = 220\ \Omega$. This resistor should not be too large otherwise the gain of this first stage would be unduly reduced and it should be large enough to accommodate the negative feedback implementation. The design starts at stage 2 where the voltage swing is greatest. For bias stability, let the voltage at the emitter or Tr_2 be about one tenth of the supply voltage giving 3 volts. Hence $R_3 = 3/2$ mA $= 1.5$ k. For maximum symmetrical swing in Tr_2, $R_4 = \frac{(30-3)/2}{2\,\text{mA}} = 6.75$ k . Using Kirchoff's voltage law around the internal feedback loop, $3 = I_{B1}R_6 + 0.7 + I_{C1}R_1$ where I_{B1} and I_{C1} are the base and collector currents of Tr_1. Using $I_{B1} = 1$ mA/100 $= 0.01$ mA, $I_{C1} = 1$ mA and $R_1 = 220\ \Omega$ we get $R_6 = 208$ k. Since the

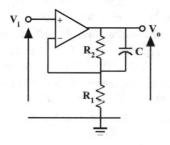

Fig. 7.48 Voltage amplifier with lead compensation

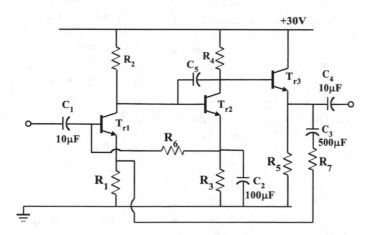

Fig. 7.49 Three-transistor feedback amplifier

emitter of Tr_2 is at 3 volts, the collector of Tr_1 is 3.7 volts. Hence $R_2 = (30 - 3.7)/1$ mA $= 26.3$ k. The voltage at the collector of Tr_2 is $3 + 13.5 = 16.5$ volts. At the output stage, the emitter of Tr_3 is $16.5 - 0.7 = 15.8$ volts. Hence $R_5 = 15.8/5$ mA $= 3$ k. The open-loop gain of the amplifier is the product of the gain of the two common emitter stages. The voltage gain of the first stage is $A_v(T_{r1}) = -R_2//h_{ie2}/(R_1 + r_{e1})$ where $h_{ie2} = h_{fe_2}/40I_{C_2} = 100/40 \times 2\text{mA} = 1.25\text{k}$ and $r_{e_1} = 1/40I_{C_1} = 25\,\Omega$ Hence $A_v(T_{r1}) = -4.8$. For the second stage, $A_v(T_{r2}) = 40I_{C_2}R_4 = -540$ from which the overall open-loop gain is $(4.8 \times 540) = 2592$. This reasonably high open-loop gain results in a closed-loop gain given by $A_{vf} = 1 + R_7/R_1$. Hence designing for a gain of 10, $10 = 1 + R_7/220$ giving $R_7 = 2$ k. Capacitor C_5 is given by $C_5 = \frac{g_m}{2\pi f_c A_V}$ where $g_m = 1/(r_e(Tr_1) + R_1) = 1/(25 + 220) = 1/245$. Using $f_c = 3$ MHz and $A_V = 10$, $C_5 = \frac{1}{245 \times 2\pi \times 3 \times 10^6 \times 10} = 22\,\text{pF}$.

7.14   Applications

Several applications of circuits involving the negative feedback principle are presented and discussed.

Three Transistor Preamplifier

The circuit in Fig. 7.50 is another three-transistor preamplifier configuration. The first stage is a common emitter amplifier using an npn transistor. The output of this stage is directly coupled to the input of the second stage which is also a common emitter configuration. This however uses a pnp transistor. The output of this second stage is also directly coupled to the output stage which is a common collector amplifier using an npn transistor. Resistor R_7 provides DC feedback that establishes quiescent current stability throughout the system. This feedback is reduced for signals since resistor R_6 is connected from the emitter of Tr_1 to ground for signals via C_3. It is evident that this feedback arrangement is voltage-series feedback and hence, for sufficiently high open-loop gain, the closed-loop gain of the system for

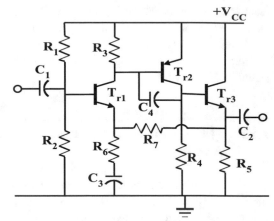

Fig. 7.50   Three-transistor preamplifier

signals is $1 + R_7/R_6$. C_4 is the Miller compensation capacitor. Because of the DC feedback loop from the output to the emitter of Tr_1, it is convenient to start the design process for this circuit at the output.

Let the currents be 0.5 mA in Tr_1, 1 mA in Tr_2 and 4 mA in Tr_3. For a 20-volt supply, let the emitter of Tr_3 be 10 V for maximum symmetrical swing. Hence, noting that the collector current of Tr_1 and Tr_3 flow into R_5, then $R_5 = 10V/(4 + 0.5)$ mA $= 2.2$ k. The voltage at the base of Tr_3 which is connected to the collector of Tr_2 is 10.7 V. Hence $R_4 = 10.7$ V/1 mA $= 10.7$ k. Since the current in Tr_1 is 0.5 mA, $R_3 = 0.7$ V/0.5 mA $= 1.2$ k. Resistor R_6 which forms part of the signal feedback loop must not be too large as it will unduly limit the gain of this stage. A value of 220 ohms is selected. If the circuit has to have a closed-loop gain of 11, then $R_7 = 2.2$ k. Since the collector current of Tr_1 flows through R_7 into R_5, the voltage drop across R_7 is 2.2 V. The base voltage of Tr_1 is then $10 + 2.2 + 0.7 = 12.9$ V. For quiescent current stability, let the current through R_1 and R_2 be 0.1 mA. Then $R_2 = 12.9$ V/0.1 mA $= 129$ k while $R_1 = (20 - 12.9)/0.1$ mA $= 71$ k. From Eq. (7.131), the value of the compensation capacitor is given by $C_4 = \frac{g_m}{2\pi f_c A_V}$ where $g_m = 1/(r_e(Tr_1) + R_6) = 1/(25 + 220) = 1/245$. Using $f_c = 3$ MHz and $A_V = 11$, $C_4 = \frac{1}{245 \times 2\pi \times 3 \times 10^6 \times 11} = 22\,\text{pF}$.

Fig. 7.51 DC voltmeter using differential amplifier

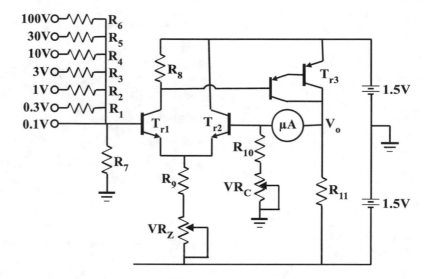

Ideas for Exploration: (i) Use the bootstrapping technique to increase the input impedance of the amplifier. (ii) Use the bootstrapping technique to increase the gain of the second stage.

DC Voltmeter Using Differential Amplifier

The circuit of a DC voltmeter is shown in Fig. 7.51. It uses a differential amplifier at the input followed by a pnp Darlington pair such as the MPSA65 in the common emitter mode. The differential amplifier provides modest gain while the Darlington pair produces higher gain. The meter is placed in a feedback loop around the amplifier such that the current through the meter is defined by the input voltage and a resistor value. Thus, the meter current I_m develops a voltage $V_x = I_m R_{10}$, $(VR_C = 0)$ at the inverting input which is subtracted from the input voltage V_i at the non-inverting input giving an error signal $(V_i - V_x)$. This error signal is multiplied by the amplifier open-loop gain A giving an output $V_o = A(V_i - V_x)$. Providing the open-loop gain A of the system is large, the difference is minimized by the feedback loop such that $I_m = V_i/R_{10}$.

Current through the Darlington pair is set at 0.5 mA giving $R_{11} = 1.5$ V/0.5 mA $= 3$ k, while that through the differential pair is set at 100uA

giving $R_9 + VR_Z \simeq 1$ V/0.1 mA $= 10$ k. Use $R_9 = 8.2$ k and potentiometer $VR_Z = 5$ k which is varied to zero the meter. Each transistor in the pair passes 50uA. These low currents ensure that the system can be operated from penlight batteries over a long period before battery replacement. The voltage across the base-emitter junction of the MPSA65 at 0.5 mA is about 1 V, and hence $R_8 = 1$ V/0.05 mA $= 20$ k. Full-scale deflection of the meter must occur for an input voltage of 100 mV. Hence 100 mV/$(R_{10} + VR_C) = 100$ µA giving $R_{10} + VR_C = 1$ k. Set $R_{10} = 680$ Ω and potentiometer $VR_C = 1$ k which is used to calibrate the meter. The upper ranges of .3 V, 1 V, 3 V, 10 V, 30 V and 100 V require input resistors $R_1 = 20k$, $R_2 = 90$ k, $R_3 = 290$ k, $R_4 = 1$ M, $R_5 = 3$ M, $R_6 = 10$ M as shown. These in conjunction with $R_7 = 10$ k will attenuate the input voltages such that 100 mV will appear at the input to the differential amplifier on each range.

Ideas for Exploration: (i) Explain the mechanism whereby VR_Z can zero the meter and VR_C can be used to calibrate the meter. (ii) Modify the circuit so that it can measure ac signals by placing the meter in a diode bridge.

RIAA Preamplifier

The signal from a phonograph disk (record) has a characteristic in which the high frequencies have

Fig. 7.52 RIAA
equalization curve

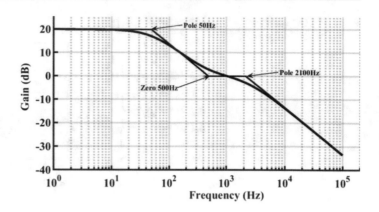

Fig. 7.53 RIAA
preamplifier

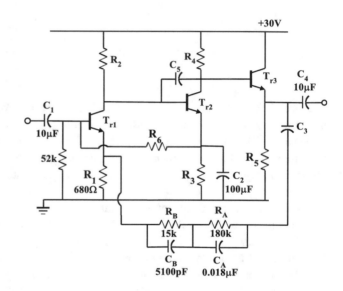

a higher amplitude than the low frequencies. Because of this, the preamplifier for such a signal must possess a frequency response characteristic that is exactly the inverse of this. This means that the amplifier must have a higher gain for low frequencies and a lower gain for high frequencies. The Record Industry Association of America (RIAA) has developed a standard for magnetic phonograph cartridges that establishes the exact nature of this equalization characteristic. This is shown in Fig. 7.52 where two poles and one zero are indicated.

The poles are at 50 Hz and 2100 Hz and the zero is at 500 Hz. A practical implementation of this is shown n Fig. 7.53. It is based on the three-transistor preamplifier of Sect. 7.13 with a CR network in the signal feedback loop which creates

the RIAA equalization characteristic. The design of the network is algebraically quite tedious. The values given in the diagram achieve close adherence to the required curve. The standard input resistance for magnetic pickups is 47 k as achieved in the diagram.

Ideas for Exploration: (i) Refer to the paper "On RIAA Equalization Networks" by Stanley Lipshitz, *Journal of the Audio Engineering Society*, *Journal of the Audio Engineering Society* 27 (6): 458–481, 1979. (ii) Use bootstrapping in the second stage to increase open-loop voltage gain.

JFET-BJT Preamplifier

A circuit for a preamplifier using a JFET at the input is shown in Fig. 7.54. This configuration is similar to the three BJT circuit except that the

Fig. 7.54 JFET hybrid
preamplifier

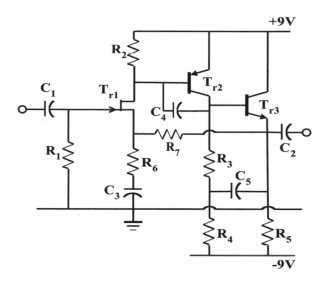

input BJT is replaced by a JFET. This gives the circuit a very high input impedance. The JFET is connected in the common source mode. The output drives Tr_2 which is a pnp BJT in the common emitter mode while the final stage is an emitter follower to buffer the second stage and also provide a low-impedance output. The circuit uses a bipolar supply which of course enhances the available output voltage swing. Resistor $R_1 = 10M$ connects the gate to ground and enables Tr_1 to be self-biased. Resistor R_7 provides DC feedback that establishes quiescent current stability throughout the system. This feedback is reduced for signals since resistor R_6 is connected from the source of Tr_1 to ground for signals via C_3. This feedback arrangement is voltage-series feedback and hence, for sufficiently high open-loop gain, the closed-loop gain of the system for signals is $1 + R_7/R_6$. C_4 is the Miller compensation capacitor. Because of the DC feedback loop from the output to the source of Tr_1, the design process for this circuit starts at the output.

Let the currents be 0.5 mA in Tr_1, 1 mA in Tr_2 and 3 mA in Tr_3. For a ± 9 V supply, let the emitter of Tr_3 be 0 volts for maximum symmetrical swing. Then, noting that the bias currents of the JFET and the emitter follower flow into R_5, the value of this resistor is given by $R_5 = 9$ V/ 3.5 mA = 2.6 k. The voltage at the base of Tr_3which is connected to the collector of Tr_2 is

0.7 V. Hence $R_3 + R_4 = 9.7/1$ mA = 9.7 k. Since the current in Tr_1 is 0.5 mA, $R_2 = 0.7$ V/ 0.5 mA = 1.2 k. From Shockley's equation, for a drain current of 0.5 mA in Tr_1, $V_{GS} = -3$ V. Hence $R_7 = 3/0.5$ mA = 6 k. If the circuit has to have a closed-loop gain of 11, then $R_6 = 600\ \Omega$. This value of resistor R_6 which forms part of the signal feedback loop limits the gain of this stage. This is overcome by bootstrapping resistor R_3 with capacitor C_5 which increases the value of R_3 for signals thereby increasing the gain of Tr_2. From Eq. (7.131), the value of the compensation capacitor is given by $C_4 = \frac{g_m}{2\pi f_c A_V}$ where $g_m = 1/ \{(1/g_m(Tr_1)) + R_6\} = 1/(100 + 600) = 1.4 \times 10^{-3}$ and $g_m(Tr1) = 10$ mA/V is used. Using $f_c = 2$ MHz and $A_V = 11$, $C_4 = \frac{1}{700 \times 2\pi \times 2 \times 10^6 \times 11} = 10$pF.

Ideas for Exploration: (i) Replace the bootstrapping arrangement in the collector of Tr_2 by a constant current source.

AC Millivoltmeter

The circuit shown in Fig. 7.55 is that of a wideband AC millivoltmeter for the measurement of signals in the millivolt range. It uses two common emitter amplifiers so that the sensitivity of the system exceeds that of the single transistor configuration discussed in Chap. 4. The input transistor Tr_1 is bootstrapped and therefore provides a high input impedance to the circuit. This stage drives a common emitter amplifier Tr_2, the

Fig. 7.55 AC
Millivoltmeter

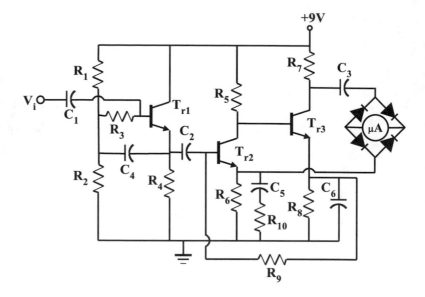

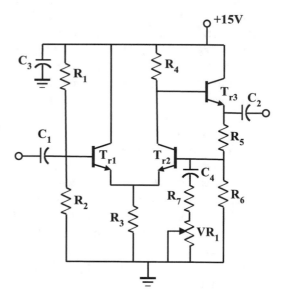

Fig. 7.56 Video amplifier

collector of which is directly coupled to a second common emitter amplifier around Tr_3. The microammeter along with a bridge rectifier is included in a feedback loop connected to the emitter of Tr_2. The first stage design follows the bootstrapped emitter follower Example 4.10. Using a 9-volt supply, let the bias voltage at the emitter of Tr_2 be 1 V. For a collector current of 0.5 mA, $R_6 = 1/0.5$ mA $= 2$ k. Since this stage precedes the output stage where there is further amplification, maximum symmetrical swing is not required. Allowing 2 V across Tr_2 for proper operation, the remaining voltage dropped across R_5 is 6 V. Hence $R_5 = 6V/0.5$ mA $= 12$ k. The voltage at the collector of Tr_2 is 3 V and therefore the voltage across R_8 is 2.3. Selecting 1 mA quiescent current in Tr_3, then $R_8 = 2.3/1$ mA $= 2.3$ k. For maximum symmetrical swing in Tr_3, the voltage across R_7 is $(9 - 2.3)/2 = 3.4$ V. Hence $R_7 = 3.4$ V/1 mA $= 3.4$ k. The voltage at the emitter of Tr_3 is 2.3 V and the voltage at the base of Tr_2 is 1.7 V. The base current of Tr_2 is 0.5 mA/100 $= 0.005$ mA. Therefore, the DC feedback resistor $R_9 = (2.3 - 1.7)/0.005$ mA $= 120$ k. Because of the AC feedback through the meter to the emitter of Tr_2, resistor $R_{10} = 10$ mV/100 μA $= 100$ Ω. Resistor R_{10} should be a potentiometer that is adjusted to give full-scale deflection for an input of 10 mV.

Ideas for Exploration: (i) Introduce an attenuator similar to that used in other meter projects to extend the range of this instrument to 100 V.

Video Amplifier
This circuit utilizes wideband stages to achieve video frequency amplification (Fig. 7.56). Thus, the input stage is a differential amplifier in which the first stage functions in the emitter follower

mode while the second is in the common base mode. The output stage is another emitter follower. In addition to these wideband stages, the system has feedback from output to the inverting input. The result is very wideband operation. With a 15 V supply, using $R_1 = 12$ k and $R_2 = 4.7$ k sets the voltage at the base of Tr_1 at 4.2 V. Let the current in each transistor in the differential pair be 1.5 mA for good high-frequency response. Then $R_3 = (4.2 - 0.7)/3$ mA $= 1.2$ k. Since the base voltage of Tr_2 is about 4.2 V, the maximum peak-to-peak swing available at Tr_3 is approximately $(15 - 4.2)/2 = 5.4$ V. Hence, for maximum symmetrical swing in Tr_2, $R_4 = 5.4/1.5$ mA $= 3.6$ k. In order to drive transmission lines which typically have characteristic impedances of between 50 ohms and 300 ohms, let the current in the emitter follower output stage be 10 mA. The voltage at the base of Tr_3 is $(15 - 5.4) = 9.6$ V, and therefore the resistance in the emitter of this transistor is 8.9/10 mA $= 890$ Ω. In order to get 4.2 V at the base of Tr_2, for $R_6 = 470$ Ω then $R_5 = 520$ Ω. The closed-loop gain of the system is set by $1 + R_5/R$ where R is the value of $VR_1 = 1$ k added to the value of $R_7 = 47$ Ω, in parallel with R_6. At a gain of 17 dB, the simulated circuit had a bandwidth of 11 MHz. All capacitors need to be large.

Ideas for Exploration: (i) Try to increase the bandwidth of the system by selecting suitable high-frequency transistors that have low junction capacitances.

Research Project 1

The circuit shown in Fig. 7.57 is that of an *audio compressor* or constant volume amplifier. It provides a near constant amplitude output signal for widely varying input signal amplitude by incorporating automatic gain control. Such a circuit is useful in reducing the dynamic range of music signals in order to enable better processing of these signals particularly in the recording industry. The first stage is an emitter follower that is biased for maximum symmetrical swing. The second stage is a JFET that is used as a voltage-controlled resistor. With a 15 V supply, the drain of this JFET is set at approximately 2 V by resistors R_4 and R_5 which form a potential divider. Resistor $R_6 = 220$ k grounds the gate of the JFET and $R_7 = 2.2$ k along with the drain voltage set the drain current which positions the operating point in the ohmic region. The drain-source resistance in conjunction with R_7 provides the dynamic adjustment that results in automatic gain control. The output at the source of the FET drives the common emitter amplifier Tr_3 while the emitter follower Tr_4 buffers the final output.

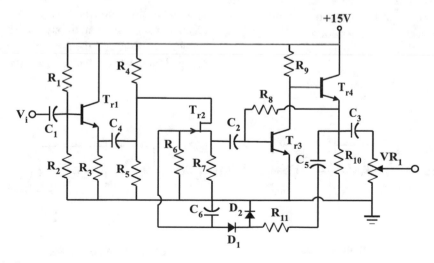

Fig. 7.57 Audio compressor

These latter two stages follow standard design procedure. The output of Tr_4 is rectified by diode D_1 on the negative half-cycle, the positive half-cycle going to ground via D_2. R_{11} and C_6 filter this rectified signal and apply a negative voltage at the gate of the JFET. Thus, for large amplitude signals, the amplitude of this negative voltage increases and this increases the drain-source resistance of the JFET, as a result of which the signal across R_7 and hence the output signal is reduced. The converse occurs for a reduced output.

Ideas for Exploration: (i) Determine the dynamic range of the circuit by applying a signal of increasing amplitude and recording the corresponding output amplitude changes. A large input signal amplitude change will produce a reduced amplitude change at the output. (ii) Experiment with different values of R_{11} and C_6 and determine the response of the circuit to these changes.

Research Project 2

A common problem when playing old records is scratches on the vinyl surface which generate high-frequency noise. This project involves the design of a *scratch filter* to remove this noise. (An introduction to active filters is presented in Chap. 11.) The basic system is shown in Fig. 7.58. It comprises two connected RC networks with one capacitor grounded and the other connected to the output of the buffer. The transfer function for this system with $R_1 = R_2 = R$ and $C_1 = C$, $C_2 = 2C$ is given by

$$\frac{V_o}{V_i} = \frac{1}{1 + \sqrt{2}j\frac{f}{f_c} + \left(j\frac{f}{f_c}\right)^2} \qquad (7.132)$$

where

$$f_c = \frac{1}{2\sqrt{2}\pi RC} \qquad (7.133)$$

This is a second-order response referred to as a Butterworth response that has the characteristic of being maximally flat. The roll-off rate is -40 dB/dec. With an appropriately placed corner frequency f_c, unwanted high frequency signals produced by vinyl scratches can be eliminated. The circuit implementing this approach is shown in Fig. 7.58 which uses a bipolar supply of ± 15 V. Here an emitter follower is the active buffer. Base bias is provided by $R_B = 120$ k and $R_E = 15$ k sets the quiescent current at about 1 mA. For an upper cut-off frequency of 10 kHz, using $C = 2.2$ nF, Eq. (7.133) gives $R = 4.7$ k.

Ideas for Exploration: (i) Use a feedback pair in place of the single transistor. This will allow an even larger value for R_B and also gain can be introduced.

Research Project 3

Using this single transistor configuration, explore the development of a *rumble filter* that removes unwanted low-frequency (less than 50 Hz) components from an audio signal. The two filter network resistors and the two network capacitors in Fig. 7.58 must be interchanged. Also, R_B would

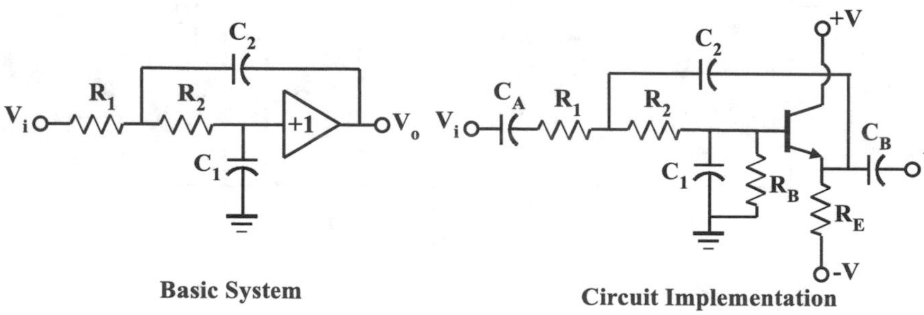

Basic System   **Circuit Implementation**

Fig. 7.58 Scratch filter

Fig. 7.59 Rumble filter

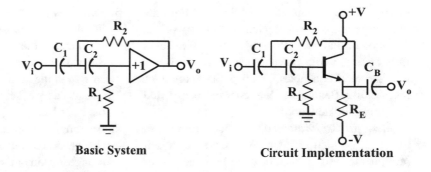

Basic System   Circuit Implementation

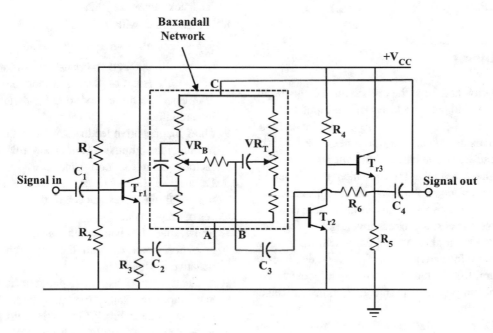

Fig. 7.60 Baxandall Tone Control

not now be necessary as base bias would be
provided by the filter network resistor R_1. The
basic system is shown in Fig. 7.59 and the
associated transfer function with $C_1 = C_2 = C$ and
$R_1 = R$, $R_2 = R/2$ is given by

$$\frac{V_o}{V_i} = \frac{\left(j\frac{f}{f_c}\right)^2}{1 + \sqrt{2}j\frac{f}{f_c} + \left(j\frac{f}{f_c}\right)^2} \quad (7.134)$$

where the corner frequency is

$$f_c = \frac{\sqrt{2}}{2\pi RC} \quad (7.135)$$

This is a Butterworth response for the high-pass
filter. For a lower cut-off frequency of 50 Hz,
using $C = 1\ \mu F$, Eq. (7.135) gives $R = 4.5$ k.

Ideas for Exploration: (i) Use a feedback pair
in place of the single transistor. This enables gain
to be introduced.

Research Project 4

A tone control circuit adjusts the amplitude of
high (treble) and low (bass) frequencies in an
audio signal in order to improve the listening
quality perception of the signal. It is usually
placed just before the power amplifier in the
audio signal chain. The most widely used circuit
for accomplishing this is the *Baxandall tone control circuit*. An implementation of this is shown in
Fig. 7.60. The stage driving the Baxandall tone
control network at A is an emitter follower while
the two-stage amplifier following the network at
B is a standard design with feedback biasing

discussed in Chap. 5. Feedback from the two-stage amplifier is applied to the network at C. Potentiometer VR$_B$ lifts/cuts the bass frequencies, while potentiometer VR$_T$ lifts/cuts the treble frequencies. The project involves researching Baxandall tone control theory and designing the network.

Ideas for Exploration: (i) Use bootstrapping to increase the gain in the second stage. This will provide increased feedback and hence lower distortion.

Problems

1. Draw and label the equivalent voltage amplifier. Indicate the levels of the input and output resistances relative to the source and load resistances and show how negative feedback can be applied around the amplifier.
2. Draw and label the equivalent current amplifier. Indicate the levels of the input and output resistances relative to the source and load resistances, and show how negative feedback can be applied around the amplifier.
3. State four advantages of negative feedback in amplifiers, and show how such feedback can be implemented around a trans-resistance amplifier.
4. Determine the amount of negative feedback in an operational amplifier having an open-loop gain of 110 dB and a closed-loop gain of 20 dB.

5. Explain what is meant by "gain stabilization" in a negative feedback amplifier and why it is important in the design of such amplifiers.
6. Show that the magnitude of the fractional change in the gain of an open-loop amplifier is reduced by a factor equal to the loop gain in the closed-loop amplifier.
7. An amplifier with gain 1500 has a gain change of 15% due to aging. Calculate the change in gain of the feedback amplifier if the feedback factor $\beta = 0.2$.
8. An amplifier with a transfer function $A = \frac{k}{1+jf/f_o}$ has a low-frequency open-loop gain of $k = 120,000$ and an open-loop bandwidth of $f_o = 1$ Hz. For a closed-loop gain of 100, calculate the closed-loop bandwidth of the amplifier.
9. Show that negative feedback reduces the distortion in the output stage of an amplifier by a factor equal to the loop gain.
10. Discuss the effect of negative feedback on the input and output resistances of the four classes of amplifiers.
11. Show that the bandwidth of a feedback amplifier increases with the application of negative feedback.
12. Evaluate the voltage gain, input resistance and output resistance for the voltage feedback amplifier in Fig. 7.61. For the transistors $h_{fe} = 120$, $h_{ie} = 5$ k, $h_{re} = h_{oe} = 0$.
13. For the current amplifier shown in Fig. 7.62, determine the closed-loop current gain, input

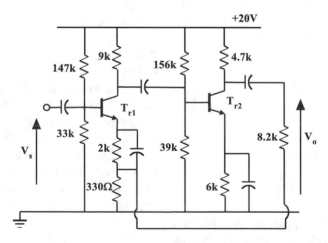

Fig. 7.61 Diagram for Question 12

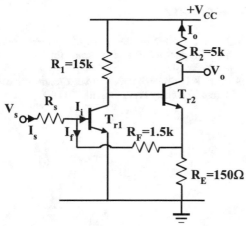

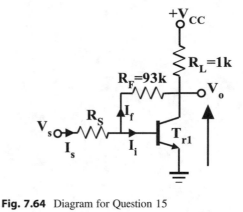

Fig. 7.64 Diagram for Question 15

Fig. 7.62 Diagram for Question 13

Fig. 7.63 Diagram for
Question 14

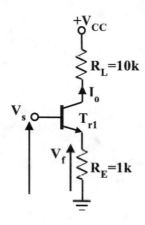

Fig. 7.65 Diagram for Question 20

resistance and output resistance. For the
transistors $h_{fe} = 100$, $h_{ie} = 2.5k$, $h_{re} - h_{oe} = 0$.
Determine R_S to give a voltage gain of 12.

14. For the transconductance amplifier of
Fig. 7.63 determine the closed-loop transconductance G_{Mf}, the closed-loop input impedance and the closed-loop output impedance.

15. For the trans-resistance amplifier of Fig. 7.64,
determine R_{Mf}, R_{if}, R_{of} and the value of R_S to
give the system a voltage gain of 25. Use
$h_{fe} = 120$ and $h_{ie} = 250$.

16. For an operational amplifier with a GBP of
8 MHz, determine the closed-loop bandwidth
if the closed-loop gain is 12.

17. Design a four-transistor feedback amplifier
using the topology in Fig. 7.40. Use a
24 volt symmetrical supply and small signal

transistors with gain of 150. Design the system for a closed-loop gain of 8.

18. Determine the open-loop voltage gain for the
circuit mentioned in Question 17.

19. Design a transistor amplifier with a gain of
8 using the configuration of Fig. 7.49. Use a
25 volt supply and transistors with current
gain of 150.

20. For the feedback amplifier shown in
Fig. 7.65, determine the current in each transistor, assuming they are all matched.

Bibliography

J. Millman, C.C. Halkias, *Integrated Electronics: Analog and Digital Circuits and Systems* (McGraw Hill, New York, 1972)

A.S. Sedra, K.C. Smith, *Microelectronic Circuits*, 6th edn. (Oxford University Press, Oxford, 2011)

Operational Amplifiers

8

The operational amplifier or op-amp is one of the most widely used linear integrated circuits today. Its great popularity arises from its versatility, usefulness, low cost and ease of use. It was introduced in the 1940s mainly for use in analog computers where it performed mathematical operations (hence the name) including addition, multiplication, integration and differentiation. The device later found ready application in a wide range of circuits and functions, many of which will be discussed in this chapter. At the end of the chapter, the student will be able to:

- Understand the operation of the device
- Derive the mathematical relations that govern its operations
- Use the device in a range of linear circuit applications

8.1 Introduction

The operational amplifier in Fig. 8.1 is a direct-coupled voltage amplifier that performs certain mathematical operations. It has differential input and a single-ended output. It responds only to the voltage difference V_D between the two inputs given by $V_D = V_1 - V_2$ and not to their common potential. It amplifies this differential input voltage by a large factor A_d which is the differential gain of the amplifier and delivers an output voltage given by

$$V_0 = A_d(V_1 - V_2) = A_d V_D \qquad (8.1)$$

A positive-going signal at the input defined as the non-inverting (+) input produces a positive-going signal at the output, while a positive-going signal at the input defined as the inverting (−) input produces a negative-going signal at the output.

The ideal operational amplifier possesses several important properties. Its voltage gain is infinite (very large), i.e. $A_d = \infty$, and is constant for all frequencies which means that the operational amplifier has infinite bandwidth. It has zero output impedance and infinite input impedance. The ideal operational amplifier also has the property that $V_O = 0$ when $V_D = 0$. In other words, no offset voltage is present at its output terminal when the differential input voltage is zero. The operational amplifier can therefore be represented by the ideal voltage-controlled voltage source shown in Fig. 8.2. Because $A_d = \infty$, the differential input voltage in an ideal operational amplifier is zero. Also, since the input resistance in either input terminal is infinite, there is no current flow into or out of either terminal.

The operational amplifier is an active device typically powered by a bipolar or dual-polarity supply $\pm V$. This enables both input and output to be referred to ground or zero potential. Depending on the technology, it is constructed from bipolar transistors and/or field-effect transistors. The operational amplifier may operate

© Springer Nature Switzerland AG 2021
S. J. G. Gift, B. Maundy, *Electronic Circuit Design and Application*,
https://doi.org/10.1007/978-3-030-46989-4_8

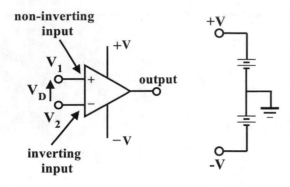

Fig. 8.1 Symbol for the op-amp with a dual power supply

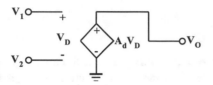

Fig. 8.2 The ideal voltage-controlled voltage source

from a $\pm 15V$ supply or a supply as low as $\pm 4V$. The output of the operational amplifier is always constrained to fall between the supply rails $+V$ and $-V$, both of which are usually equal. Usually, the details of the power supply of the operational amplifier are assumed to be standard and will henceforth be omitted unless otherwise stated.

To understand why A_d must be large, we look back at (8.1) which defines the linear behaviour of the operational amplifier. Now in order for V_o to be finite and contained within the operational amplifier power supply limits, the product A_dV_D must be finite. Under this constraint, as A_d becomes large or $A_d \rightarrow \infty$, V_D must become small or tend to zero. As a result for A_d large, $V_D \simeq 0$ or $V_1 \simeq V_2$. The role of having a large gain can also be visualized in terms of the operational amplifier transfer characteristics as shown in Fig. 8.3. Here the x-axis can be represented in terms of mV or μV depending on the DC gain of the particular operational amplifier. The y-axis shows V_o in volts up to the power supply limits. The ideal operational amplifier shown as the thick line has a hard-limiting transfer characteristic of a slope of ∞. A real operational amplifier has a "soft" characteristic as shown in Fig. 8.3. From

henceforth we will always use the ideal model of the operational amplifier unless stated otherwise.

In summary therefore the five important properties of the operational amplifier are listed below:

(i)   The voltage gain is infinite, i.e. $A_d = \infty$.
(ii) The input resistance is infinite.
(iii) The output resistance is zero.
(iv) The bandwidth is infinite.
(v)   There is zero input offset voltage, i.e. $V_O = 0$ when $V_D = 0$.

In the following section, we assume that these properties are true and examine several important configurations of the ideal operational amplifier.

8.2   The Inverting Amplifier

An amplifier which inverts an input signal while providing gain can be easily realized using the ideal operational amplifier as shown in Fig. 8.4. It is referred to as the inverting configuration. In this configuration, the non-inverting input is grounded, and the input signal is applied to the inverting input through resistor R_1, with negative feedback applied from the output through R_2. In response to an input signal V_i, a current I_1 flows into R_1, I_2 flows through R_2, and I_A flows into the inverting terminal. From Kirchhoff's current law (KCL),

$$I_1 = I_2 + I_A \tag{8.2}$$

Since the op-amp has an infinite input resistance, no current flows into the inverting input and hence $I_A = 0$. Therefore (8.2) reduces to

$$I_1 = I_2 \tag{8.3}$$

The current I_1 through R_1 is given by

$$I_1 = \frac{V_i - V_D}{R_1} \tag{8.4}$$

where V_D is the voltage at the inverting input of the op-amp. Since the amplifier gain is infinite, then from (8.1) for a finite output V_o, the voltage $V_D = V_o / A_d \rightarrow 0$ as $A_d \rightarrow \infty$. Hence from (8.4),

Fig. 8.3 Transfer
characteristics of the ideal
(bold line) op-amp and the
real op-amp

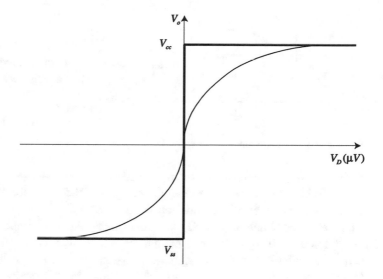

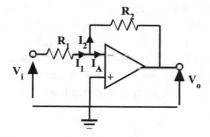

Fig. 8.4 Inverting amplifier

$$I_1 = \frac{V_i}{R_1} \quad (8.5)$$

Finally, the current I_2 in R_2 is given by

$$I_2 = \frac{V_D - V_O}{R_2} \quad (8.6)$$

Since $V_D = 0$, Eq. (8.6) becomes

$$I_2 = \frac{-V_O}{R_2} \quad (8.7)$$

It follows from Eqs. (8.3, 8.5) and (8.7) that

$$I_2 = \frac{-V_O}{R_2} = I_1 = \frac{V_i}{R_1} \quad (8.8)$$

from which we get

$$\frac{V_O}{V_i} = -\frac{R_2}{R_1} \quad (8.9)$$

Equation (8.9) is an expression for the voltage
gain of the inverting amplifier. The negative sign
indicates that the output voltage signal is inverted
relative to the input voltage signal. The input
impedance Z_i to the inverting amplifier is given
by

$$Z_i = \frac{V_i}{I_i} \quad (8.10)$$

From (8.5), $I_i = I_1 = V_i/R_1$ and therefore

$$Z_i = \frac{V_i}{V_i/R_1} - R_1 \quad (8.11)$$

In this configuration, since $V_D = 0$ and the
non-inverting terminal is at zero potential, it
follows that the inverting terminal is also a zero
potential, even though it is not directly connected
to ground. It is therefore called a **virtual earth**.
Note that if the non-inverting terminal was at a
potential different from zero, the inverting terminal will not still be a virtual earth. It in fact will
assume the potential of the non-inverting terminal. The high amplifier gain ensures that the
inverting input is always at the same potential as
the non-inverting input; the two terminals track
each other.

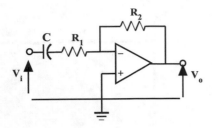

Fig. 8.5 AC inverting amplifier

$$C = 1/(2\pi f_L R_1) \qquad (8.12)$$

If the source from which the circuit is driven has a resistance R_S, then

$$C = 1/[2\pi f_L (R_1 + R_S)] \qquad (8.13)$$

Note the gain $-R_2/R_1$ is often termed the closed-loop gain since it represents the gain under feedback or in the closed-loop system. If we let $R_1 = 1\ k\Omega$ and $R_2 = 5\ k\Omega$, then the gain is -5. Of course, if $R_2 = R_1$, then the gain is -1, and the circuit produces signal inversion (corresponding to a phase shift of 180°) with unity gain.

Example 8.1

Design an operational amplifier circuit that has a gain of -2.

Solution Using the inverting configuration, let $R_1 = 10\ k$. Then from Eq. (8.9), $R_2 = 2\ R_1 = 20\ k$.

Example 8.2

If an inverting amplifier has $R_1 = 5\ k$ and $R_2 = 20\ k$, determine the gain of the system.

Solution Using (8.9), $A_V = -R_2/R_1 = -20\ k/5\ k = -4$.

In choosing R_1 and R_2, their values should not be too small as to produce loading effects or not too large which may result in parasitic capacitance effects. Values in the range $1\ k$ to $1\ M$ are suitable. Thus, for a gain of 10, for example, $R_1 = 10\ k$ and $R_2 = 100\ k$ are reasonable values.

Inverting AC Amplifier

If the circuit is used to amplify AC signals only, then a capacitor C can be added at the input as shown in Fig. 8.5. The value of C sets the lower −3 dB cut-off frequency f_L at

8.3   The Non-inverting Amplifier

Another important configuration is the non-inverting configuration shown in Fig. 8.6. It produces a gain equal to or greater than one as we will show. In this configuration, the input signal is applied to the non-inverting input, while negative feedback from the output is applied to the inverting input through resistor R_2. The inverting input is connected to ground through R_1. In response to an input signal V_i, currents I_1, I_2 and I_A that are similar to what was seen in the inverting configuration flow. From KCL,

$$I_1 = I_2 + I_A \qquad (8.14)$$

Because of the infinite input impedance of the op-amp, $I_A = 0$, and therefore (8.14) becomes

$$I_1 = I_2 \qquad (8.15)$$

From Kirchhoff's voltage law (KVL),

$$V_i = V_D + I_1 R_1 \qquad (8.16)$$

Since $V_D = 0$, it follows that

$$V_i = I_1 R_1 \qquad (8.17)$$

i.e. the input voltage is dropped across resistor R_1. From (8.17),

$$I_1 = \frac{V_i}{R_1} \qquad (8.18)$$

But the voltage at the junction of R_1 and R_2 is V_i. Hence

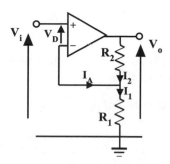

Fig. 8.6 Non-inverting amplifier

$$I_2 = \frac{V_o - V_i}{R_2} \tag{8.19}$$

Using (8.15), (8.18) and (8.19)

$$\frac{V_i}{R_1} = \frac{V_o - V_i}{R_2} \tag{8.20}$$

which reduces to

$$\frac{V_o}{V_i} = 1 + \frac{R_2}{R_1} \tag{8.21}$$

This is an expression for the voltage gain of the non-inverting amplifier. While the inverting amplifier inverts the input signal, the non-inverting amplifier does not. Further, while the minimum gain of the inverting amplifier is zero (corresponding to $R_2 - 0$), the minimum gain of the non-inverting amplifier is 1 (corresponding to $R_2 = 0$). The output then follows the input with $0°$ phase shift and with a gain greater than one provided that $R_1 < \infty$. The input impedance Z_i of the non-inverting amplifier is given by

$$Z_i = \frac{V_i}{I_i} = \infty \tag{8.22}$$

since $I_i = 0$.

Example 8.3
Design a non-inverting amplifier with a gain of 10.

Solution Using the non-inverting configuration, let $R_1 = 1$ k. Then from Eq. (8.21), $10 = 1 + \frac{R_2}{1k}$ giving $R_2 = 9$ k.

Example 8.4
A non-inverting amplifier has $R_1 = 2$ k and $R_2 = 22$ k; determine the gain of the system.

Solution Using (8.21), $A_V = 1 + R_2/R_1 = 1 + 22$ k/2 k = 12.

In a practical non-inverting amplifier, a bias current needs to be provided to the non-inverting input terminal for proper operation of the circuit. This bias may come from a direct-coupled circuit that is connected to the input. In general, however, a bias resistor R_B must be introduced as shown in Fig. 8.7 and should be as low as possible. Since the effective input impedance is now reduced to R_B, these conflicting requirements dictate that this resistor has an intermediate value between say 10 k and 1 M.

As shown above, the non-inverting amplifier is intrinsically capable of a very high input impedance since the input is directly to the non inverting terminal. This high impedance is however limited by the bias resistor R_B. This can be overcome by bootstrapping R_B as shown in Fig. 8.8 by applying a signal in phase with V_i at the lower end of R_B using capacitor C_2. The effective value of R_B is increased by a factor equal to the loop gain of the amplifier. In order to show this, from Fig. 8.8, consider the following:

For a feedback amplifier, the voltage gain A_V is given by

$$A_V = \frac{V_o}{V_i} = \frac{A}{1 + A\beta} \tag{8.23}$$

where A is the open-loop gain of the amplifier and $\beta = R_1/(R_1 + R_2)$ is the feedback factor. Using (8.23), the feedback voltage V_f is given by

$$V_f = \beta V_o = V_i \frac{A\beta}{1 + A\beta} \tag{8.24}$$

Therefore, the current I_B through R_B is given by

$$I_B = (V_i - V_f)/R_B = V_i \frac{1}{(1 + A\beta)R_B} \tag{8.25}$$

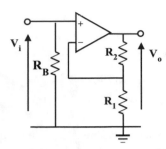

Fig. 8.7 Non-inverting amplifier with bias resistor

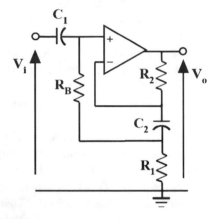

Fig. 8.8 Non-inverting amplifier with bootstrapping

Hence the input impedance $Z_i = V_i/I_B$ to the bootstrapped amplifier is given by

$$Z_i = V_i/I_B = (1 + A\beta)R_B \qquad (8.26)$$

For a 741 operational amplifier, R_B is typically 100 k with $A = 10^5$ at low frequencies. C_2 sets the lower operating frequency below which the technique becomes ineffective. Because of the availability of high input impedance op-amps with low bias current requirements, this technique may not be necessary since the value of R_B can be quite high in such devices. Also, since bootstrapping is only effective for AC signals, C_1 is introduced to eliminate DC.

Non-inverting AC Amplifier

For the amplification of AC signals, a capacitor C must be added in series at the input as shown in Fig. 8.9. Its value is set by the resulting lower cut-off frequency f_L and is given by

$$C = 1/(2\pi f_L R_B) \qquad (8.27)$$

If the circuit is driven from a source having impedance R_S, then

$$C = 1/[2\pi f_L (R_B + R_S)] \qquad (8.28)$$

In this situation, bias resistor R_B is absolutely necessary. The bootstrapped version of this circuit is of course that shown in Fig. 8.8.

8.4   Voltage Follower

A special case of the non-inverting amplifier is the unity-gain stage shown in Fig. 8.10. Here, R_1 goes to infinity (removed), and R_2 goes to zero (short-circuited). V_O is then exactly equal to V_i since $V_D = 0$ corresponding to a gain of +1. Thus,

$$\frac{V_O}{V_i} = 1 \qquad (8.29)$$

The resulting circuit shown in Fig. 8.10 has high input impedance, low output impedance and a unity-gain closed-loop transfer function. The high input impedance $Z_i = \infty$ is again reduced by any bias resistor that may be required at the input.

Again, for AC signals, a high input impedance may be restored by using bootstrapping as shown in Fig. 8.11. From Eq. (8.26), since $\beta = 1$, the input impedance is given by $Z_i = (1 + A)R_B$. For a 741, $A = 10^5$ at low frequencies, and therefore the low-frequency input impedance is given by $Z_i = (1 + 10^5) \times 100\,k = 10^3 M$ which is an extremely high value!

8.5   Summing Amplifier

As mentioned in Sect. 8.1, the operational amplifier derives its name from the fact that it is capable of performing mathematical operations. One such operation is the ability to perform the sum of input signals with the added advantage of having gains larger than one. To see how this is possible,

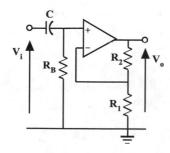

Fig. 8.9 Non-inverting AC amplifier

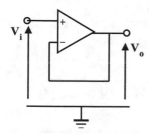

Fig. 8.10 Voltage follower

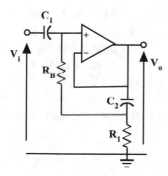

Fig. 8.11 High input impedance AC voltage follower

we examine an extension of the inverting ampli-fier circuit shown in Fig. 8.12. Here, several (n) input signals drive input resistors connected to the inverting terminal. By noting that the inverting node of the operational amplifier because of negative feedback acts as a virtual ground, then KCL dictates that

$$\frac{V_1 - 0}{R_1} + \frac{V_2 - 0}{R_2} + \frac{V_3 - 0}{R_3} + \cdots + \frac{V_n - 0}{R_n} = \frac{0 - V_O}{R_f} \quad (8.30)$$

which reduces to

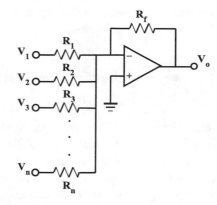

Fig. 8.12 Summing inverter

$$V_O = -\left(\frac{R_f}{R_1}V_1 + \frac{R_f}{R_2}V_2 + \frac{R_f}{R_3}V_3 + \cdots + \frac{R_f}{R_n}V_n\right) \quad (8.31)$$

Hence the output voltage is the inverted sum of the scaled input voltages. Because each input resistor is connected to a virtual earth, it follows that the input resistance seen by each source $V_k(k = 1, 2, 3, \ldots n)$ is R_k. This makes the sum-ming amplifier very useful for mixing audio signals from different sources. As in the single input inverting amplifier, the circuit performs inversion of the inputs.

The overall gain of the circuit is set by R_f, while the individual channel gains are set by the individual input resistors $R_1, R_2, R_3, \ldots, R_n$ which are also the input impedances of the respective inputs. Note that if $R_1 = R_2 = R_3 = \ldots = R_n = R_i$, then

$$V_0 = -\frac{R_f}{R_i}(V_1 + V_2 + V_3 + \ldots + V_n) \quad (8.32)$$

Further if $R_1 = R_2 = R_3 = \ldots = R_n = R_f$, then

$$V_O = -(V_1 + V_2 + V_3 + \cdots + V_n) \quad (8.33)$$

which is a true summing inverter. Finally, linear signal mixing takes place at the summing junction without interaction between inputs since all signal source resistors feed into a virtual earth. The circuit can accommodate almost any number of inputs by adding additional input resistors at the summing point, but this is limited by noise and amplifier overload considerations.

Example 8.5

Design an operational amplifier circuit to realize $V_O = -2\,V_1 - 3\,V_2 - V_3$.

Solution Comparing the requirement with Eq. (8.31), we must satisfy the following relations: $R_f/R_1 = 2$, $R_f/R_2 = 3$ and $R_f/R_3 = 1$. The obvious choice in this case is to choose R_2 with the largest gain term to be the smallest possible resistor we wish to use; in this case, let $R_2 = 1\,k\Omega$. It follows then that $R_f = 3\,k$, $R_1 = 1.5\,k$ and $R_3 = 3\,k$.

Example 8.6

Determine the output voltage V_o for the circuit shown in Fig. 8.13.

Solution $V_A = 2 \times \frac{-10}{5} = -4\,\text{V}$; $V_o = -4 \times \left(1 + \frac{2}{1}\right) = -12\,\text{V}$.

Example 8.7

Determine the output voltage V_o for the circuit shown in Fig. 8.14.

Solution $V_A = 1 \times \frac{-9}{3} = -3\,\text{V}$; $V_o = -3 \times \frac{-4}{2} = 6\,\text{V}$.

Example 8.8

Determine the output voltage V_o for the circuit shown in Fig. 8.15.

Solution $\frac{-V_A}{5} = \frac{2}{5} + \frac{1}{5} + \frac{-4}{5}$; $V_A = 1$; $V_o = 1 \times \left(1 + \frac{1}{1}\right) = 2\,\text{V}$.

Example 8.9

Determine the output voltage V_o for the circuit shown in Fig. 8.16.

Solution $\frac{5-2}{6} = \frac{2-V_A}{3}$; $V_A = 0.5\,\text{V}$; $V_o = 0.5 \times \left(1 + \frac{3}{1}\right) = 2\,\text{V}$ (Fig. 8.16).

8.6   The Differential Amplifier

A circuit capable of finding the difference of two analog signals is shown in Fig. 8.17. Note that this circuit offers different input impedances to each of its input sources. For V_1 the input resistance is R_1, while for V_2 its $R_a + R_b$. Rather than use KCL as we have done earlier, we can make use of the superposition principle to derive an

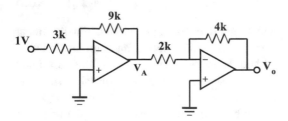

Fig. 8.13 Diagram for Example 8.6

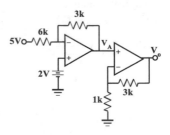

Fig. 8.15 Diagram for Example 8.8

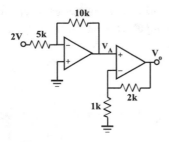

Fig. 8.14 Diagram for Example 8.7

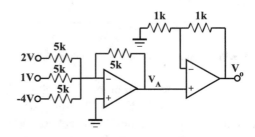

Fig. 8.16 Diagram for Example 8.9

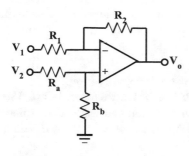

Fig. 8.17 Differential amplifier

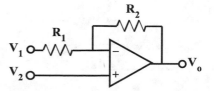

Fig. 8.18 A special case of the difference amplifier

expression for V_o. First we short V_2 keeping V_1 active and recognizing that in that configuration R_a which is in parallel with R_b has no effect on the circuit and can be replaced by a short to ground. The circuit therefore resembles the inverting amplifier configuration: That is, $V_O' = -\frac{R_2}{R_1}V_1$. Next, we short V_1 and note that the circuit resembles the non-inverting configuration this time except that the input V_2 is scaled by $\frac{R_b}{R_a+R_b}$ at the non-inverting terminal of the operational amplifier. Hence the output voltage is given by $V_O^\wedge = \frac{R_b}{R_a+R_b}\left(1+\frac{R_2}{R_1}\right)V_2$. Adding the two results from superposition $\left(V_O = V_O' + V_O^\wedge\right)$ yields the result

$$V_O = -\frac{R_2}{R_1}\left(V_1 - \frac{1+\frac{R}{R_2}}{1+\frac{R_a}{R_b}}\cdot V_2\right) \qquad (8.34)$$

Two interesting scenarios arise out of the circuit of Fig. 8.17. The first occurs when the resistance ratios are all equal. That is, $\frac{R_2}{R_1} = \frac{R_b}{R_a}$. In that case (8.34) simplifies to

$$V_O = -\frac{R_2}{R_1}(V_1 - V_2) \qquad (8.35)$$

and the output is thus proportional to the true difference of two inputs (but inverted).

The second case occurs when $R_a = 0$ and $R_b = \infty$. This is shown in Fig. 8.18. In that case V_o is given by

$$V_O = \left(1+\frac{R_2}{R_1}\right)V_2 - \frac{R_2}{R_1}V_1 \qquad (8.36)$$

The advantage of the circuit of Fig. 8.18 over Fig. 8.17 is that it uses less components and offers a single high input impedance node. Its disadvantage is that it is capable of realizing only a limited range of functions. In either case both Fig. 8.17 and Fig. 8.18 find uses in the design of instrumentation amplifiers.

Example 8.10
Design a circuit to realize the expression $V_o = V_2 - V_1$.

Solution From (8.35) this condition occurs when we have equal resistances in the circuit of Fig. 8.13. In that case choose $R_1 = R_2 = 10\ k$ and $R_a = R_b = 10\ k$.

Example 8.11
Design a circuit to realize $V_o = 3V_2 - 2\ V_1$.

Solution Right away the circuit of Fig. 8.18 and Eq. (8.36) come to mind. We will use Fig. 8.18 because of the advantages previously mentioned. Hence the design equations are $\frac{R_2}{R_1} = 2$. Note $1+\frac{R_2}{R_1} = 3$ is satisfied by default. Letting $R_1 = 1\ k$ yields $R_2 = 2\ k$ which is the desired result.

Example 8.12
For the special case of the difference amplifier shown in Fig. 8.19, show that the relationship between the output and the inputs is $V_O = -\frac{R_2}{R_1}(V_1 - V_2)$.

Solution The differential amplifier amplifies the difference between two signals to be amplified. The potential at the non-inverting terminal is not zero, but instead is set by the potential divider arrangement of R_1 and R_2 and is given by

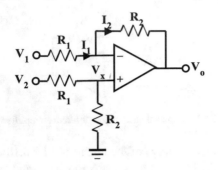

Fig. 8.19 Differential amplifier

$$V_x = \frac{R_2}{R_1 + R_2} V_2 \qquad (8.37)$$

Since the terminals of the op-amp track each other, it follows that the voltage at the inverting terminal is also V_x. Hence the current in resistor R_1 is given by

$$I_1 = \frac{V_1 - V_x}{R_1} \qquad (8.38)$$

while that in resistor R_2 is given by

$$I_2 = \frac{V_x - V_o}{R_2} \qquad (8.39)$$

But

$$I_1 = I_2 \qquad (8.40)$$

Therefore

$$\frac{V_1 - V_x}{R_1} = \frac{V_x - V_o}{R_2} \qquad (8.41)$$

Substituting for V_x using Eq. (8.37) yields

$$\frac{V_1 - \frac{R}{R_1 + R_2} V_2}{R_1} = \frac{\frac{R}{R_1 + R_2} V_2 - V_o}{R_2} \qquad (8.42)$$

After manipulation this gives

$$V_o = -\frac{R_2}{R_1}(V_1 - V_2) \qquad (8.43)$$

If $R_1 = R_2$, then (8.43) reduces to

$$V_o = -(V_1 - V_2) \qquad (8.44)$$

This means that with equal resistors throughout, the circuit functions as an inverting subtractor.

Exercise 8.1
Show that in the realization of $V_o = 2\,V_1 - 3\,V_2$, only Fig. 8.17 can be used and one solution is to choose $R_a = R_b = R_1 = 1\ k$ and $R_2 = 3\ k$.

8.7   Integrator

The integrator, sometimes referred to as the Miller integrator, is shown in Fig. 8.20. In it the feedback resistor R_2 in the inverting amplifier is replaced by a capacitor C. For this circuit,

$$I_1 = I_2 \qquad (8.45)$$

and this leads to

$$\frac{V_i}{R} = -C\frac{dV_o}{dt} \qquad (8.46)$$

Hence

$$V_o = -\frac{1}{RC}\int V_i.dt + c \qquad (8.47)$$

where c is the constant of integration. If at the start of the integration the output voltage is zero, then $c = 0$. If, for example, V_i is a constant voltage k and the capacitor is initially uncharged in which case $c = 0$, then $V_o = -kt/RC$, and the output keeps rising until the op-amp saturates. At DC, the circuit functions as an open-loop amplifier, and therefore offset currents flowing into C would result in an error voltage at the output. The magnitude of this voltage can be reduced by including a resistor R_C in parallel with C as shown in Fig. 8.21. This limits the low-frequency gain and hence minimizes the voltage output error. The circuit operates as an integrator providing $R_C > > R$ and this it does above the frequency $f = 1/2\pi R_C C$. Integrators find many users in filter design, signal processing, A/D converters and much more.

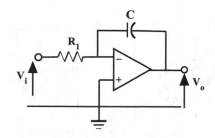

Fig. 8.20 Integrator

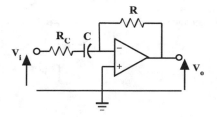

Fig. 8.23 Modified differentiator

$$V_o = -CR\frac{dV_i}{dt} \qquad (8.50)$$

If, for example, V_i is a constant voltage k, then $V_o = 0$ since $d/dt(k) = 0$. If $v_I(t) = A \sin \omega t$, the output of the differentiator will be a cosine function with amplitude $A\omega RC$, angular frequency ω and a phase shift of $180°$. The gain of this circuit increases with increasing frequency, and this makes the circuit susceptible to high-frequency noise. Moreover, the feedback resulting from this arrangement makes the circuit prone to instability. Both of these problems can be reduced by the inclusion of a resistor R_C in series with the capacitor as shown in Fig. 8.23. The circuit functions as a differentiator providing $R_C < < R$ and does so below the frequency given by $f = 1/2\pi R_C C$.

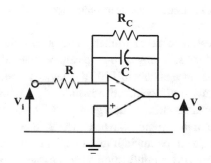

Fig. 8.21 Modified integrator

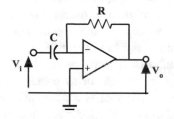

Fig. 8.22 Differentiator

8.8 Differentiator

A circuit capable of performing analog differentiation is shown in Fig. 8.22. It is called a differentiator. In this circuit the input resistor R_1 of the inverting amplifier is replaced by capacitor C.

Again,

$$I_2 = I_1 \qquad (8.48)$$

and hence

$$\frac{-V_o}{R} = C\frac{dV_i}{dt} \qquad (8.49)$$

from which

8.9 Transimpedance Amplifier

The operational amplifier can also function as a transimpedance amplifier or current-to-voltage converter. As a transimpedance amplifier, it converts an input current into an output voltage as shown in Fig. 8.24. Note that this circuit is almost identical to the inverting amplifier configuration except that R_1 is absent and the amplifier is current driven. In this case it is easy to show that according to KCL, $V_o = - I_i R_2$. The inverting terminal acts as a virtual earth as before, allowing for current driving at the inverting terminal. In addition, this terminal has a low impedance due to negative feedback. If the non-inverting terminal is at different potential from ground, say V_{ref}, then it follows that the inverting terminal will also be at that voltage since they track each other and hence $V_o = - I_i R_2 + V_{ref}$.

Fig. 8.24 A
transimpedance amplifier

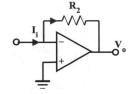

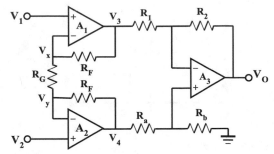

Fig. 8.26 Instrumentation amplifier

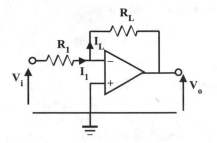

Fig. 8.25 A transconductance amplifier with a floating
load R_L

8.10 Transconductance Amplifier

A transconductance amplifier performs a voltage-
to-current conversion into a load. To accomplish
this, we consider the circuit of Fig. 8.25 and view
R_L as a load resistor that we wish to drive. Then
the current though R_1 given by $I_1 = V_i/R_1$ must be
equal to the current I_L through R_L. This current is
independent of R_L and depends only on the input
voltage V_i and R_i. The load resistor R_L is a floating
load since neither end is connected to ground.

Example 8.13

Design a transconductance amplifier which drives
a floating load $R_L = 2$ k with a transconductance
of 1 mA/V.

Solution The desired transconductance is 1 mA/
V and therefore $1/R_1 = 1$ mA/V giving $R_1 = 1$ k.

8.11 The Instrumentation Amplifier

In many applications, there is a need to have an
amplifier that can provide large amounts of gain
so as to adequately amplify μV signals at moderate frequencies. Examples may include signals
from temperature or pressure sensors (e.g. strain

gauges), medical instruments and other industrial
transducers. A single op-amp cannot generally do
the job since at gain of 1000 the device is unlikely
to have sufficient bandwidth for practical
applications. In fact, at a gain of 1000, the bandwidth of the op-amp can be expected to be 1/1000
of its gain bandwidth product. An alternative
approach to amplification is to cascade several
stages of smaller gain, but this introduces
unwanted complications such as offsets and
added noise at the final output stage. One of the
best approaches is the three op-amp instrumentation amplifier shown in Fig. 8.26.

Operation of this device can best be understood if it is thought of as consisting of three
sub-circuits. The first is a differential amplifier
A_3 as discussed in Sect. 8.6 and shown in
Fig. 8.17, and the other two are special cases of
the difference amplifier A_1 and A_2 shown previously in Fig. 8.18. In most commercial instrumentation amplifiers, the resistor R_G is typically
applied externally as shown in Fig. 8.27 with
some devices also offering R_F externally (Example: Analog Devices AD625). According therefore to Eqs. (8.36) and (8.34), the node voltages
V_3, V_4 and V_o can be expressed as (Fig. 8.27)

$$V_3 = \left(1 + \frac{R_F}{R_G}\right)V_1 - \frac{R_F}{R_G}V_y \qquad (8.51)$$

$$V_4 = \left(1 + \frac{R_F}{R_G}\right)V_2 - \frac{R_F}{R_G}V_x \qquad (8.52)$$

$$V_O = \frac{R_2}{R_1}\left(\frac{1 + \frac{R}{R_2}}{1 + \frac{R_a}{R_b}} \cdot V_4 - V_3\right) \qquad (8.53)$$

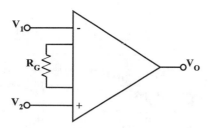

Fig. 8.27 The schematic representation of the instrumentation amplifier

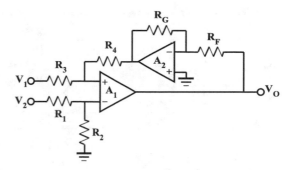

Fig. 8.28 Alternative variable gain differential amplifier

For matched resistors $R_1 = R_a$ and $R_2 = R_b$, and noting that voltages V_x and V_y are equal to V_1 and V_2, respectively, Eqs. (8.51) to (8.53) can be solved to yield an output voltage expressed as

$$V_O = \frac{R_2}{R_1}\left(1 + \frac{2R_F}{R_G}\right)(V_2 - V_1) \qquad (8.54)$$

Clearly the product of the first two terms in the right-hand side of (8.54) can yield very large gains if chosen appropriately. The ratio $V_o/(V_2 - V_1)$ can be thought of as the differential gain

$$A_D = \frac{V_O}{V_2 - V_1} = \frac{R_2}{R_1}\left(1 + \frac{2R_F}{R_G}\right) \qquad (8.55)$$

for this amplifier which can be easily changed by varying the single resistor R_G. The gain can also be changed by varying the other resistors, but for this case, several resistors of equal value would have to be changed simultaneously, and that becomes challenging in a practical sense.

Exercise 8.2

The circuit shown in Fig. 8.28 is an alternative variable gain amplifier similar to the instrumentation amplifier just examined except that the inputs are not high impedance. Show for this circuit that its output voltage V_o is given by

$$V_O = \left(\frac{R_4}{R_3}\right)\left(\frac{R_F}{R_G}\right)V_2 - \left(\frac{R_4 + R_3}{R_3}\right)$$
$$\times \left(\frac{R_2}{R_1 + R_2}\right)\left(\frac{R_F}{R_G}\right)V_1$$
$$= \left(\frac{R_2}{R_1}\right)\left(\frac{R_F}{R_G}\right)(V_2 - V_1) \qquad (8.56)$$

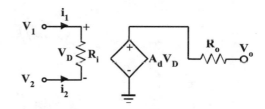

Fig. 8.29 A realistic op-amp model

when $R_2 = R_4$ and $R_3 = R_1$. Can you identify any possible draw backs of this circuit?

8.12   A Realistic Operational Amplifier

In the previous section, the model of Fig. 8.2 represented an ideal operational amplifier, and the circuits considered utilized this model. However, a practical op-amp varies somewhat from this with respect to the internal parameters of the operational amplifier such as input and output resistances. Also, while the gain was assumed to be infinite, all amplifiers have finite gain. Thus, a more realistic model that incorporates finite resistances and gain is shown in Fig. 8.29 where the input resistance R_i is not infinite and the output resistance R_o is not zero. For bipolar operational amplifiers, R_i can be of the order of tens of $M\Omega$, while for CMOS operational amplifiers, it can be around 10 $G\Omega$ which is much larger. Similarly, R_o is usually of the order of tens of ohms for both bipolar and CMOS operational amplifiers and is expected to be low. Note that in the case of bipolar operational amplifiers, a small bias current $i_1 \simeq i_2$ is present at the input terminals. For CMOS operational

Table 8.1 A comparison of various op-amp technologies with the ideal op-amp

	Bipolar	JFET	CMOS	Ideal op-amp
R_i	$6M\Omega$	$10^{13}\Omega$	$10^{12}\Omega$	∞
R_o	$<100\ \Omega$	$<100\ \Omega$	$<100\ \Omega$	0
A_d	$10^3 - 10^6$	$10^3 - 10^6$	$10^3 - 10^6$	∞
$i_1 \simeq i_2$	$\pm40\ \mu A$	$\pm1pA$	$<1pA$	0
$\pm V_{os}$	$25\ \mu V$	$5\ mV$	$\simeq0$	0

Fig. 8.30 Inverting amplifier with finite input and output impedances

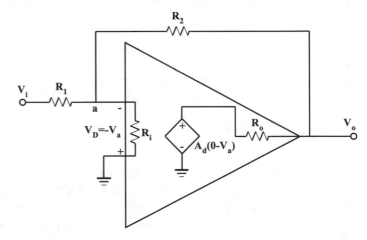

amplifiers, this current is virtually nonexistent and can be neglected altogether. Real op-amps also have offset voltages $\pm V_{os}$ not modelled in Fig. 8.29, but its effect is to cause the output to be nonzero when the differential input is zero. Finally, the open-loop voltage gain varies across op-amps but is normally large. Typically values range from 1000 to 10^7. The ubiquitous 741 has a typical voltage gain of 10^5.

In all operational amplifiers, the gain A_d is actually a frequency-dependent quantity, that is, $A_d \equiv A_d(s)$, but for the purposes of the following section, we shall assume it is constant for all frequencies. In other words, $A_d(s) = A_o$ where A_o is often referred to as the open-loop DC gain. Table 8.1 summarizes the important parameters of realistic operational amplifier (Bipolar, JFET and CMOS) with typical values and compares it to the ideal amplifier shown in Fig. 8.2.

As it turns out, the presence of R_i and R_o does not alter the results derived with the ideal op-amp model. For example, consider the practical model in the inverting configuration of an op-amp where

R_i and R_o have been included in the inverting configuration as shown in Fig. 8.30.

As before, using KCL yields

$$\frac{V_a - V_i}{R_1} + \frac{V_a - V_o}{R_2} + \frac{V_a - 0}{R_i} = 0 \quad (8.57)$$

$$\frac{V_o - (-A_o V_a)}{R_o} + \frac{V_o - V_a}{R_2} = 0 \quad (8.58)$$

Solving (8.57) and (8.58) yields

$$\frac{V_o}{V_i} = -\frac{R_2}{R_1} \cdot \frac{1}{1 + \frac{(R_o + R_2)\left(\frac{1}{R_1} + \frac{1}{R_2} + \frac{1}{R_i}\right)}{A_o - \frac{R_o}{R_2}}} \quad (8.59)$$

Letting $R_1 = 1\ k$ and $R_2 = 10\ k$ and substituting typical values of $R_o = 100\ \Omega$, $R_i = 100\ k$ and $A_o = 10^5$ yield $\frac{V_o}{V_i} = -9.998879$. This value is almost exactly equal to the result obtained using the expression $-R_2/R_1$ or Eq. (8.9) which gives $-R_2/R_1 = -10$. Thus, our assumptions used in the ideal model hold to a very good approximation for the practical case.

8.12.1 Single Supply Operation

In the case of the dual power supply operational amplifier, the output is always referenced to ground. However, in the case of the single supply operational amplifier, the output is typically referenced to mid-supply. Some special op-amps such as the LM324-Quad op-amp are specifically designed to operate on single supplies, but virtually any dual supply op-amp can be operated on a single supply if properly configured. What is needed is that one of the input terminals be referenced to mid-supply as shown in Fig. 8.31 for an inverting amplifier configuration. Here the series connected resistors R are set equal so that a voltage that is half of the supply voltage is generated. The main consideration in the choice of the values of R would be the current drain on the supply. The large capacitor C_{bp} is present to ensure the voltage at the inverting terminal is grounded for AC signals. Resistors R_1 and R_2 function in the normal feedback mode to yield an inverting gain $-R_2/R_1$. Note that raising the non-inverting input terminal to mid-supply automatically raises the inverting terminal to the same voltage though negative feedback. The feedback will also cause the output of the op-amp to be at mid-supply voltage. In order to block this DC, capacitor C_2 is added but can be removed if DC coupling at this voltage level is required. A coupling capacitor C_1 is also necessary at the input because of the DC at the inverting terminal as shown in Fig. 8.31.

To use the op-amp in a non-inverting configuration off a single supply, the circuit of Fig. 8.32 can be employed. The circuit of Fig. 8.32 is essentially the dual of the previous one shown in Fig. 8.31 and operates in the same manner except that the gain is $1 + R_2/R_1$. This circuit no longer enjoys very high input impedance due to the presence of resistors R at the non-inverting input of the op-amp. Therefore, the value of R in this configuration is typically chosen in the mega ohms range, but having large resistors at the input increases the noise contribution to the amplifier.

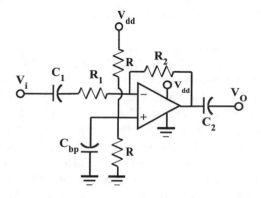

Fig. 8.31 Single supply operation of an op-amp configured as an inverting amplifier

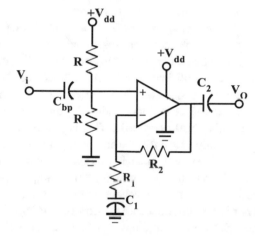

Fig. 8.32 Single supply operation of an op-amp configured as a non-inverting amplifier

Example 8.14
Design an inverting amplifier with a gain of -10 using the LM741 that operates from a 10 V power supply.

Solution For a 10 volt supply, choose $R = 10$ k, thereby allowing 0.5 mA to flow through this potential divider. For a gain of -10, let $R_2 = 100$ k and $R_1 = 10$ k. For a frequency response down to about 10 Hz, $C_1 = 1/2\pi \times 10 \times R_1 = 1.6$ μF. Use $C_1 = 2$ μF. A similar value cam be used for C_2. Finally, $C_{bp} = 1/2\pi \times 10 \times 5 \times 10^3 = 3.2$ μF. Use $C_{bp} = 10$ μF or larger.

8.13   Frequency Effects

In the previous sections, it was assumed that the large op-amp gain A_d is constant for all frequencies and effects on gain caused by internal component capacitances were ignored. In all operational amplifiers, however, the gain A_d is actually a frequency-dependent quantity, that is, $A_d \equiv A_d(s)$. In this section therefore, we examine the effect of the frequency-dependent gain $A_d(s)$ as it applies to real op-amps and its effect on overall closed-loop gain and bandwidth. The op-amp is a multi-stage voltage amplifier having several poles (and zeros) in its transfer function. This manifests itself as a roll-off rate of the frequency response plot that is greater than -20 dB/decade. Previously in Sect. 7.10 of Chap. 7 where feedback amplifiers were discussed, for such a system to be stable in the presence of feedback, we have established that the roll-off rate of the frequency response characteristic where it intersects the closed-loop gain plot must be no greater than -20 dB/decade. This is achieved by compensation such that the amplifier behaves as a single pole system, and this was also discussed in Chap. 7. As pointed out there, the op-amp is an excellent example of a voltage amplifier as it has a very high input impedance and low output impedance. Since most op-amps will find uses at unity gain whether as a buffer or an inverter, they are designed or internally compensated so that one pole dominates and hence are unity-gain stable. In this section we treat the op-amp as a single pole system and apply the theory developed earlier to the op-amp.

Thus, an op-amp is a voltage amplifier with a high input impedance and low output impedance. The input and output signals being voltages, the feedback type is voltage-series as shown in Fig. 8.33. This is the **non-inverting** configuration.

The transfer function $A_v = V_o/V_i$ of the open-loop amplifier in Fig. 8.33 (a) is represented as

$$A_v = \frac{A_o}{1 + jf/f_o} \qquad (8.60)$$

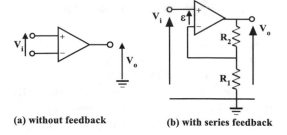

(a) without feedback   **(b) with series feedback**

Fig. 8.33 Voltage amplifier

where A_o is the low-frequency gain and f_o is the open-loop bandwidth. The transfer function $A_f = V_o/V_i$ of the closed-loop amplifier in Fig. 8.33 (b) is given by

$$A_f = \frac{A_v}{1 + \beta A_v} \qquad (8.61)$$

where $\beta = R_1/(R_1 + R_2)$ is the feedback factor. Substituting for A_v using (8.60) and manipulating, we get

$$A_f = \frac{\left(1 + \frac{R}{R_1}\right) \cdot \frac{A_o \beta}{1 + A_o \beta}}{1 + jf/f_o(1 + A_o \beta)} \qquad (8.62)$$

At low frequencies, Eq. (8.62) reduces to

$$A_f(DC) = \left(1 + \frac{R_2}{R_1}\right) \cdot \frac{A_o \beta}{1 + A_o \beta} \qquad (8.63)$$

In the operational amplifier, A_o is very large, and hence the loop gain $A_L = A_o\beta$ is such that $A_L = A_o\beta >> 1$. As a result, Eq. (8.63) reduces to

$$A_f(DC) = 1 + \frac{R_2}{R_1} = 1/\beta \qquad (8.64)$$

which is Eq. (8.21) that was previously derived. Note that from (8.64), $A_L = A_o\beta = A_o/A_f(DC)$, i.e. the loop gain is equal to the open-loop gain divided by the closed-loop gain. In general, $A_o\beta >> 1$ reduces (8.62) to

$$A_f = \frac{1 + R_2/R_1}{1 + jf/\overline{f}_o} \qquad (8.65)$$

where

$$\bar{f}_o = f_o A_o \beta \qquad (8.66)$$

is the closed-loop bandwidth.

Thus from (8.66), the closed-loop bandwidth $\bar{f}_o$ is increased by the factor of the loop gain as compared with the open-loop bandwidth f_o, directly as a result of feedback. Moreover, it is directly proportional to the loop gain $A_o\beta$. Since A_o is fixed, by expressing (8.66) in terms of $A_f(DC)$ as

$$\bar{f}_o = f_o A_o / A_f(DC) \qquad (8.67)$$

it is clear that the closed-loop bandwidth $\bar{f}_o$ is inversely proportional to the closed-loop low-frequency gain $A_f(DC)$. The loop gain $A_L = A_o\beta = A_o/A_f(DC)$ is the ratio of the open-loop gain and the closed-loop gain. When expressed in dB, it is given by

$$A_L(dB) = 20\log A_o - 20\log A_f(DC)$$
$$= \text{low-freq open-loop gain } (dB) -$$

low-freq closed-loop gain (dB) \qquad (8.68)

This represents the amount of feedback applied in the system as shown in Fig. 8.34. The maximum feedback occurs when $\beta = 1$ and is given by $A_{L(max)} = 20\log A_o$. This is the unity-gain configuration and is obtained by setting $R_1 \to \infty$ and/or $R_2 = 0$ in Fig. 8.33. The corresponding bandwidth is given by

$$\bar{f}_o = f_o A_o \qquad (8.69)$$

which is referred to as the gain bandwidth product (GBP) of the voltage amplifier.

Finally, from (8.67),

$$A_f(DC)\bar{f}_o = A_o f_o = GBP = constant \qquad (8.70)$$

This equation states: closed-loop gain x closed-loop bandwidth = GBP = constant.

Example 8.15

An operational amplifier has an open-loop bandwidth $f_o = 10$ Hz with $GBP = 1$ MHz. Determine the bandwidth for closed-loop gains (i) 10 and (ii) 100.

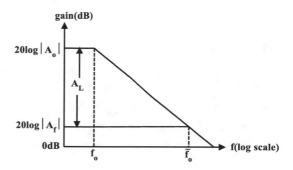

Fig. 8.34 Frequency response plot showing loop gain

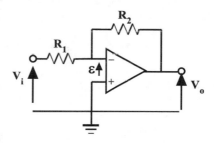

Fig. 8.35 Inverting amplifier

Solution (i) Using Eq. (8.70), the bandwidth for a closed-loop gain of 10 is $\bar{f}_o = GBP/A_f(DC) = 1\,\text{MHz}/10 = 100\,\text{kHz}$.
(ii) Similarly, the bandwidth for a closed-loop gain of 100 is $\bar{f}_o - GBP/A_f(DC) = 1\,\text{MHz}/100 = 10\,\text{kHz}$.

Inverting Configuration

The non-inverting amplifier can be converted to the **inverting** configuration by interchanging the signal input and ground points as shown in Fig. 8.35. By straightforward analysis using first principles, the transfer function $A_f \equiv V_o/V_i$ of the closed-loop amplifier is given by

$$A_f = \frac{-\frac{R}{R_1}\cdot\frac{A_o\beta}{1+A_o\beta}}{1 + jf/f_o(1+A_o\beta)} \qquad (8.71)$$

At low frequencies, Eq. (8.71) reduces to

$$A_f(DC) = -\frac{R_2}{R_1}\frac{A_o\beta}{1+A_o\beta} \qquad (8.72)$$

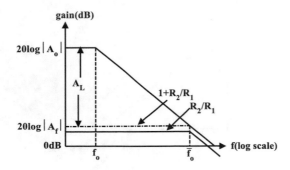

Fig. 8.36 Frequency response of inverting amplifier

which for large loop gains such that $A_o\beta >> 1$ becomes

$$A_f(DC) = -\frac{R_2}{R_1} \qquad (8.73)$$

The full Eq. (8.71) for large loop gains reduces to

$$A_v = \frac{-R_2/R_1}{1 + jf/\bar{f}_o}, A_o\beta >> 1 \qquad (8.74)$$

where

$$\bar{f}_o = f_o A_o \beta \qquad (8.75)$$

is the closed-loop bandwidth and the low-frequency closed-loop gain $A_f \equiv \frac{V_o}{V_i}(DC)$ is given by

$$A_f(DC) = -R_2/R_1 = 1 - 1/\beta \qquad (8.76)$$

The expression for the closed-loop bandwidth (8.75) of the inverting configuration is exactly the same as the expression (8.66) for the closed-loop bandwidth of the non-inverting configuration. Since the magnitude of the low-frequency gain R_2/R_1 is less than that of the non-inverting configuration $1 + R_2/R_1$, the frequency response plot is shifted downward relative to that of the non-inverting configuration as shown in Fig. 8.36.

It follows also that the closed-loop bandwidth for the inverting configuration also varies as β is varied and hence as the low-frequency gain $A_f(DC)$ is varied. Specifically, using (8.76) in (8.75) gives

Fig. 8.37 Transimpedance amplifier

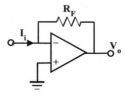

$$\bar{f}_o = f_o A_o / \left(1 - A_f(DC)\right) \qquad (8.77)$$

The maximum feedback $20 \log A_o$ which occurs for $\beta = 1$ results in $A_f(DC) = 0$. This corresponds to $R_1 \to \infty$ and/or $R_2 = 0$.

Finally, in the inverting mode, it is possible to replace R_1 and V_i by a current source such that the input signal is a current I_i as shown in Fig. 8.37. Let $R_2 = R_F$. The circuit now functions as a transimpedance amplifier (current in, voltage out), but the feedback remains voltage-series. The transfer function $R_M \equiv V_o/I_i$ is given by

$$R_M = \frac{-R_F.A_o/(1+A_o)}{1 + jf/f_o(1+A_o)} \qquad (8.78)$$

which reduces to

$$R_M = \frac{-R_F}{1 + jf/\bar{f}_o} \qquad (8.79)$$

where

$$\bar{f}_o = f_o A_o \qquad (8.80)$$

and the low-frequency closed-loop transimpedance $R_M(DC)$ is given by

$$R_M(DC) = -R_F \qquad (8.81)$$

This is the well-known voltage-to-current converter. Since the feedback mode is still voltage-series and the impedance of the current source is (ideally) infinite, it follows that $R_1 \to \infty$ and hence $\beta \to 1$. This configuration therefore corresponds to maximum voltage-series feedback with $20 \log A_o dB$, and as is evident from (8.80), the closed-loop bandwidth $\bar{f}_o$ is the GBP of the associated amplifier. It is interesting to note that because of the high impedance of the signal source, the closed-loop bandwidth $\bar{f}_o$ is independent of the closed-loop low-frequency transimpedance R_F and hence can be varied without affecting the bandwidth.

8.14 Non-ideal Effects

Apart from bandwidth issues, there are other factors or non-ideal effects that greatly influence the performance of an op-amp. These include offset voltages and currents, common mode rejection ratio, dynamic range and slewing. In considering some of these, we begin with some definitions.

Offset Voltages *(V*os*)* Offset voltages in an op-amp occur because of mismatch in the input stages which results in an output voltage even when the differential input voltage is zero. Hence an offset voltage can be viewed as the voltage that must exist between the two input terminals of the op-amp in order to make the output voltage zero. Such voltages can have values of both polarities. A typical range of values might be $\pm 10 \ \mu V$ to ± 10 mV for bipolar op-amps and as high as ± 50 mV for FET input op-amps.

Input Offset Currents *(I*os*)* Offset currents exist because there is a finite current flowing in/out of the input terminals that is required to bias the input transistors of the op-amp. By definition, I_{os} is equal to the difference of the input bias currents or

$$I_{os} = \pm |I_{B_1} - I_{B_2}| \qquad (8.82)$$

where I_{B_1} and I_{B_2} are the input bias currents which may flow into or out of the op-amp. Sometimes a manufacture may specify an op-amp's input bias current (I_B) which is the average of the two currents I_{B1} and I_{B2}. That is,

$$I_B = \frac{I_{B_1} + I_{B_2}}{2} \qquad (8.83)$$

Bias currents are worse for op-amps made with bipolar input transistors than for those made with FET transistors. The input bias current I_B is typically in the range 10 nA to 1 μA for bipolar op-amps and $<100 pA$ for FET input op-amps. Although typically in the nanoampere to microampere range, input offset currents cause problems by creating unwanted voltage drops across input resistors. The overall effect is to cause the output of the op-amp to have a small offset voltage depending on the amplifier configuration used.

Common Mode Rejection Ratio (CMRR) CMRR is the ratio of the differential gain of the op-amp to the common mode gain expressed in dB. Stated mathematically it is

$$CMRR = \left| \frac{A_d}{A_{cm}} \right| \qquad (8.84)$$

Common mode gain A_{cm} is the ratio of the output voltage of an op-amp to the input voltage applied to both inputs simultaneously, i.e. with the inputs of the op-amp electrically connected. Under these conditions, there ought to be no output voltage from the op-amp as it ideally responds only to a differential voltage. The principal reason for an output in these circumstances is that the gain at one input is slightly different from that of the other. This is an undesirable effect since the objective of having differential inputs is to cancel any common mode signals that may be present. In op-amp circuits employing negative feedback, common mode gain is usually not a problem, but in instrumentation amplifiers without overall feedback, it can become problematic. Because A_d is frequency dependent, it means that the CMRR is also frequency dependent. Manufactures usually specify only the low-frequency value of the CMRR as its value decreases with increasing frequency. Expressed in dB the CMRR is given by

$$CMRR_{dB} = 20 \log_{10} \left| \frac{A_d}{A_{cm}} \right| \qquad (8.85)$$

For a typical op-amp, CMRR is in the range 60–70 dB. More high-performance op-amps however have CMRR of the order of 120 dB.

Power Supply Rejection Ratio (PSRR) PSRR is a measure of the level of insensitivity of the op-amp input to power supply changes. Expressed mathematically it can be defined in dB as

$$PSRR = 20 \log_{10} \left| \frac{\Delta V_{\text{supply}}}{\Delta V_{in}} \right| = \frac{1}{S} \qquad (8.86)$$

where S is power supply voltage sensitivity which is the change in input voltage ΔV_{in} per unit change in supply voltage ΔV_{supply}. Because op-amps have two supplies, PSRR is usually defined for each supply.

Slew Rate (SR) Slew rate is the maximum rate of change of the output voltage V_o. It is typically specified in *volts per microsecond* and is usually stated by the manufacturer. Slew rate limiting is a non-linear characteristic that can occur in all op-amps. It occurs when the maximum rate of change of the output voltage is reached. For general-purpose op-amps, this rate is typically in the range of 0.1 to 10 *V/μs*. For more special-purpose op-amps, it can reach as high as a few hundred *V/μs* to a few thousand *V/μs*. In the subsequent subsections, we discuss each of these effects in more details and look at ways of addressing them and their implications in a practical design.

8.14.1   Offset Voltage and Currents

As mentioned previously, manufacturing differences in the input stage of an op-amp are primarily responsible for the presence of offset voltages. Also, finite input currents at the input stages in the presence of resistances give rise to offset voltages at the output of the op-amp. In order to make sense of this effect, we consider the closed-loop model in Fig. 8.38. Here we model the offset voltage by V_{os} and the two input bias currents by I_{B1} and I_{B2}. Note that it is their difference $I_{B1} - I_{B2}$ that gives rise to an input offset current. Straightforward analysis of this circuit using superposition yields an output of the form

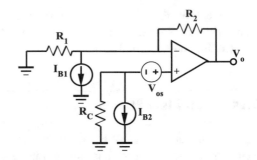

Fig. 8.38 Model used to represent offset voltage and current in an op-amp

$$V_o = I_{B_1} R_2 - I_{B_2} R_c \left(1 + \frac{R_2}{R_1} \right)$$
$$+ V_{os} \left(1 + \frac{R_2}{R_1} \right) \qquad (8.87)$$

which after manipulation can be written as

$$V_o = (I_{B_1} - I_{B_2}) R_2 + I_{B_2} \left[R_2 - R_c \left(1 + \frac{R_2}{R_1} \right) \right] + V_{os} \left(1 + \frac{R_2}{R_1} \right)$$
$$= I_{os} R_2 + I_{B_2} \left[R_2 - R_c \left(1 + \frac{R_2}{R_1} \right) \right] + V_{os} \left(1 + \frac{R_2}{R_1} \right) \right]$$
$$(8.88)$$

Equation (8.88) indicates that the output is made up of three terms: the first one due to the input offset current I_{os} and the third term due to the input offset voltage V_{os}. The middle term is due to one of the input bias currents, and it can be observed that it can be immediately cancelled if we choose $R_c = R_1 // R_2$. In other words, to reduce the middle term to zero, we can add a resistor to the non-inverting input whose value is given by $R_c = R_1 // R_2$. Of course, if the non-inverting terminal is directly connect to ground ($R_c = 0$), the size of R_2 and I_{B_2} will help determine the output offset voltage. The first term in (8.88) indicates that to reduce its contribution to offset voltage for a given op-amp, R_2 must be small. Of course, this comes at the expense of a reduction in gain. The minimum value of R_2 depends on the output signal swing and the load to which the op-amp is connected. Reducing R_2 also affects the third term in (8.88), but in this case it is the ratio R_2/R_1 that must be considered since this ratio also affects the closed-loop gain.

In some op-amps, it is possible to null the output voltage produced by the offset sources using an external voltage or as in the case of the 741 op-amp a potentiometer connected across two offset "null" points. For the 741 these are pins 1 and 5 on the 8-pin DIP package with the potentiometer wiper connected to the negative supply. Note however that temperature changes in the components will eventually cause the output voltage to change with temperature. For other op-amps laser trimming techniques are used to keep the output offset voltage to a minimum.

Example 8.16

Using the specifications for the LF351 JFET input op-amp, determine the output offset voltage for the uncompensated and compensated inverting amplifier with resistors $R_2 = 10$ k and $R_1 = 1$ k.

Solution From the data sheet of the LF351 op-amp, it lists maximum values of $V_{os} = \pm 10$ mV, $I_B = 200$ pA and $I_{os} = \pm 100$ pA all at $25\,°C$. We shall assume room temperature operation. For the uncompensated op-amp, $R_c = 0$, but I_{B2} is not specified as required by (8.88). We can, however, make a best guess estimate by using (8.82) and (8.83) to solve for I_{B2}. Solving yields two values for I_{B2} of 450pA and 350pA, but let us use the smaller of the two numbers. Substitution of the relevant numbers into (8.88) yields

$$V_{o(nc)} = 10\,k\Omega(\pm 10\,pA) + 350\,pA(10\,k\Omega)$$
$$+ (\pm 10\,mV)\left(1 + \frac{10\,k\Omega}{1\,k\Omega}\right)$$
$$= \pm 0.1\,\mu V + 3.5\,\mu V \pm 110\,mV$$

In other words, the output offset voltage lies within the range -110 mV $\leq V_{0(nc)} \leq 110$ mV. Clearly for the uncompensated amplifier, the offset voltage is the term that dominates because I_{os} and I_B are so small. With that in mind if the nominal value of $V_{os(nom)} = \pm 5$ mV is used, then the output voltage lies in the range -55 mV $\leq V_{o(nom,\ nc)} \leq 55$ mV. For the compensated amplifier, the situation does not change much since the second term in (8.88) is zero. Thus the output voltage lies in the range -110 mV $\leq V_{o\ (c)} \leq 110$ mV and with the nominal value in the range -55 mV $\leq V_{o(nom,\ c)} \leq 55$ mV. Therefore, in this case compensation can be omitted, but it will not be the case for all op-amps.

8.14.2 CMRR and PSRR

As stated earlier the CMRR is a measure of how well an op-amp amplifies differential signals and rejects common mode ones as given by the ratio

$$CMRR = \left|\frac{A_d}{A_{cm}}\right| = \left|\frac{A_o}{A_{cm}}\right| \tag{8.89}$$

where we shall assume that we are working at low frequencies so that $A_d = A_o$ can be used. An equation that fully describes the effects of both differential and common mode gain is therefore given by

$$V_o = A_d(V_1 - V_2) + A_{cm}\left(\frac{V_1 + V_2}{2}\right)$$
$$= A_d V_D + A_{cm} V_{cm}. \tag{8.90}$$

In Eq. (8.90), the factor of ½ is there because under common mode conditions when $V_1 = V_2 = V_{cm}$, then $V_o = A_{cm}V_{cm}$ which is consistent with the definition of common mode gain. Equation (8.90) can therefore be rewritten as

$$V_o = A_o\left[V_D + \frac{V_{cm}}{CMRR}\right] \tag{8.91}$$

or

$$V_o \cong A_o\left[V_1 - V_2\left(1 - \frac{1}{CMRR}\right)\right] \tag{8.92}$$

if $V_1 \simeq V_2$. Both representations are equally valid, but (8.92) indicates that we can simply add a voltage source of value $V_2/CMRR$ to the positive input on an op-amp to model its effect. Such a model is shown in Fig. 8.39.

In a similar manner, we can describe the PSRR by the equation $V_o = A_o V_D + S\Delta V_{supply}A_o$

or

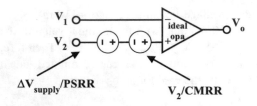

Fig. 8.39 Op-amp model that accounts for both CMRR and PSRR

$$V_o = A_o \left(V_D + \frac{\Delta V_{\text{supply}}}{PSRR} \right) \qquad (8.93)$$

As in the case of the CMRR, we can add a voltage source of value $\Delta V_{supply}/PSRR$ to the non-inverting terminal. A model that incorporates both CMRR and PSRR is shown in Fig. 8.39. Finally note that it is not uncommon to find the CMRR and the PSRR of the same order for a given op-amp.

Exercise 8.3
Show that by using the circuit of Fig. 8.18 with $V_1 = V_2 = V_i$ the CMRR of an op-amp can be measured with value $CMRR = 1 + \frac{R_2}{R_1}$.

8.14.3 CMRR of the Instrumentation Amplifier

For the instrumentation amplifier previously discussed in Sect. 8.1.10, it can be shown that its common mode gain is given by

$$A_{cm} = \frac{V_o}{V_{cm}} = \frac{R_a}{R_a + R_b} \left(1 + \frac{R_2}{R_1} \right) - \frac{R_2}{R_1} \quad (8.94)$$

Using (8.55) therefore the CMRR for this circuit becomes

$$CMRR = \left| \frac{A_{diff}}{A_{cm}} \right| = \frac{\left(1 + \frac{2R_F}{R_G} \right)}{\left(\frac{1+\frac{R}{R_2}}{1+\frac{R_a}{R_b}} - 1 \right)} \qquad (8.95)$$

Clearly if $R_1 = R_a$ and $R_2 = R_b$ as stated before, then the $CMRR \to \infty$. In practice this will not be the case because of resistance tolerances. However, it is not uncommon to easily achieve CMRRs in excess of $100\ dB$. Like

op-amps, the CMRR in instrumentation amplifiers is frequency dependent, but its frequency dependency usually comes from parasitic capacitances at its input nodes.

8.14.4 Slew Rate

Slew rate limiting occurs when an op-amp is required to change its output at a rate that the internal dynamics of the op-amp do not allow. It represents the maximum rate of change of the output voltage expressed in volts/μs. To understand this limitation, consider the simplified model of a typical three-stage op-amp in Fig. 8.40 represented by a differential transconductor stage, a second gain stage and a third stage buffer for driving low-impedance nodes. The second stage usually contains a compensation capacitor C_c, and the first-stage transconductor has a bias current I_b to bias the differential input pair. If the input voltage suddenly receives a large differential voltage, then the output of the transconductance stage will either sink current into C_c or source current from C_c depending on the polarity of the differential voltage. In either case, that current approaches the bias current value so that $I_o \simeq \pm I_b$. Assuming for the moment that I_o flows into C_c so that the positive sign can be used, then the corresponding rate of change of output voltage from amplifier A_2 is given by

$$\frac{dV_c}{dt} = \frac{I_b}{C_c} \qquad (8.96)$$

Since this voltage is directly conveyed to the output buffer, it follows that $V_o = V_c$ so that the output voltage of the op-amp experiences a rate of change of voltage which we define as the slew rate of

$$SR^+ = \frac{dV_o}{dt} = \frac{I_b}{C_c} \qquad (8.97)$$

The positive sign is included in (8.97) to show that it represents the positive slew rate. By a similar reasoning, if the polarity of the input

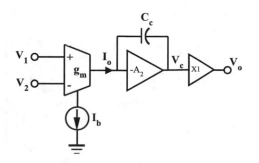

Fig. 8.40 Simplified model of a three-stage op-amp

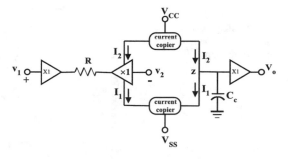

Fig. 8.41 An alternative op-amp model

differential voltage were reversed, then $-I_b$ would flow out of C_c generating a negative slew rate of

$$SR^- = \frac{dV_o}{dt} = -\frac{I_b}{C_c} \qquad (8.98)$$

The primary reason there are two different values for positive and negative slew rates is that sometimes the input stage is not completely symmetric and different values of output current I_o exist for negative and positive differential inputs. A typical value is therefore quoted by most manufacturers with a square wave or step input being the most commonly used test signal.

The topology shown in Fig. 8.40 is not the only topology used in designing op-amps. New high slew rate op-amps such as the LT1364 employ a different topology that offers higher slew rates than the approach of Fig. 8.40 and is shown in Fig. 8.41. This is achieved by sensing currents that are proportional to the difference of the differential inputs. Buffers provide high-impedance inputs to the op-amp. Resistor R limits the maximum current that can flow through the buffers. The sensed currents are then copied, and their difference is used to drive a high-impedance node labelled z in Fig. 8.41 to achieve high gain in the op-amp. The voltage at that node is buffered to preserve the high gain. The high slew rates come for the fact that C_c can be made small, while the sensed currents $I_{1, 2}$ can be large.

If the output of the op-amp is a sine wave described by the equation $V_o(t) = V_a \sin \omega t$ where V_a is the amplitude and ω is the angular frequency, then the output signal achieves its maximum slope at its zero crossings and is given by

$$\left.\frac{dV_o(t)}{dt}\right|_{max} = V_a \omega = 2\pi V_a f \qquad (8.99)$$

For linear op-amp operation, the slew rate given in (8.99) must not exceed the slew rate of the op-amp; otherwise slew rate limiting occurs. This imposes an upper bound on the frequency of operation of the op-amp for a given sinusoidal output amplitude. That bound is sometimes referred to as the *full power bandwidth* (BW$_p$) and is given by

$$BW_p = \frac{SR}{2\pi V_a}. \qquad (8.100)$$

Equation (8.100) indicates that bandwidth can be traded for output amplitude and vice versa to avoid slew rate limiting. As an example, consider an op-amp with a slew rate of $SR = 13$ V/μs and maximum peak output voltage $V_a = 14$ V being powered from a $\pm$ 15 V supply. The full power bandwidth can then be calculated to be $BW_p = 13 \times 10^6/(2\pi \times 14) = 148$ kHz. Its theoretical peak-to-peak output swing versus frequency will therefore resemble the plot shown in Fig. 8.42. Therefore in order to operate this op-amp at 1 MHz, the maximum peak output voltage V_a that can be delivered using (8.100) is given by $V_a = SR/2\pi BW_p = 13 \times 10^6/2 \times \pi \times 10^6 = 2.1$ V. Finally, manufactures often specify a plot like that in Fig. 8.41 but seldom provide BW_p in their specifications because it depends on the supply voltage, gain and load conditions.

Fig. 8.42 Peak-to-peak
output voltage versus
frequency for a slew rate-
limited op-amp

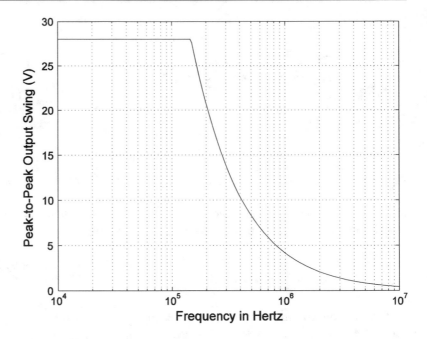

8.15 The Current Feedback Amplifier

The current feedback amplifier (CFA) is a
transimpedance amplifier. It accepts a current at
the input and delivers a voltage at the output and
has a low input impedance and a low output
impedance. It therefore employs voltage-shunt
feedback in which the output voltage is sampled
and the feedback signal is a current. Because of its
topology, it typically has larger bandwidths than
the op-amp and extremely high slew rates.
Indeed, bandwidths as high as 2GHz and slew
rates of the order of thousands of volts/μs are
possible with the CFA. However, the CFA suffers
from several disadvantages compared with the
op-amp including more noise, larger DC offsets
and certain feedback restrictions.

The operation of the CFA can best be understood by considering the idealized model shown
in Fig. 8.43. Between the two inputs is an
idealized buffer with infinite impedance on the
non-inverting input and zero impedance on the
inverting terminal. As a result of the buffer, the
inverting terminal will always follow the
non-inverting terminal much like an op-amp but

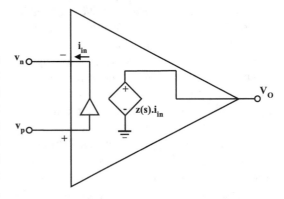

Fig. 8.43 Simplified model of the CFA

of course for a different reason. In an op-amp,
negative feedback causes this following action,
but it is the buffer in the case of the CFA. At the
zero-impedance inverting terminal, a current out
of this terminal generates an output voltage V_o
through a transimpedance $Z(s)$ given by

$$V_o = Z(s)I_i \qquad (8.101)$$

It is because of this why it is sometimes called
a transimpedance amplifier. At the output we
assume that the current-controlled voltage source

Fig. 8.44 The CFA configured as a transresistance amplifier

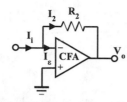

(CCVS) has zero impedance. The transimpedance $Z(s)$ is assumed to follow the one pole model given by

$$Z(jf) = \frac{Z_o}{1 + jf/f_o} \qquad (8.102)$$

where f_o is the -3 dB trans-resistance pole frequency and Z_o is the low-frequency trans-resistance.

In order to utilize the CFA in a transimpedance application, let us consider the circuit of Fig. 8.44 with feedback resistor R_2. An input current I_i is injected into the low-impedance inverting node, and we wish to determine the output voltage V_o. Because the input resistance at the inverting input is zero, the voltage at this terminal is also zero. Using KCL at this terminal yields

$$I_i = I_2 + I_\varepsilon \qquad (8.103)$$

For the device,

$$V_o = -I_\varepsilon Z \qquad (8.104)$$

and

$$I_2 = (0 - V_o)/R_2 = -V_o/R_2 \qquad (8.105)$$

Hence, (8.103) becomes

$$I_i = \frac{-V_o}{Z} + \frac{-V_o}{R_2} = \frac{-V_o}{R_2//Z} \qquad (8.106)$$

Then the transfer function V_o/I_i is given by

$$\frac{V_o}{I_i} = -R_2\left(\frac{1}{1 + R_2/Z}\right) \qquad (8.107)$$

If $Z \to \infty$, then $\frac{V_o}{I_i} = -R_2$. Considering now the frequency effects associated with the transimpedance given by $Z(jf) = \frac{Z_o}{1+jf/f_o}$, then (8.107) becomes

$$\frac{V_o}{I_i} = -R_2//Z$$

$$= -R_2 \frac{Z_o\beta/(1 + Z_o\beta)}{1 + jf/f_o(1 + Z_o\beta)} \qquad (8.108)$$

where $\beta = I_2/V_o = 1/R_2$ is the feedback factor and $Z_o\beta$ is the loop gain. In current feedback amplifiers, $Z_o > > 1$, and hence $Z_o\beta > > 1$. Equation (8.108) therefore reduces to

$$\frac{V_o}{I_i} = \frac{-R_2}{1 + jf/f_o(1 + Z_o\beta)} \qquad (8.109)$$

From this the closed-loop bandwidth is given by

$$\bar{f}_o = f_o(1 + Z_o\beta) \qquad (8.110)$$

Since $Z_o\beta > > 1$, this reduces to

$$\bar{f}_o = f_oZ_o\beta = f_oZ_o/R_2 \qquad (8.111)$$

From (8.109), the low-frequency trans-resistance is $-R_2$, and from (8.111), it can be seen that the closed-loop bandwidth varies with R_2 (Fig. 8.45).

A plot of eq. (8.109) with $Z_o = 3$ M and $f_o = 5000$ Hz showing trans-resistance (log scale) versus frequency for varying values of R_2 is shown in Fig. 8.38. For example, for $R_2 = 1$ k, the magnitude of the low-frequency trans-resistance is $R_2 = 1$ k giving $\log 10^3 = 3$ on the log scale, and from (8.111), the closed-loop bandwidth is $\bar{f}_o = f_oZ_o/R_2 = 5 \times 10^3 \times 3 \times 10^6/10^3 = 15$ MHz. Clearly as R_2 decreases in value, the bandwidth of the amplifier increases, and the curves of Fig. 8.38 closely resemble those of the closed-loop gain of the op-amp as R_2 changes. Conversely if R_2 increases, the transresistance increases, but the bandwidth decreases.

Can the value of R_2 be reduced to zero? The answer to this is no. The reason for this lies in the fact that (8.102) represents a single pole model and extra poles due to circuit design exist in the circuit. As R_2 decreases, the effect of additional circuit poles will cause peaking in the frequency response of the amplifier. If R_2 is decreased even further, it is possible that the circuit may break into oscillations. For this reason, CFA manufactures usually specify a minimum value

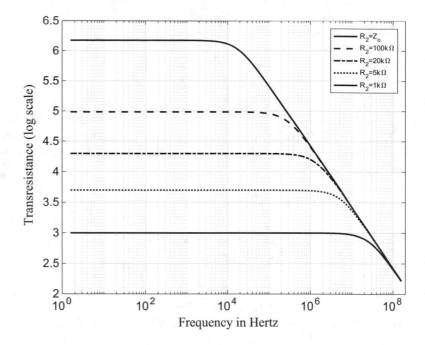

Fig. 8.45 Frequency response plot of trans-resistance with varying R_2

for R_2 below which the CFA performance suffers considerably. Typical values of R_2 fall in the few kilo-ohms range and are rarely less than 1 $k\Omega$.

Inverting Amplifier

If an input resistor R_1 is now added to the inverting terminal and a voltage source drives this amplifier, an inverting voltage amplifier results as shown in Fig. 8.46 For this configuration, noting that the input resistance at the inverting terminal is zero, then $I_i = V_i/R_1$, and therefore substituting for I_i in Eq. (8.109) yields the transfer function given by

$$\frac{V_o}{V_i} = \frac{-R_2/R_1}{1 + jf/f_o(1 + Z_o\beta)} \quad (8.112)$$

which for low frequencies becomes

$$\frac{V_o}{V_i} = -\frac{R_2}{R_1} \quad (8.113)$$

Equation (8.113) indicates that the expression $-R_2/R_1$ for the closed-loop gain of the inverting amplifier for the CFA is the same as that for the op-amp. The -3 dB closed-loop bandwidth is given by

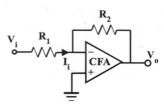

Fig. 8.46 CFA inverting amplifier

$$\bar{f}_o = f_o(1 + Z_o\beta) = f_o(1 + Z_o/R_2) \quad (8.114)$$

which is the same as the tras-resistance case in Eq. (8.110). Note that while this value is dependent on R_2, it is not dependent on R_1. Thus, the closed-loop bandwidth of the CFA in the inverting configuration is independent of resistor R_1. This fact makes the CFA very attractive compared to the op-amp whose gain and bandwidth are directly related by the *gain bandwidth product*. It follows that resistor R_1 may be changed to vary the CFAs closed-loop gain via $-R_2/R_1$ and the bandwidth will remain constant provided R_2 is fixed.

Non-inverting Amplifier

To utilize the CFA in a non-inverting application, let us consider the circuit of Fig. 8.47 with

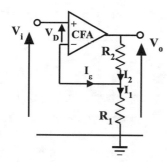

Fig. 8.47 The CFA used in a non-inverting application

feedback resistor R_2 and grounded resistor R_1. The input voltage is applied to the non-inverting high-impedance terminal. Then from KCL,

$$I_1 = I_\varepsilon + I_2 \qquad (8.115)$$

Noting that $V_D = 0$, from (8.115) we have

$$\frac{V_i}{R_1} = \frac{V_o}{Z} + \frac{V_o - V_i}{R_2} \qquad (8.116)$$

After substitution for Z and manipulation, we get

$$\frac{V_o}{V_i} = \frac{\left(1 + \frac{R}{R_1}\right) \cdot \left(\frac{Z_o \beta}{1 + Z_o \beta}\right)}{1 + jf/f_o(1 + Z_o \beta)} \qquad (8.117)$$

which for $Z_o \beta >> 1$ becomes

$$\frac{V_o}{V_i} = \frac{1 + \frac{R}{R_1}}{1 + jf/f_o(1 + Z_o \beta)} \qquad (8.118)$$

At low frequencies this reduces to

$$\frac{V_o}{V_i} = 1 + \frac{R_2}{R_1} \qquad (8.119)$$

Equation (8.119) indicates that the expression $(1 + R_2/R_1)$ for the closed-loop gain of the non-inverting amplifier for the CFA is the same as that for the op-amp. The -3 dB closed-loop bandwidth is given by

$$\bar{f}_o = f_o(1 + Z_o \beta) = f_o(1 + Z_o/R_2) \qquad (8.120)$$

which is the same as the trans-resistance and the inverting cases. Once again note that while this value is dependent on R_2, it is not dependent on R_1. Thus, the closed-loop bandwidth of the CFA

in the non-inverting configuration is independent of resistor R_1. It follows that resistor R_1 may be changed to vary the CFAs closed-loop gain by way of $(1 + R_2/R_1)$ and the bandwidth will remain constant provided R_2 is fixed. Plots of the closed-loop gain for an ideal CFA with $Z_o = 3$ M, $R_2 = 2$ k and $f_o = 9000$ Hz in the non-inverting configuration are shown in Fig. 8.41. Using (8.120), the bandwidth can be seen to remain constant at $f_o = 9000 \times 3 \times 10^6/2 \times 10^3 = 13.5$ MHz with changing gain effected by changing R_1.

Example 8.17
For an ideal CFA having $Z_o = 2$ M and $f_o = 7$ kHz, determine the closed-loop bandwidth for $R_2 = 5$ k.

Solution Using (8.106), $\bar{f}_o = f_o(1 + Z_o/R_2) = 5 \times 10^3 \times 2 \times 10^6/5 \times 10^3 = 2\,\text{MHz}$ (Fig. 8.48).

As in the case of the op-amp, if $R_1 \to \infty$, then the closed-loop gain is unity, and the bandwidth is given of course by Eq. (8.120). While in the case of the op-amp, we can go further and set $R_2 = 0$ to yield the classical buffer, this cannot be done in a CFA since R_2 determines the amount of feedback around the device and lowering its value increases the amount of feedback. The CFA will become unstable if this value is reduced below about 1 k as explained earlier. To use the CFA as a buffer, it must therefore be connected as shown in Fig. 8.42 where $R_F = 1$ k (Fig. 8.49).

Mixer
As a summing amplifier or mixer, the CFA performs particularly well since the inverting input has an already low impedance, thereby allowing the mixing of currents without interference. This low impedance is made even lower by the applied negative feedback. The bandwidth of the mixer is also not affected by the input source resistances since the CFA bandwidth is fixed by the feedback resistor R_f as shown in Fig. 8.50. Note this is in direct contrast to a mixer designed using op-amps in which the low-impedance

Fig. 8.48 CFA frequency plots in non-inverting configuration

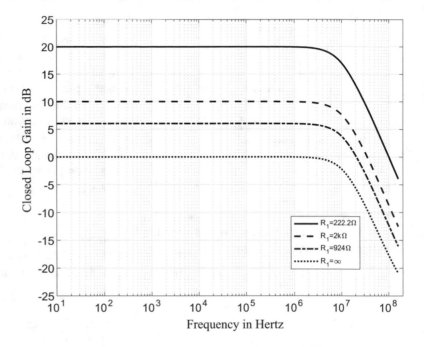

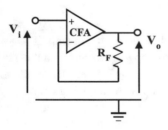

Fig. 8.49 The CFA connected as a buffer

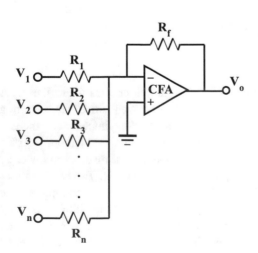

Fig. 8.50 The CFA used as a mixer

inverting node is only made low by negative feedback. The op-amp also suffers from the fact that its bandwidth as a mixer is dependent on the input resistors R_i, $i = 1, 2\ldots n$.

Differential Amplifier

The CFA can also be used as a differential amplifier in much the same way as its op-amp counterpart. In that case the circuit shown in Fig. 8.51 can be used, and the output is given by

$$V_o = -\frac{R_2}{R_1}(V_1 - V_2) \qquad (8.121)$$

The important consideration when using this amplifier configuration is that resistor R_2 sets the bandwidth of the amplifier and the remaining resistors are defined around this value depending on the gains required. Finally, note the special case of the difference amplifier governed by Eq. (8.36) with $R_1 = 0$ and $R_2 = \infty$ driven by V_2 also apply to this circuit.

Integrator

The most notable exception to the CFA's general use is as an integrator. The problem that the CFA has in the integrator configuration is that the

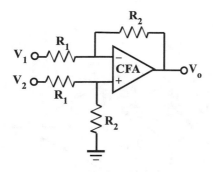

Fig. 8.51 The CFA as a differencing amplifier

impedance of the feedback capacitor C_f is frequency dependent and can fall below the minimum value of R_2 set for stable operation. This configuration should therefore be avoided.

Further Bandwidth Considerations

In real CFAs however, the bandwidth situation is never quite as Fig. 8.41 indicates and as Eq. (8.120) predicts. The bandwidth unfortunately does vary with changing closed-loop gain as a result of variation in resistor R_1. The main reason for this is the finite input impedance at the input to the buffer at the input of the device. If we include impedances r_x and r_o in the CFA model as illustrated in Fig. 8.52, then it can be shown that the new closed-loop bandwidth using the non-inverting amplifier is given by

$$f_{cl} \simeq \frac{f_o Z_o}{R_2}$$

$$\times \left(\frac{1}{r_x \left(\frac{1}{R_1} + \frac{1}{R_2} + \frac{r_o}{R_1 R_2} \right) + \frac{r_o}{R_2} + 1} \right)$$

(8.122)

Since $r_o < < R_2$, Eq. (8.122) reduces to

$$f_{cl} \simeq \frac{f_o Z_o}{R_2}$$

$$\times \left(\frac{1}{r_x \left(\frac{1}{R_1} + \frac{1}{R_2} + \frac{r_o}{R_1 R_2} \right) + 1} \right)$$

(8.123)

Equation (8.123) predicts that the bandwidth will decrease as R_1 decreases although the change

is not as marked as that for an op-amp. Of course if $r_x = 0$, Eq. (8.123) reverts to Eq. (8.120). For the previous case where $R_2 = 2$ k, $Z_o = 3$ M and $f_o = 9$ kHz that LED to Fig. 8.41, the bandwidth changes if the input impedances $r_x = 50 \ \Omega$ and $r_o = 100 \ \Omega$ are included as shown in Fig. 8.53. For the values of R_1 given, the bandwidths are 10.1 MHz, 11.7 MHz, 12 MHz and 12.4 MHz in the order of increasing R_1. Finally, it should be noted that the input impedance r_x at the inverting input is not purely resistive but rather has an inductive component associated with it. Hence its impedance eventually rises with frequency. For most applications this is not a problem since the shunt feedback reduces it even further from its open-loop value.

8.16 Applications

In this section, we discuss several circuits based on the op-amp. Most of them employ the 741 op-amp which is perhaps the most popular device. Typical parameters for this and three other devices are presented in Table 8.2.

In all of these applications, the power supply to the op-amp should be decoupled by connecting 1 microfarad capacitors at the power supply pins of the op-amp to ground.

DC Non-inverting Amplifier

The circuit in Fig. 8.54 is a non-inverting amplifier that can amplify DC. The gain is $1 + \frac{R_2}{R_1} = 1 + \frac{100 \text{k}}{1 \text{k}} = 101$. If $R_1 = 1.01$ k, then the gain is 100. If the source is another op-amp circuit with a direct connection, then the bias needed will be supplied by that circuit. In that case the (low-frequency) input impedance of the circuit is the open-loop resistance multiplied by the loop gain which can be very high. If a direct path to the driving circuit is not available, then a resistor $R_B = 100$ k must be connected between the non-inverting input and ground to provide the bias. In such a case, the input impedance is now 100 k.

Fig. 8.52 Small-signal model of the CFA to include finite impedances

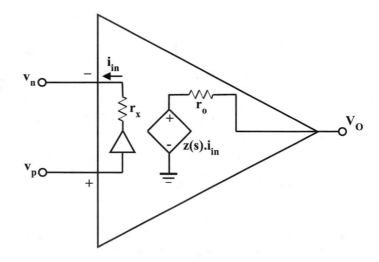

Fig. 8.53 Frequency response plot for varying gains with finite impedances

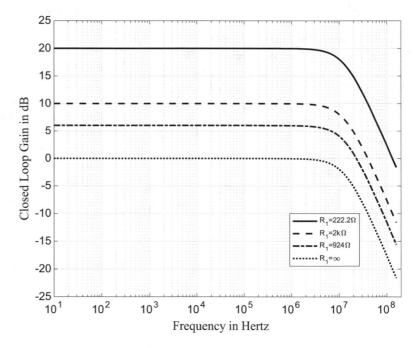

Ideas for Exploration (i) Convert the circuit into a variable gain amplifier by replacing resistor R_2 by a 100 k potentiometer with the feedback connection to the inverting terminal coming from the wiper of the potentiometer. With this arrangement the gain can be varied from 1 to 101. (ii) Introduce **offset nulling** in order to minimize offset voltage at the output. Consult the manufacturer's literature on implementing this.

AC Non-inverting Amplifier

The circuit in Fig. 8.55 is an AC non-inverting amplifier. A bias resistor $R_B = 100$ k is included as there is no other DC path to provide a bias current. The gain is $1 + \frac{100k}{1k} = 101$. For a low-frequency response down to 20 Hz, the input coupling capacitor value is given by $C = 1/2 \times \pi \times 20 \times 10^5 = 0.08 \ \mu F$. A value of $C = 0.1 \ \mu F$ is used. A coupling capacitor at the output may be necessary if the amplifier is

Table 8.2 Characteristics of several unity-gain stable op-amps

Symbol	Parameter	741	OPA134	OPA452	LF351
A_o	Open-loop voltage gain	100 dB	120 dB	110 dB	106 dB
Z_i	Input impedance	1 M	10G	10G	1G
I_b	Input bias current	200 nA	±100 pA	±100 pA	±100 pA
V_S	Maximum supply voltage	±18 V	±18 V	±40 V	±18 V
V_{imx}	Maximum input voltage	±13 V	±15 V	±34.5 V	±15 V
V_{omx}	Maximum output voltage	±14 V	±15.5 V	±34.5 V	±12.5 V
I_{omx}	Maximum output current	±25 mA	±35 mA	±50 mA	±30 mA
V_{io}	Input offset voltage	2mV	±3 mV	±3 mV	±5 mV
CMRR	Common mode rejection ratio	90 dB	90 dB	94 dB	86 dB
GBP	Gain bandwidth product	1 MHz	8 MHz	1.8 MHz	4 MHz

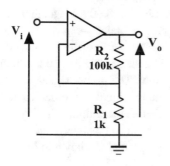

Fig. 8.54 DC non-inverting amplifier

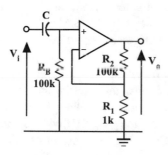

Fig. 8.55 AC non-inverting amplifier

expected to connect to a circuit which has DC that may be affected. A similar value capacitor connected at the output will suffice.

Ideas for Exploration (i) Convert the circuit into a variable gain amplifier by replacing resistor R_2 by a 100 k potentiometer with the feedback

connection to the inverting terminal coming from the wiper. With this arrangement the gain can be varied from 1 to 101. (ii) Introduce a capacitor C_2 as shown Fig. 8.56 in order to introduce 100% DC feedback that will enhance DC stability.

Bootstrapped AC Amplifier
In the circuit in Fig. 8.56, by moving the resistor R_B to the position shown in Fig. 8.57, this resistor is effectively bootstrapped. As a result, the low-frequency input impedance goes from $Z_i = 100$ k to $Z_i = 100$ k $\times$ *loopgain* $= 100$ k $\times 10^3 = 100$ M where loop gain is given by open-loop gain/closed-loop gain. Capacitor C_2 is calculated by ensuring that its reactance at the lowest frequency under consideration is small compared with R_B. Using 1 Hz we have $C_2 = 1/2 \times \pi \times 1 \times 10^5 = 1.6$ µF. The gain of the system continues to be 101.

Ideas for Exploration (i) Convert the circuit into a variable gain amplifier by replacing resistor R_2 by a 100 k potentiometer with the feedback connection to the inverting terminal coming from the wiper. With this arrangement the gain can be varied from 1 to 101.

DC Voltage Follower
This circuit in Fig. 8.58 is that of a voltage follower which responds down to DC. It is derived from the non-inverting amplifier by setting $R_2 = 0$

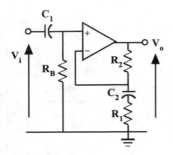

Fig. 8.56 AC non-inverting amplifier with 100% DC feedback

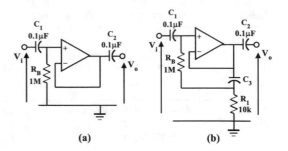

(a)   **(b)**

Fig. 8.59 (a) AC voltage follower; (b) bootstrapped voltage follower

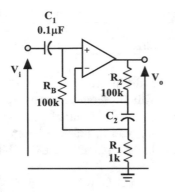

Fig. 8.57 Bootstrapped AC amplifier

Fig. 8.58 DC voltage follower

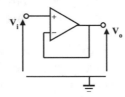

and removing R_1. The output follows the input to within a few millivolts, and this can be further improved by using offset nulling. Since this constitutes 100% feedback, i.e. $\beta = 1$, the input impedance is at its highest value. Specifically, the open-loop input impedance of the 741 which is 1 M is increased by the factor of the loop gain $A\beta = 10^5$ to 1 M $\times$ $10^5 = 10^{11}$ Ω!

Ideas for Exploration (i) Modify the DC voltage follower to create an AC voltage follower as shown in Fig. 8.59 (a). Here a larger value of R_B is used since the input impedance to the non-inverting input is because of the feedback

so very high. (ii) Modify the AC circuit to increase the input impedance as shown in Fig. 8.59 (b).

Inverting Amplifier

This circuit in Fig. 8.60 (a) is an inverting amplifier. It has a gain $-R_2/R_1 = -$ 100 k/ 10 k $= -$ 10. Unlike the non-inverting case, the input impedance is $R_1 = 10$ k which is comparatively low. The output signal is out of phase with the input signal. In order to reduce any voltage offset at the output, a resistor $R_3 = R_1//R_2 = 9$ k is placed between the non-inverting terminal and ground as shown in Fig. 8.60 (b).

Ideas for Exploration (i) Convert the circuit into a variable gain amplifier by replacing resistor R_2 by a 1 M potentiometer in series with a 10 k resistor as shown in Fig. 8.61. With this arrangement the gain can be varied from -1 to -100.

Modified Inverting Amplifier

A modification of the inverting amplifier is shown in Fig. 8.62. Here the voltage feedback signal is obtained from the potential divider comprising R_3 and R_4. The voltage gain for this circuit is given by $-\frac{R_2}{R_1}\left(1 + \frac{R_4}{R_3}\right) = -101$. Because of the presence of the potential divider R_3 and R_4, larger resistor values can be used for R_1 and R_2, thereby enabling a larger input impedance $R_1 = 1$ M.

Ideas for Exploration (i) Convert the circuit into a variable gain amplifier by replacing resistor R_4 by a variable 100 k resistor. With this arrangement the gain can be varied from -1 to -101.

Fig. 8.60 Inverting amplifier

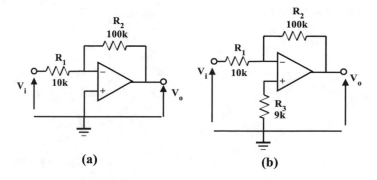

(a)

(b)

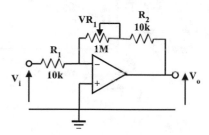

Fig. 8.61 Variable gain inverting amplifier

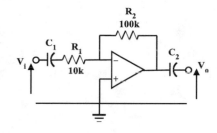

Fig. 8.63 AC inverting amplifier

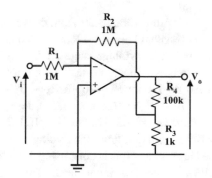

Fig. 8.62 Modified inverting amplifier

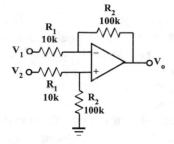

Fig. 8.64 Differential amplifier

Note however that at low gains, R_3 will be a load on the op-amp and hence may have to be increased.

AC Inverting Amplifier

This inverting amplifier processes AC by the inclusion of coupling capacitors C_1 and C_2 as shown in Fig. 8.63. These capacitors block any DC and ensure that only AC passes through the amplifier. Using a lower cut-off frequency of 20 Hz and noting the input impedance to the amplifier is $R_1 = 10$ k, then capacitor C_1 is

given by $C_1 = 1/2 \times \pi \times 20 \times 10^4 = 0.8$ μF. A value of 1 μF is suitable for both capacitors.

Ideas for Exploration (i) Convert the circuit into a variable gain inverting amplifier by replacing resistor R_2 by a 1 M potentiometer in series with a 10 k resistor as shown in Fig. 8.67 for the DC case. With this arrangement the gain can be varied from -1 to -100.

Differential Amplifier

The circuit in Fig. 8.64 is that of a differential amplifier. It has a gain $-R_2/R_1 = -100$ k/ 10 k $= -10$ and an output given by $V_o = -10$

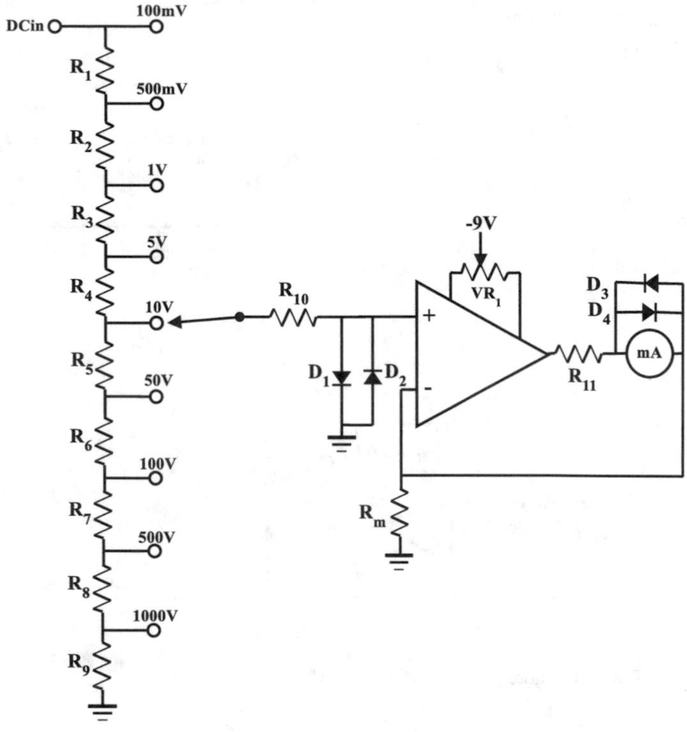

Fig. 8.65 DC voltmeter

$(V_1 - V_2)$. If all resistors are made equal say 100 k, then the circuit would have unity gain and would be that of a subtractor with output given by $V_o = V_2 - V_1$.

DC Voltmeter

This circuit shown in Fig. 8.65 is that of a DC voltmeter. It uses the 741 op-amp in a voltage-to-current configuration since the meter is included in a current-series feedback loop. Thus, the current I_m through the meter M is sampled by resistor R_m across which a feedback voltage is developed that is applied in series with the input voltage V_i. The result is that the ratio I_m/V_i is stabilized because of the feedback and the current I_m through the meter is simply $I_m = V_i/R_m$. The input impedance to the non-inverting input is already high without feedback (1 M) and is increased by the feedback. With $R_m = 100\ \Omega$, full-scale deflection with a 1 mA milliammeter occurs with $V_i = 100$ mV. Diodes D_1 and D_2 along with resistor $R_{10} = 10$ k protect the input of the op-amp from excessive voltage, while diodes D_3 and D_4 in conjunction with resistor

$R_{11} = 3.3$ k protect the meter from current overload. The resistor chain R_1 to R_9 forms an attenuator that provides a voltage range 0.1 V to 1000 V in nine ranges. These values are as follows: $R_1 = (680$ k $+ 120$ k$), R_2 = 100$ k$, R_3 = (68$ k $+ 12$ k$), R_4 = 10$ k$, R_5 = (6.8$ k $+ 1.2$ k$), R_6 = 1$ k$, R_7 = (680\ \Omega + 120\ \Omega), R_8 = 100\ \Omega$ and $R_9 = 100\ \Omega$. Finally, any offset voltage must be reduced to zero, and this is done using potentiometer VR_1, while the input leads are shorted together.

Ideas for Exploration (i) Convert the circuit into a current meter by connecting a resistor R_C across the input terminals as shown in Fig. 8.66. The current to be measured is passed through this resistor, and the choice of R_C is given by $R_C = 0.1$ V/I_{mx} where I_{mx} is the maximum current. Thus, for a 1 mA range, $R_C = 0.1/1$ mA $= 100\ \Omega$, or for a 100uA range, $R_C = 0.1/100$ uA $= 1$ k.

High-Impedance DC Voltmeter

The circuit shown in Fig. 8.67 is a high-impedance voltmeter using a MOSFET op-amp

Fig. 8.66 Modification for conversion to current meter

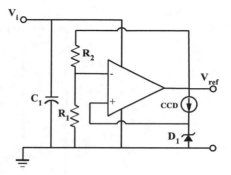

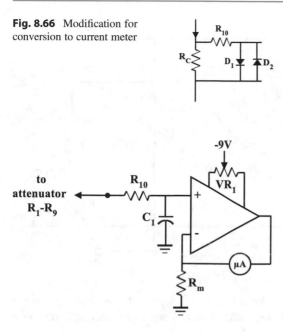

Fig. 8.67 High-impedance DC voltmeter

the CA3140E. It uses essentially the same basic configuration as that in Fig. 8.65. Thus using $I_m = V_i/R_m$, for $V_i = 0.1$ V and $I_m = 100$ μA, then $R_m = V_i/I_m = 0.1$ V/100 μA $= 1$ k. No meter protection circuitry is used in this design. The resistor chain R_1 to R_9 of Fig. 8.65 forms an attenuator that provides a voltage range 0.1 V to 1000 V in nine ranges. The resistor values are as follows: $R_1 = (6.8$ M $+ 1.2$ M$)$, $R_2 = 1$ M, $R_3 = (680$ k $+ 120$ k$)$, $R_4 = 100$ k, $R_5 = (68$ k $+ 12$ k$)$, $R_6 = 10$ k, $R_7 = (6.8$ k $+ 1.2$ k$)$, $R_8 = 1$ k and $R_9 = 1$ k. Also, any offset voltage must be reduced to zero, and this is done using potentiometer $VR_1 = 10$ k, while the input leads are shorted together. Resistor $R_{10} = 1$ M and capacitor $C_1 = 10$ nF form a low-pass filter to remove high-frequency noise from the signal.

Ideas for Exploration (i) Convert the circuit into a current meter by connecting a resistor R_C across the input terminals as shown in Fig. 8.66. The current to be measured is passed through this resistor, and the choice of R_C is given by $R_C = 0.1V/I_{mx}$ where I_{mx} is the maximum current. Thus, for a 1 mA range, $R_C = 0.1/1$ $mA = 100$ Ω, or for a 100uA range, $R_C = 0.1/100$ $uA = 1$ k.

Fig. 8.68 Stable voltage reference

Stable Voltage Reference

A stable reference voltage can be realized using an op-amp and a 5.6 V Zener. In Zener diodes below about 5.6 V, the Zener effect discussed in Chap. 1 is predominant, as a result of which the device exhibits a negative temperature coefficient. For devices above this voltage, the avalanche effect also discussed in Chap. 1 is the major operating effect, and this is marked by a positive temperature coefficient. In a 5.6 V diode, the two effects approximately cancel each other with the result that the device exhibits a temperature-stable voltage. This device is used in creating a very stable voltage reference as shown in Fig. 8.68. The voltage from the 5.6 V Zener diode D_1 is applied at the non-inverting input of the op-amp which is connected in the non-inverting configuration with a modest gain giving an output of $5.6(1 + R_2/R_1) = 6.6$ V. Feedback ensures that the output voltage is accurate and stable. Current to the Zener is supplied from the stabilized output of the op-amp through a constant current diode (CCD). This enhances system performance since the supply is stable and any residual ripple is reduced by the constant current diode. The voltage output needs to be higher than the Zener voltage so that there is a minimum voltage across the CCD. Line regulation for this circuit is better than 0.005%, and the ripple rejection is greater than 80 dB.

Ideas for Exploration (i) Use the LM336 reference diode from Texas Instruments instead of a 5.6 V Zener in a voltage reference circuit. This diode provides 2.5 V, has a low temperature

Fig. 8.69 Linear scale
ohmmeter

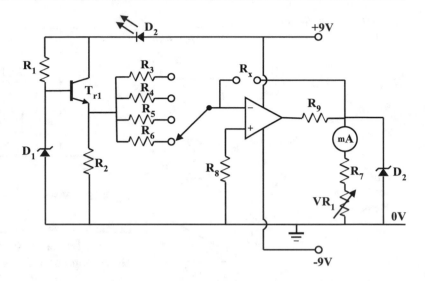

coefficient and operates from 400 µA to 10 mA
with a low dynamic resistance.

Linear Scale Ohmmeter

The circuit in Fig. 8.69 is that of an ohmmeter that
has a linear scale. In this circuit, a stable reference
voltage from the 5.6 V Zener D_1 (with current
supplied by $R_1 = 2.2$ k) is developed via the
transistor across resistor $R_2 = 1$ k. This voltage
which is approximately 5 V drives the inverting
amplifier through range resistors $R_3 = 1$ k,
$R_4 = 10$ k, $R_5 = 100$ k and $R_6 = 1$ M. These
along with the feedback resistor R_x which is the
resistor to be measured set the gain of the amplifier. The output of the op-amp drives a 1mA meter
in series with $R_7 = 2.7$ k and $VR_1 = 5$ k. The
output of the op-amp is given by $V_o = 5\frac{R_x}{R_R}V$
where R_R represents one of the range resistors. It
follows that the output voltage is proportional to
R_x the resistance to be measured. In order to
calibrate the instrument, resistor R_x is set equal
to the value of the selected range, for example,
$R_x = 10$ k on the 10 k range ($R_4 = 10$ k selected).
Then the op-amp will be set for unity gain and
will therefore output approximately 5 V. VR_1 is
adjusted for full-scale deflection on the 1 mA
meter. If high precision resistors (2% tolerance)
are used for the range resistors, then all the ranges
will be calibrated following this procedure on one
range. Resistor $R_8 = 1$ k reduces offset voltage of
the op-amp. Zener diode D_2 with voltage 5.6 V

along with $R_9 = 1$ k protects the meter by limiting
the voltage across the meter during over-range
conditions.

Ideas for Exploration (i) The CA3140 op-amp
is a direct replacement for the 741 and can operate
on voltages down to 4 V either single or dual
supply. Use one of these op-amps to replace the
741 and use another to design a split supply to
enable operation from a single 9 V battery.

AC Voltmeter

The circuit shown in Fig. 8.70 is that of a simple
AC voltmeter. The meter is placed in a diode
bridge D_1 to D_4 to allow the measurement of
AC signals. Diode D_5 protects the meter from
over-current, while capacitor $C_2 = 10$ µF
provides filtering to remove ripples from the current through the meter. Offset nulling may not be
necessary in this circuit because of the presence
of the diodes. Resistors R_{m1} to R_{m4} enable variation of voltage ranges. Thus using $I_m = V_i/R_m$, for
$V_i = 0.1$ V and $I_m = 1$ mA, then $R_{m1} = V_i/I_m = 0.1$
V/1 mA = 100 Ω. Similarly, $R_{m2} = 1$ k for $V_i = 1$
V, $R_{m3} = 10$ k for $V_i = 10$ V and $R_{m4} = 100$ k for
$V_i = 100$ V. Resistor $R_B = 100$ k provides bias to
the non-inverting input of the op-amp, and capacitor $C_1 = 1$ µF couples the AC signal to the input.

Ideas for Exploration (i) Convert this circuit to
an AC/DC instrument by removing C_1 and

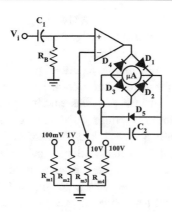

Fig. 8.70 AC voltmeter

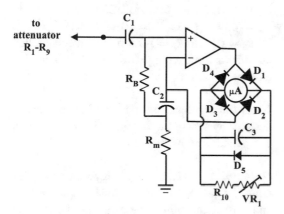

Fig. 8.71 High-impedance AC voltmeter

applying the input voltage directly to the non-inverting terminal.

High-Impedance AC Voltmeter

The circuit of Fig. 8.71 uses a 741 op-amp in the non-inverting configuration to realize a high-impedance AC voltmeter. The 100 μA microammeter is placed in a diode bridge D_1 to D_4 to allow the measurement of AC signals, while diode D_5 protects the meter from over-current. These diodes can be 1 N914 small-signal silicon diodes. Capacitor $C_3 = 10$ μF provides filtering to remove ripples form the current through the meter. Potentiometer $VR_1 = 5$ k is used for calibration of the meter, and $R_{10} = 1$ k limits the minimum resistance across the microammeter. Using $I_m = V_i/R_m$, for $V_i = 0.1$ V and $I_m = 100$ μA, then $R_m = V_i/I_m = 0.1$ V/100 μA $= 1$ k. Bias

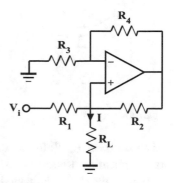

Fig. 8.72 Howland current pump

resistor $R_B = 100$k is bootstrapped, thereby significantly increasing the input impedance of the instrument. The resistor chain R_1 to R_9 from the DC voltmeter project (Fig. 8.65) but with values $R_1 = (6.8$ M $+ 1.2$ M), $R_2 = 1$ M, $R_3 = (680$ k $+ 120$ k), $R_4 = 100$ k, $R_5 = (68$ k $+ 12$ k), $R_6 = 10$ k, $R_7 = (6.8$ k $+ 1.2$ k), $R_8 = 1$ k and $R_9 = 1$ k forms an attenuator that provides a voltage range 0.1 V to 1000 V in nine ranges. Capacitor $C_1 = 1$ μF is a coupling capacitor to remove DC from the input signal, while capacitor $C_2 = 100$ μF provides the bootstrapping action.

Howland Current Pump

The Howland current pump (Fig. 8.72) is a voltage-controlled current source that delivers a current that is independent of the load into which the current is being delivered and is dependent on the controlling voltage to the system. The system comprises an op-amp with voltage-series feedback to its inverting terminal and a resistive feedback connection to the non-inverting terminal. The load R_L is connected to the non-inverting terminal. For the balance condition $R_4/R_3 = R_2/R_1$, the output current I is given by $I = V_i/R_1$. The output impedance is given by $Z_T = R_1//R_2\left(1 + \frac{A}{1+R_2/R_1}\right)$. The low-frequency output impedance is $Z_T(DC) = R_1//R_2\left(1 + \frac{A_o}{1+R_2/R_1}\right)$. For example, with $R_1 = R_2 = R_3 = R_4 = 1$k and using an op-amp with $A_o = 10^5$, then $Z_T(DC) = 10^3//10^3(1 + 10^5/2) \simeq 25$ M.

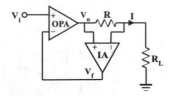

Fig. 8.73 Alternative current pump

Ideas for Exploration (i) Refer to the Texas Instruments Application Report AN-1515 "A Comprehensive Study of the Howland Current Pump", April 2013.

Alternative Current Pump

A current pump that does not require a balance condition and achieves a higher output impedance than the Howland current pump is shown in Fig. 8.73. It comprises an op-amp whose output current I to the load flows through a resistor R, thereby developing a voltage IR across this resistor. This floating voltage is detected by a unity-gain instrumentation amplifier at its differential inputs and delivered as feedback to the inverting input of the op-amp. Noting that $V_o = (V_i - V_f) A$, $V_o = I(R + R_L)$ and $V_f = IR$, then the output current I is given by $I = \frac{V_i/R}{1+R_L/AR}$. As $A \to \infty$, $I = \frac{V_i}{R}$. Thus, the high gain of the op-amp ensures that the output current I is independent of the load R_L and dependent only on the input voltage and the resistor R. By grounding the input and applying a voltage source at the output, the output impedance Z_T of the circuit is found to be $Z_T = R(1 + A)$. The low-frequency output impedance $Z_T(DC) = R(1 + A_o)$. For example, with $R = 1$ k and using an op-amp such as the 741 with $A_o = 10^5$, then $Z_T(DC) = 10^3(1 + 10^5) = 100$ M. This output impedance for this circuit is higher than that achievable in the Howland current pump which (for comparable resistor values) is 25 M.

Ideas for Exploration (i) Introduce gain in the instrumentation amplifier and examine the effect on the performance of the circuit.

AC/DC Universal Voltmeter

The circuit of Fig. 8.74 is a voltmeter that can measure both AC and DC signals. It also uses a MOSFET op-amp the CA3140E in the same basic configuration as that in Fig. 8.67. Here however the meter is placed in a diode bridge D_1 to D_4 to allow the processing of AC signals. Diode D_5 protects the meter from over-current, while capacitor $C_1 = 10$ μF provides filtering to enable a smooth meter response. Potentiometer $VR_1 = 5$ k along with $R_{10} = 1$ k is used to calibrate the instrument. No offset nulling is necessary in this circuit because of the presence of the diodes. Thus using $I_m = V_i/R_m$, for $V_i = 0.1$ V and $I_m = 100$ μA, then $R_m = V_i/I_m = 0.1$ V/ 100 μA $= 1$ k. The resistor chain R_1 to R_9 from Fig. 8.65 with values $R_1 = (6.8$ M $+ 1.2$ M), $R_2 = 1$ M, $R_3 = (680$ k $+ 120$ k), $R_4 = 100$ k, $R_5 = (68$ k $+ 12$ k), $R_6 = 10$ k, $R_7 = (6.8$ k $+ 1.2$ k), $R_8 = 1$ k and $R_9 = 1$ k forms an attenuator that provides a voltage range 0.1 V to 1000 V in nine ranges.

Ideas for Exploration (i) Introduce a polarity detector by taking the output of the op-amp to a comparator driving two leds.

Op-Amp with Transformer-Coupled Output Op-amps have specified output voltage and current capability. For example, the 741 can deliver a peak output voltage of about 12 V and a peak output current of about 25 mA. It therefore cannot directly drive a low-impedance load such as an 8 Ohm speaker since the maximum peak current demand would be 12 V/8 Ω $= 1.5$ A which exceeds the maximum current rating of the device. One method of addressing this is to use a step-down transformer between the output of the op-amp and the speaker as shown in Fig. 8.75. Let the peak output voltage of the op-amp be V_{opk} and the peak voltage into the speaker from the transformer be V_{spk}. Then $V_{opk}/V_{spk} = n$ where n is the transformer turns ratio. From transformer principles, $I_{opk}/I_{spk} = 1/n$ where I_{opk} and I_{spk} are the peak output current of the op-amp and the peak current into the speaker from the transformer, respectively. Let $V_{opk}/I_{opk} = R'_L$ and $V_{spk}/I_{spk} = R_L$. Then, $R'_L/R_L = n^2$ and $n = \sqrt{R'_L/R_L}$. With this turns ratio, an impedance R_L at the output of the transformer secondary is reflected as R'_L at the input of

Fig. 8.74 AC/DC universal voltmeter

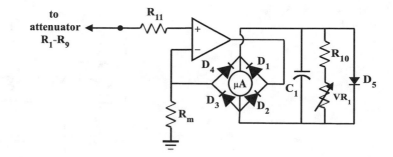

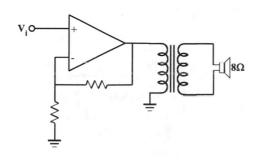

Fig. 8.75 Transformer-coupled op-amp

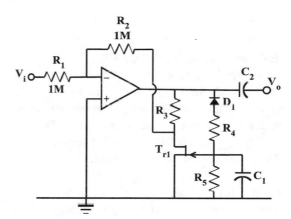

Fig. 8.76 Constant volume amplifier

the transformer primary. This limits the output current of the op-amp to its rated output current when delivering its rated output voltage. Thus, for the 741, $R'_L = 12\,\text{V}/25\,\text{mA} = 480\,\Omega$ and $R_L = 8\,\Omega$ giving $n = \sqrt{480/8} = 7.7$. Use $n = 8$. Therefore, a step-down transformer with a turns ratio $n = 8$ will enable the full voltage and current capacity of the 741 op-amp to be delivered to the 8 Ω speaker. The average power to the speaker is then $P = \frac{V_{opk}}{\sqrt{2}} \frac{I_{opk}}{\sqrt{2}} = 12 \times 25\,\text{mA}/2 = 150\,\text{mW}$. If the speaker is connected directly to the output of the op-amp, then the maximum power that can be delivered to the load is limited by the maximum current available from the op-amp and is given by $\left(I_{opk}/\sqrt{2}\right)^2 \times R_L = \left(0.025/\sqrt{2}\right)^2 \times 8 = 2.5\,\text{mW}$.

Research Project 1

The circuit shown in Fig. 8.76 is that of a **constant volume amplifier**. It provides an approximately constant amplitude output signal for varying input signal amplitude by incorporating a system of automatic gain control. This circuit is useful in reducing the dynamic range of music

and other signals without introducing significant distortion. Resistor $R_5 = 1$ M grounds the gate of the JFET and $R_3 = 330$ k along with the drain voltage set the drain current such that the operating point is in the ohmic region. The drain-source resistance in conjunction with $R_3 = 330$ k provides the dynamic adjustment of thec op-amp feedback that results in automatic gain control. The JFET operates as a voltage-controlled resistor with the control voltage being derived from the output signal of the op-amp. This signal is rectified by diode D_1 on the negative half-cycle, filtered by R_4 and C_1 and this negative voltage applied at the gate of the JFET. Thus, for low-level signals, the amplitude of this negative voltage is low, and hence the drain-source resistance is a few hundred ohms. This means that the feedback signal at point A is low and therefore amplifier gain is high and the output of the op-amp is at a reasonable level. When the signal amplitude is large, the amplitude of this negative voltage increases and this increases the

Fig. 8.77 Baxandall tone
control

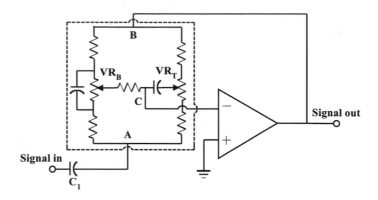

drain-source resistance of the JFET, as a result of
which the feedback signal at point A is high
thereby reducing the circuit gain and across R_3
and hence the output of the op-amp is reduced to a
reasonable level. Capacitor C_2 eliminates DC offset Fig. 8.75 set voltages from the output signal.
The project is to design the circuit.

Ideas for Exploration (i) Determine the
dynamic range of the circuit by applying a signal
of increasing amplitude and recording the
corresponding output amplitude changes. A
large input signal amplitude change will produce
a reduced amplitude change at the output.
(ii) Experiment with different values of R_3 and
C_1 and determine the response of the circuit to
these changes.

Research Project 2
This project involves the design of a **Baxandall
tone control circuit** that was introduced in Chap. 7
using discrete transistors. Here the project is to
design this system around the 741 op-amp, as
shown in Fig. 8.77, using the theory of the
Baxandall tone control employed in Chap. 7. The
signal driving the Baxandall tone control network
enters at A and may come from an op-amp pre-amplifier. The output of the network at B enters the
inverting input of the op-amp. Feedback from the
output of the op-amp is applied to the network at
C. Potentiometer VR_B lifts/cuts the bass
frequencies, while potentiometer VR_T lifts/cuts the
treble frequencies. The project involves
researching Baxandall tone control theory and
designing the network.

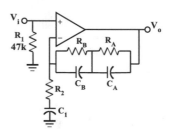

Fig. 8.78 RIAA pre-amplifier

Ideas for Exploration (i) Design a pre-amplifier
circuit using a 741 op-amp with a suitable gain to
drive the tone control circuit. (ii) Refer to the
article The Texan 20 + 20 W IC Stereo Amplifier
by Richard Mann, Parts 1 to 4, Practical Wireless,
May 1972, p 48 to August 1972, p318.

Research Project 3
This project requires the design of an **RIAA
pre-amplifier** using an op-amp as shown in
Fig. 8.78. The signal from a phonograph disk
(record) has a characteristic in which the high
frequencies have a higher amplitude than the
low frequencies. Because of this, the
pre-amplifier for such a signal must possess a
frequency response characteristic that is exactly
the inverse of this. The standard developed by
The Record Industry Association of America
(RIAA) for magnetic phonograph cartridges that
establishes the exact nature of this equalization
characteristic was discussed in Chap. 7. The characteristic comprises poles at 50 Hz and 2100 Hz
and a zero at 500 Hz. A practical implementation
of this using a 741 op-amp is shown in Fig. 8.78.

It uses the op-amp in a non-inverting configuration with a CR network in the signal feedback loop which creates the RIAA equalization characteristic. The standard input resistance for magnetic pickups is 47 k as shown in the diagram.

Ideas for Exploration (i) Refer to the paper "On RIAA Equalization Networks" by Stanley Lipshitz, Journal of the Audio Engineering Society, Volume 27 Issue 6 pp. 458–481, June 1979. (ii) Use a different CR network to achieve the RIAA equalization.

Research Project 4

A common problem when playing old records is scratches on the vinyl surface which generate high-frequency noise. This project involves the design of a **scratch filter** to remove this noise as was discussed in Chap. 7, but here using a 741 op-amp. The system is shown in Fig. 8.79. It comprises two connected RC networks with one capacitor grounded and the other connected to the output of the buffer. The transfer function for this system with $R_1 = R_2 = R$ and $C_1 = C$, $C_2 = 2C$ is given by

$$\frac{V_o}{V_i} = \frac{1}{1 + \sqrt{2}j\frac{f}{f_c} + \left(j\frac{f}{f_c}\right)^2} \tag{8.124}$$

where

$$f_c = \frac{1}{2\sqrt{2}\pi RC} \tag{8.125}$$

This is a second-order response referred to as a Butterworth response that has the characteristic of being maximally flat. With an appropriately placed corner frequency f_c, unwanted high-frequency signals produced by vinyl scratches can be eliminated. For an upper cut-off frequency of 10 kHz, using $C = 2.2$ nF, Eq. (8.125) gives $R = 4.7$ k.

Ideas for Exploration (i) Minimize DC offset by introducing a resistor in the feedback connection between the output of the op-amp and the inverting terminal.

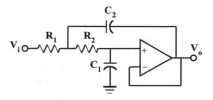

Fig. 8.79 Scratch filter using op-amp

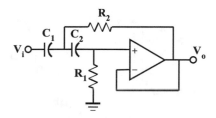

Fig. 8.80 Rumble filter using op-amp

Research Project 5

Using the voltage follower op-amp configuration, explore the development of a **rumble filter** that removes unwanted low-frequency (less than 50 Hz) components from an audio signal as discussed in Chap. 7. The two filter network resistors and the two network capacitors in Fig. 8.79 must be interchanged. Resistor R_1 which is part of the filter network also provides bias to the non-inverting input of the op-amp. The op-amp filter is shown in Fig. 8.80 and the associated transfer function with $C_1 = C_2 = C$ and $R_1 = R$, $R_2 = R/2$ is given by

$$\frac{V_o}{V_i} = \frac{\left(j\frac{f}{f_c}\right)^2}{1 + \sqrt{2}j\frac{f}{f_c} + \left(j\frac{f}{f_c}\right)^2} \tag{8.126}$$

where the corner frequency is

$$f_c = \frac{\sqrt{2}}{2\pi RC} \tag{8.127}$$

This is a Butterworth response for the high-pass filter.

Ideas for Exploration (i) Minimize DC offset by introducing a resistor in the feedback

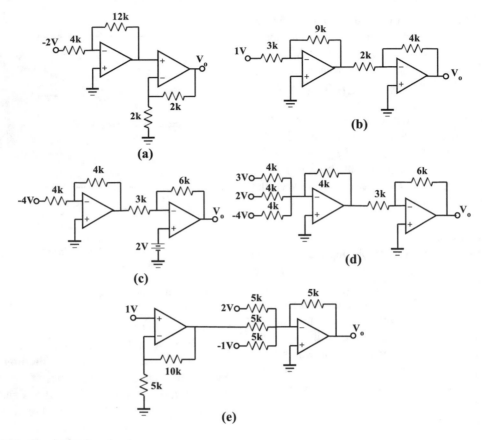

Fig. 8.81 Circuits for Question 1

connection between the output of the op-amp and the inverting terminal.

Problems

1. For each of the circuits shown in Fig. 8.81, determine V_o.
2. Design an inverting amplifier with a voltage gain of -5.
3. Design an inverting amplifier with a voltage gain of -8 and an input resistance of 10 k.
4. Design a non-inverting amplifier with a voltage gain of 8.
5. Design a non-inverting amplifier with a voltage gain of 8 and an input resistance of 150 kΩ.

6. A non-inverting amplifier has $R_1 = 5$ k and $R_2 = 25$ k. Determine the gain of the system.
7. Configure an op-amp circuit to realize the operation $V_O = 5\,V_1 - 2\,V_2$. If a third input was added to the differential amplifier such that $V_O = 5\,V_1 - 2\,V_2 + 3\,V_3$, realize such a circuit using a single op-amp.
8. Design a non-inverting amplifier with a gain of 12 using an op-amp that operates from a single-ended 15 V supply.
9. Design an inverting amplifier with a gain of -9 using an op-amp that operates from a single-ended 18 V supply.
10. Show that the closed-loop bandwidth of a unity-gain stable op-amp is equal to the open-loop bandwidth multiplied by the loop gain. Such an operational amplifier has a

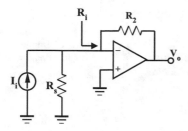

Fig. 8.82 Circuit for Question 14

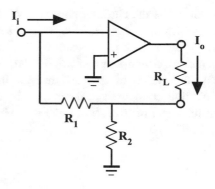

Fig. 8.83 Circuit for Question 15

GBP = 20 MHz. Determine the closed-loop bandwidth for gains of (i) 10 and (ii) 50.

11. A (unity-gain stable) op-amp has a *GBP* = 8 MHz. Determine the gain for a bandwidth of 1 MHz.

12. A (unity-gain stable) op-amp has an open-loop bandwidth of 20 Hz and a low-frequency open-loop gain of 120 dB. Determine the gain bandwidth product of the device.

13. A particular op-amp has an input resistance $R_i = 1$ M, $R_o = 100\ \Omega$ and open-loop DC gain $A_o = 2000$. Estimate the error in the gain of this op-amp if used as an inverting amplifier, when $R_1 = 1$ k and $R_2 = 10$ k. Repeat for the case $R_1 = 1$ M and $R_2 = 10$ M. Which choice of resistors results in a lower gain error while maintaining the same gain?

14. For the transimpedance amplifier of Fig. 8.82, show that the input impedance as seen to the right of the source is given by $R_i = R_2/(1 + A_o)$. If $R_2 = 10$ k and $A_o = 1000$, determine the range of R_s such that the output voltage V_o deviates from its ideal value by 1%.

15. The circuit shown in Fig. 8.83 contains a floating load and has the characteristics of an ideal current amplifier. Show that for this circuit i_o and i_i are related by $i_o = -\left(1 + \frac{R_1}{R_2}\right)i_i$. Determine the input impedance as seen at the inverting terminal of this amplifier.

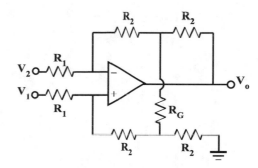

Fig. 8.84 Circuit for Question 16

16. Consider the variable gain difference amplifier shown in Fig. 8.84 where resistor R_G is used to change the gain. Show for this amplifier, the output voltage is given by

$$V_o = \frac{2R_2}{R_1}\left(1 + \frac{R_2}{R_G}\right)(V_2 - V_1)$$

17. An ideal CFA has $Z_o = 5$ M and $f_o = 1$ kHz. Determine the closed-loop gain and closed-loop bandwidth in the inverting and non-inverting configurations for $R_1 = 1$ k and $R_2 = 10$ k.

18. An op-amp has a slew rate of 5 V/μs and a maximum output voltage swing of ±8 V. Determine the full power bandwidth BW_p. What is the maximum voltage output the device can deliver at 500 kHz?

19. A CFA has the characteristics $f_o = 7.5$ kHz, $Z_o = 4.5\ M$ and $r_x = 50\ \Omega$. Determine the value of the feedback resistor R_2 in order to achieve a closed-loop bandwidth of 10 MHz. For this value of feedback resistor in the inverting configuration, determine the value of the input resistor R_1 in order to achieve a gain of -10.

Bibliography

R. Coughlin, F. Driscoll, *Operational Amplifiers and Linear Integrated Circuits*, 5th edn. (Prentice Hall, New Jersey, 1998)

Power Amplifiers

9

An amplifier is a device for receiving a signal at its input and delivering a larger signal at its output. They are used in a wide range of applications including music systems, in driving industrial loads, and elsewhere. In small-signal amplifiers, the active devices are operated such that the voltage and current changes are small and the devices are operated in their approximately linear regions. In such amplifiers, the main factors are amplifier linearity, gain and noise performance.

Large-signal amplifiers, also called power amplifiers, experience relatively large changes in current and voltage and are usually expected to deliver large amounts of power to an external load. This class of amplifiers usually range from hundreds of milliwatts to hundreds of watts. The features that characterize such amplifiers are power output, gain, load rating and total harmonic distortion. In this chapter, we consider the operation of power amplifiers and how to design working power amplifier circuits. At the end of the chapter the student will be able to:

- Explain the operation of transistor power amplifiers
- Explain the operation of different classes of power amplifiers
- Design transistor power amplifiers using several topologies
- Utilize power amplifier integrated circuits

9.1 Amplifier Classes

Amplifiers may be classified by class, i.e. the extent the active output devices of the amplifiers conduct while delivering an output. In a *class A* amplifier, the output device(s) conducts for the complete wave form cycle, i.e., 360°. In a *class AB* amplifier, the output devices conduct for more than half of the signal cycle but less than the complete cycle, i.e. more than 180° but less than 360°. Therefore, the output device turns off for a brief period during the cycle in which case, another device must conduct and deliver the output signal. In a *class B* amplifier, the output device conducts for half of the signal cycle, i.e. 180°, and turns off for the other half of the cycle. In order to reproduce a complete signal, two such amplifiers must be connected to the load. These will conduct alternatively so that the full cycle is reproduced. A class C amplifier is one in which the output device conducts for less than half-cycle, i.e. 180°. It is difficult to fully reproduce a complete wave cycle with such an amplifier and distortion is the result. A class D amplifier is one in which the output device is alternatively fully on or fully off.

9.2 Fixed-Bias Class A Amplifier

The common emitter fixed-bias amplifier discussed in Chap. 2 actually operates in class A since the device never turns off during the signal

© Springer Nature Switzerland AG 2021
S. J. G. Gift, B. Maundy, *Electronic Circuit Design and Application*,
https://doi.org/10.1007/978-3-030-46989-4_9

cycle. This same arrangement may be used as a power amplifier by replacing the small signal transistor by a power transistor as shown in Fig. 9.1.

A signal at the input of the amplifier causes the collector current of the transistor to change, and hence an output voltage is developed across R_L the load resistor. Class A operation follows from the fact that a DC bias is established in the transistor with the device operating along the DC "load line" as shown in Fig. 9.2. The Q point represents the DC voltage and current in the transistor when no signal is present. As is evident, the

transistor is always on as movement up and down the load line occurs in response to an input signal.

The signal power in R_L, P_L is given by

$$P_L(AC) = \frac{1}{T} \int_0^T i_C^2 R_L dt \qquad (9.1)$$

The peak value of the maximum collector AC current is given by

$$i_{Cpk}(\max) = \frac{V_{cc}}{2R_L} \qquad (9.2)$$

and occurs when the transistor is symmetrically biased. Under these conditions, the maximum AC power to the load becomes

$$P_L(AC)_{\max} = \left(\frac{i_{Cpk}(\max)}{\sqrt{2}}\right)^2 R_L$$

$$= \left(\frac{V_{cc}}{2\sqrt{2}R_L}\right)^2 R_L = \frac{V_{cc}^2}{8R_L} \qquad (9.3)$$

In low-power audio systems, a loudspeaker is sometimes the load resistor R_L, and the low impedance (8 ohms) gives the load line a very steep slope. However, while the system quiescent current will be permanently passing through the loudspeaker, changes in the value of this current as a result of signal input to the transistor result in

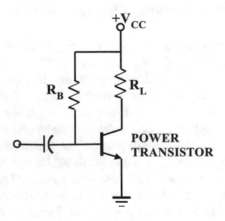

Fig. 9.1 Class A common emitter power amplifier

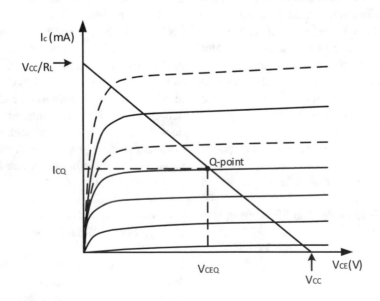

Fig. 9.2 Transistor DC load line

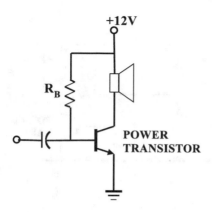

+12V

R_B

POWER
TRANSISTOR

Fig. 9.3 Class A common emitter power amplifier with speaker load

audio output from the speaker. Consider the common emitter amplifier in Fig. 9.3 where an 8-ohm loudspeaker has replaced the load resistor in Fig. 9.3 and a moderate quiescent current of say $I_C = 100$ mA flows in the transistor collector. This current results in a voltage drop across the speaker of $I_C R_L = 100$ mA $\times$ 8 $\Omega = 0.8$ V which, with a supply voltage of 12 volts, cannot result in maximum symmetrical swing. For maximum symmetrical swing, the load resistance would need to be 6 V/100 mA $= 60$ Ω. It is possible to achieve this with the 8-ohm speaker by using a transformer, and this is discussed in the next section. For the current arrangement, the maximum symmetrical change ΔI_C in the collector current is $\Delta I_{Cpk} = 100$ mA, i.e. a minimum I_C of zero and a maximum I_C of 200 mA. While this causes only a small change in the collector voltage of the transistor, power is delivered to the speaker and is given by $P_L(AC) = \left(\frac{\Delta I_{Cpk}}{\sqrt{2}}\right)^2 R_L = \left(\frac{0.1}{\sqrt{2}}\right)^2 \times 8 = 40$ mW . There is however significant power dissipation in the transistor which may have to be mounted on a heatsink in order to prevent overheating. This approach was explored in a research project in Chap. 5, and we can now outline the design process.

Example 9.1
Using the configuration in Fig. 9.3 and a transistor with a current gain of 50, design a low-power

amplifier to deliver 75 mW into an 8-ohm load using a 9-volt supply.

Solution Using Equation (9.3), the peak collector current change required to produce 75 mW in an 8-ohm speaker is given by $\left(\frac{\Delta I_{Cpk}}{\sqrt{2}}\right)^2 \times 8 = 75 \times 10^{-3}$. This yields $\Delta I_{Cpk} = 137$ mA and therefore the quiescent current in the power transistor needs to be about 150 mA. The value of R_B required is given by $9 = I_B R_B + 0.7$ where $I_B = I_C/50 = 0.15$ A/50 $= 3 \times 10^{-3}$. Hence $R_B = (9 - 0.7)/I_B = 8.3/3 \times 10^{-3} = 2.8$ k.

Example 9.2
If the 8-ohm speaker in the above amplifier is replaced by a 15-ohm speaker, calculate the output power.

Solution In this circuit with a 15-ohm speaker, the power output is given by $\left(\frac{\Delta I_{Cpk}}{\sqrt{2}}\right)^2 \times R_L = \left(\frac{137 \text{ mA}}{\sqrt{2}}\right)^2 \times 15 = 141$ mW . Thus, if the 8-ohm speaker is replaced by a 15-ohm speaker, the power out increases from 75 mW to 141 mW.

This basic power amplifier system is subject to an effect referred to as thermal runaway. As the system operates, the transistor heats up and the collector current increases. This causes increased heat dissipation in the transistor which causes further increase in collector current. This self-reinforcing cycle may eventually result in the failure of the transistor. Various strategies are available to prevent this phenomenon from occurring. A simple method is to use collector-base feedback biasing which requires that the bias resistor R_B be returned to the collector of the transistor instead of the power supply. The result is that any increase in collector current with temperature will cause an increased voltage drop across the speaker which in turn results in a reduction in the collector voltage of the transistor. The result is that the base current flowing through R_B is reduced thereby reducing the collector current. Higher values of speaker resistance enhance this collector current stabilizing action.

Fig. 9.4 Power amplifier
using a power Darlington
pair

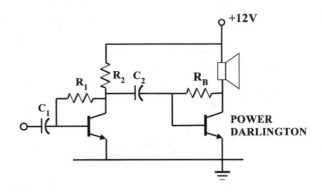

Further improvement of this circuit can be realized by replacing the power transistor with a power Darlington or power MOSFET (which does not suffer from thermal runaway) that results in a higher input impedance and the addition of a preamplifier stage to increase the sensitivity of the overall system. This enhanced arrangement using a power Darlington is shown in Fig. 9.4.

Because of the low resistance of the speaker, the voltage drop across it is low ($0.15 \times 8 = 1.2$ V), and therefore most of the supply voltage would be dropped across the transistor. This does not allow for maximum (symmetrical) swing of the collector voltage. This can be corrected by the use of a step-down transformer in the collector circuit such that a greater impedance is reflected in the collector circuit of the output transistor. This will be discussed under transformer-coupled amplifiers.

9.2.1   Efficiency Calculations

Power amplifiers act as power converters as they deliver large amounts of power to a load. They are driven by an input signal and convert power available from a power supply (input or DC power) to useful signal output power in the load (AC power). The efficiency of the amplifier η is a measure of the effectiveness of this conversion and is given by

$$\eta = \frac{P_{AC}}{P_{DC}} \times 100\% \qquad (9.4)$$

where P_{AC} is the output power delivered to the load and P_{DC} is the input DC power from the supply. In general, not all of P_{DC} is converted to P_{AC}, i.e. $\eta < 100\%$, and the component of the input power that is not converted to AC output power is dissipated as heat in the active output devices.

Example 9.3
A power amplifier has an efficiency of 20% and delivers an output power of 25 W to a resistive load. Calculate the power dissipated and the DC input power to the amplifier.

Solution $\eta = \frac{P_{AC}}{P_{DC}} \times 100$. $\therefore P_{DC} = \frac{P_{AC}}{\eta} \times 100 = \frac{25 \times 00}{20} = 125\,\text{W}$. Hence power dissipated P_{dis} is given by $P_{dis} = 125\,\text{W} - 25\,\text{W} = 100\,\text{W}$.

Example 9.4
A large signal amplifier delivers an output power of 5 W and consumes 20 W from the associated DC supply. Calculate its efficiency.

Solution

$$\eta = \frac{P_{AC}}{P_{DC}} \times 100 = \frac{5}{20} \times 100 = 25\%.$$

Power considerations are extremely important in large-signal amplifiers because of the large voltages and currents involved. For example, a power transistor will be destroyed if its power dissipation capacity is exceeded. In small-signal amplifiers, it was noted that the Q-point must lie below the maximum dissipation rating if the transistor is not to be destroyed. Here we explore new rules to ensure safe operation in power amplifiers.

The average power dissipated by any device is given by

$$P_{AVG} = \frac{1}{T} \int_o^T V(t)i(t)dt \qquad (9.5)$$

where $V(t)$ is the voltage across the device, $i(t)$ is the current through it, and T is the period of the time-varying portion of V or i. For the *fixed-bias Class A amplifier*, the average power P_{DC} delivered by the power supply, neglecting the small power associated with the bias, is given by

$$\begin{aligned} P_{DC} &= \frac{1}{T} \int_o^T V_{cc} i_C dt \\ &= \frac{1}{T} \int_o^T V_{cc}(I_c + i_c(t))dt \\ &= V_{cc} I_c \end{aligned} \qquad (9.6)$$

where i_c is the signal component and I_c the DC component of i_C. The power dissipated in the transistor P_c is calculated using

$$\begin{aligned} P_C &= P_{DC} - P_{AC} \\ &= P_{DC} - \frac{1}{T} \int_o^T i_C^2 R_L dt \\ &= P_{DC} - \frac{1}{T} \int_o^T (I_c + i_c)^2 R_L dt \\ &= P_{DC} - I_c^2 R_L - \frac{1}{T} \int_o^T i_c^2 R_L dt \end{aligned} \qquad (9.7)$$

From this equation, it can be seen that maximum power dissipation in the transistor occurs when the signal current i_c is zero which corresponds to no signal present, i.e.

$$P_{C(max)} = V_{CEq} I_C \qquad (9.8)$$

The peak voltage across the load is $V_{CC}/2$, and hence the maximum power dissipated in the load is $P_{AC\,max} = \frac{(V_{cc}/2\sqrt{2})^2}{R_L}$. Therefore, the maximum efficiency η_{max} is given by

$$\begin{aligned} \eta_{max} &= \frac{P_{AC\,max}}{P_{DC}} \times 100 \\ &= \frac{V_{CC}^2/8R_L}{V_{cc} I_C} \times 100 \end{aligned} \qquad (9.9)$$

Since $I_C = \frac{V_{cc}/2}{R_L}$,

$$\eta_{max} = \frac{V_{cc}^2}{8R_L} \Big/ \frac{V_{cc}^2}{2R_L} \times 100 = 25\% \qquad (9.10)$$

This means that the maximum efficiency of the class A common emitter amplifier is 25%.

9.3 Transformer-Coupled Class A Amplifier

In order to increase the efficiency, the DC power dissipated in the load R_L may be reduced by coupling the load to the transistor using a step-down transformer as shown in Fig. 9.5.

Assuming the resistance of the transformer primary is zero, then the DC load line is vertical, passing through V_{cc} on the V_{CE} axis as shown in Fig. 10.4. Because of the transformer, the load $\overline{R_L}$ seen by the transistor collector is given

$$\overline{R_L} = a^2 R_L \qquad (9.11)$$

where a is the turns ratio of the transformer. With this arrangement, a larger output voltage swing for the same collector current change can be

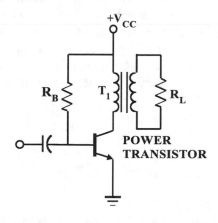

Fig. 9.5 Simple transistor amplifier using a transformer

Fig. 9.6 AC load line for
amplifier using output
transformer

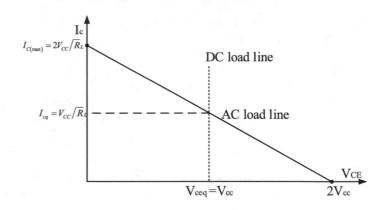

realized as compared with the case where R_L was
placed in the collector circuit. In fact, the AC load
line intercepts V_{CE} axis at a value greater than V_{cc}.
The value of $\overline{R}_L$ determines the slope of the AC
load line. If $\overline{R}_L$ is chosen such that $V_{CEpk} = 2V_{CC}$
and $I_{Cpk} = 2I_{Cq}$, then maximum symmetrical
swing is possible. Then, input power is given by
$V_{CC}I_{Cq}$ and output power (AC) is given by $\frac{V_{cc}}{\sqrt{2}} \times$
$\frac{I_{Cq}}{\sqrt{2}}$. Hence

$$\eta_{\max} = \frac{V_{cc}I_{Cq}}{2V_{cc}I_{Cq}} \times 100 = 50\% \qquad (9.12)$$

The output voltage in this arrangement is able
to exceed the supply voltage because of the
inductive effect of the transformer primary as
shown in the AC load line of Fig. 9.6. Thus, the
maximum efficiency of a transformer-coupled
amplifier is 50%, which is twice as good as the
efficiency of the class A amplifier with the load in
the collector where there was DC power dissipation in the load. Again, for the transformer-
coupled case, the maximum transistor dissipation
occurs when there is no signal present and is
given by

$$P_C = V_{CEq}I_{Cq} = \frac{V_{cc}^2}{R_L} \qquad (9.13)$$

since $V_{CEq} = V_{CC}$ and $I_{Cq} = V_{CC}/\overline{R}_L$.

Example 9.5

Using the configuration shown in Fig. 9.5, design
a transformer-coupled amplifier delivering 1 W

into 8 ohms and operating from a 20-volt supply.
Use a power Darlington with $\beta = 1000$.

Solution The power into the load is given by
$P_L = \left(\frac{V_{pk}}{\sqrt{2}}\right)^2 /R_L$ where V_{pk} is the peak voltage
across the speaker from the output of the trans-
former secondary. Thus, for 1 W output into
8 ohms, $1 = \left(\frac{V_{pk}}{\sqrt{2}}\right)^2 /8$ which yields $V_{pk} = 4$ V
and $I_{pk} = 4$ V/8 = 0.5 A. Since the power supply
is 20 volts, the peak change in collector voltage is
$\Delta V_{CEpk} = 20$ V. This is the input signal voltage to
the transformer primary in the collector. Hence
the transformer turns ratio a to realize this is
$a = \Delta V_{CEpk}/V_{pk} = 20/4 = 5$. (Alternatively, the
power supply voltage may be varied to accommodate a different transformer turns ratio.) The peak
signal current into the transformer primary is
$\Delta I_{Cpk} = I_{pk}/a = 0.5$ A/5 = 100 mA. Therefore
$I_{Cq} = \Delta I_{Cpk} = 100$ mA. Hence, noting that the
Darlington transistor has a base-emitter voltage of
1.4 V, the fixed-bias resistor is calculated from
$20 = R_B I_C/\beta + 1.4 = R_B \times 0.1$ A/30 + 1.4 giving
$R_B = 186$ k. Note that the power supplied to the
system is $V_{CC}I_{Cq} = 20$ V $\times$ 0.1 A = 2 W which
is double the output power of 1 W corresponding
to 50% efficiency.

A preamplifier can also be used with this system to increase sensitivity as shown in Fig. 9.7. In
this configuration, resistor R_2 ensures bias stability in the output stage and R_3 provides feedback
bias to Tr_1.

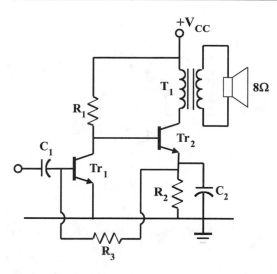

Fig. 9.7 Transformer-coupled power amplifier with direct coupled preamplifier

Example 9.6

Using the configuration shown in Fig. 9.7, design a transformer-coupled amplifier delivering 1 W into 8 ohms and operating from a 20-volt supply. Use a power Darlington with $\beta = 1000$.

Solution From Example 9.6 for 1 W output into 8 ohms, $V_{pk} = 4$ V and $I_{pk} = 4$ V/8 $= 0.5$ A, where V_{pk} is the peak voltage across the speaker from the output of the transformer secondary. Since the power supply is 20 volts, if we allow 2 volts at the emitter of Tr_2 for bias stability in the output stage, then the peak change in collector voltage is $\Delta V_{CEpk} = (20 - 2) = 18$ V. This is the input signal voltage to the transformer primary in the collector. Hence the transformer turns ratio a to realize this is $a = \Delta V_{CEpk}/V_{pk} = 18/4 = 4.5$. (Alternatively, the power supply voltage may be varied to accommodate a different transformer turns ratio.) The peak signal current into the transformer primary is $\Delta I_{Cpk} = I_{pk}/a = 0.5$ A/ $4.5 = 111$ mA. Therefore $I_{Cq} = \Delta I_{Cpk} = 111$ mA. Resistor R_2 is given by $R_2 = 2$ V/111 mA $= 18$ Ω. The base current of the Darlington is 111 mA/ $1000 = 0.11$ mA. This must be supplied by the collector of Tr_1. Choosing the collector current I_{C_1} in Tr_1 to be greater than 10 times this current, we use $I_{C_1} = 2$ mA. Noting that the Darlington transistor has a base-emitter voltage of 1.4 V, the

voltage at the collector of Tr_1 is 3.4 V. Therefore resistor R_1 is found from $R_1 = (20 - 3.4)/$ 2 mA $= 8.3$ k. The base current of Tr_1 is given by $I_{C_1}/100 = 2$ mA/100 $= 20$ μA. Therefore, the bias resistor R_3 is calculated from $R3 = (2 - 0.7)/$ 20 μA $= 65$ k.

9.4   Class B Push-Pull Amplifier

In the class A amplifier considered earlier, the maximum efficiency was 50%. Part of the reason for this is the fact that in class A, the active device (s) is always on, even when no signal is present. Hence there is continuous heat dissipation in this device. In a class B amplifier, the active device conducts for only half of the signal cycle, and therefore greater efficiency is possible. In order to provide full cycle conduction to the load in such circumstances, it is necessary to have at least two active devices, which will conduct for alternative half-cycles. Such an arrangement is traditionally called push-pull based on circuits used in the past. A modern approach to implementing such a design is by the use of a complimentary symmetry emitter follower as shown in Fig. 9.8.

The action of the circuit is as follows: when the input voltage V_i is at zero, then V_o is zero. When V_i goes positive, the emitter follower action of Tr_1 causes V_o to go positive thereby resulting in a flow of current into R_L from Tr_1. Tr_2 remains off during this positive half-cycle. When V_i goes negative, V_o goes negative by emitter follower action of Tr_2 and current flows out of the load into Tr_2. Tr_1 is off during this half-cycle. Thus Tr_1 supplies the load current during the positive half-cycle, while Tr_2 sinks the load current during the negative half-cycle.

Each transistor therefore operates for only approximately half of the signal cycle – class B. When the input signal is at zero, both transistors are off since $V_{BE} = 0$ for both and only a tiny leakage current flows. Therefore, as V_i goes positive, Tr_1 must first turn on, i.e. the input voltage must exceed 0.7 volts the threshold voltage. As a result, for the early part of the

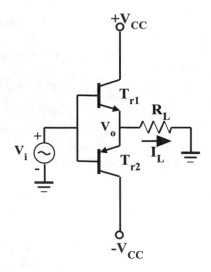

Fig. 9.8 Complimentary symmetry emitter follower

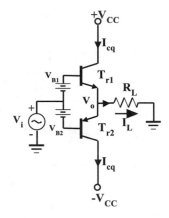

Fig. 9.10 Complimentary Symmetry Emitter Follower with Biasing

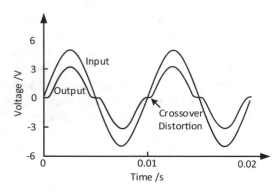

Fig. 9.9 Crossover distortion

positive half-cycle, V_o remains zero until $V_i \geq 0.7$ such that Tr$_1$ turns on. After turn-on, V_o then follows V_i. The effect is that part of the input voltage is not reproduced. This occurs when V_i is increasing above zero and decreasing to zero.

Similarly, for the negative half-cycle, for |V_i| < 0.7, $V_o = 0$ and again part of the input signal is lost. The total effect is shown in Fig. 9.9, and the resulting distortion is called crossover distortion since it occurs as the delivery of the signal output V_o crosses over from one transistor to the other. This distortion can be substantially reduced by biasing the transistors on as shown in Fig. 9.10 (class AB) such that when $V_i = 0$, both transistors are on with a small current I_{cq}. In such a case, as V_i goes positive since Tr$_1$ is already on, no turn-on is required and V_o immediately follows V_i. A

similar action takes place on the negative half-cycle with Tr$_2$ already on. The overall effect is that the crossover distortion is eliminated as a smooth transition from one transistor to the other occurs.

A simple method of forward biasing Tr$_1$ and Tr$_2$ is the use of forward-biased diodes as shown in Fig. 9.11. The resistors R provide current to the diodes as well as the transistor bases. The diodes provide the bias voltage for the transistors and provide temperature stability for I_{cq}. For a transistor, V_{BE} decreases with increasing temperature according to

$$\frac{\Delta V_{BE}}{\Delta T} \simeq -2.5 \,\mathrm{mV}/{}^\circ\mathrm{C} \qquad (9.14)$$

Since the voltage across a diode varies in approximately the same way, by placing D_1 and D_2 in physical contact with Tr$_1$ and Tr$_2$, I_{cq} can be maintained approximately constant. By decreasing R, the voltage across the diodes is increased, and this in turn increases the base emitter voltage of the transistors. I_{cq} therefore increases. This increase in I_{cq} moves the amplifier into class AB and usually reduces the amount of cross over distortion. The optimum value of I_{cq} is normally found experimentally by adjusting I_{cq} for minimum distortion. Since increase in I_{cq} increases the heat dissipated in the transistors, a minimum I_{cq} for minimum harmonic is sought.

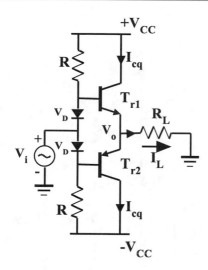

Fig. 9.11 Complimentary symmetry emitter follower with diode biasing

For the push-pull class B circuit, the power delivered by the supply V_{cc}, is

$$P_{cc_1} = V_{cc} \frac{1}{T} \int_0^t i_{c_1}(t) dt \qquad (9.15)$$

If $i_{c_1}(t)$ during the positive half-cycle is given by $i_{c_1}(t) = I_{c\,max} \sin \omega t$, then (9.15) becomes

$$P_{cc1} = V_{cc1} \frac{I_{c\,max}}{\pi} \qquad (9.16)$$

Assuming symmetrical supplies, then $P_{cc_1} = P_{cc_2}$ and the total power P_{cc} supplied by the two supplies becomes

$$P_{cc} = \frac{2}{\pi} V_{cc} I_{c\,max} \qquad (9.17)$$

where $|+V_{cc}| = |-V_{cc}| = V_{cc}$. The power delivered to R_L is

$$P_L = \frac{1}{2} I_{c\,max}^2 R_L \qquad (9.18)$$

The maximum power delivered to the load P_{Lmax} is

$$P_{L\,max} = \frac{1}{2} \left(\frac{V_{cc}}{R_L} \right)^2 R_L = \frac{V_{cc}^2}{2R_L} \qquad (9.19)$$

The power P_D dissipated in the two transistors Tr_1 and Tr_2 is the difference between the power delivered by the power supply and the power delivered to the load, i.e.

$$P_D = P_{cc} - P_L$$
$$= \frac{2}{\pi} V_{cc} I_{c\,max} - \frac{1}{2} R_L (I_{c\,max})^2 \qquad (9.20)$$

In order to determine the value of I_{cmax} for maximum dissipation, P_D is differentiated with respect to I_{cmax} and the result set to zero giving i.e.

$$I_{c\,max} = \frac{2}{\pi} \frac{V_{cc}}{R_L} \qquad (9.21)$$

Substituting (9.21) into (9.20), we get

$$P_D = \frac{2}{\pi} V_{cc} \frac{2}{\pi} \frac{V_{cc}}{R_L} - \frac{1}{2} R_L \left(\frac{2}{\pi} \frac{V_{cc}}{R_L} \right)^2$$
$$= \frac{1}{2} R_L \left(\frac{2}{\pi} \frac{V_{cc}}{R_L} \right)^2 = \frac{2V_{cc}^2}{\pi^2 R_L} \qquad (9.22)$$

Hence each transistor must dissipate half of this energy, i.e. $P_T = V_{cc}^2/\pi^2 R_L$. We note that

$$\frac{P_{L\,max}}{P_T} = \frac{V_{cc}^2}{2R_L} \bigg/ \frac{V_{cc}^2}{\pi^2 R_L} = \frac{\pi^2}{2} = 4.93 \simeq 5 \quad (9.23)$$

Therefore, $P_T \approx \frac{1}{5} P_{L\,max}$ which means that in a class B amplifier, the maximum power dissipated in each transistor is 1/5 the power dissipated in the load. The efficiency η is given by

$$\eta = \frac{P_L}{P_{cc}} \times 100 = \frac{1}{2} \frac{I_{c\,max} R_L}{\frac{2}{\pi} V_{cc} I_{c\,max}} \times 100$$
$$= \frac{\pi}{4} \frac{I_{c\,max} R_L}{V_{cc}} \times 100 \qquad (9.24)$$

This is a maximum when $I_{c\,max} = \frac{V_{cc}}{R_L}$ giving $\eta = 78.5\%$.

There are several characteristics which define the performance of a power amplifier. Some of these are discussed below:

Power rating in watts is a measure of the power output that can be delivered by the amplifier into a particular load, usually 8 Ω or 4 Ω for domestic amplifiers. The formula $P = V_{rms}^2/R_L$ gives the continuous average power sometimes (erroneously) referred to as RMS power. The output power into 8 Ω for a typical domestic power amplifier is of the order of 100 W. It is interesting to note that for the same amplifier output voltage, the power output into 4 Ω is twice that into 8 Ω since the current into the lower resistance increases. The *Gain* of an amplifier is the ratio of the output voltage amplitude to the input voltage amplitude and is usually measured in decibels. Thus the *gain G* in decibels is given by $GdB = 20 \log (V_{out}/V_{in})$ where V_{in} is the amplitude of the input voltage and V_{out} is the amplitude of the output voltage. The gain of many amplifier circuits is between 10 and 50. The sensitivity is the input signal level in volts required to drive the amplifier to its full rated power output and for an amplifier can be about 1 V. The *frequency response* of an amplifier is a measure of the range of signal frequencies that the amplifier will reproduce at the rated gain within certain gain limits. Hi-fi amplifiers typically have a frequency response of 20 Hz to 20 kHz ±3 dB. *Total harmonic distortion* or THD is a measure of the amount of distortion of the input signal arising as a result of the addition by the amplifier of harmonics of the input signal to the output signal. It is usually given as a percentage and can range from about 0.001% to about 0.1% for high fidelity amplifiers, the lower the better. Finally *signal-to-noise ratio* or SNR is the ratio of the desired signal level to the noise level at the output of the amplifier. This ratio is normally expressed in decibels.

9.5   Low-Power Amplifier Design

The low-power amplifier configuration shown in Fig. 9.12 is our first example of a practical power amplifier. It utilizes complimentary symmetry at the output based on transistor pairs in the emitter follower configuration and a common emitter stage. Transistor Tr_1 is connected in a common emitter configuration with its output developed across R_2. Because of the heavy current demand, medium power transistors are used in the emitter follower configuration to couple the voltage at the collector of Tr_1 to the load. The complementary symmetry arrangement enables the circuit to

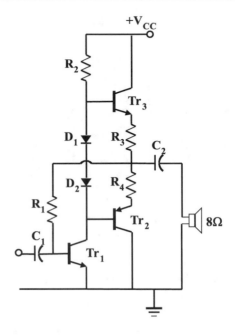

Fig. 9.12 Low-power amplifier

function in class B with each output transistor delivering current to the load on alternative half-cycles. Two diodes D_1 and D_2 connected in series provide the bias voltage that turns on transistors Tr_2 and Tr_3. This results in 1.4 V appearing across the base-emitter junctions of Tr_2 and Tr_3 thereby turning these transistors on and as a result reducing cross over distortion. Resistors R_3 and R_4 are small high wattage resistors that help to stabilize the quiescent current I_{Cq} in the output stage. Biasing for the voltage amplifier Tr_1 is set by the resistor R_1 which also provides signal feedback from the output to the input. A single-ended supply is employed, and hence a voltage of $V_{CC}/2$ sits at the output of the amplifier. This must be therefore coupled to the speaker load via a large electrolytic capacitor C_2. The input coupling capacitor C_1 is necessary to preserve the DC bias at the base of Tr_1. It is important to note that in this circuit, while on the negative half-cycle, Tr_1 sinks base current from Tr_2, while on the positive half-cycle, R_1 supplies base current to Tr_3. Thus this resistor must be carefully chosen to ensure that adequate current can be supplied.

Example 9.7

Using the configuration in Fig. 9.12 design a low-power amplifier that delivers 250 mW into an 8 Ω speaker. All the transistors are silicon.

Solution For $P = 250$ mW into 8 Ω, using $P = V^2/2R_L$, $V_{pk} = \sqrt{P \times 2 \times R_L} = \sqrt{0.25 \times 2 \times 8} = 2$ V. $I_{pk} = 2/8 = 250$ mA. In order to allow for transistor saturation losses, use a 9-volt supply. This can be conveniently supplied by a 9-volt battery. For maximum symmetrical output, the quiescent voltage at the output of the power amplifier is $V_{CC}/2 = 9/2 = 4.5$ V. When the peak voltage across the load is $+2$ V corresponding to the amplifier driving current into the speaker, the amplifier's DC output voltage is $4.5 + 2 = 6.5$ V. Ignoring the small voltage drop across R_3 the voltage at the base of Tr_3 is $6.5 + 0.7 = 7.2$ V. Hence using a 9 V supply the voltage drop across R_2 is $9 - 7.2 = 1.8$ V. At that point, the maximum current $I_{pk} = 250$ mA flows into the load, and hence $I_B(Tr_3)$ is at its maximum and given by $I_b(Tr_3) = 0.25$ A/$100 = 2.5$ mA where 100 is the approximate current gain of the transistor Tr_3. Since this current is supplied by R_2, it follows that $R_2 = 1.8$ V/2.5 mA $= 720$ Ω. However, in order to ensure that R_2 can always supply the necessary, the peak output current requirement is increased by a factor of 1.5 to $I_{pk} = 375$ mA. Hence R_2 now becomes $R_2 = 1.8$ V/3.75 mA $= 480$ Ω. Resistors R_3 and R_4 must be chosen to maximize bias current stability (large), while minimizing power dissipation from signal current (small). A value between 2 Ω and 3 Ω for each of these resistors is reasonable. The quiescent current in Tr_1 is given by $I_{Cq}(Tr_1) = (9 - 4.5 - 0.7)/480$ $\Omega \approx 8$ mA. Hence, taking the current gain of Tr_1 to be 100, the base current $I_b(Tr_1) = 8$ mA/$100 = 80$ μA. This gives the base current in R_1 as 80 μA and hence $R_1 = \frac{(9/2-0.7)}{80 \, \mu A} = 47.5$ k . Because the value of R_1 depends so directly on the current gain of Tr_1, its value may have to be adjusted in order to set the amplifier quiescent output voltage at 4.5 volts for maximum symmetrical output swing. Capacitor C_2 must be sufficiently large in order to achieve a satisfactory low-frequency amplifier response. For a 100 Hz low-frequency response, C_2 is given by $C_2 = 1/2\pi \times 100 \times 8 = 199$ μF. A value of 200 μF is therefore chosen. The input impedance of the amplifier is approximately h_{ie} for Tr_1 which is $h_{ie} = h_{fe}/40I_C = 100/(40 \times 8$ mA$) \approx 300$ Ω. Therefore, the value of the input capacitor C_1 is given by $C_1 = 1/2\pi \times 100 \times 300 = 5.3$ μF. A value of the at least 10 μF is chosen. Transistors Tr_2 and Tr_3 must be medium-power transistors having a collector current rating of at least 250 mA the peak load current, a collector emitter voltage BV_{CEO} of at least 9 V and power rating of 1/5 the amplifier output power. In practice, significant over rating is employed in order to ensure adequate safety margins. Transistors MPSA05 and MPSA55 are very suitable. Transistor Tr_1 dissipates only very modest power. Also, the maximum current is about twice the quiescent current giving 16 mA. Almost any general purpose small-signal transistor such as the 2 N3904 will suffice. Finally, diodes D_1 and D_2 can be 1 N4148 silicon diodes.

A technique for improving the distortion performance of this amplifier is shown in Fig. 9.13. It involves the *bootstrapping* technique applied to the resistor R_2 such that the resistance seen by the collector of Tr_1 is increased, and this results in an increase in the open-loop gain of the amplifier.

This resistor is split into two resistors R_{2a} and R_{2b} and a capacitor C_3 connected from the amplifier output to the junction of these two resistors. The effect is that bootstrapping action by the emitter follower Tr_3 around R_{2b} via C_3 increases the effective value of this resistor for signals. Thus R_{2b} becomes $R_{2b}/(1 - \rho)$ where ρ is the voltage gain of Tr_3 given by $\rho = \frac{g_m R_E}{1+g_m R_E}$ and R_E is the resistance being driven by the transistor emitter which is 8 ohms. For a current in the output of 20 mA say, then $\rho = \frac{40 \times 20 \times 8}{1+40 \times 20 \times 8} = 0.86$, $R_{2b} \rightarrow R'_{2b} = R_{2b}/(1 - 0.86) = 7.4R_{2b}$. Using $R_{2b} = 240$ Ω then $R'_{2b} = 7.4 \times 240 = 1.8$ k. Note that resistor R_{2a} is part of the load of the amplifier output and is in parallel with speaker.

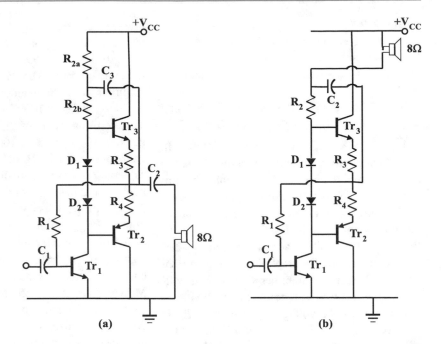

Fig. 9.13 Low-power amplifier with bootstrapping

(a)   (b)

It is possible to eliminate capacitor C_3 by connecting the speaker in place of resistor R_{2a} as shown in Fig. 9.13 and making resistor R_{2b} equal to R_2. This popular configuration has the disadvantage of DC current always flowing through the speaker.

Power Supply

These and other power amplifiers to be discussed in this chapter can generally be powered by unregulated power supplies which are fully discussed in Chap. 10. These supplies can either be single-ended or bipolar. However the low-power systems may be powered from a battery supply.

9.6   Medium-Power Amplifier Design

The power amplifier of Fig. 9.12 is restricted to power of less than about 1 W. At higher-power levels, the increased current demand at the bases of the two output transistors necessitates a large bias current in transistor Tr_1 which is undesirable. For increased power output, the basic

configuration can be modified as shown in Fig. 9.14. It utilizes complimentary Darlington pairs instead of normal transistors at the output to couple the voltage at the collector of Tr_1 to the load. The high current gain of these devices reduces the quiescent current level in Tr_1 while still enabling high output current. The *bootstrapping* technique involving resistor R_2 in the previous low-power amplifier may be applied here to good effect because of the higher input impedance presented by the Darlington pairs. Instead, a constant current source involving Tr_4 and its associated components is used. This circuit ensures that all the quiescent current in Tr_1 is available to drive Tr_3 during the positive half-cycle. More importantly, the combination of a current-source load and the high input impedance presented by the Darlington output pair results in a higher open-loop gain for the circuit. Significant voltage-shunt feedback is then applied via R_2 while R_1 ensures that this feedback is not short-circuited to ground by the input source.

Bias voltage for the Darlington pair may be a string of four diodes connected in series so that these on voltages appear across the four base-emitter junctions of the Darlington pair. This

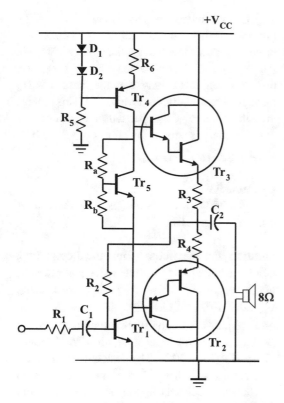

Fig. 9.14 Medium-power amplifier using Darlington pairs

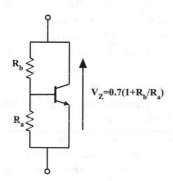

$$V_Z = 0.7(1 + R_b/R_a)$$

Fig. 9.15   Variable Zener

turns on the pair thereby ensuring that cross over distortion is reduced. A diode string unfortunately does not permit easy variation of the bias voltage and the setting of the quiescent current in the output stage. A better approach is to use a V_{BE} multiplier as shown in Fig. 9.15. This circuit acts as a variable Zener diode, producing reasonably wide variation in the voltage V_Z across its terminals by varying R_b. In practice, V_Z would

be varied until the quiescent current in the output transistors is between 10 mA and 100 mA or alternatively until cross over distortion is minimized.

Example 9.8
Design a 10 W power amplifier using the configuration in Fig. 9.14 along with the variable Zener for biasing.

Solution For maximum symmetrical output, the quiescent voltage at the output of the power amplifier is $V_{CC}/2$. For $P = 10$ W into 8 Ω, using $V_{pk} = \sqrt{2PR_L}$, $V_{pk} = \sqrt{2 \times 10 \times 8} = 12.6$ V. In order to allow for transistor saturation losses, the peak voltage swing is set at $12.6 + 5 = 17.6$ V. Hence $V_{CC} = 2 \times 17.6 = 35$ V. The peak output current is given by $I_{pk} = 12.6/8 = 1.6$ A. Using a current gain of 1000 for each Darlington pair, the maximum output current results in maximum base currents for the Darlington of $I_B(Tr_{2,\ 3})$ given by $I_b(Tr_{2,\ 3}) = 1.6$ A/1000 $= 1.6$ mA. This current flows into the collector of Tr_1 during negative-going half-cycles or is supplied from the constant current flowing out of the collector of Tr_4 during positive-going half-cycles. However, in order to ensure that Tr_4 can always supply the necessary current, the constant current is set at 5 mA which is more than three times the requirement for the base current to the Darlington pairs. Resistors R_3 and R_4 are chosen to maximise bias current stability while minimising power dissipation from signal current. Since higher output currents are involved, the power dissipation is a greater problem than before. These resistors would therefore have to be smaller than in the low-power amplifier and a value between 0.2 Ω and 0.3 Ω for each of these resistors is reasonable. The quiescent current in Tr_1 is equal to the current in the current source which is 5 mA. Hence, taking the current gain of Tr_1 to be 100, the base current $I_b(Tr_1) = 5$ mA/100 $= 50$ μA. This gives the current in R_2 as 50 μA and hence $R_2 = \frac{(17.5 - 0.7)}{50\,\mu A} = 336$ k . This value may need to be adjusted to set the output at 17.5 V. Resistor R_6 is calculated from $R_6 = 0.7/5$ mA $= 140$ Ω. To allow about 3 mA through D_3 and D_4, $R_5 = (35 - 1.4)/3$ mA $= 11.2$ k. For a 20 Hz low-frequency response, C_2 is given by $C_2 = 1/2\pi \times 20 \times 8 = 995$ μF. A value of 1000 μF is therefore chosen. The input impedance at the base of Tr_1 is approximately h_{ie} for Tr_1 which is $h_{ie} = h_{fe}/40I_C = 100/(40 \times 5$ mA$) = 500$ Ω. The feedback reduces it further. Therefore the minimum value of the input capacitor C_1 is given by $C_1 = 1/2\pi \times 20 \times 500 = 15.9$ μF. A value of at least 20 μF is chosen. The presence of R_1 further reduces this value. Darlingtons Tr_2 and Tr_3 must be power transistors having a collector current rating of at least 1.6 A the peak load current, a collector emitter voltage BV_{CEO} of at least 35 V and power rating of 1/5 the amplifier output power. Complimentary Darlington transistors TIP 121 and TIP126 (80 V, 5 A, 65 W) are suitable. Transistors Tr_1 and Tr_4 can be small signal transistors but must have a breakdown voltage of greater than 35 volts. MPSA05 and MPSA55 are adequate. The V_{BE} multiplier must provide at least 4 V_{BE} drops across Tr_2 and Tr_3. Let $I_{Ra} = 1$ mA. Then $R_a(Tr_5) = 0.7$ V/ 1 mA $= 700$ Ω.

Hence $1 + \frac{R_b}{R_a} = \frac{2.4}{0.7} = 3.4$, $\frac{R_a}{R_b} = 2.4$, $R_b = 2.4$ k. A 5 k potentiometer can be used for R_b to enable adjustment of the quiescent current in Tr_2 and Tr_3. A 2 N3904 transistor can be used. Resistor R_1 at the input sets the overall voltage gain of the power amplifier. For a voltage gain of 20, $R_1 \approx R_2/10 = 16.8$ k (see Chap. 8). Because of the moderate open-loop gain of the circuit, R_1 may have to be reduced somewhat in order to achieve the desired gain of 10. Under test, the circuit performed as expected delivering 15 volts peak into an 8-ohm load which is equivalent to 14 W. Resistor R_2 was adjusted to 496 k in order to set the DC output voltage at 17.5 V. The 5 k potentiometer was adjusted to set the bias current in the output stage at about 14 mA. Finally the circuit gain was set at -20 by adjusting $R_1 = 16$ k. The frequency response was measured at 100 Hz to 20 kHz.

An improved non-inverting version of the power amplifier in Fig. 9.14 is shown Fig. 9.16. An additional common emitter amplifying stage

Tr_6 is introduced in the circuit. This stage provides further gain and allows the application of voltage-series feedback at the emitter of Tr_6 via R_2, R_1 and C_1. The Zener regulator D_3, D_4 provides regulated bias for the base of Tr_6 through potential divider R_9 and R_{10}. This configuration requires a compensation capacitor, C_4 whose value is given by $C_4 = \frac{g_m}{2\pi f_c A_V}$ where $g_m = 1/(r_e(Tr_6) + R_1)$.

Example 9.9

Design a 25 W power amplifier using the configuration in Fig. 9.16.

Solution For maximum symmetrical output, the quiescent voltage at the output of the power amplifier is $V_{CC}/2$. For $P = 25$ W into 8 Ω, using $V_{pk} = \sqrt{2PR_L}$, $V_{pk} = \sqrt{2 \times 25 \times 8} = 20$ V. In order to allow for transistor saturation losses, the peak voltage swing is set at $20 + 5 = 25$ V. Hence $V_{CC} = 2 \times 25 = 50$ V. The peak output current is given by $I_{pk} = 20/8 = 2.5$ A. Using a current gain of 1000 for each Darlington pair, the maximum output current results in maximum base currents for the Darlington of $I_B(Tr_{2, 3})$ given by $I_b(Tr_{2, 3}) = 2.5$ A/1000 $= 2.5$ mA. This current flows into the collector of Tr_1 during negative-going half-cycles or is supplied from the constant current flowing out of the collector of Tr_4 during positive-going half-cycles. However, in order to ensure that Tr_4 can always supply the necessary current, the constant current is set at 5 mA which is twice the requirement for the base current to the Darlington pairs. Resistors R_3 and R_4 are chosen to maximize bias current stability in the output stage while minimizing power dissipation from signal current. Since higher output currents are involved, the power dissipation is a greater problem than before. These resistors would therefore have to be smaller than in the low-power amplifier and a value between 0.2 Ω and 0.3 Ω for each of these resistors is reasonable. The quiescent current in Tr_1 is equal to the current in the current source which is 5 mA. Hence, taking the current gain of Tr_1 to be 100, the base current $I_b(Tr_1) = 5$ mA/ 100 $= 50$ mA. Since this current is supplied by Tr_6 the quiescent current in this transistor is set at

Fig. 9.16 Improved
medium-power amplifier
using Darlington pairs

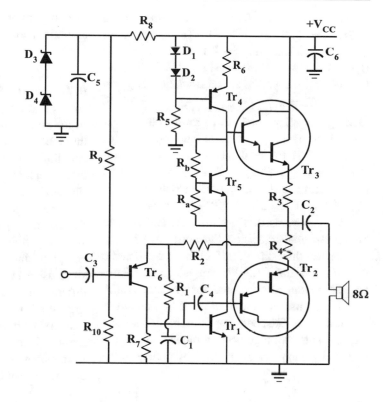

say 0.5 mA. Hence $R_7 = 0.7/0.5$ mA $= 1.4$ k.
Resistor R_6 is calculated from $R_6 = 0.7/$
5 mA $= 140\ \Omega$. To allow about 3 mA through
diodes D_1 and D_2, $R_5 = (50 - 1.4)/3$ mA $= 16.2$ k.
For a 20 Hz low-frequency response into the
8 Ω load, C_2 is given by $C_2 = 1/2\pi \times 20 \times 8 = 995$
μF. A value of 1000 μF is therefore chosen. The
V_{BE} multiplier must provide at least $4V_{BE}$ drops
across Tr_2 and Tr_3. Let $I_{Ra} = 1$ mA. Then
$R_a(Tr_5) = 0.7V/1$ mA $= 700\ \Omega$. Hence $1 + \frac{R_b}{R_a} =$
$\frac{2.4}{0.7} = 3.4\frac{R_a}{R_b} = 2.4, R_b = 1.7$ k. A 5 k potentiometer
for R_b can be used to allow adjustment of the
quiescent current in the output stage. Resistor R_2
along with resistor R_1 provides feedback to the
emitter of the input transistor Tr_6. Let $R_2 = 4.7$ k
say. Then the voltage at the emitter of Tr_6 is
$V_E(Tr_6) = 25 - 0.5$ mA$(4.7$ k$) = 22.7$ V. This
sets the voltage at the base of Tr_6 at
$V_B(Tr_6) = 22.7 - 0.7 = 22$ V. Using a current of
0.25 mA through resistors R_9 and R_{10} to ensure
bias voltage stability for Tr_6, then $R_{10} = 22/$
0.25 mA $= 88$ k. Using two 18 V Zener diodes

for D_3 and D_4, the Zener regulator would provide 36 V. Hence $R_9 = (36 - 22)/0.25$ mA $= 56$
k. Allowing 2 mA into D_3 and D_4 for proper
operation, then $R_8 = (50 - 36)/2$ mA $= 7$ k.
The input impedance of the amplifier is less than
$R_9//R_{10}$ which is 56 k//88 k $= 32$ k. Therefore
the value of the input capacitor C_3 can be found
using $C_3 = 1/2\pi \times 20 \times 32$ k $= 0.25\ \mu$F. A
value of 1 μF is then chosen since the input
impedance at the base of Tr_6 while high because
of voltage-series feedback introduced by R_2 and
R_1 reduces the overall input impedance below
32 k. The gain of the amplifier is given by
$1 + R_2/R_1$. Since $R_2 = 4.7$ k then for a gain of
20, $R_1 = 247\ \Omega$. A small filter capacitor C_6 of
about 0.1 μF can be used to reduce the effect of
long power supply leads, while $C_5 = 1\ \mu$F
improves the regulation of the Zener diodes.
For an adequate low-frequency response, capacitor C_1 must be chosen such that its reactance is
equal to R_1 at 20 Hz. Thus $C_1 = 1/$
$2 \times \pi \times 20 \times 247 = 32\ \mu$F. Capacitor C_4
introduces Miller compensation for circuit

stability. For Tr_6, $I_C = 0.5$ mA and hence $r_e(Tr_6) = 1/40 I_C = 50 \,\Omega$. Therefore using $f_c = 500$ kHz, $A_V = 20$ and $g_m = 1/(r_e(Tr_6) + R_1) = 1/(50 + 247) = 1/297$ in equation (9.37), the values for capacitor C_4 is given by $\qquad C_4 = \frac{g_m}{2\pi f_c A_V} = \frac{1}{297 \times 2\pi \times 5 \times 10^5 \times 20} = 54 \,\mathrm{pF}$. Darlingtons Tr_2 and Tr_3 must be power transistors having a collector current rating of at least 2.5 A the peak load current, a collector emitter voltage BV_{CEO} of at least 50 V and power rating of 1/5 the amplifier output power. Transistors TIP142 and TIP147 (100 V, 10 A, 125 W) can be used. Transistors Tr_1 and Tr_4 can be small-signal transistors but must have BV_{CEO} of greater than 50 V. MPSA06 and MPSA56 (80 V, 500 mA, 1.5 W) are adequate. These transistors can also be used for Tr_5 and Tr_6. A simple but useful modification of this circuit is the use of a Cascode amplifier in the second stage. This arrangement removes distortion caused by early effect and thereby enhances the overall harmonic distortion performance of the amplifier particularly at high frequencies.

9.7 High-Power Amplifier Design

In the power amplifiers considered thus far, neither the input nor the output stages permit direct coupling. An enhanced power amplifier circuit that overcomes these problems is shown in Fig. 9.17. It comprises three amplifying stages: the input stage made up of the differential amplifier Tr_1 and Tr_2, the intermediate common emitter stage Tr_3 with current source load Tr_4 and the complimentary symmetry emitter follower output stage using Darlington pairs Tr_5 and Tr_6. In the differential amplifier input stage, two pnp transistors Tr_1 and Tr_2 are connected at their emitters with resistor R_2 supplying current to these emitters from a Zener-regulated supply fed from the supply rail. Each transistor is basically in a common emitter mode with resistor R_3 in the

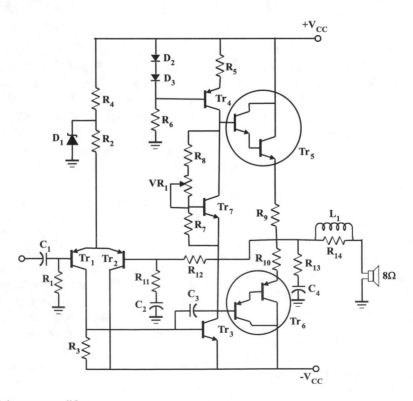

Fig. 9.17 High-power amplifier

collector of Tr_1. This first stage produces modest voltage gain. A signal applied at the base of amplifier Tr_1 is amplified and passed via R_3 to the second stage Tr_3. The common emitter amplifier second stage with active load Tr_4 provides most of the voltage gain.

The complimentary symmetry emitter follower output stage of Darlington pairs Tr_5 and Tr_6 provides a high-impedance load to the collector of Tr_3 and low-impedance output drive to the speaker as before. The use of high-gain Darlington pairs reduces the current demands on the second stage and thereby reduces the required bias current in Tr_3. Also, the increased input impedance of these Darlington transistors along with the active load results in a larger open-loop amplifier gain. The active load comprises transistor Tr_4 and the associated components, while Tr_7 is an amplified Zener that provides output stage bias. The input resistor R_1 both supplies bias current to the base of Tr_1 and establishes a near-zero potential at the input of the amplifier.

This enables direct coupling at the input of the circuit. Resistors R_{11} and R_{12} introduce negative feedback around the amplifier. Capacitor C_1 is for input coupling, while C_2 enables the minimization of DC offset. This will be discussed in the actual design to follow. Capacitor C_3 provides Miller compensation. It ensures that the amplifier has only one dominant pole and therefore is stable when feedback is applied. Its value can be determined analytically or experimentally. Capacitor C_4 along with R_{13} ensures stable amplifier operation into inductive loads. Inductor L_1 and resistor R_{14} enable stable operation into capacitive loads.

In analysing this circuit, we need to determine the open-loop gain, i.e. the gain before feedback is applied. In order to do so, the feedback resistors must be disconnected. As we have discussed in Chap. 8, there are rules which govern how this should be done. However, in this circuit their effect is minimal.

Input Stage

The gain A_{V1} of the first stage Tr_1 with collector current I_{C1} is given by

$$A_{V1} = -\frac{h_{fe}R_{L1}}{2h_{ie}} \approx -\frac{R_{L1}}{2r_e} \qquad (9.25)$$

where

$$r_e = \frac{0.025}{I_{C1}} \qquad (9.26)$$

In the circuit of Fig. 9.17 the load R_{L1} is resistor R_3 in parallel with the input impedance h_{ie} to transistor Tr_3. Thus, the voltage gain of the input stage of Fig. 9.17 is given by

$$A_{V1} = -\frac{R_3//h_{ie3}}{2r_e} \qquad (9.27)$$

where

$$h_{ie3} = \frac{h_{fe}}{40I_{C3}} \qquad (9.28)$$

Intermediate Stage

The second stage of the amplifier consists of transistor Tr_3 in a common emitter amplifier configuration with an active load realized using Tr_4. This stage amplifies the output from the differential amplifier stage and provides most of the voltage gain of the overall amplifier. The gain of the stage is given by

$$A_{v2} = -40I_{C3}R_{L3} \qquad (9.29)$$

where I_{C3} is the current in transistor T_{r3} and R_{L3} is the load made up of the active load in parallel with the input impedance of the complimentary Darlington pairs. Since the Darlington pairs are in the emitter follower configuration, the input impedance is $h_{fe}(D)8$ where $h_{fe}(D)$ is the current gain of the Darlington pair. Since the impedance of the active load is high, it follows that $R_{L3} \approx h_{fe}(D)8$.

Output Stage

The output stage of the amplifier consists of a complimentary emitter follower whose gain is approximately one. It presents a low output impedance and provides the amplifier with the ability to deliver current to the speaker load without loading down the amplifier and reducing the open-loop gain. The use of the Darlington pair gives the stage a high current gain of approximately the product of the current gain of each of the two transistors. For typical transistor gains, this value can be of the order of one to ten thousand.

The overall voltage gain A_V of the open-loop amplifier is the product of the open loop gain of the three stages and is given by

$$A_V = 40 I_{C3} R_{L3} \cdot \frac{R_2 // h_{ie3}}{2r_e} \qquad (9.30)$$

The closed-loop amplifier shown in Fig. 9.17 involves voltage-series negative feedback being applied as in the non-inverting operational amplifier. For a finite open-loop gain A_V, the closed-loop gain A_f is given by

$$A_f = \left(1 + \frac{R_{12}}{R_{11}}\right) \left(\frac{1}{1 + 1/A_V \beta}\right) \qquad (9.31)$$

where

$$\beta = \frac{R_{11}}{R_{11} + R_{12}} \qquad (9.32)$$

If $A_V \gg 1$, the closed-loop gain (9.31) reduces to

$$A_f = 1 + \frac{R_{12}}{R_{11}} \qquad (9.33)$$

This is the well-known expression for the voltage gain of a non-inverting operational amplifier circuit. It is therefore desirable to have a large open-loop gain so that equation (9.33) applies. Assuming large open loop gain, the effect of capacitor C_2 in the transfer function can be found from $A_f = \frac{1+sC_2(R_{11}+R_{12})}{1+sC_2R_{11}}$ which means that the circuit has a low-frequency pole at

$$f_p = 1/2\pi C_2 R_{11} \qquad (9.34)$$

Compensation Capacitor

It was shown in Chap. 7 that Miller capacitor compensation succeeds in realizing a dominant pole system. The closed-loop system has a gain of A_f and the closed-loop graph intersects the open-loop plot at frequency f_c. In order to determine the value of this capacitor, we note that at f_c,

$$A_V = A = A_{V1} \times A_{V2} \approx -\frac{R_{L1}}{2r_e} \times A_{V2} \qquad (9.35)$$

At f_c, $R_{L1} \to X_{CM}$ where X_{CM} is the reactance of the effective Miller capacitance at the input of Tr_3 and is given by $X_{CM} = 1/2\pi f_c A_{V2} C_3$. Hence

$$A_V = \frac{X_{CM}}{2r_e} \times A_{V2}$$

$$= g_M \times \frac{1}{2\pi f_c A_{V2} C_3} \times A_{V2}$$

$$= \frac{g_M}{2\pi f_c C_3} \qquad (9.36)$$

where $g_M = 1/2r_e$ is the transconductance of the input stage. From this the value of capacitor C_3 is given by

$$C_3 = \frac{g_M}{2\pi f_c A_V} \qquad (9.37)$$

Example 9.10

Design a power amplifier that delivers 100 W average power into an 8 ohm load using the configuration in Fig. 9.17.

Solution In order to produce 100 W into 8 ohms, the peak output voltage V_{opk} is given by $\frac{(V_{opk}/\sqrt{2})^2}{R_{spk}} = 100$ where $R_{spk} = 8\ \Omega$. This gives $V_{opk} = 40$ V. The corresponding peak output current is 40/8 = 5 A. Resistors R_9 and R_{10} provide output stage quiescent current stability. However since they are in the output current path there will be voltage losses across them therefore they cannot be too large. If we allow a voltage drop of 1 V peak, then R_9 and R_{10} should be 1/5 = 0.2 Ω. On the positive half-cycle, this current flows through resistor R_9 resulting in a 1 V drop across R_9. This and the base-emitter voltages of Darlington Tr_5 sum to 40 + 1 + 1.4 = 42.4 V. If we allow about 5 V across the active load in order that it is fully operational, then the positive supply voltage needs to be about 50 V. On the negative half-cycle, resistor R_{10} and the base-emitter voltages of Darlington Tr_6 again sum to about 42.4 V. For symmetry, we choose a negative supply voltage of −50 V. This allows at least 7 V across Tr_3

during its operation. For good transistor operation, let the quiescent current in each transistor of the differential pair be 1 mA. The total current through R_2 is therefore 2 mA. This current is provided by the Zener diode D_1 which is selected to be about half of the supply voltage. A Zener with voltage of 24 V is available. The base of Tr_1 is at an approximately zero potential and hence R_2 is given by $R_2 = (24 - 0.7)/2$ mA $= 11.7$ kΩ. Resistor R_4 supplies current to the Zener of about 5 mA as well as the 2 mA to the differential pair. Hence $R_4 = (50 - 24)/7$ mA $= 3.7$ kΩ. The base-emitter voltage of Tr_3 appears across R_3 and therefore $R_3 = 0.7/1$ mA $= 700$ Ω. Resistor R_1 supplies base current to Tr_1. It should be low in order to minimize the voltage drop arising from the base current flow while high enough to provide a high input impedance to the circuit. A value of about 10 k or higher is a reasonable compromise. For a peak output voltage of 40 V, the peak output current is 40/8 = 5 A. Using Darlington current gain of 1000, the maximum peak base current of each Darlington is 5/1000 = 5 mA. Since this current is sourced from the collector current of Tr_3, we therefore choose a quiescent current of 10 mA for this transistor. The collector current of Tr_4 must also be set at 10 mA. This is accomplished by noting that the voltage across diode D_2 is dropped across R_5 through which the 10 mA also flows giving $R_5 = 0.7/10$ mA $= 70$ Ω. Resistor R_6 is chosen such that about 5 mA flows through diodes D_2 and D_{13} to turn them fully on. Hence $R_6 = (50 - 1.4)/5$ mA $= 9.7$ k. In order to minimize DC offset at the output, feedback resistor R_{12} is chosen to be the same value 10 k as the input resistor R_1. Resistor R_{11} then sets the closed-loop gain. A value of 520 Ω gives a closed-loop gain of 20. For the V_{BE} multiplier, the current in this stage is the 10 mA collector current of Tr_3. Because of the base-emitter voltage of Tr_7 across R_7, a value of 1 k results in a current through this resistor of $0.7/1$ k $= 0.7$ mA. This is approximately seven times larger than the base current of Tr_7 which is 10 mA/100 = 0.1 mA. If we set $R_8 = 1$ k, then a 5 k potentiometer for VR_1 will produce a maximum

voltage across Tr_7 of 0.7 mA × (1 k + 1 k + 5 k) = 4.9 volts. This is more than the 2.8 volts required to turn on the Darlington pairs. The minimum voltage across Tr_7 (corresponding to $VR_1 = 0$) is 0.7 mA × (3.3 k) = 2.3 V. Therefore, the potentiometer can be used to set the bias current in the output stage to a value that minimizes cross-over distortion. This value is likely to lie in the range 10 mA to 100 mA. The capacitor C_2 influences the low-frequency performance of the amplifier. A value of 100 µF sets a pole at $f_p = 1/(2\pi 100 \times 10^{-6} \times 520) = 3$ Hz. The capacitor C_1 also influences low-frequency roll-off. Using the input resistance as 10 k (since feedback will increase the input impedance into Tr_1), a value of 10 µF for this capacitor will result in a pole at $f_p = 1/2\pi 10 \times 10^{-6} \times 10000 = 1.59$ Hz. Capacitor C_3 is determined using equation (9.37). The choice of frequency f_c is critical. It must be chosen high enough such that adequate feedback is available at all frequencies of interest. However, setting this frequency too high will increase the risk of instability arising from increased phase shift around the feedback loop. For audio amplifiers, f_c is likely to be in the range 200 kHz to 2 MHz using $f_c = 500$ kHz and $g_m = 1/2r_e = 1/50$, where $r_e = 1/40I_C$, we get $C_3 = \frac{1}{50} \times \frac{1}{2\pi \times 5 \times 10^5 \times 20} = 318$ pF. The final value of C_3 can be experimentally determined by adjusting for minimum square wave overshoot.

Suggested transistors are as follows: Power Darlingtons MJ11021/MJ11022 (250 V, 15 A, 175 W); Tr_3/Tr_4 MJE340/MJE350 (300 V, 500 mA, 20 W); Tr_1/Tr_2 2 N5400 pnp (140 V, 600 mA, 625 mW).

This amplifier can be improved in several ways. The first improvement is to replace the resistor R_2 supplying current to the differential pair by a constant current source as shown in Fig. 9.18. The effect of this is to improve the ripple rejection capability of the amplifier so that power supply changes do not get into the amplifier via the emitters of Tr_1 and Tr_2. A second improvement is the inclusion of small resistors R_e in the emitters of Tr_1 and Tr_2. These resistors reduce the distortion produced in this stage,

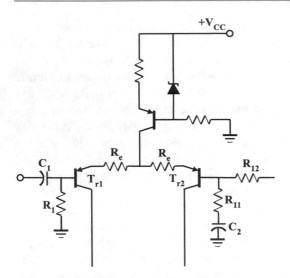

Fig. 9.18 Introduction of constant current source

increase the bandwidth and improve the current equalization between the transistor pair. If this is done, the compensation capacitor can be reduced, and hence the amplifier slew rate increases. Note that g_m becomes $g_m = 1/2(r_e + R_e)$. A third significant improvement is the utilization of a Cascode amplifier in place of Tr_3 as shown in Fig. 9.19. This reduces distortion arising from modulation of the collector base junction capacitor of Tr_3 as a result of the large voltage swing occurring in the output stage. Finally, the Darlington pairs in the output stage may be assembled using discrete transistors or may be replaced by power MOSFETS which have extremely high input impedance. A fully symmetrical design using these latter devices is the subject of the next section.

Example 9.11

The amplifier circuit shown in Fig. 9.20 uses feedback pairs in the output stage and can deliver 60 W into 8 ohms. Analyse the circuit to determine the quiescent currents in the input and intermediate stages as well as the closed-loop gain. Investigate the purpose of the 100 µF capacitor connected between the output and the junction of the two 2.7 k resistors.

Solution The voltage at the emitters of the input transistors is approximately +0.7 V. Noting that

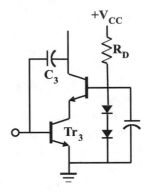

Fig. 9.19 Cascode amplifier

the other end of the 6.8 k resistor supplying current to these transistors is supplied by a 15 V Zener diode regulator, the current through the 6.8 k resistor is $(15 - 0.7)/6.8$ k = 2.1 mA. This is the total current flowing into the emitters of the two input transistors. The first of these transistors has a 680 Ω resistor in its collector circuit. This in turn is connected across the base-emitter junction of the intermediate stage. Hence the current through this resistor is $0.7/680 = 1$ mA. Therefore, the current in Tr_2 is $2.1 - 1 = 1.1$ mA. The current through R_3 is $(35 - 15)/2.2$ k = 9 mA. The current through the second stage transistor is $(35 - 0.7)/(2.7 + 2.7)$ k = 6.4 mA. Finally, the closed-loop gain is 1 + 20 k/1 k = 21. The 100 uF capacitor is used to bootstrap resistor R_6 thereby producing a quasi-constant current source. This makes more drive current available to drive the base of the feedback pair for positive output swing.

9.8 High-Power MOSFET Amplifier

A high-quality power amplifier may be designed using power MOSFETS instead of power transistors in the output stage. Power MOSFETS became available in the 1970s and enable the exploitation of the high input impedance of the field effect transistor while being able to handle significant power levels. They can be used to replace the Darlington Pairs in Fig. 9.17 in the output stage. These devices respond to very high

Fig. 9.20 60 W power
amplifier circuit

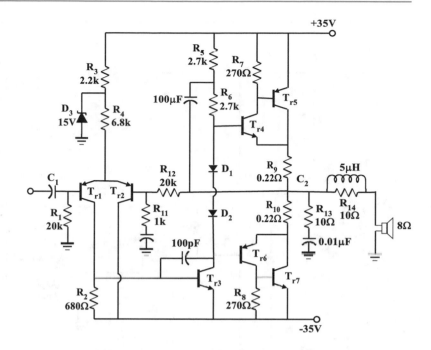

frequencics and arc more easily biased. Another advantage of these devices is their extremely high input impedance completely isolates the second stage transistor collector from the effects of the amplifier load. A completely symmetrical amplifier design using power MOSFETS is shown in Fig. 9.21. Note that four MOSFETS are used, two n-channel and two p channel devices each pair sharing the load current on alternatively positive and negative signal cycles. Along with power MOSFETS at the output, this circuit uses a complimentary symmetry differential amplifier at the input and complimentary Cascode intermediate voltage gain stages. The result is a fully symmetrical and very linear circuit that produces low levels of harmonic distortion. The dominant pole is set by the capacitance at the gate-drain junctions of the MOSFETS with all other poles at high frequencies.

Example 9.12

Design a power amplifier that delivers 150 W average power into an 8-ohm load using the configuration in Fig. 9.21.

Solution In order to produce 150 W into 8 ohms, the peak output voltage V_{opk} is given by

$\frac{\left(V_{opk}/\sqrt{2}\right)^2}{R_{spk}} - 150$ where $R_{spk} = 8\ \Omega$. This gives $V_{opk} = 49$ V. The corresponding peak output current is 49/8 = 6.13 A. Resistors R_{20} to R_{23} provide output stage quiescent current stability. As before, since they are in the output current path they cannot be large. Values of 0.22 Ω will be used and, because the current is approximately shared, will result in a peak voltage drop of less than 1 V. On the peak positive half-cycle with maximum current flowing into the load, the voltage drop in R_{20} (or R_{21}) added to the gate-source FET voltage of 2 V gives the voltage at the collector of Tr_7 as 49 + 1 + 2 = 52 V. If we allow about 5 V across the active load in order that it is fully operational, then the positive supply voltage needs to be about +60 V. Similarly, we choose a negative supply voltage of −60 V. The FET gate resistors $R_{16} - R_{19}$ are necessary for MOSFET stability and each needs to be about 220 Ω. Diodes D_5, D_7, D_6 and D_8 protect the gate source junctions from over-voltage with Zeners D_5 and D_6 limiting the voltage to 12 V. Potentiometer V_{R1} is adjusted in order to set the FET bias current. Note that a V_{be} multiplier is not necessary. Suitable power MOSFETS are 2SK1058/2SJ162 (160 V, 7 A, 100 W).

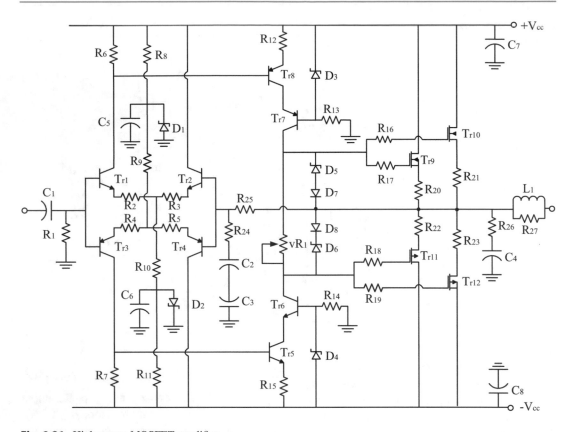

Fig. 9.21 High-power MOSFET amplifier

At the input, the complimentary symmetry differential amplifier has a reduced current flowing in the bias resistor R_1 since Tr_1 and Tr_3 approximately provide each other with the necessary base current. This resistor can therefore assume a higher value while still keeping the transistor bases at an approximately zero potential. Suitable transistors are 2 N5551/2 N5401 (160 V, 600 mA). We let $R_1 = R_{25}$ for minimum DC offset and set the value at 33 k. Hence with $R_{24} = 1$ k, the overall closed-loop gain is 34. This gives a sensitivity of about 1 V rms for full output into 8 ohms. The input transistors all have emitter degeneration resistors R_2 to R_5. These resistors balance the current in Tr_1–Tr_2 and Tr_3–Tr_4, improve the frequency response and reduce distortion. However, the gain of this stage is reduced. Values of 100 ohms represent a reasonable compromise. Diodes D_1 and D_2 provide regulated supplies for the currents into Tr_1–Tr_2 and Tr_3–Tr_4. If we set the currents in these

transistors to be 0.5 mA, then using diodes D_1 and D_2 at 24 V, it follows that $R_9 = R_{10} = (24 - 0.7)/1$ mA $= 23.3$ k. R_8 and R_{11} supply current to the Zener diodes. This current needs to be about 5 mA. Hence $R_8 = R_{11} = (60 - 24)/(5 + 1) = 6$ k. The Cascode amplifiers Tr_5–Tr_6 and Tr_8–Tr_7 using MJE340/MJE350 high-voltage transistors for Tr_6 and Tr_7 must operate with enough current to charge the input (gate-drain) capacitance of the MOSFETS which varies with drain-source voltage. Using gate capacitance of 500 pF and a current of 15 mA, the slew rate will be $SR = \frac{2 \times 15\,\text{mA}}{500\,\text{pF}} = 60\,\text{V}/\mu\text{s}$. This is sufficient to avoid slewing distortion for peak output signals at 20 kHz which is $SR = 2\pi \times 49 \times 20,000 = 6.2$ V/μs. Resistors R_{12} and R_{15} introduce local feedback in the Cascode amplifiers thereby improving linearity. A value for these resistors of 82 ohms results in a voltage drop across it of 15 mA $\times$ 82 $= 1.23$ V. Therefore, the voltage between the base of Tr_5

and the negative supply rail is 1.23 + 0.7 = 1.93 V. This is the voltage across resistor R_7 which, because it is carrying the Tr_3 collector current of 0.5 mA, must be $R_7 = 1.93/0.5$ mA = 3.8 k. Resistor R_6 is calculated in a similar manner to give the same value. Diodes D_3 and D_4 set the voltage at the bases of Tr_6 and Tr_7. Their values are chosen to be 3.9 volts in order to allow sufficient voltage across the cascade arrangement. Current for these diodes is delivered by resistors R_{14} and R_{13} which are given by $R_{13} = R_{14} = (60 - 3.9)/5$ mA = 11.2 k where 5 mA is the value of this current. Capacitor C_1 is an input coupling capacitor whose value is set to ensure an adequate low-frequency response. Since the input impedance of the circuit is approximately $R_1 = 33$ k, for a low-frequency cut-off of 20 Hz, $C_1 = 1/2\pi \times 20 \times 33,000 = 0.24$ µF. We chose a value of 1 uF. Capacitors C_5 and C_6 are filter capacitors chosen to reduce the ripple at the Zener diodes. A value of 10 uF will give sufficiently low impedance at ripple frequency of 120 Hz from a bridge rectifier. Capacitors C_7 and C_8 perform a similar function and need to be about 100 uF. Capacitors C_2 and C_3 are connected as a non-polarized capacitor and minimize the dc offset at the output of the amplifier. This capacitor along with R_{24} also influences the low-frequency cut-off frequency and is given by $C_T = 1/2\pi \times 20 \times 1000 = 8$ µF. A value of 100 uF is chosen and therefore $C_2 = C_3 = 200$ µF.

9.9   IC Power Amplifiers

The power amplifiers discussed thus far all employ discrete components. An alternative approach to power amplification is the use of power ICs. These are specifically designed to deliver high voltages and currents, unlike low-power operational amplifier ICs. A few devices selected from the large number available are presented.

Low-Power IC Amplifier: LM386
The first example of a power amplifier IC is the LM386 from National Semiconductor shown in a

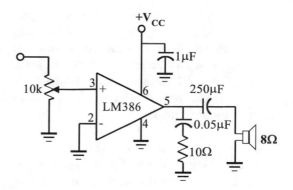

Fig. 9.22 LM386 power amplifier

typical application in Fig. 9.22. This device has been optimized such that it requires a minimum number of supplementary components and can deliver up to 1 W average power into an 8 ohm load. It operates from a single-ended power supply that can vary from 4 to 12 V. It can therefore be powered from three 1.5 V batteries, the 5 V supply from a computer, a 9 V battery or a 12 V car battery. Internal circuitry provides bias for the IC such that the output terminal is at half of the supply voltage thereby allowing for maximum symmetrical swing. The output terminal must therefore be coupled to the load through a capacitor C resulting in a low-frequency cut-off frequency given by $f_L = 1/2\pi R_{load}C$. Even though the IC operates from a single-ended supply, the inputs are referenced to ground. Direct coupling at the input is therefore possible. Overall negative feedback is applied within the package. The input signal is applied to the non-inverting input with the inverting input connected to ground. The gain is 20 but can be adjusted using IC pins. The 10 k potentiometer limits the input signal amplitude to the maximum rated value of + − 0.4 volts. With a 9–12 volt supply, the IC can deliver a 6 volt peak-peak output which corresponds to about 0.5 W rms into 8 ohms.

In order to increase the output power, two LM386 ICs can be operated in a bridge configuration shown in Fig. 9.23. Here one IC is connected in the non-inverting configuration (signal applied to the non-inverting input with inverting input grounded), while the other is connected in the inverting configuration (signal applied to the inverting input with non-inverting input grounded). The load is connected across the IC outputs which are out of phase. Thus, while the output of one IC swings positive towards the supply voltage, the output of the other swings negative towards ground.

USB-Powered Amplifier: LM4871

The LM4871 uses the bridge configuration in Fig. 9.23 in its internal structure to deliver 3 W into a 3-ohm load from a 5-volt supply such as is available from a computer USB port. It requires few external components not including output coupling capacitor. It possesses thermal shutdown protection and is unity gain stable. The amplifier gain can be set externally and is

intended primarily for use in portable devices. Fig. 9.24 shows it in a typical application.

Medium-Power IC Amplifier: LM3886

For higher output power levels than can be delivered by the LM386, the LM3886 shown in Fig. 9.25 is available. This is a high performance audio power amplifier that can deliver 56 W continuous power into an 8 ohm load with low harmonic distortion between 20 Hz and 20 kHz. It operates from ± 10 V to ± 42 V and can supply up to 4 A. The device is fully protected against over-temperature, over-current and voltage transients. It operates essentially as a power op-amp, and therefore the gain is given by $A_V = 1 + \frac{R_2}{R_1}$. The peak output voltage is about 5 V less than the supply voltage.

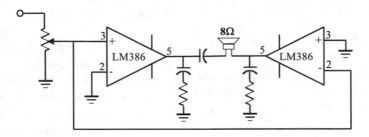

Fig. 9.23 Bridge configuration

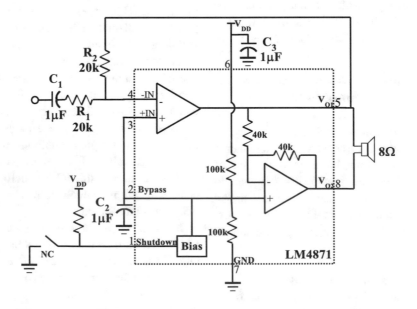

Fig. 9.24 Power amplifier using the LM4871

Example 9.13

Configure the LM3886 to produce 50 W into 8 ohms with reasonable sensitivity.

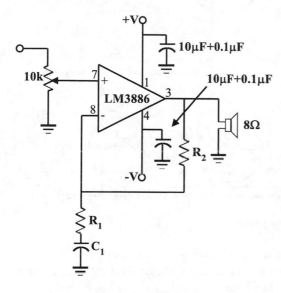

Fig. 9.25 LM3886 power amplifier

Solution For 50 W into 8 ohms, the peak output voltage is given $V_{opk} = \sqrt{50 \times 2 \times 8} = 28.3\,\text{V}$. The peak output current is then $I_{opk} = 28.3/8 = 3.54$ A. Therefore, the required supply voltage is given by $\pm V_{CC} = 28.3 + 5 = \pm 33.3$ V, where 5 volts is the dropout from the IC. If the device is powered from an unregulated supply, then adequate allowance must be introduced such that the minimum supply voltage is 33.3 volt. In order to produce the full output from an input of say 1 V, the amplifier gain must be $\frac{28.3/\sqrt{2}}{1} = 20$. Therefore $20 = 1 + \frac{R_f}{R_i}$. If $R_i = 1$ k then $R_f = 19$ k. Note that with readily available ± 15 V supplies, this IC can deliver 10 W into an 8 ohm load.

High-Power IC Amplifier Driver: LME49811
The LME49811 is a power IC driver from Texas Instruments that, working in conjunction with power output transistors, is able to deliver up to 500 W output into an 8 Ω load at extremely low levels of harmonic distortion. The basic system is shown in Fig. 9.26. It operates from ± 20 V to ± 100 V with the power output increasing with

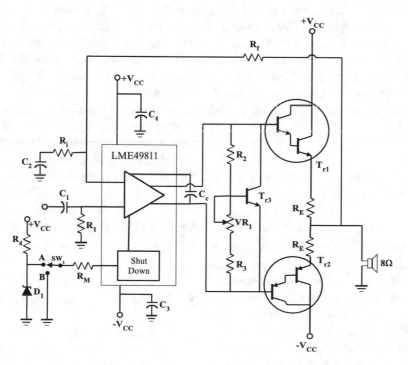

Fig. 9.26 Power Amplifier using LME49811

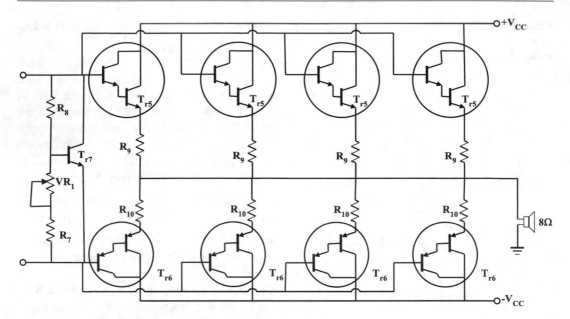

Fig. 9.27 Parallel combination of complimentary Darlington pairs

the supply voltage. The IC is used to deliver drive current (source and sink) to complimentary Darlington pairs Tr_1 and Tr_2 such as the MJ11021 and MJ11022 (250 V, 15 A, 175 W). Transistor Tr_3 is a V_{BE} multiplier that sets the bias current in Tr_1 and Tr_2. The resistors R_f and R_i set the closed-loop gain at $A_V = 1 + R_f/R_i$. For stability, A_V must be greater than about 20. The capacitor C_C provides compensation and must be optimized experimentally. The IC driver has a shutdown mode and is thermally protected. Power supply decoupling of the IC using C_3 and C_4 is essential.

Example 9.14

Configure the LME49811 IC to produce 200 W into 8 ohms with a gain of 21.

Solution In order to produce 200 W into 8 ohms, the peak output voltage is given by $V_{Opk} = \sqrt{2 \times 8 \times 200} = 56.6$ V. The peak output current is then $I_{Opk} = 56.6/8 = 7.1$ A. The specifications for the LME49811 indicate that in order to deliver an output voltage of ± 56.6 V, the supply voltage must be ± 80 V. The components of the V_{BE} multiplier are determined as done previously. For a gain of 21, R_i is set to 1 k and hence R_f is

set to 20 k. Resistor R_1 is also set at 20 k. From the data sheet, the shutdown resistor R_M is calculated using $R_M = (V_{ref} - 2.9)/I_{SD}$. With $V_{ref} = 6.8$ V and $I_{SD} = 1.5$ mA, $R_M = 2.6$ k. Capacitors, C_1 and C_2, are selected to be 10 μF for adequate low-frequency response. The compensation capacitor C_c is set at a nominal value of 30 pF with the final value determined experimentally. The power supply pins to the IC must be decoupled with $C_3 = C_4 = 0.1$ μF capacitors. In order to deliver this level of power to the speaker, the Darlington output stage needs to comprise a parallel arrangement of four sets of complimentary Darlington pairs MJ11021/MJ11022 as shown in Fig. 9.27. Each pair will deliver a portion of the output load current with all four pairs being driven by the LME49811. Adequate heatsinking of these transistors as well as the LME49811 is necessary for successful operation of this high-power system.

9.10 Amplifier Accessories

A power amplifier is often used to drive inductive and capacitive loads such as loud speakers and significant lengths of connecting cable. If the amplifier is to remain stable under such conditions, appropriate networks need to be connected at the output of the power amplifier. The two networks normally used are a shunt *Zobel* network for maintaining stability into inductive loads and a parallel connected inductor and resistor in series with the output for preserving stability when driving capacitive loads.

A capacitive load connected to the output of a power amplifier can result in the amplifier going unstable. This occurs since the output resistance of the amplifier in conjunction with the capacitor form what is effectively a low-pass filter at the output of the amplifier. This network introduces a pole into the transfer function of the amplifier which causes phase shift that pushes the closed-loop system towards instability. In order to prevent this, an inductor is placed between the amplifier output and the capacitive load such that the capacitance is isolated from the amplifier. Its value is chosen such that the upper cut-off frequency created by the inductor working in conjunction with the lowest load resistance presented by the speaker load is above the highest frequency to be amplified. Additionally, a small resistor is placed in parallel with this inductor in order to provide damping. This network is referred to as a *Thiele* network. Typical values are 2 μH in parallel with a 1 Ω resistor of suitable wattage. (See L_1 and R_{27} in Fig. 9.21)

The voice coil of a loud speaker presents an inductive load to a power amplifier that can excite instability at high frequencies as the inductive reactance increases. This problem can be resolved by placing a *Zobel* network across the speaker. This network comprises a resistor in series with a capacitor, and its effect is to cause the inductive speaker load to appear resistive across the frequency of operation of the amplifier. The resistor is usually in the range 4.7 Ω – 10 Ω, while the capacitor is almost always 0.1 μF. (See R_{26} and C_4 in Fig. 9.21).

It may be necessary to provide protection for the output transistors in a power amplifier. This is to prevent transistor burn-out due to current overload. A simple mechanism to achieve this is the circuit shown in Fig. 9.28. This is incorporated in the output stage of the amplifier. A large current flow through resistors R_E produces a voltage drop that turns on transistors Tr_a and Tr_b. These in turn reduce the drive current to the base of the output transistors and as a result limit the current in these devices.

9.10.1   Heatsink Design

In the class B amplifier, almost 25% of the input power from the power supply is converted into heat, most of which is dissipated in the output stage. As a result, during operation power amplifiers dissipate a large amount of heat in the power transistors. A characteristic of transistors is that as the temperature is increased, their power-handling capability decreases. A graphical illustration of this is presented in Fig. 9.29. Here the curve ABCD defines the safe operating area of the transistor. Portion AB represents the maximum current that the transistor can handle, portion CD is set by the maximum collector-emitter voltage of the transistor and portion BC is the hyperbola defined by $V_{CE}I_C = P_D$ where P_D is the transistor power rating at room temperature (25°C).

As the temperature of the case of the transistor and hence the collector junction temperature increases, the power rating of the transistor decreases. The result is that the hyperbola represented by curve BC moves towards the axes as shown thereby decreasing the safe operating area of the transistor. The transistor is then said to be de-rated. A curve showing the transistor de-rating as a function of case temperature is shown in Fig. 9.30. In order to prevent this transistor de-rating, the dissipated heat must be conducted away such that the transistor junction temperature is kept low. To aid in heat conduction away from the transistor, the device must be mounted on a piece of metal called a heatsink. This along with the metal case of the transistor provides a large surface area from which the heat can be radiated. This results in a lower transistor junction temperature as compared with the transistor operating without the heatsink. The heat path is from the transistor junction to the case, from the case to the heatsink and finally from the heatsink to the surrounding atmosphere.

Fig. 9.28 Electronic amplifier protection

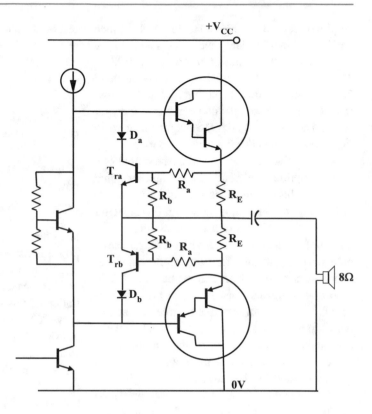

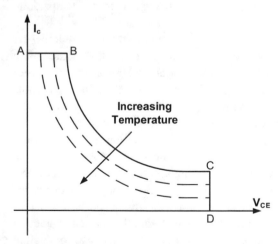

Fig. 9.29 Change of the safe operating area with temperature for power transistors

Information regarding transistor de-rating is contained in the equation

$$P_D(T) = P_D(T_o) - (T - T_o)DF \qquad (9.38)$$

where $P_D(T)$ is the maximum power dissipation at temperature T, T_o is the temperature above

which de-rating begins with $P_D(T_o)$ the maximum power dissipation rating of the transistor and DF (Watts/degree of temperature) is the de-rating factor given in manufacturer specification sheet.

Example 9.15
Determine the maximum allowable dissipation at 100°C in a power transistor rated at 125 W at 25°C if de-rating occurs above this temperature by a de-rating factor of 0.5 W/°C.

Solution Using equation (10.35) the maximum power dissipation for the transistor at 100°C is $P_D(100°C) = P_D(25°C) - (100°C - 25°C)$ $(0.5 \text{ W/°C}) = 125 \text{ W} - 75°C(0.5 \text{ W/°C}) = 87.5$ W. This means that at 100°C this 125 W power transistor can only dissipate 87.5 W.

In order to determine the heatsink required for a particular application, it is necessary to consider the thermal conditions associated with the transistor heatsink combination. In particular the junction temperature for a heatsink mounted transistor is given by

Fig. 9.30 Typical
de-rating curve for silicon
power transistors

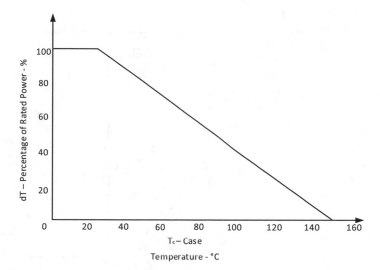

$$T_J = T_A + P_D(\theta_{JC} + \theta_{CS} + \theta_{SA}) \quad (9.39)$$

where P_D is the power to be dissipated, T_J is the transistor junction temperature, T_A is the ambient temperature in the immediate vicinity of the transistor, $\theta_{JC}(°C/W)$ is the transistor thermal resistance between junction and case, θ_{CS} is the insulator thermal resistance between case and heatsink and θ_{SA} is the heatsink thermal resistance between heatsink and atmosphere. θ_{JC} is set by the transistor manufacturer, while θ_{CS} is determined by the transistor mounting arrangements. If a good thermal compound is used between the transistor case and the heatsink, $\theta_{CS} \approx 0.2°C/W$. It is however often necessary to isolate the transistor case from the heatsink by introducing a mica wafer between them. This electrically isolates the transistor but raises the thermal resistance to typically $\theta_{CS} \approx 2°C/W$. The last parameter θ_{SA} is the main heatsink specification and is determined using Equation (9.39).

Example 9.16

In a power amplifier design, a power transistor is expected to dissipate 50 W. Using the transistor in the previous example, determine the thermal resistance of the heatsink required for safe operation of this transistor. Use $\theta_{JC} = 0.4°C/W$ and $\theta_{CS} \approx 2°C/W$.

Solution From the previous example, the transistor was able to dissipate 87.5 W at a junction temperature of 100°C. Thus, at this temperature the transistor can safely dissipate 50 W. From Equation (10.37), $\theta_{SA} = (T_J - T_A)/P_D - \theta_{JC} - \theta_{CS} = (100 - 25)/50 - 0.4 - 0.2 = 0.9°C/W$.

9.11 Applications

A variety of power amplifiers and associated systems are presented below.

1.5 W Low-Power Amplifier

The circuit shown in Fig. 9.31 is an improved non-inverting version of the low-power amplifier with bootstrapping in Fig. 9.13a and is similar to the configuration in Fig. 9.16 (improved medium power). An additional common emitter amplifying stage Tr_4 is introduced in the circuit. This stage provides further gain and allows the application of voltage-series feedback at the emitter of Tr_4. The resistors R_1 and R_2 provide a fixed voltage for the base of Tr_4 and resistor R_8 and capacitor C_5 provide filtering. This configuration requires a compensation capacitor C_4 whose value is given by $C_4 = \frac{g_m}{2\pi f_c A_V}$ where $g_m = 1/(r_e(Tr_4) + R_9)$ and A_V is the closed-loop gain of the amplifier.

Fig. 9.31 1.5 W
low-power amplifier

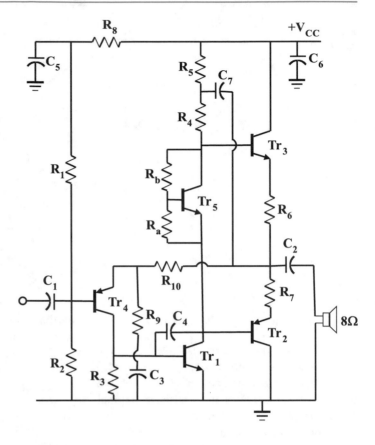

For maximum symmetrical output, the quiescent voltage at the output of the power amplifier is $V_{CC}/2$. For $P = 1.5$ W into 8 Ω, using $V_{pk} = \sqrt{2PR_L}$, $V_{pk} = \sqrt{2 \times 1.5 \times 8} = 4.9$ V. In order to allow for transistor saturation losses, the peak voltage swing is set at 6 V, and therefore $V_{CC} = 2 \times 6 = 12$ V. The peak output current is given by $I_{pk} = 4.9/8 = 0.6$ A. Using a current gain of 100 for each output transistor, the maximum output current results in maximum base currents for each given by $I_b(Tr_{2, 3}) = 0.6$ A/$100 = 6$ mA. This current flows into the collector of Tr_1 during negative-going half-cycles or is supplied by the bootstrapped resistor R_4 during positive-going half-cycles. In order to ensure that this current is always available, the collector current of Tr_1 is set at 12 mA which is twice the requirement for the base current to the output transistors. This gives $R_4 + R_5 = 6$ V/ 12 mA $= 500$ Ω. We use $R_4 = R_5 = 220$ Ω. Resistors R_6 and R_7 are chosen to maximize bias

current stability in the output stage while minimizing power dissipation from signal current. Since only moderate output currents are involved, the power dissipation is the less serious issue, and therefore these resistors can be selected for bias stability. A value of about 1 Ω for each of these resistors is reasonable. Taking the current gain of Tr_1 to be 100, the base current $I_b(Tr_1) = 12$ mA/100 $= 120$ μA. Since this current is supplied by Tr_4 the quiescent current in this transistor is set at say 0.5 mA. Hence $R_3 = 0.7/0.5$ mA $= 1.4$ k. For a 20 Hz low-frequency response into the 8 Ω load, C_2 is given by $C_2 = 1/2\pi \times 20 \times 8 = 995$ μF. A value of 1000 μF is therefore chosen. The V_{BE} multiplier must provide at least $2V_{BE} = 1.4$ V drops across the base-emitter junctions of Tr_2 and Tr_3. Let $I_{Ra} = 1$ mA. Then $R_a(Tr_5) = 0.7$ volts/ 1 mA $= 700$ Ω. Hence $1 + \frac{R_b}{R_a} = \frac{1.4}{0.7} = 2$. $\frac{R_b}{R_a} = 1$, $R_b = 700$ Ω. A 2 k potentiometer can be used for R_b to allow adjustment of the quiescent current in the output stage. Resistor R_{10} along with resistor R_9 provides feedback to the emitter of the input transistor Tr_4. Let $R_{10} = 2.7$ k say. Then the voltage at the emitter of Tr_4 is $V_E(Tr_4) = 6 - 0.5$ mA(2.7 k) = 4.7 V. This sets the voltage at the base of Tr_6 at $V_B(Tr_4) = 4.7 - 0.7 = 4$ V. Using a current $I_R = 0.05$ mA through resistors $R_8 - R_1 - R_2$ which is one tenth of the collector current of Tr_4 to ensure bias voltage stability for Tr_4, then $R_8 + R_1 + R_2 = 12$ V/0.05 mA = 240 k and $R_2 = 4/0.05$ mA = 80 k. Choosing $R_1 = 80$ k then $R_8 = 240$ k $- 80$ k $- 80$ k = 80 k. The input impedance of the amplifier is less than $R_1//R_2$ which is 80 k//80 k = 40 k. Therefore the value of the input capacitor C_1 can be found using $C_1 = 1/2\pi \times 20 \times 40$ k = 0.2 μF. A value of 1 μF is then chosen since the input impedance at the base of Tr_4 while high because of voltage-series feedback introduced by R_{10} and R_9 reduces the overall input impedance below 40 k. The gain of the amplifier is given by $1 + R_{10}/R_9$. Since $R_{10} = 2.7$ k then for a gain of 25, $R_9 = 112$ Ω. A filter capacitor C_6 of about 100 μF can be used to reduce the effect of long power supply leads, while $C_5 = 100$ μF forms a low-pass filter with R_8 to remove ripples from the supply going to the base of Tr_4. For an adequate low frequency response, capacitor C_3 must be chosen such that its reactance is equal to R_9 at 20 Hz. Thus $C_3 = 1/2 \times \pi \times 20 \times 112 = 71$ μF. Choose $C_3 = 100$ μF. Capacitor C_4 introduces Miller compensation for circuit stability and is determined using equation (9.37). For Tr_4, $I_C = 0.5$ mA and hence $r_e(Tr_4) = 1/40 I_C = 50$ Ω. Therefore using $f_c = 500$ kHz, $A_V = 25$ and $g_m = 1/(r_e(Tr_4) + R_9) = 1/(50 + 112) = 1/162$ in equation (9.37), the values for capacitor C_4 is given by $C_4 = \frac{g_m}{2\pi f_c A_V} = \frac{1}{162 \times 2\pi \times 5 \times 10^5 \times 25} = 79$ pF

. Finally, bootstrapping capacitor C_7 must have a reactance at 20 Hz that is lower than the resistance of R_4. Hence $C_7 = 1/2 \times \pi \times 20 \times 220 = 36$ μF. Choose $C_7 = 100$ μF.

Ideas for Exploration (i) Redesign the input stage of this circuit, using a differential amplifier to replace Tr_4. (ii) Employ a Cascode amplifier in place of Tr_1.

50 W MOSFET Amplifier Using Bootstrapping

This circuit is that of a 50 Watt MOSFET amplifier using a differential input stage and a bootstrapped driver stage (Fig. 9.32). Each transistor in the differential pair has about 1 mA. For $P = 50$ W into 8 Ω, using $V_{pk} = \sqrt{2PR_L}$, $V_{pk} = \sqrt{2 \times 50 \times 8} = 28.2$ V. In order to allow for transistor saturation losses, the supply is set at 35 V and therefore $\pm V_{CC} = \pm 35$ V. The MOSFETs being used are the IRF530 and the IRF9530 which together have input capacitance of 150 pF. The driver stage Tr_3 must have adequate current to drive the input capacitances of the MOSFETs. Using a current of 6 mA, the slew rate is $SR = \frac{6\,\text{mA}}{150\,\text{pF}} = 40$ V/μs which will prevent slew rate limiting on audio signals. Let $R_5 = R_6$. Then $R_5 + R_6 = 35$ V/6 mA = 5.8 k giving $R_5 = R_6 = 2.9$ k. For a current of 1 mA in Tr_1, $R_2 = 0.7/1$ mA = 700 Ω. With 1 mA through Tr_2 also, then $R_4 + R_3 = 35/2$ mA = 17.5 k. Let $R_4 = 2.5$ k then $R_3 = 15$ k. Let $R_1 = R_8 = 47$ k. Setting these resistors equal minimizes DC offset at the output of the amplifier. For a gain of 20, $R_7 = 2.1$ k. The capacitor values are calculated to give adequate low-frequency response as well as filtering of ripple.

Ideas for Exploration (i) Replace the resistor R_3, R_4 and capacitor C_5 by a constant current source supplied from the positive supply rail. (ii) Employ a Cascode amplifier in place of Tr_3. This will lower the distortion produced by the circuit.

Amplifier Clipping Indicator

The circuit in Fig. 9.33 gives an indication when clipping occurs in a power amplifier. It comprises a diode bridge connected to the output of the amplifier being monitored which ensures that signals of either polarity can be monitored. The signal drives two transistors connected as a silicon-controlled rectifier (see Chap. 14) such that if the voltage at the base-emitter junction of Tr_1 equals about 0.7 volts, this transistor is turned on and thereby turns on Tr_2, the action of which further turns on Tr_1 latching both transistors fully

Fig. 9.32 50 W MOSFET
amplifier

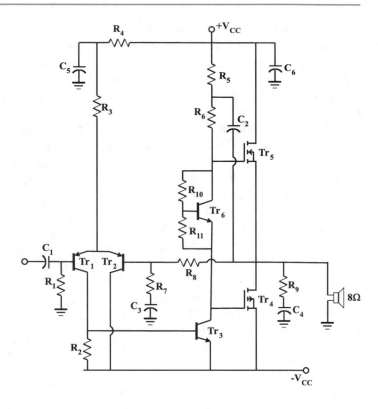

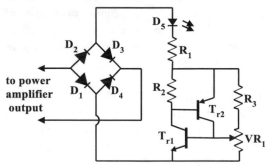

Fig. 9.33 Amplifier Clipping Indicator

example, for $P = 100$ W, $R_1 = 40\sqrt{100} = 400\,\Omega$. The potential divider formed by resistor R_3 and potentiometer $VR_1 = 1$ k sets the voltage to be applied to the base-emitter junction of Tr_1 (noting that R_1 is small). We allow for a 1 volt drop across the LED and set the potentiometer at a value of say 700 ohms when triggering occurs for a given $V_{pk} = 4\sqrt{P}$. Then $\frac{4\sqrt{P}-1}{R_3+1000} = \frac{0.7}{700}$ from which $R_3 = 2000(2\sqrt{P} - 1)$. For example, for $P = 100$ W, $R_3 = 2000(2\sqrt{100} - 1) = 38$k . $R_2 = 10$ k assists in turning on Tr_2.

on. They remain in this condition until the voltage across the arrangement goes to zero or the current through them goes to zero. When the transistors are turned on, current flows through the LED which then illuminates. Resistor R_1 limits the current through the LED to about 100 mA. Thus $R_1 = V_{pk}/100$ mA where V_{pk} is the maximum peak output voltage of the amplifier. For an amplifier delivering output power P into an 8-ohm speaker load, $V_{pk} = 4\sqrt{P}$. Therefore $R_1 = V_{pk}/100\,\text{mA} = 4\sqrt{P}/0.1 = 40\sqrt{P}$. For

Ideas for Exploration (i) Investigate the possibility of including a green LED in series with R_3 such that it illuminates when there is no clipping but goes off when clipping occurs since then the current through R_3 goes to zero.

Power Operational Amplifiers
A very simple method of realizing a power amplifier is to use an operational amplifier and boost the output current as shown in Fig. 9.34. The output of the operational amplifier drives a

complimentary emitter follower with a resistor R_3 connected between the output of the operational amplifier and the emitters of the two transistors. If the output current is less than 0.7 V/R_3, then the transistors remain off, and the output current is supplied by the operational amplifier. As the current exceeds 0.7 V/R_3, the voltage drop across R_3 exceeds 0.7 V and the transistors turn on Tr_1 for positive-going signals and Tr_2 for negative-going signals and supply the current. Crossover distortion is suppressed by the operation of resistor R_3 and the inclusion of the booster transistors in the feedback loop of the operational amplifier ensures that distortion introduced by these transistors is significantly reduced. For example, with $R_3 = 100\ \Omega$ the limiting current is 0.7 V/ $100 = 7$ mA above which the transistors take over. It is important to note that while the bulk of the output current is supplied by the transistors, base current to the transistors must be supplied by the operational amplifier. This limits the amount of current that can be delivered to about 1 ampere since with a transistor current gain of say 100, the base current demand for 1 A output is 10 mA which is the maximum current typically available from an operational amplifier. Further enhancement of the output current capacity can be achieved by the use of complimentary feedback pairs in place of the two transistors. The current gain for these pairs is better than 1000 in which case as much as 10 A can be delivered using this approach. In the system, $R_B = 100$ k provides bias to the non-inverting input of the operational amplifier and R_1 and R_2 set the gain at $1 + R_2/R_1$. For a 741 operational amplifier operating from ±15 V supplies, the peak output voltage is about 12 V giving a power of $P = \frac{(12/\sqrt{2})^2}{8} = 9$ W into 8 ohms and 18 W into 4 ohms.

Another operational amplifier current booster circuit is shown in Fig. 9.35. The transistors are driven by current flowing into the power leads of the operational amplifier in response to a signal.

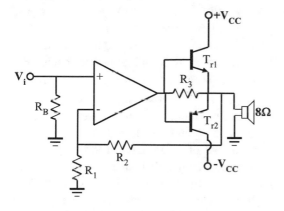

Fig. 9.34 Operational amplifier power system 1

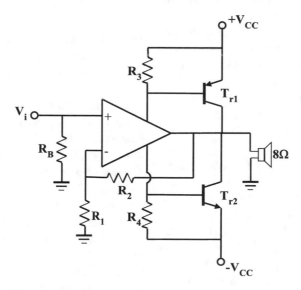

Fig. 9.35 Operational amplifier power system 2

One advantage of this circuit over the previous current booster is R_3 and R_4 can be selected such that a quiescent current flows in the transistors. The result is that there is little or no crossover distortion.

Using a 741 op-amp, $R_3 = R_4 = 400 \ \Omega$ produced a quiescent current in the transistors of 20 mA. Setting $R_B = 100$ k, $R_2 = 9$ k and $R_1 = 1$ k, a 1 V input resulted in a 10 V output. Like the first current booster circuit, enhancement of the output current capacity can be achieved by the use of complementary power Darlington pairs such as the MJ3001/MJ2501 (80 V, 10 A) or the TIP142/TIP147 (100 V, 10 A) in place of the two transistors. The current gain for these Darlington pairs can exceed 1000 enabling the circuit to deliver currents approaching 10 A. In the system, R_B provides bias to the non-inverting input of the operational amplifier, and R_1 and R_2 set the gain at $1 + R_2/R_1$. Using this circuit, an OPA452 operational amplifier operating from ± 35 V supply can deliver 50 W into 8 ohms and 100 W into 4 ohms.

Ideas for Exploration (i) While both the operational amplifier and the complimentary emitter follower can be operated from the same bipolar supply, it is important that the supply is decoupled at the transistors and at the power pins to the operational amplifier. Also, separate ground connectors must be used for the speaker load and the operational amplifier ground. This prevents "ground loops" from creating spurious results. (ii) Explore the use of the bridge configuration in order to increase the power output of the systems. (iii) Reconfigure this circuit to operate in the inverting configuration and examine its performance using simulation.

75 W Power Amplifier Using Feedback Pairs
This power amplifier in Fig. 9.36 utilizes complementary feedback pairs in the output stage. The design follows the procedures previously discussed. Resistors $R_{13} = 560 \ \Omega$ and $R_{14} = 560 \ \Omega$ provide current to Tr_6 and Tr_7 and also assist in the proper operation of the output transistors Tr_5 and Tr_8. For $P = 75$ W into 8 Ω,

using $V_{pk} = \sqrt{2PR_L}$, $V_{pk} = \sqrt{2 \times 75 \times 8} = 34.6V$. Allowing for transistor saturation and other losses, the supply is selected to be $\pm V_{CC} = \pm 40$ V.

Ideas for Exploration (i) Replace the bootstrap arrangement in the collector of Tr_3 by a transistor constant current source; (ii) Introduce a transistor constant current source to supply the emitters of the differential pair Tr_1 and Tr_2. What improvement does this provide?

Research Project 1
This power amplifier is intended to deliver *extremely high-power* output. It must produce 250 W into 8 ohms or 500 W into 4 ohms. The project requires the scaling up of the power MOSFET amplifier described in Example 9.12 in order to produce these power levels. For $P = 250$ W into 8 Ω, using $V_{pk} = \sqrt{2PR_L}$, $V_{pk} = \sqrt{2 \times 250 \times 8} = 63.2$ V . Allowing for transistor saturation and other losses, the supply is set at 75 V and therefore $\pm V_{CC} = \pm 75$ V. The number of MOSFETS in the output stage must at least be doubled compared to the 150 W circuit therefore giving eight devices: four n-channel and four p-channel. These can be Renesas 2SK2221 and 2SJ352 or Magnatec BUZ901 and BUZ906. This is a major project that should be undertaken with great care.

Ideas for Exploration (i) Refer to the article "High-Power Audio Frequency Amplifier", Elektor Electronics, May 1986, p60.

Research Project 2
Peter Walker and the Acoustical Manufacturing Company introduced the *current dumping amplifier* in 1975. It uses the system described in the op-amp power amplifier project in Fig. 9.34 where transistors are turned on to "dump" current into an external load. However, while the op-amp power amplifier uses negative feedback to reduce the distortion arising from the dumping action, the system developed by this company is claimed to cancel this distortion completely. The project requires the reader to implement the scheme in an

Fig. 9.36 75 W Power
Amplifier Using Feedback
Pairs

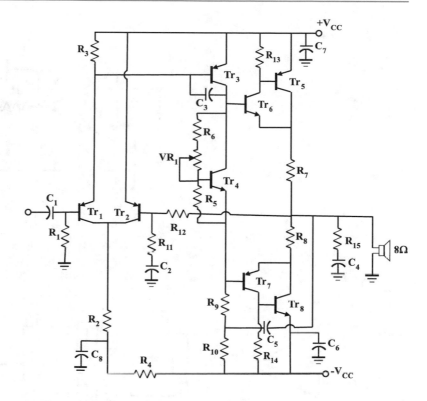

amplifier and verify by experiment that distortion
cancellation does indeed occur. The basic system
is shown in Fig. 9.37. It can be derived from the
power op-amp circuit 1 of Fig. 9.34 by splitting
resistor R_3 of that circuit into R_3 and R_4 and taking
the output to the speaker load R_L at the junction of
these resistors. Note that as $R_4 \rightarrow 0$, the current
dumping circuit reduces to the power op-amp
circuit 1 of Fig. 9.34.

Using standard circuit analysis, for the system

$$V_1 = A(V_i - \beta V_2) \qquad (9.40)$$

where A is the gain of the low-power linear amplifier and $\beta = R_1/(R_1 + R_2)$. The second equation is

$$\frac{V_o}{R_L} = \frac{V_1 - V_o}{R_3} + \frac{V_2 - V_o}{R_4} \qquad (9.41)$$

which reduces to

$$\frac{V_o}{R_P} = \frac{V_1}{R_3} + \frac{V_2}{R_4} \qquad (9.42)$$

where $R_P = R_L//R_3//R_4$. The third defining equation is

$$V_2 = BV_1 \qquad (9.43)$$

where B is the voltage gain of the current dumping transistors which is highly non-linear primarily as a result of crossover distortion. Eliminating
V_1 and V_2 from equations (9.40), (9.42) and
(9.43), we get

$$V_o = \frac{R_P}{R_3} \frac{A}{1 + \beta AB} \left(1 + \frac{R_3}{R_4}B\right) V_i \qquad (9.44)$$

If the condition

$$\frac{R_3}{R_4} = \beta A \qquad (9.45)$$

is satisfied, then (9.44) reduces to

$$V_o = \frac{R_P}{R_3} A V_i \qquad (9.46)$$

Thus, from equation (9.46), the output signal
V_o driving the speaker is independent of the
non-linear dumper characteristic B and depends
only on the gain A of the linear low-power amplifier. Therefore, the crossover distortion of

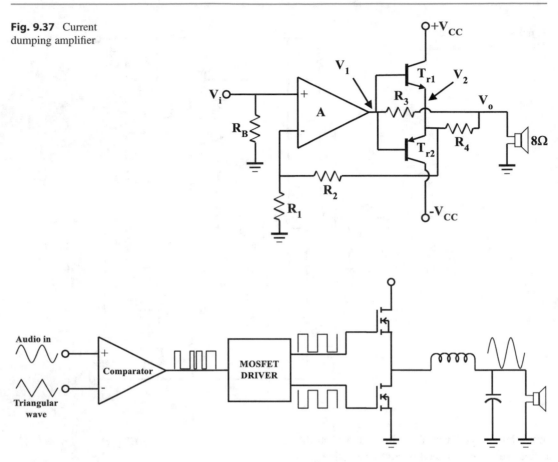

Fig. 9.37 Current dumping amplifier

Fig. 9.38 Class D amplifier

transistors Tr_1 and Tr_2 is not present in the output signal.

Ideas for Exploration (i) Refer to the articles "Current Dumping Audio Amplifier", P.J. Walker, Wireless World, December 1975, p560 and Current Dumping Analysis by H.S. Malvar, Wireless World, March 1981, p69. (ii) Examine other distortion cancelling schemes used by amplifier manufacturers or published in the audio engineering literature.

Research Project 3

This project involves the design of a *class D amplifier*. This is a class of amplifiers that operates by switching on and off the output to the external load. When on, the voltage across the switching device is low and when off, the current through the switching device is very low,

therefore there is little power loss in the switching devices and as a result such amplifiers have very high efficiencies (better than 90%). As a consequence, they require only small heatsink area since relatively low power is dissipated. The basic scheme is shown in Fig. 9.38. The input signal is compared with a triangular wave of a higher frequency of about 200 kHz using a comparator. The output of the comparator is a pulse width modulated waveform whose pulse widths vary according to the signal amplitude. This waveform then drives the push-pull MOSFET output stage, switching it fully on or fully off for each pulse. The output of this stage is applied to the loudspeaker through a low-pass LC filter to remove the unwanted switching components and recover the amplified analog signal.

Ideas for Exploration (i) Refer to International Rectifier Application Note AN-1071, "Class D Amplifier Basics" by Jun Honda and Jonathan Adams. (ii) Consider using the International Rectifier IRS2092 Class D Audio Amplifier Controller IC that contains the control and switching elements for driving the power MOSFETS.

Problems

1. Using the configuration in Fig. 9.39 and a transistor with a current gain of 50, design a low-power amplifier to deliver 60 mW into an 8-ohm load using a 6-volt supply (Fig. 9.39).

2. If the 8-ohm speaker in the above amplifier is replaced by a 15-ohm speaker, calculate the output power.

3. A power amplifier has an efficiency of 30% and delivers an output power of 15 W to a resistive load. Calculate the power dissipated and the DC input power to the amplifier.

4. A large signal amplifier delivers an output power of 25 W and consumes 75 W from the associated DC supply. Calculate its efficiency.

5. Using the configuration shown in Fig. 9.40, design a transformer-coupled amplifier delivering 1.5 W into 8 ohms and operating from a 16-volt supply. Use a power Darlington with $\beta = 2000$.

6. Using the configuration shown in Fig. 9.41, design a transformer-coupled amplifier delivering 2 W into 8 ohms and operating from a 24-volt supply. Use a power Darlington with $\beta = 1000$.

7. Explain the phenomenon of crossover distortion and the reason it occurs. Suggest one method of overcoming this problem.

8. Design a low-power amplifier using the topology of Fig. 9.12 and a 12-volt supply.

9. For the amplifier circuit shown in Fig. 9.42, determine the quiescent current and the base voltage in transistor Tr_3 if the output voltage is 4.5 volts. Develop a design procedure for this configuration.

10. What would be the effect on amplifier performance of increasing the input and output capacitors? Discuss the use of a variable Zener circuit to adjust the output bias current.

11. Design a 500 mW amplifier using the circuit of Fig. 9.43. Introduce bootstrapping to improve the performance of this circuit.

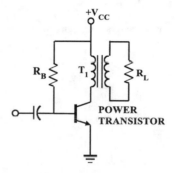

Fig. 9.40 Circuit for Question 5

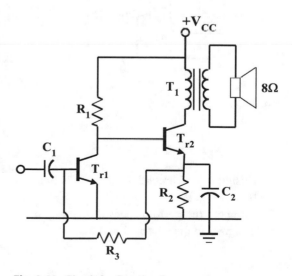

Fig. 9.41 Circuit for Question 6

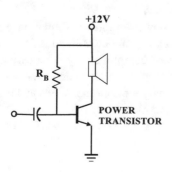

Fig. 9.39 Circuit for Question 1

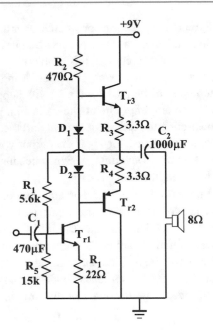

Fig. 9.42 Circuit diagram for Question 9

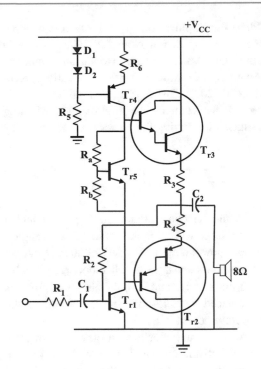

Fig. 9.44 Diagram for Question 12

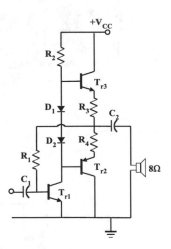

Fig. 9.43 Circuit for Question 11

12. Design a 5 W amplifier using the configuration of Fig. 9.44.

13. Compare the performances of the amplifiers designed in questions 11 and 12.

14. Identify the type of feedback that is used in the amplifier in Fig. 9.44 as well as the number of voltage gain stages.

15. Design a 20 W amplifier driving an 8 ohm speaker using the configuration in Fig. 9.16.

16. Design a 60 W amplifier driving a 4 ohm speaker using the configuration in Fig. 9.17. (i) Discuss the effect of including emitter resistors in the input stage. (ii) Design and introduce a constant current source in the differential amplifier stage and indicate how this improves amplifier performance. (iii) Design short-circuit protection for the amplifier.

17. The power amplifier shown in Fig. 9.45 utilizes an output configuration referred to as quasi-complimentary symmetry since the two power transistors are of the same polarity. Analyse the circuit shown and investigate the operation of the output stage.

18. Design a 75 W power MOSFET amplifier using the circuit in Fig. 9.21.

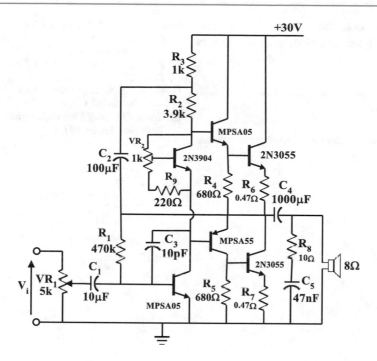

Fig. 9.45 Diagram for Question 17

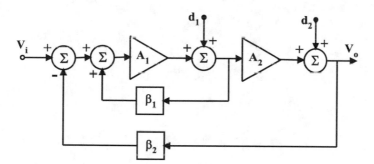

Fig. 9.46 Diagram for Question 21

19. In a power amplifier design a power transistor is expected to dissipate 75 W. Using a transistor that can dissipate 100 W at a junction temperature of 100°C determine the thermal resistance of the heatsink required for safe operation of this transistor. Use $\theta_{JC} = 0.4°C/W$ and $\theta_{CS} \approx 2°C/W$.

20. As a research project, modify the configuration in Fig. 9.45 to include another transistor stage at the input as in Fig. 9.16.

21. An audio power amplifier can be represented in a simplified model as an amplifier A_1 with relatively low distortion d_1 driving an output stage A_2 with significant distortion d_2 (Fig. 9.46). Positive feedback is applied around A_1 via β_1 and negative feedback is applied around the system through β_2. Show that when $A_1\beta_1 = 1$, the output is given by $V_o = \frac{V_i}{\beta_2} + \frac{d_1}{A_1\beta_2}$. This means that in the amplifier output signal V_o, the already low

distortion d_1 is further reduced by the factor $A_1 \beta_2$ and the major distortion d_2 completely vanishes!

Bibliography

B. Cordell, *Designing Audio Power Amplifiers* (Mc Graw Hill, New York, 2011)

Motorola Product Application Note AN-483B, Complementary Darlington Output Transistor in Audio Amplifiers, April 1974

D. Self, *Audio Power Amplifier Design*, 6th edn. (Focal Press, 2013)

G. Slone, *Randy, High-Power Audio Amplifier Construction Manual* (McGraw-Hill, New York, 1999)

M. Yunik, *Design of Modern Transistor Circuits* (Prentice Hall, New Jersey, 1973)

Power Supplies

10

An electrical power source is essential for the operation of most electronic equipment. Some portable equipment such as radio receivers and pocket calculators may be battery operated, but most electronic equipment requires an electrical supply. In this chapter, we examine the operation and design of linear power supplies that convert an AC voltage from the public mains supply into a stable DC voltage. At the end of the chapter, the student will be able to:

- Explain the operation of linear regulators
- Design several forms of linear regulated power supplies
- Utilize IC regulators in regulated power supplies
- Design short-circuit protection circuits

10.1 Basic System

A regulated power supply consists of several sub-systems, a transformer, a rectifier, a filter and a voltage regulator, as shown in Fig. 10.1. The transformer converts the high AC mains voltage to a lower usable voltage while also providing isolation from the mains. The rectifier converts the low AC voltage from the transformer to a varying DC voltage, while the filter removes the ripple from the rectified voltage, thereby converting it from high ripple content to low ripple content. Finally the regulator converts the

unregulated output of the filter to a regulated one where the DC voltage is very stable and ripple-free. The power supply comprising only of transformer, rectifier and filter is referred to as an unregulated supply. When a regulator is added, the system is referred to as a regulated power supply.

10.2 Rectification

The principle of rectification has been introduced in Chap. 1 where diode applications were discussed. In this chapter, more quantitative analysis is done so that design issues may be considered. The circuit shown in Fig. 10.2 is that of a half-wave rectifier with a transformer to reduce the mains voltage to a more desirable level. The diode permits current flow in one direction only. The voltage V_S represents the secondary voltage of the transformer and is given by $V_S = V_m \sin \omega t$ where $\omega = 2\pi f$. For $f = 60$ Hz, $\omega = 377$ rad/s. The output waveform is shown in Fig. 10.3. Its average value is the ratio of the area under one cycle of the half-wave rectified voltage waveform and the waveform period $T = 1/60$ given by

$$V_{AV} = \frac{\int_0^T V_m \sin 377t \, dt}{1/60} = \frac{V_m}{\pi} \qquad (10.1)$$

Note that the area under half of the output voltage cycle is zero. The supply output is

© Springer Nature Switzerland AG 2021
S. J. G. Gift, B. Maundy, *Electronic Circuit Design and Application*,
https://doi.org/10.1007/978-3-030-46989-4_10

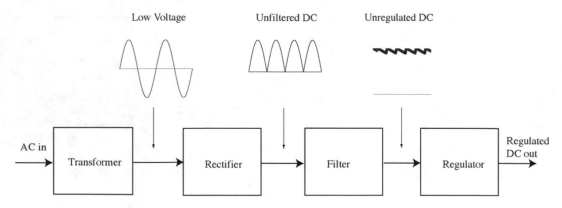

Fig. 10.1   Basic power supply system

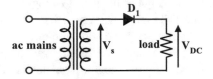

Fig. 10.2   Half-wave rectifier

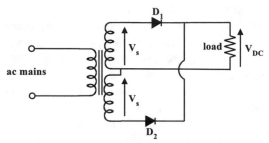

Fig.   10.4 Full-wave   rectifier   using   centre-tapped transformer

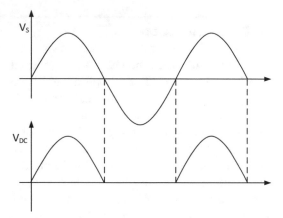

Fig. 10.3   Output waveform of half-wave rectifier

waveform is shown in Fig. 10.6. For this waveform, the average value increases to

$$V_{AV} = \frac{\int_0^T V_m \sin 377t \, dt}{1/120} = \frac{2V_m}{\pi} \quad (10.2)$$

where $T = 1/120$ since the waveform frequency is doubled. This value is twice as large as the value for the half-wave rectifier since both half-cycles are available in the output. The AC component still has a peak value of V_m. By definition, the effective or root mean square (*rms*) value of a periodic function of time is the square root of the ratio of the area under the square of one cycle of the waveform and the waveform period and for a sinusoidal waveform is given by

$$V_{rms} = \sqrt{\frac{1}{T} \int_0^T V_m^2 \sin^2 377t \, dt} \quad (10.3)$$

For a half-wave rectifier, Equation (10.3) leads to $V_{rms} = V_m/2$, while the value for a full-wave

comprised of two components: a DC component V_{DC}, which is essentially the average value of the output voltage, and an AC component V_{AC} which determines the ripple content of the output voltage, i.e. $V_{AV} = V_{DC} + V_{AC}$. It is therefore desirable to reduce this ripple to a minimum.

The   circuits   shown   above   are   full-wave rectifiers using a centre-tapped transformer in Fig. 10.4 and a bridge rectifier in Fig. 10.5. The diodes conduct such that there is current flow in only one direction and both half-cycles of the waveform   are   utilized.   The   resulting   output

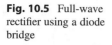

Fig. 10.5 Full-wave rectifier using a diode bridge

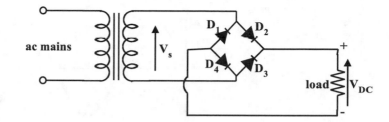

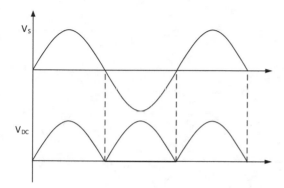

Fig. 10.6 Output waveform of the full-wave rectifier

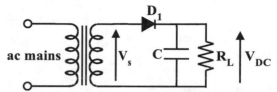

Fig. 10.7 Half-wave rectifier with capacitive filter

rectifier is the same as that for a sinusoidal waveform, i.e. $V_{rms} = V_m/\sqrt{2}$.

10.3 Filtering

In order to convert the half- and full-wave rectifiers into viable supplies, the ripple component (V_{AC}) in the output needs to be reduced. This process is referred to as filtering, and one straightforward method of doing this is to connect a "reservoir" capacitor C across the output of the rectifier. If the capacitor is sufficiently large, it acts as a short-circuit for the AC output voltage component, thereby removing it from the output signal. The resulting circuit consisting of transformer, rectifier and filter is called an unregulated supply.

The arrangement for the half-wave rectifier is shown in Fig. 10.7 with a load R_L. As the transformer voltage that is applied to the diode rises, the diode conducts, and the capacitor is charged up to the peak voltage at point A in Fig. 10.8. When the transformer voltage falls such that the cathode voltage of the diode set by the capacitor

is greater than the anode voltage set by the transformer, the diode turns off, and the output load is supplied by the capacitor which discharges through the load to point B in Fig. 10.8. The capacitor in effect prevents the output voltage from going to zero at any point in the cycle. The discharge rate depends on the value of the capacitor and the load current, and this voltage is exponential in form with time constant CR_L seconds. The instantaneous voltage across the capacitor and load is given by

$$V_L = \left(V_{pk} - V_D\right)e^{-t/CR_L} \qquad (10.4)$$

where V_D is the diode voltage drop. When the transformer voltage rises again above the capacitor voltage, the diode turns on the capacitor is recharged and the cycle is repeated. The resulting voltage waveform is shown in Fig. 10.8.

If the time constant CR_L is chosen to be significantly larger than the period of the waveform, then the capacitor voltage will not fall significantly before being recharged through the diode, and therefore, the amplitude of the ripple voltage will be reduced. However, the conduction time Δt_D of the diode is quite short compared with the period T of the waveform as shown in Fig. 10.9. During this time, the current pulse delivered by the transformer and diode must supply the charge delivered to the load. Since $I = dQ/dt$ if I_{DC} the

Fig. 10.8 Output voltage waveform of the half-wave unregulated supply

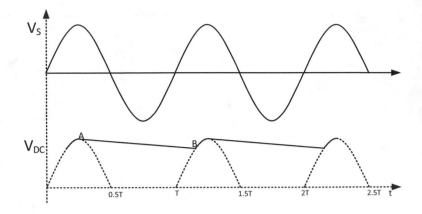

Fig. 10.9 Voltage and current waveforms for a half-wave rectifier

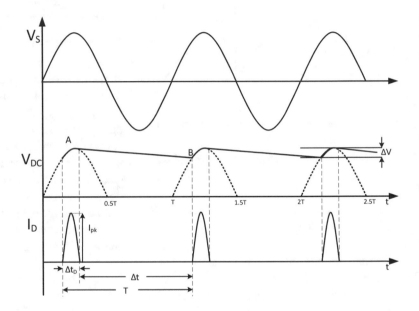

load current is approximately constant, then the average current pulse I_{pk} through the diode is given by

$$I_{pk}\Delta t_D = I_{DC}T \qquad (10.5)$$

from which

$$I_{pk} = \frac{T}{\Delta t_D}I_{DC} \qquad (10.6)$$

Since $T \gg \Delta t_D$, the magnitude of the current pulse is quite large. Moreover because Δt_D decreases as capacitor size increases, then the peak diode current increases. If the capacitor is too large, the diode peak current rating could be exceeded causing damage to the diode (and/or the transformer). Voltage and current waveforms are shown in Fig. 10.9. Because of the capacitor voltage, the diode is subjected to a maximum reverse voltage of $2V_s$, and therefore the diode PIV rating must be $PIV = 2V_s$.

The filtering of full-wave rectifiers, either that using the centre-tapped transformer or the bridge rectifier version, can be similarly accomplished using a reservoir capacitor as shown in Figs. 10.10 and 10.11. Because of the availability of both cycles in the rectifier output, the capacitor is recharged twice as frequently as in the half-wave rectifier, and therefore the ripple content is approximately half of that in the half-wave rectifier.

Also, since the capacitor discharge is reduced, the diode current flow during recharge is also reduced. The peak inverse voltage for the full-wave rectifier using the centre-tapped transformer

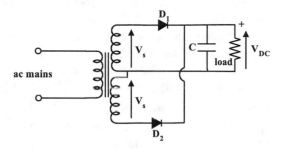

Fig. 10.10 Full-wave rectifier with capacitive filter

is $2V_s$, while that for the bridge rectifier is V_s. The output voltage and current waveforms for these circuits are shown in Fig. 10.12.

10.4   **Average DC Output Voltage**

In order to calculate the peak-to-peak value of the output ripple and the average output voltage, certain approximations need to be made. Specifically, from Fig. 10.8 for the half-wave rectifier, the diode is off for most of the period T. During this time, the capacitor delivers a load current I_{DC} and therefore discharges in the process. The decay is exponential but can be approximated as a linear decay. Also, the load current is assumed to be approximately constant during the discharge. For a capacitor, the relationship $I = CdV/dt$ holds. This can be approximated by $I \approx C\Delta V/\Delta t$ where

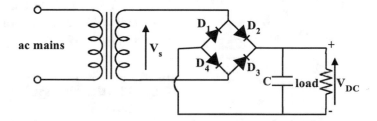

Fig. 10.11 Full-wave rectifier using diode bridge and capacitive filter

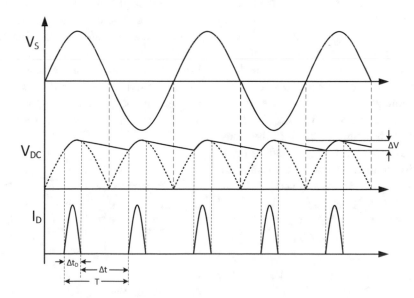

Fig. 10.12 Voltage and current waveforms for full-wave rectifiers

ΔV is a small change in capacitor voltage and Δt is the associated small time interval over which this voltage change occurs. For the half-wave rectifier,

$$I_{DC} = C \frac{\Delta V}{\Delta t} = C \frac{\Delta V_R}{T_H} \qquad (10.7)$$

where ΔV_R is the change in voltage of the reservoir capacitor and T_H is the period of the half-wave rectified waveform. From this, the peak-peak output ripple is given by

$$\Delta V_R = \frac{I_{DC} T}{C} = \frac{I_{DC}}{Cf} \qquad (10.8)$$

where f is the frequency of the input voltage. For a full-wave rectifier, the period T_F of the rectified waveform is half that of the half-wave rectifier and therefore $T_F = 1/2f$. Hence for the full-wave rectifier, the peak-peak output ripple is given by

$$\Delta V_R = \frac{I_{DC} T}{C} = \frac{I_{DC}}{2Cf} \qquad (10.9)$$

From this, the average DC output voltage is

$$V_{DC} = V_{pk} - \Delta V/2 \qquad (10.10)$$

Example 10.1

Using the circuit in Fig. 10.7 (without the load), design a half-wave unregulated power supply that can deliver 12V DC at 1A with less than 2.5 V peak-to-peak ripple.

Solution Since the peak transformer voltage is greater than the root mean square value by a factor of 1.414, it is reasonable to choose the transformer voltage V_{rms} to be close to the required DC output voltage, i.e. $V_{rms} = 10$ V. Also, to prevent transformer heating, the transformer secondary current rating should be larger than the maximum load current by at least a factor of 1.5, i.e. $I_{rms}(\text{sec}) = 1.5 I_{DC}(\text{max}) = 1.5$ A. Choose a diode with $I_{AV} \geq I_{DC}(\text{max}) = 1$ A and $PIV > V_{pk} = 12\sqrt{2} = 16.8\,\text{V}$. Capacitor C is given by $C = \frac{I_{DC}}{f\Delta V} = \frac{1}{60 \times 2.5} = 6666\,\mu\text{F}$. A rough indication of the output voltage under full-load conditions is given by $V_{DC} = V_{pk} - \Delta V/2 = 14 - 2.5/2 \approx 13$.

Example 10.2

Design a full-wave unregulated power supply delivering 15 volts DC at 1 A with less than 1.5 V peak-to-peak ripple. Use a centre-tapped transformer as in Fig. 10.10 (without the load).

Solution Since the peak transformer voltage is greater than the root mean square value by a factor of 1.414, we choose the transformer voltage V_{rms} to be close to the required DC output voltage, i.e. $V_{rms} = 12$ V. Also, to prevent transformer heating, the transformer secondary current rating should be larger than the maximum load current by at least a factor of 1.5, i.e. $I_{rms}(\text{sec}) = 1.5 I_{DC}(\text{max}) = 1.5$ A. Choose a diode with $I_{AV} \geq I_{DC}(\text{max}) = 1$ A and $PIV > V_{pk} = 15\sqrt{2} = 21.2\,\text{V}$. Capacitor C is given by $C = \frac{I_{DC}}{2f\Delta V} = \frac{1}{2 \times 60 \times 1.5} = 5555\,\mu\text{F}$. A rough indication of the output voltage under full-load conditions is given by $V_{DC} = V_{pk} - \Delta V/2 = 17 - 1.5/2 \approx 16$. In both designs, diode voltage drops and transformer winding resistance reduce the output voltage.

10.5   Bipolar Unregulated Power Supplies

Many electronic devices, especially operational amplifiers, require both positive and negative supply voltages. These voltages are defined relative to a common ground terminal. Such a supply is referred to as a bipolar supply and is shown in Fig. 10.13. In this circuit, a centre-tapped transformer is used to drive four diodes with the centre tap serving as the common ground. Diodes D_1 and D_2 provide positive full-wave rectification

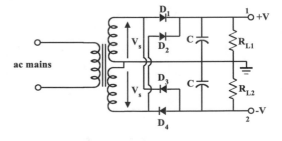

Fig. 10.13 Bipolar unregulated power supply

Fig. 10.14 Bipolar unregulated power supply using half-wave rectifiers

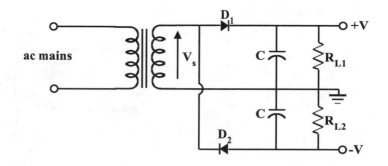

Fig. 10.15 Voltage doubler

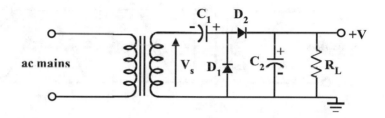

and make terminal 1 positive with respect to the centre tap, while diodes D_3 and D_4 provide negative full-wave rectification and make terminal 2 negative relative to the centre tap. Capacitors C_1 and C_2, respectively, filter the positive and negative voltages. It is possible to realize a bipolar unregulated supply using a single winding transformer. This elegant circuit is shown in Fig. 10.14. Each section is effectively a half-wave rectifier and therefore would in general produce twice as much ripple as the circuit of Fig. 10.13. However, it does not require a centre-tapped transformer, and this is its main advantage. Moreover, this circuit can function as a voltage doubler (as seen in the Zener diode example of Fig. 1.69 in Chap. 1) by moving the ground to either the +V or the −V terminal.

Example 10.3
Design a full-wave unregulated bipolar power supply delivering ±50 V DC at 5 A with less than 5 V peak-to-peak ripple to supply the 100 W power amplifier in Example 9.10 of Chap. 9.

Solution The basic system is shown in Fig. 10.13 (excluding loads R_{L1} and R_{L2}). In order to produce 50 V, a 40 V transformer

(80 V centre tap) is used. When rectified the peak output will be $40\sqrt{2} = 56.6\,\text{V}$. The maximum peak-to-peak ripple is 5 V, and allowing for diode voltage drop and transformer resistance will give an average output voltage of about ±50 VDC. The transformer current rating is chosen as 5 A which is sufficient to deliver the 5 A peak current requirement into 8 ohms. The filter capacitors are given by $C = \frac{I}{2f\Delta V} = \frac{5\,\text{A}}{2\times 60\times 5} = 8333\,\mu\text{F}$. Use $C = 10,000\,\mu\text{F}$ with 75 VDC working voltage. A diode bridge with piv 200 V and 5 A rating is used to provide the rectification.

10.6   Voltage Multipliers

It is possible to produce a high DC voltage using a low-voltage transformer. This can be done using a combination of diodes and capacitors along with the transformer. Such circuits rely on capacitor charge storage and are therefore not able to deliver high currents. In the circuit of Fig. 10.15, when the AC input is negative-going such that diode D_1 is forward-biased, capacitor C_1 is charged up to the peak transformer voltage $\sqrt{2}V_s$ with the polarity shown. During this time D_2 is reverse-biased. When the

Fig. 10.16   Voltage tripler

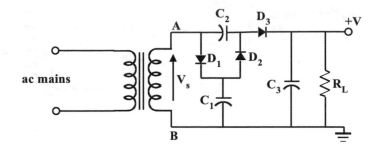

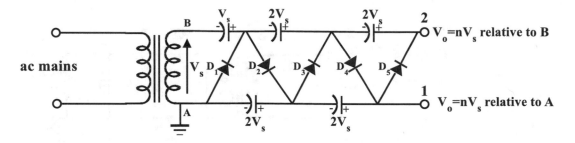

Fig. 10.17   General voltage multiplier

transformer input reverses and is positive going, D_1 is reverse-biased, but D_2 is now forward-biased allowing capacitor C_2 to charge up to the sum of the transformer voltage $\sqrt{2}V_s$ and the voltage $\sqrt{2}V_s$ on C_1, both of which are in series. After a few cycles, capacitor C_2 charges up to $2\sqrt{2}V_s$ making this circuit a voltage doubler. When a load is connected, however, C_2 will be constantly discharging, and therefore, the output voltage will in general be less than $2\sqrt{2}V_s$ and will have high ripple content. Increasing the size of the capacitors will reduce but not eliminate the problem.

The principle of the voltage doubler may be extended to voltage tripling, quadrupling and beyond. In the voltage tripler circuit of Fig. 10.16, during the half-cycle when terminal A is positive relative to terminal B, diode D_1 turns on, and C_1 is charged to $\sqrt{2}V_s$. When the transformer voltage reverses such that terminal B is positive relative to terminal A, diode D_2 conducts and the transformer voltage in series with the voltage across C_1 together charge capacitor C_2 to $2\sqrt{2}V_s$. When A is again positive relative to B, diode D_3 conducts, and the series combination of the transformer voltage and the voltage

$2\sqrt{2}V_s$ across capacitor C_2 charges capacitor C_3 to $3\sqrt{2}V_s$. After a few cycles, the output voltage across C_3 rises to approximately $3\sqrt{2}V_s$.

Any number of multiplications of the basic input voltage is realized in the multiplier circuit of Fig. 10.17. The peak transformer secondary voltage $\sqrt{2}V_s$ is multiplied by the number of cascaded stages resulting in an output voltage at terminal 1 of $n\sqrt{2}V_s$ relative to terminal A where n corresponds to the number of diodes in the circuit path. As can be seen from the circuit, n is always an even number. If point B is grounded instead of A, then the output is taken at terminal 2 and is still $n\sqrt{2}V_s$, but now n is an odd number. Thus, any multiplication of $\sqrt{2}V_s$ is possible using this circuit. The diode and capacitor voltage ratings must be twice the transformer output voltage $\sqrt{2}V_s$. Again, the capacitor values and diode current ratings depend on the output current.

10.7   Voltage Regulators

Unregulated power supplies have certain undesirable characteristics. Firstly, the DC output voltage decreases, and the AC ripple increases as load current increases. Secondly, the output voltage changes significantly in response to a changing input voltage resulting from changing line voltage. These disadvantages can be minimized by adding a voltage regulator to the unregulated supply. The resulting supply is called a regulated power supply. The function of the regulator is therefore to maintain a constant output voltage under conditions of changing load current and changing line voltage. There are two types of voltage regulators: (a) the shunt regulator in which a control element is in parallel (shunt) with the load and (b) the series regulator in which a control element is in series with the load.

In a *shunt regulator*, the regulating device is placed in parallel with the load and works in conjunction with a resistor in series with the load and the unregulated supply. The regulator operates by varying the current through the control element depending on the load current. This results in a varying voltage drop across the series resistor such that the load voltage remains constant. One example of a shunt regulator is the Zener diode regulator which was introduced in Chap. 1. In a *series regulator*, the regulating device is placed in series with the load and the unregulated supply, and the voltage across the control element is adjusted so that the load voltage remains constant. Examples of these types of regulators will be discussed in this chapter.

10.7.1   Ripple and Regulation

The performance of a voltage regulator can be specified by several parameters. One of these gives an indication of the ripple content of the DC output. The output of a rectifier contains considerable ripple, which would have to be reduced by the regulator in order to make the DC voltage useable. The quality of the resulting output may be expressed using *percent ripple R* which is defined as

$$R = \frac{\text{Ripple voltage (rms)}}{\text{Average or DC output voltage}} \times 100\%$$

$$(10.11)$$

This parameter is however difficult to calculate because of the difficulty in determining the *rms* ripple voltage in the output. A ripple measure that is more easily evaluated is the *ripple factor r* given by

$$r = \frac{\text{Peak to peak ripple in output voltage}}{\text{Average or DC output voltage}}$$

$$(10.12)$$

The peak-to-peak output voltage ripple involved in this specification is easily determined using an oscilloscope. Both *percent ripple* and *ripple factor* measures should be as low as possible. Another factor of importance in a power supply is *load regulation* which specifies the change in the DC output voltage arising from changing load current. This factor is given by

$$\text{Load regulation} = \frac{V_{DC}(\text{no load}) - V_{DC}(\text{full load})}{V_{DC}(\text{full load})} \times 100\%$$

$$(10.13)$$

where V_{DC} (no load) is the average value of the output voltage when the external load resistance is removed and V_{DC} (full load) is the average value of the output voltage when the external load resistance is at a minimum. In practical circuits, diode resistance and transformer winding resistance will result in voltage drops that cause the output voltage to be reduced with increasing load. It is desirable to reduce this drop and thereby have a low load regulation. *Line regulation* specifies the change in the DC output voltage resulting from a change in the line or input voltage. It is a measure of effectiveness of the regulation (output voltage stability) in the presence of changing input voltage. It is given by

$$\text{Line regulation} = \frac{\Delta V_{out}}{\Delta V_{in}} \times 100\% \quad (10.14)$$

where ΔV_{out} is the change in the output voltage of the regulator and ΔV_{in} is the associated change in the input voltage to the regulator with constant load. Ideally, both line and load regulation should be zero, and in practice, the values are less than 0.01% for good regulators. The ratio $\Delta V_{out}/\Delta V_{in}$ is sometimes referred to as *stability factor*.

10.7.2 Zener Diode Regulator

As discussed in Chap. 1, the Zener diode is essentially a semiconductor diode designed to operate in the reverse-biased region. In the forward bias region, the device operates as a normal diode, while in the reverse bias region, it again operates like a diode for voltages less than the breakdown voltage. For voltages exceeding the breakdown (Zener) voltage, the Zener diode conducts and operates along that part of the I/V characteristic having a steep slope. Providing the current is prevented from rising too high, the voltage across the Zener in this condition is approximately constant at the breakdown or Zener voltage, and the dynamic resistance r_d is quite low, ranging from a few ohms to about 50 ohms. In this region, the voltage across the Zener is approximately constant, and when operated here, the device provides a constant voltage output.

The current through the Zener in this region is limited by the power rating P_D of the diode and must be operated such that $V_z I_z < P_D$. If $I_z(max) = P_D/V_z$ is exceeded, the diode will be destroyed. Also, I_z must not fall below the minimum value $I_z(min)$; otherwise, the Zener voltage will fall. This corresponds to the knee of the curve in Fig. 10.18. Zener diodes are available in voltages ranging from about 2.4 V to about 200 V. The power ratings range from about 1/4 w to over 100 W.

The Zener diode operates as an effective voltage regulator. The basic system is shown in Fig. 10.19. A (varying) voltage V_i is applied to the Zener diode through a resistor R that limits the current. The voltage V_Z across the Zener is then applied to a load R_L. If V_i increases, say, more current flows through the Zener but based on its characteristic, V_Z stays approximately constant. Similarly, if V_i decreases, the current though the Zener drops but again V_Z remains constant. Thus, the voltage V_Z is approximately constant despite changes in the input voltage or load current. The load R_L therefore sees a regulated (constant) voltage providing $I_{zmin} \leq I_z \leq I_{zmax}$. This design of a Zener-regulated power supply involves the determination of R. In order to do so, the minimum and maximum values of V_i as well as the maximum

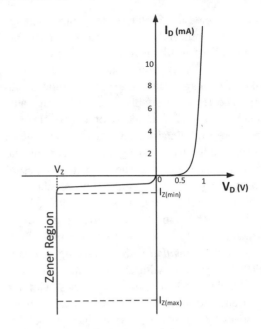

Fig. 10.18 Zener diode characteristic

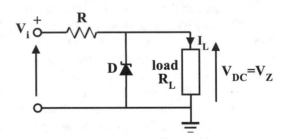

Fig. 10.19 Zener diode regulator

value of I_L must be known. When I_L is at maximum corresponding to minimum R_L, and V_i at a minimum, the value of R must be such that $I_z \geq I_{zmin}$.

Thus,

$$\frac{V_{i\min} - V_Z}{R} - I_{L\max} \geq I_{Z\min} \qquad (10.15)$$

i.e.

$$R \leq \frac{V_{i\min} - V_Z}{I_{L\max} + I_{Z\min}} \qquad (10.16)$$

When V_i rises and I_L goes to zero (load removed), the current through the Zener is at its maximum, and the required power rating P_D of the diode must satisfy

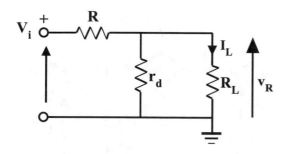

Fig. 10.20 Equivalent circuit of Zener diode regulator

$$P_D > V_Z \frac{V_{i\,max} - V_Z}{R} \qquad (10.17)$$

Zener-regulated power supplies are useful for low-power applications involving less than a few watts since under the condition where no load current is being supplied, the Zener must accommodate the full load current.

In order to determine the ripple at the output of the Zener diode regulator, the dynamic resistance of the Zener must be considered. The dynamic resistance is the reciprocal of the slope of the diode characteristic in the operating region. It is an indication of the resistance of the diode to small voltage and current changes. The value is typically in the range 5 Ω to 50 Ω. The actual circuit under small-signal conditions is equivalent to that shown Fig. 10.20. For input ripple voltage v_i, the output ripple v_R is given by

$$v_R = v_i \frac{r_d//R_L}{R + r_d//R_L} \qquad (10.18)$$
$$= v_i \frac{r_d}{R + r_d}$$

where $r_d \ll R_L$. Hence to minimize v_R, R should be as large as possible, for example, by using a large resistor or a constant current diode. Hence in (10.16), equality is used with I_{zmin} set at a value determined by the specification sheet. If the load is variable and V_i is fixed, then $V_{imin} = V_i$. If V_i is variable and the load is fixed, then at no time will current flowing through R all flow into the Zener since some current will always be flowing into the load. Hence (10.17) becomes

$$P_D > V_Z \left(\frac{V_{i(max)} - V_Z}{R} - I_L \right) \qquad (10.19)$$

This circuit has the advantage of short-circuit protection provided by the presence of resistor R which will act to limit excessive current to the load.

Example 10.3
A voltage from an unregulated supply varies between a minimum value of 16 volts and a maximum value of 20 volts. Using this unregulated supply, design a Zener-regulated supply that delivers 10 volts DC with a current capacity of 10 mA.

Solution The first step is to choose a 10 volt Zener for the circuit of Fig. 10.19. The power rating will be determined later. To determine R we consider the case where Zener current is at a minimum. This occurs when the input voltage is at a minimum, in this case 16 volts, and the load current is at a maximum at 10 mA. We determine from diode specifications the minimum diode current for proper Zener operation. Assume a value of 5 mA. The voltage across R is (16–10) volts and the current through R is (10 + 5) mA. R is therefore given by $R = \frac{(16-10)}{(10+5)} \times 10^3 = 400\,\Omega$. Note that if the load were suddenly removed, then the current through D would rise to 15 mA when the input voltage is at 16 volts. To determine the diode power rating, we calculate the maximum current through the diode. This occurs when the unregulated input voltage is at a maximum of 20 volts and the load current is at a minimum (zero) corresponding to a removal of the load. The voltage across R is then $(20 - 10) = 10$ V, and hence, the current I_D through D is given by $I_D = \frac{10}{400} = 25\,mA$. This current flows through R, and hence the required Zener power rating P_D is given by $P_D = I_D \cdot V_D = 25$ mA $\times$ 10 V $= 0.25$ W. We choose a 1/2 W diode to provide a safety margin and for longer diode life. In this design, the input ripple voltage is $(20 - 16) = 4$ V pk-to-pk. Hence, assuming $r_d = 10\,\Omega$, the output ripple voltage is $v_R = \frac{10}{400+10} \times 4 = 0.1\,V$. The ripple factor r is then $r = \frac{v_R}{V_{DC}} = \frac{0.1}{10} = 0.01$.

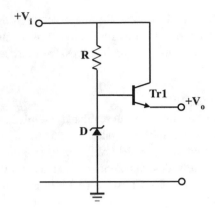

Fig. 10.21 Single transistor regulator

10.7.3   Simple Series Transistor Regulator

One method of improving the current capacity of the Zener regulator is the use of a transistor in the emitter follower configuration. A simple single transistor regulator of this type is shown in Fig. 10.21. It utilizes the Zener diode regulator and an npn transistor connected as an emitter follower. The transistor collector is supplied from the unregulated voltage source. Thus, the Zener voltage is provided at the transistor emitter giving an output of $V_o = V_z - 0.7$ volts. The transistor effectively increases the load current capacity of the Zener by a factor of the transistor current gain. Alternatively, since the transistor base current is significantly less than the load current ($I_B < \; < I_L$) for large transistor current gain, the changes in base current with large changes in load current are small. Therefore, the Zener current does not change by very much. Hence the output voltage is quite stable for large load current changes. The circuit output impedance is given by

$$Z_O = \frac{r_d + h_{ie}}{1 + h_{fe}} \qquad (10.20)$$

The voltage V_i must always be larger than V_z (by a few volts) to ensure that the transistor is properly biased. In order to determine the ripple at the output, consider the equivalent circuit of the regulator shown in Fig. 10.22. Then

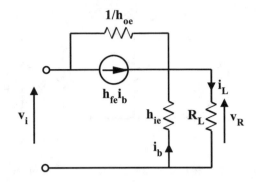

Fig. 10.22 Equivalent circuit of single transistor regulator

$$\frac{v_R}{R_L} = (v_i - v_R)h_{oe} + (1 + h_{fe})i_b \qquad (10.21)$$

$$i_b = \frac{-v_R}{h_{ie}} \qquad (10.22)$$

giving

$$\frac{v_R}{v_i} = \frac{h_{oe}}{\frac{1}{R_L} + \frac{1 + h_{fe}}{h_{ie}} + h_{oe}} \frac{1}{1 + \frac{r_{oe}}{r_e} + \frac{r_{oe}}{R_L}}$$

$$= \frac{1}{1 + r_{oe}\left(\frac{1}{r_e} + \frac{1}{R_L}\right)} \qquad (10.23)$$

where $r_{oe} = \frac{1}{h_{oe}}$ and $r_e = \frac{h_{ie}}{1 + h_{fe}}$. Now $r_{oe}\left(\frac{1}{R_L} + \frac{1}{r_e}\right) \gg 1$. Hence,

$$\frac{v_R}{v_i} = \frac{R_L//r_e}{r_{oe}} \qquad (10.24)$$

r_{oe} is of the order $10^5\ \Omega$ and r_e is of the order $10^2\ \Omega$, hence $\frac{v_R}{v_i} \approx 10^{-3}$. Thus, the ratio is quite small indicating that the ripple output of the emitter follower regulator is quite low.

Example 10.4
Design a single transistor regulated supply using a Zener diode capable of delivering 12 V at 100 mA. Use a half-wave rectifier to drive the regulator.

Solution For a 12 volt output, in order to allow for sufficient voltage across the transistor, we choose a transformer with secondary voltage V_S

of 15 volts. This gives a peak output voltage of $15\sqrt{2} = 21$ V. If we set a maximum peak-to-peak ripple of the unregulated supply of 1 volt, then using (10.8), the required capacitor C is given by $C = \frac{I_{DC}}{\Delta V_{Rf}} = \frac{0.1\,A}{1 \times 60} = 1667\,\mu F$. In order to realize a 12 volt output, we choose a Zener having a voltage $V_Z = 12 + 0.7 = 12.7$ V. The closest available voltage is 13 volts. Let the current through the Zener be 5 mA chosen for good Zener operation. A transistor with a current gain of 50 when delivering the full current of 100 mA will have a base current of 100 mA/50 = 2 mA. This must be supplied by the Zener. Then from (10.16), $R = (V_S(\text{min}) - V_Z)/(2 + 5)\text{mA} = (15\sqrt{2} - 1 - 13)/(2 + 5)\text{mA} = 1\,k\Omega$. Note that the minimum voltage drop across the transistor is $V_S(\text{min}) - V_O = 21 - 1 - 12 = 8$ V. This circuit can be modified to deliver higher currents by using a Darlington pair instead of a single transistor. In such a case, the design proceeds as before but using the gain of the Darlington pair.

Example 10.5

Design a regulated supply using a Darlington pair and a Zener diode that delivers 15 V at 1 A. Use a full-wave rectifier to drive the regulator.

Solution For a 15 volt output, in order to allow for sufficient voltage across the transistor, we choose a transformer with secondary voltage V_S of say 18 volts. This gives a peak output voltage of $18\sqrt{2} = 25$ V. If we set a maximum peak-to-peak ripple of the unregulated supply of 1.5 volts, then using (10.8), the required capacitor C is

given by $C = \frac{I_{DC}}{\Delta V_{Rf}} = \frac{1\,A}{1.5 \times 120} = 5555\,\mu F$. In order to realize a 15 volt output, we choose a Zener diode having a voltage $V_Z = 15 + 1.4 = 16.4$ V. The closest available voltage is 16 volts. Let the current through the Zener be 5 mA chosen for good Zener operation. A Darlington pair with a current gain of 1000 when delivering the full current of 1 A will have a base current of $1A/1000 = 1$ mA. This must be supplied by the Zener. Then, $R = (V_S(\text{min}) - V_Z)/(1 + 5)\text{mA} = (18\sqrt{2} - 1.5 - 16)/(1 + 5)\text{mA} = 1.3\,k\Omega$. Note that the minimum voltage drop across the transistor is $V_S(\text{min}) - V_O = 25 - 1.5 - 15 = 9$ V. The full circuit is shown in Fig. 10.23.

This configuration utilizing a Darlington transistor is especially well-suited for introducing a variable voltage output. This is because the high input impedance of the Darlington allows the employment of a potentiometer as shown in Fig. 10.24. The loading on the potentiometer is minimal, and accurate voltage variation can be realized.

Short-circuit protection of the basic single transistor regulator can be realized by the introduction of a fuse at the output. Electronic short-circuit protection is easily realized as shown in Fig. 10.25. The protection circuit comprises Tr_2 and R_X. When there is a current overload and the protection transistor Tr_2 turns on, the base current to the pass transistor is limited, and the Zener voltage falls. This circuit is simple and works very well. For example, to limit current to 100 mA, $R_X = 0.7$ V/0.1 A $= 7\ \Omega$.

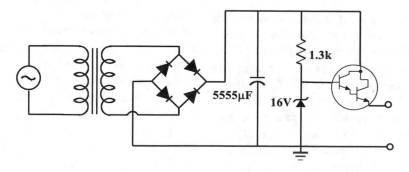

Fig. 10.23 Regulated power supply using Darlington pair

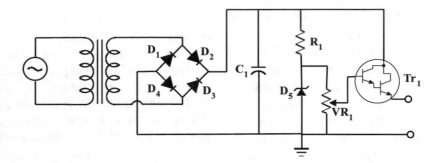

Fig. 10.24 Transistor regulator with variable output voltage

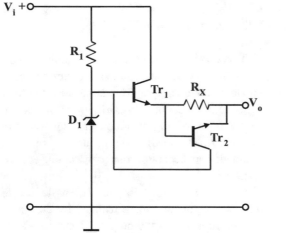

Fig. 10.25 Transistor regulator with electronic protection

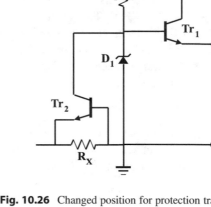

Fig. 10.26 Changed position for protection transistor

This circuit has the disadvantage of causing a change in the regulated output voltage as current flows through R_X. This problem can be overcome by placing the protection transistor in the position shown in Fig. 10.26. In this position the regulated output is unaffected by the voltage drop across R_X.

The second transistor introduced to provide electronic protection can provide that protection in a different manner as shown in Fig. 10.27. The introduced transistor is a pnp transistor along with resistor R_2 and signal diode D_2. The collector current I_{C1} of Tr_1 flows out of the base of Tr_2 into R_2. This current results in a collector current I_{C2} in Tr_2 which flows through diode D_2 also into R_2. The current flowing in R_2 is therefore made up of the collector currents of the two transistors and is given by $(V_Z - 0.7)/R_2$. As load current increases, the current through D_2 decreases with a consequent fall in the voltage across D_2. Eventually D_2 turns off and all of the current through R_2 is supplied by Tr_1. This results in a maximum current through Tr_2 into the load given by $I_{L(max)} = h_{fe2}(V_Z - 0.7)/R_2$ where h_{fe2} is the current gain of Tr_2. The circuit now converts from a constant voltage source to a constant current source, and as a result any further decrease in load resistance results in a fall in the output voltage. This may be described as a *constant current* characteristic.

Under short-circuit conditions, the maximum current $I_{L(max)}$ continues to flow into through the short circuit. It is possible to reduce this short-circuit current by the introduction of another signal diode D_3 as shown in Fig. 10.28. When the maximum output current flows and the regulator

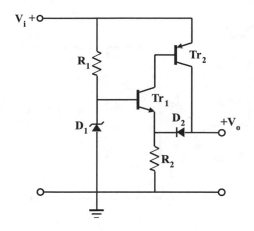

Fig. 10.27 Transistor regulator with alternative electronic protection

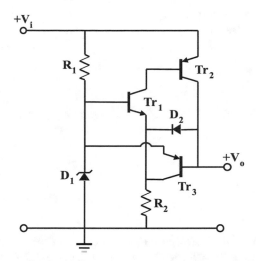

Fig. 10.29 Transistor regulator with shutdown electronic protection

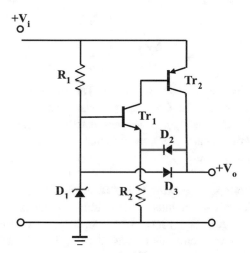

Fig. 10.28 Transistor regulator with enhanced electronic protection

is in the constant current mode, as load resistance decreases, diode D_2 becomes reverse-biased, and the additional diode D_3 becomes forward-biased. This causes current to flow from resistor R_1 through D_3 into the load. This reduces the current into the Zener diode, thereby causing the Zener diode voltage to fall. The result is decrease in both the output voltage and load current. Under short-circuit conditions, the load current will be reduced to near zero with zero output voltage. This is the *fold-back* characteristic.

A further improvement in the protection scheme is possible by the replacement of diode

D_3 by another pnp transistor Tr_3 as shown in Fig. 10.29. When the maximum output current flows and the regulator is in the constant current mode, as load resistance decreases, diode D_2 becomes reverse-biased, and the base-emitter junction of Tr_3 becomes forward-biased. This causes current to flow from resistor R_1 into the emitter of Tr_3 and out through the collector into resistor R_2. This has two effects: (i) it reduces the current into the Zener diode, thereby causing the Zener diode voltage to fall, and (ii) it reduces the collector current through Tr_1 and hence the maximum current available from Tr_2. The result is a rapid switch off of the supply when the maximum load current is exceeded. This may be referred to as the *switch-off* characteristic. In order to reset the supply after the overload condition is removed, the connection from the collector of Tr_3 to R_2 must be interrupted. This can be done with a normally closed push-button switch in the collector circuit of Tr_3.

10.7.4 Series Feedback Voltage Regulators

In the single transistor voltage regulator, while the output ripple is low, there are still variations in the output voltage arising from the V_{BE}/I_C characteristic of the transistor. Thus, as load

Fig. 10.30 Basic series regulator

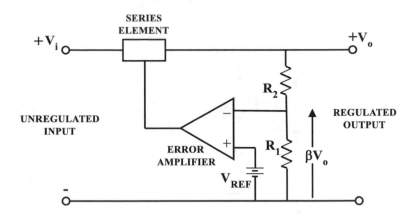

current increases, V_{BE} increases, and hence $V_o = V_z - V_{BE}$ decreases. In order to reduce the magnitude of this change, feedback can be used to sense the change and effect a correction at the output. The basic elements of a series feedback regulator circuit are shown in Fig. 10.30. It consists of an error amplifier, a sampling circuit, a voltage reference and a series element. During operation, the sampling circuit R_2R_1 samples the regulator output voltage and sends it to the error amplifier. This amplifier compares the sample with a reference voltage and generates a signal proportional to the difference. This error signal is used to drive the series pass element, which varies the output voltage such that the error is reduced and the output voltage regulated.

Considering the basic feedback system, the voltage across the series element is $(V_i - V_o)$, while the input voltage to the error amplifier is $(\beta V_o - V_{REF})$ where

$$\beta = \frac{R_1}{R_1 + R_2} \qquad (10.25)$$

The output voltage across the series element may be viewed as an amplified version of the input voltage, i.e.

$$V_i - V_o = A(\beta V_o - V_{REF}) \qquad (10.26)$$

where A is the gain of the error amplifier. Consider changes in the input and output voltages resulting in

$$\Delta V_i - \Delta V_o = A(\beta \Delta V_o - \Delta V_{REF}) \qquad (10.27)$$

Since V_{REF} is constant, then

$$\Delta V_i = (1 + A\beta)\Delta V_o \qquad (10.28)$$

$$\Delta V_o = \Delta V_i/(1 + A\beta) \qquad (10.29)$$

Hence,

$$\frac{\Delta V_o}{\Delta V_i} = \frac{1}{1 + A\beta} \qquad (10.30)$$

which is the stability factor. The ripple voltage component of the output is also reduced by a factor $(1 + A\beta)$. It follows that the higher the loop gain $A\beta$, the better the regulator performance.

Voltage Regulator Using Discrete Transistors
A transistor regulator circuit utilizing feedback is shown in Fig. 10.31. Transistor Tr_2 is the series pass transistor, and transistor Tr_1 is the error amplifier. The overall system operates according to the block diagram in Fig. 10.30. The regulator in Fig. 10.31 accepts an unregulated DC input voltage V_i and delivers a regulated DC output voltage V_O. It operates by sampling the output voltage V_O via R_1 and R_2 and comparing the sampled voltage which appears at the base of Tr_1 with a reference voltage V_Z set at the emitter of Tr_1 by the Zener diode D. The resulting error voltage across the base-emitter junction of Tr_1 establishes a current in R_3 that sets the output voltage V_O. If V_O increases, the sampled voltage and hence the voltage at the base of Tr_2 increase. Since the emitter voltage of Tr_1 is held fixed by the Zener, the base-emitter voltage increases. This further turns on Tr_1, increasing its collector current. The resulting increased voltage drop across

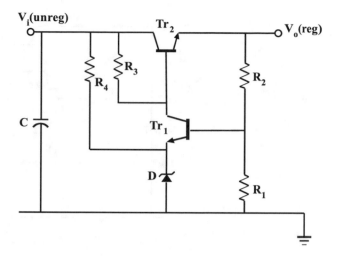

Fig. 10.31 Regulator using discrete transistors

R_3 lowers the collector voltage of Tr_1, which then lowers the output voltage V_O by the emitter follower action of Tr_2. If the output voltage decreases, then the base-emitter voltage of Tr_1 is reduced, thereby reducing its collector current. The voltage at the collector of Tr_1 increases which then increases the output voltage V_O. The capacitor C provides filtering action for the output of the associated rectifier.

The design procedure involves first choosing a current to be passed into the Zener through R_4. Let this value be I_z. Then R_4 must be calculated to enable this current to flow when V_i is at its minimum. Hence,

$$R_4 = \frac{V_{i\,min} - V_z}{I_z} \qquad (10.31)$$

where $V_{i\,min} = V_s\sqrt{2} - \Delta V$ and ΔV is the peak-to-peak ripple voltage at the input to the regulator. Resistor R_3 is calculated by noting that it must supply collector current I_{C1} to Tr_1 as well as base current I_{B2} to Tr_2. It must do so under worst-case conditions, i.e. when V_i is a minimum and I_{B2} is a maximum. Noting that the voltage at the collector of Tr_1 is $V_{out} + V_{BE}$, then

$$R_3 = \frac{V_{i\,min} - (V_{out} + V_{BE})}{I_{B2}(\max) + I_{C1}} \qquad (10.32)$$

where

$$I_{B2}(\max) = \frac{I_{L\,max}}{h_{fe}(Tr_2)} \qquad (10.33)$$

In order to determine R_1 and R_2, we note that $V_B(Tr_1) = V_Z + V_{BE}$. Assuming that the base current of Tr_1 does not load the R_2, R_1 potential divider, it follows that

$$\frac{V_z + V_{BE}}{R_1} = \frac{V_{out}}{R_2 + R_1} \qquad (10.34)$$

giving

$$V_{out} = \left(1 + \frac{R_2}{R_1}\right)(V_z + V_{BE}) \qquad (10.35)$$

To prevent loading, choose current $I(R_2, R_1)$ through R_2 and R_1 such that $I(R_2, R_1) \geq 10 I_B (Tr_1)$

Example 10.6
Design a regulated supply using the topology in Fig. 10.31 rated at 9 volts and 100 mA. Power for the regulator must come from an unregulated supply using a full-wave bridge rectifier and a 12 volt transformer having a ripple voltage of 1 V peak-to-peak at maximum load current. Use transistors with current gain of 100.

Solution For $V_O = 9$, choose an intermediate value of Zener diode voltage of say 5.6 volts

Fig. 10.32 Transistor feedback regulator using Darlington pair

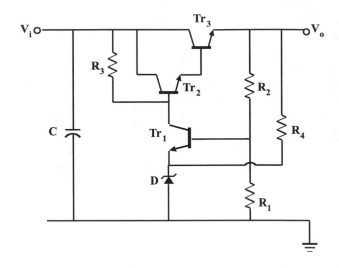

(why?). Let the minimum current through the Zener be 5 mA for good Zener performance. Also, choose $I_{C1} = 1$ mA as a reasonable operating current for Tr_1. Hence, $R_4 = \frac{V_i(\min) - V_z}{5\,\text{mA}} = \frac{12\sqrt{2} - 1 - 5.6}{5\,\text{mA}} = 2\,\text{k}$. Also, $R_3 = \frac{V_i(\min) - (9 + 0.7)}{I_L/h_{fe} + I_C(Tr_1)} = \frac{12\sqrt{2} - 1 - 9.7}{100\,\text{mA}/100 + 1\,\text{mA}} = 3\,\text{k}2$. The current flowing through R_2 and R_1 must be at least ten times the base current of Tr_2. The base current I_{B1} of Tr_1 is approximately 1 mA/100 = 10 μA. Hence, $\frac{9}{R_1 + R_2} \geq 10 \times 10\,\mu\text{A}$, $R_1 + R_2 \leq 90$ kΩ, $V_o = (V_Z + 0.7)\left(1 + \frac{R_2}{R_1}\right)$ and $\frac{R_2}{R_1} = \frac{9}{6.3} - 1 = 0.43$. Thus for $R_1 = 10$ kΩ, $R_2 = 4.3$ kΩ and note that for these values $R_1 + R_2 \leq 90$ kΩ as required. Finally, $C = \frac{I_{L\max}}{120\Delta V} = \frac{100\,\text{mA}}{120 \times 1} = 833\,\mu\text{F}$.

There are several ways to improve the performance of this circuit. A great improvement can be realized by supplying the Zener from the regulated output instead of the unregulated input by connecting resistor R_4 to the regulated output. This reduces the ripple across the Zener diode and hence at the output. A second improvement is to place a capacitor across the Zener diode, again reducing the ripple. The value of this capacitor is that value that has an impedance comparable to or lower than the dynamic resistance of the Zener. A third method of performance improvement is to replace R_4 by

a constant current source. An excellent method of implementing this scheme is the use of the constant current diode made using a JFET. This is a two-terminal device made, for example, by Siliconix that can simply replace the resistor feeding the Zener. Finally, a small capacitor placed across the circuit output removes output noise and reduces the supply impedance at high frequencies. In order to handle a large load current, a second transistor Tr_3 can be used along with Tr_2 in a Darlington arrangement as shown in Fig. 10.32. This reduces the pass transistor base current demand that may otherwise cause resistor R_3 to be too low in value and thereby result in too large a standing current in the error transistor Tr_1. The calculation would now involve the combined transistor current gain $h_{fe2}h_{fe3}$. Further ripple reduction can be achieved by applying bootstrapping to the load resistor R_3 of the error transistor Tr_1. This increases the error transistor gain which increases the applied feedback.

Example 10.7

Using the circuit shown in Fig. 10.32, design a +20 volt, 1.5 A regulator that is driven by an unregulated supply having 2 volt peak-peak input ripple. Use a 22 volt transformer and a bridge rectifier to provide the unregulated input. Assume the power transistor has a gain of 25 and the other transistors have gains of 150. Justify all design steps.

Solution Choose $V_Z = 10$ V for good feedback and reasonable value of R_4, i.e. not too low. Choose a Zener current for good Zener operation say 5 mA. The minimum unregulated input voltage is $V_i(\min) = 22\sqrt{2} - 2 = 29.1\,\text{V}$. Hence, $R_4 = \frac{20-10}{5\,\text{mA}} = \frac{10}{5\,\text{mA}} = 2\,\text{k}\Omega$. Let $I_C(\text{Tr}_1) = 1$ mA for good transistor operation. Then, $R_3 = \frac{29.1-(20+1.4)}{1.5\,\text{A}/(150\times25)+1\,\text{mA}} = \frac{7.7}{1.4\,\text{mA}} = 5.5\,\text{k}\Omega$. In order that the base of Tr_1 does not load the voltage divider R_2, R_1, $\frac{V_o}{R_1+R_2} > 10 \times \frac{1\,\text{mA}}{150} = 66\,\mu\text{A}$, i.e. $R_1 + R_2 < \frac{V_o}{66\,\mu\text{A}} = \frac{20}{66\,\mu\text{A}} \approx 300\,\text{k}\Omega$. Now $\frac{V_Z+0.7}{R_1} = \frac{V_o}{R_1+R_2}$, i.e. $\frac{10.7}{R_1} = \frac{20}{R_1+R_2}$ from which $\frac{R_2}{R_1} =$ 0.87 . For $R_1 = 10$ k, $R_2 = 8.7$ k. Note $R_1 + R_2 < 300$ k. Since the unregulated input is from a full-wave rectifier, $C = \frac{I}{2\Delta Vf} = \frac{1.5}{2\times2\times60} = 6250\,\mu\text{F}$.

Example 10.8

Introduce bootstrapping of resistor R_3 in Example 10.7, and add a suitable filter capacitor across the Zener diode as shown in Fig. 10.33

Solution In order to apply bootstrapping to R_3, this resistor is split into two resistors. The exact values are not critical, and they can be made equal. Hence, R_3 is split into $R_{3a} = 2.2$ k and $R_{3b} = 2.2$ k. A capacitor C_3 is chosen to have a value such that its reactance at the ripple frequency is lower than R_{3a}. Hence, $C_3 = 1/(2\pi fR_{3a}) = 1/(2 \times \pi \times 120 \times 2.2 \times 10^3) = 0.6\,\mu\text{F}$. A value of 10 µF is chosen. Choose $C_2 = 100$ µF with a reactance of 13 Ω at 120 Hz.

A particularly useful version of this transistor series regulator that can be adapted to a wide range of output voltages and currents is shown in Fig. 10.34. The Zener diode is moved from between the emitter of Tr_1 and ground and connected between the base of Tr_1 and the regulated output. Resistor R_1 across the base-emitter junction of Tr_1 establishes a near constant current in the Zener diode of $0.7/R_1$. The output voltage is $V_o = (V_Z + 0.7)$, and this arrangement constitutes 100% negative feedback as compared with the original system. As before, Tr_2 could be

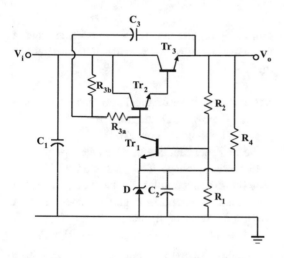

Fig. 10.33 Transistor feedback regulator with bootstrapping

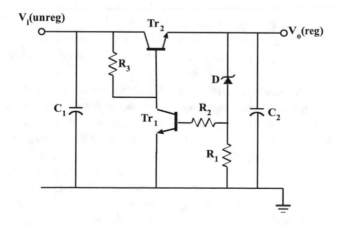

Fig. 10.34 Alternative transistor feedback regulator

either a single transistor or a Darlington pair for higher currents. Resistor R_2 provides protection for Tr_1 in the event of component failure. The application of bootstrapping to resistor R_3 will increase the amount of available feedback and hence improve the performance of the power supply.

Example 10.9 Using the configuration in Fig. 10.34, design a +15 V, 100 mA regulator that is driven by an unregulated supply having 1.5 V peak-peak input ripple. Use a 20 V transformer and a half-wave rectifier to provide the unregulated input.

Solution For $V_O = 16$ V, choose a value of Zener diode with a voltage that is close to the desired voltage. Thus, a diode with $V_Z = 15$ V will result in a regulator voltage of 15.7 V. Let the minimum current through the Zener be 5 mA for good Zener performance. Then $R_1 = 0.7/5$ mA $= 140\ \Omega$. Also, choose $I_{C1} = 1$ mA as a reasonable operating current. Hence, $R_3 = \frac{V_i(\min)-(15+0.7)}{I_L/h_{fe}+I_C(Tr_1)} = \frac{20\sqrt{2}-1.5-15.7}{100\,\text{mA}/100+1\,\text{mA}} = 5\text{k}5$. In order to provide some protection to Tr_1, let $R_2 = 47\ \Omega$. Finally, $C_1 = \frac{I_{L\max}}{60\Delta V} = \frac{100\,\text{mA}}{60\times1.5} = 1111\,\mu\text{F}$. Capacitor $C_2 = 1\ \mu\text{F}$ provides further filtering of the regulated output voltage.

Example 10.10
Introduce bootstrapping of resistor R_3 in Example 10.9.

Solution In order to apply bootstrapping to R_3, this resistor is split into two resistors. The exact values are not critical and equal values will do. Hence, R_3 is split into $R_{3a} = 2.2$ k and $R_{3b} = 2.2$ k. A capacitor C_3 is chosen to have a value such that its reactance at the ripple frequency is lower than R_{3b}. Hence, $C_3 = 1/(2\pi f R_{3a}) = 1/(2\times\pi\times 60\times 2.2\times 10^3) = 1.2\ \mu\text{F}$. A value of 10 μF is chosen (Fig. 10.35).

Example 10.11
Using the basic configuration in Fig. 10.34, design a +12 V, 5 A regulator that is driven by an unregulated supply having 2 V peak-peak

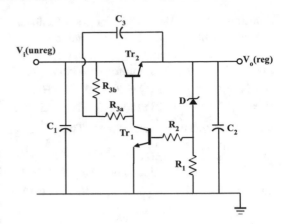

Fig. 10.35 Alternative feedback regulator with bootstrapping

input ripple. Use an 18 V transformer and a full-wave rectifier to provide the unregulated input.

Solution Because of the large current requirement, a power Darlington is used having $\beta = 1000$ is used. For $V_O = 12$ V, choose a value of Zener diode with a voltage that is close to the desired voltage. Thus, a diode with $V_Z = 11$ V will result in a regulator voltage of 11.7 V. Let the minimum current through the Zener be 5 mA for good Zener performance. Then $R_1 = 0.7/5$ mA $= 140\ \Omega$. Also, choose $I_{C1} = 1$ mA as a reasonable operating current. Hence, $R_3 = \frac{V_i(\min)-(11+0.7)}{I_L/h_{fe}+I_C(Tr_1)} = \frac{18\sqrt{2}-2-11.7}{(5/10^3)\times 10^3\,\text{mA}+1\,\text{mA}} = 2\text{k}$. In order to provide some protection to Tr_1, let $R_2 = 47\ \Omega$. Finally, $C_1 = \frac{I_{L\max}}{60\Delta V} = \frac{5\,\text{A}}{60\times2} = 41,666\ \mu\text{F}$. Capacitor $C_2 = 1\ \mu\text{F}$ provides further filtering of the regulated output voltage.

Op-Amp Series Regulator
In order to further improve the regulator performance, the loop gain $A\beta$ needs to be increased. A simple method of doing this is to replace the transistor error amplifier by an operational amplifier which has a much higher open-loop gain. This is shown in Fig. 10.36. The operational amplifier compares the reference voltage of the Zener with the feedback voltage sampled by resistors R_1 and R_2. Using basic op-amp theory,

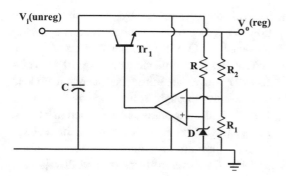

Fig. 10.36 Op-amp series regulator

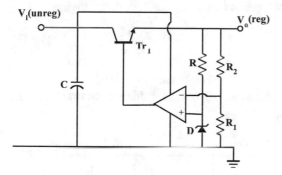

Fig. 10.37 Improved op-amp voltage regulator

$$V_o = V_z \left(1 + \frac{R_2}{R_1} \right) \qquad (10.36)$$

Note that the op-amp must be supplied from the unregulated supply. (Why?) Similar to the improved transistor case, the circuit can be improved by supplying the Zener diode from the regulated output (Fig. 10.37). All the other improvement techniques (except bootstrapping) are applicable here. The current limiting techniques are also applicable.

Example 10.12

Using the circuit shown in Fig. 10.38, design a +15 volt, 2 A regulator that is driven by an unregulated supply having 2 volt peak-peak input ripple. Use a 20 volt transformer and a full-wave rectifier to provide the unregulated input. Assume the power transistor has a gain of 150.

Solution Choose $V_Z = 7.5$ V. Set $I_Z = 5$ mA for good Zener operation. Hence, $R_3 = (15 - 7.5)/$

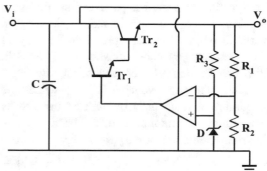

Fig. 10.38 Op-amp voltage regulator with Darlington pair

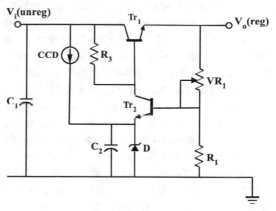

Fig. 10.39 Variable transistor voltage regulator

5 mA $= 1.5$ kΩ. From (10.36), $7.5\left(1 + \frac{R_2}{R_1}\right) =$ 15 . Hence, $R_1 = 10$ k $= R_2$. $C = \frac{I}{2\Delta Vf} = \frac{2}{2\times2\times60} = 8333\,\mu$F.

Variable Output Voltage

The basic feedback voltage regulator can be modified to produce a variable output voltage. In such a situation since the regulated output will now be variable, the Zener diode which provides the reference voltage will have to be supplied from the unregulated input voltage and not the regulated output if its current supply is to be maintained.

In the case of the transistor feedback regulator, one approach is shown in Fig. 10.39. Here the Zener diode is supplied from the unregulated voltage through a constant current diode to minimize ripple across the Zener diode. Additionally, a large filter capacitor is placed in parallel with the Zener diode to further reduce ripple. The variable output voltage is produced by varying the feedback through adjustment of VR_1 in Fig. 10.39 which replaced feedback resistor R_2.

The minimum voltage is V_Z, while the maximum voltage must be at least three volts less than the minimum unregulated voltage in order to ensure that the pass transistor is properly biased. Thus, $\left(V_Z + V_{BE}\right)\left(1 + \frac{VR_1}{R_1}\right) = \left(V_{mn}(unreg) - 3\right)$ is the design equation.

With respect to the op-amp feedback regulator, one simple approach to achieving variable regulated voltage is shown in Fig. 10.40. Here again the Zener diode is supplied from the unregulated voltage through a constant current diode to minimize ripple, and a large filter capacitor C_1 is placed in parallel with the Zener diode. The variable output voltage is produced by varying the reference voltage through adjustment of VR_1 in Fig. 10.40. This approach has the advantage that the feedback around the system is unaltered, thereby maintaining the system characteristics as the voltage is varied. It is important to note that the output of this system cannot go to zero because of saturation voltage loss within the op-amp. Once again, the maximum voltage must be set at about 3 V below the minimum unregulated voltage.

10.7.5 Protection Circuits

In this linear regulator, if the output is short-circuited to ground, the pass transistor will be immediately destroyed by excessive current flow. Various protection schemes can be used to prevent this. One circuit that can be used is a current limiting circuit shown in Fig. 10.41. Transistor Tr_3 and R_5 are introduced such that the load current flows through R_5 and thereby develops a voltage across the base-emitter junction of Tr_3. For a sufficiently large load current, this voltage exceeds the turn-on voltage of the transistor (0.7 V). Tr_3 therefore turns on and diverts base current away from the base of Tr_1, thereby limiting the load current. The design equations are simple: Let the maximum load current be $I_{L(max)}$. Then at turn on $I_{L(max)}R_5 = 0.7$ V giving

$$R_5 = \frac{0.7}{I_{L(max)}} \qquad (10.37)$$

This current limiting circuit at switch-on converts the regulator from a constant voltage

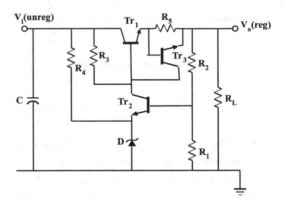

Fig. 10.41 Transistor voltage regulator with electronic protection

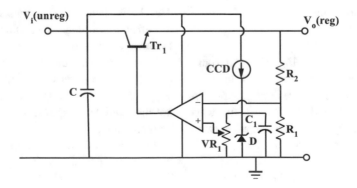

Fig. 10.40 Variable op-amp voltage regulator

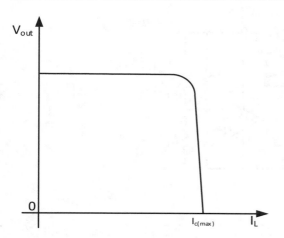

Fig. 10.12 Voltage characteristic of regulator with constant current protection

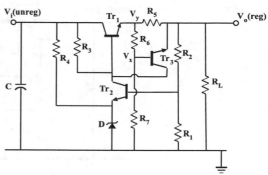

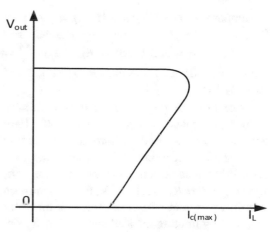

Fig. 10.43 Transistor regulator with fold-back protection

Fig. 10.44 Fold-back characteristic

output to a constant current output as shown in Fig. 10.42. Hence during its operation, as the external load resistance R_L decreases, the constant current $I_{L(max)}$ develops a voltage across R_L given by

$$V_o = I_{L(max)}R_L \qquad (10.38)$$

Therefore, as R_L goes to zero, so does the output voltage.

For the circuit in Fig. 10.41 in the presence of a short circuit, a significant current $I_{L(max)}$ flows into a short circuit. This can result in heating from residual circuit resistance. It is therefore desirable that the load current $I_{L(max)}$ be also reduced. In the fold-back limiting circuit shown in Fig. 10.43, reduction of both the output voltage and load current in an overload condition is achieved. The fold-back action shown in Fig. 10.44 is introduced by the potential divider R_6 and R_7. The base of Tr_3 is connected to the junction of these resistors. Transistor Tr_3 is turned on when the voltage V_x at the base of Tr_3 is 0.7 volts higher than the output voltage V_o. Initially, with no load current I_L, V_y is equal to V_o, and V_x is lower than V_y making V_x less than V_o. Tr_3 is therefore off. As the load current increases, $V_y = V_o + I_L R_5$ increases, and hence V_x also increases. Eventually when $V_x - V_o = 0.7$ volts, Tr_3 turns on, thereby diverting current form Tr_1 and hence limiting I_L. The circuit is now in the

constant current mode. As the load resistor R_L is reduced further, V_o decreases. But because of the potential divider R_6 and R_7, V_x decreases by a smaller amount, and therefore to maintain $V_x - V_o = 0.7$, $I_L R_5$ must decrease, i.e. I_L must fall. The design equations are the following: At turn-on,

$$0.7 = V_x - V_o = V_y - V_o$$
$$= \rho(V_o + I_L R_5) - V_o \qquad (10.39)$$

where

$$\rho = \frac{R_7}{R_6 + R_7} \qquad (10.40)$$

This gives

$$0.7 = (\rho - 1)V_o + \rho I_L R_5 \qquad (10.41)$$

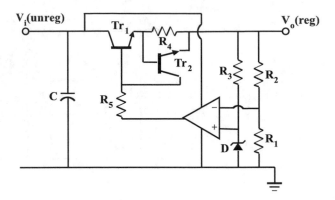

Fig. 10.45 Op-amp regulator with electronic protection

Suppose $V_o = 12$ V and $I_L(\text{max}) = 1$ A; then for $R_5 = 1\ \Omega$, $\rho = \frac{0.7+12}{12+1} = \frac{12.7}{13} = 0.9692$. For $R_L = 0$, $V_o = 0$ and hence $I_L(R_L = 0) = \frac{0.7}{\rho R_5} = \frac{0.7}{0.9692 \times 1} = 0.619$ A. From this, $R_7 = 10$ kΩ and $R_6 = 317\ \Omega$.

Protection for the op-amp regulator follows similar principles as seen in Fig. 10.45. Here transistor Tr_2 is used to introduce electronic short-circuit protection as in the discrete transistor case. When load current flows such that the voltage drop across R_4 equals 0.7 V, then Tr_2 turns on and draws current away from the pass transistor Tr_1, thereby limiting the load current drawn from the supply. Resistor R_5 ensures that the output voltage of the op-amp decreases as current through Tr_2 increases, thereby reducing the output voltage of the regulated supply.

10.7.6 IC Voltage Regulators

Instead of building regulators using transistors and operational amplifiers, IC voltage regulators that greatly simplify the process are available. These ICs contain all the elements of the basic block including reference voltage, error amplifier, series pass device and overload protection circuitry. They provide excellent voltage regulation in three basic formats, fixed positive voltage, fixed negative voltage and adjustable output voltages. The overall power supply consists of an unregulated supply providing the input voltage to the IC regulator. A wide range of output

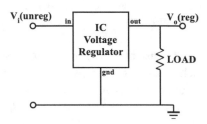

Fig. 10.46 Three-terminal IC regulator

voltages and currents are available. The design of a fully regulated power supply using an IC regulator is quite straightforward.

Three-Terminal Voltage Regulator

The simplest of the IC regulators is probably the three-terminal device shown in Fig. 10.46. An unregulated voltage is supplied to the input terminal, a regulated output is available at the output terminal, and the third terminal is grounded. Both positive and negative voltage devices are available.

Fixed Positive Voltage Regulators

Fixed positive voltages are available in the 7800 series of regulators. The voltages range from 5 volts to 24 volts as shown in Table 10.1. As an example, the 7815 IC regulator is shown in Fig. 10.47. The output voltage is +15 volts. A positive rectified input voltage is supplied at V_i with capacitor C_1 providing filtering (reservoir capacitor). The output is filtered by capacitor C_2 which is a smaller capacitor, designed to remove mostly high-frequency noise. The ground terminal is connected to ground. If the supply transformer has an output voltage of V_s rms, then the peak voltage into the regulator is $\sqrt{2}V_s$, and the minimum voltage is $V_{mn} = \sqrt{2}V_s - \Delta V$ where ΔV is the peak-to-peak ripple.

Based on the maximum load current I_{Lmx}, the reservoir capacitor C_1 is given by

$$C_1 = \frac{I_{L(\max)}}{2f\Delta V} \qquad (10.42)$$

Note that the minimum value of the unregulated input voltage must exceed the minimum

input voltage specification for the regulator. Some specifications include the following:

Output Voltage: Regulated Output Voltage

Line Regulation: Typically, 0.1%

Load Regulation: Typically, 0.1%

Current Limit: The maximum current that flows at the output of the regulator into a short circuit.

Drop Out Voltage: This is the minimum voltage across the input and output terminal of the regulator that must be maintained if regulating action is to be sustained.

Ripple Rejection Ratio: Ratio in dB of input ripple to output ripple.

Output Resistance: The DC output resistance of the regulator.

Table 10.1 7800 series positive voltage regulators

IC	Output voltage (V)	Minimum input voltage (V)
7805	+5	7.3
7806	+6	8.3
7808	+8	10.5
7810	+10	12.5
7812	+12	14.6
7815	+15	17.7
7818	+18	21.0
7824	+24	27.1

Fixed Negative Voltage Regulators

Similar to the 7800 series positive voltage regulators, the 7900 series provides negative regulated voltages. Similar specifications apply with the list of available voltages shown in Table 10.2. The connection of a −5 volt regulator is shown in Fig. 10.48.

Adjustable Voltage Regulators

Adjustable voltage regulators allow the user to change the output voltage of the regulator to a desired level. The LM317 is an example of such a regulator. Its output voltage can be adjusted over the range 1.2 volts to 37 volts. The circuit connection is shown in Fig. 10.49. The output voltage V_o is given by

$$V_o = V_{ref}\left(1 + \frac{R_2}{R_1}\right) + I_{adj}R_2 \qquad (10.43)$$

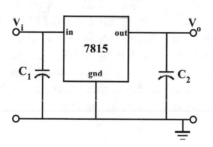

Fig. 10.47 Three-terminal positive fixed voltage regulator

Table 10.2 7900 series negative voltage regulators

IC No.	Output voltage (V)	Minimum input voltage (V_i)
7905	−5	−7.3
7906	−6	−8.4
7908	−8	−10.5
7909	−9	−11.5
7912	−12	−14.6
7915	−15	−17.7
7818	−18	−20.8
7824	−24	−27.1

where $R_1 = 240 \, \Omega$ and $R_2 = 5 \, \text{k}$ is a potentiometer.

Typical values are $V_{ref} = 1.25$ V and $I_{adj} = 100$ μA. The corresponding adjustable negative regulator is the LM337. These devices are able to provide output current of up to 1.5 A. For the safe operation of this regulator protection diodes are required and the reader is referred to the device datasheet. A typical application for a supply that can be varied from 1.25 to 13.5 is shown in Fig. 10.50. Capacitor $C_2 = 10$ μF improves

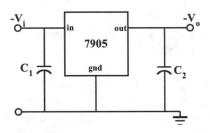

Fig. 10.48 Three-terminal negative fixed voltage regulator

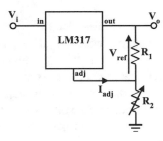

Fig. 10.49 Adjustable IC voltage regulator

ripple rejection. Capacitor $C_3 = 1$ μF improves transient response of the supply. Diodes D_1 and D_2 provide protection to the IC against capacitor discharge.

Example 10.13
Design a variable voltage regulated power supply using the LM1084 adjustable regulator IC.

Solution The LM1084 regulator IC is a low dropout voltage regulator that can deliver high current at a positive voltage. It has the same pin-out configuration as the LM317. It is available with fixed output voltages of 3.3 V, 5.0 V and 12.0 V as well as variable output. The device is current limited and thermally protected and can deliver 5 A at a line regulation of 0.015% and line regulation of 0.1%. A typical application as a variable voltage regulator is shown in Fig. 10.51. For this circuit the output voltage is given by $V_{DC} = 1.25\left(1 + \frac{R_2}{R_1}\right)$. The unregulated part of the supply follows the established design procedure. It utilizes a BR34 bridge rectifier which contains four diodes. Resistor R_1 is set at 120 ohms and R_2 is a potentiometer. Using $R_2 = 5$ k, then the output voltage varies from 1.25 volts to $V_{DC} = 1.25\left(1 + \frac{5000}{120}\right) = 15 \, \text{V}$. Resistor $R_3 = 1.5$ k and the red LED provide indication that the supply is on and also enable discharge of capacitor $C_1 = 3300$ μF when the supply is turned off. A small capacitor $C_2 = 10$ μF at the output removes any high-frequency noise from the supply.

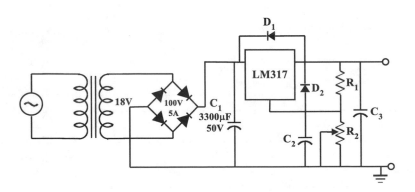

Fig. 10.50 Adjustable regulated power supply using the LM317 IC regulator

Fig. 10.51 Variable
regulator using
the LM1084 IC

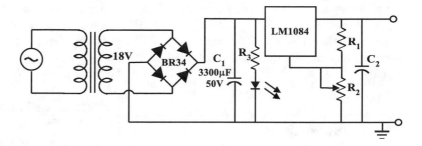

Fig. 10.52 Simple
regulated power supply

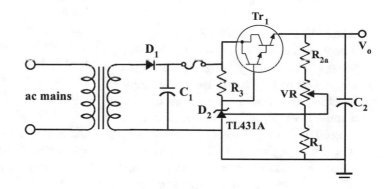

10.7.7   Simple Approach to Regulated Power Supplies

The TL431A integrated circuit is a three-terminal device that operates essentially as a programmable Zener diode whose effective voltage V_Z can be varied from 2.5 V to 36 V using two external resistors R_1 and R_2 as shown in Fig. 1.70 of Chap. 1. The Zener voltage is determined by $V_Z = \left(1 + \frac{R_2}{R_1}\right) V_{ref}$ where $V_{ref} = 2.5$ V. The Zener current range is 1 mA to 100 mA, the typical dynamic resistance is 0.22 ohms, and the temperature coefficient is a low 0.4% at room temperature. It was used in Chap. 1 in Zener-regulated power supplies. The versatility of the TL431A is based on it containing both a variable reference voltage and an error amplifier that enables the application of corrective feedback. In this application, its performance is enhanced by the introduction of a pass transistor Tr_1 (Darlington) to boost the current capacity, thereby implementing the basic series regulator of Fig. 10.30. The simple system is shown in Fig. 10.52 where the transistor is within the

feedback loop of the TL431A and therefore the output voltage is given by $V_o = \left(1 + \frac{R_2}{R_1}\right) 2.5\,\text{V}$. Resistor $R_2 = R_{2a} + VR$ comprises fixed resistor R_{2a} and variable resistor VR, thereby allowing easy variation of the output voltage. It can be adapted to a wide range of applications and power supply requirements at a very low cost.

Example 10.14
Using the configuration in Fig. 10.52 and a bridge rectifier, design a regulated supply delivering 20 V at a current of 2 A.

Solution   A 24 volt transformer is used to ensure adequate voltage drop across the pass transistor. Allowing a 2 volt drop in the voltage across reservoir capacitor C_1, the capacitor value is given by $C_1 = 2A/(2 \times 2 \times 60) = 8333\ \mu\text{F}$. The MJ3001 Darlington pair has a current gain of 1000; hence for a maximum current of 2 A, the base current is 2 A/1000 = 2 mA. The minimum value of the input voltage is $24\sqrt{2} - 2 = 32\,\text{V}$. Allowing a current of 10 mA into the TL431A Zener gives $R_3 = (32 - 20)/10$ mA = 1.2 k. For a

20 volt output, $20 = (1 + R_2/R_1) \times 2.5$. Using $R_1 = 2$ k, then $R_2 = 14$ k. Capacitor $C_2 = 100$ μF reduces ripple across the Zener, and $C_3 = 1$ μF removes any high-frequency noise and residual ripple.

10.8 Applications

In this section, several regulated power supply designs are presented.

Variable Voltage Regulated Supply
We here explore the design of a variable voltage regulated power supply using the basic op-amp configuration of Fig. 10.37. The system must be adjustable down to zero and up to about 15 volts and must have adjustable current limiting from about 100 mA to 1 A. In order to secure a 15 volt output from a regulator, a 30 volt centre-tapped transformer rated at twice the required current is used. This would enable the design of a full-wave rectifier using two rectifying diodes that provides sufficient voltage to ensure pass transistor operation. One system topology is shown in Fig. 10.53. It is an enhancement of the circuit of Fig. 10.37. In order to achieve variable output, the Zener diode D_4 is supplied from the unregulated input through a constant current diode D_6 for improved ripple reduction. The Zener voltage supplies a potentiometer VR_1 which in turn supplies a

variable reference voltage to the reference input of the op-amp. In order that zero output voltage be possible, the negative supply terminal of the op-amp cannot go to zero (why?) but must be taken to a negative voltage. A negative voltage of -6.8 volts is furnished by Zener diode D_5 which is powered via R_3 by the half-wave rectifier $D_3 - C_2$. The total voltage between the power supply terminals of the op-amp is $15\sqrt{2} + 6.8 = 28$ V which is less than the maximum rated supply voltage of $2 \times 22 = 44$ V for the LM741. A Darlington pair comprising Tr_1 and Tr_2 is used and can be made up using a 2 N3053 and a 2 N3055, respectively, or a MJ000 power Darlington. Potentiometer VR_2 permits variation of the current limit. Tr_3 can be a small-signal transistor 2 N3904. Rectifying diodes D_1 and D_2 must have piv ratings of better than $2 \times 15 \times \sqrt{2} = 42$ V and current rating higher than 1 A. The 1 N5401 diode is rated at 100 V and 3 A and can be used. Since $I = 1$ A and $f = 60$ Hz, in order to realize a peak-to-peak ripple voltage of about 2.5 volts, capacitor C_1 must be $C_1 = \frac{I}{2f\Delta V} = \frac{1}{2 \times 60 \times 2.5} = 3333$ μF. The negative supply to the op-amp must deliver at least 2 mA to the op-amp. Hence using $I = 0.002$ A and $f = 60$ Hz, in order to realize a pk-to-pk ripple voltage of about 2.5 volts, capacitor C_2 must be $C_2 = \frac{I}{f\Delta V} = \frac{0.002}{60 \times 1} = 33$ μF. Use $C_2 = 100$ μF. Diode D_3 can be an 1 N4002 with piv $= 100$ V and

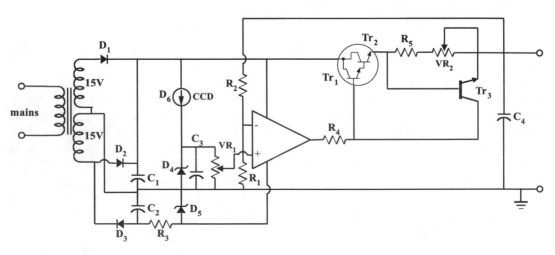

Fig. 10.53 Variable voltage regulated power supply

current rating of 1 A. Resistor R_3 must supply the Zener and the op-amp with current. Allowing 10 mA through D_5 in order to accomplish this, then $R_3 = (15\sqrt{2} - 6.8)/10\,\text{mA} = 1.4\,\text{k}$. A 5.6 V Zener is used for the reference Zener D_4. This value will be amplified by the op-amp to produce the desired output voltage. VR_1 is a 10 k potentiometer to provide the voltage variation. Capacitor C_3 should have a reactance that is comparable to the dynamic resistance of the Zener which is of the order of tens of ohms. A 10 uF capacitor has a reactance at $f = 120$ Hz of $1/2\pi 120 \times 10 = 133\ \Omega$ which is acceptable. The constant current diode D_6 is a Siliconix J509 which supplies about 3 mA to D_4. The high impedance of D_6 and the low impedance of D_4 in parallel with C_3 ensure that the ripple content at the non-inverting input at the op amp is very low. A choice of $R_1 = 10$ k and $R_2 = 18$ k means that the maximum output of the system is $5.6\left(1 + \frac{18}{10}\right) = 15.7\,\text{V}$ just above the desired output voltage level of 15. Resistor R_5 sets the maximum current of say 1.1 A and is given by $R_5 = 0.7/1.1\ \text{A} = 0.6\ \Omega$. The 10 ohm potentiometer VR_2 allows variation of the current limit to a minimum value of $I_{\text{min}} = 0.7/10.6 = 66$ mA. The resistor $R_4 = 330\ \Omega$ assists in lowering the output voltage during a current overload condition. Finally, capacitor $C_4 = 0.01\ \mu\text{F}$ is intended to reduce high-frequency noise at the output of the power supply.

Ideas for Exploration (i) Re-design the system to deliver a higher voltage using the OPA452

(80 V, 50 mA) operational amplifier to replace the LM741. (ii) Utilize the LM338A regulator IC in the design of a variable regulated power supply.

Variable Power Supply with Zener Stabilization

This regulated supply utilizing a 741 op-amp includes variable output voltage from 5 V to 20 V at 1 A while supplying the Zener from the regulated output. This is achieved by varying the resistance supplying the 3.9 V Zener while also varying the applied feedback. The circuit is shown in Fig. 10.54. The essence of the approach is to vary both the amount of feedback and the value of the resistor supplying the Zener. This latter action will ensure that as the output voltage changes, the current through the Zener will remain approximately constant. Because of the saturation loss of the op-amp, the output is not allowed to fall below 5 V. With an input unregulated voltage of 30 V, the maximum voltage is 20 V. The output of the Zener is passed through a low-pass filter $R_4 - C_2$ to reduce any noise, and a capacitor C_3 is implemented across the output also for purposes of filtering. Both can be about 10 μF. A Darlington pair comprising a 2 N3053 medium power transistor and a 2 N3055 power transistor or an MJ3000 is used as the series device.

Ideas for Exploration (i) Re-design the system to deliver a higher voltage using the OPA452

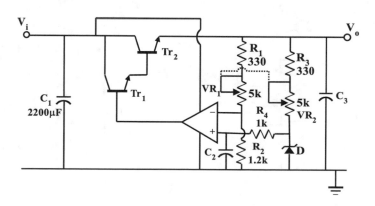

Fig. 10.54 Variable power supply with Zener stabilization

(80 V, 50 mA) operational amplifier to replace the 741.

Discrete Voltage Regulator

This Zener diode regulator enhanced by an emitter follower enables the adjustment of the Zener voltage to compensate for the voltage drop caused by emitter follower action. The follower is really a feedback pair with modest gain where the second transistor is a pnp Darlington pair such as the MJ2501. Thus, with a 15 V Zener, the voltage that would be available at the output of Tr_2 would be 14.3 V because of the V_{BE} drop. Thus if $R_1 = 12$ k and $R_2 = 600 \, \Omega$, then $V_o = (1 + 600/12000) \times 14.3 = 15$ V. The JFET is connected as a constant current diode supplying the Zener diode. Capacitor $C_1 = 10 \, \mu F$ provides filtering for the Zener (Fig. 10.55).

Ideas for Exploration (i) Re-design the system to provide a negative voltage with respect to ground. (ii) Explore the introduction of short-circuit protection.

Bipolar Zener

The circuit in Fig. 10.56 is essentially two complimentary feedback pairs connected as V_{BE} multipliers with voltage adjustment by a single resistor. Since variable resistor R_V is across the base-emitter junctions of Tr_1 and Tr_2, it follows that the current I through R_V is $I = \frac{1.4V}{R_V}$.

Therefore, the positive Zener voltage with respect to the zero potential is $V_Z = IR_1 + 0.7 = 1.4R_1/R_V + 0.7$, and the negative Zener voltage with respect to the zero reference is $-V_Z = -(IR_1 + 0.7) = -(1.4R_1/R_V + 0.7)$. Thus, a symmetrical supply is available with easy adjustment using R_V. Variable resistor R_V should comprise a variable component R_{Va} and a fixed component R_b such that R_V has a minimum value $R_V = R_b$. For example, let $R_1 = 6.8$ k, $R_{Va} = 10$ k and $R_b = 1$ k. Then noting that $R_{Vmax} = (10 \text{ k} + 1 \text{ k}) = 11$ k and $R_{Vmin} = 1$ k, then $V_{Zmin} = (1.4 \times 6.8 \text{ k}/11 \text{ k}) + 0.7 = 1.6$ V and $V_{Zmax} = (1.4 \times 6.8 \text{ k}/1 \text{ k}) + 0.7 = 10.2$ V. This means that the bipolar Zener voltages are variable from ±1.6 V to ±10.2 V using R_V.

Ideas for Exploration (i) Use this configuration to design an adjustable Zener-regulated bipolar power supply.

Simple Feedback Power Supply

This feedback regulator delivers 6 V at 100 mA from a 12 V supply. It does not use a reference voltage Zener but instead uses the error transistor base-emitter junction as the reference voltage. The circuit is shown in Fig. 10.57. The defining equation is $\frac{0.7}{R_1} = \frac{V_o}{R_1+R_2}$ where V_o is the regulated

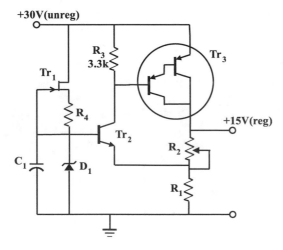

Fig. 10.55 Discrete voltage regulator

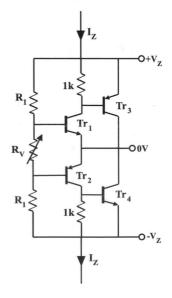

Fig. 10.56 Bipolar Zener

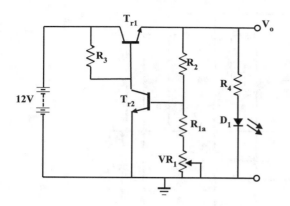

Fig. 10.57 Simple feedback power supply

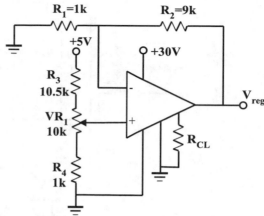

Fig. 10.58 Regulated power supply using OPA549 power op-amp

output voltage and $R_1 = R_{1a} + VR_1$. For $V_o = 6$ V, if $R_2 = 4.7$ k then $R_1 = 620\ \Omega$. A good approach is to set $R_{1a} = 470\ \Omega$ and potentiometer $VR_1 = 1$ k in series with R_{1a} in order to adjust the output voltage to the desired voltage. A load current of 100 mA produces a base current requirement in Tr_1 of 100 mA/50 = 2 mA where $\beta = 50$ is used. With a collector current in Tr_2 of 2 mA, then $R_3 = (12 - 6 - 0.7)/(2\ \text{mA} + 2\ \text{mA}) = 1.3$ k. This circuit can be adapted to a variety of applications using an unregulated input. If an unregulated supply is used, R_3 can be bootstrapped to realize a higher gain in Tr_2 and therefore greater ripple-reducing feedback. $R_4 = 1$ k provides about 5 mA to the LED to indicate the on-condition.

Ideas for Exploration (i) Add electronic protection to this circuit. (ii) Use this configuration to design an adjustable Zener-regulated bipolar power supply. (ii) Use a 20 V transformer and an unregulated half-wave rectifier to supply the circuit.

High-Current Variable Power Supply Using OPA549 Power Op-Amp
The OPA549 is a high-voltage, high-current operational amplifier that can be used as a variable regulated power supply. It is simply connected as an amplifier with a gain of 10 with DC at the input and the amplified DC at the output. The op-amp is powered from a 30 V unregulated supply. This is shown in Fig. 10.58, where a variable DC voltage of 0.1 V to 2.5 V at the non-inverting input results

in a regulated output of 1 V to 25 V at a maximum current of 8 A. Current limit is set by resistor $R_{CL} = \left(\frac{75}{I_{LM}} - 7.5\right)$ kΩ where I_{LM} is the current limit in amperes. For example, for a current limit of 3 A, $R_{CL} = \left(\frac{75}{3} - 7.5\right)$k $= 17.5$ k.

Ideas for Exploration (i) Re-design the system to deliver a negative voltage relative to ground.

Dual Tracking Power Supply
This project involves the design of the dual tracking power supply shown in Fig. 10.59. It has one positive and one negative output of equal voltage. The positive voltage is adjustable from 0 to 20 V, and the negative supply automatically tracks the positive voltage. The supply can deliver up to 1 A. The basic system for each voltage polarity is essentially the op-amp system discussed earlier in this chapter with certain modifications. A 24 V transformer is used to supply each regulator. A low-cost high-voltage replacement for the 741 op-amp is used: It is the OPA452. This op-amp operates from up to ±40 V, can deliver 50 mA, has a GBP of 1.8 MHz and is unity-gain stable. The pass transistors are MJ3001 and MJ2501 complementary power Darlington which are 80 V, 10 A, 150 W devices with $\beta = 1000$. These ensure that the current drawn from each op-amp is no more than 1 A/ 1000 = 1 mA. The feedback components

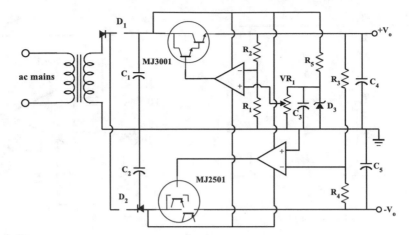

Fig. 10.59 Dual tracking power supply

$R_2 = 15$ k and $R_1 = 5.1$ k for the positive regulator set the gain at 4. Hence a 0 to 5 V at the non-inverting input will produce 0 to 20 V at the output. The input reference voltage is supplied by the 5 V Zener and potentiometer $VR_1 = 5$ k. On the negative side, the non-inverting input of the op-amp is connected to ground and the inverting input connected to the junction of equal resistors $R_3 = R_4 = 10$ k which are connected between the positive and negative supplies. Hence as the positive voltage changes in response to the changing potentiometer, the negative output will track it as the feedback ensures that the inverting potential is held at zero potential as is the non-inverting terminal. Resistor $R_5 = 5.6$ k passes about 5 mA into the Zener diode D_3. Allowing a 2 V peak-to-peak ripple voltage, capacitor C_1 and C_2 are given by $C_1 = C_2 = 1$ A$/2 \times 60 = 8333$ μF. Capacitor C_3 is set at $C_3 = 100$ μF to ensure low ripple across the reference Zener. $C_4 = C_5 = 1$ μF are placed at the outputs to provide further filtering.

Ideas for Exploration (i) Replace R_5 by a constant current diode to improve performance. (ii) Introduce electronic protection. (iii) Modify the system to allow for switching between tracking and independent voltage adjustment of the two supplies.

Research Project 1

The *transistor feedback regulator* in Fig. 10.60 considered previously was enhanced by

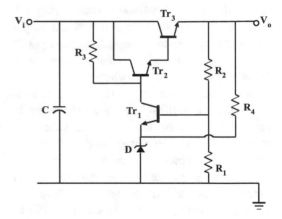

Fig. 10.60 Regulated supply for Research Project 1

bootstrapping resistor R_3. Another approach to performance enhancement is the use of a constant current diode in place of the resistor R_3. In the existing circuit, ripples from the unregulated supply will cause variation in the current in R_3 which must be accommodated by the error transistor. This therefore shows up in the regulated output. With a constant current diode instead of resistor R_3, there will be significantly reduced variation in the current resulting from ripples in the unregulated supply and hence reduced ripple in the regulated output. The project requirement is to design the regulator using a constant current diode instead of resistor R_3. The current considerations are the same as before in that current through the CCD must supply both the collector current of the error transistor and the base current of the pass transistor.

Ideas for Exploration (i) Simulate this system and that using bootstrapping and compare the load regulation and line regulation performances of the two systems.

Research Project 2

This project involves the development of a *precision voltage divider* circuit that converts 30 V into ± 15 V with current boosting for powering operational amplifiers. The circuit is shown in Fig. 10.61 and uses the op-amp current booster circuit discussed in Chap. 9, Fig. 9.34 where resistors R_1 and R_2 set the voltage ratio at the output. Feedback via R_4 ensures that the ratio set by R_1 and R_2 is maintained as current is drawn from the output. A 741 is suitable for this application.

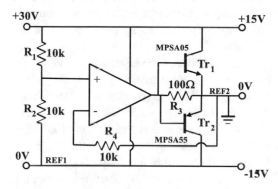

Fig. 10.61 Precision voltage divider

Ideas for Exploration (i) Examine whether resistor R_3 is necessary in this application. (ii) Use an OPA452 to produce higher dual voltages.

Research Project 3

This project examines a *regulated power supply* configuration that is slightly different from those already considered. The basic circuit is shown in Fig. 10.62. where a 15 V, 1 A transformer drives a bridge rectifier and is filtered by capacitor $C_1 = 1000$ µF. This results in an unregulated voltage of about 20 V. This voltage is applied to the 13 V Zener diode in series with current setting resistor R_1. Setting $R_1 = 1$ k results in a Zener diode current of $(20 - 13)/1$ k $= 7$ mA. The stabilized voltage produced by the Zener is applied to a voltage divider string of resistors $R_2 = 2.7$ k, $R_3 = 820$ Ω, $R_4 = 1.5$ k and $R_5 = 1.5$ k. This (in conjunction with the output transistor) provides four reference voltages: 4.5 V, 6 V, 9 V and 12 V. The voltage at the positive terminal is the zero reference, and that on the negative terminal is the variable voltage. This is different from the previous regulators where the positive rail is usually the variable one. Transistors Tr_2 and Tr_3 form a pnp feedback pair, and Tr_1 provides short-circuit protection. With $R_6 = 1.5$ Ω, load currents in excess of about 450 mA causes Tr_1 to turn on and limit

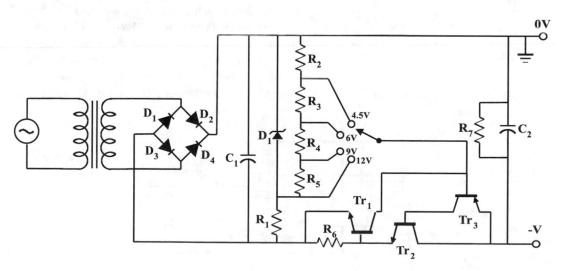

Fig. 10.62 Alternative regulated power supply

the current to the base of the feedback pair, thereby converting the supply from a constant voltage supply into a constant current supply. Under short-circuit conditions, the full unregulated supply voltage is across Tr_2 and with 450 mA flowing through the device, some 10 W of power would be dissipated in this device. It must therefore be mounted on a suitable heatsink. Capacitor $C_2 = 100$ μF provides additional filtering at the output, and resistor $R_7 = 1$ k discharges this capacitor when the supply is turned off.

Ideas for Exploration (i) Introduce an LED in series with R_7 to indicate the ON condition. (ii) Convert the regulated supply into a fully variable supply by replacing the resistor string by a potentiometer of value about 10 k. The wiper of the potentiometer then goes to the base of Tr_3. (iii) Introduce fold-back current limiting to this circuit.

Research Project 4

This research project involves the investigation of a *switch-mode power supply*. Such a supply converts one voltage level to another by switching devices on and off at high frequency

and using energy storage devices such as an inductor and/or a capacitor. Its main advantage over the linear regulators discussed in this chapter is its high efficiency which is better than 90%. One class of switch-mode power supply is the step down or *buck converter* shown in Fig. 10.63. The operation can be divided into two phases: a charge phase and a discharge phase. In the charge phase, the switch S is closed, and energy is stored in the inductor L. In the discharge phase, the switch is open, and the energy stored in the inductor is transferred to the output capacitor C and load through the diode D. The capacitor maintains the load voltage during the charging phase of the inductor which is central to the energy transfer process. Since there is no build-up of current in the inductor, it follows that the change in inductor current ΔI_C during the charging phase must be equal to the change in inductor current ΔI_D during the discharge phase, i.e. $|\Delta I_C| = |\Delta I_D|$. From this it can be shown that the relation between the input voltage V_i and the output voltage V_o is given by $V_o = DV_i$ where $D = T_{ON}/T_S$, T_{ON} is the on-time of the switch and T_S is the switching period. Since $D < 1$, then $V_o < V_i$ which represents a reduction of the input voltage as generally occurs with the series regulators discussed in this chapter.

A simple implementation of this system is shown in Fig. 10.64. A variable duty cycle square wave is generated by the 555 timer, and this is used to switch on and off a p-channel MOSFET IRF4905. Potentiometer $VR_1 = 10$ k varies the duty cycle of the square wave and hence the output voltage. The output voltage in this circuit varies with changing load since no feedback is present. An improved circuit is shown in

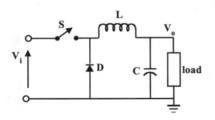

Fig. 10.63 Step down buck converter

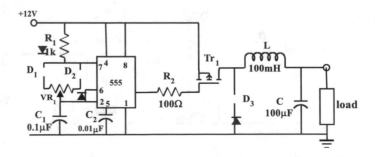

Fig. 10.64 Buck converter using 555 timer

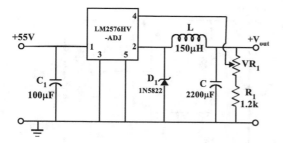

Fig. 10.65 Buck converter using the LM2576T-ADJ

Fig. 10.65. It uses a Texas Instruments LM2576HV-ADJ integrated circuit that contains the pulse generation and switching circuitry necessary to realize a buck converter. Importantly, it contains feedback circuitry that changes the pulse width and thereby maintains a constant output voltage. Potentiometer $VR_1 = 50$ k varies the amount of feedback and hence the output voltage. This version of the IC enables a variable output voltage from 1.2 to 50 V at 3 A.

Ideas for Exploration (i) Investigate the operation of a *boost converter* that increases the voltage supplied to the system.

Problems

1. Describe the operation of a full-wave rectifier and derive the expression for the average and rms output voltages.
2. Explain how the inclusion of a filter capacitor reduces the ripple in the output voltage and derive an expression for the peak-to-peak value of this ripple voltage.
3. Design an unregulated power supply to deliver 15 volts at 1 ampere. Use a 15 volt transformer with a bridge rectifier and allow for a maximum ripple voltage of 1.5 volts peak-to-peak.
4. Re-design the circuit of Question 3 using a half-wave rectifier.
5. A voltage from an unregulated supply varies between a minimum value of 12 volts and a maximum value of 18 volts. Using this unregulated supply, design a Zener-regulated

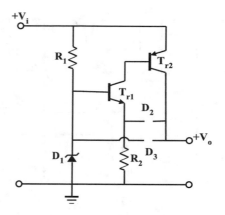

Fig. 10.66 Circuit for Question 8

supply that delivers 9 volts DC with a current capacity of 15 mA.

6. Design a single transistor regulated supply using a Zener diode capable of delivering 15 V at 120 mA. Use a half-wave rectifier to drive the regulator. Show how this circuit can be equipped with short-circuit protection.
7. Design a regulated supply using a Darlington pair and a Zener diode that delivers 9 V at 1.5 A. Use a full-wave rectifier to drive the regulator.
8. Explain the operation of the regulator shown in Fig. 10.66. Using this circuit, design a regulated supply to deliver 9 volts at 0.5 A.
9. Explain the operation of the regulator circuit shown in Fig. 10.67, describing the circuit corrective action for unwanted increases or decreases in the output and giving an indication of the action of capacitor C.
10. Using the circuit shown in Fig. 10.67, design a +16 volt, 1.2 A regulator that is driven by an unregulated supply having 2 volt peak-peak input ripple. Use a 20 volt transformer and a bridge rectifier to provide the unregulated input. Assume the power transistor has a gain of 20 and the other transistors have gains of 125. Justify all your design steps.
11. Show how electronic protection can be added to this circuit and describe its operation.
12. Introduce bootstrapping of resistor R_3 in the design of problem 10 and describe the effect of this bootstrapping.

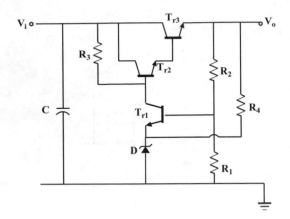

Fig. 10.67 Circuit for Question 9

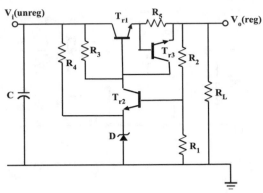

Fig. 10.69 Circuit for Question 14

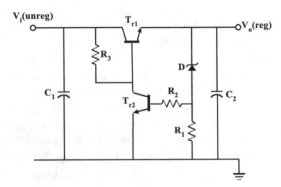

Fig. 10.68 Circuit for Question 13

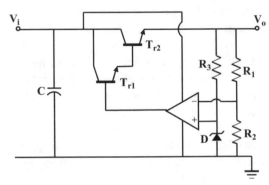

Fig. 10.70 Circuit for Question 15

13. Using the circuit shown in Fig. 10.68, design a +16 volt, 1.2 A regulator that is driven by an unregulated supply having 2 volt peak-peak input ripple. Use a 20 volt transformer and a bridge rectifier to provide the unregulated input. Assume the power transistor has a gain of 20 and the other transistors have gains of 125. Justify all your design steps.

14. Design a regulated supply using the basic topology in Fig. 10.69 rated at 10 volts and 150 mA. Power for the regulator must come from an unregulated supply using a half-wave bridge rectifier and a 12 volt transformer having a ripple voltage of 1.5 volts peak-to-peak at maximum load current. Use a pass transistor with current gain of 150. Include short-circuit protection in your design.

15. Using the circuit shown in Fig. 10.70, design a +20 volt, 1.5 A regulator that is driven by an unregulated supply having 2.5 volt peak-peak input ripple. Use a 30 volt transformer and a full-wave rectifier to provide the unregulated input. Assume the power transistor has a gain of 50 and the other transistor has a gain of 150. Justify all design steps.

16. Discuss the manner in which output voltage variation can be introduced in this circuit.

17. Design a 9 volt regulator using the 7809 regulator IC.

18. Design a single transistor regulator to power a small radio from the mains supply. The required voltage is 6 volts and the maximum current demand is 50 mA.

19. How can a constant current diode be used to improve the performance of a regulated power supply?

Fig. 10.71 Circuit for
Question 24

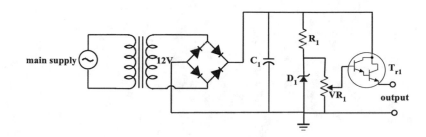

20. Indicate other methods of improving the performance of a regulator.
21. Design a +12 V regulated supply using regulator IC from the 7800 series.
22. Design a ±15 V regulated bipolar supply using regulator ICs from 7800 and 7900 series.
23. Using the circuit of Fig. 10.51, design a variable voltage regulated supply that can deliver 0–9 V at 100 mA.
24. Design a variable voltage bench power supply using a Darlington pair and the topology of Fig. 10.71.

25. Outline the operation of a switch-mode buck regulator.

Bibliography

J. Millman, C.C. Halkias, *Integrated Electronics: Analog and Digital Circuits and Systems* (Mc Graw Hill Book Company, New York, 1972)

R. Coughlin, F. Driscoll, *Operational Amplifiers and Linear Integrated Circuits*, 5th edn. (Prentice Hall, New Jersey, 1998)

Active Filters

11

A filter is an electrical network that passes signals within a specified band of frequencies while attenuating those signals that fall outside of this band. Passive filters utilize passive components, namely, resistors, capacitors and inductors, while active filters contain passive as well as active components such as transistors and operational amplifiers. Passive filters have the advantage of not requiring an external power supply as do active filters. However, they often utilize inductors which tend to be bulky and costly, whereas active filters utilize mainly resistors and capacitors. Additionally, active filters can produce signal gain and have high input and low output impedances which allow simple cascading of systems with little or no interaction between stages. This chapter discusses the principles of active filter operation and treats with several types and configurations of active filters. At the end of the chapter, the student will be able to:

- Understand the principles of operation of the basic types of filters
- Explain the operation of a range of active filters
- Design a range of active filters

11.1 Introduction to Filters

There are five basic types of filters: low-pass, high-pass, band-pass, band-stop (or notch) and all-pass filters. A low-pass filter passes signals with constant amplitude from DC up to a frequency f_c referred to as the cut-off frequency. The signal output above f_c is attenuated. This is shown in Fig. 11.1 where the magnitude of the output voltage of the ideal (solid line) and practical (dashed line) responses are plotted against the frequency. The ideal response shows the pass-band, the range of transmitted frequencies, and the stop-band, the range of attenuated frequencies and the cut-off or break frequency f_c. In the practical response, f_c is the frequency at which the magnitude of the output voltage falls to 0.707 of its low-frequency value which is -3 dB. Here $f_c = 1$ Hz for the normalized response.

A high-pass filter attenuates signals from DC up to a frequency f_c while passing all signals with constant amplitude above f_c. The normalized response is shown in Fig. 11.2 where the ideal and practical responses with pass-band and stop-band are indicated. A band-pass filter passes a band of frequencies, while frequencies outside of that band are attenuated. This is shown in Fig. 11.3 where the ideal and practical responses are indicated. A band-stop filter stops a band of frequencies while passing all frequencies outside that band. It performs in a manner that is exactly opposite to the band-pass filter. The band-stop response is shown in Fig. 11.4 with ideal and practical responses. Finally, an all-pass filter passes all frequencies with a constant amplitude output but with a changing phase. This is shown in Fig. 11.5.

© Springer Nature Switzerland AG 2021
S. J. G. Gift, B. Maundy, *Electronic Circuit Design and Application*,
https://doi.org/10.1007/978-3-030-46989-4_11

Fig. 11.1 Low-pass filter
response

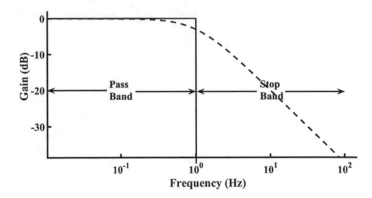

Fig. 11.2 High-pass filter
response

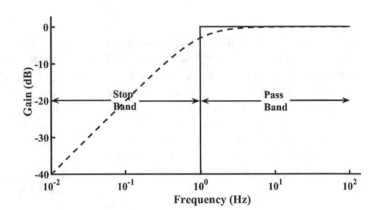

Fig. 11.3 Band-pass filter
response

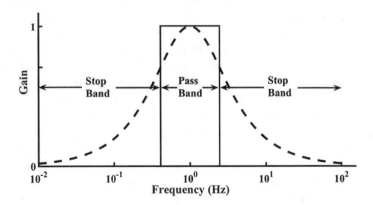

11.2   Basic First-Order LP Filter

We first consider a low pass of the simplest kind
shown in Fig. 11.6(a). It is a non-inverting
low-pass active filter using an RC network feeding
into a unity-gain operational amplifier. A resistor
R_{os} is sometimes included in the feedback loop of
the op-amp in order to minimize DC offset at the
output as shown in Fig. 11.6(b). For the circuit,

$$A_f = \frac{V_o}{V_i} = \frac{1/sC}{R + 1/sC} = \frac{1}{1 + j\omega RC}$$

$$= \frac{1}{1 + jf/f_c} \quad (11.1)$$

where

$$f_c = 1/2\pi RC \quad (11.2)$$

Fig. 11.4 Band-stop filter response

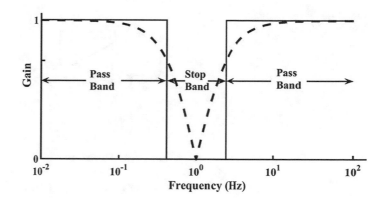

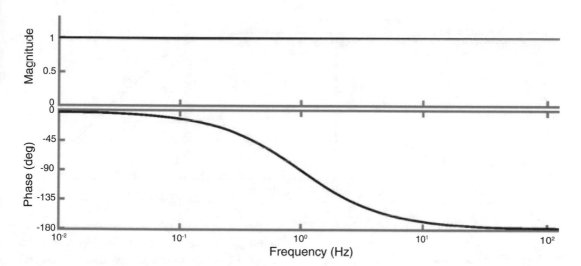

Fig. 11.5 Magnitude and phase responses of first-order all-pass filter

Note that for $f \ll f_c$, $|A_f| \rightarrow 1$ and for $f \gg f_c$, $|A_f| \rightarrow 0$. The frequency response plot for this circuit is shown in Fig. 11.7 which is clearly a low-pass response. The frequency $f_c = 1/2\pi RC$ is the cut-off frequency. The system has one pole at $f = f_c$ and is therefore first-order. At this frequency, the filter transfer function becomes

$$A_f(f_c) = \frac{1}{1 + jf_c/f_c} = \frac{1}{1+j} \qquad (11.3)$$

This reduces to

$$A_f(f_c) = \frac{1}{\sqrt{2}\angle 45^\circ} = 0.707\angle -45^\circ \qquad (11.4)$$

Hence

$$|A_f|_{f_c} = \frac{1}{\sqrt{2}} = 0.707 \qquad (11.5)$$

with

$$|A_f|_{f_c}(\text{dB}) = 20\log 0.707 = -3 \text{ dB} \qquad (11.6)$$

The signal phase is

$$\angle A_f(f_c) = -45^\circ \qquad (11.7)$$

Equation (11.2) is used to design the filter.

Example 11.1
Design a low-pass non-inverting filter with unity gain and a break frequency of 1 kHz.

Solution Choose $C = 0.01\ \mu\text{F}$. Then from Eq. (11.2), $R = 1/2\pi \times 10^3 \times 0.01 \times 10^{-6} = 16$ k. If low DC offset is required, then $R_{os} = R = 16$ k.

Fig. 11.6 Basic first-order low-pass filter circuit (unity gain)

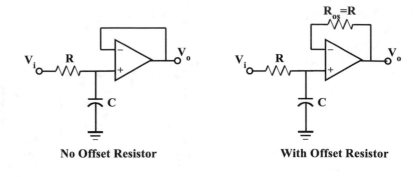

No Offset Resistor

With Offset Resistor

(a)

(b)

Fig. 11.7 First first-order low-pass filter response

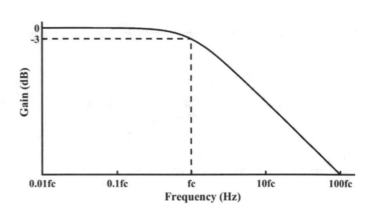

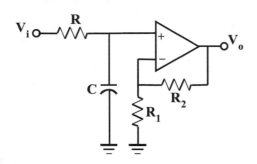

Fig. 11.8 First-order non-inverting low-pass filter with gain

11.2.1 Low-Pass Filter with Gain

The low-pass filter with gain is achieved by reducing the feedback around the op-amp as shown in Fig. 11.8. The transfer function is given by

$$\frac{V_o}{V_i} = \frac{k}{1 + jf/f_c} \qquad (11.8)$$

where $f_c = 1/2\pi RC$ is the break frequency and $k = 1 + \frac{R_2}{R_1}$ is the low-frequency gain. For low DC offset, $R = R_2//R_1$.

An alternative configuration for a first-order low-pass filter is shown in Fig. 11.9(a). It is an inverting amplifier configuration with a capacitor C across the feedback resistor R_2. From Fig. 11.9(a),

$$\frac{V_i}{R_1} = \frac{-V_o}{R_2//1/sC} = -V_o \Big/ \frac{R_2}{1 + sCR_2} \qquad (11.9)$$

which yields

$$A_f = -\frac{R_2}{R_1} \cdot \frac{1}{1 + sCR_2} = \frac{-k}{1 + jf/f_c} \qquad (11.10)$$

where $f_c = 1/2\pi R_2 C$ is the break frequency and $k = R_2/R_1$ is the low-frequency gain. For reduced DC offset, a resistor $R_{os} = R_1//R_2$ can be inserted between the non-inverting terminal and ground as shown in Fig. 11.9(b).

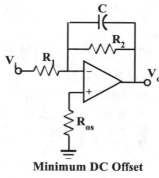

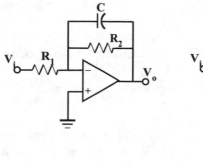

(a)   **(b)**

Fig. 11.9 First-order inverting low-pass filter with gain

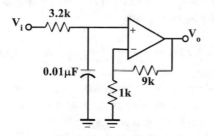

Fig. 11.10 Solution for Example 11.2

Example 11.2
Design a first-order non-inverting low-pass filter
with a gain of 10 and a cut-off frequency of 5 kHz
using Fig. 11.8.

Solution Choose $C = 0.01\ \mu F$. Then from
$f_c = 1/2\pi RC$, resistor R is found to be $R = 1/$
$(2 \times \pi \times 5 \times 10^3 \times 0.01 \times 10^{-6}) = 3.2$ k. Since
$1 + R_2/R_1 = 10$, for $R_1 = 1$ k, then $R_2 = 9$ k. The
solution is shown in Fig. 11.10.

Design Procedure
In preparation for the discussion on higher-order
filters and their realization, we wish to develop a
more ordered procedure for designing first-order
filters. Thus, the basic first-order system can be
written as

$$A(s) = \frac{A_o}{1 + a_1(s/\omega_c)} \qquad (11.11)$$

where $A_o = k$ is the low-frequency gain, $\omega_c = 2\pi f_c$
is the cut-off frequency and here $a_1 = 1$. For

higher-order filters considered later, the condition
$a_1 \neq 1$ will arise.

For the first-order *non-inverting* low-pass filter,
the transfer function is given by equation (11.12):

$$A_f(s) = \frac{1 + \frac{R_2}{R_1}}{1 + RCs} \qquad (11.12)$$

Comparing coefficients in (11.11) and (11.12)
gives

$$A_o = 1 + \frac{R_2}{R_1} \qquad (11.13)$$

$$a_1/2\pi f_c = RC \qquad (11.14)$$

The design steps are as follows:
1. Select C usually between 100 pF and 1 μF.
2. Set the coefficient $a_1 = 1$.
3. Determine R using $R = a_1/2\pi f_c C$.
4. Select R_1 and find R_2 using $R_2 = R_1(A_o - 1)$.

For the first-order **inverting** low-pass filter, the
basic first-order system can be written as

$$A(s) = -\frac{A_o}{1 + a_1(s/\omega_c)} \qquad (11.15)$$

The transfer function for the actual system is
given by equation (11.16):

$$A_f(s) = -\frac{\frac{R_2}{R_1}}{1 + R_2Cs} \qquad (11.16)$$

Comparing coefficients in (11.15) and (11.16)
gives

$$A_o = \frac{R_2}{R_1} \qquad (11.17)$$

$$a_1/2\pi f_c = R_2 C \qquad (11.18)$$

The design steps are as follows:
1. Select C usually between 100 pF and 1 μF.
2. Set the coefficient $a_1 = 1$.
3. Determine R_2 using $R_2 = a_1/2\pi f_c C$.
4. Determine R_1 using $R_1 = R_2/A_o$.

11.3   Low-Pass Second-Order Filter

It is possible to cascade two first-order low-pass filters with the same cut-off frequency f_c each with -20 dB/dec roll-off rate to get a second-order filter with a roll-off rate of -40 dB/dec. This is shown in Fig. 11.11. Such an approach has several disadvantages. The first is that the cut-off frequency of the overall filter is lower than the cut-off frequency f_c of the individual first-order filters. The result is that the pass-band gain begins to fall well before the cut-off frequency f_c. Secondly, the transition from the pass-band to the stop-band lacks sharpness as is desired in a filter. Thirdly, it can be shown that the phase response is not linear with the effect that signals distortion results. A low-pass filter can be optimized to meet at least one of the following three requirements: (i) maximum flatness in the pass-band; (ii) maximum sharpness in the transition from pass-band to stop-band; and (iii) linear phase response.

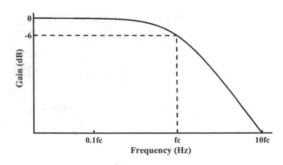

Fig. 11.11 Two cascaded first-order filters

In order to allow optimization of a filter, the transfer function of the filter must contain complex poles and hence, for a second-order filter, must take the following form:

$$A(s) = \frac{A_o}{\left(1 + a_1(s/\omega_c) + b_1(s/\omega_c)^2\right)} \qquad (11.19)$$

where A_o is the low-frequency gain in the pass-band and a_1 and b_1 are positive real filter coefficients. These coefficients define the location of the complex poles of the filter and thereby determine the characteristics of the filter transfer function. The angular frequency $\omega_c = 2\pi f_c$ represents the cut-off frequency of the filter. This transfer function can be normalized by setting $\omega_c = 1$ giving

$$A(s) = \frac{A_o}{(1 + a_1 s + b_1 s^2)} \qquad (11.20)$$

Several filter types are available: (i) *Butterworth* response, maximum pass-band flatness; (ii) *Chebyshev* response, sharpened transition from pass-band to stop-band; (iii) *Bessel* response, linearized phase response within the pass-band; (iv) *inverse Chebyshev* response, flat pass-band and ripples in the stop-band; and (v) *elliptic (Cauer)* response, optimal approximation to the ideal low-pass filter with ripples in the pass-band and the stop-band. These filter responses occur for second- and higher-order and not for first-order where complex poles are not possible, and therefore all filter types have identical responses for a first-order low-pass filter (corresponding to $a_1 = 1$ in (11.11)). Only the Butterworth, Chebyshev and Bessel are considered in this chapter.

The amplitude responses of first-, second- and third-order *Butterworth* low-pass filters using the normalized characteristic are shown in Fig. 11.12. The filter provides maximum flatness in the pass-band, and as the order increases, the pass-band flatness occurs for higher frequencies before the break frequency. This type of filter is therefore ideal for audio systems where high-frequency noise needs to be removed with maximum pass-

Fig. 11.12 Low-pass filter response plots for Butterworth filters

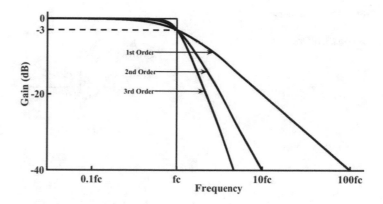

Fig. 11.13 Gain responses of fourth order low-pass filters

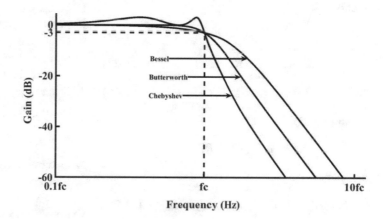

band flatness in order to preserve signal quality. The *Chebyshev* low-pass filter provides increased roll-off rate as compared with the Butterworth filter. The price paid for this is constant amplitude ripples in the pass-band. The amplitude of these ripples for a filter of given order may be varied by changing the filter coefficients; the higher the ripple amplitude, the steeper the roll-off rate. The *Bessel* low-pass filters have a linear phase response as compared with the Butterworth and the Chebyshev filters resulting in constant group delay. The pass-band gain however is not at flat as the Butterworth though it is flatter than the Chebyshev and the roll-off rate is less than both the Chebyshev and the Butterworth. A graphical comparison of these three filter responses is presented in Fig. 11.13.

The second-order low-pass filter transfer function (11.19) can be realized around a single active device using two topologies, the Sallen-Key and the multiple feedback.

11.3.1   Sallen-Key or Voltage-Controlled Voltage Source (VCVS) Topology

The circuit shown in Fig. 11.14 is one implementation of a second-order low-pass filter referred to as the Sallen-Key or VCVS topology. Using KCL, the node equations at junctions A and B are

$$\frac{V_i - V_A}{R_1} = \frac{V_A - V_B}{R_2} + (V_A - V_o)sC_2 \quad (11.21)$$

$$\frac{V_A - V_B}{R_2} = V_B sC_1 \quad (11.22)$$

where $V_B = V_o$. Eliminating V_A and V_B yields

$$A_f = \frac{V_o}{V_i}$$

$$= \frac{1}{1 + sC_1(R_1 + R_2) + s^2 R_1 R_2 C_1 C_2} \quad (11.23)$$

Fig. 11.14 Sallen-Key
second-order unity-gain
low-pass filter

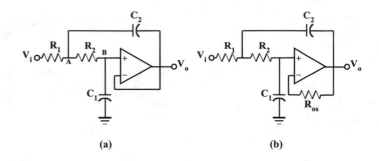

(a)   (b)

Comparing coefficients for (11.23) and (11.19)
yields

$$A_o = 1 \qquad (11.24)$$

$$a_1/\omega_c = C_1(R_1 + R_2) \qquad (11.25)$$

$$b_1/\omega_c^2 = R_1 R_2 C_1 C_2 \qquad (11.26)$$

From (11.25),

$$R_2 = a_1/\omega_c C_1 - R_1 \qquad (11.27)$$

and hence

$$b_1/\omega_c^2 = R_1 C_1 C_2(a_1/\omega_c C_1 - R_1) \qquad (11.28)$$

After manipulation, this leads to a quadratic
equation in R_1 given by

$$\omega_c^2 C_1 C_2 R_1^2 - \omega_c a_1 C_2 R_1 + b_1 = 0 \qquad (11.29)$$

Solving for R_1 we get

$$R_1 = \frac{a_1 C_2 + \sqrt{a_1^2 C_2^2 - 4b_1 C_1 C_2}}{4\pi f_c C_1 C_2} \qquad (11.30)$$

where the positive sign from the square root is
taken and

$$C_2 \geq 4b_1 C_1/a_1^2 \qquad (11.31)$$

Substituting (11.30) for R_1 in (11.27) gives

$$R_2 = \frac{a_1 C_2 - \sqrt{a_1^2 C_2^2 - 4b_1 C_1 C_2}}{4\pi f_c C_1 C_2} \qquad (11.32)$$

We can now state a design procedure for a specific cut-off frequency:

Design Procedure
1. Select C_1 usually between 100 pF and 1 μF.
2. For the specific type of filter (Butterworth,
   Chebyshev, Bessel) and the specific filter
   requirements, obtain the coefficients a_1 and
   b_1 from the relevant table.
3. Determine C_2 using $C_2 \geq 4b_1 C_1/a_1^2$.
4. Find R_1 and R_2 using $R_1 = \frac{a_1 + \sqrt{a_1^2 - 4b_1 C_1/C_2}}{4\pi f_c C_1}$
   and $R_2 = \frac{a_1 - \sqrt{a_1^2 - 4b_1 C_1/C_2}}{4\pi f_c C_1}$.
5. Determine $R_{os} = R_1 + R_2$ for minimum offset
   voltage.

Example 11.3
Design a VCVS second-order unity-gain
low-pass Butterworth filter with a cut-off frequency of 10 kHz.

Solution For this case, $f_c = 10$ kHz. Select
$C_1 = 0.001$ μF. For second-order Butterworth,
$a_1 = \sqrt{2}, b_1 = 1$. Then $C_2 \geq (4 \times 1 \times 0.01 \times 10^{-6})/$
$\left(\sqrt{2}\right)^2 = 0.02\,\mu F$. Select $C_2 = 0.05$ μF. Hence
$R_1 = \frac{1.4142 + \sqrt{2 - \left(4 \times 1 \times 0.01 \times 10^{-6}/0.05 \times 10^{-6}\right)}}{4\pi \times 10 \times 10^3 \times 0.01 \times 10^{-6}} = 2$ k
and $R_2 = \frac{1.4142 - \sqrt{2 - \left(4 \times 1 \times 0.01 \times 10^{-6}/0.01 \times 10^{-6}\right)}}{4\pi \times 10 \times 10^3 \times 0.01 \times 10^{-6}} =$
254 Ω. Finally, for minimum offset,
$R_{os} = 2000 + 254 = 2254\Omega$.

Based on the design Eqs. (11.30) and (11.32),
the resistor and capacitor values can be scaled in
order to get more appropriate values. Specifically,
$R_1, R_2 \rightarrow \alpha R_1, \alpha R_2$ where α may be referred to as
an impedance scaling factor (ISF) and C_1,
$C_2 \rightarrow C_1/\alpha$, C_2/α will leave the filter

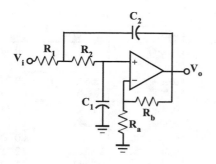

Fig. 11.15 Sallen-Key second-order low-pass filter with gain

characteristics unaffected. This can be seen by simply substituting αR_1, αR_2 and C_1/α, C_2/α in Eqs. (11.30) and (11.32) to see that the equations are unchanged. Also, the system can be designed for gain as shown in Fig. 11.15 where

$$k = 1 + R_b/R_a \qquad (11.33)$$

The process of developing the associated design procedure follows that used for the unity-gain case.

Example 11.4
Develop a design procedure for the VCVS Butterworth second-order low-pass filter with gain for the special case when $R_1 = R_2$ and $C_1 = C_2$.

Solution Using nodal analysis, the transfer function for the filter with gain is given by

$$A_f = \frac{V_o}{V_i}$$

$$= \frac{k}{1 + s(C_1 R_1 + C_1 R_2 + C_2 R_1 - kC_2 R_1) + s^2 R_1 R_2 C_1 C_2} \qquad (11.34)$$

For the case where $R_1 = R_2 = R$ and $C_1 = C_2 = C$, then (11.34) reduces to

$$A_f = \frac{V_o}{V_i} = \frac{k}{1 + s(3 - k)CR + s^2 R^2 C^2} \qquad (11.35)$$

Comparing coefficients with (11.19)

$$A(s) = \frac{A_o}{\left(1 + a_1(s/\omega_c) + b_1(s/\omega_c)^2\right)} \qquad (11.36)$$

yields

$$A_o = k \qquad (11.37)$$

$$a_1/\omega_c = (3 - k)CR \qquad (11.38)$$

$$b_1/\omega_c{}^2 = R^2 C^2 \qquad (11.39)$$

For a Butterworth filter, $a_1 = \sqrt{2}$ and $b_1 = 1$. Hence from (11.39),

$$\omega_c = 1/RC, f_c = 1/2\pi RC \qquad (11.40)$$

and from (11.38)

$$\sqrt{2} = 3 - k \qquad (11.41)$$

giving

$$k = 3 - \sqrt{2} = 1.6 \qquad (11.42)$$

For minimum offset

$$R_1 + R_2 = R_a//R_b = \frac{R_a R_b}{R_a + R_b} = \frac{R_b}{k} \qquad (11.43)$$

Hence

$$R_b = k(R_1 + R_2) = 3.2R \qquad (11.44)$$

Also

$$R_1 + R_2 = R_a//R_b = \frac{R_a R_b}{R_a + R_b} = R_a \frac{(k - 1)}{k} \qquad (11.45)$$

Hence

$$R_a = \frac{k(R_1 + R_2)}{k - 1} = 3.2R/0.6 = 5.3R \qquad (11.46)$$

These equations enable the determination of R_a and R_b to satisfy both the gain requirement and the minimum offset requirement. We can now state a design procedure for a specific cut-off frequency:

Table 11.1   Butterworth, Bessel and Chebyshev (0.5 dB) coefficients

n	i	a_i(Bu)	b_i(Bu)	a_i(Be)	b_i(Be)	a_i(.5 dB)	b_i(.5 dB)
1	1	1.0000	0.0000	1.000	0.0000	1.0000	0.0000
2	1	1.4142	1.0000	1.3617	0.6180	1.3614	1.3827
3	1	1.0000	0.0000	0.7560	0.0000	1.8636	0.0000
	2	1.0000	1.0000	0.9996	0.4772	0.0640	1.1931
4	1	1.8478	1.0000	1.3397	0.4889	2.6282	3.4341
	2	0.7654	1.0000	0.7743	0.3890	0.3648	1.1509
5	1	1.0000	0.0000	0.6656	0.0000	2.9235	0.0000
	2	1.6180	1.0000	1.1402	0.4128	1.3025	2.3534
	3	0.6180	1.0000	0.6216	0.3245	0.2290	1.0833
6	1	1.9319	1.0000	1.2217	0.3887	3.8645	6.9797
	2	1.4142	1.0000	0.9686	0.3505	0.7528	1.8573
	3	0.5176	1.0000	0.5131	0.2756	0.1589	1.0711
7	1	1.0000	0.0000	0.5937	0.000	4.0211	0.0000
	2	1.8019	1.0000	1.0944	0.3395	1.8729	4.1795
	3	1.2470	1.0000	0.8304	0.3011	0.4861	1.5676
	4	0.4450	1.0000	0.4332	0.2381	0.1156	1.0443
8	1	1.9616	1.0000	1.1112	0.3162	5.1117	11.9607
	2	1.6629	1.0000	0.9754	0.2979	1.0639	2.9365
	3	1.1111	1.0000	0.7202	0.2621	0.3439	1.4206
	4	0.3902	1.0000	0.3728	0.2087	0.0885	1.0407
9	1	1.0000	0.0000	0.5386	0.0000	5.1318	0.0000
	2	1.8794	1.0000	1.0244	0.2834	2.4283	6.6307
	3	1.5321	1.0000	0.8710	0.2636	0.6839	2.2908
	4	1.0000	1.0000	0.6320	0.2311	0.2559	1.3133
	5	0.3473	1.0000	0.3257	0.1854	0.0695	1.0272
10	1	1.9754	1.0000	1.0215	0.2650	6.3648	18.3695
	2	1.7820	1.0000	0.9393	0.2549	1.3582	4.3453
	3	1.4142	1.0000	0.7815	0.2351	0.4822	1.9440
	4	0.9080	1.0000	0.5604	0.2059	0.1994	1.2520
	5	0.3129	1.0000	0.2883	0.1665	0.0563	1.0263

Design Procedure

1. Select $C_1 = C_2 = C$ usually between 100 pF and 1 μF.
2. Determine $R = 1/2\pi f_c C$.
3. Set $R_1 = R_2 = R$.
4. Find R_a and R_b using (11.46) and (11.44).

Simplified Design Procedure for Second-Order Low-Pass Unity-Gain Butterworth Filter

A simplified design procedure for the unity-gain second-order Butterworth filter can be developed. Consider the transfer function (11.33) for a low-pass second-order filter:

$$A(s) = \frac{A_o}{\left(1 + a_1(s/\omega_c) + b_1(s/\omega_c)^2\right)} \quad (11.47)$$

For unity gain, $A_o = 1$ and for a Butterworth response from Table 11.1, $a_1 = \sqrt{2}$ and $b_1 = 1$. Then setting $s = j\omega$, (11.10) becomes

$$A(jf) = \frac{1}{1 + j\sqrt{2}\frac{f}{f_c} + \left(j\frac{f}{f_c}\right)^2} \quad (11.48)$$

In the general second-order transfer function (11.23) of Fig. 11.14, if we let $R_1 = R_2 = R$ and $C_1 = C_2/2 = C$, then for $s = j\omega$, the transfer function (a) reduces to

$$\frac{V_o}{V_i}(j\omega) = \frac{1}{1 + 2RCj\omega + 2R^2C^2(j\omega)^2} \quad (11.49)$$

Substituting $\omega = 2\pi f$ and

$$f_c = 1/2\sqrt{2}\pi RC \quad (11.50)$$

yields

$$\frac{V_o}{V_i}(jf) = \frac{1}{1 + j\sqrt{2}\frac{f}{f_c} + \left(j\frac{f}{f_c}\right)^2} \quad (11.51)$$

Table 11.2 Chebyshev (1 dB, 2 dB, 3 dB) coefficients

n	i	$a_i(1\text{ dB})$	$b_i(1\text{ dB})$	$a_i(2\text{ dB})$	$b_i(2\text{ dB})$	$a_i(3\text{ dB})$	$b_i(3\text{ dB})$
1	1	1.0000	1.0000	1.0000	0.0000	1.0000	0.0000
2	1	1.3022	1.5515	1.1813	1.7775	1.0650	1.9305
3	1	2.2156	0.0000	2.7994	0.0000	3.3496	0.0000
	2	0.5442	1.2057	0.4300	1.2036	0.3559	1.1923
4	1	2.5904	4.1301	2.4025	4.9862	2.1853	5.5339
	2	0.3039	1.1697	0.2374	1.1896	0.1964	1.2009
5	1	3.5711	0.0000	4.6345	0.0000	5.6334	0.0000
	2	1.1280	2.4896	0.9090	2.6036	0.7620	2.6530
	3	0.1872	1.0814	0.1434	1.0750	0.1172	1.0686
6	1	3.8437	8.5592	3.5880	10.4648	3.2721	11.6773
	2	0.6292	1.9124	0.4925	1.9622	0.4077	1.9873
	3	0.1296	1.0766	0.0995	1.0826	0.0815	1.0861
7	1	4.9520	0.0000	6.4760	0.0000	7.9064	0.0000
	2	1.6338	4.4899	1.3258	4.7649	1.1159	4.8963
	3	0.3987	1.5834	0.3067	1.5927	0.2515	1.5944
	4	0.0937	1.0432	0.0714	1.0384	0.0582	1.0348
8	1	5.1019	14.7608	4.7743	18.1510	4.3583	20.2948
	2	0.8916	3.0426	0.6991	3.1353	0.5791	3.1808
	3	0.2806	1.4334	0.2153	1.4449	0.1765	1.4507
	4	0.0717	1.0432	0.0547	1.0461	0.0448	1.0478
9	1	6.3415	0.0000	8.3198	0.0000	10.1759	0.0000
	2	2.1252	7.1711	1.7299	7.6580	1.4585	7.8971
	3	0.5624	2.3278	0.4337	2.3549	0.3561	2.3651
	4	0.2076	1.3166	0.1583	1.3174	0.1294	1.3165
	5	0.0562	1.0258	0.427	1.0232	0.0348	1.0210
10	1	6.3634	22.7468	5.9618	28.0376	5.4449	31.3788
	2	1.1399	4.5167	0.8947	4.6644	0.7414	4.7363
	3	0.3939	1.9665	0.3023	1.9858	0.2479	1.9952
	4	0.1616	1.2569	0.1233	1.2614	0.1008	1.2638
	5	0.0455	1.0277	0.0347	1.0294	0.0283	1.0304

Clearly, transfer function (11.51) is identical to transfer function (11.48), the transfer function of a second-order unity-gain Butterworth filter. Hence a simplified design procedure for a unity-gain second-order Butterworth filter is as follows:

1. Choose the desired cut-off frequency f_c.
2. Choose C_1, usually an available value between about 100 pF and 1 μF.
3. Let $C_2 = 2C_1$.
4. Determine R using $R = 1/2\sqrt{2}\pi f C_1$ and then set $R_1 = R_2 = R$.
5. For minimum offset use $R_{os} = 2R$.

Example 11.5
Using the configuration shown in Fig. 11.14, re-design the second-order Butterworth filter of Example 11.5 using the simplified approach.

Solution Choose $C_1 = 0.01$ μF and then $C_2 = 0.02$ μF. Using (11.36), $R = 1/2\sqrt{2}\pi \times 10 \times 10^3 \times 0.01 \times 10^{-6} = 1.1\text{k}$ and hence $R_1 = R_2 = 1.1$ k. $R_{os} = 1.1$ k $\times 2 = 2.2$ k.

Example 11.6
Design a VCVS second-order unity-gain low-pass Chebyshev filter with a cut-off frequency of 10 kHz and a ripple width $RWdB = 2$ dB.

Solution For this case, $f_c = 10$ kHz. Select $C_1 = 0.01$ μF. From Table 11.2, the coefficients for $RWdB = 2$ dB are $a_1 = 1.1813$, $b_1 = 1.7775$. Then from (11.31), $C_2 \geq (4 \times 1.7775 \times 0.01 \times 10^{-6})/(1.1813)^2 = 0.05$ μF. Select $C_2 = 0.1$ μF. Hence from (11.30) $R_1 = $

$$\frac{1.1813+\sqrt{(1.1813)^2-\left(4\times1.7775\times0.01\times10^{-6}/0.1\times10^{-6}\right)}}{4\pi\times10\times10^3\times0.01\times10^{-6}}=$$

1598 Ω and from (11.32) R_2 =

$$\frac{1.1813-\sqrt{(1.1813)^2-\left(4\times1.7775\times0.01\times10^{-6}/0.1\times10^{-6}\right)}}{4\pi\times10\times10^3\times0.01\times10^{-6}}=$$

282 Ω. Finally, for minimum offset, $R_{os}=1598+282=1880$ Ω.

Example 11.7
Design a VCVS second-order unity-gain low-pass Bessel filter with a cut-off frequency of 10 kHz.

Solution For this case, f_c = 10 kHz. Select C_1 = 0.01 μF. From Table 11.1, the Bessel coefficients are a_1 = 1.3617, b_1 = 0.6180. Then $C_2 \geq (4 \times 0.6180 \times 0.01 \times 10^{-6})/(1.3617)^2$ = 0.01333 μF. Select C_2 = 0.02 μF. Hence R_1

$$=\frac{1.3617+\sqrt{(1.3617)^2-\left(4\times0.6180\times0.01\times10^{-6}/0.02\times10^{-6}\right)}}{4\pi\times10\times10^3\times0.01\times10^{-6}}$$

=   1700   Ω   and R_2 =

$$\frac{1.3617-\sqrt{(1.3617)^2-\left(4\times0.6180\times0.01\times10^{-6}/0.02\times10^{-6}\right)}}{4\pi\times10\times10^3\times0.01\times10^{-6}}=$$

458 Ω. Finally, for minimum offset, $R_{os}=1700+458=2158$ Ω.

11.3.2 Low-Pass Multiple Feedback Topology

Another available topology for implementation of second-order low-pass filters is the multiple feedback (MFB) topology shown in Fig. 11.16. In this circuit there are multiple feedback paths, and the active element is operating in a high-gain mode. The amplifier introduces 180° of phase shift in addition to that introduced by the network elements. Assuming an ideal op-amp as the active element, then for nodes a and b,

$$\frac{V_i - V_a}{R_1} = V_a s C_2 + \frac{V_a - V_b}{R_3} + \frac{V_a - V_o}{R_2} \tag{11.52}$$

$$\frac{V_a - V_b}{R_3} = (V_b - V_o)sC_1 \tag{11.53}$$

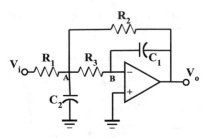

Fig. 11.16 Multiple feedback low-pass filter

Noting that $V_b = 0$ because node b is a virtual earth and eliminating V_a, we get

$$\frac{V_o}{V_i} = -\frac{\frac{R_2}{R_1}}{1 + C_1\left(R_2 + R_3 + \frac{R_2 R_3}{R_1}\right)s + C_1 C_2 R_2 R_3 s^2} \tag{11.54}$$

Comparing the coefficients of (11.54) with the general second-order low-pass transfer function

$$A(s) = -\frac{A_o}{1 + a_1(s/\omega_c) + b_1(s/\omega_c)^2} \tag{11.55}$$

(where a negative sign is introduced to account for the inversion produced by the MFB configuration) yields

$$A_o = R_2/R_1 \tag{11.56}$$

$$\frac{a_1}{\omega_c} = C_1\left(R_2 + R_3 + \frac{R_2 R_3}{R_1}\right) \tag{11.57}$$

$$\frac{b_1}{\omega_c^2} = C_1 C_2 R_2 R_3 \tag{11.58}$$

Solving for R_1, R_2, R_3 in terms of C_1 and C_2, we get

$$R_1 = \frac{a_1 + \sqrt{a_1^2 - 4b_1(1 + A_o)C_1/C_2}}{4\pi f_c C_1 A_o} \tag{11.59}$$

In order to ensure a positive value under the square root and hence a real solution for R_1, then

$$C_2 \geq C_1 \frac{4b_1(1 + A_o)}{a_1^2} \tag{11.60}$$

Using (11.59) in (11.56) and (11.58), we get

$$R_2 = A_o R_1 \qquad (11.61)$$

and

$$R_3 = \frac{b_1}{4\pi^2 f_c^2 C_1 C_2 R_2} \qquad (11.62)$$

Design Procedure

1. Select C_1 usually between 100 pF and 1 μF.
2. For the specific type of filter (Butterworth, Chebyshev, Bessel) and the specific filter requirements, obtain the coefficients a_1 and b_1 from the relevant table.
3. Determine C_2 to satisfy $C_2 \geq C_1 \frac{4b_1(1+A_o)}{a_1^2}$.
4. Find R_1, R_2 and R_3 using

$R_1 = \frac{a_1 + \sqrt{a_1^2 - 4b_1(1+A_o)C_1/C_2}}{4\pi f_c C_1 A_o}$, $R_2 = A_o R_1$ and

$R_3 = \frac{b_1}{4\pi^2 f_c^2 C_1 C_2 R_2}$.

5. Scale values if necessary.

Example 11.8
Design a MFB second-order low-pass Butterworth filter with a gain of 10 and a cut-off frequency of 10 kHz.

Solution For this case, $f_c = 10$ kHz. Select $C_1 = 0.001$ μF. From Table 11.1, the Butterworth coefficients are $a_1 = 1.4142$, $b_1 = 1$. Then $C_2 \geq C_1 \frac{4b_1(1+A_o)}{a_1^2}$

$C_2 \geq 0.001 \times 10^{-6} \times 4 \times 1 \times (1 + 10)/ (\sqrt{2})^2 = 0.022$ μF. Select $C_2 = 0.1$ μF. Hence

$R_1 = \frac{1.4142 + \sqrt{(\sqrt{2})^2 - (4 \times 1 \times 11 \times 0.001 \times 10^{-6}/0.1 \times 10^{-6})}}{4\pi \times 10 \times 10^3 \times 0.001 \times 10^{-6} \times 10}$
$= 2.12$ k, $R_2 = A_o R_1 = 2.12$ k and $R_3 = \frac{1}{4 \times \pi \times (10^4)^2 \times 0.001 \times 10^{-6} \times 0.1 \times 10^{-6} \times 21.2 \times 10^3} = 119.5$ Ω.

11.4   Higher-Order Low-Pass Filters

We wish to consider those higher-order filters in which the transfer function has a denominator polynomial that is of degree n corresponding to the order of the filter and a numerator polynomial with degree $m < n$. There are two general methods for realizing these higher-order transfer functions. The first is to synthesize the filter around a single active element such as an op-amp. This works well up to about $n = 3$ but problems arise for filters of higher order. The second general method is to factor the higher-order transfer function $A(s)$ into first- and second-order transfer functions $A_1(s)$ and $A_2(s)$, each of which can be synthesized and then cascaded. This approach has the advantage of simplicity, and the resulting first- and second-order transfer functions are those already analysed and used in filter implementation. In this chapter, the factorizing approach is pursued in the development of higher-order filters.

Therefore, the approach adopted here in realizing higher-order low-pass filters is to cascade first- and/or second-order filters in order to approximate the higher-order filter. Thus, in order to realize a fourth-order filter, two second-order filters are cascaded. In order to realize an odd-order filter, at least one first-order must be cascaded with one or more second-order filters. For example, in order to synthesize a third-order filter, one first-order and one second-order filters are cascaded. Similarly, if a fifth-order filter is required, then one first-order and two second-order filters must be cascaded.

The transfer function of one stage is given by

$$A(s) = \frac{A_o}{\left(1 + a_1(s/\omega_c) + b_1(s/\omega_c)^2\right)} \qquad (11.63)$$

In a first-order transfer function, the coefficient b_1 is zero giving the transfer function

$$A(s) = \frac{A_o}{1 + a_1(s/\omega_c)} \qquad (11.64)$$

seen earlier. Therefore, the general transfer function representing this cascade of second- and first-order transfer functions is given by

$$A(s) = \frac{A_o}{\prod_i \left(1 + a_i(s/\omega_c) + b_i(s/\omega_c)^2\right)}$$

$$= \frac{A_o}{\left(1 + a_1(s/\omega_c) + b_1(s/\omega_c)^2\right)\left(1 + a_2(s/\omega_c) + b_2(s/\omega_c)^2\right)\ldots\left(1 + a_n(s/\omega_c) + b_n(s/\omega_c)^2\right)}$$

$$(11.65)$$

The multiplication of the factors in the denominator of the transfer function yields an nth-order polynomial where n represents the order of the filter. The filter coefficients a_i and b_i determine the filter characteristics such as Butterworth, Chebyshev and Bessel, the types being considered in this discussion. These are tabulated in Table 11.1 up to order 10. This cascading of

first- and second-order stages is shown diagrammatically in Fig. 11.17 up to the sixth order. As can be seen, a filter with even order comprises only second-order stages, while a filter with odd order has a first-order stage included.

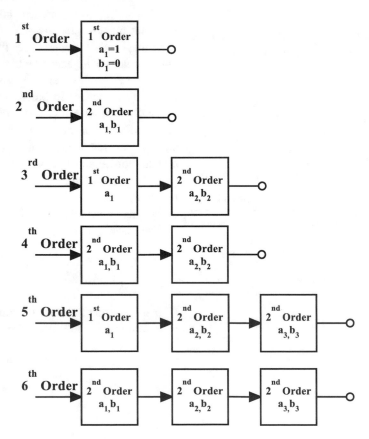

Fig. 11.17 Cascading first- and second-order stages to realize higher-order filters

11.4.1 Third-Order Low-Pass Unity-Gain Filter

In order to develop a third-order unity-gain low-pass filter, we cascade first- and second-order filters such that

$$A(s) = \frac{A_o}{\left(1 + a_1(s/\omega_c)\right)\left(1 + a_2(s/\omega_c) + b_2(s/\omega_c)^2\right)}$$

(11.66)

Note that $b_1 = 0$ for the first-order part of the transfer function. For the unity-gain system under consideration, $A_o = 1$. Hence (11.66) can be written as

$$A(s) = \frac{1}{1 + a_1(s/\omega_c)} \cdot \frac{1}{\left(1 + a_2(s/\omega_c) + b_2(s/\omega_c)^2\right)}$$

(11.67)

The circuit of this system is shown in Fig. 11.18, comprising a unity-gain non-inverting first-order system cascaded with a unity-gain non-inverting second-order system. The design process is as follows:

Partial Filter 1
The first-order system is

$$A(s) = \frac{1}{1 + a_1(s/\omega_c)}$$

(11.68)

where $\omega_c = 2\pi f_c$ is the cut-off frequency. The transfer function for the first-order *non-inverting* unity-gain low-pass filter in Fig. 11.19 is given by

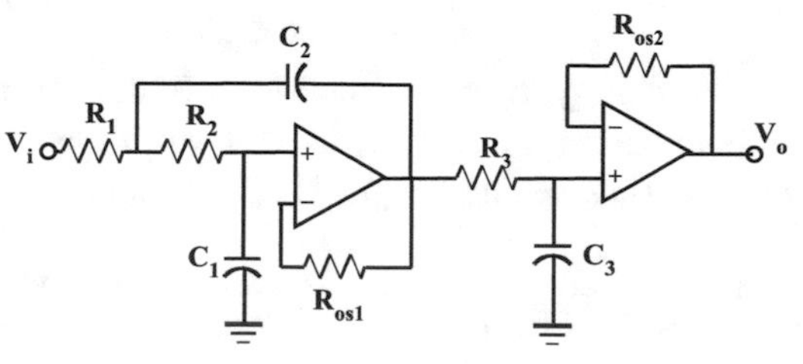

$$A_f(s) = \frac{1}{1 + R_3 C_3 s}$$

(11.69)

Comparing coefficients in (11.68) and (11.69) gives

$$a_1/2\pi f_c = R_3 C_3$$

(11.70)

The design steps are as follows:
1. Select C_3 usually between 100 pF and 1 μF.
2. For the specific type of filter (Butterworth, Chebyshev, Bessel) and the specific filter requirements, obtain the coefficient $a_1(b_1 = 0)$ from the relevant table.
3. Determine R using $R_3 = a_1/2\pi f_c C$.
4. Set $R_{os2} = R$ for minimum offset.

Partial Filter 2
The second-order system is

$$A(s) = \frac{1}{\left(1 + a_2(s/\omega_c) + b_2(s/\omega_c)^2\right)}$$

(11.71)

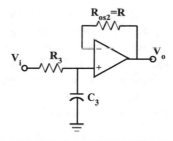

Fig. 11.19 First-order low-pass partial filter

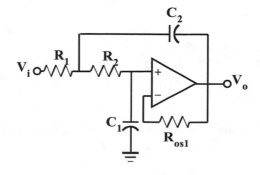

Fig. 11.18 Third-order low-pass unity-gain filter

Fig. 11.20 Second-order low-pass partial filter

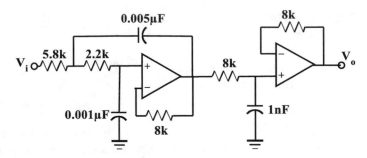

Fig. 11.21 Completed third-order low-pass Butterworth filter for Example 11.21

The transfer function for the second-order *non-inverting* unity-gain low-pass filter in Fig. 11.20 is given by

$$A_f = \frac{1}{1 + sC_1(R_1 + R_2) + s^2R_1R_2C_1C_2}$$
(11.72)

Comparing coefficients for (11.63) and (11.62) yields

$$a_2/\omega_c = C_1(R_1 + R_2)$$
(11.73)

$$b_2/\omega_c^2 = R_1R_2C_1C_2$$
(11.74)

From this, the design steps are as follows:
1. Select C_1 usually between 100 pF and 1 μF.
2. For the specific type of filter (Butterworth, Chebyshev, Bessel) and the specific filter requirements, obtain the coefficients a_2 and b_2 from the relevant table.
3. Determine C_2 using $C_2 \geq 4b_2C_1/a_2^2$.
4. Find R_1 and R_2 using $R_1 = \frac{a_2 + \sqrt{a_2^2 - 4b_2C_1/C_2}}{4\pi f_c C_1}$ and $R_2 = \frac{a_2 - \sqrt{a_2^2 - 4b_2C_1/C_2}}{4\pi f_c C_1}$.
5. Determine $R_{os1} = R_1 + R_2$ for minimum offset voltage.

Example 11.9
Design a VCVS third-order unity-gain low-pass Butterworth filter with a cut-off frequency of 20 kHz.

Solution The design of each of the two partial filters is considered separately.

Partial Filter 1

For this case, $f_c = 20$ kHz. Select $C_3 = 1$ nF. From Table 11.1, $a_1 = 1$, $b_1 = 0$. Hence $R_3 = a_1/2\pi f_c C_3 = 1/2\pi \times 20 \times 10^3 \times 10^{-9} = 8$ k.

Partial Filter 2
For this case, $f_c = 20$ kHz. Select $C_1 = 0.001$ μF. From Table 11.1, $a_2 = 1$, $b_2 = 1$. Then $C_2 \geq 4b_2C_1/a_2^2 = 4 \times 1 \times 0.001 \times 10^{-6} = 0.004$ μF. Select $C_2 = 0.005$ μF. Hence $R_1 = \frac{1 + \sqrt{1 - (4 \times 1 \times 0.001 \times 10^{-6}/0.005 \times 10^{-6})}}{4\pi \times 20 \times 10^3 \times 0.001 \times 10^{-6}} = 5.8$ k and $R_2 = \frac{1 - \sqrt{1 - (4 \times 1 \times 0.001 \times 10^{-6}/0.005 \times 10^{-6})}}{4\pi \times 20 \times 10^3 \times 0.001 \times 10^{-6}} = 2.2$ k.
Finally, for minimum offset, $R_{os} = 5.8$ k + 2.2 k = 8 k.

The completed filter is shown in Fig. 11.21.

Simplified Design Procedure for Third-Order Low-Pass Unity-Gain Butterworth Filter
A simplified design procedure for the unity-gain third-order Butterworth filter can be developed. Consider the transfer function (11.75) for a low-pass third-order filter:

$$A(s) = \frac{A_o}{(1 + a_1(s/\omega_c))(1 + a_2(s/\omega_c) + b_2(s/\omega_c)^2)}$$
(11.75)

For unity gain, $A_o = 1$ and for a Butterworth response from Table 11.1, $a_1 = 1$ and $a_2 = 1$, $b_2 = 1$. Then setting $s = j\omega$, (11.10) becomes

$$A = \frac{1}{1 + jf/f_c} \cdot \frac{1}{1 + jf/f_c + (jf/f_c)^2}$$
(11.76)

Fig. 11.22 Third-order
low-pass unity-gain
Butterworth filter

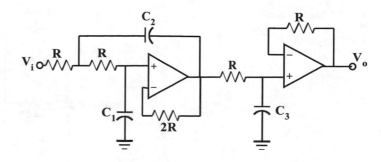

This transfer function can be implemented using the configuration shown in Fig. 11.18 which comprises cascaded first- and second-order systems. For this system, the transfer function is given by

$$A_f = \frac{1}{1 + sR_3C_3} \cdot \frac{1}{1 + sC_1(R_1 + R_2) + s^2C_1C_2R_1R_2} \quad (11.77)$$

Let $R_1 = R$, $R_2 = R$, $R_3 = R$ and $C_1 = C_3/2$ and $C_2 = 2C_3$. Then Fig. 11.18 becomes Fig. 11.22 and (11.77) becomes

$$A_f = \frac{1}{1 + sRC_3} \cdot \frac{1}{1 + sC_3R + s^2(C_3R)^2}$$

$$= \frac{1}{1 + jf/f_c} \cdot \frac{1}{1 + jf/f_c + (jf/f_c)^2} \quad (11.78)$$

where

$$f_c = 1/2\pi RC_3 \quad (11.79)$$

which is the cut-off frequency. Clearly, transfer function (11.78) is identical to transfer function (11.76). Hence a simplified design procedure for a unity-gain third-order low-pass Butterworth filter is as follows:

1. Choose the desired cut-off frequency f_c.
2. Choose C_3, usually an available value between about 100 pF and 1 μF.
3. Let $C_1 = 1/2C_3$ and $C_2 = 2C_3$.
4. Determine R using $R = 1/2\pi f_c C_3$. Set $R_1 = R_2 = R_3 = R$.

Example 11.10
Using the circuit of Fig. 11.22 and the simplified design procedure, design a third-order low-pass

Butterworth filter with a cut-off frequency of 9 kHz.

Solution Choose $C_1 = 0.005$ μF and then $C_3 = 2C_1 = 0.01$ μF and $C_2 = 2C_3 = 0.02$ μF. Using (11.70), $R = 1/2\pi f_c C_3 = 1/2\pi \times 9 \times 10^3 \times 0.01 \times 10^{-6} = 1.8$ k. Hence $R_1 = R_2 = R_3 = 1.8$ k.

Example 11.11
Design a VCVS fourth-order unity-gain low-pass Chebyshev filter with a cut-off frequency of 20 kHz and 1 dB ripple width.

Solution The design of each of the two partial filters is considered separately.

Partial Filter 1
For this case, $f_c = 20$ kHz. Select $C_1 = 0.001$ μF. From Table 11.2, $a_1 = 2.5904$, $b_1 = 4.1301$. Then $C_2 \geq 4b_1C_1/a_1^2 = 4 \times 4.1301 \times 0.001 \times 10^{-6}/(2.5904)^2 = 0.0025$ μF. Select $C_2 = 0.005$ μF. Hence R_1

$$= \frac{2.5904 + \sqrt{(2.5904)^2 - (4 \times 4.1301 \times 0.001 \times 10^{-6}/0.005 \times 10^{-6})}}{4\pi \times 20 \times 10^3 \times 0.001 \times 10^{-6}}$$

$$= 9.9 \quad \text{k} \quad \text{and} \quad R_2 =$$

$$\frac{2.5904 - \sqrt{(2.5904)^2 - (4 \times 4.1301 \times 0.001 \times 10^{-6}/0.005 \times 10^{-6})}}{4\pi \times 20 \times 10^3 \times 0.001 \times 10^{-6}} =$$

2.6 k. Finally, for minimum offset, $R_{os} = 9.9$ k + 2.6 k = 12.5 k.

Partial Filter 2
For this case, $f_c = 20$ kHz. Select $C_1 = 0.001$ μF. From Table 11.1, $a_2 = 0.3039$, $b_2 = 1.1697$. Then $C_2 \geq 4b_2C_1/a_2^2 = 4 \times 1.1697 \times 0.001 \times 10^{-6}/(0.3039)^2 = 0.05$ μF. Select $C_2 = 0.1$ μF. Hence R_1

Fig. 11.23 Completed third-order low-pass filter for Example 11.13

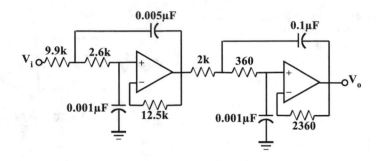

Fig. 11.24 Completed fifth-order low-pass filter for Example 11.14

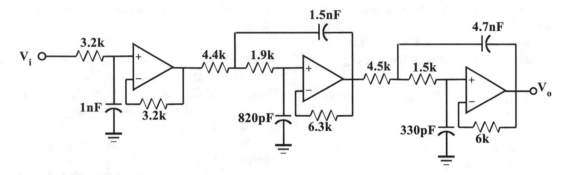

$$= \frac{0.3039 + \sqrt{(0.3039)^2 - \left(4 \times 1.1697 \times 0.001 \times 10^{-6}/0.1 \times 10^{-6}\right)}}{4\pi \times 20 \times 10^3 \times 0.001 \times 10^{-6}}$$
$$= \quad 2 \qquad \text{k} \qquad \text{and} \qquad R_2 \quad =$$
$$\frac{0.3039 - \sqrt{(0.3039)^2 - \left(4 \times 1.1697 \times 0.001 \times 10^{-6}/0.1 \times 10^{-6}\right)}}{4\pi \times 20 \times 10^3 \times 0.001 \times 10^{-6}} =$$
360 Ω. Finally, for minimum offset, $R_{os} = 2$ k + 360 Ω = 2360 Ω.

The completed filter is shown in Fig. 11.23.

Example 11.12
Design a VCVS fifth-order unity-gain low-pass Butterworth filter with a cut-off frequency of 50 kHz.

Solution The design of each of the three partial filters is considered separately.

Partial Filter 1
For this case, $f_c = 50$ kHz. Select $C = 1$ nF. From Table 11.1, $a_1 = 1$, $b_1 = 0$. Hence $R = a_1/2\pi f_c C = 1/2\pi \times 50 \times 10^3 \times 10^{-9} = 3.2$ k.

Partial Filter 2

For this case, $f_c = 50$ kHz. Select $C_1 = 820$ pF. From Table 11.1, $a_2 = 1.6180$, $b_2 = 1$. Then $C_2 \geq 4b_2 C_1/a_2^2 = 4 \times 1 \times 820 \times 10^{-12}/(1.618)^2 = 1.26$ nF. Select $C_2 = 1.5$ nF. Hence
$$R_1 = \frac{1.618 + \sqrt{(1.618)^2 - \left(4 \times 1 \times 0.820 \times 10^{-9}/1.5 \times 10^{-9}\right)}}{4\pi \times 50 \times 10^3 \times 0.820 \times 10^{-9}}$$
$$= \quad 4.4 \qquad \text{k} \qquad \text{and} \qquad R_2 \quad =$$
$$\frac{1.618 - \sqrt{(1.618)^2 - \left(4 \times 1 \times 0.820 \times 10^{-9}/1.5 \times 10^{-9}\right)}}{4\pi \times 50 \times 10^3 \times 0.820 \times 10^{-9}} = 1.9 \text{ k}.$$
Finally, for minimum offset, $R_{os} = 4.4$ k + 1.9 k = 6.3 k.

Partial Filter 3
For this case, $f_c = 50$ kHz. Select $C_1 = 330$ pF. From Table 11.1, $a_3 = 0.6180$, $b_3 = 1$. Then $C_2 \geq 4b_3 C_1/a_3^2 = 4 \times 1 \times 330 \times 10^{-12}/(0.6180)^2 = 3.5$ nF. Select $C_2 = 4.7$ nF. Hence
$$R_1 = \frac{0.618 + \sqrt{(0.618)^2 - \left(4 \times 1 \times 0.330 \times 10^{-9}/4.7 \times 10^{-9}\right)}}{4\pi \times 50 \times 10^3 \times 0.330 \times 10^{-9}}$$
$$= \quad 4.5 \qquad \text{k} \qquad \text{and} \qquad R_2 \quad =$$
$$\frac{1.618 - \sqrt{(0.618)^2 - \left(4 \times 1 \times 0.330 \times 10^{-9}/4.7 \times 10^{-9}\right)}}{4\pi \times 50 \times 10^3 \times 0.330 \times 10^{-9}} = 1.5 \text{ k}.$$

Finally, for minimum offset, $R_{os} = 4.5$ k $+ 1.5$ k $= 6$ k.

The completed filter is shown in Fig. 11.24.

11.5   High-Pass First-Order Filter-Butterworth Response

High-pass filters can be generated from low-pass filters by changing resistors to capacities and capacities to resistors in the filter network. This corresponds to the transformation $s \rightarrow 1/s$ or $j(f/f_c) \rightarrow \frac{1}{j(f/f_c)}$. Thus, the first-order low-pass filter given by $\frac{1}{1+jf/f_c}$ is transformed such that $\frac{1}{1+jf/f_c} \rightarrow \frac{1}{1+\frac{1}{jf/f_c}} = \frac{jf/f_c}{1+jf/f_c}$. A basic first-order high-pass filter is shown in Fig. 11.25(a). It employs a CR network to provide the high-pass filtering, the output of which feeds into a unity-gain buffer. It is obtained from the first-order low-pass filter by interchanging R and C in the filter network. A resistor $R_{OS} = R$ may be

included in the feedback loop in order to reduce DC offset at the output as shown in Fig. 11.25(b). For the circuit

$$A_f = \frac{V_o}{V_i}(s) = \frac{R}{R + 1/sC} = \frac{sCR}{1 + sCR}$$

$$= \frac{jf/f_c}{1 + jf/f_c} \qquad (11.80)$$

where

$$f_c = 1/2\pi RC \qquad (11.81)$$

Note that for $f \gg f_c$, $|A_f| \rightarrow 1$ and for $f \ll f_c$, $|A_f| \rightarrow 0$. The frequency response plot for this circuit is shown in Fig. 11.26, where the cut-off frequency is $f_c = 1/2\pi RC$. At $f = f_c$, the filter transfer function becomes

$$A_f(f_c) = \frac{jf/f_c}{1 + jf/f_c} = \frac{j}{1+j} \qquad (11.82)$$

Hence, $\quad |A_f(f_c)| = \left|\frac{j}{1+j}\right| = \frac{1}{\sqrt{2}} = 0.707 = -3$ dB. Thus, at the break frequency $f = f_c$ the magnitude response falls by 3 dB from its mid-band value as shown in Fig. 11.26.

Example 11.13

Design an RC high-pass first-order Butterworth filter with a cut-off frequency of 1 kHz.

Solution Choose $C = 0.01$ μF. Then from (11.81), $R = 1/2\pi f_c C = 1/2\pi \times 10^3 \times 0.01 \times 10^{-6} = 15.9$ k. Choose $R_f = R = 15.9$ k for low DC offset.

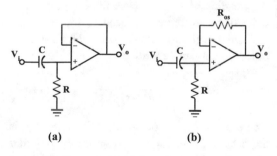

(a)   (b)

Fig. 11.25 High-pass first-order filter

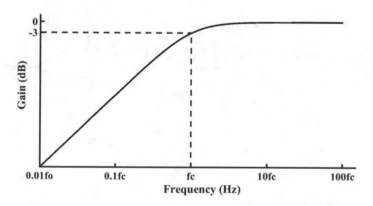

Fig. 11.26 First-order high-pass filter response

11.5.1 First-Order High-Pass Filter with Gain

The high-pass first-order filter with gain is shown in Fig. 11.27 where the active device is modified to produce a gain greater than unity. The transfer function is given by

$$A_f = \frac{V_o}{V_i} = \frac{kjf/f_c}{1 + jf/f_c} \qquad (11.83)$$

where $k = 1 + R_2/R_1$ and $f_c = 1/2\pi RC$. The system has a high-frequency gain of $k = 1 + R_2/R_1$ and a break frequency of $f_c = 1/2\pi RC$. For low DC offset, $R = R_1//R_2$.

An alternative configuration for a first-order high-pass filter is shown in Fig. 11.28. It is an inverting configuration with a capacitor C in series with the input resistor R_1. From Fig. 11.28, we have

$$\frac{V_i}{R_1 + 1/sC} = \frac{-V_o}{R_2} \qquad (11.84)$$

Hence the transfer function is given by

$$A_f = -\frac{R_2}{R_1} \cdot \frac{sCR_1}{1 + sCR_1} = -k\frac{jf/f_c}{1 + jf/f_c} \qquad (11.85)$$

where $k = R_2/R_1$ is the gain and

$$f_c = 1/2\pi R_1 C \qquad (11.86)$$

is the break frequency.

Example 11.14

Design a first-order high-pass Butterworth non-inverting filter using Fig. 11.27 with a gain of 10 and a cut-off of 1 kHz.

Solution Choose $C = 0.01$ μF. Then from (11.86), $R = 1/2\pi fC$ giving $R = 1/(2 \times \pi \times 10^3 \times 0.01 \times 10^{-6}) = 15.9$ k. Since $1 + R_2/R_1 = 10$ and $R = R_1//R_2$, therefore $R_2 = 159$ k and $R_1 = 17.7$ k.

Design Procedure

As we did for the low-pass filters, we wish to write down an ordered procedure for designing first-order high-pass filters. Thus, the basic first-order system can be written as

$$A(s) = \frac{A_\infty}{1 + a_1/(s/\omega_c)} \qquad (11.87)$$

where $A_\infty = k$ is the pass-band gain, $\omega_c = 2\pi f_c$ is the cut-off frequency and $a_1 = 1$ for first-order filters. For higher-order filters considered later,

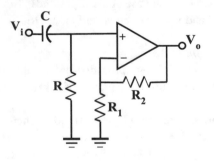

Fig. 11.27 First-order high-pass filter with gain

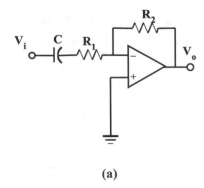

(a)

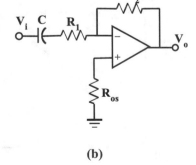

(b)

Fig. 11.28 Inverting first-order high-pass filter

the condition $a_1 \neq 1$ will arise. Equation (11.87) can be written as

$$A(s) = \frac{A_\infty s}{s + a_1 \omega_c} \qquad (11.88)$$

For the first-order *non-inverting* high-pass filter, the transfer function is given by (see earlier)

$$A_f(s) = \frac{\left(1 + \frac{R_2}{R_1}\right)}{s + 1/RC} \qquad (11.89)$$

Comparing coefficients in (11.89) and (11.88) gives

$$A_\infty = 1 + \frac{R_2}{R_1} \qquad (11.90)$$

$$a_1 \omega_c = \frac{1}{RC} \qquad (11.91)$$

The design steps are as follows:
1. Select C usually between 100 pF and 1 μF.
2. Set the coefficient $a_1 = 1$.
3. Determine R using $R = 1/a_1 2\pi f_c C$.
4. Select R_1 and find R_2 using $R_2 = R_1(A_\infty - 1)$.

For the first-order *inverting* low-pass filter, the transfer function is given by (see earlier)

$$A(s) = -\frac{\frac{R_2}{R_1} s}{s + 1/R_1 C_1} \qquad (11.92)$$

Comparing coefficients in (11.92) with the general transfer function

$$A(s) = -\frac{A_\infty s}{s + a_1 \omega_c} \qquad (11.93)$$

gives equation (11.94):

$$A_\infty = \frac{R_2}{R_1} \qquad (11.94)$$

$$a_1 \omega_c = \frac{1}{R_1 C_1} \qquad (11.95)$$

The design steps are as follows:
1. Select C_1 usually between 100 pF and 1 μF.
2. Set the coefficient $a_1 = 1$.
3. Determine R_1 using $R_1 = 1/a_1 2\pi f_c C_1$.

4. Determine R_1 using $R_1 = R_2/A_\infty$.

11.6 High-Pass Second-Order Filter

As in the case of second-order low-pass filters, a high-pass filter can be optimized to meet at least one of the following three requirements: (i) maximum flatness in the pass-band; (ii) maximum sharpness in the transition from pass-band to stop-band; and (iii) linear phase response. In order to allow this optimization of the filter, the transfer function must contain complex poles and hence, for a second-order filter, can take the following form:

$$A(s) = \frac{A_\infty s^2}{s^2 + a_1 \omega_c s + b_1 \omega_c^2} \qquad (11.96)$$

where A_∞ is the low-frequency gain in the pass-band and a_1 and b_1 are positive real filter coefficients that define the location of the complex poles of the filter and hence the characteristics of the filter. The angular frequency $\omega_c = 2\pi f_c$ represents the cut-off frequency of the filter. This transfer function can be normalized by setting $\omega_c = 1$ giving

$$A(s) = \frac{A_\infty s^2}{s^2 + a_1 s + b_1} \qquad (11.97)$$

Only the **Butterworth, Chebyshev** and **Bessel** responses are considered. These filter responses occur for second (and higher) order and not for first-order where complex poles are not possible and therefore all three filter types have identical responses for a first-order high-pass filter. The amplitude responses of first-, second- and higher-order high-pass *Butterworth* filters using the normalized characteristic are shown in Fig. 11.29. The second-order high-pass filter transfer function (11.97) can be realized around a single active device using two topologies, the Sallen-Key and the multiple feedback.

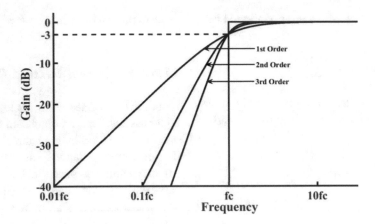

Fig. 11.29 High-pass filter response plots for Butterworth filters

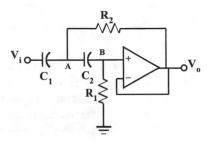

Fig. 11.30 Second-order high-pass unity-gain filter

11.6.1   Sallen-Key or Voltage-Controlled Voltage Source (VCVS) Topology

The circuit shown in Fig. 11.30 is the VCVS or Sallen-Key implementation of a second-order high-pass unity-gain filter. Comparing this with the second-order low-pass filter, it can be seen that the resistors and capacitors of the filter network are interchanged. Using KCL at nodes A and B, we have

$$(V_i - V_A)sC_1 = (V_A - V_B)sC_2 + (V_A - V_o)/R_2 \tag{11.98}$$

$$(V_A - V_B)sC_2 = V_B/R_1 \tag{11.99}$$

where $V_B = V_o$. These equations yield

$$A(s) = \frac{s^2 C_1 C_2 R_1 R_2}{1 + sR_2(C_1 + C_2) + s^2 R_1 R_2 C_1 C_2} \tag{11.100}$$

In order to simplify the development, let $C_1 = C_2 = C$. Then (11.100) becomes

$$A_f(s) = \frac{s^2}{s^2 + \frac{2}{R_1 C}s + \frac{1}{C^2 R_1 R_2}} \tag{11.101}$$

Comparing coefficients for (11.101) and (11.96) yields

$$A_\infty = 1 \tag{11.102}$$

$$a_1 \omega_c = \frac{2}{R_1 C} \tag{11.103}$$

$$b_1 \omega_c^2 = \frac{1}{R_1 R_2 C^2} \tag{11.104}$$

From (11.103),

$$R_1 = \frac{1}{a_1 \pi f_c C} \tag{11.105}$$

and from (11.104),

$$R_2 = \frac{a_1}{b_1 4\pi f C} \tag{11.106}$$

We can now state a design procedure for a specific cut-off frequency:

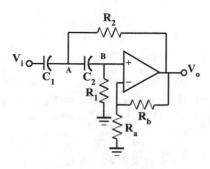

Fig. 11.31 Sallen-Key second-order high-pass filter with gain

Design Procedure

1. Select $C_1 = C_2 = C$ usually between 100 pF and 1 μF.
2. For the specific type of filter (Butterworth, Chebyshev, Bessel) and the specific filter requirements, obtain the coefficients a_1 and b_1 from the relevant table.
3. Find R_1 and R_2 using $R_1 = \frac{1}{\pi f_c a_1 C}$ and $R_2 = \frac{a_1}{4\pi f b_1 C}$.
4. Determine $R_{os} = R_1$ for minimum offset voltage if required.

Example 11.15

Design a VCVS second-order unity-gain high-pass Butterworth filter with a cut-off frequency of 10 kHz.

Solution For this case, $f_c = 10$ kHz. Select $C_1 = C_2 = 0.01$ μF. From Table 11.1, $a_1 = \sqrt{2}, b_1 = 1$. Then $R_1 = \frac{1}{\pi \times 10 \times 10^3 \times \sqrt{2} \times 0.01 \times 10^{-6}} = 2.3$ k and $R_2 = \frac{\sqrt{2}}{4\pi \times 10 \times 10^3 \times 0.01 \times 10^{-6}} = 1.1$ k. Finally, for minimum offset, $R_{os} = 2.3$ k.

The system can be designed for gain as shown in Fig. 11.31. where

$$k = 1 + \frac{R_b}{R_a} \qquad (11.107)$$

The process of developing the associated design procedure follows that used for the unity-gain case.

Example 11.16

Develop a design procedure for the VCVS Butterworth second-order high-pass filter with gain for the special case when $R_1 = R_2$ and $C_1 = C_2$.

Solution Using nodal analysis, the transfer function for the filter with gain is given by

$$A_f = \frac{V_o}{V_i}$$

$$= \frac{ks^2}{s^2 + s\left[\left(\frac{1}{C_1} + \frac{1}{C_2}\right) + \frac{1}{C_1 R_2}(1-k)\right] + \frac{1}{R_1 R_2 C_1 C_2}} \qquad (11.108)$$

For the case where $R_1 = R_2 = R$ and $C_1 = C_2 = C$, then (11.108) reduces to

$$A_f = \frac{V_o}{V_i} = \frac{ks^2}{s^2 + s\frac{1}{RC}(3-k) + \frac{1}{R^2 C^2}} \qquad (11.109)$$

Comparing coefficients with

$$A(s) = \frac{A_\infty s^2}{s^2 + a_1 \omega_c s + b_1 \omega_c^2} \qquad (11.110)$$

yields

$$A_\infty = k \qquad (11.111)$$

$$a_1 \omega_c = (3-k)CR \qquad (11.112)$$

$$b_1 \omega_c^2 = R^2 C^2 \qquad (11.113)$$

For a Butterworth filter, $a_1 = \sqrt{2}$ and $b_1 = 1$. Hence from (11.113),

$$\omega_c = 1/RC, f_c = 1/2\pi RC \qquad (11.114)$$

and from (11.112)

$$\sqrt{2} = 3 - k \qquad (11.115)$$

giving

$$k = 3 - \sqrt{2} = 1.6 \qquad (11.116)$$

For minimum offset

$$R = R_a//R_b = \frac{R_a R_b}{R_a + R_b} = \frac{R_b}{k} \quad (11.117)$$

Hence

$$R_b = kR = 1.6R \quad (11.118)$$

Also

$$R = R_a//R_b = \frac{R_a R_b}{R_a + R_b} = R_a \frac{(k-1)}{k} \quad (11.119)$$

Hence

$$R_a = \frac{kR}{k-1} = 1.6R/0.6 = 2.7R \quad (11.120)$$

We can now state a design procedure for a specific cut-off frequency:

Design Procedure

1. Select $C_1 = C_2 = C$ usually between 100 pF and 1 μF.
2. Determine $R = 1/2\pi f_c C$.
3. Set $R_1 = R_2 = R$.
4. Find R_a and R_b using (11.120) and (11.118).

Simplified Design Procedure for Second-Order High-Pass Unity-Gain Butterworth Filter
A simplified design procedure for the unity-gain second-order Butterworth filter can be developed. Consider the transfer function (11.121) for a high-pass second-order filter:

$$A(s) = \frac{A_\infty s^2}{s^2 + a_1 \omega_c s + b_1 \omega_c^2} \quad (11.121)$$

For unity gain, $A_\infty = 1$ and for a Butterworth response from Table 11.1, $a_1 = \sqrt{2}$ and $b_1 = 1$. Then setting $s = j\omega$, (11.121) becomes

$$A(jf) = \frac{(jf/f_c)^2}{1 + \sqrt{2}jf/f_c + (jf/f_c)^2} \quad (11.122)$$

The transfer function for a second-order high-pass filter using the VCVS or Sallen-Key topology in Fig. 11.30 is given by

$$A_f = \frac{s^2 C_1 C_2 R_1 R_2}{1 + sR_2(C_1 + C_2) + s^2 R_1 R_2 C_1 C_2} \quad (11.123)$$

In order to obtain the maximally flat Butterworth response, let $C_1 = C_2 = C$, $R_1 = R$, $R_2 = R/2$. Then, (11.123) reduces to

$$A_f = \frac{s^2 C_2 R_2/2}{1 + sCR + s^2 R_2 C_2/2}$$

$$= \frac{(jf/f_c)^2}{1 + \sqrt{2}jf/f_c + (jf/f_c)^2} \quad (11.124)$$

where

$$f_c = \sqrt{2}/2\pi RC \quad (11.125)$$

This transfer function is identical to the second-order Butterworth high-pass filter in (11.122). The design procedure is as follows:
1. Determine the desired cut-off frequency f_c.
2. Choose a value of C between 100 pF and 1 μF.
3. Set $C_2 = C_1 = C$.
4. Determine R using $R = \sqrt{2}/2\pi f_c C$.

Example 11.17
Using the configuration shown in Fig. 11.30, re-design the second-order high-pass Butterworth filter of Example 11.15 using the simplified approach.

Solution For $f_c = 10$ kHz, choose $C_1 = C_2 = 0.01$ μF. Using (11.125), $R = \sqrt{2}/2\pi \times 10 \times 10^3 \times 0.01 \times 10^{-6} = 2.3$k and hence $R_1 = 2.3$ k, $R_2 = 1.1$ k and $R_{os} = 2.3$ k.

Example 11.18
Using $C_1 = C_2 = 0.1$ μF, determine the cut-off frequency in the high-pass filter if $R_1 = 1$ k and $R_2 = 500$ Ω.

Solution From Eq. (11.125), $f_c = \sqrt{2}/2\pi \times 10^3 \times 0.1 \times 10^{-6} = 2.3$ kHz.

Example 11.19

Design a VCVS second-order unity-gain high-pass Chebyshev filter with a cut-off frequency of 10 kHz and a ripple width $RWdB = 1\ dB$.

Solution For this case, $f_c = 10$ kHz. Select $C_1 = C_2 = 0.001\ \mu F$. From Table 11.2, the coefficients for $RWdB = 1$ dB are $a_1 = 1.3022$, $b_1 = 1.5515$. Then from (11.105) and (11.106), $R_1 = \frac{1}{\pi \times 10 \times 10^3 \times 1.3022 \times 0.001 \times 10^{-6}} = 24$ k and $R_2 = \frac{1.3022}{4\pi \times 10 \times 10^3 \times 1.5515 \times 0.001 \times 10^{-6}} = 6.7$ k. Finally, for minimum offset, $R_{os} = 24$ k.

Example 11.20
Design a VCVS second-order unity-gain high-pass Bessel filter with a cut-off frequency of 10 kHz.

Solution For this case, $f_c = 10$ kHz. Select $C_1 = C_2 = 0.001\ \mu F$. From Table 11.1, the Bessel coefficients are $a_1 = 1.3617$, $b_1 = 0.6180$. Then $R_1 = \frac{1}{\pi \times 10 \times 10^3 \times 1.3617 \times 0.001 \times 10^{-6}} = 23.4$ k and $R_2 = \frac{1.3617}{4\pi \times 10 \times 10^3 \times 0.6180 \times 0.001 \times 10^{-6}} = 17.5$ k. Finally, for minimum offset, $R_{os} = 23.4$ k.

11.6.2 High-Pass Second-Order Multiple Feedback Filter

Another available topology for implementation of second-order high-pass filters is the multiple feedback (MFB) topology shown in Fig. 11.32. In this circuit the resistors and capacitors are interchanged compared to the corresponding low-pass MFB circuit.

Assuming an ideal op-amp, then for nodes a and b,

$$(V_i - V_a)sC_1 = (V_a - V_o)sC_2 + (V_a - V_b)$$
$$sC_3 + \frac{V_a}{R_1} \tag{11.126}$$

$$(V_a - V_b)sC_3 = \frac{(V_b - V_o)}{R_2} \tag{11.127}$$

Noting that $V_b = 0$ because node b is a virtual earth and eliminating V_a, we get

$$A_f(s) = -\frac{\frac{C_1}{C_2}s^2}{s^2 + \left(\frac{C_1 + C_2 + C_3}{R_2 C_2 C_3}\right)s + \frac{1}{R_1 R_2 C_2 C_3}} \tag{11.128}$$

For simplicity of the analysis, let $C_1 = C_3 = C$. Then (11.128) reduces to

$$A_f(s) = -\frac{\frac{C}{C_2}s^2}{s^2 + \left(\frac{2C + C_2}{R_2 C_2 C}\right)s + \frac{1}{R_1 R_2 C_2 C}} \tag{11.129}$$

Comparing the coefficients of (11.129) with the general second-order high-pass transfer function

$$A(s) = -\frac{A_\infty s^2}{s^2 + a_1 \omega_c s + b_1 \omega_c^2} \tag{11.130}$$

(where a negative sign is introduced to account for the inversion produced by the MFB configuration) yields

$$A_\infty = \frac{C}{C_2} \tag{11.131}$$

$$a_1 \omega_c = \frac{C_2 + 2C}{R_2 C_2 C} \tag{11.132}$$

$$b_1 \omega_c^2 = \frac{1}{R_1 R_2 C_2 C} \tag{11.133}$$

Solving for R_1, R_2 in terms of C and C_2, we get

$$R_1 = \frac{a_1}{b_1 2\pi f_c (C_2 + 2C)} \tag{11.134}$$

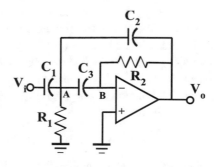

Fig. 11.32 High-pass second-order multiple feedback filter

$$R_2 = \frac{(C_2 + 2C)}{a_1 2\pi f_c C_2 C} \qquad (11.135)$$

Design Procedure
1. Select C usually between 100 pF and 1 μF.
2. Determine C_2 using $C_2 = C/A_\infty$.
3. For the specific type of filter (Butterworth, Chebyshev, Bessel) and the specific filter requirements, obtain the coefficients a_1 and b_1 from the relevant table.
4. Find R_1 and R_2 using $R_1 = \frac{a_1}{b_1 2\pi f_c (C_2 + 2C)}$ and $R_2 = \frac{(C_2 + 2C)}{a_1 2\pi f_c C_2 C}$.
5. Scale values if necessary.

Example 11.21
Design a MFB second-order high-pass Butterworth filter with a gain of 10 and a cut-off frequency of 1 kHz.

Solution For this case, $f_c = 1$ kHz. Select $C_1 = C_3 = C = 0.01$ μF. Then $C_2 = 0.01$ μF/10 $= 0.001$ μF. From Table 11.1, the Butterworth coefficients are $a_1 = 1.4142$, $b_1 = 1$. Then $R_1 = \frac{a_1}{b_1 2\pi f_c (C_2 + 2C)} = \frac{1.4142}{2 \times \pi \times 10^3 \times (0.001 + 0.02) \times 10^{-6}} = 10.7$ k and $R_2 = \frac{(C_2 + 2C)}{a_1 2\pi f_c C_2 C} = \frac{(0.001 + 0.02) \times 10^{-6}}{1.4142 \times 2 \times \pi \times 10^3 \times 0.001 \times 10^{-6} \times 0.01 \times 10^{-6}} = 236$ k.

11.7   Higher-Order High-Pass Filters

We come now to higher-order filters in which the transfer function has a denominator polynomial that is of degree n. We again use the method of factoring the higher-order transfer function $A(s)$ into first- and second-order transfer functions $A_1(s)$ and $A_2(s)$, each of which is then synthesized and cascaded.

The transfer function of one stage is given by

$$A(s) = \frac{A_\infty s^2}{s^2 + a_1 \omega_c s + b_1 \omega_c{}^2} \qquad (11.136)$$

In a first-order transfer function, the coefficient b_1 is zero giving the transfer function

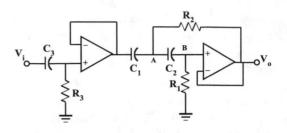

Fig. 11.33 Third-order high-pass filter

$$A(s) = \frac{A_\infty s}{s + a_1 \omega_c} \qquad (11.137)$$

Therefore, the general transfer function representing this cascade of second- and first-order transfer functions is given by

$$\begin{aligned} A(s) &= A_\infty \prod_i \left(\frac{s^2}{s^2 + a_i \omega_c s + b_i \omega_c{}^2} \right) \\ &= A_\infty \left(\frac{s^2}{s^2 + a_1 \omega_c s + b_1 \omega_c{}^2} \right) \left(\frac{s^2}{s^2 + a_2 \omega_c s + b_2 \omega_c{}^2} \right) \\ &\quad \cdots \left(\frac{s^2}{s^2 + a_n \omega_c s + b_n \omega_c{}^2} \right) \end{aligned} \qquad (11.138)$$

11.7.1   Third-Order High-Pass Unity-Gain Filter

In order to develop a third-order unity-gain high-pass filter, we cascade first- and second-order filters such that

$$A(s) = A_\infty \frac{s}{s + a_1 \omega_c} \cdot \frac{s^2}{(s^2 + a_2 \omega_c s + b_2 \omega_c{}^2)} \qquad (11.139)$$

Note again that $b_1 = 0$ for the first-order part of the transfer function. For the unity-gain system under consideration, $A_\infty = 1$. Hence (11.139) can be written as

$$A(s) = \frac{s}{s + a_1 \omega_c} \cdot \frac{s^2}{(s^2 + a_2 \omega_c s + b_2 \omega_c{}^2)} \qquad (11.140)$$

The circuit of this third-order system is shown in Fig. 11.33. It comprises a unity-gain non-inverting first-order system cascaded with a unity-gain non-inverting second-order system. The design process is as follows:

Partial Filter 1

The first-order system is

$$A(s) = \frac{s}{s + a_1\omega_c} \qquad (11.141)$$

where $\omega_c = 2\pi f_c$ is the cut-off frequency. The transfer function for the first-order *non-inverting* unity-gain high-pass filter in Fig. 11.33 is given by

$$A_f(s) = \frac{R_3 C_3 s}{1 + R_3 C_3 s} = \frac{s}{s + 1/R_3 C_3} \qquad (11.142)$$

Comparing coefficients in (11.142) and (11.141) gives

$$a_1\omega_c = \frac{1}{R_3 C_3} \qquad (11.143)$$

The design steps are as follows:

1. Select C_3 usually between 100 pF and 1 μF.
2. For the specific type of filter (Butterworth, Chebyshev, Bessel) and the specific filter requirements, obtain the coefficient $a_1(b_1 = 0)$ from the relevant table.
3. Determine R using $R_3 = 1/a_1 2\pi f_c C_3$.

Partial Filter 2

The second-order system is

$$A(s) = \frac{s^2}{(s^2 + a_2\omega_c s + b_2\omega_c^2)} \qquad (11.144)$$

The transfer function for the second-order *non-inverting* unity-gain low-pass filter in Fig. 11.33 is given by ($C_1 = C_2 = C$)

$$A_f(s) = \frac{s^2}{s^2 + \frac{2}{R_1 C}s + \frac{1}{C_2 R_1 R_2}} \qquad (11.145)$$

Comparing coefficients for (11.145) and (11.144) yields

$$a_2\omega_c = \frac{2}{R_1 C} \qquad (11.146)$$

$$b_2\omega_c^2 = \frac{1}{R_1 R_2 C^2} \qquad (11.147)$$

From (11.146),

$$R_1 = \frac{1}{a_2 2\pi f_c C} \qquad (11.148)$$

and from (11.147),

$$R_2 = \frac{a_1}{b_2 4\pi f_c C} \qquad (11.149)$$

We can now state a design procedure for a specific cut-off frequency:

Design Procedure

1. Select C usually between 100 pF and 1 μF. Set $C_1 = C_2 = C$.
2. For the specific type of filter (Butterworth, Chebyshev, Bessel) and the specific filter requirements, obtain the coefficients a_2 and b_2 from the relevant table.
3. Find R_1 and R_2 using $R_1 = \frac{1}{\pi f_c a_2 C}$ and $R_2 = \frac{a_2}{4\pi f_c b_2 C}$.

Example 11.22

Design a VCVS third-order unity-gain high-pass Butterworth filter with a cut-off frequency of 20 kHz.

Solution The design of each of the two partial filters is considered separately.

Partial Filter 1

For this case, $f_c = 20$ kHz. Select $C_3 = 1$ nF. From Table 11.1, $a_1 = 1$, $b_1 = 0$. Hence $R_3 = 1/2\pi f_c a_1 C_3 = 1/2\pi \times 20 \times 10^3 \times 10^{-9} = 8$ k.

Partial Filter 2

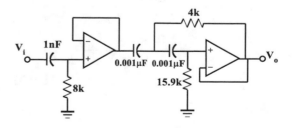

Fig. 11.34 Solution to Example 11.24

For this case, $f_c = 20$ kHz. Select $C = 0.001$ μF and select $C_1 = C_2 = C$. From Table 11.1, $a_2 = 1$, $b_2 = 1$. Hence $R_1 = \frac{1}{\pi f_c a_2 C} = \frac{1}{\pi \times 20 \times 10^3 \times 0.001 \times 10^{-6}} =$ 15.9 k and $R_2 = \frac{a_2}{4\pi f_c b_2 C} = \frac{1}{4\pi \times 20 \times 10^3 \times 0.001 \times 10^{-6}} =$ 4 k. The completed filter is shown in Fig. 11.34.

Simplified Design Procedure for Third-Order High-Pass Unity-Gain Butterworth Filter
A simplified design procedure for the unity-gain high-pass third-order Butterworth filter can be developed. Consider the transfer function for a high-pass second-order filter:

$$A(s) = A_\infty \frac{s}{s + a_1 \omega_c} \cdot \frac{s^2}{(s^2 + a_2 \omega_c s + b_2 \omega_c{}^2)}$$
(11.150)

For unity gain, $A_\infty = 1$ and for a Butterworth response from Table 11.1, $a_1 = 1$ and $a_2 = 1$, $b_2 = 1$. Then setting $s = j\omega$, (11.150) becomes

$$A = \frac{(jf/f_c)^2}{1 + jf/f_c + (jf/f_c)^2} \cdot \frac{jf/f_c}{1 + jf/f_c}$$
(11.151)

where f_c is the cut-off frequency. This transfer function can be implemented using the configuration shown in Fig. 11.33 which comprises cascaded first- and second-order systems. For this system, the transfer function is given by

$$A_f = \frac{s^2 C_1 C_2 R_1 R_2}{1 + sR_2(C_1 + C_2) + s^2 C_1 C_2 R_1 R_2} \cdot \frac{sC_3 R_3}{1 + sR_3 C_3}$$
(11.152)

Let $C_1 = C_2 = C_3 = C$ and $R_1 = 2R_3$ and $R_2 = R_3/2$. Then (11.152) becomes

$$A_f = \frac{s^2 (CR_3)^2}{1 + sR_3 C + s^2 (CR_3)^2} \cdot \frac{sCR_3}{1 + sCR_3}$$

$$= \frac{(jf/f_c)^2}{1 + jf/f_c + (jf/f_c)^2} \cdot \frac{jf/f_c}{1 + jf/f_c}$$
(11.153)

where

$$f_c = 1/2\pi CR_3$$
(11.154)

which is the cut-off frequency. This transfer function (11.153) is the same as (11.151). We can therefore write down a simplified design procedure:

1. Choose the desired cut-off frequency f_c.
2. Choose C, usually an available value between about 100 pF and 1 μF. Let $C_1 = C_2 = C_3 = C$.
3. Determine R_3 using $R_3 = 1/2\pi f_c C$. Set $R_1 = 2R_3$ and $R_2 = R_3/2$.

Example 11.23
Using the simplified design procedure, design a third-order high-pass Butterworth filter with a cut-off frequency of 9 kHz.

Solution Choose $C = 0.01$ μF and then $C_1 = C_2 = C_3 = 0.01$ μF. Using (11.154), $R_3 = 1/2\pi f_c C = 1/2\pi \times 9 \times 10^3 \times 0.01 \times 10^{-6} = 1.8$ k. Hence $R_1 = 2R_3 = 3.6$ k and $R_2 = R_3/2 = 1.8$ k/2 $= 900$ Ω.

Example 11.24
Design a VCVS third-order unity-gain Bessel high-pass filter using Fig. 11.33 with a cut-off frequency of 1 kHz.

Solution The design of each of the two partial filters is considered separately.

Partial Filter 1
For this case, $f_c = 1$ kHz. Select $C_3 = 0.1$ μF. From Table 11.1, $a_1 = 0.756$, $b_1 = 0$. Hence $R_3 = 1/2\pi f_c a_1 C = 1/2\pi \times 10^3 \times 0.756 \times 0.1 \times$

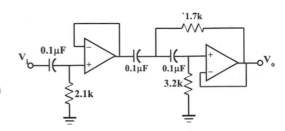

Fig. 11.35 Solution to Example 11.24

$10^{-6} = 2.1$ k.

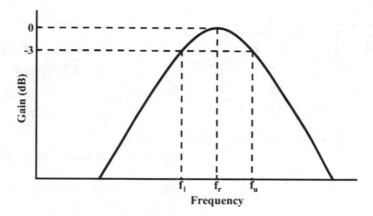

Fig. 11.36 Band-pass filter response

Fig. 11.37 Normalized band-pass filter response for varying values of Q

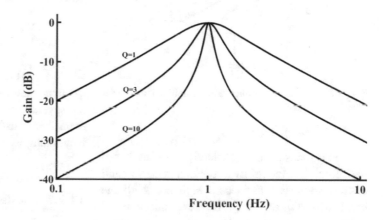

Partial Filter 2

For this case, $f_c = 1$ kHz. Select $C = 0.1$ μF and select $C_1 = C_2 = C$. From Table 11.1, $a_2 = 0.9996$, $b_2 = 0.4772$. Hence $R_1 = \frac{1}{\pi f_c a_2 C} = \frac{1}{\pi \times 10^3 \times 0.9996 \times 0.1 \times 10^{-6}} = 3.2$ k and $R_2 = \frac{a_2}{4\pi f_c b_2 C} = \frac{1}{4\pi \times 10^3 \times 0.4772 \times 0.1 \times 10^{-6}} = 1.7$ k. The completed filter is shown in Fig. 11.35.

11.7.2   Band-Pass Filter

A band-pass filter passes frequencies within a particular frequency band and attenuates all frequencies outside that band. This is shown in Fig. 11.36. This filter has a maximum gain at a frequency f_R referred to as the resonant or centre frequency. The frequencies f_L and f_U where the output falls to 0.707 of its maximum value are the lower cut-off frequency and upper cut-off frequency, respectively. The bandwidth B is the frequency difference given by

$$B = f_U - f_L \qquad (11.155)$$

The relationship between f_U, f_L and the resonant frequency f_R is $f_R = \sqrt{f_L f_U}$. Band-pass filters are generally classified as being either narrow-band or wide-band. A narrow-band filter is one having a bandwidth less than one-tenth of the resonant frequency, otherwise the filter is classified as wide-band. The quality factor Q of the circuit is the ratio of resonant frequency to bandwidth given by

$$Q = f_R/B \qquad (11.156)$$

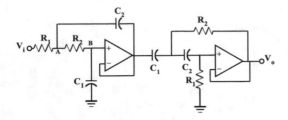

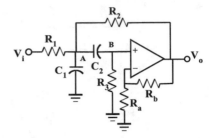

Fig. 11.38 Band-pass filter using cascaded low-pass and high-pass filters

Fig. 11.40 Sallen-Key band-pass filter

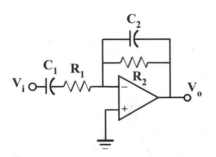

Fig. 11.39 Band-pass filter using combined first-order filters

It is a measure of the selectivity of the band-pass filter, i.e. its ability to select a specific band of frequencies while rejecting signals outside the band. The higher the Q, the more selective the band-pass filter as shown in Fig. 11.37. From (11.156), $B < 0.1f_R$ is equivalent to $Q > 10$ which corresponds to a narrow-band filter, and $B > 0.1f_R$ is equivalent to $Q < 10$ which corresponds to a wide-band filter.

In general, a band-pass filter can be realized by cascading a low-pass filter and a high-pass filter of the same order and gain with the respective cut-off frequencies suitably separated with appropriate overlap. Such a circuit is shown in Fig. 11.38.

The lower cut-off frequency is set by the high-pass filter, and the upper cut-off frequency is set by the low-pass filter. The pass-band gain will be the gain of each filter and will be constant for a band of frequencies. The roll-off rate on each side of the frequency response characteristic would be determined by the order of the filters. In the case of first-order filters, the low-pass and high-pass

filters can be combined in an inverting amplifier as shown in Fig. 11.39.

In narrow-band filters, $B < 0.1f_R$. This type of filter has a general second-order transfer function given by

$$\frac{V_o}{V_i} = \frac{s\omega_o \frac{G}{Q}}{s^2 + s\frac{\omega_o}{Q} + \omega_o^2} \qquad (11.157)$$

where ω_o is the centre frequency, G is the centre frequency gain and Q is the quality factor. There are several topologies that can be used to implement this transfer function including the Sallen-Key, multiple feedback and Wien.

11.7.3 Sallen-Key Band-Pass Filter

The circuit shown in Fig. 11.40 is a second-order Sallen-Key (VCVS) band-pass filter. Node analysis at A and B yields the equations

$$V_A\left(\frac{1}{R_1} + \frac{1}{R_2} + s(C_1 + C_2)\right) = \frac{V_i}{R_1} + sC_2V_B + \frac{V_o}{R_2}$$
$$(11.158)$$

$$V_B\left(\frac{1}{R_3} + sC_2\right) = sC_2V_A \qquad (11.159)$$

$$V_B = \frac{V_o}{k} \qquad (11.160)$$

where

$$k = 1 + \frac{R_b}{R_a} \qquad (11.161)$$

This gives

$$\frac{V_o}{V_i}(s) =$$

$$\frac{\frac{k}{C_1 R_1}s}{s^2 + s\frac{1}{C_1}\left(\frac{1}{R_1} + \frac{1}{R_3} + \frac{1}{R_2}(1-k) + \frac{C_1}{C_2 R_3}\right) + \frac{R_1+R_2}{C_1 C_2 R_1 R_2 R_3}} \tag{11.162}$$

Let $C_1 = C_2 = C$ and $R_1 = R_2 = R$ and $R_3 = 2R$. Then (11.162) reduces to

$$\frac{V_o}{V_i}(s) = \frac{\frac{k}{CR}s}{s^2 + s\frac{(3-k)}{RC} + \frac{1}{C^2 R^2}} \tag{11.163}$$

Comparing coefficients in Eqs. (11.163) and (11.157) yields

$$\omega_o^2 = \frac{1}{R^2 C^2} \tag{11.164}$$

$$\frac{\omega_o}{Q} = \frac{(3-k)}{RC} \tag{11.165}$$

$$\frac{G\omega_o}{Q} = \frac{k}{RC} \tag{11.166}$$

From this,

$$f_o = \frac{1}{2\pi RC} \tag{11.167}$$

$$Q = \frac{1}{3-k} \tag{11.168}$$

$$G = \frac{k}{3-k} \tag{11.169}$$

In the design process, a value of C is selected, and then using (11.167), R is determined for a given f_o. Following this, either a particular Q using (11.168) or G using (11.169) can be designed for by selecting k. Note that as Q increases, the value of $(3-k)$ becomes smaller indicating that k approaches $3(k < 3)$ and hence the circuit gain G in (11.169) changes.

Example 11.25
Design a Sallen-Key second-order band-pass filter with a centre frequency of 5 kHz and a quality factor of 5.

Solution We select $C = 0.01\ \mu F$. Then using (11.167), $R = 1/2\pi \times 5 \times 10^3 \times 0.01 \times 10^{-6}$ $= 3.2$ k. Using (11.168) (11.143), for $Q = 5$ we

have $k = 3 - 1/Q = 3 - 1/5 = 2.8$. Since $k = 1 + R_b/R_a$, for $R_a = 1$ k then $R_b = 1.8$ k. Note that the resulting value for G using (11.169) is $G = 2.8/(3 - 2.8) = 14$.

Thus, this design process allows designing for a particular f_o and either Q or G. If greater design freedom is desirable for specified values of f_o, Q and G, then the conditions $C_1 = C_2 = C$ and $R_1 = R_2 = R$, $R_3 = 2R$ would need to be relaxed.

11.7.4   Multiple Feedback Band-Pass Filter

A simple circuit to implement the band-pass transfer function is shown in Fig. 11.41. It is a multiple feedback configuration referred to as the Delyiannis-Friend circuit. Setting the capacitor values equal such that $C_1 = C_2 = C$ simplifies the analysis without compromising the filter. The node equations are given by

$$\frac{V_i - V_A}{R_1} = \frac{V_A}{R_2} + (V_A - V_o)sC + V_A sC \tag{11.170}$$

$$V_A sC = -\frac{V_o}{R_3} \tag{11.171}$$

Eliminating V_A yields

$$\frac{V_o}{V_i} = -\frac{\frac{s}{CR_1}}{s^2 + \frac{2s}{CR_3} + \frac{R_1+R_2}{C^2 R_1 R_2 R_3}} \tag{11.172}$$

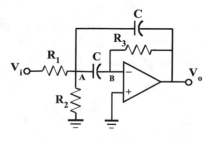

Fig. 11.41 Multiple feedback band-pass filter

Comparing terms in Eqs. (11.172) and

$$\frac{V_o}{V_i} = -\frac{s\omega_o \frac{G}{Q}}{s^2 + s\frac{\omega_o}{Q} + \omega_o^2} \qquad (11.173)$$

yields

$$\omega_o = \frac{1}{C\sqrt{R_1//R_2.R_3}}, \quad f_o = \frac{1}{2\pi C\sqrt{R_1//R_2.R_3}} \qquad (11.174)$$

$$Q = \frac{1}{2}\sqrt{\frac{R_3}{R_1//R_2}} \qquad (11.175)$$

$$G = \frac{1}{2}\frac{R_3}{R_1} \qquad (11.176)$$

Solving for R_1, R_2, R_3 yields

$$R_1 = \frac{Q}{2\pi f_o GC} \qquad (11.177)$$

$$R_3 = \frac{2Q}{\omega_o C} \qquad (11.178)$$

$$G_2 = \frac{1}{R_2} = 4\pi f_o CQ\left(1 - \frac{G}{2Q^2}\right) \qquad (11.179)$$

where G_2 is conductance. This must be positive and therefore the following realizability condition must be satisfied:

$$\left(1 - \frac{G}{2Q^2}\right) \geq 0 \Rightarrow \frac{G}{2Q^2} \leq 1 \qquad (11.180)$$

If the realizability condition is satisfied such that $G/2Q^2 = 1$, then $G_2 = 0$ which means $R_2 \rightarrow \infty$(open circuit). In these circumstances, resistor R_2 is removed from the circuit.

Example 11.26
Design a second-order MFB band-pass filter with $f_0 = 1$ kHz, $Q = 7$ and $G = 10$.

Solution Checking the realizability condition, we have $G/2Q^2 = 10/2 \times 7^2 = 0.1 < 1$. Hence the condition is satisfied. Select $C_1 = C_2 = C = 0.01$ μF. Then

$$R_1 = \frac{Q}{2\pi f_o GC}$$
$$= 7/2\pi \times 10^3 \times 10 \times 0.01 \times 10^{-6}$$
$$= 11.1\,\text{k};$$

$$R_3 = \frac{2Q}{\omega_o C}$$
$$= 2 \times 7/(2 \times \pi \times 10^3 \times 0.01 \times 10^{-6})$$
$$= 222\,\text{k};$$

$$R_2 = 1/4\pi f_o CQ\left(1 - \frac{G}{2Q^2}\right)$$
$$= 1/4\pi \times 10^3 \times 0.01 \times 10^{-6} \times 7 \times (1 - 0.1)$$
$$= 1.1\,\text{k}$$

(In the case where the centre frequency gain G is not specified, then G can be selected to ensure that the realizability condition is satisfied with equality. Then resistor R_2 can be removed from the circuit.) If the realizability condition is not met, then the requirement $C_1 = C_2$ would have to be relaxed.

11.7.5   Wien Band-Pass Filter

A third band-pass filter is shown in Fig. 11.42. It utilizes a Wien network which has a band-pass response. At $f_o = 1/2\pi RC$, the output of the network is at a maximum of 1/3 of the input voltage. By placing this network in a positive feedback loop, the band-pass response is sharpened. The transfer function is given by

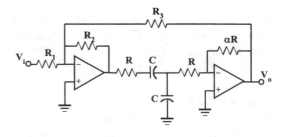

Fig. 11.42 Wien band-pass filter

$$\frac{V_o}{V_i} = \frac{\alpha \frac{R_2}{R_1} \omega_o s}{s^2 + \left(3 - \alpha \frac{R_2}{R_1}\right) \omega_o s + \omega_o^2} \quad (11.181)$$

where

$$\omega_o = \frac{1}{RC}, \ f_o = \frac{1}{2\pi RC} \quad (11.182)$$

$$Q = \frac{1}{3 - \alpha \frac{R_2}{R_3}} \quad (11.183)$$

$$G = \alpha \frac{R_2}{R_1} Q \quad (11.184)$$

In designing this filter, C is first chosen and then R is selected to achieve the desired centre frequency using (11.182). Following this, R_2, R_3 and α are determined to give the desired Q using (11.183) and then R_1 is adjusted to give the desired centre frequency gain using (11.184). This procedure is called orthogonal tuning.

Example 11.27
Design a second-order Wien band-pass filter with $f_0 = 1$ kHz, $Q = 7$ and $G = 10$.

Solution Select $C = 0.01 \ \mu F$ and then $R = 1/2\pi \times 10^3 \times 0.01 \times 10^{-6} = 15.9$ k. To get the specified Q, choosing $\alpha = 1$ in (11.183) gives $7 = \frac{1}{3 - \frac{R_2}{R_3}}$. Selecting $R_3 = 10$ k we get $R_2 = 8.6$ k. Finally, using (11.184), in order to achieve the desired gain, we get $10 = \frac{8.6 k}{R_1} \times 7$ which yields $R_1 = 6$ k.

11.8 Band-Stop Filter

The general transfer function of a second-order band-stop or notch filter is given by

$$\frac{V_o}{V_i} = \frac{s^2 + \omega_o^2}{s^2 + s\frac{\omega_o}{Q} + \omega_o^2} \quad (11.185)$$

where $\omega_o = 2\pi f_o$ is the notch frequency and Q is the quality factor. The normalized response for this filter is shown in Fig. 11.43 for varying values of Q.

11.8.1 Twin-T Notch Filter

A very popular notch filter is the twin-T circuit shown in Fig. 11.44. It utilizes a twin-T network at the input to create the notch at frequency f_o and sharpens the notch with positive feedback. The feedback here is applied to both legs of the T network, and the level of this feedback is

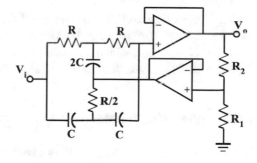

Fig. 11.44 Twin-T notch filter

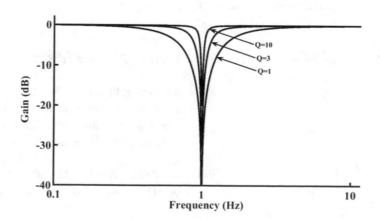

Fig. 11.43 Normalized band-stop filter response for varying values of Q

adjustable with the potential divider R_1, R_2. Both op-amps provide buffering. The transfer function is given by

$$\frac{V_o}{V_i} = \frac{s^2 + \omega_o{}^2}{s^2 + s\frac{\omega_o}{Q} + \omega_o{}^2} \qquad (11.186)$$

where

$$\omega_o = \frac{1}{RC}, f_o = \frac{1}{2\pi RC} \qquad (11.187)$$

$$Q = \frac{1}{4(1 - \rho)} \qquad (11.188)$$

$$\rho = \frac{R_1}{R_1 + R_2} \qquad (11.189)$$

With this notch filter, Qs up to about 50 are achievable.

Example 11.29
Using the circuit in Fig. 11.44, design a twin-T notch filter to have a notch frequency of 60 Hz and a Q of 20.

Solution Select $C = 1$ μF. Then using (11.187), $R = 1/2\pi \times 60 \times 10^{-6} = 2.7$ k. Using (11.188) $20 = 1/4 \times (1 - \rho)$ from which $\rho = 0.9875$. From (11.189) with $R_1 = 10$ k, then $R_2 = 127$ Ω.

11.8.2   Wien Notch Filter

The notch filter shown in Fig. 11.45 is easier to tune than the twin-T filter, and higher Qs can be achieved. It utilizes the Wien network in a feedback loop to increase Q. The transfer function is given by

$$\frac{V_o}{V_i} = k\frac{s^2 + \omega_o{}^2}{s^2 + s\frac{\omega_o}{Q} + \omega_o{}^2} \qquad (11.190)$$

where

$$\omega_o = \frac{1}{RC}, f_o = \frac{1}{2\pi RC} \qquad (11.191)$$

$$Q = \frac{1}{3}\left(1 + \frac{R_2}{R_3}\right) \qquad (11.192)$$

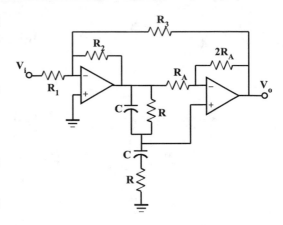

Fig. 11.45 Wien notch filter

$$k = -\frac{R_2//R_3}{R_1} \qquad (11.193)$$

where k is the low-frequency gain. Thus, R and C are selected to give a particular notch frequency, and R_2 and R_3 are selected to give the desired Q after which R_1 is chosen to achieve a particular gain.

Example 11.30
Using the circuit in Fig. 11.45, design a Wien notch filter to have a notch frequency of 60 Hz, a Q of 20 and a gain of 10.

Solution Select $C = 1$ μF. Then using (11.191), $R = 1/2\pi \times 60 \times 10^{-6} = 2.7$ k. Using (11.192), $20 = \frac{1}{3}\left(1 + \frac{R_2}{R_3}\right)$, and therefore for $R_3 = 10$ k, $R_2 = 590$ k. From (11.193) and noting that $R_2//R_3 = 9.8$ k, we have $10 = \frac{R_2//R_3}{R_1} = \frac{9.8\,k}{R_1}$. Hence $R_1 = 980$ Ω.

11.8.3   All-Pass Filters

An all-pass filter is one in which the output signal amplitude is constant for all frequencies but whose phase varies with frequency. Because of these properties, all-pass filters are used to correct spurious signal phase shifts and to introduce signal delay. The all-pass filter needs to have a constant group delay over the relevant frequency band, in order that transmitted signal suffer minimum distortion. The group delay refers to the time by which signal frequencies within the band are delayed. It is defined as

$$t_{gr} = -\frac{d\phi}{d\omega} \qquad (11.194)$$

where ϕ is the phase difference between output and input.

11.8.4   First-Order All-Pass Filter Realization

The transfer function of a first-order all-pass filter is given by

$$\frac{V_o}{V_i}(s) = \frac{1 - \frac{s}{\omega_o}}{1 + \frac{s}{\omega_o}} \qquad (11.195)$$

The amplitude response is given by

$$\left|\frac{V_o}{V_i}\right| = \left|\frac{1 - s/\omega_o}{1 + s/\omega_o}\right| = \frac{\sqrt{1 + (\omega/\omega_o)^2}}{\sqrt{1 + (\omega/\omega_o)^2}} = 1 \qquad (11.196)$$

which is constant for all frequencies. Note that at low frequencies,

$$\frac{V_o}{V_i}(DC) = 1 \qquad (11.197)$$

The phase response is given by

$$\sphericalangle\left(\frac{V_o}{V_i}\right) = \sphericalangle(1 - s/\omega_o) - \sphericalangle(1 + s/\omega_o)$$

$$= -\tan^{-1}\omega/\omega_o - \tan^{-1}\omega/\omega_o$$

$$= -2\tan^{-1}\omega/\omega_o \qquad (11.198)$$

This system produces zero phase at low frequencies with no inversion and $-180°$ at high frequencies. The magnitude and phase responses are shown in Fig. 11.46. At $f = f_o$, $\sphericalangle(V_o/V_i) = -90°$. Using (11.194), the group delay produced by this filter is given by

$$t_{gr} = -\frac{d\phi}{d\omega} = -\frac{d}{d\omega}\left(-2\tan^{-1}\omega/\omega_o\right)$$

$$= \frac{1/\pi f_o}{1 + (f/f_o)^2} \qquad (11.199)$$

This has a value

$$t_{gr} = 1/\pi f_o, f \ll f_o \qquad (11.200)$$

which is the maximum group delay occurring at low frequencies. A plot of t_{gr} against f is shown in Fig. 11.47.

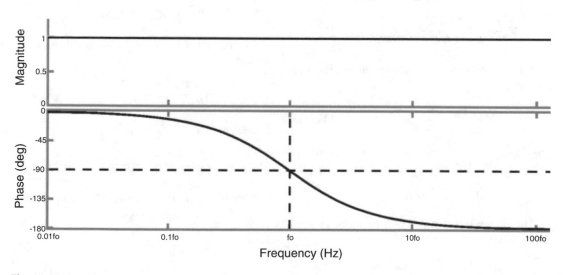

Fig. 11.46 Magnitude and phase response of first-order all-pass filter

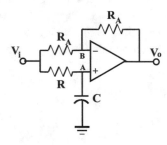

Fig. 11.47   Group delay for first-order all-pass filter (lag)

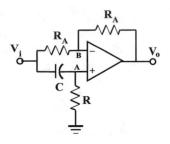

Fig. 11.49   First-order all-pass filter (lead)

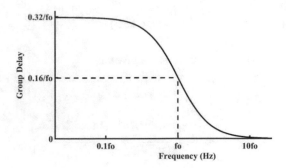

Fig. 11.48   First-order all-pass filter (lag)

First-Order Al- Pass Filter Realization

A first-order all-pass filter which produces this response is shown in Fig. 11.48. It is based on a single op-amp with negative feedback to the inverting terminal but where the non-inverting terminal is fed through an RC network. For this system at the non-inverting terminal,

$$V_A = \frac{1/sC}{R + 1/sC} = \frac{1}{1 + sCR} \qquad (11.201)$$

Noting that at the two input terminals, $V_A = V_B$, we have

$$\frac{V_i - V_B}{R_A} = \frac{V_B - V_o}{R_A} \qquad (11.202)$$

Therefore

$$\frac{V_o}{V_i} = \frac{1 - sCR}{1 + sCR} \qquad (11.203)$$

This is the transfer function of an all-pass lag network with

$$\omega_o = 1/RC, f_o = 1/2\pi RC \qquad (11.204)$$

This system produces zero phase at low frequencies with no inversion and $-180°$ at high frequencies. At $f = f_o = 1/2\pi RC$, $\angle(V_o/V_i) = -90°$. Using (11.194), the group delay produced by this filter is given by

$$t_{gr} = -\frac{d\phi}{d\omega} = -\frac{d}{d\omega}\left(-2\tan^{-1}\omega RC\right)$$

$$= 2\frac{RC}{1 + \omega^2 R^2 C^2} \qquad (11.205)$$

This has a value $t_{gr} = 2RC$, $\omega \ll 1/RC$ which is the maximum group delay occurring at low frequencies.

A first-order all-pass filter which produces a lead in the phase is shown in Fig. 11.49. In this circuit, R and C are interchanged as compared to the first-order all-pass lag circuit. For this circuit

$$V_A = \frac{R}{R + 1/sC} = \frac{sCR}{1 + sCR} \qquad (11.206)$$

Noting that at the two input terminals, $V_A = V_B$, we have

$$\frac{V_i - V_B}{R_A} = \frac{V_B - V_o}{R_A} \qquad (11.207)$$

Therefore

$$\frac{V_o}{V_i} = \frac{-(1 - sCR)}{1 + sCR} \qquad (11.208)$$

This is the transfer function of an all-pass lead network. At low frequencies,

$$\frac{V_o}{V_i}(DC) = -1 \qquad (11.209)$$

The amplitude response is given by

$$\left|\frac{V_o}{V_i}\right| = \left|\frac{-(1 - sCR)}{1 + sCR}\right| = \frac{\sqrt{1 + (\omega CR)^2}}{\sqrt{1 + (\omega CR)^2}} = 1$$

$$(11.210)$$

which is constant for all frequencies. The phase response is given by

$$\measuredangle\left(\frac{V_o}{V_i}\right) = \measuredangle(-1) + \measuredangle(1 - sCR) - \measuredangle(1 + sCR)$$

$$= 180° - \tan^{-1}\omega RC - \tan^{-1}\omega RC$$

$$= 180° - 2\tan^{-1}\omega RC \qquad (11.211)$$

This system produces 180° phase shift at low frequencies which is inversion and zero phase shift at high frequencies. At $f = f_o = 1/2\pi RC$, $\measuredangle(V_o/V_i) = +90°$.

11.9 State Variable Filter

The state variable filter also called the KHN (Kerwin, Huelsman, Newcomb) filter is another versatile filter realization. Because of its great flexibility, its good performance and low sensitivity to component variation, it is one of the most widely used active filters. The filter name arises from the state variable method used to solve differential equations that is used here to develop the filter topology. In outlining the method, consider the second-order low-pass transfer function

$$\frac{V_o}{V_i} = \frac{1}{s^2 + a_1 s + a_o} \qquad (11.212)$$

In this equation, an arbitrary function $X(s)/s^2$ is introduced such that (11.212) becomes

$$\frac{V_o}{V_i} = \frac{\frac{X(s)}{s^2}}{\frac{X(s)}{s^2}(s^2 + a_1 s + a_o)}$$

$$= \frac{\frac{X(s)}{s^2}}{X(s) + a_1\frac{X(s)}{s} + a_o\frac{X(s)}{s^2}} \qquad (11.213)$$

Equating numerator and denominator, we get

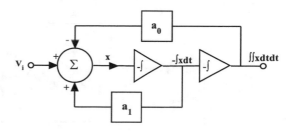

Fig. 11.50 Block diagram of state variable filter system

$$V_o = \frac{X(s)}{s^2} \qquad (11.214)$$

and

$$V_i = X(s) + a_1\frac{X(s)}{s} + a_o\frac{X(s)}{s^2} \qquad (11.215)$$

Expressing Eqs. (11.214) and (11.215) in the time domain (by taking the inverse Laplace transform) yields

$$x(t) = V_i(t) - a_1\int x(t)dt - a_o\int\left(\int x(t)dt\right)dt$$

$$(11.216)$$

$$V_o(t) = \int\left(\int x(t)dt\right)dt \qquad (11.217)$$

where the quantities $x(t)$, $\int x(t)dt$ and $\int(\int x(t)dt)dt$ are referred to as state variables and hence the name of the filter. A block diagram of this system is shown in Fig. 11.50, and an implementation of this using op-amps is shown in Fig. 11.51. Here op-amp 1 provides the summing action, while op-amp 2 and op-amp3 are used as integrators. For this system, the output V_{o3} of op-amp 3 represents the low-pass output V_o as given in (11.212), i.e. $V_{o3} = V_o = V_{LP}$. Also, the output V_{o2} of op-amp 2 is $V_{o2} = -sV_{o3} = -V_{BP}$, and the output V_{o1} of op-amp 1 is $V_{o1} = -sV_{o2} = V_{HP}$. The actual transfer functions are given by setting $R_4 = R_5 = R$, $C_1 = C_2 = C$ and $R_2 = R_3 = R_x$, then

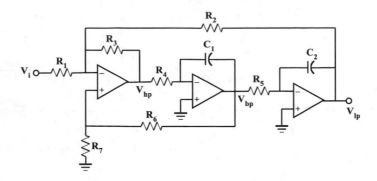

Fig. 11.51 Circuit implementation of state variable filter system

$$\frac{V_{HP}}{V_i} = \frac{-\frac{R_x}{R_1}s^2}{s^2 + \frac{R_7(2R_1+R_x)}{CRR_1(R_6+R_7)}s + \frac{1}{C^2R^2}} \quad (11.218)$$

of general form

$$\frac{V_{HP}}{V_i} = -\frac{ks^2}{s^2 + s\frac{\omega_o}{Q} + \omega_o^2} \quad (11.219)$$

$$\frac{V_{BP}}{V_i} = \frac{\frac{R_x}{R_1}\frac{1}{CR}s}{s^2 + \frac{R_7(2R_1+R_x)}{CRR_1(R_6+R_7)}s + \frac{1}{C^2R^2}} \quad (11.220)$$

of general form

$$\frac{V_{BP}}{V_i} = \frac{s\omega_o\frac{G}{Q}}{s^2 + s\frac{\omega_o}{Q} + \omega_o^2} \quad (11.221)$$

$$\frac{V_{LP}}{V_i} = \frac{-\frac{R_2}{R_1}\frac{1}{C^2R^2}}{s^2 + \frac{R_7(2R_1+R_x)}{CRR_1(R_6+R_7)}s + \frac{1}{C^2R^2}} \quad (11.222)$$

of general form

$$\frac{V_{LP}}{V_i} = -\frac{k\omega_o^2}{s^2 + s\frac{\omega_o}{Q} + \omega_o^2} \quad (11.223)$$

From these for $\beta = R_7/(R_6 + R_7)$,

$$f_0 = 1/2\pi RC \quad (11.224)$$

$$k = R_x/R_1 \quad (11.225)$$

$$Q = \frac{R_1}{\beta(2R_1 + R_x)} = \frac{1}{\beta(2 + k)} \quad (11.226)$$

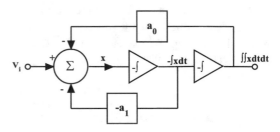

Fig. 11.52 Block diagram of modified state variable filter system

$$G = \frac{R_x}{\beta(2R_1 + R_x)} = kQ \quad (11.227)$$

Example 11.31

Using the state variable filter, design a band-pass filter with a centre frequency $f_o = 10$ kHz, a quality factor $Q = 50$ and a centre frequency gain $G = 10$.

Solution Choosing $C = 0.001$ μF, then $R = 1/2\pi \times 10 \times 10^3 \times 0.001 \times 10^{-6} = 15.9$ kΩ. $G = kQ$ giving $10 = 50\ k$ from which $k = 1/5$. This yields $R_x = 10$ kΩ and $R_1 = 50$ kΩ. Finally, $Q = 1/\beta(2 + k)$ which becomes $50 = 1/\beta(2 + 0.2)$ which yields $\beta = 1/110$. For $R_7 = 1$ kΩ then $R_6 = 109$ kΩ.

11.9.1 Modified State Variable Filter

We wish to exploit the excellent summing properties of the virtual earth of an op-amp in the inverting configuration. Thus, in the state variable

Fig. 11.53 Circuit
implementation of modified
state variable filter system

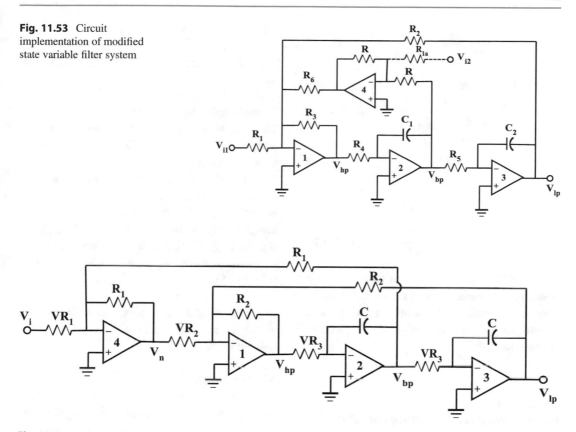

Fig. 11.54 Universal filter

filter configuration in Fig. 11.50, instead of adding the term $-a_1 \int x(t)dt$ derived from the band-pass output at the non-inverting terminal of the op-amp, the term $-\int x(t)dt$ is inverted (sign changed) using an inverting amplifier and the term $a_1 \int x(t)dt$ added at the inverting terminal using the normal summing process via the virtual earth. The resulting modified block diagram is shown in Fig. 11.52, and the modified circuit is shown in Fig. 11.53.

The inverting amplifier is op-amp 4 with the output signal being applied to the virtual earth of the input op-amp 1 via R_6. This modified topology achieves the same results as the original circuit. However, the presence of op-amp 4 allows the input to be moved from op-amp 1 to op-amp 4 such that the notch filter response becomes available at the output of op-amp 4. This is so since in general, $V_i - V_{BP} = V_N$ with the summing action resulting in $-V_N$. The fully redrawn circuit is shown in Fig. 11.54 and may be considered a universal filter as it provides band-stop,

high-pass, band-pass and low-pass outputs. Moreover, the availability of quad op-amp packages such as the LM148 series or the superior OPA4134 facilitates the implementation of this system.

For this system, the transfer functions are as follows:

$$\frac{V_N}{V_i} = \frac{-\frac{R_1}{VR_1}\left(s^2 + \frac{1}{VR_3{}^2 C^2}\right)}{s^2 + \frac{R_2}{VR_2 VR_3 C}s + \frac{1}{VR_3{}^2 C^2}} \quad (11.228)$$

$$\frac{V_{HP}}{V_i} = \frac{\frac{R_1 R_2}{VR_1 VR_2}s^2}{s^2 + \frac{R_2}{VR_2 VR_3 C}s + \frac{1}{VR_3{}^2 C^2}} \quad (11.229)$$

$$\frac{V_{BP}}{V_i} = \frac{-\frac{R_1 R_2}{VR_1 VR_2}\frac{1}{CVR_3}s}{s^2 + \frac{R_2}{VR_2 VR_3 C}s + \frac{1}{VR_3{}^2 C^2}} \quad (11.230)$$

$$\frac{V_{LP}}{V_i} = \frac{\frac{R_1 R_2}{VR_1 VR_2} \frac{1}{C^2 VR_3{}^2}}{s^2 + \frac{R_2}{VR_2 VR_3 C} s + \frac{1}{VR_3{}^2 C^2}} \qquad (11.231)$$

For this system,

$$f_o = \frac{1}{2\pi VR_3 C} \qquad (11.232)$$

$$Q = \frac{VR_2}{R_2} \qquad (11.233)$$

$$G = \frac{R_1}{VR_1} \qquad (11.234)$$

$$k = \frac{G}{Q} \qquad (11.235)$$

From these equations it can be seen that f_o is continuously variable using a variable resistor for VR_3, Q is continuously variable using a variable resistor for VR_2, and G is variable using a variable resistor for VR_1.

Example 11.32
Design a universal filter with a cut-off frequency of 20 kHz, a quality factor of 100 and a centre frequency gain of 10.

Solution Using Eq. (11.232) and choosing $C = 0.001\ \mu F$, we get $VR_3 = 7957\ \Omega$. For $Q = 100$ from Eq. (11.233), we get $VR_2 = 100\ k$ and $R_2 = 1\ k$. Finally for $G = 10$ from Eq. (11.234), we have $VR_1 = 10\ k$ and $R_1 = 100\ k$.

11.10 Applications

Several filter applications are presented in this section.

Second-Order Speech Filter
This system is a straightforward application of a wide-band filter comprising a second-order low-pass filter with a cut-off frequency of 300 Hz and a second-order high-pass filter with cut-off frequency 3.4 kHz. Both filters are Sallen and Key types which employ a simplified design procedure such that the cut-off frequency is given by $f_c = 1/2\pi RC$ and the circuit gain is set at 1.6 by resistors $R_a = 39\ k$ and $R_b = 22\ k$ (Fig. 11.55).

Ideas for Exploration (i) Compare the performance of the 741 op-amp with the OPA134 op-amp in this application.

All-Pass Notch
Figure 11.56 shows a notch filter made using two first-order all-pass filters. The output of the second all-pass filter is $V_A = \left(\frac{1-Ts}{1+Ts} \cdot \frac{1-Ts}{1+Ts}\right) V_i =$

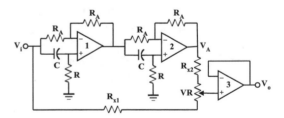

Fig. 11.56 All-pass notch filter

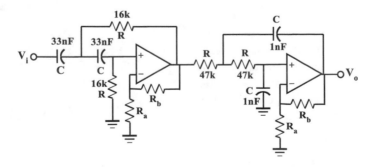

Fig. 11.55 Second-order speech filter

$\left(\frac{1-2Ts+T^2s^2}{1+2Ts+T^2s^2}\right)V_i$ where $T = RC$. This output is applied to one end of the potential divider comprising two series connected equal value resistors $R_{x1} = R_{x2}$ (Assume initially $VR = 0$). The input signal is applied at the other end of these two resistors and op-amp 3 connected as a buffer takes its input from the junction of these two resistors. The voltage across these resistors is therefore $V_A - V_i = \left(\frac{1-2Ts+T^2s^2}{1+2Ts+T^2s^2} - 1\right)V_i = -\left(\frac{4Ts}{1+2Ts+T^2s^2}\right)V_i$. Half of this is dropped across each resistor R_{x1} and is given by $V_{R_{x1}} = -\left(\frac{2Ts}{1+2Ts+T^2s^2}\right)V_i$. Therefore, the voltage V_o at the output V_o of op-amp 3 is given by $V_o = V_i + V_{R_{x1}} = \left(\frac{1+2Ts-2Ts+T^2s^2}{1+2Ts+T^2s^2}\right)V_i = \left(\frac{1+T^2s^2}{1+2Ts+T^2s^2}\right)V_i.$

This goes to zero at $f_o = 1/2\pi RC$ and is non-zero at all other frequencies. This represents a notch filter with notch frequency $f_o = 1/2\pi RC$ and $Q = 1/2$. We can choose $R_A = 10$ k and

$R_{x1} = R_{x2} = 10$ k. In the circuit the potentiometer $VR = 1$ k has been added to enable an adjustment for the deepest possible notch. Finally, for a notch frequency $f_o = 1$ kHz, choose $C = 0.01$ μF and therefore $R = 1/2\pi \times 10^3 \times 0.01 \times 10^{-6} = 15.9$ k.

Ideas for Exploration (i) Replace fixed resistors R by ganged variable resistors in series with fixed resistors such that the notch frequency can be continuously varied.

Variable Wide-Band Filter
The circuit in Fig. 11.57 is a variable wide-band filter whose cut-off frequencies are made variable by including variable resistors in series with fixed resistors. Each filter uses equal value resistors and capacitors and unity gain for simplicity. In order to realize a Butterworth response, the gain of each op-amp must be set to 1.6. Such a response is especially useful in audio systems as a scratch and rumble filter to remove unwanted high and low frequencies. The Butterworth response ensures minimum amplitude distortion in the pass-band.

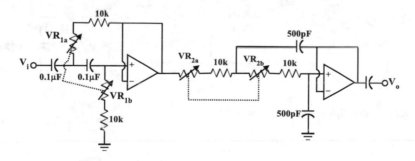

Fig. 11.57 Variable wide-band filter

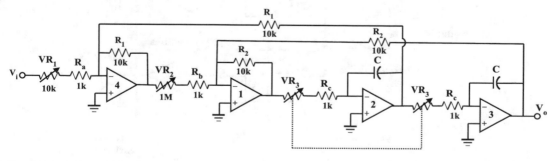

Fig. 11.58 Universal filter system

The dual-ganged potentiometers are 100 k linear type.

Ideas for Exploration (i) Adjust the gains to 1.6 for Butterworth response. (ii) Compare the performance of the 741 op-amp with the OPA134 op-amp in this application.

Universal Filter
The circuit in Fig. 11.58 is a *universal filter* based on the modified sate variable filter. With the values selected ($C = 0.01$ μF and the ganged potentiometer $VR_3 = 100$ k), the resonant frequency ($f_o = 1/2\pi CVR_3$) is variable from about 160 Hz to about 16 kHz, the quality factor ($Q = (VR_2 + 1 \text{ k})/R_2$) is variable from 0.1 to 100, and the centre frequency gain ($G = R_1/(VR_1 + 1 \text{ k})$) is variable from about

1 to 10. The circuit is assembled around a quad op-amp such as the OPA2134.

Ideas for Exploration (i) Use a bank of switched capacitors to increase the frequency range.

Research Project 1
This research project involves the analysis and optimization of the *variable frequency band-pass filter* in Fig. 11.59. It uses cascaded all-pass filters in a feedback loop such that at the frequency where the phase shift is $180°$, the feedback around the input op-amp goes to zero such that the gain is at its maximum. At all other frequencies, the gain is reduced by the feedback. The loop gain is kept less than unity to ensure there is no oscillation. Potentiometer $VR_1 = 20$ k

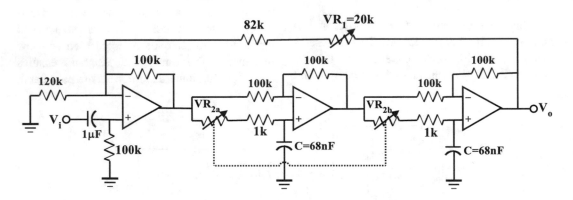

Fig. 11.59 Variable frequency band-pass filter

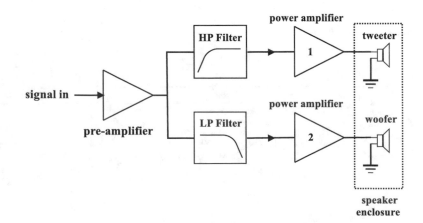

Fig. 11.60 Active cross-over system

adjusts the quality factor, while the ganged potentiometers VR_2 vary the centre frequency of the band-pass filter.

Ideas for Exploration (i) Introduce a bank of switched capacitors for C in order to realize a wide frequency range.

Research Project 2
This project involves research into and the design of an active cross-over system used in audio speaker systems. A diagrammatic representation of such a system is shown in Fig. 11.60. This system separates the high frequencies and the low frequencies of a line level audio signal coming from a preamplifier using a high-pass filter and a low-pass filter, respectively. The output of the high-pass filter drives a power amplifier which delivers power to the tweeter in the speaker enclosure, while the output of the low-pass filter drives a separate power amplifier which drives the woofer in the speaker system. Such an approach allows the signal separation to take place at low signal levels, thereby avoiding the use of the expensive components found in passive cross-over filters used in speaker systems.

Ideas for Exploration (i) Examine the introduction of an active band-pass filter that separates the mid-band frequencies and used to drive a third power amplifier that delivers power to a mid-range speaker.

Problems

1. Design a first-order low-pass non-inverting filter with unity gain and a break frequency of 5 kHz.
2. Design a first-order non-inverting low-pass filter with gain of 8 and a cut-off frequency of 17 kHz.
3. Design a first-order low-pass filter with gain of 12 and a cut-off frequency of 25 kHz using the inverting configuration.

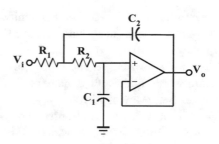

Fig. 11.61 Circuit for Question 7

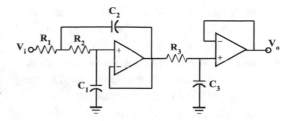

Fig. 11.62 Circuit for Question 12

4. Show that in a first-order low-pass filter, the amplitude response is down 3 dB at the cut-off frequency and determine the phase angle of the output relative to the input at that frequency.
5. Using the Sallen-Key configuration, design a second-order low-pass Butterworth filter with unity gain and a cut-off frequency of 50 kHz. Ensure that offset voltage is minimized.
6. Show that in the Sallen-Key second-order low-pass Butterworth filter, for the case where the network resistors are equal and the network capacitors are equal, the gain must be 1.6. Hence repeat the design of Question 5 using this modified approach.
7. Show that in the Sallen-Key second-order unity-gain low-pass filter configuration shown in Fig. 11.61, if $R_1 = R_2$ and $C_2 = 2C_1$, then the transfer function reduces to a second-order Butterworth response.
8. Design a VCVS second-order unity-gain low-pass Chebyshev filter with a cut-off frequency of 22 kHz and a ripple width $RWdB = 1$dB.

9. Design a VCVS second-order unity-gain low-pass Bessel filter with a cut-off frequency of 550Hz.

10. Design a MFB second-order low-pass Butterworth filter with a gain of 8 and a cut-off frequency of 3 kHz.

11. Design a VCVS third-order unity-gain low-pass Butterworth filter with a cut-off frequency of 33 kHz. What is the ultimate roll-off rate of such a filter?

12. Show that in the Sallen-Key third-order unity-gain low-pass filter configuration shown in Fig. 11.62, if $R_1 = R_2 = R_3$ and $C_2 = 2C_3$ and $C_1 = C_3/2$, then the transfer function reduces to a third-order Butterworth response. Hence repeat Question 11 using this simplified design procedure.

13. Design a VCVS fourth-order unity-gain low-pass Chebyshev filter with a cut-off frequency of 12 kHz and 3 dB ripple width.

14. Design a VCVS fifth-order unity-gain low-pass Butterworth filter with a cut-off frequency of 1200 Hz.

15. Design a first-order high-pass non-inverting filter with unity gain and a break frequency of 5 kHz.

16. Design a first-order non-inverting high-pass filter with gain of 8 and a cut-off frequency of 17 kHz.

17. Design a first-order high-pass filter with gain of 15 and a cut-off frequency of 10 kHz using the inverting configuration.

18. Show that in a first-order high-pass filter, the amplitude response is down 3 dB at the cut-off frequency and determine the phase angle of the output relative to the input at that frequency.

19. Using the Sallen-Key configuration, design a second-order high-pass Butterworth filter with unity gain and a cut-off frequency of 5 kHz.

20. Show that in the Sallen-Key second-order high-pass Butterworth filter, for the case where the network resistors are equal and the network capacitors are equal, the gain must be 1.6. Hence repeat the design of Question 19 using this modified approach.

21. Show that in the Sallen-Key second-order unity-gain high-pass filter configuration shown in Fig. 11.63, if $R_1 = 2R_2$ and $C_1 = C_2$, then the transfer function reduces to a second-order high-pass Butterworth response.

22. Design a VCVS second-order unity-gain high-pass Chebyshev filter with a cut-off frequency of 14 kHz and a ripple width $RWdB = 1$dB.

23. Design a VCVS second-order unity-gain high-pass Bessel filter with a cut-off frequency of 5 kHz.

24. Design a MFB second-order high-pass Butterworth filter with a gain of 6 and a cut-off frequency of 2 kHz.

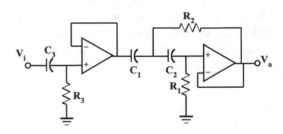

Fig. 11.64 Circuit for Question 26

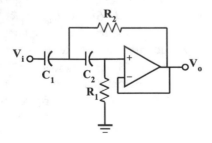

Fig. 11.63 Circuit for Question 21

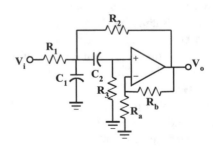

Fig. 11.65 Circuit for Question 29

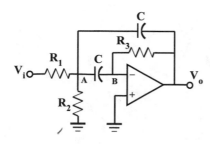

Fig. 11.66 MFB band-pass filter for Question 31

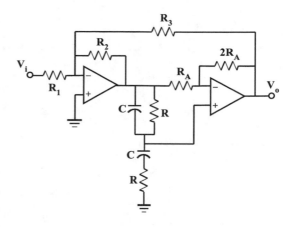

Fig. 11.67 Circuit for Question 35

25. Design a VCVS third-order unity-gain high-pass Butterworth filter with a cut-off frequency of 900 Hz. What is the ultimate roll-off rate of such a filter?

26. Show that in the Sallen-Key third-order unity-gain low-pass filter configuration shown in Fig. 11.64, if $C_1 = C_2 = C_3$ and $R_1 = 2R_3$ and $R_2 = R_3/2$, then the transfer function reduces to a third-order Butterworth response. Hence repeat Question 25 using this simplified design procedure.

27. Design a VCVS fourth-order unity-gain high-pass Chebyshev filter with a cut-off frequency of 1200 Hz and 2 dB ripple width.

28. Design a VCVS fifth-order unity-gain high-pass Bessel filter with a cut-off frequency of 6 kHz.

29. Using the VCVS band-pass filter shown in Fig. 11.65, design a Sallen-Key second-order band-pass filter having $C_1 = C_2$, $R_1 = R_2$ with a centre frequency of 13 kHz and a quality factor of 8.

30. For the VCVS band-pass filter shown in Fig. 11.65, develop design equations for R_1, R_2 and R_3 if $C_1 = C_2 = C$ and $1 + R_b/R_a = 2$. Hence, design a VCVS band-pass filter having $f_o = 20$ kHz, $Q = 6$ and $G = 10$.

31. Design a second-order MFB band-pass filter with $f_0 = 2$ kHz, $Q = 9$ and $G = 5$. Use the circuit shown in Fig. 11.66.

32. For the second-order MFB band-pass filter shown in Fig. 11.66, develop design equations without imposing the condition

$C_1 = C_2$. Hence, design a second-order MFB band-pass filter having $f_o = 1$ kHz, $Q = 5$ and $G = 10$.

33. Design a second-order Wien band-pass filter with $f_0 = 60$ Hz, $Q = 10$ and $G = 2$.

34. Design a twin-T notch filter having a notch frequency of 2.5 kHz and $Q = 25$.

35. Using the circuit shown in Fig. 11.67, design a Wien notch filter to have a notch frequency of 200 Hz, a Q of 25 and a gain of 2.

36. By using a ganged potentiometer for R and two banks of switched capacitors for C, convert the Wien notch filter in Fig. 11.67 to a variable frequency circuit and specify the frequency range.

37. Develop the transfer function for a Tow-Thomas biquadratic filter.

Bibliography

R. Schaumann, M. Van Valkenburg, *Design of Analog Filters*, 1st edn. (Oxford University Press, 2001)

T. Kugelstadt, in *Active filter design techniques in op amps for everyone*, ed. by R. Mancini, (Texas Instruments, 2002)

A. Waters, *Active filter design* (Mc Graw Hill, 1991)

Oscillators

12

Oscillators deliver an essentially sinusoidal output waveform without input excitation. They are used in a wide range of applications including testing, communication systems and computer systems. Frequencies range from about 10^{-3} Hz to 10^{10} Hz, and in many applications a low distortion is required. There are two main classes of oscillators; RC oscillators in which the frequency determining elements are resistors and capacitors and LC oscillators in which the frequency-determining elements are inductors and capacitors. RC oscillators operate from very low frequencies up to about 10 MHz, making them very useful for audio frequency applications, while LC oscillators are useful for frequencies above about 100 kHz, and they are usually used in communications applications. This chapter explores the basic principles governing this class of circuits. The conditions required for oscillation are investigated and frequency and amplitude stability studied. At the end of the chapter, the student will be able to:

- Explain the principle of operation of an oscillator and the criterion for oscillation
- Explain the operation of a wide range of oscillators
- Design a wide range of oscillators
- Create new oscillator systems

12.1 Conditions for Oscillation

The diagram in Fig. 12.1 shows a closed-loop positive feedback system consisting of an amplifier A, a feedback network β and a summing junction $\sum$. The transfer function for this system is given by

$$\frac{X_o}{X_i} = \frac{A(s)}{1 - A(s)\beta(s)} \qquad (12.1)$$

where X_o is the output signal, X_i is the input signal, X_f is the feedback signal and X_e is the error signal resulting from a summation of the feedback signal with the input signal. $A\beta(s)$ is the loop gain. In negative feedback systems, the feedback signal X_f is subtracted from the input signal. In positive feedback systems, X_f is added as it is here, resulting in an increase in the error signal X_e. If the loop gain goes to unity then, the closed-loop transfer function goes to infinity implying that the system can sustain a finite output signal with no input signal. In such a circumstance, the output waveform need not be sinusoidal, and the amplifier need not be linear. All that is required is that $A\beta = 1$ which means that the instantaneous value of X_i is at all times equal to the value of X_f. This condition is called the Barkhausen criterion and is expressed as

$$A(s)\beta(s) = 1 \qquad (12.2)$$

This condition is really equivalent to two conditions:

© Springer Nature Switzerland AG 2021
S. J. G. Gift, B. Maundy, *Electronic Circuit Design and Application*,
https://doi.org/10.1007/978-3-030-46989-4_12

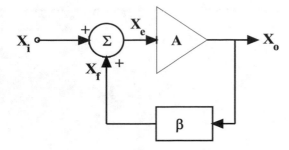

Fig. 12.1 Oscillator system

$$|A(s)||\beta(s)| = 1 \qquad (12.3)$$

i.e. the magnitude of the loop gain must be unity and

$$\angle A(s)\beta(s) = 2\pi n \qquad (12.4)$$

where $n = 0,1,2,\ldots$, i.e. the total phase shift through the amplifier and feedback network must be zero or multiples of 2π.

If the feedback network (or amplifier) contains reactive elements and the entire oscillator is approximately linear, then the only periodic waveform which will preserve its form is the sinusoid. What this means in effect is that since the reactive element introduces phase shift that changes with frequency, then the Barkhausen criterion will only be met at a particular frequency, and hence a sustained sinusoidal output at that frequency is produced. A general principle for oscillators can therefore be stated:

> The frequency of operation of an oscillator is that frequency at which the total phase shift around the system is zero (or multiples of 2π) and the loop gain is one (or greater).

When these conditions are satisfied, a signal X_i entering the system returns such that X_f is equal to or greater than X_i. This results in an equal or larger signal X_i entering the system, which will again reappear as a still larger signal X_f. This will continue and the output signal X_o will increase indefinitely. It is limited by the saturation of the amplifier delivering the output signal X_o. If the loop gain is less than one, i.e. $|A\beta| < 1$, then removal of the external signal will result in the cessation of oscillations. If the loop gain is greater than one, i.e. $|A\beta| > 1$, the amplitude of the oscillations will

increase until limited by amplifier saturation. Since it is impossible to set the loop gain exactly equal to one, i.e. $|A\beta| = 1$, in practice, the loop gain is usually set somewhat larger than unity in which case the amplitude of the oscillations is limited by the onset of non-linearity. This onset of non-linearity through saturation of the oscillator amplifier is a general feature of practical oscillators. However, it results in distortion of the output signal, which increases as the loop gain is made greater than one. As a practical matter, it is therefore important that the loop gain be made only slightly greater than one, say by about 5% or so, so that oscillation is maintained while distortion is minimized. It is possible to manually set the loop gain close to unity but this is not a reliable approach. Amplitude stabilization by electronic means is possible and is discussed in Sect. 12.6. We will now apply the basic criterion to the design of several types of sinusoidal oscillators.

12.2   RC Oscillators

From the analysis in Sect. 12.1, an oscillator that produces sinusoidal waveforms can be realized by combining a linear amplifier with a frequency-dependent network in its feedback loop. Oscillators that utilize networks comprising resistors and capacitors are referred to as *RC* oscillators.

12.2.1   Wien Bridge Oscillator

An RC oscillator circuit that works on the basis of the Barkhausen criterion is the Wien bridge oscillator shown in Fig. 12.2a. It consists of an amplifier A with a Wien network comprising *RC* elements connected in its feedback loop as shown. The input impedance of the amplifier is assumed to be high, while the output impedance is assumed to be low. These conditions are easily met by using an operational amplifier as the amplifying element. The term Wien bridge derives from the fact that the Wien network and the amplifier gain-setting resistors R_A and R_B as

Fig. 12.2 Wien bridge oscillator. (**a**) Basic system; (**b**) implemented using op-amp

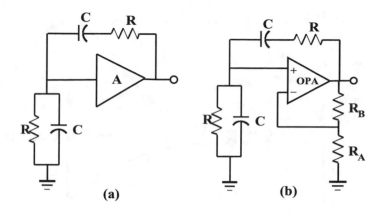

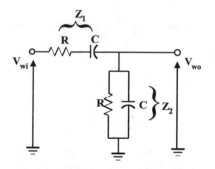

Fig. 12.3 Wien network

shown in Fig. 12.2b form a bridge across which a null occurs at f_o.

In analysing this system, consider the Wien network as shown in Fig. 12.3. The transfer function V_{wo}/V_{wi} for this network is given by

$$\frac{V_{wo}}{V_{wi}} = \frac{Z_2}{Z_1 + Z_2}$$

$$= \frac{R\frac{1}{sC}/\left(R + \frac{1}{sC}\right)}{R + \frac{1}{sC} + R\frac{1}{sC}/\left(R + \frac{1}{sC}\right)} \quad (12.5)$$

This reduces to

$$\frac{V_{wo}}{V_{wi}} = \frac{sCR}{1 + 3sCR + s^2C^2R^2} \quad (12.6)$$

This response is called a band pass response and has centre frequency $\omega_o = 2\pi f_o$ and pass band gain G given by

$$\omega_0 = \frac{1}{CR}, f_o = \frac{1}{2\pi CR} \quad (12.7)$$

and

$$G = \frac{1}{3} \quad (12.8)$$

The magnitude and phase responses can be determined analytically by setting $s = j\omega$ in (12.6) giving

$$\frac{V_{wo}}{V_{wi}}(j\omega) = \frac{j\omega CR}{1 + 3CRj\omega + (j\omega)^2C^2R^2}$$

$$= \frac{jf/f_o}{1 + 3jf/f_o + (jf/f_o)^2} \quad (12.9)$$

The responses are shown in Fig. 12.4. Clearly at the frequency f_o, the phase response shows that there is no phase shift between V_{wo} and V_{wi}. Also, at this frequency, $|V_{wo}/V_{wi}| = 1/3$.

From (12.9) for V_{wo} to be in phase with V_{wi}, since the numerator is imaginary, then the denominator must also be imaginary. Hence, the real part of the denominator must be zero, that is

$$1 - \omega^2C^2R^2 = 0 \quad (12.10)$$

i.e.

$$\omega_0 = \frac{1}{CR} \quad \text{or} \quad f_0 = \frac{1}{2\pi CR} \quad (12.11)$$

This gives

$$\frac{V_{wo}}{V_{wi}}(\omega_0) = \frac{j\omega CR}{3j\omega CR} = \frac{1}{3} \quad (12.12)$$

Thus, a signal V_{wi} entering the network produces a signal V_{wo} at the output and at the frequency f_o;

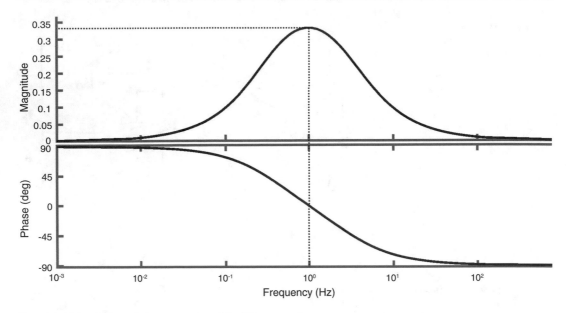

Fig. 12.4 Magnitude and phase response of the Wien network

both signals are in phase with the output signal being attenuated relative to the input by a factor of 3. When this signal passes through the amplifier A, if the amplifier has a gain 3, then the Barkhausen criterion $A\beta = 1$ is satisfied and oscillation results.

The conditions for oscillation in the Wien bridge oscillator can be more directly determined by applying the Barkhausen criterion to the system. Thus for the system in Fig. 12.2a,

$$A\beta = A\frac{j\omega CR}{1 + 3j\omega CR + (j\omega CR)^2} = 1 \quad (12.13)$$

where β is the transfer function of the Wien network given in (12.9) and A is the amplifier gain. From (12.13), we have

$$j\omega CRA = 1 + 3j\omega CR - (\omega CR)^2 \quad (12.14)$$

from which

$$1 - (\omega CR)^2 + j\omega CR(3 - A) = 0 \quad (12.15)$$

Hence equating real and imaginary parts to zero, we have $A - 3 = 0$ and $1 - \omega^2 C^2 R^2 = 0$ giving

$$A = 3 \quad (12.16)$$

$$f_o = \frac{1}{2\pi RC} \quad (12.17)$$

Thus, the amplifier gain for oscillation at $f_o = 1/2\pi RC$ in the Wien bridge oscillator is $A = 3$. A practical implementation of this method is shown in Fig. 12.5 where an operational amplifier is used as the amplifying element with $R_A = 1$ k and $R_B = 2$ k.

Example 12.1
Design a Wien bridge oscillator to operate at a frequency of 1 kHz.

Solution Using the configuration in Fig. 12.5, we first choose a reasonable value for C say $C = 0.01$ μF. Then, for oscillation at $f = 1$ kHz using Eq. (12.17) the value of R is given by $R = \frac{1}{2\times\pi\times10^3\times0.01\times10^{-6}} = 15.9$k. Also, from (12.16), the gain of the op-amp is given by $1 + \frac{R_B}{R_A} = 3$. If we choose $R_A = 1$ k then $R_B = 2$ k. In order to sustain oscillation, we make R_2 slightly greater than 2 k say 2.01 k

This circuit was simulated in the laboratory using an LF351 operational amplifier and a ±15-volt supply. The frequency of the resulting sinusoidal waveform was 968 Hz and the

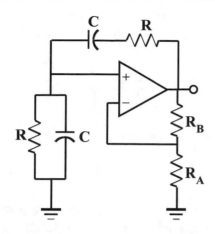

Fig. 12.5 Complete circuit of Wien bridge oscillator

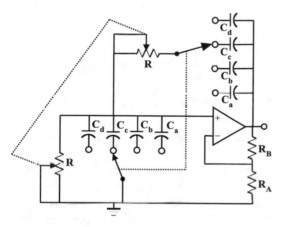

Fig. 12.6 Wien bridge oscillator with wide-range tuning

amplitude 13.3 volts peak. Better accuracy in the frequency can be achieved by using an operational amplifier with a higher gain bandwidth product. For amplitude stabilization, R_B can be a thermistor with a negative temperature coefficient. This is discussed in Sect. 12.6. An advantage of the Wien bridge oscillator is its ability to cover a wide frequency range by using ganged variable resistors for R for continuous variation of f_o over a given band and switched banks of capacitors for C to vary the frequency band as shown in Fig. 12.6.

Example 12.2

Derive the transfer function for the RC network shown in Fig. 12.7, and determine the frequency

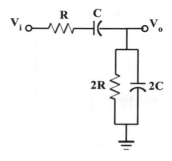

Fig. 12.7 RC network

at which the output is in phase with the input. If the network is used as the basis of a sinusoidal oscillator, determine the minimum voltage gain required from the associated amplifier, and design the system to oscillate at 10 kHz.

Solution For the network, $\frac{V_o}{V_i} = \frac{2R \cdot \frac{1}{2sC}/\left(2R + \frac{1}{2sC}\right)}{R + \frac{1}{sC} + \frac{2R.1/2sC}{2R + 1/2sC}} = \frac{2sCR}{1 + 7sCR + 4s^2C^2R^2} = \frac{2sCR}{1 - 4\omega^2R^2C^2 + 7sCR}$ where $s = j\omega$. For V_o to be in phase with V_i, $1 - 4\omega^2R^2C^2 = 0$ which gives $\omega_o{}^2 = 1/4C^2R^2$ or $f_o = 1/4\pi CR$. Hence, $\frac{V_o}{V_i}(f_o) = \frac{2}{7}$. An oscillator using this network would replace the Wien network in Fig. 12.5 by this one. Since the network attenuation at the oscillating frequency is 2/7, for oscillation, based on the Barkhausen criterion, it follows that the amplifier gain must be 7/2, i.e. $1 + \frac{R_B}{R_A} = \frac{7}{2}$. If we choose $R_A = 1$ k, then $R_B = 2.5$ k. For $f_o = 10$ kHz $= 1/4\pi CR$, choose $C = 0.001$ μF. Then $R = \frac{1}{4 \times \pi \times 0.001 \times 10^{-6} \times 10^4} = 8$ k . Hence $2R = 16$ k and $2C = 0.002$ μF.

12.2.2 Phase Shift Oscillator-Lead Network

The phase shift oscillator is shown in Fig. 12.8 and consists of an inverting amplifier A driving three cascaded RC combination, the output of the last RC combination being returned to the input of the amplifier. If the loading of the phase shift network by the amplifier is neglected, the amplifier shifts the signal appearing at its input by 180° and the RC network shifts the signal phase by an

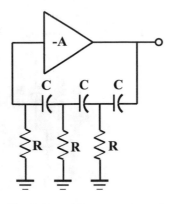

Fig. 12.8 Phase shift oscillator system

additional 180°. At the frequency f_o at which the phase shift introduced by the RC network is precisely 180°, the total phase shift around the system will be zero. This frequency f_o will be the frequency at which the circuit will oscillate, provided that the magnitude of the loop gain is sufficiently large, i.e. greater than or equal to one. For the lead phase shift network shown in Fig. 12.9, the transfer function V_o/V_i is given by

$$\frac{V_o}{V_i} = \frac{V_o}{V_B}\frac{V_B}{V_A}\frac{V_A}{V_i}$$

$$= \frac{R}{R - jX_C}\frac{Z_B}{Z_B - jX_C}\frac{Z_A}{Z_A - jX_C} \quad (12.18)$$

where

$$X_C = \frac{1}{2\pi f C} \quad (12.19)$$

$$Z_B = \frac{R(R - jX_C)}{2R - jX_C} \quad (12.20)$$

$$Z_A = \frac{R(Z_B - jX_C)}{R + Z_B - jX_C} \quad (12.21)$$

After substitution and extensive manipulation, the transfer function becomes

$$\frac{V_o}{V_i} = \frac{1}{1 - 5\alpha^2 - j(6\alpha - \alpha^3)} \quad (12.22)$$

where

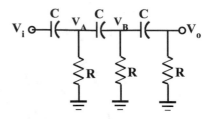

Fig. 12.9 Lead RC network

$$\alpha = \frac{1}{\omega CR} \quad (12.23)$$

In order that the phase shift network contributes 180° at f_o, the imaginary term in (12.22) must be zero, i.e.

$$6\alpha - \alpha^3 = 0 \quad (12.24)$$

giving

$$\omega_0 = \frac{1}{\sqrt{6}CR} \text{ or } f_0 = \frac{1}{2\pi\sqrt{6}CR} \quad (12.25)$$

By substituting $\omega_o = 2\pi f_o$ in (12.22), we get

$$\frac{V_o}{V_i}(f_o) = -\frac{1}{29} \quad (12.26)$$

In order to satisfy the Barkhausen criterion, then $A = -29$, i.e. the magnitude of the gain of the associated amplifier must be 29 or greater.

A practical implementation of the phase shift oscillator using the lead network is shown in Fig. 12.10. An inverting op-amp is used to provide 180° phase shift as well as the gain of −29. Note that the input resistor R of the op-amp is one of the resistors of the phase shift network. This is possible since the inverting input of the op-amp is a virtual earth. The phase shift oscillator is not very useful as a variable frequency oscillator since frequency variation requires variation of three components simultaneously. It can however be used as a fixed frequency oscillator.

Example 12.3
Design a phase shift oscillator to operate at a frequency of 1 kHz.

Solution Using the circuit of Fig. 12.10, we first design the phase shift network. Choosing

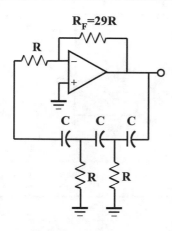

Fig. 12.10 Practical phase shift oscillator

$C = 0.01\ \mu F$, Eq. (12.25) yields $R = \frac{1}{2\pi 10^3 \sqrt{6} \times 0.01 \times 10^{-6}} = 6497\ \Omega \cong 6.5\text{k}$. Hence $R_f = 29 \times R = 29 \times 6.5\text{ k} = 188.5$ k. Use $R_f = 190$ k to ensure oscillation. This circuit was simulated using an LF351 op-amp and ± 15-volt supply. The measured frequency was 958 Hz, and the measured peak amplitude was 13.1 volts.

12.2.3 Phase Shift Oscillator-Lag Network

The phase shift oscillator may be constructed using three lag RC sections instead of the lead RC sections used previously. The basic system is shown in Fig. 12.11.

Neglecting amplifier loading the inverting amplifier shifts the signal by 180°, while the lag network shifts negatively by an additional factor. At a frequency f_o at which this shift is $-180°$, with a loop gain of one or greater, oscillation results. For the phase shift lag network shown in Fig. 12.12, the transfer function V_o/V_i is given by

$$\frac{V_o}{V_i} = \frac{V_o}{V_B}\frac{V_B}{V_A}\frac{V_A}{V_i}$$

$$= \frac{-jX_C}{R - jX_C}\frac{Z_B}{Z_B + R}\frac{Z_A}{Z_A + R} \quad (12.27)$$

where

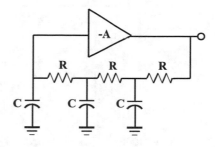

Fig. 12.11 Phase shift oscillator system

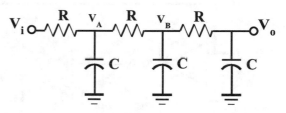

Fig. 12.12 RC lag network

$$X_C = \frac{1}{2\pi f C} \quad (12.28)$$

$$Z_B = \frac{-jX_C(R - jX_C)}{R - 2jX_C} \quad (12.29)$$

$$Z_A = \frac{-jX_C(R + Z_B)}{R + Z_B - jX_C} \quad (12.30)$$

After substitution and considerable manipulation, the transfer function becomes

$$\frac{V_o}{V_i} = \frac{1}{1 - 5\alpha^2 + j(6\alpha - \alpha^3)} \quad (12.31)$$

where

$$\alpha = \omega CR \quad (12.32)$$

For a 180° phase shift contribution at ω_o, the imaginary term in (12.31) must be zero, i.e.

$$6\alpha - \alpha^3 = 0 \quad (12.33)$$

giving

$$\omega_0 = \frac{\sqrt{6}}{CR} \text{ or } f_0 = \frac{\sqrt{6}}{2\pi CR} \quad (12.34)$$

Substituting ω_o in (12.31) gives

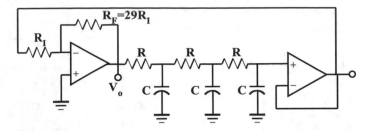

Fig. 12.13 Phase shift lag oscillator with buffered network output

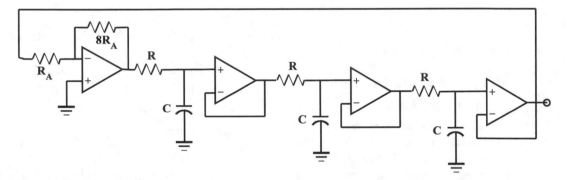

Fig. 12.14 Buffered phase shift oscillator

$$\frac{V_o}{V_i}(f_0) = -\frac{1}{29} \qquad (12.35)$$

In order to satisfy the Barkhausen criterion, $A = -29$ which means that the associated amplifier must have a gain whose magnitude is 29 or greater. In order to prevent the inverting amplifier from loading the network and thereby affect its oscillating frequency, the phase shift oscillator is implemented using the configuration shown in Fig. 12.13 in which a buffer is included at the output of the third RC section. Again, an inverting operational amplifier is used to provide 180° phase shift and −29 gain. A dual op-amp such as the LF353 is convenient for the implementation of this circuit.

Example 12.4
Design a phase shift lag oscillator with an oscillating frequency of 1 kHz.

Solution Using the circuit of Fig. 12.13, for the phase shift network, choose $C = 0.1$ μF. Then

Eq. (12.34) yields $R = \frac{\sqrt{6}}{2\pi 10^3 \times 0.1 \times 10^{-6}} = 4\,\text{k}$. For $R_I = 2.5$ k, then $R_F = 72.5$ k. This circuit based on the LF351 was simulated and produced a sinusoidal waveform of frequency 965 Hz and amplitude 1.22 volts peak. An interesting feature of this circuit is that the three RC low-pass sections filter out harmonics from the signal and hence the distortion at the output of the voltage follower is low.

12.2.4 Buffered Phase Shift Oscillator

The buffered phase shift oscillator shown in Fig. 12.14 involves buffering all RC sections of the phase shift lag network such that there is no interaction between the sections. The effect of this is the simplification of the analysis since there is no interaction between successive RC sections. The transfer function for each section is given by $1/(1 + sRC)$, and hence the Barkhausen criterion gives

$$A\beta = \frac{1}{(1 + sCR)^3} = 1 \qquad (12.36)$$

This reduces for $s = j\omega$ to

$$A = (1 - 3\omega^2 C^2 R^2) + j\omega CR(3 - \omega^2 C^2 R^2) \qquad (12.37)$$

For A real, it follows that the imaginary component on the right must be zero, i.e.

$$3 - \omega^2 C^2 R^2 = 0 \qquad (12.38)$$

giving

$$\omega_0 = \frac{\sqrt{3}}{CR} \text{ or } f_0 = \frac{\sqrt{3}}{2\pi CR} \qquad (12.39)$$

This is the frequency of oscillation. Substituting this in (12.37) gives $A = -8$, i.e. each RC section produces an attenuation of 1/2 at f_o resulting in a total attenuation of 1/8. This circuit is easily implemented using a quad operational amplifier. It is particularly useful for low distortion oscillation as each RC network filters the harmonics from the signal. Also, since the amplifier gain is only 8, more feedback can be employed around this amplifier than for the case where the gain is 29.

Example 12.5

Design a buffered phase shift oscillator to oscillate at 5 kHz. Use an LF347 quad op-amp and a ±15-volt power supply.

Solution Using the circuit in Fig. 12.14, for oscillation at 5 kHz, we choose $C = 0.01$ μF.

Then Eq. (12.39) gives $R = \frac{\sqrt{3}}{2\pi \times 5 \times 10^4 \times 0.01 \times 10^{-6}} = 5.5$ k. The gain of the amplifier is given by $\frac{R_2}{R_1} = 8$. We choose $R_1 = 5$ k giving $R_2 = 40$ k. We use $R_2 = 42$ k in order to sustain oscillation. This circuit was simulated, and it oscillated at a frequency of 4.92 kHz with amplitude of 1.9 volts peak.

12.2.5   Multiphase Sinusoidal Oscillator

Another type of phase shift oscillator is the multiphase sinusoidal oscillator shown in Fig. 12.15. This circuit uses active phase shift first-order elements to produce 360° around the loop. Each of the elements is identical, and therefore three equal amplitude outputs are produced each shifted in phase from the other by 120°. This kind of oscillator is particularly useful in communications systems. The transfer function for each section is given by $-k/(1 + sRC)$, and applying the Barkhausen criterion gives

$$A\beta = \left(\frac{-R/R_1}{1 + sRC}\right)^3 = 1 \qquad (12.40)$$

For $s - j\omega$, this reduces to

$$(1 + j\omega\tau)^3 + k^3 = 0 \qquad (12.41)$$

where $k = R/R_1$ and $\tau = RC$. This becomes

$$(1 + k^3 - 3\omega^2\tau^2) + j\omega(3 - \omega^2\tau^2) = 0 \qquad (12.42)$$

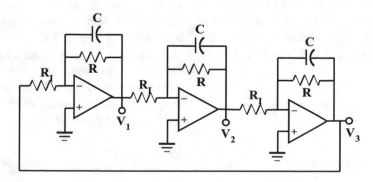

Fig. 12.15 Multiphase sinusoidal oscillator

In order that (12.42) be satisfied, the imaginary part must be zero, i.e.

$$3 - \omega^2 \tau^2 = 0 \qquad (12.43)$$

from which

$$\omega_o = \frac{\sqrt{3}}{RC} \text{ or } f_o = \frac{\sqrt{3}}{2\pi RC} \qquad (12.44)$$

Substituting this in (12.42) gives $k^3 = 8$, and therefore $k = 2$, i.e. each stage must have a gain of 2.

Example 12.6
Design a multiphase sinusoidal oscillator using the circuit of Fig. 12.15 to give a sinusoidal output of frequency 2 kHz.

Solution Using the circuit of Fig. 12.15, we select $C = 0.01\,\mu\text{F}$. Hence Eq. (12.44) gives $R = \frac{\sqrt{3}}{2\pi \times 2 \times 10^3 \times 0.01 \times 10^{-6}} = 13.8\,\text{k}$. The gain of the amplifier is given by $\frac{R}{R_1} = 2$. For $R = 13.8$ k, then $R_1 = 6.9$ k. The simulated circuit oscillated at a frequency of 1.96 kHz and an amplitude of 13.3 volts peak.

12.2.6   Quadrature Oscillator

The quadrature oscillator achieves oscillation by cascading sections that shift the phase of the signal 360° at a loop gain of unity. Three RC sections are cascaded such that two of the sections produce 45° of phase shift while the third in conjunction with the inversion of the op-amp produces 270°. The outputs of the two op-amps are 90° out of phase and hence are referred to as sine and cosine, i.e. quadrature. The basic circuit is shown in Fig. 12.16. The loop gain $A\beta$ is given by

$$A\beta = -\frac{1}{R_1 C_1 s} \cdot \frac{1}{1 + R_2 C_2 s} \cdot \frac{1 + R_3 C_3 s}{R_3 C_3 s} \qquad (12.45)$$

For

$$R_1 C_1 = R_2 C_2 = R_3 C_3 \qquad (12.46)$$

Equation (12.45) reduces to

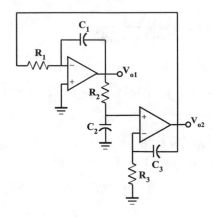

Fig. 12.16 Quadrature oscillator

$$A\beta = -\frac{1}{R^2 C^2 s^2} = \frac{1}{R^2 C^2 \omega^2} \qquad (12.47)$$

For oscillation, $A\beta = 1$, and hence (12.47) yields the frequency of oscillation.

$$\omega_o = \frac{1}{RC} \text{ or } f_o = \frac{1}{2\pi RC} \qquad (12.48)$$

By inspection of the second and third terms in the transfer function (12.45), it can be seen that at f_o, the second term corresponding to the passive section $R_2 C_2$ contributes 45° phase lag while the third term corresponding to the elements $R_3 C_3$ and the operational amplifier contributes 45° phase lag. Hence the output V_{o1} is 90° out of phase with V_{o2}. These outputs are therefore labelled sine and cosine or quadrature. One advantage of this circuit is that the circuit offsets tend to balance each other. Dual operational amplifiers such as the 1458 are convenient for the implementation of this oscillator. R_3 is usually made slightly less than R to ensure that the circuit oscillates. In such a scenario, the op-amps will eventually saturate, and it may be necessary to use amplitude limiting Zener diodes across the capacitor in the Miller integrator. As is the case with all of these circuits, the maximum oscillating frequency is determined by the GBP of the operational amplifiers involved.

Example 12.7
Design a quadrature oscillator to oscillate at 1 kHz.

Solution Using the configuration in Fig. 12.16, for C = 0.01 µF and f_o = 1 kHz, Eq. (12.48) gives $R = \frac{1}{2\pi \times 10^3 \times 10^{-8}} = 15.9\,\mathrm{k}$. This circuit oscillated at a frequency of 1 kHz and amplitude 13.5 V peak. R_3 was set at 15.5 k in order to start oscillation.

12.2.7 Another Quadrature Oscillator

The final RC oscillator to be discussed is another quadrature oscillator. It utilizes two Miller integrators, each producing a phase shift of 90° along with an inverting amplifier, which produces a further phase shift of −180° such that the total phase shift is 0°. The circuit is shown in Fig. 12.17. The loop gain $A\beta$ is given by

$$A\beta = \frac{-1}{sC_1R_1} \cdot \frac{-1}{sC_2R_2} \cdot \frac{-R_4}{R_3} \qquad (12.49)$$

Since $A\beta = 1$ for oscillation, then, letting $C_1 = C_2 = C$, $R_1 = R_2 = R$ and $R_3 = R_4 = R_A$, then (12.49) gives

$$\omega_o = \frac{1}{RC} \text{ or } f_o = \frac{1}{2\pi RC} \qquad (12.50)$$

This circuit has the disadvantage that offset voltages must be precisely balanced; otherwise, they will cause a build-up of voltage across the capacitors of the two integrators and consequently dampen the oscillations. The circuit in Fig. 12.16 automatically balances its offsets and hence is superior.

12.3 LC Oscillators

RC oscillators require at least two RC sections in order to produce the required phase shift to realize oscillation. Any resulting signal loss is compensated for by amplification. If the amplifier gain is excessive, the distortion level is higher since less negative feedback is available. LC oscillators are able to produce the necessary phase shift with two components and the frequency of oscillation is the resonant frequency of the LC circuit. Moreover, tuning is usually via one component. Finally, LC oscillators are often used at higher frequencies than RC oscillators since the amplifying element can be a single transistor having a higher frequency response than an op-amp. In this section we consider the theory and design of LC oscillators.

12.3.1 LC Resonant Oscillator

The first of the LC oscillators for producing sinusoidal waveforms to be considered is the LC resonant oscillator shown in Fig. 12.18. It utilizes an LC frequency-selective circuit in the drain of a JFET. A secondary winding couples signal from the coil back to the JFET input. The JFET in the common source configuration produces 180° phase shift while an appropriate connection of the transformer (see dots) produces a further 180° phase shift. For the FET, the voltage gain A_v is given by

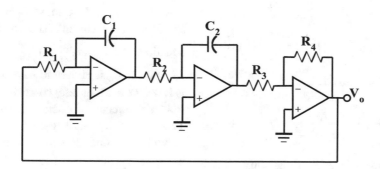

Fig. 12.17 Another quadrature oscillator

Fig. 12.18 LC resonant oscillator

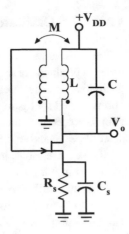

$$A_V = \frac{V_o}{V_i} = -g_m r_d \| jX_L \| -jX_C \qquad (12.51)$$

where r_d is the output resistance of the FET. The high FET input impedance prevents transformer loading. Now, for the circuit,

$$V_o = L\frac{di_1}{dt} + M\frac{di_2}{dt} \approx L\frac{di_1}{dt} \qquad (12.52)$$

$$V_f = -L\frac{di_2}{dt} - M\frac{di_1}{dt} \approx -M\frac{di_1}{dt} \qquad (12.53)$$

since $i_2 \approx 0$ and where M is the mutual inductance between the two windings, i_1 is the current in the transformer primary and i_2 is the current in the transformer secondary. From (12.52) to (12.53)

$$\beta = \frac{V_f}{V_o} = \frac{-M}{L} \qquad (12.54)$$

Hence

$$A\beta = -g_m r_d \| jX_L \| - jX_C \cdot \frac{-M}{L} \qquad (12.55)$$

$$A\beta = \frac{M}{L}g_m \frac{r_d}{1 + \frac{r}{jX_L \| -jX_C}}$$

$$= \frac{M}{L}g_m \frac{r_d}{1 + r_d \frac{j(X_L - X_C)}{X_L X_C}} \qquad (12.56)$$

For the Barkhausen criterion to be satisfied, we must have zero phase shift. From (12.56), this occurs when the imaginary component of $A\beta$ is zero, i.e.

$$X_L = X_C \qquad (12.57)$$

giving

$$f_o = \frac{1}{2\pi\sqrt{LC}} \qquad (12.58)$$

The system will sustain oscillations if $A\beta = 1$, i.e.

$$\frac{M}{L}g_m r_d = 1 \qquad (12.59)$$

If the coil has resistance r_s which is effectively in series with the inductor L, then the equivalent parallel resistance R_{eff} is given by

$$R_{eff} = r_s Q^2 \qquad (12.60)$$

where Q is the quality factor of the inductor given by

$$Q = \frac{\omega_o L}{R} \qquad (12.61)$$

The equations are adjusted by replacing r_d by $r_d \| R_{eff} = \overline{r_d}$. The frequency of oscillation is therefore unaffected while the loop gain condition (12.59) becomes

$$\frac{M}{L}g_m\overline{r_d} = 1 \qquad (12.62)$$

Example 12.8
Design an LC resonant oscillator to operate at 31.8 kHz. Use a 2N3819 JFET and a 15-volt supply.

Solution For the 2N3819, $V_P = -3$ V, $I_{DSS} = 10$ mA and $r_d = 50$ k. Choose $I_D = 1$ mA for the JFET operating current. Then using Shockley's equation, $R_S = \left(\sqrt{\frac{I_D}{I_{DSS}}} - 1\right)\frac{V_P}{I_D} = \left(\sqrt{\frac{1}{10}} - 1\right) \times \frac{-3}{1} = 2.1\,\text{k}$, C_S is chosen to have a reactance that is low (one tenth or less) compared with R_S at the lowest frequency of operation. Use $C_S = 1\,\mu\text{F}$. The transconductance g_m of the FET at this current is using Shockley's equation given by $g_m = \frac{\partial I_D}{\partial V_{GS}} = -2\frac{I_{DSS}}{V_P}\left(\frac{I_D R_S}{V_P} + 1\right) = 2.2\,\text{mA/V}$. For $C = 1000$ pF, (12.58) gives $L = 25$ mH. Hence $\frac{M}{L} \geq \frac{1}{g_m r_d} = 0.009$. In order to ensure oscillation in the simulation, M was set at 10 mH and the gain of the amplifier reduced by bypassing only half of R_S. The system oscillated at 29.3 kHz with amplitude of 24.2 volts peak to peak.

The LC resonant oscillator may be implemented using a special current feedback amplifier, the AD844, in which access to the compensation node is available. This is shown in Fig. 12.19. The Barkhausen criterion gives

$$A\beta = \frac{r_z\|jX_L\|-jX_C}{R_3 + r_x}\rho$$
$$= \frac{\rho}{R_3 + r_x}\frac{r_z}{1 + r_z\frac{j(X_L - X_C)}{X_L X_C}} = 1\angle 0^o \quad (12.63)$$

where

$$\rho = \frac{R_1}{R_1 + R_2} \quad (12.64)$$

and r_x and r_z are parameters of the AD844. From (12.63), it follows that the imaginary component of $A\beta$ must be zero, i.e.

$$X_L = X_C \quad (12.65)$$

giving

$$f_o = \frac{1}{2\pi\sqrt{LC}} \quad (12.66)$$

At the frequency of oscillation, $A\beta(f_o) = 1$, i.e.

$$\frac{R_1}{R_1 + R_2}\frac{r_z}{R_3 + r_x} = 1 \quad (12.67)$$

In this implementation, the amplifier produces zero phase shift at f_o, and, hence, no further signal inversion is required. The signal is fed back to the non-inverting input of the current feedback amplifier.

Example 12.9

Design an LC resonant oscillator using the AD844 to operate at 1 kHz.

Solution For a frequency of 1 kHz choose $C = 0.005$ μF. Then using (12.66), we get $L = 5$ H. Condition (12.67) gives $\frac{R_1}{R_1 + R_2} \cdot \frac{r_z}{R_3 + r_x} = 1$. For the AD844, $r_z = 3 \times 10^6$ Ω and $r_x = 50$ Ω. Choose $R_1 = 1$ kΩ, $R_2 = 2$ kΩ and hence $R_3 = 500$ kΩ. This circuit was simulated and the frequency of oscillation was measured at 980 Hz. The primary advantage of this implementation is that a coupling transformer is not required.

12.3.2 Colpitts and Hartley Oscillators

The Colpitts and Hartley oscillators can be represented in the general form shown in Fig. 12.20. The active device may be a FET or operational amplifier with high input impedance. Its gain with no load is A_v and its output impedance R_o. The equivalent circuit is shown. The load impedance Z_L consists of Z_2 in parallel with a series combination Z_1 and Z_3. Feedback to the

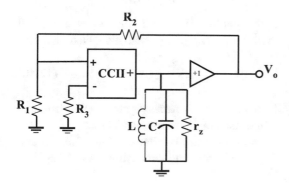

Fig. 12.19 Resonant LC oscillator using the AD844 current feedback amplifier

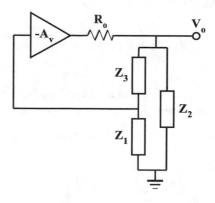

Fig. 12.20 Generalized Colpitts-Hartley oscillator

amplifier is taken from the junction of Z_1 and Z_3. The gain A of the forward loop is given by

$$A = \frac{-A_v Z_L}{Z_L + R_o} \quad (12.68)$$

where Z_L is the load impedance. The feedback factor β is given by

$$\beta = \frac{Z_1}{Z_1 + Z_3} \quad (12.69)$$

Hence the loop gain $A\beta$ is

$$A\beta = \frac{-A_v Z_L}{Z_L + R_o} \cdot \frac{Z_1}{Z_1 + Z_3} \quad (12.70)$$

where

$$Z_L = Z_2 \| (Z_1 + Z_3) = \frac{Z_2(Z_1 + Z_3)}{Z_1 + Z_2 + Z_3}$$

$$= \frac{Z_2(Z_1 + Z_3)}{\Sigma Z} \quad (12.71)$$

therefore

$$A\beta = \frac{-A_v \cdot \dfrac{Z_2(Z_1 + Z_3)}{\Sigma Z}}{R_o + \dfrac{Z_2(Z_1 + Z_3)}{\Sigma Z}} \cdot \frac{Z_1}{Z_1 + Z_3}$$

$$= \frac{-A_v \cdot Z_2(Z_1 + Z_3)}{R_o(Z_1 + Z_2 + Z_3) + Z_2(Z_1 + Z_3)} \cdot \frac{Z_1}{Z_1 + Z_3}$$

$$= \frac{-A_v Z_1 Z_2}{R_o(Z_1 + Z_2 + Z_3) + Z_2(Z_1 + Z_3)}$$

$$(12.72)$$

kIf the load impedance are pure reactances (either inductive or capacitive), then $Z_1 = jX_1$, $Z_2 = jX_2$ and $Z_3 = jX_3$. For an inductor, $X = \omega L$, and for a capacitor $X = -1/\omega C$. Then

$$A\beta = \frac{A_v X_1 X_2}{jR_o(X_1 + X_2 + X_3) - X_2(X_1 + X_3)}$$

$$(12.73)$$

For zero phase shift around the loop, the imaginary component in $A\beta$ must go to zero, i.e.

$$X_1 + X_2 + X_3 = 0 \quad (12.74)$$

giving

$$A\beta = \frac{A_v X_1 X_2}{-X_2(X_1 + X_3)} = \frac{-A_v X_1}{X_1 + X_3} \quad (12.75)$$

From $\Sigma X = 0$, it follows that the circuit will oscillate at the resonant frequency of the series combination of X_1, X_2 and X_3. Using $\Sigma X = 0$,

$$A\beta = A_v \frac{X_1}{X_2} \quad (12.76)$$

Since $A\beta$ must be positive and at least unity in magnitude, then X_1 and X_2 must have the same sign. This means that they must be the same kind of reactance, either both inductive or both capacitive. Then $X_3 = -(X_1 + X_2)$ must be inductive if X_1 and X_2 are capacitive or vice versa. If X_1 and X_2 are capacitors and X_3 is an inductor, the circuit is called a Colpitts oscillator. If X_1 and X_2 are inductors and X_3 is a capacitor, the circuit is called a Hartley oscillator. These circuits can be implemented using BJT technology, but the low input impedance makes the analysis more difficult.

For the Colpitts oscillator in Fig. 12.21 where X_1 and X_2 are capacitors of equal value, the condition (12.74) gives

$$\frac{-1}{\omega C} + \frac{-1}{\omega C} + \omega L = 0 \quad (12.77)$$

yielding

$$f_o = \frac{\sqrt{2}}{2\pi\sqrt{LC}} \quad (12.78)$$

Hence

$$A\beta = A_v \geq 1 \quad (12.79)$$

If the amplifier is a FET, then $A_v = g_m R_D$ and therefore $g_m R_D \geq 1$ for oscillation. Note that at f_o,

$$\beta = \frac{Z_1}{Z_1 + Z_3} = -1 \quad (12.80)$$

Inclusion of R_o is necessary to enable identification of the condition for oscillation.

For the Hartley oscillator in Fig. 12.22, X_1 and X_2 are inductors and the condition (12.74) gives

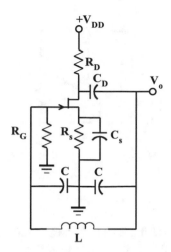

Fig. 12.21 Colpitts oscillator

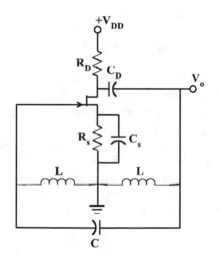

Fig. 12.22 Hartley oscillator

$$\omega L + \omega L - \frac{1}{\omega C} = 0 \qquad (12.81)$$

giving

$$f_o = \frac{1}{2\pi\sqrt{2LC}} \qquad (12.82)$$

and $A_v \geq 1$ for oscillation.

Example 12.10

Design a Hartley oscillator to operate at 35.6 kHz using a 2N3819 JFET. Use a supply voltage of 15 volts.

Solution Using a 2N3819 n-channel JFET, we choose a 1 mA drain current. Then, as in Example 12.8, $R_S = 2.1$ kΩ and C_S is chosen to be 1 μF. Hence for $V_{DD} = 15$ V and $V_{RS} = 2.1$ V, then for maximum symmetrical swing, $V_{DS} = (15 - 2.1)/2 = 6.45$ V giving $R_D = 6.45$ k. Use a 5.6 k resistor. Note that no biasing resistor is needed since the gate is grounded through L. The coupling capacitor C_D is chosen to be 1 μF so that its impedance compared to R_D is small. Using (12.82), for $f = 35.6$ kHz and $C = 0.001$ μF, $L = 10$ mH. Since $g_m R_D = 2.2 \times 5.6 = 12.3 \geq 1$ the circuit will oscillate. Under simulation the circuit oscillated at 32 kHz with amplitude of 8 V peak to peak. For effective oscillation in the simulation, the gain of the amplifier was reduced by introducing local feedback at the source of the FET.

Example 12.11

Design a Colpitts oscillator to oscillate at 31.8 kHz using a 2N3819 JFET. Use a supply voltage of 15 volts.

Solution For $I_D = 1$ mA, Shockley's equation gives $R_S = 2.1$ k. $C_S = 1$ μF is chosen. Also $R_G = 1$ M is used to ground the gate while giving it a high input impedance. Using (12.78), for $C = 0.005$ μF, then $L = 10$ mH. Since $g_m R_D = 12.3 \geq 1$, the circuit will oscillate. $C_D = 1$ μF is a coupling capacitor. The circuit was simulated in the laboratory, and f_o was measured at 31 kHz. In order to reduce the distortion, the FET gain may be reduced by only bypassing a part of R_S.

12.3.3 Clapp Oscillator

The Clapp oscillator shown in Fig. 12.23 is a variation of the Colpitts oscillator in which the tuning element is an inductor L in series with a capacitor C_T across the normal dual capacitor arrangement of the Colpitts.

Thus, Z_1 and Z_2 are capacitors, and Z_3 is an inductor in series with a capacitor. For this circuit, the loop gain is given by

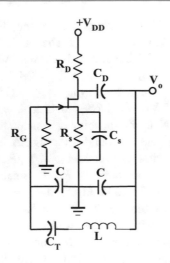

Fig. 12.23 Clapp oscillator

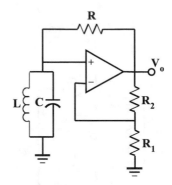

Fig. 12.24 Simple LC oscillator

$$A\beta = \frac{-A_V Z_1 Z_2}{R_o(Z_1 + Z_2 + Z_3) + Z_2(Z_1 + Z_3)} \tag{12.83}$$

where Z_1 and Z_2 are capacitors C and Z_3 is an inductor L in series with a capacitor C_T. Substituting $Z_1 = jX_C$, $Z_2 = jX_C$ and $Z_3 = j(X_L + X_{CT})$ in (12.83) results in

$$A\beta = \frac{A_V X_C X_C}{jR_o(2X_C + X_L + X_{CT}) - X_C(X_C + X_L + X_{CT})} \tag{12.84}$$

For zero phase shift around the loop, then

$$2X_C + X_L + X_{CT} = 0 \tag{12.85}$$

Substituting this in (12.84) results in the condition $A_v \geq 1$. Note that (12.85) yields

$$f_o = \frac{1}{2\pi} \frac{\sqrt{\frac{C}{C_T} + 2}}{\sqrt{LC}} \tag{12.86}$$

The capacitor C_T enables easy tuning of the oscillator.

Example 12.12
Evaluate the frequency of oscillation of the resulting Clapp oscillator by the introduction of a capacitor $C_T = 0.01$ μF in the Colpitts oscillator of Example 12.11.

Solution Using Eq. (12.86), $C_T = 0.01$ μF along with $C = 0.005$ μF and $L = 10$ mH gives $f_o = 35.6$ kHz. The simulated circuit oscillated at 34 kHz with amplitude 7.3 V peak to peak.

12.3.4 Simple LC Oscillator

The LC oscillator shown in Fig. 12.24 utilizes an LC tank circuit in the positive feedback loop of an amplifier. At the resonant frequency of the circuit, the impedance becomes (i) very large such that positive feedback is maximum and (ii) resistive such that the phase shift of the feedback signal is zero. Hence, since the phase shift produced by the amplifier is zero, the conditions for oscillation can be met. The loop gain is given by

$$A\beta = \frac{Z_{LC}}{R + Z_{LC}} \cdot \frac{R_1 + R_2}{R_1} \tag{12.87}$$

where

$$Z_{LC} = \frac{jX_L(-jX_C)}{jX_L - jX_C} \tag{12.88}$$

At the resonant frequency ω_o

$$jX_L - jX_C = 0 \tag{12.89}$$

giving

$$\omega_o = \frac{1}{\sqrt{LC}} \text{ or } f_o = \frac{1}{2\pi\sqrt{LC}} \tag{12.90}$$

At this frequency, from (12.87) the loop gain goes to

$$A\beta = 1 + \frac{R_2}{R_1} \geq 1 \qquad (12.91)$$

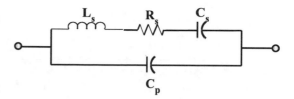

Example 12.13

Design a simple LC oscillator to oscillate at a frequency of 1 kHz, using an LF351 op-amp.

Fig. 12.25   Equivalent circuit of crystal

Solution Using an LF351 op-amp, we choose $L = 5$ H. Then, (12.90) gives $C = 0.005$ μF with $R = 1$ kΩ. Then based on (12.91), we select $R_2 = 1$ kΩ and $R_1 = 500$ kΩ. This circuit was simulated and oscillated at 1 kHz with amplitude of 13.5 volts peak. Note that the loop gain was set close to one to avoid undue waveform clipping.

There are many variations of LC-type oscillators that can be explored. Many of these can be implemented using BJT technology, and the reader is encouraged to explore some of these circuits.

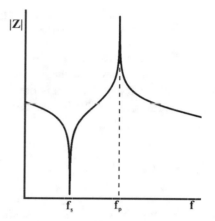

Fig. 12.26   Impedance characteristic of crystal

12.4   Crystal Oscillators

Apart from *RC* and *LC* circuits, crystals also represent a frequency-dependent circuit. A small crystal, normally made using quartz, is mounted under tension in a mechanical frame. Stress is applied to create a natural mechanical resonant frequency of vibration, which depends on both crystal properties and characteristics of the mount. Quartz possesses the piezoelectric property, i.e. it generates electricity when the material is subjected to pressure. Thus when a piezoelectric crystal is mechanically deformed, a voltage appears across its faces. Inversely, when a voltage is applied to the faces of the crystal, the crystal will mechanically deform.

Because quartz is piezoelectric in a circuit, a mechanical-electrical interaction can occur with the crystal. These characteristics provide the device with an equivalent circuit as shown in Fig. 12.25. It comprises a series *RLC* circuit in parallel with a capacitor. The reactance characteristic is shown in Fig. 12.26. It has both series resonance frequency f_s and parallel resonance frequency f_p. At the series resonance, the

inductive reactance cancels the capacitive reactance, and the resulting impedance is resistive and at a minimum. The parallel resonant frequency occurs a few kilohertz higher, and the crystal impedance is at a maximum. The fundamental frequency or frequency of operation and the load capacitance in the case of parallel resonance must be specified. The fundamental frequency is usually between 500 kHz and 20 MHz and is usually parallel resonance. It is difficult to make crystals operating at higher frequencies. Below about 500 kHz, the fundamental frequency is usually series resonance, which is more stable.

The electrical characteristics of the crystal can be determined by considering the equivalent circuit of the crystal. The two parameters L_s and C_s are set by the mechanical properties of the crystal. The damping is represented by R_s which is small while the capacitance of the electrodes is represented by C_p. Typical values for these parameters in a 4 MHz crystal are as follows: $L_s = 100$ mH, $C_s = 0.015$ pF, $R_s = 100$ Ω, $C_p = 5$ pF and $Q = 25,000$. The impedance of the circuit in Fig. 12.25 is given by

$$Z_{cry} = \left(\frac{1}{sC_s} + R_s + sL_s\right) \Big\| \frac{1}{sC_p}$$

$$= \frac{1 - \omega^2 C_s L_s + j\omega C_s R_s}{-R_s C_s C_p + j\omega\left(C_s + C_p - \omega^2 C_s C_p L_s\right)}$$

$$(12.92)$$

Since R_s is small, this reduces to

$$Z_{cry} = \frac{j}{\omega} \cdot \frac{\left(\omega^2 L_s C_s - 1\right)}{\left(C_s + C_p - \omega^2 C_p C_s L_s\right)} \quad (12.93)$$

From Eq. (12.92), the crystal has two resonant frequencies. The series-resonant frequency ω_s corresponds to $Z_{cry} = 0$ which is obtained by setting the numerator in (12.93) to zero. This gives

$$\omega_p = \frac{1}{\sqrt{L_s C_s}} \text{ or } f_p = \frac{1}{2\pi\sqrt{L_s C_s}} \quad (12.94)$$

The parallel resonance frequency ω_p corresponds to $Z_{cry} = \infty$ which is obtained by setting the denominator in (11.89) to zero. This results in

$$\omega_p = \frac{1}{\sqrt{L_s C_s}}\sqrt{1 + \frac{C_s}{C_p}} \text{ or }$$

$$f_p = \frac{1}{2\pi\sqrt{L_s C_s}}\sqrt{1 + \frac{C_s}{C_p}} \quad (12.95)$$

Note that $f_p > f_s$. From (12.94), it is clear that the series-resonant frequency is determined solely by the parameters L_s and C_s, which are well defined by the crystal. From (12.95), however, the parallel resonant frequency is dependent on the electrode capacitance C_p that is susceptible to variations.

12.4.1 Crystal Oscillator Using an Op-Amp

In the circuit shown in Fig. 12.27, the crystal, along with resistor R, provides positive feedback around the operational amplifier. The loop gain is given by

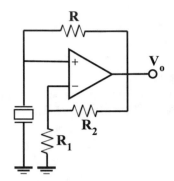

Fig. 12.27 Crystal oscillator

$$A\beta = \left(1 + \frac{R_2}{R_3}\right) \cdot \frac{Z_{cry}}{R + Z_{cry}} \quad (12.96)$$

Near parallel resonance, which is set for a given crystal, $|Z_{cry}|$ is large resulting in $\frac{Z_{cry}}{R+Z_{cry}} \approx 1$, and the phase shift is zero. At other frequencies, $|Z_{cry}|$ falls and $A\beta = 1$ is not met. Hence, for oscillation at the crystal resonant frequency f_o

$$A\beta \approx \left(1 + \frac{R_2}{R_3}\right) \geq 1 \quad (12.97)$$

The operation of this circuit is similar to the simple LC oscillator; the crystal effectively replaces the LC tank circuit of that oscillator.

Example 12.14
Design a crystal oscillator to oscillate at 100 kHz using a suitable operational amplifier.

Solution Using a suitable operational amplifier and a 100 kHz crystal, the circuit was configured using $R_1 = 1$ kΩ, $R_2 = 1$ kΩ and $R_3 = 1$ kΩ. Note that the amplifier gain $A = 2$ ensures that $A\beta > 1$ and hence that oscillation occurs.

12.4.2 Miller Oscillator

A simple crystal oscillator making use of the natural oscillations of the crystal is the Miller oscillator shown in Fig. 12.28. The output of the crystal is used to drive a tuned amplifier. The load of the amplifier consists of an *LC* circuit tuned approximately to the crystal frequency. The design procedure is quite straightforward. At the resonant frequency of the tank circuit

$$jX_L - jX_C = 0 \qquad (12.98)$$

giving

$$\omega_o = \frac{1}{\sqrt{LC}} \text{ or } f_o = \frac{1}{2\pi\sqrt{LC}} \qquad (12.99)$$

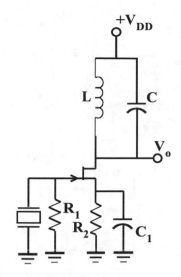

Fig. 12.28 Miller oscillator

Example 12.15

Design a Miller oscillator using a 2N3918 JFET to oscillate at 100 kHz. Use a 15-volt supply.

Solution The bias resistor R_2 as well as the bypass capacitor C_1 can be the same values as those in the *LC*-tuned oscillator. The drain current is therefore 1 mA. The gate bias resistor R_1 is chosen to be 1 MΩ. Using (12.99), for $L = 1$ mH, $C = 2.53$ nF. A 100 kHz crystal must be used at the input of the circuit. In order that oscillations begin, this circuit relies on Miller feedback capacitance between the drain and gate of the FET.

12.4.3 Clapp Oscillator with Crystal Control

The resonant frequency of an *LC* oscillator can be stabilized by the appropriate inclusion of a crystal. In such an application, it is preferable to operate the crystal in its series resonance rather than parallel resonance mode as this provides better frequency stability. The circuit of Fig. 12.29 illustrates the use of a crystal to improve the frequency stability of a Clapp oscillator discussed in Sect. 12.3.3.

12.4.4 Pierce Crystal Oscillator

This oscillator, shown in Fig. 12.30, involves the incorporation of a crystal in the *LC* resonant oscillator to improve its stability. The crystal

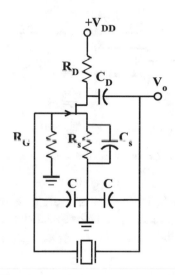

Fig. 12.29 Clapp oscillator with crystal control

permits feedback only at the series-resonant frequency when the impedance is near zero.

12.4.5 AD844 Crystal Oscillator

The *LC* resonant oscillator of Fig. 12.19 involving the AD844 may be implemented using a crystal instead of the *LC* tank circuit. This circuit is shown in Fig. 12.31. The design equations of the original circuit apply with the parallel resonant frequency of the crystal being operational.

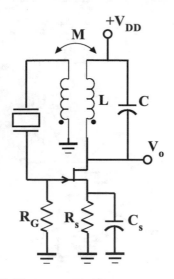

Fig. 12.30 Pierce crystal oscillator

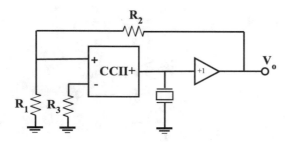

Fig. 12.31 Crystal oscillator using current feedback amplifier

As a result, the output of the circuit will be a maximum at the parallel resonant frequency of the crystal.

12.5   Frequency Stability

The frequency of an oscillator will tend to drift from its design frequency as a result of aging, temperature changes and other factors. The frequency stability of an oscillator is a measure of its ability to maintain its frequency over time. The frequency of oscillation is usually determined by a small number of components, and this enables reasonable levels of stability to be achieved. In these frequency-determining components, the ratio $d\theta/d\omega$ is a measure of the phase shift as a function of frequency change. In a circuit where

this parameter is large, then if phase changes introduced by other circuit elements, as the frequency changes to satisfy the Barkhausen criterion, the phase shift introduced by the frequency-determining components as a result of the frequency change is large, and hence only a small shift in frequency results. Thus, for frequency-determining components with large $d\theta/d\omega$, the frequency stability is high. LC-tuned circuits and crystals tend to have large $d\theta/d\omega$ at f_o and hence excellent frequency stability.

12.6   Amplitude Stabilization

The Barkhausen criterion includes the condition that the loop gain must be equal to unity for oscillation, i.e. $A\beta = 1$. If $A\beta < 1$, the circuit ceases to oscillate. Since $A\beta = 1$ is not practical, $A\beta > 1$ has to be set in order to ensure that oscillation is sustained. The extent to which the loop gain is greater than unity largely determines the amplitude of the output waveform. Without appropriate amplitude control, this amplitude will increase indefinitely thereby producing extreme levels of distortion as the amplifier saturates. In such circumstances, the output progressively changes from a sinusoidal waveform to a square wave as the loop gain exceeds unity.

For a value of loop gain marginally above the critical value of unity, the output amplitude increases. However, as the non-linear region of the amplifier is entered, the amplifier gain drops, and the output is reduced, causing the amplifier to return to the linear region, this in turn causing an increase in amplifier gain. A dynamic point may be reached where the non-linearity reduces the gain such that oscillation is sustained with a stable amplitude and low distortion. Adjusting the loop gain manually to achieve this is extremely difficult. Automatic gain control can be introduced by using a thermally sensitive non-linear device as one of the gain-setting components of the associated amplifier. Thus, amplitude increase would cause heating and hence resistance increase (sensistor) or decrease (thermistor) that can be used to regulate oscillator output

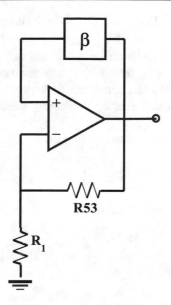

Fig. 12.32 Thermistor amplitude stabilization

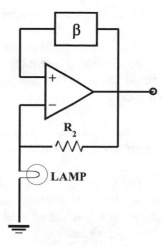

Fig. 12.33 Sensistor amplitude stabilization

amplitude. Examples of this amplitude control are shown in the Wien bridge oscillator.

Figure 12.32 illustrates amplitude control in a Wien bridge oscillator using the R53 thermistor. This device, though now rare, has suitable negative temperature coefficient characteristics for use in this application. In the particular case where $A > 3$, R_1 needs to be between 600 Ω and 1 kΩ. Thus, as the amplitude of the output signal increases, more current flows into R53 resulting in heating. As a result, the resistance of the thermistor decreases, and hence the amplifier gain also decreases. The overall effect is a reduction in the amplitude of the output signal. Conversely, a falling output signal causes a decrease of the heating in R53 and hence an increase in its resistance. The overall effect is an increase in the amplifier gain resulting in the increase of the amplitude of the output signal. A dynamic equilibrium is reached where the output signal is approximately constant.

Figure 12.33 illustrates amplitude control in a Wien bridge oscillator using a sensistor, specifically a low-voltage incandescent lamp, a cheaper alternative to the R53. Note that the lamp replaces R_1 in the negative feedback loop. Thus, if the output signal increases, the lamp is heated and its resistance increases. As a result, the amplifier gain is reduced thereby reducing the amplitude of the output signal. The converse occurs for a decrease in signal amplitude and the overall effect is the stabilization of the output signal amplitude. Typically, the lamp is a low-voltage medium-current device. The value of R_2 depends on the characteristics of the specific lamp but is likely to be in the range 250 Ω to 1 kΩ.

A third amplitude stabilization technique is the use of an electronically variable resistor such as a FET, in conjunction with R_1 and R_2 to set the amplifier gain. The scheme is shown in Fig. 12.34. It utilizes the fact that over a limited range of drain-source voltage, a JFET behaves as a variable resistor. It operates as follows: At turn-on, the gate resistor R_G biases the FET on (zero gate-source voltage) such that with a drain-source voltage less than pinch-off, the drain-source resistance is low. The gain of the oscillator amplifier is therefore set by R_1 and R_2 giving a gain greater than three; oscillation therefore begins. The sinusoidal output signal is amplified by the control amplifier A then half-wave rectified by diode D_1.

This signal is then filtered by C_1, and the resulting negative DC voltage is applied to the gate of the JFET to reverse bias it. This increases the channel resistance of the FET and as a result reduces the gain of the oscillator amplifier. The amplitude of the negative DC voltage is proportional to the amplitude of the sinusoidal signal, and the actual value that is applied to the gate of the FET is adjusted until a stable output signal is obtained. This circuit works very well and maintains constant output amplitude over a wide frequency range. For proper operation the amplitude of the signal across the FET must maintain it in the ohmic region.

12.7   Oscillator Creation

Having discussed the principle of operation of an oscillator and the Barkhausen criterion, it is possible to create "new" oscillator structures by employing these principles.

Oscillator 1 A simple illustration of this is the use of an all-pass filter network to realize an oscillator structure. Each of the cascaded stages

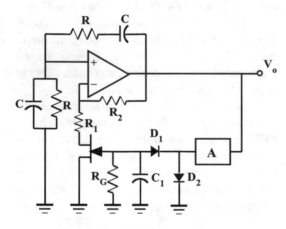

Fig. 12.34 Feedback amplitude stabilization

in Fig. 12.35 is an all-pass filter. Its transfer function is given by $H(s) = \frac{1-Ts}{1+Ts}$ where $T = RC$. It has the unique feature of unity gain and varying phase with changing frequency. The phase varies from $0°$ to $180°$, and hence the Barkhausen criterion can be easily met by combining three all-pass stages as shown.

The Barkhausen criterion gives

$$A\beta = \left(\frac{1 - Ts}{1 + Ts}\right)^3 = 1 \qquad (12.100)$$

which is equivalent to

$$|A\beta(j\omega_o)| = 1 \qquad (12.101)$$

and

$$\angle(A\beta(j\omega_o)) = -2\pi \qquad (12.102)$$

Since $|H(s)|=1$ in the all-pass network, condition (12.101) is satisfied. From condition (12.102)

$$3\left(-2\tan^{-1}(\omega_o T)\right) = -2\pi \qquad (12.103)$$

from which the oscillation frequency is given by

$$f_o = \frac{1}{2\pi RC} \tan\left(\frac{\pi}{3}\right) = \frac{\sqrt{3}}{2\pi RC} \qquad (12.104)$$

This circuit was tested using LF351 operational amplifiers. All resistors were set to 10 kΩ and C was set at 0.01 μF. From (12.104), we get $f_o = \frac{\sqrt{3}}{2\pi \times 10 \times 10^3 \times 0.01 \times 10^{-6}} = 2757\,\text{Hz}$. The system was tested in the laboratory and oscillated at 2756 Hz the calculated frequency.

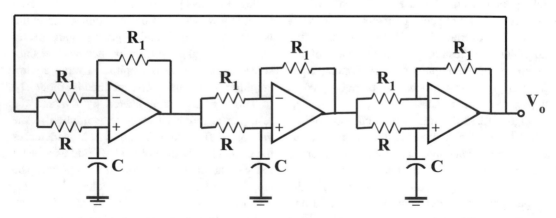

Fig. 12.35 Oscillator using all-pass filters

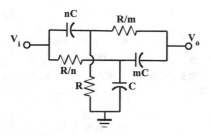

Fig. 12.36 RC network

Exercise 12.1

Show that for four all-pass networks, the frequency of oscillation is $f_o = \frac{1}{2\pi RC}$.

Oscillator 2 Another oscillator can be created using the network shown in Fig. 12.36. For this network, using nodal analysis the transfer function is given by

$$\frac{V_o}{V_i}(s) = \frac{nsCR(m+n+1)}{n(sCR)^2 + sCR(mn+n^2+m+1)+n}$$

(12.105)

Setting $s = j\omega$, then at a frequency $\omega_o = 1/RC$ corresponding to $f_o = 1/2\pi RC$, the transfer function becomes

$$\frac{V_o}{V_i}(f_o) = \frac{n(1+m+n)}{1+m+mn+n^2}$$

(12.106)

Since the transfer function in (12.106) is real, this indicates that the phase shift introduced by the network is zero at f_o. If $n = 3$ and $m = 0.2$, then

$$\frac{V_o}{V_i}(f_o) = \frac{n(1+m+n)}{1+m+mn+n^2}$$

$$= \frac{3(1+0.2+3)}{1+0.2+0.2\times3+3^2}$$

$$= 1.16$$

(12.107)

This means that the network produces an output that is greater than the input at the frequency f_o at which the phase shift is zero. This network can therefore be used with an emitter follower to produce oscillation providing the gain of the emitter follower is greater than $1/1.16 = 0.862$ which ensures that the Barkhausen criterion for

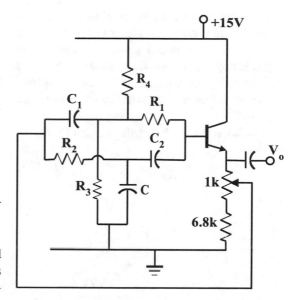

Fig. 12.37 Emitter follower oscillator

oscillation is satisfied. A simple application is shown in Fig. 12.37. The output of the network drives the input of the emitter follower while the output of the emitter follower drives the input of the network. At the single frequency f_o the signal phase shift is zero while the loop gain is greater than unity and therefore oscillation occurs. The potentiometer sets the loop gain for minimum distortion while sustaining oscillation. The load at the output of the emitter follower must be high so that its gain is not reduced below 0.862. The transistor is biased using $R_4 = 16.8$ k and $R_3 = 16.8$ k through $R_1 = 42$ k. Since resistor R_4 is grounded through the power supply, resistor R of the network comprises $R = R_3//R_4 = 8.4$ k. Hence $R_2 = R/n = 8.4$ k/3 = 2.8 k and $R_1 = R/m = 8.4$ k/0.2 = 42 k. For $C = 0.01$ μF, then $C_1 = nC = 3 \times 0.01 = 0.03$ μF and $C_2 = mC = 0.2 \times 0.01 = 0.002$ μF. The resulting frequency of oscillation is $f_o = 1/2\pi RC = 1/2\pi \times 8.4 \times 10^3 \times 0.01 \times 10^{-6} = 1895$ Hz.

Exercise 12.2

Design an oscillator using $n = 2$ and $m = 0.5$.

Oscillator 3 The oscillator shown in Fig. 12.38 is based on the Colpitts circuit but here using an emitter follower which has gain less than unity instead of an inverting amplifier with gain greater than unity. The model shown in Fig. 12.39 representing this oscillator now contains a non-inverting A_v instead of an inverting amplifier as discussed in the Colpitts-Hartley system in Fig. 12.20.

Thus, the system loop gain is given by

$$A\beta = \frac{A_v X_1}{X_1 + X_3} = -\frac{A_v X_1}{X_2} \quad (12.108)$$

where

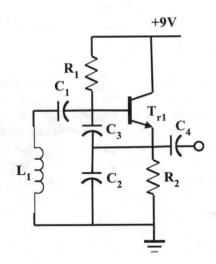

Fig. 12.38 Oscillator 3

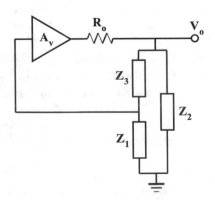

Fig. 12.39 Equivalent circuit

$$X_1 + X_2 + X_3 = 0 \quad (12.109)$$

In order to satisfy the Barkhausen criterion for oscillation, $A\beta > 1$, and therefore X_1 and X_2 must have different signs. Let $X_1 = \omega L_1$ and $X_2 = -1/\omega C_2$. Then,

$$A\beta = -A_v \frac{\omega L_1}{-1/\omega C_2} = A_v \omega^2 L_1 C_2 > 1 \quad (12.110)$$

Since A_v is only slightly less than unity, this condition reduces to

$$\omega^2 L_1 C_2 > 1 \quad (12.111)$$

From (12.109), $X_3 = -(X_1 + X_2)$. For $X_3 = -1/\omega C_3$, then $-1/\omega C_3 = -\omega L_1 + 1/\omega C_2$, and therefore the frequency of oscillation is given by

$$\omega_o = \frac{1}{\sqrt{L_1 C_T}}, \quad f_o = \frac{1}{2\pi\sqrt{L_1 C_T}} \quad (12.112)$$

where $C_T = \frac{C_2 C_3}{C_2 + C_3}$. Using $R_1 = 470$ k and $R_2 = 4.7$ k to bias the transistor, with coupling capacitors $C_1 = C_4 = 0.1$ μF, then for $C_2 = C_3 = 1$ nF and $L_1 = 1$ mH, we get $f_o = \frac{1}{2\pi\sqrt{10^{-3} \times 0.5 \times 10^{-9}}} = 225$ kHz. The frequency of oscillation using simulation software was measured at 215 kHz. Note that using (12.112), (12.111) becomes $\omega^2 L_1 C_2 = \frac{L_1 C_2}{L_1 C_T} = \frac{C_2}{C_T} > 1$. Therefore, the Barkhausen criterion is generally satisfied.

Exercise 12.3 Show that if X_1 and X_3 are capacitors and X_2 is an inductor, while X_1 and X_2 have different signs as required, the system will not oscillate.

12.8　Applications

Oscillators producing sinusoidal outputs are used in a wide range of applications including laboratory testing, communications and reference standards. In laboratory testing, a sinusoidal generator is a standard piece of equipment. It is used for determining the frequency and phase response of amplifiers and other signal processing circuits. This enables the transfer function of the circuit to be determined and also gives an indication of the stability of the circuit, particularly under varying loads. The sine-wave generator is also used for troubleshooting as it can be injected into a circuit at different points and monitored at others.

Another extremely important application of sine-wave generators is in communication systems. In radio transmitters, information to be transmitted is superimposed on a sinusoidal signal called a carrier. The resulting modulated carrier signal is then amplified and radiated through radio antennas as radio waves. In radio receivers, this radio wave induces an electrical signal in a receiving antenna. It is then mixed with a sinusoidal signal produced by a local oscillator and the resulting lower-frequency signal processed by the receiver circuitry to extract the information. This mixing technique has long formed the basis of super heterodyne receivers. Sinusoidal oscillators are also used in standard frequency sources and in the production of synthesized music.

Wide-Range Wien Bridge Oscillator
The Wien bridge oscillator circuit in Fig. 12.40 uses the LF351 JFET input op-amp and is intended for laboratory use. It covers the frequency range 15 Hz − 150 kHz in four bands. It is a simple system and the frequency is easily varied over a wide frequency range using the switched banks of capacitors and the ganged variable potentiometer. The components of the Wien network giving the frequency ranges shown are selected using Eq. (12.11). Fixed resistors $R_a = 1$ k are used to set the minimum value of R, and a dual 10 k linear potentiometer is used for VR_a as the remaining part of R. This arrangement provides continuous variation within a band set by a pair of equal capacitors C. Values of C of 1 μF, 0.1 μF, 0.01 μFand 0.001 μF establish four bands. By switching in different equal pairs of C using the two-pole multi-throw switch, four frequency bands can be realized, each allowing a decade change in frequency by varying the potentiometer. These bands are 15 Hz − 150 Hz, 150 Hz − 1.5 kHz, 1.5 kHz − 15 kHz and 15 kHz − 150 kHz.

The gain of the oscillator amplifier is set by R_1 and R_2 along with the channel resistance of the JFET. From its specifications, the 2N3819 for $V_{DS} < V_P$ has a channel resistance r_d that varies between 100 Ω and 3.6 k for V_{GS} varying from 0 V to −2.5 V. Therefore $R_2 = 20$ k and $R_1 = 8$ k result in gain variation from 3.4 when $r_d = 100$ Ω to 2.7 when $r_d = 3.6$ k, inclusive of the critical value 3. The gate-biasing resistor R_5 is set at 1 M. Capacitor C_1 is chosen to provide adequate filtering of the output ripple and thereby produces a DC voltage. Since R_5 is large, this capacitor need not be very large. It cannot be too large otherwise the time constant R_5C_1 will prevent a quick response to varying amplitude changes. A compromise value of 2 μF is used. The diodes are small signal devices such as 1N4148. The output of the Wien bridge oscillator is low and the output amplifier opa3 is used to boost this signal by a factor of two by setting $R_3 = R_4 = 10$ k. The control amplifier opa2 is set at a gain of 6 with $R_7 = 10$ k and $R_6 = 2$ k in order to provide sufficient drive for the half-wave rectifier. The potentiometer $VR_2 = 10$ k provides a variable signal level to the control amplifier. The circuit is operated from a ±15 V supply. Resistors are 2% tolerance and all capacitors are polystyrene.

Ideas for Exploration: (i) Introduce another potentiometer in order to provide a variable output signal amplitude from the oscillator.

Wien Bridge Oscillator Using a 741 Op-Amp
The circuit in Fig. 12.41 uses the 741 op-amp in a simple Wien bridge oscillator. It uses the Wien network components as the previous case. An R53 thermistor (if available) and resistor $R_1 = 470$ Ω produce the necessary gain of three with amplitude stabilization as a result of the action of the thermistor. Note however that the GBP of the 741 limits the upper frequency of operation of the circuit to less than 1 MHz/ 3 = 333 kHz.

Ideas for Exploration: (i) The R53 thermistor is difficult to find and appears to be no longer available. Use another approach to amplitude stabilization such as the low power incandescent lamp as shown in Fig. 12.42. In operation,

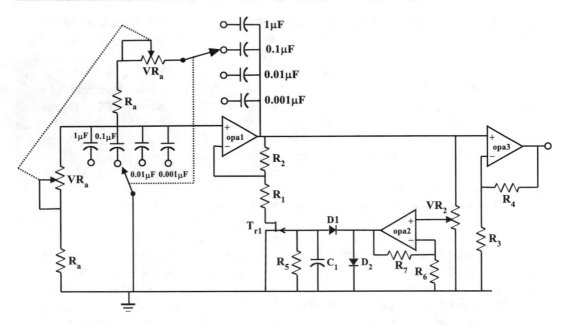

Fig. 12.40 Wide-band Wien bridge oscillator

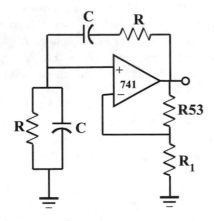

Fig. 12.41 Wien bridge oscillator using a 741 op-amp

potentiometer VR1 is adjusted for a maximum signal output of about 2 V. (ii) Examine a third approach to amplitude stabilization as shown in Fig. 12.43. Here the potentiometer is adjusted such that the output is just above zero. As signal amplitude build-up takes place, eventually the diodes begin to conduct thereby limiting the amplifier gain and therefore constraining the signal amplitude.

Oscillator Using a JFET Input Amplifier

The Wien bridge oscillator shown in Fig. 12.44 uses discrete components with a MPF102 JFET at the input stage of the amplifier and is one of many configurations that can be used. The JFET presents a high input impedance to the Wien network thereby reducing damping on the Wein network and enabling the use of a high-value dual potentiometer for wide frequency variation (20 Hz to 20 kHz) without capacitor switching. A ganged potentiometer $VR_a = 500$ k is employed along with $R_a = 820\ \Omega$ which sets a minimum value of R in the Wien network. Capacitor $C = 0.01\ \mu F$ in the Wien network gives a minimum frequency of 31 Hz and maximum frequency of 19.4 kHz. Potentiometer $VR_2 = 20$ k sets the current in the JFET to a value that minimizes the distortion, and $C_4 = 100\ \mu F$ bypasses this potentiometer for signals. In the absence of an R53 thermistor, amplitude stabilization is achieved using diodes in parallel with the feedback resistor.

Ideas for Exploration: (i) Design the oscillator using the JFET input amplifier of Fig. 7.54 in Chap. 7.

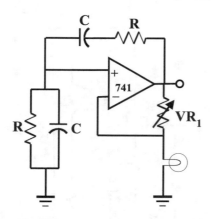

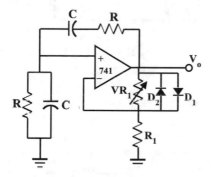

Fig. 12.42 Amplitude stabilization using a low-power incandescent lamp

Fig. 12.43 Amplitude stabilization using signal diodes

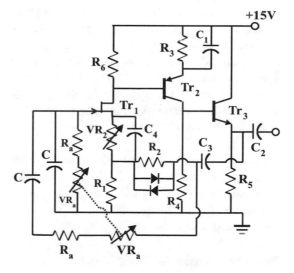

Fig. 12.44   JFET Wien bridge oscillator

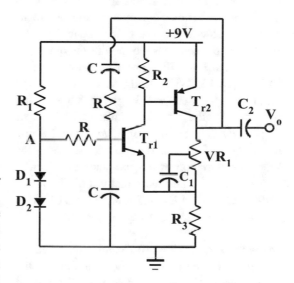

Fig. 12.45 Wien bridge oscillator using feedback pair

Wien Bridge Oscillator Using Two-Transistor Amplifier

The oscillator shown in Fig. 12.45 is based on the two-transistor feedback pair with gain discussed in Chap. 5. The design follows the principles discussed there and application of the Wien bridge oscillator principles. Transistors Tr_1 and Tr_2 constitute the feedback pair with the input at the base of Tr_1 and the output at the collector of Tr_2. Potentiometer $VR_1 = 10$ k along with bypass capacitor $C_1 = 100$ μF and resistor $R_3 = 1.5$ k sets the gain of the feedback pair at just over 3 for oscillation. The Wien network from output back to non-inverting input comprises capacitors $C = 0.01$ μF and resistor $R = 15$ k which give frequency of oscillation $f_o = 1/2\pi \times 15 \times 10^3 \times 0.01 \times 10^{-6} = 1061$ Hz. Note that point A is a signal ground. The voltage there is 1.4 V with current of about 1.4 mA supplied to the diodes by $R_1 = 5.6$ k. This is the approximate voltage

applied at the base of Tr_1 with bias current to this transistor through one of the resistors $R = 15$ k of the Wien network. (The base current through this resistor is small, and hence the voltage drop across this resistor is small.) The voltage at the emitter of Tr_1 is therefore 0.7 V. The current through Tr_1 is given by $0.7/R_2$. A value of $R_2 = 6.8$ k gives a current through Tr_1 of 0.1 mA. The current through resistor $R_3 = 1.5$ k (made up of currents of both transistors) is given by 0.7 V/1.5 k = 0.47 mA. Therefore, the current through Tr_2 is $0.47 - 0.1 = 0.37$ mA. Noting that only the current of Tr_2 flows through the potentiometer, the voltage drop across the potentiometer is 0.37 mA $\times$ 10 k $= 3.7$ V. Hence the DC voltage at the output is $0.7 + 3.7 = 4.4$ V which allows for maximum symmetrical swing. The low transistor currents in this circuit allow for operation from a 9 V battery.

Ideas for Exploration: (i) Introduce variable amplitude by including a 10 k potentiometer after capacitor C_2 to ground with the output at the wiper; (ii) implement the circuit using the four-transistor amplifier of Chap. 7.

Phase Shift Oscillator Using a JFET
The phase shift oscillator shown in Fig. 12.46 uses a JFET in the common source mode at the input and an emitter follower at the output to provide a low-impedance drive capability. The JFET uses the self-bias mode with the quiescent current set at 1 mA by appropriate selection of R_3 with $VR_1 = 1$ k. With a load resistor $R_1 = 6$ k, the drain voltage is set at 9 V giving 8.3 V at the output of the emitter follower. Resistor $R_2 = 4.7$ k sets a current of $8.3/4.7$ k $= 1.8$ mA in the emitter follower. The potentiometer $VR_1 = 1$ k enables variation of the gain such that oscillation results with minimum distortion. The phase shift network uses $R = 6.5$ k and $C = 0.01$ μF to give an oscillating frequency of 1 kHz.

Ideas for Exploration: (i) Examine the effect on distortion by introducing bootstrapping of the load resistor R_1. This would increase the gain of the JFET, and therefore the potentiometer would have to be adjusted to allow greater levels of local feedback in order to reduce the gain to just above 29.

Hartley Oscillator Using an Op-Amp
The circuit shown in Fig. 12.47 is a Hartley oscillator using a LF351 JFET op-amp in the inverting configuration. It is a simple application of the principle presented in Sect. 12.3.2. Here the input resistor R_1 is chosen to be 1 M to ensure a high input impedance as assumed in the analysis of the oscillator. The feedback resistor

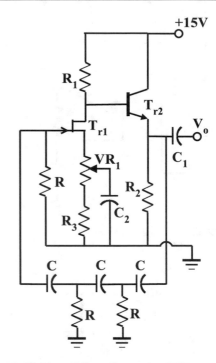

Fig. 12.46 Phase shift oscillator using JFET

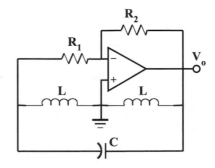

Fig. 12.47 Hartley oscillator using op-amp

$R_2 \simeq 1$ M is chosen slightly greater than 1 M in order to enable oscillation. Using $f = 1/2\pi\sqrt{2LC}$, then for $L = 10$ mH and $C = 0.001$ μF, the frequency of oscillation is given by $f = 1/2\pi\sqrt{2 \times 10 \times 10^{-3} \times 0.001 \times 10^{-6}} = 35.6$ kHz.

Ideas for Exploration: (i) Explore the use of amplitude stabilization techniques in this circuit.

Colpitts Oscillator Using an Op-Amp

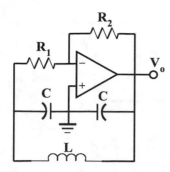

Fig. 12.48 Colpitts oscillator using op-amp

The circuit shown in Fig. 12.48 is a Colpitts oscillator using a LF351 JFET op-amp in the inverting configuration. It also is a simple application of the principle presented in Sect. 12.3.2. The input resistor R_1 is chosen to be 1 M to ensure a high input impedance as assumed in the theory of the oscillator and the feedback resistor $R_2 \simeq 1$ M is chosen slightly greater than 1 M to ensure that oscillation results. Using $f = \sqrt{2}/2\pi\sqrt{LC}$, then for $L = 10$ mH and $C = 0.005$ μF, the frequency of oscillation is given by $f = \sqrt{2}/2\pi\sqrt{10 \times 10^{-3} \times 0.005 \times 10^{-6}} = 31.8$ kHz.

Ideas for Exploration: (i) Explore the use of amplitude stabilization techniques in this circuit.

LC Resonant Oscillator Using a BJT
The circuit shown in Fig. 12.49 is an LC resonant oscillator implemented using a BJT. The lower input impedance of the BJT compared with the JFET means that the analysis of Sect. 12.3.1 needs to be modified to take this into account. Here the transistor is biased by $R_1 = 47$ k and $R_2 = 22$ k which set the voltage at the base of the transistor at 4.7 V. This along with $R_3 = 1$ k sets the quiescent current in the BJT at 4 mA. With $L = 100$ mH and $C = 100$ pF, the resonant frequency is given by $f = 1/2\pi\sqrt{LC} = 1.59$ MHz. The high transconductance of the BJT compared with the JFET ensures that oscillation occurs. Note that coupling capacitor C_2 and decoupling capacitors C_1 and C_3 can be smaller (typically

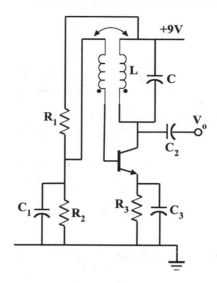

Fig. 12.49 LC resonant oscillator using BJT

0.1 μF and less) because of the high frequency of operation.

Ideas for Exploration: (i) Replace C by a variable trimmer capacitor and investigate the frequency variation that can be achieved.

Hartley Oscillator Using a BJT
The circuit shown in Fig. 12.50 is a Hartley oscillator implemented using a BJT. Again, the lower input impedance of the BJT compared with the JFET means that the analysis of Sect. 12.3.2 should be adjusted to accommodate this. Here the transistor is biased using collector-base feedback biasing with $R_1 = 470$ k and $R_2 = 4.7$ k. Capacitor $C_1 = 0.1$ μF couples the output of the LC network to the input of the transistor while capacitor $C_2 = 0.1$ μF couples the output of the transistor to the input of the network. With $L = 10$ mH and $C = 1$ nF, the oscillating frequency is given by $f = 1/2\pi\sqrt{2LC} = 32$ kHz.

Ideas for Exploration: (i) Replace the network capacitor C by a variable trimmer capacitor and investigate the frequency variation that can be achieved.

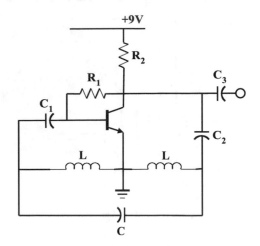

Fig. 12.50 Hartley oscillator using BJT

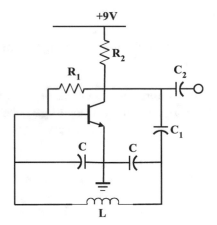

Fig. 12.51 Colpitts BJT

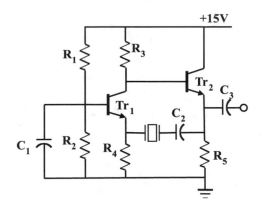

Fig. 12.52 Crystal oscillator using common base and common emitter amplifiers

inductor. (ii) Introduce a capacitor in series with the inductor thereby converting the system into a *Clapp oscillator*. Compare the performance of the two systems. (iii) Replace the inductor by a series-resonant crystal (*crystal-controlled Clapp oscillator*), and examine the performance.

Crystal Oscillator 1

The circuit in Fig. 12.52 is an oscillator that employs a series-resonant crystal in a positive feedback loop around two transistor stages. The first stage around Tr_1 is a common base amplifier with the output of the crystal feeding into the transistor emitter. The output at the collector of Tr_1 is fed into the emitter follower stage Tr_2, the output of which feeds into the series-resonant crystal. Thus, at resonance the impedance of the crystal is at a minimum, and positive feedback is at a maximum. At all other frequencies, the impedance is higher and therefore oscillation occurs at the series-resonant frequency of the crystal. Resistors $R_1 = 47$ k and $R_2 = 12$ kset the base voltage of Tr_1 at 3 V. Therefore with $R_4 = 1.2$ k, the current in the first stage is 2.3/1.2 k = 1.9 mA. With $R_3 = 2.2$ k, the voltage at the base of the emitter follower is $15 - (1.9$ mA $\times 2.2$ k$) = 10.8$ V. Resistor $R_5 = 2.2$ k sets the current in Tr_2 at 10.1/2.2 k = 4.6 mA. Capacitor C_2 works in conjunction with the crystal and is typically about 100 pF. This circuit operates with a wide range of crystals from 100 kHz to 10 MHz.

Colpitts Oscillator Using a BJT

The circuit shown in Fig. 12.51 is a Colpitts oscillator implemented using a BJT. Here the transistor is again biased using collector-base feedback biasing with $R_1 = 470$ k and $R_2 = 4.7$ k. Capacitor $C_1 = 0.1$ μF couples the output of the transistor amplifier to the input of the LC network. The output of the network is directly connected to the input of the transistor with no coupling capacitor being necessary. With $L = 10$ mH and $C = 5$ nF, the oscillating frequency is given by $f = \sqrt{2}/2\pi\sqrt{LC} = 31$ kHz.

Ideas for Exploration: (i) Examine the possibility of frequency variation by using a variable

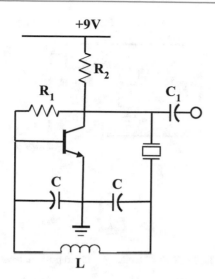

Fig. 12.53 Colpitts oscillator with series resonance crystal

Ideas for Exploration: (i) Try a range of crystals to test the functionality of this circuit.

Crystal Oscillator 2

The circuit in Fig. 12.53 is a Colpitts oscillator that employs a series-resonant crystal working into the Colpitts network. The tuned circuit is designed to resonate at the series-resonant frequency of the crystal, thereby ensuring stable oscillation at a single frequency. Here the biasing arrangement is the same as before with $R_1 = 470$ k and $R_2 = 4.7$ k. The output of the transistor amplifier drives the crystal which is connected to the input of the LC network. The output of the network is directly connected to the input of the transistor. The crystal frequency f_c is used in choosing L and C according to $f_c = \sqrt{2}/2\pi\sqrt{LC}$.

Ideas for Exploration: (i) Try a range of crystals to test the functionality of this circuit.

Phase Shift Oscillator Using a BJT

The circuit of Fig. 12.54 demonstrates a phase shift oscillator using a BJT. The BJT is in the common emitter mode and is biased by $R_1 = 43$ k and $R_2 = 10$ k, these setting a base voltage of 1.7 V. This base voltage in conjunction with $VR_1 = 1$ k sets the quiescent current at 1 mA.

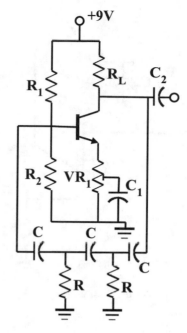

Fig. 12.54 Phase shift oscillator using BJT

Collector resistor $R_L = 3.9$ k gives the circuit close to maximum symmetrical swing. The output at the collector drives the phase-shift network. Note that the last resistor of the network is made up of R_2 in parallel with R_1 and the input impedance at the transistor base. The potentiometer is adjusted for minimum visual distortion of the signal when viewed on an oscilloscope. This adjustment changes the gain of the amplifier so that it is just greater than 29.

Ideas for Exploration: (i) Introduce a direct coupled emitter follower at the output in order to drive low-impedance loads and not affect the gain of the system. Drive the phase shift network from the output of the emitter follower. This should improve the accuracy of the frequency of oscillation given by $f_c = 1/2\pi\sqrt{6}CR$.

Research Project 1: Sinusoidal Oscillator Using the Twin-T Network

The twin-T oscillator shown in Fig. 12.55 utilizes a twin-T network comprising parallel networks $R_1 - R_2 - C_3$ and $C_1 - C_2 - R_3$. Under the conditions $R_1 = R_2 = 2R_3 = R$ and $C_1 = C_2 = C_3/2 = C$, straightforward circuit analysis yields the

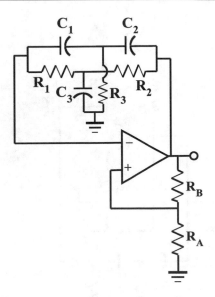

Fig. 12.55 Twin-T oscillator using op-amp

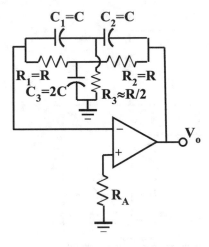

Fig. 12.56 Twin-T oscillator with no positive feedback

transfer function for this network given by $\frac{V_o}{V_i} = \frac{1+(sCR)^2}{1+4sCR+(sCR)^2}$. By setting $s = j\omega$, this can be written as $\frac{V_o}{V_i}(j\omega) = \frac{1-(\omega CR)^2}{1-(\omega CR)^2+4j\omega CR}$. At $\omega = \omega_o$ where $\omega_o = 1/RC$, i.e. $f_o = 1/2\pi RC$, the magnitude of this transfer function is given by $\left|\frac{V_o}{V_i}(j\omega)\right|_{\omega=\omega_o} = \frac{1-1}{\sqrt{(1-1)^2+4^2}} = 0$. Thus at $\omega = 1/RC$ corresponding to $f = 1/2\pi RC$, the signal transmission through the network is zero. Therefore, at this single frequency, there is no negative feedback around the op-amp, and the positive feedback around the system produces oscillation at this frequency. At all other frequencies, the network produces an output such that negative feedback signal is applied to the amplifier. Thus the condition $A\beta \geq 1$ around the positive feedback loop is operational at $f_o = 1/2\pi RC$ where A is the amplifier gain at f_o and $\beta = R_A/(R_B + R_A)$. Using $R = 2$ k and $C = 0.01$ μF, the oscillating frequency is given by $f_o = 1/2\pi RC = 1/2\pi \times 2000 \times 0.01 \times 10^{-6} = 8$ kHz. For sustained oscillation, $R_A = 1$ k and $R_B = 10$ k giving $\beta = 1$ k/((1 k + 10 k) = 0.09. This value may have to be adjusted, but increasing values of β will generally result in increased distortion of the signal.

Ideas for Exploration: (i) Investigate amplitude stabilization techniques for this circuit; (ii) When there is a slight imbalance in the network resulting from R_3 slightly less than $R/2$, there is a small but nonzero output from the network with phase inverted relative to the input signal. In these circumstances it is possible to sustain a sinusoidal output from the amplifier without any positive feedback. This occurs since the finite and inverted output signal from the network into the inverting terminal of the amplifier results in zero phase shift around the system and with amplification satisfies $A\beta \geq 1$. Use the circuit shown in Fig. 12.56 where there is no positive feedback with adjustable R_3 in order to produce oscillation.

Research Project 2: Sinusoidal Oscillator Using an Inverter

The oscillator shown in Fig. 12.57 satisfies the Barkhausen criterion by using a NAND gate connected as an inverter producing 180° phase shift along with amplification. The output of the NAND gate drives a phase-lag network that produces the additional 180° phase shift. The result is continuous oscillation. The square wave out of the inverter is filtered by the three-stage RC phase shift network thereby removing harmonics and producing a sine-wave output. The frequency of the sine wave is $f = \sqrt{6}/2\pi RC$, and the amplitude of this signal is given by $V_{\sin} = 4V_{sq}/$

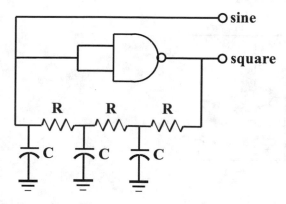

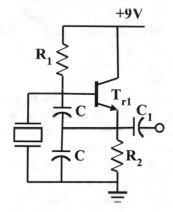

with the crystal provides the necessary signal gain to ensure oscillation. $R_1 = 470$ k and $R_2 = 4.7$ k will adequately bias the transistor which can be a 2N3904.

Ideas for Exploration: (i) Try a range of crystals to test the functionality of this circuit; (ii) use this circuit to design a *crystal tester* as shown in Fig. 12.59. Capacitors $C_3 = 1$ nF and $C_4 = 0.0047$ µF along with diodes D_1 and D_2 form a voltage multiplier. If the crystal is in good condition, the multiplied output drives transistor T_{r3} on and the LED lights. The diodes should be germanium or Schottky diodes since these have low turn-on voltages.

Fig. 12.57 Sine-wave oscillator using inverter

Fig. 12.58 Crystal oscillator using emitter follower

29π which for 5 V logic gate is $V_{\sin} = 4 \times 5$ V/ $29 \times \pi = 220$ mV.

Ideas for Exploration: (i) Replace the gate by a single transistor common emitter amplifier circuit and examine the operation; (ii) reduce the gain of the circuit to close to 29 by introducing local feedback such that the output at the collector becomes a low-distortion sine wave (phase shift oscillator).

Research Project 3: Crystal Oscillator Using Emitter Follower

The requirement is to investigate the crystal oscillator circuit in Fig. 12.58. It is a Colpitts oscillator implemented in an emitter follower that employs a parallel-resonant crystal in place of an inductor. The idea is that at the oscillating frequency, the circuit gives zero phase shift and the tuned circuit

Problems

1. State the Barkhausen criterion and discuss its meaning with respect to producing oscillation in a system.

2. For the network shown in Fig. 12.60 determine the frequency at which phase shift through the network is zero and the magnitude of the associated transfer function at that frequency. Using this network design an oscillator to oscillate at a frequency of 15 kHz.

3. Design a Wien bridge oscillator to oscillate at 10 kHz, and explain how a low-power incandescent lamp can be used to achieve amplitude stabilization.

4. For the circuit shown in Fig. 12.61, find the frequency at which the output signal V_o is in phase with the input signal V_i and the transfer function value at this frequency. Hence design an oscillator using this circuit to have an oscillating frequency of 7 kHz.

5. Draw the circuit of a phase shift lead oscillator, and describe its operation, referring to the Barkhausen criterion.

6. Design a phase shift lead oscillator to oscillate at 5 kHz, and explain why this circuit is not suitable for a variable frequency oscillator.

7. For the circuit shown in Fig. 12.62, determine showing all analysis the ratio R_B/R_A in order to achieve oscillation and the frequency of the

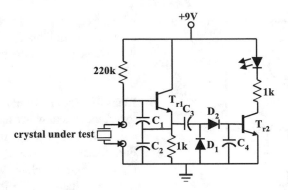

Fig. 12.59 Crystal tester

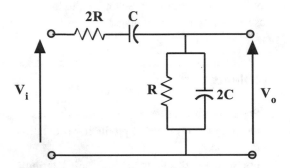

Fig. 12.60 Circuit for Question 1

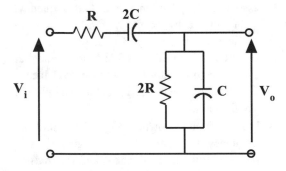

Fig. 12.61 Circuit for Question 4

oscillation. Hence design an oscillator using this circuit to have an oscillating frequency of 15 kHz.

8. For the circuit shown in Fig. 12.63, determine the inverting amplifier low-frequency gain in order to produce oscillation and the frequency of that oscillation. Hence design an oscillator using this circuit to have an oscillating frequency of 9 kHz.

9. Draw the circuit of the Meacham oscillator, and briefly explain its operation. Using an operational amplifier, design a Meacham oscillator having an oscillating frequency of 6 kHz, and explain how oscillation can be maintained.

10. Draw the circuit of a crystal oscillator using an operational amplifier as the active element. If the crystal is replaced by an inductor and capacitor in parallel, determine the condition for oscillation and the oscillating frequency of the circuit.

11. For the circuit shown in Fig. 12.64 determine the ratio R_B/R_A for oscillation and the frequency of that oscillation. Using this circuit, design an oscillator with an oscillating frequency of 18 kHz.

12. For the circuit shown in Fig. 12.65 determine the ratio R_2/R_1 in order to produce oscillation and the frequency of that oscillation. Hence design an oscillator using this circuit to oscillate at a frequency of 15 kHz.

13. Design an LC resonant oscillator to operate at 50 kHz. Use a 2N3819 JFET and a 20-volt supply.

14. Design an LC resonant oscillator using the AD844 to operate at 15 kHz.

15. Explain the operation of the Hartley oscillator. Design a Hartley oscillator to operate at 50 kHz using a 2N3819 JFET. Use a supply voltage of 24 volts.

16. Explain the operation of the Colpitts oscillator. Design a Colpitts oscillator to oscillate at

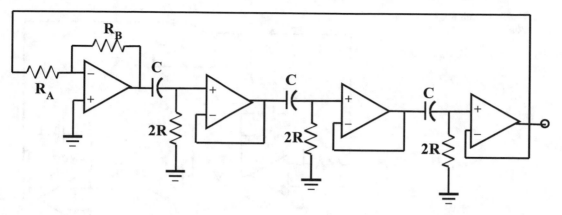

Fig. 12.62 Circuit for Question 7

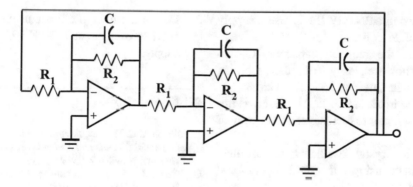

Fig. 12.63 Circuit for Question 8

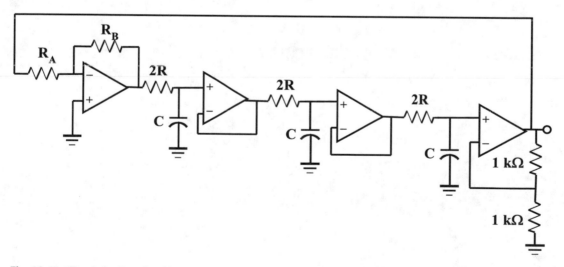

Fig. 12.64 Circuit for Question 11

Fig. 12.65 Circuit for
Question 12

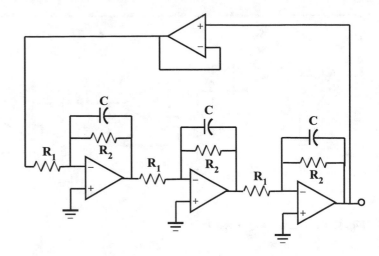

50 kHz using a 2N3819 JFET. Use a supply
voltage of 24 volts.

17. Determine the required capacitor, C_T, to
adjust the frequency of the Colpitts
oscillators in Question 16 to 55 kHz.

18. Using the circuit in Fig. 12.24, design a simple LC oscillator to oscillate at a frequency of
5 kHz.

19. Design a crystal oscillator to oscillate
at 75 kHz using an OP42 operational
amplifier.

20. Design a Miller oscillator using a 2N3918
JFET to oscillate at 110 kHz. Use a 25-volt
supply.

Bibliography

J. Millman, C.C. Halkias, *Integrated Electronics: Analog
and Digital Circuits and Systems* (Mc Graw Hill Book
Company, New York, 1972)

U. Tietze, C. Schenk, *Advanced Electronic Circuits*
(Springer-Verlag, Berlin, 1978)

Waveform Generators and Non-linear Circuits

In this chapter, we consider circuits that involve non-linear operation of the operational amplifier. These can be used to realize waveform generators which are circuits that produce a variety of non-sinusoidal waveforms. They are fundamentally instrumentation building blocks used for signal generation and test and measurement. Typically, waveform generators produce square waves, triangular waves, pulses and in some cases arbitrary waveforms over a range of frequencies with constant amplitude. Here we discuss comparators, op-amp-based free-running (astable) and one-shot (monostable) multivibrator circuits as well as precision rectifiers and other non-linear circuits. At the end of the chapter, the student will be able to:

- Explain the operation of several non-linear circuits
- Explain the operation of a wide range of waveform generators
- Design a wide range of non-linear circuits and waveform generators

13.1 The Comparator

A comparator is very similar to an op-amp except that the differential gain is made larger with the addition of internal positive feedback circuitry. As a result, the output has two states V_{omx} and V_{omn} whose value typically approaches the comparator supply rails V_{CC} and $-V_{SS}$, respectively. The symbol for a comparator is also identical to an op-amp. In fact, because of its high open-loop gain, an op-amp can operate as a comparator by omitting feedback. To understand the operation of a comparator, consider the circuit shown in Fig. 13.1.

Here V_i the input voltage is applied at the inverting input, while the reference voltage V_R is applied at the non-inverting input of the comparator. If $V_i < V_R$, then the differential voltage $V_D = V_i - V_R$ is negative and hence the output of the comparator is positive, i.e. $V_o = V_{omx}$, and if $V_i > V_R$, then $V_D = V_i - V_R$ is positive and the output of the comparator is negative, i.e. $V_o = V_{omn}$. This comparator would be operating in the inverting mode. If the input terminals are reversed such that V_i is applied at the non-inverting input while the reference voltage V_R is applied at the inverting input of the comparator, then if $V_i < V_R$, then the differential voltage $V_D = V_i - V_R$ is negative and hence the output of the comparator is negative, i.e. $V_o = V_{omn}$, and if $V_i > V_R$, then $V_D = V_i - V_R$ is positive and the output of the comparator is positive, i.e. $V_o = V_{omx}$. This comparator would be operating in the non-inverting mode.

If $V_R = 0$, then the comparator is referred to as an inverting or non-inverting zero-crossing detector. This action for $V_R = 0$ for a triangular wave input is shown in Fig. 13.2 for a non-inverting comparator and in Fig. 13.3 for an inverting comparator.

© Springer Nature Switzerland AG 2021
S. J. G. Gift, B. Maundy, *Electronic Circuit Design and Application*,
https://doi.org/10.1007/978-3-030-46989-4_13

Fig. 13.1 A comparator with input V_i and Reference voltage V_R

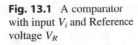
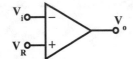

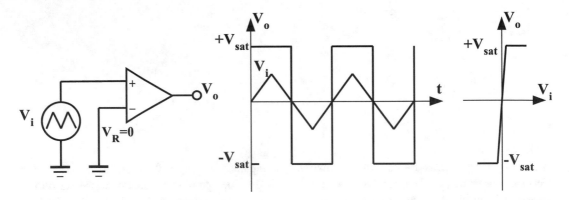

Fig. 13.2 Non-inverting zero-crossing detector

Fig. 13.3 Inverting zero-crossing detector

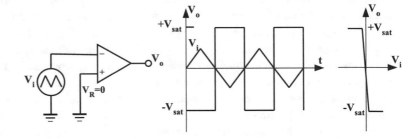

Example 13.1

An op-amp is connected as a comparator with the inverting terminal held at 1 V. Describe the action of the circuit when the non-inverting input is below and above 1 V.

Solution With the inverting input at 1 V, when the non-inverting input is less than 1 V, then the high amplification causes the output to saturate negatively giving $V_o = -V_{sat}$. When the voltage at the non-inverting input is greater than 1 V, the high amplification causes the output to saturate positively giving $V_o = +V_{sat}$.

Example 13.2

An op-amp is connected as a comparator with the non-inverting terminal held at 1 V. Describe the action of the circuit when the inverting input is below and above 1 V.

Solution With the non-inverting input at 1 V, when the inverting input is less than 1 V, then the high amplification causes the output to saturate positively giving $V_o = +V_{sat}$. When the voltage at the inverting input is greater than 1 V, the high amplification causes the output to saturate negatively giving $V_o = -V_{sat}$.

One straightforward but useful application of the comparator is in an over-temperature alarm system. The basic system is shown in Fig. 13.4. Here an LM35 temperature sensor produces an output voltage that is proportional to the ambient temperature given by $V_o = 10\,T\,mV$ where T is Celsius temperature. This output is connected to the non-inverting input of an op-amp. The inverting input is supplied with a reference voltage via a potential divider which sets the reference threshold. The op-amp can be powered by a

Fig. 13.4 Over-temperature alarm system

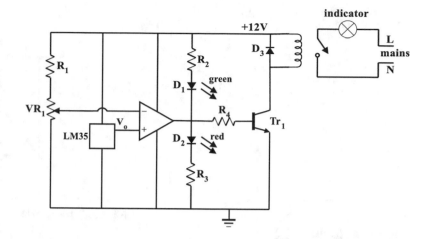

single-ended power supply since both inputs will have voltages at values between ground and the supply voltage. Under normal temperature conditions, the output voltage of the temperature transducer falls below the reference voltage, and the op-amp output is therefore saturated at close to zero. When the temperature rises above some specific value, the output of the sensor exceeds the reference voltage, and the op-amp output saturates at close to the supply voltage. This output is used to switch on an indicator using a relay.

Example 13.3
Using the basic circuit shown in Fig. 13.4, design a system to turn on a fan when the ambient temperature in a room exceeds 35 °C.

Solution Here an op-amp is connected as a comparator. The output of the LM35 temperature sensor drives the non-inverting input of the op-amp, while the inverting terminal is connected to the output of a potential divider arrangement comprising $R_1 = 10$ k and $VR_1 = 2$ k. With a 12 V supply, this means that the wiper of the potentiometer can be varied from 0 to 2 V. For a threshold temperature of 35 °C, the potentiometer is set to 10 mV × 35 = 350 mV. When the temperature is below 35 °C, the output of the sensor will be less than 350 mV, and therefore the output of the op-amp will be low. Therefore, the green LED will be on with current limited by $R_2 = 1$ k to about 10 mA, and the red LED will be off. The base of the transistor will also be at a low

voltage, and therefore the transistor will be off. When the temperature is above 35 °C, the output of the op-amp will be high. Therefore, the red LED will be on with current limited by $R_3 = 1$ k to about 10 mA, and the green LED will be off. The high op-amp output will turn on the transistor thereby causing the relay to switch on power to the fan. Resistor $R_4 = 10$ k ensures transistor saturation.

If noise is present on the input signal, this can lead to spurious or false switching of the comparator. In order to overcome this and other potential problems, the comparator's performance can be enhanced if positive feedback is added in the form of resistors R_a and R_b as shown in Fig. 13.5.

For this circuit the output voltage V_o in terms of the input V_i can be determined by writing the equation

$$V_o = A_d \left(\frac{R_a}{R_a + R_b} V_o - V_i \right) \qquad (13.1)$$

where A_d is the differential gain of the comparator. Solving for V_o gives

$$V_o = \frac{A_d}{A_d \beta - 1} V_i \qquad (13.2)$$

where β a constant is defined as $\beta = \frac{R_a}{R_a + R_b}$. Since β is always less than one, it follows that the denominator of (13.2) could be less than one or greater than one. For the case $0 \leq A_d \beta \leq 1$, the gain of the comparator is negative and larger than

Fig. 13.5 A comparator with positive feedback

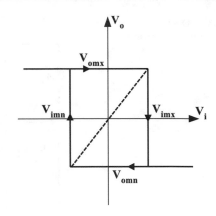

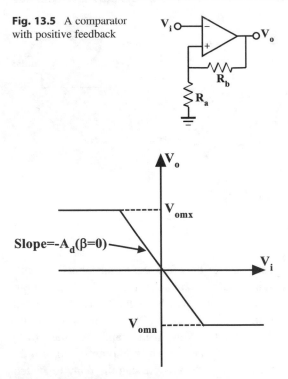

Fig. 13.6 Transfer characteristic of the comparator when $\beta = 0$

Fig. 13.7 Transfer characteristic of the Schmitt trigger

$$V_{imx} = \frac{A_d\beta - 1}{A_d} V_{omx} \qquad (13.3)$$

and

$$V_{imn} = \frac{A_d\beta - 1}{A_d} V_{omn} \qquad (13.4)$$

A_d which corresponds to the case of the simple comparator in Fig. 13.1. When $A_d\beta = 0$ which corresponds to setting $R_a = 0$ or $R_b \rightarrow \infty$, $V_R = 0$ and the comparator has the transfer characteristic shown in Fig. 13.6. For values of $0 \leq A_d\beta \leq 1$ the slope of the transfer characteristic is $-\frac{A_d}{1-\beta A_d} > -A_d$.

For $A_d\beta > 1$ the situation changes as the slope of the transfer characteristic becomes positive and less than A_d. Under this condition the comparator is referred to as a Schmitt trigger with a transfer characteristic shown in Fig. 13.7. From Fig. 13.7, the transfer characteristic has a unique shape in that for $V_i > V_{imx}$ the output voltage is $V_o = V_{omn}$. However, if V_i starts off being greater than V_{imx} and is decreased to V_{imx} and less, the output remains at V_{omn} until $V_i < V_{imn}$ is reached. When this occurs, $V_o = V_{omx}$. The converse is true, i.e. if we start with V_i being less than V_{imn} and increasing to V_i and greater, the output will eventually switch from V_{omx} to V_{omn}. Thus, the Schmitt trigger has two distinct switching points, V_{imx} and V_{imn} given by

Note that if a voltage V_R is applied to the grounded end of resistor R_a by removing it from ground, the transfer characteristic of Fig. 13.7 shifts to the right by an amount V_R if $V_R > 0$ and to the left by an equal amount if $V_R < 0$. The new switching points can be found by adding V_R to Eqs. (13.3) and (13.4). Finally, an op-amp can function as a comparator or a Schmitt trigger. However, its switching times between output states will typically not be as good as a dedicated comparator.

Example 13.4

Design a Schmitt trigger circuit with a switching threshold of ± 5 V using the LF411 op-amp powered by ± 15 V supplies. For the op-amp assume $|V_{omx}| = |V_{omn}| = 14.2$ V and $A_d = 200{,}000$. Verify your result using PSPICE with an input signal $V_i = 10 \sin (2\pi 10^3 t) V$.

Solution To find β we use Eq. (13.3) with $V_{imx} = 5$ V. Solving yields $\beta = 0.352$. Choosing $R_a = 1$ k yields $R_b = 1.84$ k. The PSPICE simulation result is shown in Fig. 13.8 verifying that the circuit triggers at ± 5 V.

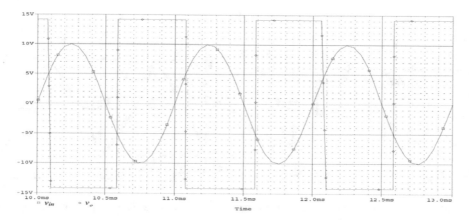

Fig. 13.8 PSPICE simulation

13.2   Square-Wave Generation

A simple square-wave generator is shown in Fig. 13.9. It consists of a single amplifier A_1 which may be an op-amp or a comparator configured as a Schmitt trigger. Because of the positive feedback, upon switch on the output of the op-amp saturates to either the positive rail $+V_{sat}$ or the negative rail $-V_{sat}$. Assuming that on applying power the system goes to the positive rail $V_o = +V_{sat}$, then the voltage at the inverting input terminal is therefore $+\beta V_o$ where the resistive feedback factor β is given by $\beta = \frac{R_a}{R_a+R_b}$. This voltage acts as a threshold voltage. If the capacitor C_1 is initially uncharged, then it begins to charge from zero volts with its voltage V_c increasing towards $+V_{sat}$ with a time constant $\tau = R_1 C_1$. Eventually V_c at the inverting terminal reaches and slightly exceeds the threshold voltage $+\beta V_o$ set at the non-inverting terminal and the output of A_1 switches to the negative saturation voltage $-V_{sat}$. The capacitor C_1 will now begin to charge towards the negative rail, and once again the output will switch to the positive rail when the new threshold $-\beta V_o$ is exceeded. The process is then repeated indefinitely as shown in Fig. 13.9b.

In practice, the value of V_{sat} is imprecise and therefore leads to inaccuracy in the calculation of the waveform characteristics which depend on its value. In order to improve the accuracy, the output V_o of A_1 is clamped by two back-to-back Zener diodes so that instead of V_o being

$V_o = \pm V_{sat}$, we have $V_o = +(V_z + V_D)$ where V_z is the Zener voltage and $V_D = 0.7$ V. Resistor R limits the current through the Zener diodes. To determine the period of oscillation, we may arbitrarily assign $t = 0$ as the starting point at which the capacitor voltage is $-\beta V_o$. The time domain response of the capacitor voltage from basic circuit theory is then given by

$$V_{C_1}(t) = -\beta V_o + (1+\beta)V_o\left[1 - e^{-\frac{t}{\tau}}\right] \quad (13.5)$$

for $0 \le t \le T/2$. At $t = T/2$ the half-cycle is completed as $V_{c1}(T/2) = +\beta V_o$ is reached. Substituting in (13.5) yields

$$+\beta V_o = -\beta V_o + (1+\beta)V_o\left[1 - e^{-\frac{t}{\tau}}\right] \quad (13.6)$$

and solving for the period T gives the result

$$T = \frac{1}{f_{osc}} = 2\tau \ln\left(\frac{1+\beta}{1-\beta}\right)$$
$$= 2\tau \ln\left[1 + 2\left(\frac{R_a}{R_b}\right)\right] \quad (13.7)$$

Note that the period T is independent of the supply voltage and the output voltage V_o and is comprised of two equal periods T_1 and T_2. If an unsymmetrical waveform is desired, the network of Fig. 13.10 can be used in place of R_1. It consists of two resistors of unequal value. The diodes ensure that only one resistor is enabled in the circuit during a given cycle. Hence during the positive half cycle as $V_c(t)$ increases, D_2 is on

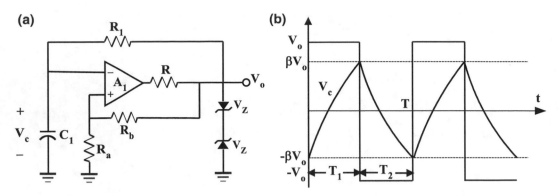

Fig. 13.9 (**a**) A square-wave generator with a clamped output, (**b**) output voltage V_o and capacitor voltage V_c waveforms

Fig. 13.10 A method to obtain an unsymmetrical square wave using diodes

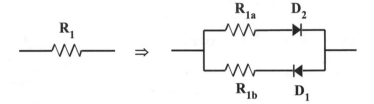

while D_1 is off. For this half-cycle the time constant $\tau_1 = R_{1a}C_1$. Conversely for the negative half-cycle D_1 is on while D_2 is off, and the time constant is now $\tau_2 = R_{1b}C_1$. The expression for the periods T_1 and T_2 are therefore given by

$$T_1 = \tau_1 \ln\left(\frac{1+\beta}{1-\beta}\right)$$
$$= \tau_1 \ln\left[1 + 2\left(\frac{R_a}{R_b}\right)\right] \qquad (13.8)$$

$$T_2 = \tau_2 \ln\left(\frac{1+\beta}{1-\beta}\right)$$
$$= \tau_2 \ln\left[1 + 2\left(\frac{R_a}{R_b}\right)\right] \qquad (13.9)$$

The only disadvantage to using the diodes in this manner is that the output voltage is reduced by one diode drop. As a general rule of thumb, the square-wave generator is useful to frequencies of about 1/10 of the GBP of the amplifier used. As the frequency is increased beyond that slew rate effects and the capacitance seen at the output of A_1 due to Zener diodes alters the output rise and fall times. The effect of slew rate limitations can

be minimized by using a comparator for A_1 in place of an op-amp. Since the comparator is designed for high gain and not in a feedback configuration, its slew rates can be very high. Also, as a general rule of thumb, if the output of the square-wave generator is required to have a particular rise time t_r, then it is related to the amplifiers slew rate by

$$t_r = 0.8 \times \frac{2V_o}{SR} \qquad (13.10)$$

Here rise, time is defined as the time it takes the output to get from 10% to 90% of the asymptote. To minimize the capacitance, two back-to-back diode-connected transistors can be employed in place of the Zener diodes. The effective Zener voltage is then set by the emitter-base breakdown voltage of the transistors used, and the nodal capacitance is considerably less than that of a commercial Zener diodes.

An alternative method of generating unsymmetrical square waves is to remove the lower end of resistor R_a from ground and apply a voltage V_R to the lower end of resistor R_a. For this configuration, show that the new periods T_1 and T_2 are given by

$$T_1 = \tau \ln \left[\frac{(1+\beta) - \alpha(1-\beta)}{(1-\beta) - \alpha(1-\beta)} \right] \quad (13.11)$$

and

$$T_2 = \tau \ln \left[\frac{(1+\beta) + \alpha(1-\beta)}{(1-\beta) + \alpha(1-\beta)} \right] \quad (13.12)$$

where $\alpha = V_R/V_o$ is a constant whose magnitude is less than one but greater than zero. Note that for $\alpha = 0$, the equation above revert back to Eq. (13.7) and for $0 < \alpha < 1$, $T_1 > T_2$ and for $-1 < \alpha < 0$, $T_2 > T_1$.

Example 13.5

Design a square-wave oscillator to have an oscillation frequency of 100 kHz, with an output of ± 5 V using a ± 12 V bipolar supply.

Solution To clamp the outputs to ± 5 V, we begin by choosing appropriate Zener diodes. Let us choose the 1N437A Zener diode which has a $V_Z = 4.3$ V. The reason for choosing this value is due to the fact that $V_o = V_Z + V_D$. With $V_D = 0.7$ V the solution to this equation yields $V_Z = 4.3$ V. To satisfy the frequency of oscillation, let us choose $R_a = 10$ k and $R_b = 39$ k yielding $\beta = 10/39$. If we let $C_1 = 1200$ pF, then from (13.7) $R_1 = 10$ k. For the amplifier we may choose an op-amp or a dedicated comparator. We choose the LM111 comparator for its

reasonable response time of 200 ns and its operation over a wide voltage range. The final circuit is shown in Fig. 13.11. Note the presence of the 4 kΩ resistor which is there because the LM111 has an open collector output. A PSPICE simulation of the output waveform is shown in Fig. 13.12.

Example 13.6

Modify the previous design so that the circuit now operates from a single 12 V supply and generates a 5 V TTL output at the same frequency.

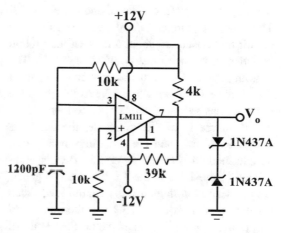

Fig. 13.11 Circuit for Example 13.5

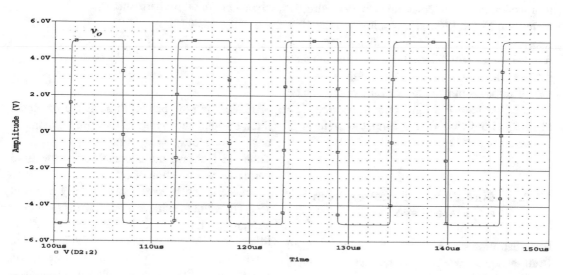

Fig. 13.12 PSPICE simulation for Example 13.5

Fig. 13.13 Circuit for Example 13.6

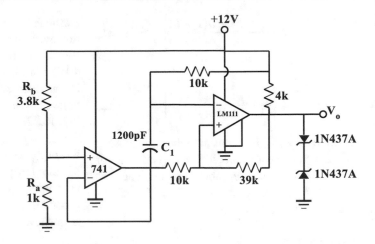

Solution To accomplish this we recognized that the original circuit oscillates about zero volts. Therefore, for single supply operation, the new circuit must oscillate about a new virtual ground which in this case must be $V_o/2$ or 2.5 V. The bottom plate of C_1 and the bottom end of R_a must therefore be supplied with 2.5 V. At this point the designer has several options open to them. If the 2.5 V source is readily available, the design is complete. If not, then one must improvise. Once simple method is to use a potential divider comprising of the 1 k and the 3.8 k resistor to generate the desired voltage that is then buffered. Such a solution is shown Fig. 13.13. If the number of components is an important factor, the circuit of Fig. 13.13 will also work if the op-amp is removed and the potential divider directly connected to the 1200 pF capacitor and the 10 k resistor, but virtual "2.5 V" will oscillate. Low resistor values will reduce the oscillation, but they increase the power consumption of the circuit. If the buffer is employed, the resistor values can be increased to reduce the power consumption. The PSPICE simulation of the output waveform is shown in Fig. 13.14.

13.2.1 Sine Wave Generation from a Square-Wave Input

Low distortion sine waves can be easily generated from a square-wave input by using a simple bandpass filter to extract the fundamental

frequency. To understand why, we recall from *Fourier* series that a square wave is made up of the sum of sinusoidal waves. That is, the output voltage $V_o(t)$ of Fig. 13.9 can be described by

$$V_o(t) = \frac{4A}{\pi}\left(\sin\omega_0 t + \frac{1}{3}\sin 3\omega_0 t + \frac{1}{5}\sin 5\omega_0 t + \frac{1}{7}\sin 7\omega_0 t + \cdots \right)$$

(13.13)

where A is the peak amplitude of the square wave. Examining Eq. (13.13) closely shows us that the third harmonic has an attenuation $\alpha_{vo}(3\omega_o) = 9.54$ dB greater than the fundamental. We only need to consider the third harmonic since the fifth and seventh harmonics, etc. have greater attenuation. For a unity gain second-order bandpass filter tuned to ω_o, the attenuation for a given Q to the third harmonic is

$$\alpha_{BP}(3\omega_0) = -20\log_{10}\left(\frac{3}{\sqrt{9 + 64Q^2}} \right)$$

(13.14)

If $V_o(t)$ is now passed through the bandpass filter, the total attenuation of the third harmonic is now $\quad \alpha_T(3\omega_0) = \alpha_{v_o}(3\omega_0) + \alpha_{BP}(3\omega_0)$. Table 13.1 shows the attenuation for the third harmonic as a function of Q.

Hence it can be seen that only a Q of about 10 is sufficient to achieve almost 40 dB of attenuation for the third harmonic. Thus, a state variable filter, Tow-Thomas or Antonio's GIC, could be used to generate a low distortion sine wave. Finally, note that the resulting sine wave will

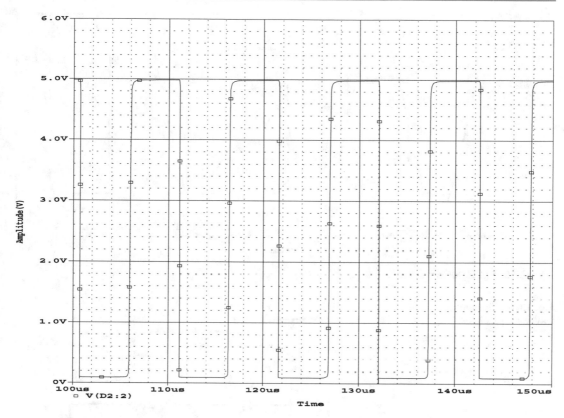

Fig. 13.14 PSPICE simulation of circuit

Table 13.1 Attenuation of the third-order harmonic in a square wave to various values of the bandpass filter Q

Q	$\alpha BP(3\omega o)$	$uvo(3wv)$	$\alpha_T(3\omega_o)$
10	28.25	9.54	38.07
15	32.04	9.54	41.60
20	34.54	9.54	44.08

have an amplitude that is $4/\pi$ times that of the input signal. An example of the filtering action of the second-order bandpass filter is shown in the MATLAB simulation Fig. 13.15 where a square wave of input frequency 1 kHz and 1 V peak-to-peak is the input and the output is a sine wave of amplitude 1.27 V.

13.3   Triangular Wave Generation

In the previous section we noted that the capacitor voltage was exponential with a time constant. To generate a triangular waveform which has constant slope, the current charging or discharging of the capacitor must be constant so that the voltage developed across the capacitor is the integral of a constant. Hence a current source whose current is a function of the output will generate a triangular output waveform. An example of this is shown in Fig. 13.16. Here transistors Q_1 and Q_2 which could be bipolar or CMOS act as current sources and current sinks. The operation of this circuit is therefore conceptually similar to that of Fig. 13.9. That is, during the positive upswing diode, D_2 is forward-biased (D_1 is reversed biased) so that transistor Q_2 is effectively off. Transistor Q_1 therefore acts as a current source charging C_1 with a constant current $I_{C_1} \simeq (V_{Z_3} - V_{BE_1})/R_1$. When the capacitor voltage $V_c(t)$ reaches the

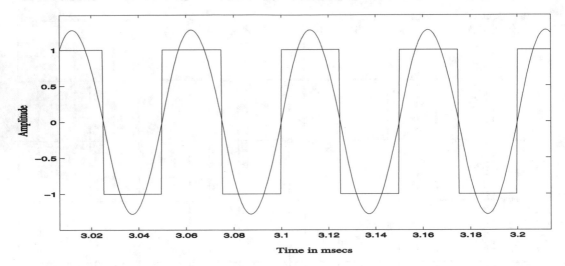

Fig. 13.15 MATLAB simulation

Fig. 13.16 Triangular
waveform generator

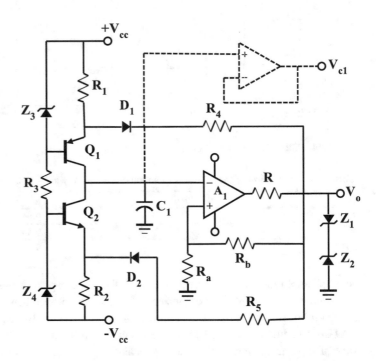

threshold voltage $+\beta V_o$, the output of A_1 goes
negative and D_1 becomes forward-biased with
D_2 now reversed-biased. Transistor Q_1 therefore
switches off and Q_2 switches on acting as a current sink connected to C_1 with constant current
$I_{C_2} \simeq (V_{Z_4} - V_{BE_2})/R_2$. Eventually, the capacitor
voltage $V_c(t)$ reaches the threshold voltage $-\beta V_o$
and the output of the amplifier A_1 goes positive
and the cycle repeats as shown in Fig. 13.17. The

role of diodes D_1 and D_2 is therefore to steer the
current source or sink depending on the output
state. Resistor R serves to limit the current
through Zener diodes Z_1 and Z_2. To compute the
period of oscillation, we need to evaluate each
half-cycle period. Assigning $t = 0$ as the starting
point at which the capacitor voltage is $-\beta V_o$, the
voltage across the capacitor $V_c(t)$ is given by,

Fig. 13.17 Output
waveform of Fig. 13.16

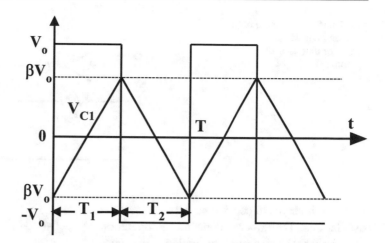

$$V_c(t) = \frac{1}{C} \int_0^t I_{C_1} dt \qquad (13.15)$$

for $0 \leq t \leq T_1$. At $t = T_1$ the half-cycle is
completed as $V_{c1}(T_1) = +\beta V_o$ is reached. But
from (13.15)

$$V_c(T_1) = \frac{T_1}{C} \left(\frac{V_{Z_3} - V_{BE_1}}{R_1} \right) - \beta V_o \quad (13.16)$$

Solving for T_1 yields $T_1 = 2C\beta V_o \left(\frac{R_1}{V_{Z_3} - V_{BE}} \right)$.
In a similar manner half-cycle period T_2 can be
solved as $T_2 \simeq 2C\beta V_o \left(\frac{R_2}{V_{Z_4} - V_{BE_2}} \right)$. The total
period $T = T_1 + T_2$ and hence the frequency of
oscillation is therefore given by

$$T = \frac{1}{f_{osc}}$$

$$= 2C\beta V_o \left(\frac{R_1}{V_{Z_3} - V_{BE_1}} + \frac{R_2}{V_{Z_4} - V_{BE_2}} \right)$$

$$(13.17)$$

If $V_{Z_3} = V_{Z_4}, V_{BE_1} \simeq V_{BE_2}$ and $R_1 = R_2$, then

$$f_{osc} = \frac{V_{Z_3} - V_{BE}}{4C\beta V_o R_1} \qquad (13.18)$$

If $T_1 \neq T_2$ either by setting $R_1 \neq R_2$ or $V_{Z_4} \neq V_{Z_3}$, a sawtooth waveform can be generated. Note
that buffering at the capacitor node is required to
drive any load so as not to disturb the nodal
currents as illustrated by the dashed op-amp.

Example 13.7
Design a square-wave/sawtooth oscillator that
oscillates at a frequency of 3.2 kHz.

Solution We begin, by choosing a typical
op-amp, the TL082 which has a bandwidth of
4 MHz. Next we choose $V_{z3} = V_{z4} = 4.7$ V
using the 1N750 Zener diodes and $R_1 = R_2 = 1$ k.
Since the average Zener diode current is about
1A, we can choose $R_3 = 2$ k, which limits the
Zener diode current to approximately 10 mA.
Diodes D_1 and D_2 can be regular 1N4148 diodes,
and we can set R_4 and R_5 to both be 1 k. Next we
choose $C_1 = 0.22$ μF, and with $V_o \simeq 13.4$ V, it
follows that $\beta = \frac{V_{z3} - V_{BE}}{4CV_o R_1 f_{osc}}$ or $\beta \simeq 0.1$. Since $\beta = \frac{R_a}{R_a + R_b}$, let us choose $R_a = 1$ k which results in
$R_b = 9$ k. To observe the triangular waveform, we
add a simple buffer to the capacitor voltage to
allow it to drive an output. The value of that
output is βV_o or 1.38 V.

A circuit that is capable of driving low impedance loads and simultaneously generating triangular or square waves is shown in Fig. 13.18.
Here amplifier A_2, a dedicated op-amp functions
as a lossless integrator to replace the Zener diode/
transistor arrangement of Fig. 13.16.
Conceptually Figs. 13.16 and 13.18 are the same
as op-amp A_2 integrates the input voltage V_{o1} to
yield a constant slope triangular waveform. V_R
and V_S represent control voltages. For the
moment let us assume that $V_S = 0$. Later we

Fig. 13.18 A variation on
the triangular wave
generator that uses all
op-amps

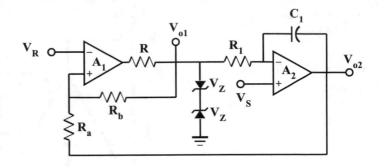

shall see it serves to control the duty cycle of the
oscillations. To further understand the operation
of this circuit, we need to look at the non-
inverting node of amplifier A_1. By superposition
the voltage V_x at this node is given by

$$V_x = -V_{o1}\beta + V_{o2}(1-\beta) \qquad (13.19)$$

where β was previously defined as $R_a/(R_a + R_b)$.
At a switching point, the voltage V_x must be equal
to the reference voltage V_R so that the op-amp
output voltage V_{o2} can be computed from (13.19)
as

$$V_{o2} = \frac{V_R}{1-\beta} + V_{o1}\frac{\beta}{1-\beta} \qquad (13.20)$$

However, the output voltage V_{o1} alternates
between $\pm V_o$, and therefore (13.20) can be
rewritten as

$$V_{o2} = V_R\frac{R_a + R_b}{R_b} \pm \frac{R_a}{R_b}V_o \qquad (13.21)$$

From (13.21) it can be seen that the integrator
output swings between $\pm\frac{R_a}{R_b}V_o$ with an average
value of $V_R(R_a + R_b)/R_b$. This point is illustrated
in Fig. 13.19. Hence by controlling V_R we can
control the offset of the triangular waveform,
while the output amplitude can be controlled by
varying the ratio R_a/R_b. To determine the fre-
quency of operation, we note that $V_{o2} = -\frac{1}{R_1C_1}\int V_{o1}dt$. Using $t = 0$ as the starting point,
we note that at $t = T_1$ or at the end of the first half-
cycle

$$V_{o2}(T_1) = -\frac{1}{R_1C_1}\int_0^{T_1} V_{o1}(t)dt$$

$$+ V_{o2}(0)$$

$$= -\frac{1}{R_1C_1}[V_{o1}(T_1)T_1]$$

$$+ V_{o2}(0) \qquad (13.22)$$

But $V_{o2}(T_1) = -\frac{R_a}{R_b}V_o + V_R\frac{R_a+R_b}{R_b}$, $V_{o1}(0) = \frac{R_a}{R_b}V_o + V_R\frac{R_a+R_b}{R_b}$, and $V_{o1}(T_1) = V_o$. Substituting
in (13.22) and solving for T_1 yields,

$$T_1 = \frac{2R_1C_1R_a}{R_b} \qquad (13.23)$$

Similar analysis applied to the second period
T_2 will also yield

$$T_2 = \frac{2R_1C_1R_a}{R_b} \qquad (13.24)$$

With $T_1 = T_2 = T/2 = 1/2f_{osc}$, the frequency of
oscillation is therefore given by

$$f_{osc} = \frac{R_b}{4R_1C_1R_a} \qquad (13.25)$$

The resulting composite waveforms are shown
in Fig. 13.19. Clearly from (13.25), the frequency
of oscillation can be controlled by varying R_1
(or C_1) and is independent of V_o. Usually decades
of frequency change are obtained by changing
C by a factor of 10 and within a decade by
varying R continuously. In terms of high frequency limitations, this waveform generator is
limited by either the slew rate of the op-amp A_2
used or its maximum output current.

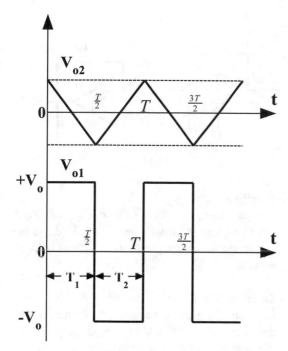

Fig. 13.19 Triangular and square-wave output waveforms of Fig. 13.18

Example 13.8

Design a sawtooth/square oscillator using the alternative structure of Fig. 13.18 that oscillates at $f_{osc} = 15$ kHz.

Solution Let us begin the design of this oscillator by setting $V_R = V_S = 0$ V and by choosing $C_1 = 0.1$ μF and $R_1 = 1$ k. This implies from (13.25) that $\frac{R_b}{R_a} = 6$. Choosing $R_a = 1$ k yields $R_b = 6$ k. Note that the integrator output swings between $\pm\frac{1}{6}V_O$ or roughly 2.2 V with an average value of zero volts.

Exercise 13.1

An alternative method of frequency control in this waveform generator is to use the potentiometer arrangement as shown in Fig. 13.20. If $\alpha V_{o1}(\alpha \le 1)$ represents the portion of the output signal V_{o1} that supplies the integrator, show that the frequency of oscillation of this circuit is given by

$$
f_{osc} = \frac{\alpha R_b}{4C_1 R_a [R_1 + \alpha(1 - \alpha)R_C]}
$$

$$
\cong \frac{\alpha R_b}{4R_1 C_1 R_a} \tag{13.26}
$$

if $R_1 \gg R_C$ where R_C is the total potentiometer resistance.

13.3.1 Duty Cycle Modulation

In the previous section, a 50% duty cycle square-wave/triangular wave generator was introduced. Here we define duty cycle δ as the ratio $T_1/T \times 100\%$. To be able to vary the duty cycle of the circuit of Fig. 13.18, we must set $V_S \ne 0$. To see why we first note that with $V_S \ne 0$, $V_{o2} = \frac{1}{R_1 C_1} \int (V_S - V_{o1})dt + V_S \delta(-\infty)$. That is, the slope of the integrator is now $(V_S \pm V_o)/R_1 C_1$ depending on the value of the output $V_{o1}(= \pm V_o)$ in the cycle. The impulse function is present in the above equation only for mathematical completeness and disappears after power-up. Assigning $t = 0$ once again as an arbitrary starting point and following the logic used earlier in deriving (13.22), (13.23), and (13.24), the periods T_1 and T_2 are now given by

$$
T_1 = \frac{2R_1 C_1 R_a}{R_b} \frac{V_o}{(V_S - V_o)} \tag{13.27}
$$

$$
T_2 = \frac{2R_1 C_1 R_a}{R_b} \frac{V_o}{(V_S + V_o)} \tag{13.28}
$$

Therefore the duty cycle δ of the square wave or the triangular wave is given by

$$
\delta \equiv \frac{T_1}{T} = \frac{T_1}{T_1 + T_2} = \frac{1}{2}\left(1 + \frac{V_S}{V_o}\right) \tag{13.29}
$$

Hence the duty cycle varies from 0 when $V_S = -V_o$ to 50% when $V_S = 0$ to 100% when $V_S = +V_o$. In practice the extremities of the duty cycle are never achieved. Note that the frequency of oscillation can be easily deduced to be

Fig. 13.20 An alternative
method of frequency
control of a waveform
generator

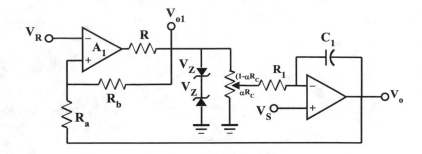

$$f_{osc} = \frac{R_b}{4R_1C_1R_a}\left[1 - \left(\frac{V_S}{V_O}\right)^2\right] \qquad (13.30)$$

which is a non-linear function of frequency.

Example 13.9
Determine the frequency of oscillation of
Fig. 13.18 if the control voltage changes from
$V_S = 0$ V, 2 V, 3 V. You may assume that
$C_1 = 0.1$ μF, $R_1 = 1$ k and $\frac{R_b}{R_a} = 6$.

Solution With $Vo \cong 13.4$ V, $f_{osc} =$
$\frac{6}{4\times1000\times0.1\times10^{-6}}\left[1 - \frac{V_s}{13.4}\right]$. For $V_s = 0$ V,
$f_{osc} = 15$ kHz. For $V_s = 2$ V, $f_{osc} =$
$\frac{6}{4\times1000\times0.1\times10^{-6}}\left[1 - \frac{2}{13.4}\right] = 14.66$ kHz . Finally,
for $V_s = 3$ V, $f_{osc} = \frac{6}{4\times1000\times0.1\times10^{-6}} \times$
$\left[1 - \frac{3}{13.4}\right] = 14.25$ kHz.

13.3.2 Sawtooth Generation

A natural extension of the triangular wave generator is that of a sawtooth waveform. Sawtooth
waveforms are used where a repeatable ramp
function whose single positive or negative slope
has a predefined period T. In fact, the sawtooth
waveform can be regarded as a triangular waveform with one slope of the triangular waveform
being set at $\pm\infty$. That is either $T_1 = 0$ or $T_2 = 0$.
As such, the triangular wave generator of
Fig. 13.18 can be used, but its duty cycle must
be set to near 0% for a positive-going sawtooth
waveform and near 100% for a negative-going
sawtooth waveform.

Example 13.10
Design a triangular wave generator to generate a
positive-going sawtooth wave of 5 V pk-pk
amplitude with a frequency of 5 kHz of a 15 V
supply.

Solution We first note that for a positive ramp
function, VS must be negative. Since we require
the output $V_o = \pm 5$ V we can employ 1N437
Zener diodes for D_1 and D_2 as before to clamp the
output to the desired value. Furthermore let us
choose $V_S = -4.8$ V so that according to
(13.29), $\delta = 2\%$ and hence $T_1 < < T_2$. Employing
Eq. (13.30) with $f_{osc} = 5$ kHz, let us choose
$R_a = R_b = 10$ k. Hence, we can solve for the
time constant $R_1C_1 = 3.92$ μs. Choosing
$C_1 = 1000$ pF yields $R_1 = 3.92$ k. We can use a
3.9 k resistor. The last step involves the choice of
amplifiers. If the design application is not critical,
then a general purpose op-amps such as the
LF411 op-amp can be employed for A_1 and A_2.
The final circuit and PSPICE simulation results
are shown in Figs. 13.21 and 13.22. Note that the
output of the generator has a negative DC offset.
Why?

Another means of altering the duty cycle is to
employ the circuit of Fig. 13.23 which uses the
circuit of Fig. 13.10 to alter the time constants.
Here we choose $R_{1b} > > R_{1a}$ so that the integrator
time constant during the positive cycle is more.

13.3.3 Voltage-Controlled Oscillators

A voltage-controlled oscillator (VCO) is a circuit
whose frequency of oscillation can be directly

Fig. 13.21 Final circuit for Example 13.10

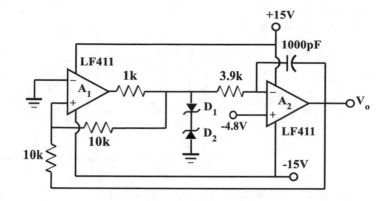

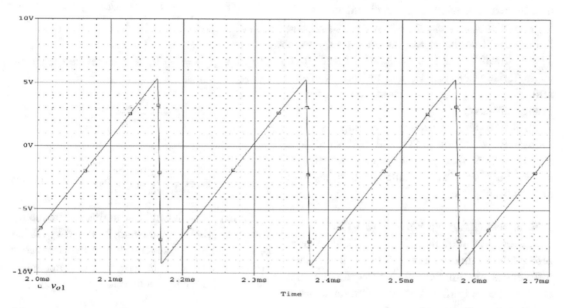

Fig. 13.22 PSPICE simulation for Example 13.10

Fig. 13.23 An alternative sawtooth waveform generator

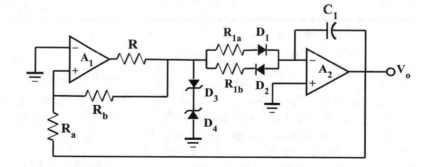

controlled by a voltage. Note in some cases the controlling signal can be a current. VCOs can be of two types: those that produce square and/or triangular waveforms and those that produce sinusoidal waveforms. In regards to multivibrators that produce square/triangular waveforms, the circuit of Fig. 13.20 can be considered a VCO since its frequency is dependent

Fig. 13.24 A square/
triangular wave voltage-
controlled oscillator

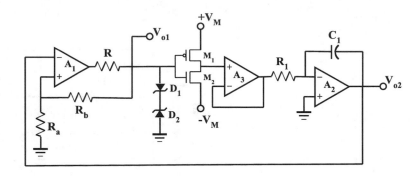

on V_S albeit non-linearly. Sinusoidal VCOs are usually realized using a tuned circuit in a feedback connection or using a sinewave shaper. The tuned circuit could be RC, LC or crystal based. The sinewave shaper exploits the logarithmic characteristic between V_{be} and the collector current in a bipolar transistor to smooth a triangular wave.

In this section we describe a simple square/triangular waveform VCO and two sinusoidal VCOs. VCOs that generate square waveforms can be built using ring oscillators, but they will not be discussed here.

A Square/Triangular Waveform VCO

Consider the circuit of Fig. 13.24 which represents a modification of Fig. 13.18. Two differences are immediately observed. The first is that the amplifier is now configured as a Schmitt trigger by grounding R_a. Secondly, a CMOS version of a single pole double throw (SPDT) switch has been introduced between the comparator and the integrator. The effect of the SPDT is to provide the integrator with a sweep speed of $\pm V_o/RC$ depending on the sign of V_{o1}. Hence when $V_{o1} = +V_o$, M_2 is on while M_1 is off, and the integrator receives a voltage of $-V_M$. Conversely, when $V_{o1} = -V_o$, M_1 is on while M_2 is off, and the input voltage to the integrator is now $+V_M$. The Schmitt trigger functions as before, switching when V_{o2} exceeds the variable threshold $\pm\beta V_o$. The waveforms produced by the circuit of Fig. 13.24 are therefore identical to those shown in Fig. 13.17 with identical half cycles. Using Fig. 13.17 as a guide, during the positive sweep

$$V_{o2}(t) = -\frac{1}{R_1 C_1} \int_0^{\frac{T}{2}} (-V_M)dt - \beta V_o \quad (13.31)$$

But $V_{o2}\left(\frac{T}{2}\right) = +\beta V_o$ and therefore the period can be solved as.

$$T = \frac{4\beta V_o R_1 C_1}{V_M} \quad (13.32)$$

The frequency of oscillation is therefore

$$f_{osc} = \frac{R_a + R_b}{4 R_a R_1 C_1} \cdot \frac{V_M}{V_o} \quad (13.33)$$

From (13.33) it can be seen that the frequency varies linearly with the control voltage V_M. It has been reported that several decades of frequency are available with this type of VCO making it useful for frequency modulation.

Example 13.11

Design an oscillator whose frequency of oscillation is variable from $f_{osc} = 5.3$ kHz to $f_{osc} = 10.63$ kHz. You may assume that $C_1 = 0.1$ μF, $R_1 = R_a = 1$ k and $R_b = 4.7$ k.

Solution Solution: With $V_o \cong 13.4$ V, $f_{osc} = \frac{1 \text{k}+4.7 \text{k}}{4\times 1 \text{k}\times 1000\times 0.1\times 10^{-6}} \frac{V_M}{13.4} = 1063.43\ V_M$. Hence when $f_{osc} = 5.3$ kHz, $V_M = 5$ V and when $f_{osc} = 10.63$ kHz, $V_M = 10$ V.

A Sinusoidal VCO

The circuit of Fig. 13.24 can be used to generate sinewaves and hence a sinusoidal VCO if we use its triangular waveform output to drive the sinewave shaper circuit shown in the bottom half of Fig. 13.25. This circuit functions by

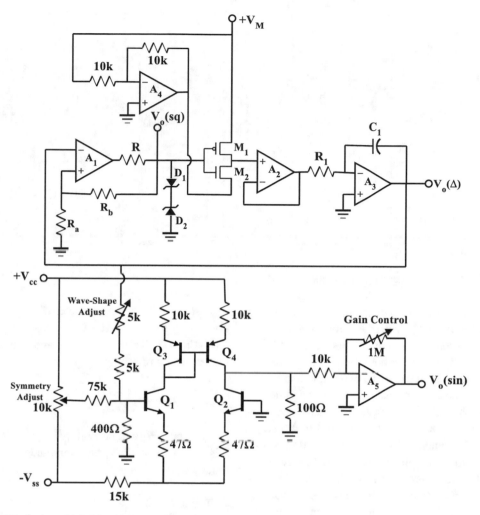

Fig. 13.25 A sinusoidal/triangular/square-wave VCO

exploiting the I_C/V_{BE} characteristics of transistors Q_1 and Q_2 to approximate the sine function over a limited input range. It can be recognized that transistors $Q_1 - Q_4$ form a differential amplifier with the inverting terminal grounded and a significantly reduced input signal applied to Q_1. Here the 47 Ω resistors act as degeneration resistors, and transistors $Q_{3,4}$ form a current mirror that copies the collect current of Q_1 and subtracts it from the collector current of Q_2. The 100 Ω resistor forces the node labeled a to be low impedance and converts the output current to a voltage that is then amplified by amplifier A_5. Two variable resistors are included that offer symmetry adjustment and wave shaping. The value of the wave-

shaping resistor is crucial because it determines the distortion levels of the sine wave produced. Note in Fig. 13.25 amplifier A_4 generates $-V_M$ for MOS switches M_1 and M_2. It is equally possible to use JFETs instead of MOS transistors with the addition of some bias circuitry. Figure 13.26 shows the outcome of a PSPICE simulation of the circuit of Fig. 13.25 using LF411 op-amps, $R_1 = 10$ k, $C_1 = 12$ nF and transistors $Q_{1,2}$ and $Q_{3,4}$ as general purpose 2N3904/2N3906 transistors. Input voltage V_M is varied from 1 to 10 V.

A direct alternative means to construct a linear sinusoidal VCO requires designing a system

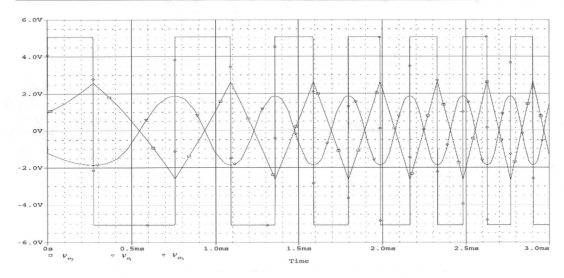

Fig. 13.26 PSPICE output waveforms for Fig. 13.25

whose imaginary axis poles move linearly with a control voltage V_c. Most such commercial VCOs are made using transconductors connected in a filter configuration whose poles are controlled via a transconductance. To better understand this suppose the characteristic equation of a system or a bandpass filter was of the form $s^2 + a_1 s + a_0$. Next we arrange it so that $a_1 = 0$ and then set $a_0 = KV_c^2$ where K is a constant. Setting $a_1 = 0$ ensures that the poles are on the $j\omega$ axis, resulting in a frequency of oscillation $f_{osc} = \sqrt{KV_C}$. A simple circuit that accomplishes this is shown in Fig. 13.27 consisting of an op-amp and two multipliers. Each multiplier is assumed to have a multiplier constant K_i, $i = 1, 2$. The two multipliers are needed to generate the term V_c^2. If only one multiplier was used, then the frequency of oscillation would vary with the square root of V_c. To determine the frequency of oscillation, we must compute the loop gain and set it equal to one. Breaking the loop at the point indicated by the $\times$ the loop gain of this circuit can be computed as

$$LG = \frac{K_1 K_2 V_C^2}{C_1 C_2 R_1 R_2 s^2 + \left(C_1 R_1 R - C_2 R^2\right)s + R_1 - R}$$

$$= 1 \tag{13.34}$$

If $R_1 = R$ then the characteristic equation is given by $C_1 C_2 R^2 s^2 + R(C_1 - C_2)s - K_1 K_2 V_C^2 = 0$. Setting $C_1 = C_2$ and solving for the frequency of oscillation yields

$$f_{osc} = \frac{1}{R}\sqrt{\frac{K_1 K_2}{C_1 C_2} \cdot V_C} \tag{13.35}$$

In practice R_1 will be slightly less than R to get this circuit to oscillate, and amplitude control is required.

Example 13.12
The variable frequency oscillator of Fig. 13.27 is set up using a TL082 op-amp and two multipliers (AD534) whose $K_1 = K_2 = 0.1$. You may assume that $C_1 = C_2 = 1$ nF and $R = 1$ k. Determine the frequency of oscillation if V_c is varied from 1 to 0.5 V.

Solution Using Eq. (13.35), the frequency of oscillation is given as $f_{osc} = \frac{1}{1000}\sqrt{\frac{0.1 \times 0.1}{1 \times 10^{-18}} V_c}$, which for $V_c = 1$ V yields $f_{osc} = 100$ kHz and for $V_c = 0.5$ V yields $f_{osc} = 70.7$ kHz.

13.4 Monostable Multivibrators

A monostable multivibrator produces a single pulse for a fixed period of time T. Unlike the

Fig. 13.27 Variable
frequency oscillator

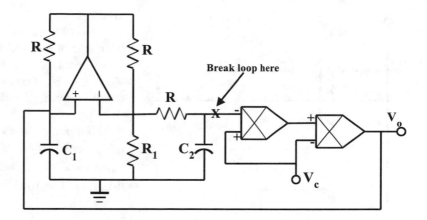

Fig. 13.28 Monostable
multivibrator

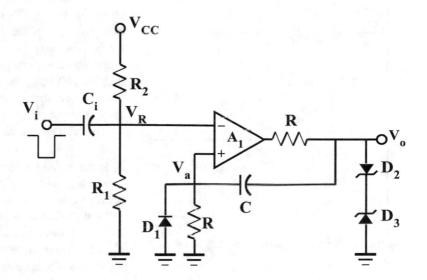

free-running or astable multivibrators discussed
earlier on, the monostable multivibrator has one
stable state and a quasi-stable state. It is during
the quasi-stable state that a pulse is generated for
a time T before returning to the stable state. To
generate the pulse, however, requires that a trigger input be applied to the circuit whose period
must be less than the desired period T. One such
simple but fairly accurate monostable pulse generator and its associated waveforms are shown in
Figs. 13.28 and 13.29, respectively.

The purpose of R_1 and R_2 is to establish a
positive reference voltage $V_R = \frac{R_1}{R_1+R_2} V_{cc}$ at the
inverting terminal of amplifier A_1. In the stable
state, the voltage V_a at the non-inverting terminal
is zero because the capacitor is open circuit. Note

since the voltage at the inverting terminal is
greater than zero, the output is initially at $-V_o$
and so is the voltage across the capacitor C. If a
negative going pulse of sufficient amplitude is
now applied to V_i, V_R subsequently follows it
going negative first then positive. When this
happens the output of A_1 switches to $+V_o$ and
the capacitor voltage changes to V_o with the
node voltage $V_a = 2V_o$. At once the C begins to
discharge through R with time constant $\tau = RC$ as
D_1 is open. Eventually V_a falls to V_R and the
output switches back to $-V_o$ at which point V_a
tries to follow but is clamped by D_1 to -0.6 V.
With the amplifier output at V_o and V_a at -0.6 V,
C again discharges through R until V_a reaches
zero. Since the capacitor discharges during

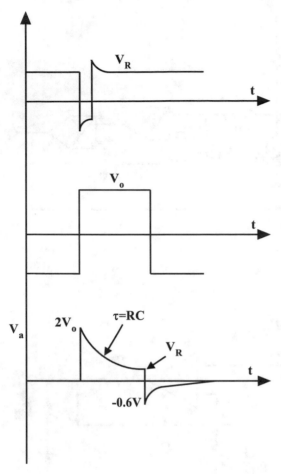

Fig. 13.29 Associated waveforms for Fig. 13.28

T from voltage $2V_o$ to V_R during the period T, it can be easily shown that T can be computed as

$$T = RC \ln \left(\frac{2V_O}{V_R} \right) \qquad (13.36)$$

While this circuit is a very useful monostable multivibrator, there are two practical concerns in its design. The first relates to the fact that the time constant of the input circuit $\tau_i = (R_1 // R_2)C < < \tau$. This way the input circuit can recover long before the period T is complete. Secondly, the amplitude $V_R < V_O$. If this is not observed, false triggering can result in the operation of this circuit. Note that V_R is a function of V_{cc} and therefore changes in power supply voltage will affect T.

Example 13.13

A monostable multivibrator is constructed using the circuit of Fig. 13.28. The values chosen were $R_1 = 1$ k, $R_2 = 2$ k, $R = 1$ k, $C = 0.1$ μF and $V_{cc} = 15$ V. Determine the period of this multivibrator if $V_z = 10$ V.

Solution From the specifications, we can determine that $V_R = \frac{1\,k}{1\,k+2k} \cdot 15$ V $= 5$ V and $V_o = V_Z = 10$ V. The period T from (13.36) is therefore $T = 1000 \times 0.1 \times 10^{-6} \ln \left(\frac{2 \times 10}{5} \right)$ or $T = 138.6$ μs.

A second circuit capable of producing pulses is shown in Fig. 13.30 having certain advantages over the circuit of Fig. 13.28. Here a JFET transistor is effectively used to short a capacitor that originally is fully charged to V_{cc}. We shall assume that the JFET being used has sufficiently low resistance r_{on} so that C is effectively shorted. The associated waveforms are shown in Fig. 13.31.

The voltage at the non-inverting terminal $V_R = \frac{R_1}{R_1+R_2} V_{cc} = \beta V_{cc}$. Resistor R_a ensures that J_1 is off in the stable state in the absence of an input pulse since the gate of J_1 is negative. If a positive-going pulse of amplitude greater than V_B and pulse width of at least $5r_{on}C_1$ is now applied to V_i, J_1 is turned on and V_a goes to zero. Since $V_R > V_a$ the output of the comparator goes to V_o. As V_i returns to zero J_1 turns off and C_1 begins to charge towards V_{cc}. Eventually V_a exceeds βV_{cc} and the amplifier output reverts back to $-V_o$ and V_a continues on to $+V_{cc}$. If the pulse width is sufficiently small so that the capacitor C starts charging immediately, the capacitor voltage is defined by $V_c(t) = V_{cc}\left(1 - e^{-\frac{t}{RC}}\right)$. Switching, however, occurs when $V_c(T) = +\beta V_{cc}$. Solving for the period T yields,

$$T = RC \ln \left(1 + \frac{R_1}{R_2} \right) \qquad (13.37)$$

Note unlike the circuit of Fig. 13.28, the period of the circuit of Fig. 13.30 is completely independent of V_{cc} and $-V_B$. In addition, this circuit can be re-triggered immediately, this despite the fact that a full pulse period occurs only when the

Fig. 13.30 Improved
multivibrator

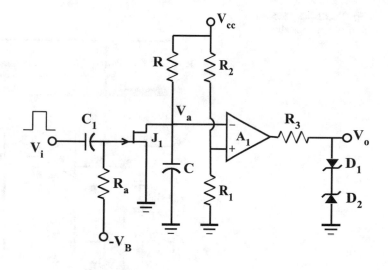

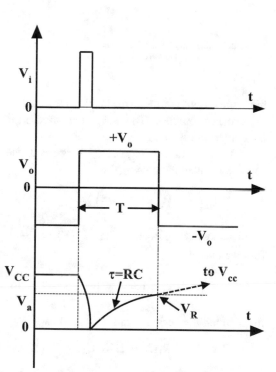

Fig. 13.31 Associated waveforms for Fig. 13.30

capacitor voltage exceeds βV_{cc}. Finally, if desired
the JFET transistor can be replaced by an NMOS
transistor. In that case the input pulse needs to
exceed the threshold voltage of the transistor to
turn it on.

Example 13.14
A monostable multivibrator is constructed using
the circuit of Fig. 13.30. The values chosen were
$R_1 - 1.5$ k, $R_2 = 3$ k, $R - 1$ k and $C = 0.1$ μF
with $V_{cc} = 15$ V. Determine the period of this
multivibrator if $V_z = 10$ V.

Solution From the specifications we can determine that $V_o = 10$ V. The period T from (13.37)
is   therefore $T = 10^3 \times 0.1 \times 10^{-6} \times$
$\ln\left(1 + \frac{3\,k}{1.5\,k}\right)$ giving $T = 109.86$ μs.

13.4.1   The 555 Timer

Any discussion about multivibrators, astable or
monostable, would be incomplete without the
subject of the NE555 timer. Originally introduced
by Philips Semiconductors under the name
NE555, this versatile device has been employed
in a variety of applications enjoying immense
popularity throughout the years. Today
low-power CMOS versions are available through
many manufacturers.

   The basic chip consists of two comparators, a
bistable flip-flop, a discharge transistor and an
internal resistive network. The resistive network
consists of three equal value $R_2 = 5$ k resistors
that set the voltages at the comparators to $A = \frac{V_{cc}}{3}$
and $B = \frac{2V_{cc}}{3}$. The flip-flop is an $R - S$ bistable

Fig. 13.32 The 555 timer used as a monostable multivibrator

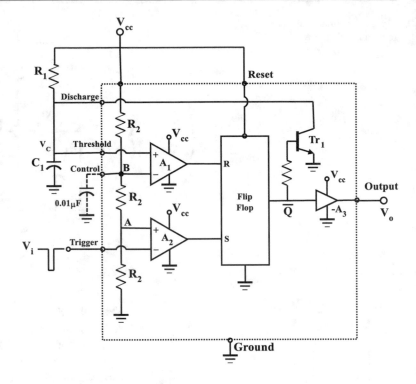

flip-flop whose state can be forced to reset by a reset pin. The output of the flip-flop is taken from $\overline{Q}$ where it is fed to an amplifier A_3 that inverts $\overline{Q}$ to provide Q and be able to drive a load. The block diagram of the 555 timer is shown in Fig. 13.32 being used as a monostable multivibrator. In the stable state, S is low because the trigger input V_i is larger than $\frac{V_{cc}}{3}$. The input R is also low because the control voltage which is equal to $\frac{2V_{cc}}{3}$ is larger than the threshold voltage V_c at the capacitor terminal. The output of the flip-flop $\overline{Q}$ is subsequently high forcing Q_1 on so that C is shorted to ground. At this point the main output V_o is approximately zero volts. Assuming now that a trigger pulse comes along and its value is less than $\frac{2V_{cc}}{3}$ for a short period of time. When this occurs, comparator A_2 changes state, and S goes high forcing $\overline{Q}$ low and turning off the discharge transistor Q_1. Capacitor C_1 immediately begins to charge towards V_{cc} as the output V_o remains high. Upon V_c exceeding $\frac{2V_{cc}}{3}$, the output of the comparator A_1 changes state forcing R to go high and subsequently setting $\overline{Q}$ high which shorts out the capacitor C_1 once more via discharge transistor Q_1. Immediately as $\overline{Q}$ goes

high, the output returns to the stable low state and the cycle is complete. The associated waveforms are shown in Fig. 13.33.

If the saturation voltage of Q_1 is ignored, the capacitor voltage is governed by

$$V_c(t) \cong V_{cc}\left(1 - e^{-\frac{t}{RC}}\right) \qquad (13.38)$$

Since switching occurs when $V_c(T) = \frac{2V_{cc}}{3}$, it follows that

$$\frac{2V_{cc}}{3} \cong V_{cc}\left(1 - e^{-\frac{T}{RC}}\right) \qquad (13.39)$$

from which the period T can be solved as

$$T \simeq \ln(3)R_1C_1 = 1.1R_1C_1 \qquad (13.40)$$

Practical pulse widths can range from a few microseconds to several seconds with the 555 possessing good temperature stability. The propagation time from an input pulse to when the output is generated is in the order of tens of nanoseconds to a few hundred nanoseconds depending on the technology and manufacturer. Note in the circuit of Fig. 13.32, an optional 0.01 μF capacitor C is present to ensure that the

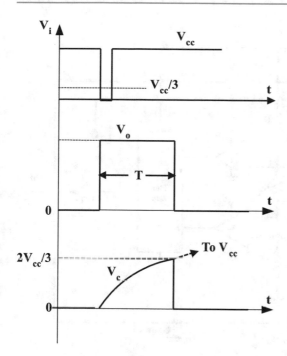

Fig. 13.33 Associated waveforms for Fig. 13.32

control node is not affected by transients during the switching periods as the basic 555 timer can sink or source up to 200 mA. In low-power CMOS versions of the 555 timer, typically C is not needed in the monostable mode.

To use the 555 timer in the astable mode as shown in Fig. 13.34, an extra resistor is required and the trigger and threshold points must be connected together. At power up the capacitor would be initially discharged so that $S = 1$ or a logic high and $R = 0$. $\overline{Q}$ is therefore low, and hence V_o is high with the discharge transistor Q_1 being off. Eventually C starts to charge through R_a and R_b towards V_{cc} with time constant $\tau_1 = (R_a + R_b)C$. Note during the time that the capacitor voltage is between one third and two thirds of V_{cc}, S is set to 0 and V_o is high. When the capacitor voltage reaches $\frac{2V_{cc}}{3}$ during a period T_1 which starts from V_c exceeding $\frac{V_{cc}}{3}$, $R = 1$ setting $\overline{Q} = 1$ and turning discharge transistor Q_1 on. Since Q_1 shorts R_b to ground C begins to discharge through R_b setting up a second astable period T_2 with time constant $\tau_2 = R_bC$ and the output V_o low. When the capacitor voltage reaches $\frac{V_{cc}}{3}$, changes to 1, setting $\overline{Q}$ to zero and

hence V_o high and C once again charges through $R_a + R_b$ repeating the cycle. Since the capacitor voltage alternates between $\frac{V_{cc}}{3}$ and $\frac{2V_{cc}}{3}$, it can be shown that the periods T_1 and T_2 are given by

$$T_1 = 0.69(R_a + R_b)C \qquad (13.41)$$

$$T_2 = 0.69R_bC \qquad (13.42)$$

and $T = T_1 + T_2 = 0.69(R_a + 2R_b)C$. The frequency of oscillation is therefore given by

$$f_{osc} = \frac{1.44}{(R_a + 2R_b)C} \qquad (13.43)$$

The associated waveforms are shown in Fig. 13.35. Note that because $T_1 > T_2$ the duty cycle is always greater than 50%. To achieve a duty cycle of exactly 50%, diode D_1 indicated by dashed lines can be added to the circuit and the resistors arranged so that $R_a = R_b = R$. The effect of the diode is to short R_b during the positive cycle when $\overline{Q}$ is low (V_o is high) and transistor Q_1 is off, so that $T_1 = 0.69R_a$. During the negative half-cycle when $\overline{Q}$ is high (V_o is low) the diode is reversed-biased so that C discharges normally through $R_b = R$. Note that if $R_b > R_a$ duty cycles of less than 50% can be achieved. Likewise, if $R_b < R_a$ duty cycle greater than 50% can be achieved. From typical manufacturer specifications with D_1 in place, duty cycles can range from less than 5% to greater than 95% provide that a minimum of $R_b = 3$ k is maintained. Note for the astable version of the 555 timer, the 0.01 μF on the control voltage is typically required to nullify switching transients.

Example 13.15

A monostable 555 timer is required to produce a time delay within a circuit. If a timing capacitor $C_1 = 10$ μF is used, calculate the value of the resistor required to produce a minimum output time delay of 500 ms.

Solution 500 ms is the same as 0.5 s, so by rearranging the formula above, we get the calculated value for the resistor R_1 as $R_1 = \frac{t}{1.1C_1} = \frac{0.5}{1.1 \times 10 \times 10^{-6}} = 45.5$ k.

Fig. 13.34 The 555 timer used as an astable multivibrator

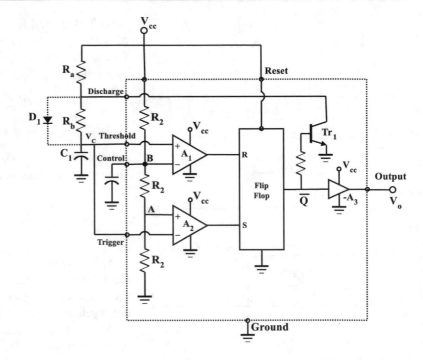

Fig. 13.35 Associated waveforms for Fig. 13.34

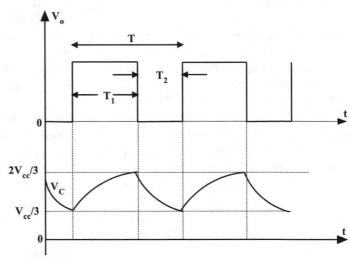

Example 13.16

An astable 555 oscillator is constructed using the following components, $R_a = 1$ k, $R_b = 2$ k and capacitor $C_1 = 0.1$ μF. Calculate the output frequency from the 555 oscillator and the duty cycle of the output waveform.

Solution From Eq. (13.43) the frequency of oscillation is $f_{osc} = \frac{1.44}{(1\,k+2\times2\,k)\times0.1\times10^{-6}} = $

2.88 kHz. The duty cycle D of the output waveform is $D = \frac{R_a+R_b}{R_a+2R_b} = \frac{1000+2000}{1000+2\times2000} = 0.6$ or 60%.

13.5   Precision Rectifiers

A non-linear op-amp function that is of considerable importance is a precision rectifier. This function arises since ordinary silicon diodes cannot rectify signals of amplitude less than 0.7 volts, the diode threshold voltage. Using the op-amp, a circuit that functions like an ideal diode that is capable of rectifying signals of a few millivolts can be realized. Such a circuit is called a precision rectifier and can be classified as either a linear half-wave rectifier or a full-wave rectifier. The linear half-wave rectifier delivers an output signal that depends on the amplitude and polarity of the input signal. It is also called a precision half-wave rectifier. The precision full-wave rectifier delivers an output signal that depends on the amplitude but not the polarity of the input signal. It is sometimes referred to as an absolute value circuit.

13.5.1   Linear Half-Wave Rectifier

Linear half-wave rectifier circuits deliver one-half cycle of a signal and eliminate the other half-cycle while holding the output at zero volts. A circuit for achieving this is shown in Fig. 13.36. It involves an inverting amplifier in which two diodes are added thereby converting the amplifier into a precision half-wave rectifier. The operation of the circuit is straightforward. During a positive half-cycle of the input signal, V_i goes positive and therefore the voltage V_{o1} at the output of the op-amp goes negative. This causes diode D_1 to conduct as a result of which the V_{o1} is limited to -0.7 V, the anode of the diode being connected to the virtual earth at the inverting terminal. Diode D_2 is reverse-biased by V_{o1}. The input current $I = V_i/R$ flows into the input resistor R through D_1 into the output of the op-amp. The output voltage V_o is therefore at zero potential. During

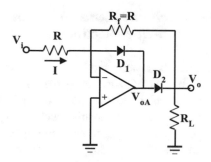

Fig. 13.36 Inverting Linear Half-Wave Rectifier: Positive Output

the negative half-cycle of the input signal, V_i goes negative, and therefore the voltage V_{o1} goes positive. This causes diode D_2 to be forward-biased. Since the anode of diode D_1 is connected to the inverting terminal and is therefore at ground potential, the effect of V_{o1} going positive is to reverse-bias this diode. The input current $I = V_i/R$ therefore flows through R_f resulting in an output voltage $V_o = -V_i$ since $R_f = R_i$. The op-amp effectively operates like an inverter. Note that since during this half-cycle V_i is negative, the inversion results in a positive-going signal. The diode turn-on voltage of D_2 is virtually eliminated since small voltages at the input of the op-amp results in amplified voltage V_{o1} at the output of the op-amp which turns on diode D_2. Because this diode is within the feedback loop of the op-amp, voltage V_{o1} assumes the necessary level required to turn on this diode. The result is that signals at the level of millivolts can be rectified. For the case where both diodes are reversed, then positive-going input signals are transmitted and inverted, while the output is zero for all negative-going signals.

13.5.2   Signal Polarity Separator

An interesting variation of the half-wave rectifier is the signal polarity separator shown in Fig. 13.37. In this circuit, signals of different polarities are conveniently separated. Thus, when the input signal V_i is positive, the output V_{oa} of the op-amp goes negative thereby forward-biasing diode D_1 and reverse-biasing diode D_2. This causes current V_i/R to flow through the input resistor R, through the feedback resistor $R_{F1} = R$ and through forward-biased diode D_1 into the output of the op-amp. This results in an output voltage $V_{o1} = -V_i$ with $V_{o2} = 0$ since D_2 is off. When the input signal V_i goes negative, the output V_{oa} of the op-amp goes positive thereby forward-biasing diode D_2 and reverse-biasing diode D_1. This causes current V_i/R to flow out of the output of the op-amp, through forward-biased diode D_2, through feedback resistor $R_{F2} = R$ and through the input resistor R into the input signal source. This results in an output voltage $V_{o2} = -V_i$ with

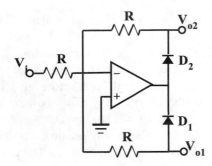

Fig. 13.37 Signal polarity separator

$V_{o1} = 0$ since D_1 is off. The result is that the signals of different polarities are separated into V_{o1} for positive-going V_i and V_{o2} for negative-going V_i. Note that each output is inverted relative to the input signal.

13.5.3 Precision Rectifiers: The Absolute Value Circuit

The precision full-wave rectifier provides full-wave rectification of an alternating input signal with a signal of one polarity being delivered at the output. The action of the op-amps effectively reduces the threshold voltage of the diodes and enables the rectification of signals with amplitudes of millivolts. The first precision full-wave rectifier circuit is shown in Fig. 13.38. It employs two op-amps with equal value resistors and an input impedance $R_1 = R$.

When V_i goes positive such that $V_i = k$ (where $k > 0$), the output voltage of op-amp A goes negative resulting in the forward-biasing of diode D_p and the reverse-biasing of diode D_n. Both op-amps have their inverting terminals at ground potential, and, hence, a current k/R flows into R_1 at the input, through the resistor R_2 into the output of op-amp a via diode D_p. As a result, a voltage $V_A = -k$ is developed at point A. Since point B is at zero potential (it tracks the non-inverting terminal of op-amp b which is at a virtual earth), a current k/R flows from point B to A through R_3 and D_p into the output of op-amp a. This current flows from the output of op-amp b through R_4. Since point B is at zero potential, it follows that the output of op-amp b is at $V_o = k = V_i$

which is positive. For negative input voltages such that $V_i = -k$, the output of op-amp a goes positive and hence diode D_n conducts while diode D_p turns off. Current flows out of op-amp a through D_n and R_5 into R_1 to the source, while current flows out of op-amp b through $R_4 - R_3 - R_2$ into R_1 to the source. Noting that $V_C = V_B = V_x$, it follows that the current I flowing into the source at the input $I = k/R$ is made up of current out of op-amp a given by V_x/R and current out of op-amp b given by $V_x/2R$. Thus the input current $I = k/R$ flowing through R_1 into the source is made up of $\frac{2}{3}I$ from op-amp a and $\frac{1}{3}I$ from op-amp b. Hence $V_o = \frac{1}{3}I(R_5 + R_4 + R_3) = \frac{1}{3}I \times 3R = k$. Thus, the output voltage V_o of op-amp b is $V_o = k = -V_i$ which is positive. Therefore, the output voltage is positive for either polarity of input voltage V_i and is equal to the absolute value of V_i. If the diodes are reversed, then the output will provide negative signals.

Example 13.17
Using the circuit in Fig. 13.38, design a precision full-wave rectifier to operate at 1 kHz and below.

Solution Using the 741 op-amp, choose $R = 10$ k and use 1N4148 signal diodes. The value of resistor R determines the current levels in the circuit. Note that the input impedance of this circuit is $R_1 = 10$ k. The low slew rate of this op-amp will limit the operation of this circuit to about 1 kHz. If higher frequency of operation is desired, then an op-amp with a higher slew rate such as the LF351 may be employed.

13.5.4 High-Impedance Precision Full-Wave Rectifier

A second precision rectifier is shown in Fig. 13.39. It utilizes an op-amp in the non-inverting mode at the input to obtain high input impedance. For V_i positive such that $V_i = k$ (where $k > 0$), the output of op-amp a goes positive, and diode D_p conducts with D_n off as a result. Since the inverting input of op-amp b is also at V_i, then both inverting terminals have the

Fig. 13.38 Absolute value circuit

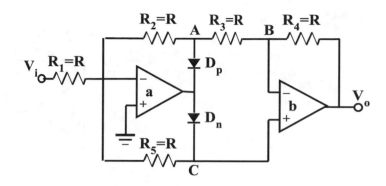

Fig. 13.39 High-impedance full-wave rectifier

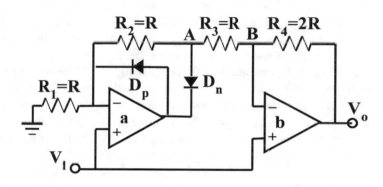

same potential V_i, and therefore no current flows through resistors R_2, R_3 and hence also R_4. From this, for positive input voltages $V_o = V_i$ which is positive. When V_i goes negative such that $V_i = -k$, the output of op-amp a goes negative turning off D_p and turning on D_n. With V_i also at the inverting terminal of op-amp a, a current k/R flows up from earth through resistor R_1 through R_2. The potential at point A is then $-2k$. Since the potential V_B at B is $V_B = V_i = -k$, then the voltage drop across R_3 is given by $V_{BA} = -k - (-2k) = k$. This results in a current k/R through R_3 through diode D_n into the output of op-amp a. (Note that the current through R_2 also flows through D_n into the output of op-amp a.) The current through $R_4 = 2R$ is therefore k/R giving a voltage drop V_{R_3} across this resistor of $V_{R_3} = 2R \times k/R = 2k$. Therefore, the output voltage V_o is given by $V_o = V_B + V_{R_3} = -k + 2k = k$. This gives $V_o = -V_i$ which is positive. Therefore V_o is positive for both polarities of the input signal V_i.

Example 13.18
Using the circuit in Fig. 13.39, design a high input impedance precision full-wave rectifier.

Solution The 741 is again used, and we choose $R = 10$ k. This means that all resistors except R_4 are of this value while $R_4 = 2R = 20$ k. The input impedance of the 741 at the non-inverting terminal is about 1 M. However, a bias resistor of say 100 k may need to be connected between the non-inverting terminals and ground. Therefore, this resistor will set the input impedance. If a JFET input op-amp such as the LF351 is used, the bias resistor can be 1 M or higher. The slew rate limitation applies to this configuration as well.

13.5.5 AC to DC Converter

The circuit in Fig. 13.40 is a third full-wave precision rectifier circuit. This particular configuration is able to provide the average value of a

Fig. 13.40 AC to DC
converter

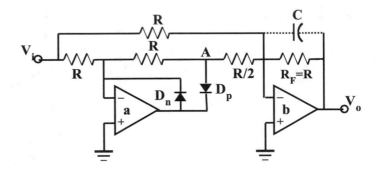

rectified AC signal. As a result, it is sometimes referred to as a mean-absolute-value circuit.

The basic system comprises a precision half-wave rectifier realized around op-amp a, the output of which is summed with the original signal using op-amp b as a summing amplifier. Note that the summing resistor for the half-wave rectifier is $R/2$ while that for the input signal is R. For positive inputs where $V_i = k$ ($k > 0$), op-amp a inverts V_i giving an output $V_A = -k$. Op-amp b sums the output of op-amp a and V_i to give $-\frac{V_o}{R} = \frac{V_A}{R/2} + \frac{V_i}{R} = \frac{-2k}{R} + \frac{k}{R} = -\frac{k}{R}$. From this, $V_o = k = V_i$ which is positive. For negative inputs where $V_i = -k$, op-amp a gives zero output, and op-amp b inverts V_i giving $V_o = k = -V_i$ which is positive. Thus, the circuit output V_o is positive and is the rectified or absolute value of the input.

This circuit produces the average value of the rectified output of op-amp b if a large-value capacitor C is placed in parallel with the feedback resistor in op-amp b. Several input voltage cycles are required for the capacitor to provide the mean value. The circuit can give the mean absolute value of sinusoidal, triangular, square and other periodic waveforms. For a full-wave rectified sinusoidal wave of peak amplitude k, the rms value is $k/\sqrt{2}$ and the average value is $k2/\pi$. Hence the form factor k_f is the ratio of these two values and is given by $k_f = \frac{k/\sqrt{2}}{2k/\pi} = \frac{\pi}{2\sqrt{2}} \simeq 1.11$. Hence for this circuit to be used to measure rms of a sinusoidal signal, the resistor R_F has to be scaled such that $R_F = 1.11R$ which will give an output voltage V_o that is the numerical value of the rms of the amplitude of the sine wave being measured.

Example 13.19
Using the circuit in Fig. 13.40, design a precision full-wave rectifier.

Solution For this circuit we use the AD844 current feedback amplifier. It has a very high slew rate (2000 μV/s) and as a current feedback amplifier is ideally suited to implement the inverting mode of the two amplifying elements in the circuit. If we choose $R = 2$ k, then the possible bandwidth of this circuit is 30 MHz. The completed circuit is shown in Fig. 13.41. If the circuit is used to measure rms, then form-factor correction must be applied by setting $R_F = 1.11 \times 2$ k = 2.22 k. Also, capacitor C must be chosen to be sufficiently large to filter out the ripple while enabling the system to track a changing voltage. A value of $C = 10$ μF will have a reactance of 16 Ω at 1 kHz and will ensure that most of the ripple does not appear across R_F.

13.6   Applications

In this section, a variety of circuits involving non-linear operations and waveform generation are discussed.

Battery Monitor
The circuit shown in Fig. 13.42 monitors the voltage at the terminals of a 12 V car battery. It comprises four op-amps connected as comparators and powered by the battery voltage. The inverting terminals are held at fixed voltages established by the 5.6 V Zener diode D_1 along with the resistor chain $R_2 - R_6$. The values

Fig. 13.41 Completed
circuit for Example 13.19

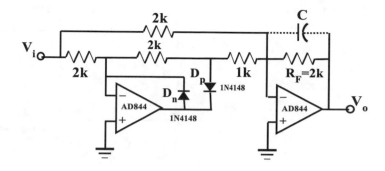

Fig. 13.42 Car Battery
Monitor

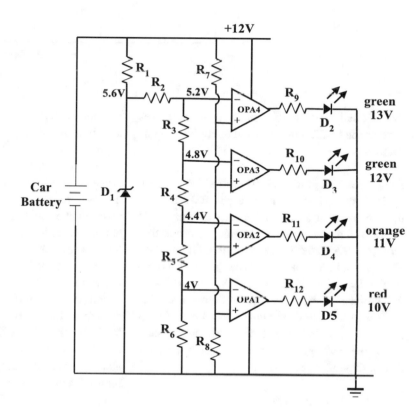

$R_2 = R_3 = R_4 = R_5 = 680\ \Omega$ and $R_6 = 6.8$ k
create reference voltages 5.2 V, 4.8 V, 4.4 V and
4 V at the inverting inputs as shown. $R_1 = 1$ k
sets a nominal Zener current of $(12 - 5.6)/$
1 k $= 6.4$ mA. The potential divider formed
with $R_7 = 15$ k and $R_8 = 10$ k monitors the
battery voltage and provides a fraction $V_M = \beta V_B$
to the input of the non-inverting terminals of the
op-amps where $\beta = R_8/(R_7 + R_8) = 10$ k/
$(10$ k $+ 15$ k$) = 0.4$ and V_B is the voltage of the
battery. With $V_B = 10$ V, $V_M = 4$ V and op-amp
1 is high turning on the red LED. All others are off.

With $V_B = 11$ V, $V_M = 4.4$ V and op-amp 2 and
op-amp 1 are high turning on the red and orange
leds. With $V_B = 12$ V a satisfactory battery voltage, $V_M = 4.8$ V and op-amp 3, op-amp 2 and
op-amp 1 are high turning on the red, orange and
one green leds. With $V_B = 13$ V, $V_M = 5.2$ V and
all op-amps are high turning on the red, orange,
and the two green leds. Therefore, a battery voltage at 12 volts or higher will result in one or two
green leds being lit corresponding to good battery
condition. Resistors $R_9 = R_{10} = R_{11} = R_{12} = 1$ k
limit the current in the leds to about 10 mA. Note

Fig. 13.43 Sound-activated switch

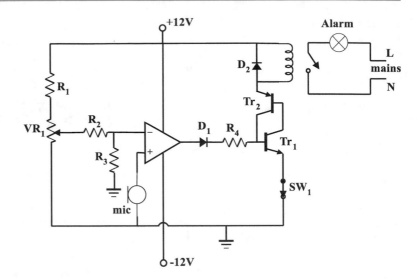

that a quad op-amp IC containing four devices such as the LM324 may be used to implement this system.

Ideas for Exploration (i) Introduce a fifth op-amp comparator driving a green LED with the inverting terminal connected to the 5.6 V Zener reference voltage and the non-inverting terminal connected to the junction of R_7 and R_8. This will add a fifth battery voltage level indication of $V_B = V_M/\beta = 5.6$ V/0.4 = 14 V; (ii) replace R_6 by a 10 k potentiometer so that the threshold battery voltages can be varied if desired; (iii) implement the original system using the 339 quad comparator which contains four independent comparators.

Sound-Activated Switch

The circuit of a sound activated switch is shown in Fig. 13.43. The output of a microphone drives the non-inverting input of an op-amp, while the inverting input is supplied with a reference voltage which can be varied. Resistor $R_1 = 2$ k and the potentiometer $VR_1 = 10$ k are connected to a 12 V supply. This enables the voltage at the wiper of the potentiometer to be variable between 0 and 10 V. This voltage is further reduced by another potential divider arrangement comprising $R_2 = 100$ k and $R_3 = 1$ k the junction of which is connected to the inverting input of the op-amp. Because of the attenuation produced by R_2 and

R_3, the voltage from the junction into the inverting terminal here can be varied from approximately $0 - 100$ mV corresponding to the $0 - 10$ V variation at the wiper of the potentiometer. This level is comparable to the output of the microphone. With a suitable voltage at the inverting terminal, if the output of the microphone exceeds this level, the output of the op-amp goes high switching on transistor Tr1 via $R_4 = 10$ k. This in turn switches on Tr2 which further turns on Tr1, and the regenerative action rapidly turns on both devices thereby activating the relay. These transistors can be MPSA05 and MPSA55 complimentary pair. The activated relay turns on the mains supply to an alarm. Should the output of the microphone be reduced causing the op-amp output to go low, the transistors remain on. They can only be turned off by opening the reset switch SW1 which interrupts the current flow through the transistors.

Ideas for Exploration This connection of transistors constitutes a thyristor which is discussed in Chap. 14. Explore the use of an actual thyristor in place of the two transistors.

High-Speed Half-Wave Rectifier

The half-wave precision rectifier of Fig. 13.44 is implemented here using the AD844 current feedback amplifier (CFA). This device has a high slew rate of 2000 V/µs and a high closed-loop

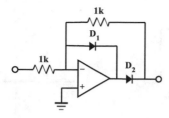

Fig. 13.44 Fast-action half-wave rectifier

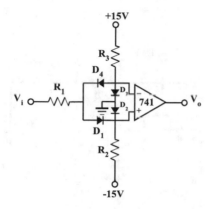

Fig. 13.45 Window comparator

bandwidth of 60 MHz. The diodes employed are 1N5711 Schottky diodes which have a low threshold voltage and fast switching action. The use of 1 k resistors gives the CFA wide bandwidth. The high slew rate of the CFA and the fast diode switching action result in good circuit performance up to high frequencies.

Ideas for Exploration Use this circuit with the Schottky diodes in the full-wave precision rectifier of Example 13.19, and compare the performance with the original circuit.

Window Comparator

The circuit in Fig. 13.45 is a modified comparator using an op-amp. Resistors R_2 and R_3 along with diodes D_2 and D_3 set voltages -0.7 V and $+0.7$ V at the non-inverting and inverting terminals. With the input voltage $V_i = 0$ applied at the input resistor R_1, the inverting terminal is at a higher potential than the non-inverting terminal and hence the output of the op-amp saturates negatively. As V_i increases with diode D_1 on and diode D_4 off, the potential at the non-inverting input

increases from -0.7 V. When V_i just exceeds a positive threshold voltage V_P such that the potential at the non-inverting input just exceeds the $+0.7$ V at the inverting terminal, the output of the op-amp saturates positively. Similarly, as V_i decreases with diode D_4 on and diode D_1 off, the potential at the inverting input decreases from $+0.7$ V. When V_i becomes just less than a negative threshold voltage V_N such that the potential at the inverting input becomes just less than the -0.7 V at the non-inverting terminal, the output of the op-amp saturates positively. Thus V_o goes positive for $V_i > V_P$ and $V_i < V_N$, while V_o goes negative for $V_N < V_i < V_P$. The equation to determine the positive threshold voltage V_P is given by

$$\frac{V_P + 15 - 0.7}{R_1 + R_2} = \frac{15 + 0.7}{R_2} \qquad \text{from} \qquad \text{which} \qquad V_P =$$

$15.7 \left(1 + \frac{R_1}{R_2}\right) - 14.3$. Using $R_1 = 15$ k and $R_2 = 33$ k, we get $V_P = 8.5$ V. The equation to determine the negative threshold voltage V_N is given by $\frac{15 - V_N - 0.7}{R_1 + R_3} = \frac{15 + 0.7}{R_3}$ from which $V_N = 14.3 - 15.7 \left(1 + \frac{R_1}{R_3}\right)$. Using $R_1 = 15$ k and $R_3 = 33$ k, we get $V_N = -8.5$ V.

Ideas for Exploration (i) Redesign the system to switch at $V_P = 5$ V and $V_N = -5$ V; (ii) investigate the standard window comparator involving two comparators and two diodes and compare the two systems.

Non-inverting Schmitt Trigger

The circuit in Fig. 13.46 is that of a non-inverting Schmitt trigger. (The inverting Schmitt trigger was discussed in Sect. 13.1.) This circuit can be implemented using an op-amp with positive feedback through R_2 to the non-inverting input with the input signal applied to this same terminal through R_1. The presence of positive feedback causes the output to saturate. If V_o is at $+V_{sat}$, then for switching to occur to $-V_{sat}$, the voltage at the non-inverting terminal must go to zero. This occurs when $\frac{+V_{sat}}{R_2} = \frac{-V_i}{R_1}$ giving $V_i = -\frac{R_1}{R_2}(+V_{sat})$. The input V_i must fall below this threshold value for switching to occur. Once switching has occurred such that $V_o = -V_{sat}$,

Fig. 13.46 Non-inverting
Schmitt trigger

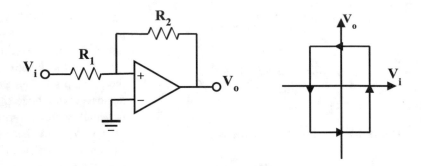

Fig. 13.47 Transistor
tester

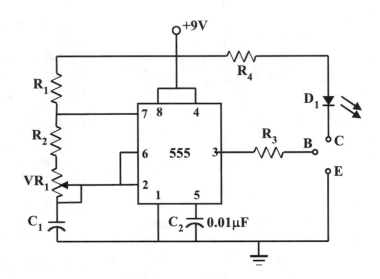

in order that the system switch to $+V_{sat}$, the voltage at the non-inverting terminal must go to zero. This occurs when $-\frac{(-V_{sat})}{R_2} = \frac{V_i}{R_1}$ giving $V_i = -\frac{R_1}{R_2}(-V_{sat})$, and therefore the input V_i must rise above the threshold value $V_i = \frac{R_1}{R_2}V_{sat}$ for switching to occur. As in the inverting Schmitt trigger, there is a switching band centered on zero with switching levels at $\pm\frac{R_1}{R_2}V_{sat}$.

Ideas for Exploration (i) Apply a bias voltage to the inverting input in order to shift the switching band to the left or right on the diagram.

Transistor Tester
The circuit shown in Fig. 13.47 is used to test npn BJTs. Here, using the design information presented in this chapter, the 555 timer is configured as an astable multivibrator. With $R_1 = 1$ k, $R_2 = 100\ \Omega$, $C_1 = 100\ \mu F$ and $VR_1 = 10$

k, then $f = 1/0.693(R_1 + 2(R_2 + VR_1))C_1$ yields oscillation between 1 and 12 Hz as VR_1 is varied. Thus, at the output on pin 3, a square-wave voltage signal is delivered to the base of the transistor through resistor $R_3 = 10$ k. This forces a base current into the transistor which, if the transistor is in good working order, will cause the flow of collector current that results in the turning on of the LED. Resistor $R_4 = 1$ k limits the current through the LED to 9 V/1 k = 9 mA.

Ideas for Exploration (i) Introduce a switch such that transistors of both polarities can be easily tested.

Sawtooth Wave Generator
The circuit shown in Fig. 13.48 uses a 555 timer to generate a sawtooth waveform. It utilizes the 555 timer in the astable mode but charges the capacitor using a constant current source. This

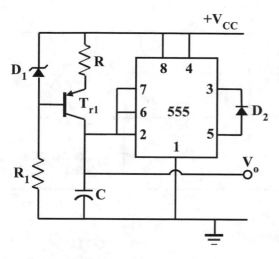

Fig. 13.48 Sawtooth wave generator

results in a linear rise in voltage across the capacitor which is eventually discharged by the 555 when the upper threshold voltage $2V_{CC}/3$ is reached giving the sawtooth wave. When the capacitor voltage falls below a lower threshold voltage V_L, charging of the capacitor restarts. The result is a sawtooth waveform whose amplitude varies between V_L and $2V_{CC}/3$. In the absence of diode D_2, the lower threshold voltage is $V_L = V_{CC}/3$. However, at the start of the capacitor discharge, the output of the 555 goes low thereby turning on D_2. The effect is to pull down the control voltage at pin 5 from $2V_{CC}/3$ to about 0.8 V which is made up of the diode voltage ($\simeq 0.7$ V) and the low-state output voltage of the 555 ($\simeq 0.1$ V). The lower threshold voltage for the capacitor discharge is half of this value, i.e. $V_L = 0.4$ V. Zener diode $D_1(2.7$ V) and the BJT set the constant current at $2/R$. Therefore, using simple integration, the frequency of the waveform is given by $f = \frac{2}{V_{pp}RC}$ where V_{pp} is the peak-peak voltage of the sawtooth waveform across the capacitor given by $V_{pp} = \frac{2}{3}V_{cc} - 0.4$. Thus for $V_{CC} = 9$ V, $V_{pp} = 6 - 0.4 = 5.6$ V and for $R = 1$ k and $C = 0.1$ μF, we get $f = \frac{2}{V_{pp}RC} = \frac{2}{5.6 \times 10^3 \times 0.1 \times 10^{-6}} = 3.6 \text{kHz}$. Resistor $R_1 = 1.8$ k sets the current through the Zener diode at $(9 - 2.7)/1.8$ k $= 3.5$ mA.

Ideas for Exploration (i) Introduce a suitable unity-gain buffer to prevent loading of the capacitor.

Zener Diode Tester
The circuit of Fig. 13.49 is a Zener diode tester. It consists of an astable multivibrator which with $R_1 = 1.8$ k, $R_2 = 82$ k and $C_1 = 3.3$ nF oscillates at about 2.6 kHz. It drives a 115-9 step-down transformer connected in reverse. Capacitor $C_2 = 3.3$ μF acts as a coupling capacitor and has a reactance of 19 Ω at the operating frequency. The output of the transformer is rectified by diode D_1 and filtered by capacitor $C_3 = 2.2$ μF to produce a DC voltage. Capacitor C_3 needs to have a working voltage of greater than 150 V. This DC voltage is applied to the Zener diode under test through one of two resistors $R_3 = 10$ k or $R_4 = 22$ k that set different current levels through the Zener diode. The Zener voltage is then measured by a voltmeter across points X and Y.

Ideas for Exploration (i) Design a suitable voltmeter circuit to be used to measure the Zener voltage.

Lamp Dimmer
The circuit in Fig. 13.50 is a light dimmer. It is made up of an astable multivibrator configured using a 555 and oscillating at about 2.8 kHz as set by the timing components $R_1 = 1$ k, $R_2 = 1$ k, $VR_1 = 50$ k and $C_1 = 0.01$ μF. The output of the 555 drives a Darlington pair such as the MPSA13, switching it on and off in response to the square-wave output. Adjusting VR_1 adjusts the duty cycle such that the brightness of the lamp is varied. Diode D_1 bypasses the lower half of the potentiometer during the capacitor charging cycle as a result of which the frequency of operation is independent of the duty cycle. Assuming the lamp L_1 is 5 W, then it requires 5 W/ 12 V = 417 mA for its operation. Using $\beta = 1000$ for the Darlington pair, then the required base current is 417 mA/ 1000 = 417 μA. Therefore resistor $R_3 \simeq 10$ V/

Fig. 13.49 Zener Diode
Tester

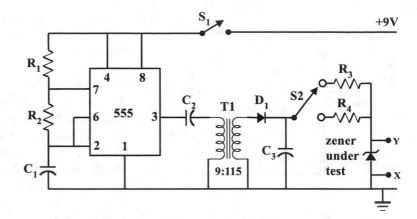

Fig. 13.50 Lamp dimmer

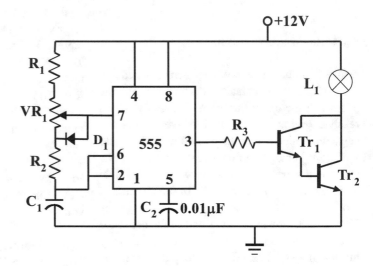

417 µA = 24 k. Use R_3 = 10 k to ensure that the transistor turns on fully.

Ideas for Exploration (i) Replace the Darlington pair by a MOSFET such as the IRF511.

Voltage Doubler

The circuit in Fig. 13.51 converts a DC voltage to twice its value with the same polarity. The output of the 555 astable multivibrator with frequency 7 kHz (as set by R_1 = 100 Ω, R_2 = 10 k, C_1 = 0.01 µF) drives a complimentary pair of transistors (such as the MPSA05 and the MPSA55) connected in the emitter follower configuration. With the arrangement of diodes D_1 and D_2 along with capacitors C_4 = 10 µF and C_5 = 10 µF, as the output goes low, transistor

T_{r2} turns on and capacitor C_4 charges via diode D_1 to the supply voltage. When the output goes high, transistor T_{r1} turns on, diode D_1 turns off, and diode D_2 turns on. The series combination of the 12 V at the output of the emitter followers and the charged capacitor C_4 charges up capacitor C_5 to twice the supply voltage or 24 V. A 75 mA load current will cause the voltage across capacitor C_5 to fall by $75 \times 10^{-3}/10 \times 10^{-6} \times 7 \times 10^3 = 1$ V before being recharged. Resistor R_3 = 220 Ω limits the base current to the transistors while ensuring that they are fully turned on.

Ideas for Exploration (i) Rearrange components D_1, D_2 so that the voltage across C_5 is negative thereby realizing ±12 V suitable for powering an op-amp. The resulting circuit would then be a positive to negative DC converter.

Fig. 13.51 Voltage doubler

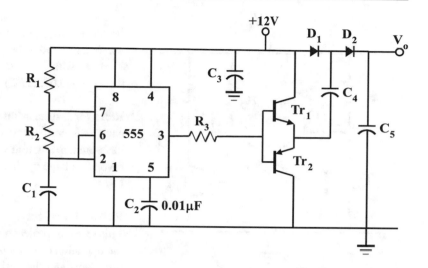

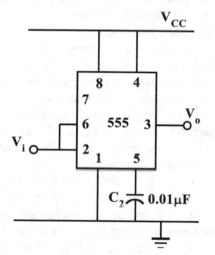

Fig. 13.52 Simple Schmitt trigger

(ii) Convert the circuit into a DC to AC inverter by removing D_1, D_2, C_4 and C_5 and using the output of the complimentary emitter followers to drive the input of a 9 V to 115 V transformer (a 115 V to 9 V step down transformer reverse connected).

Schmitt Trigger Using 555

The circuit in Fig. 13.52 is a simple Schmitt trigger using the 555. Inputs 2 and 6 are inputs to comparators where the threshold voltages are, respectively, $V_{CC}/3$ and $2V_{CC}/3$ which for $V_{CC} = 9$ V are 3 V and 6 V. Hence when the input voltage is low, the output is high. As the input increases from zero to greater than 6 V, the output goes

low. When the output decreases below 3 V, the output goes high.

Ideas for Exploration (i) Use this circuit to convert sine waves at the input to square waves at the output. In order to ensure reliable operation, introduce a two-resistor potential divider between the supply and ground with the junction of the two resistors connected to the 555 input and the sinewave coupled to the input via a coupling capacitor. Making the two resistors equal (100 k) biases the two internal comparators (with pins 2 and 6 connected) at $V_{CC}/2$. Since the upper comparator (pin 6) triggers at $2V_{CC}/3$ and the lower comparator (pin 2) triggers at $V_{CC}/3$, the bias establishes a voltage that is between these two levels. Thus, an input sine wave of sufficient amplitude will trigger both comparators and thereby produce a square-wave output. (ii) Make one of the resistors variable, and this makes the symmetry of the resulting square wave adjustable.

Research Project 1

This research project involves the investigation of the operation of an *astable multivibrator* using transistors and the design of an operating circuit. The basic configuration is shown in Fig. 13.53. It is essentially an AC-coupled amplifier comprising two common emitter stages with the output of one coupled back to the input of the other. The result is overall positive feedback that causes the

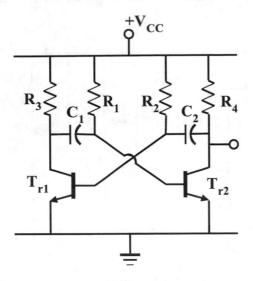

Fig. 13.53 Astable multivibrator using transistors

system to oscillate continuously as it switches back and forth between two quasi-stable states. Suppose Tr1 is on. Just prior while Tr1 was off, capacitor C_1 would have charged up to the supply voltage through R_3 and the base-emitter junction of Tr2 which would have been forward-biased. With Tr1 on, the collector of Tr1 would go to almost zero volts and the voltage on C_1 would be applied to the base-emitter junction of Tr2 taking it to $-V_{CC}$ thereby turning Tr2 off. As Tr2 turns off, its collector voltage rises at a rate limited by the need for C_2 to charge through R_4. During this time C_2 charges up to $+V_{CC}$ through R_4 and C_1 charges through R_1 until it reaches 0.7 V causing Tr2 to turn on. As Tr2 turns on, the collector of Tr2 would go to almost zero volts and the voltage on C_2 would be applied to the base-emitter junction of Tr1 taking it to $-V_{CC}$ thereby turning Tr1 off. As Tr1 turns off, its collector voltage rises at a rate limited by the need for C_1 to charge through R_3. During this time C_1 charges up to $+V_{CC}$ through R_3 and C_2 charges through R_2 until it reaches 0.7 V causing Tr1 to turn on and the cycle repeats itself. The signal out at either transistor collector is a square wave. The time T_1 that Tr1 is off can be shown to be $T_1 = 0.693R_1C_1$ and time T_2 that Tr2 is off is $T_2 = 0.693R_2C_2$. Hence the frequency of the square-wave output from

either collector is given by $f = 1/0.693$ $(R_1C_1 + R_2C_2)$. For the case where the square wave has a 50% duty cycle corresponding to $T_1 = T_2$ with $R_1 = R_2 = R$ and $C_1 = C_2 = C$, then $f = 1/0.693(2RC) = 0.72/RC$.

Ideas for Exploration (i) Introduce diodes from each emitter to ground in order to protect the base-emitter junction of each transistor from the reverse voltage applied by the capacitors C_1 and C_2.

Research Project 2

This research project involves the investigation of the operation of a *monostable multivibrator* using transistors and the design of an operating circuit. The basic configuration is shown in Fig. 13.54. It comprises two common emitter stages with AC and DC coupling such that there is one stable state to which the system always reverts and one quasi-stable state into which the system can be triggered by an external pulse. The stable state is Tr2 on and Tr1 off. If a negative-going pulse is applied to the base of Tr2 turning it off, the collector voltage rises driving base current into Tr1 thereby turning it on. Capacitor C_1, which was charged through R_3 to voltage V_{CC} while Tr1 was off and Tr1 on, applies $-V_{CC}$ to the base-emitter junction of Tr2 ensuring that it turns off. C_1 now charges through R_1 until its voltage gets to 0.7 V turning on Tr2. The collector voltage of Tr2 then falls and the base current drive through R_2 to the base of Tr1 goes to zero and Tr1 turns off. The system remains in this state until it is triggered again. Note that a positive-going signal at the base of Tr1 will also trigger the system to move from the stable state into the quasi-stable state from which it returns after a fixed time $T = 0.693R_1C_1$.

Ideas for Exploration (i) Introduce a speed-up capacitor across R_2 that reduces the turn-off time of Tr1 by rapidly removing charge stored in the base; (ii) convert this system into a bistable multivibrator by removing C_1 and connecting R_1 in place of C_1 as was done at Tr2 with C_1 and R_1 to produce the monostable multivibrator.

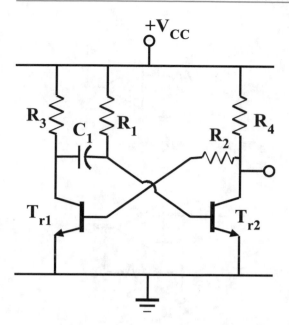

Fig. 13.54 Monostable multivibrator using transistors

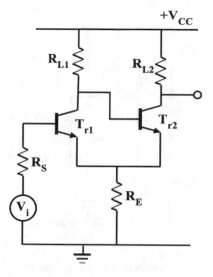

Fig. 13.55 Schmitt trigger using transistors

Research Project 3

This research project involves the investigation and design of a *Schmitt trigger* using transistors. The basic configuration is shown in Fig. 13.55. The system has two stable states in which alternate transistors are on and off. With $V_i = 0$, Tr1 is off and Tr2 is on by base drive through R_{L1}. When V_i is increased beyond an upper threshold voltage V_{UT}, Tr1 turns on and consequentlyTr2 turns off since base drive through R_{L1} is reduced. When V_i is reduced below a lower threshold voltage $V_{LT} < V_{UT}$, Tr1 turns off and Tr2 turns back on. Note that to switch states, the input voltage must either exceed V_{UT} to switch on Tr1 and switch off Tr2 or fall below V_{LT} to switch off Tr1 and switch on Tr2. The upper threshold voltage is given by $V_{UT} = V_{CC} \cdot R_E/(R_E + R_{L2})$ with $I_E(T_{r2}) = V_{UT}/R_E$ and $R_{L2} = (V_{CC} - V_{UT})/I_E(T_{r2})$. The lower threshold voltage is given by $V_{LT} = V_{CC} \cdot R_E/(R_E + R_{L1})$ with $I_E(T_{r1}) = V_{LT}/R_E$ and $R_{L1} = (V_{CC} - V_{LT})/I_E(T_{r1})$.

Ideas for Exploration (i) Use this circuit to convert sine waves at the input to square waves at the output; (ii) in order to make the symmetry of the resulting square wave adjustable, introduce a two-resistor potential divider between the

supply and ground with the junction connected to the base of Tr1 and the sine wave coupled to the input via a coupling capacitor. Make one of the resistors variable.

Research Project 4

This research project involves the investigation of the *566 voltage-controlled oscillator IC*. This system can generate triangular waves and square waves whose frequency is linearly related to the control voltage. The basic system is shown in Fig. 13.56. The frequency of oscillation is given by $f_o = \frac{2.4(V^+ - V_C)}{R_1 C_1 V^+}$.

Ideas for Exploration (i) Design a square-wave and triangular wave generator using the IC. Use a variable voltage to provide continuous frequency variation and a switched bank of capacitors to extend the frequency range.

Problems

1. An op-amp is connected as a comparator with the inverting terminal held at 3 V. Describe the action of the circuit when the non-inverting input is below and above 3 V.
2. An op-amp is connected as a comparator with the non-inverting terminal held at 2 V.

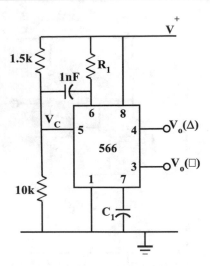

Fig. 13.56 566 voltage-controlled oscillator IC

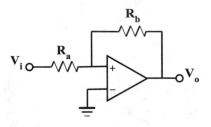

Fig. 13.57 Circuit for Question 7

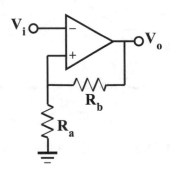

Fig. 13.58 Circuit for Question 7

Describe the action of the circuit when the inverting input is below and above 2 V.

3. Using the basic circuit shown in Fig. 13.4, design a system to turn on an alarm when the ambient temperature in a room exceeds 70 °C.

4. Investigate the use of a thermistor for temperature control.

5. Design a Schmitt trigger circuit with a switching threshold of ± 7.5 V using an op-amp powered by ± 15 V supplies. For the op-amp, assume $|V_{omx}| = |V_{omn}| = 13.2$ V and $A_d = 150,000$.

6. Using an op-amp, design a square-wave oscillator to have an oscillation frequency of 25 kHz, with an output of ± 9 V using a ± 15 V split supply.

7. For the Schmitt trigger circuit shown in Fig. 13.57 determine its threshold voltages and sketch its transfer characteristic. You may assume the output voltage is limited to the range $V_{omn} \le V_o \le V_{omx}$. What are the differences between this circuit and the one shown in Fig. 13.58?

8. If an input voltage V_{ref} is now applied to the inverting input, repeat Question 7.

9. For the collection of circuits in Fig. 13.59, determine their threshold voltages and sketch their transfer characteristics. The output level

of the comparator is ± 10 V. For Fig. 13.59, you may assume $\alpha = 0.3$, 0.5 and 0.8.

10. In the circuit of Fig. 13.60, let $R_1 = 220$ kΩ, $C_1 = 2.2$ nF, $R_a = 10$ kΩ, $R_b = 20$ kΩ and the circuit be supplied with ± 15 V supplies. For this circuit determine f_{osc}.

11. An alternative method of duty cycle control for the free-running multivibrator is shown in Fig. 13.61. Determine an expression for f_{osc} and the duty cycle, D, expressed as $D = \frac{T_1}{T_1 + T_2} \times 100\%$ in terms of v_{ctrl}.

12. Design a square-wave/sawtooth oscillator that oscillates at a frequency of 1 kHz.

13. Design a triangular wave generator to generate a positive-going sawtooth wave of 8 V pk-pk amplitude with a frequency of 500 Hz with a ± 15 V supply.

14. Describe the operation of a square/triangular voltage-controlled oscillator and derive an expression for the frequency of operation.

15. The variable frequency oscillator of Fig. 13.62 is set up using a TL082 op-amp and two multipliers (AD534) whose $K_1 = K_2 = 0.2$. You may assume that

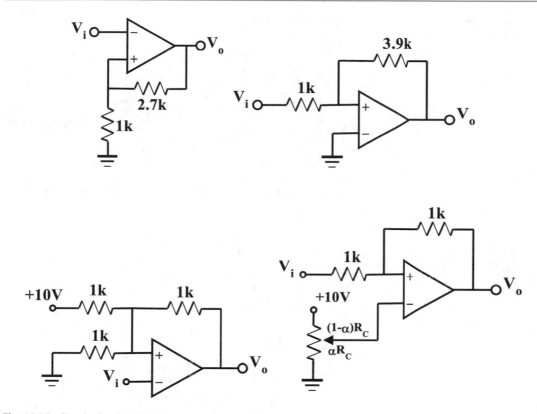

Fig. 13.59 Circuits for Question 9

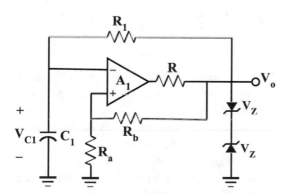

Fig. 13.60 Circuit for Question 10

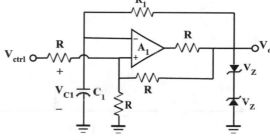

Fig. 13.61 Circuit for Question 11

$C_1 = C_2 = 0.003$ μF and $R_2 = 2.5$ k. Determine the frequency of oscillation if V_c is varied from 1 to 0.4 V.

16. A monostable multivibrator is constructed using the circuit of Fig. 13.63. The values chosen were $R_1 = 2$ k, $R_2 = 3$ k, $R = 1$ k, $C = 0.3$ μF and $V_{cc} = 15$ V. Determine the

period of this multivibrator if the voltage of each Zener diode is $V_Z = 8.2$ V.

17. Show how a JFET can be used to improve the operation of the monostable multivibrator in Fig. 13.63.

18. Describe the operation of the 555 timer and its use as an astable multivibrator. Hence design an astable multivibrator using the 555 to oscillate at 10 kHz. Show how a diode can be used to achieve 50% duty cycle.

Fig. 13.62 Circuit for
Question 15

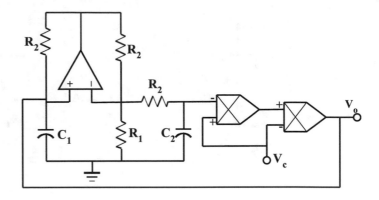

Fig. 13.63 Circuit for
Question 16

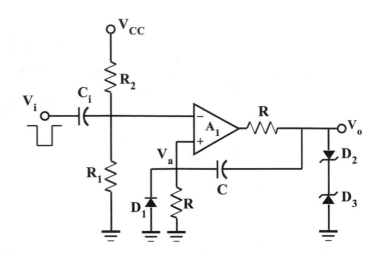

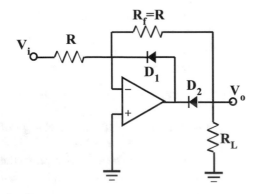

Fig. 13.64 Circuit for Question 22

19. An astable 555 oscillator is constructed using
    $R_a = 5$ k, $R_b = 5$ k and capacitor
    $C = 0.01$ μF. Calculate the output frequency
    of the square wave delivered by the 555 oscil-
    lator and the duty cycle of the output
    waveform.

20. Discuss the use of the 555 timer as a
    monostable multivibrator.

21. A monostable 555 timer is required to pro-
    duce a specified time delay in a circuit. If a
    3.3 μF timing capacitor is used, calculate the
    value of the resistor required to produce a
    minimum output time delay of 120 ms.

22. Explain the operation of the half-wave recti-
    fier shown in Fig. 13.64, and determine the
    polarity of the output signal.

23. For the precision full-wave rectifier shown in
    Fig. 13.65, describe the operation of the sys-
    tem and determine the polarity of the output
    signal.

Fig. 13.65 Circuit for
Question 23

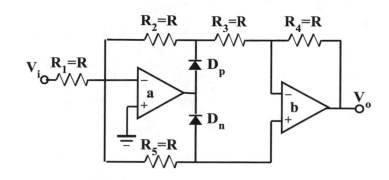

Fig. 13.66 Circuit for Question 25

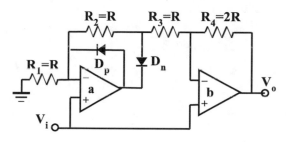

Fig. 13.67 Circuit for
Question 26

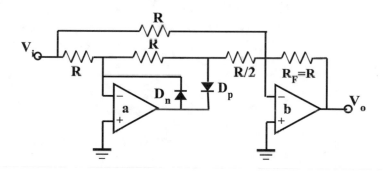

24. Using the circuit in Fig. 13.65, design a precision full-wave rectifier to operate at 10 kHz and below.
25. Using the circuit in Fig. 13.66, design a high input impedance precision full-wave rectifier.
26. Using the circuit in Fig. 13.67, design a precision full-wave rectifier.

Bibliography

R. Coughlin, F. Driscoll, *Operational Amplifiers and Linear Integrated Circuits*, 5th edn. (Prentice Hall, Upper Saddle River, 1998)

Special Devices

14

In previous chapters we discussed several semiconductor devices including the diode, the BJT, the FET, the operational amplifier and the current feedback amplifier. There are several other devices that are available to the designer and we will discuss some of these in this chapter. At the end of the chapter the student will be able to

- Demonstrate knowledge of a variety of special devices
- Use these devices in a range of circuits

14.1 Light-Dependent Resistor

A light-dependent resistor (LDR) or photoresistor is a resistor that is sensitive to light. Specifically, the device lowers its resistance significantly in response to exposure to light. It is sometimes referred to as a photoconductor, cadmium sulphide cell or photocell. It is made up of a semiconductor of high resistance. Energy absorption from incident light stimulates bound electrons into the conduction band such that the resulting electron-hole pairs increase the conductivity of the material and consequently lower the resistance. They are used in many applications including camera light metres, street lights, clock radios, security alarms and outdoor clocks. An example of an LDR is the ORP12. This device has a dark resistance of about 1 M which falls to about 5 k in bright light. The symbol for the LDR is shown in Fig. 14.1.

One specification of the device is the sensitivity which is the resistance at a specified illumination level. A typical curve containing this information is shown in Fig. 14.2 where resistance is plotted against light intensity expressed in lux. For applications involving on-off responses such as fire detectors, cells with steep slopes are suitable since then a small change in illumination results in a large change in resistance.

For applications involving signals of varying levels such as exposure control for cameras, cells with low slopes are more appropriate. Another specification is the dark resistance – the resistance of the cell under zero illumination. This establishes the maximum current that can be expected for a given voltage applied to the cell. This parameter is typically in the range 500 k to 20 M. The power rating of the cell in Watts is yet another specification, and maximum cell voltage lies between 100 and 300 V. This voltage must never be exceeded.

Typical applications of the photoconductive cell are shown in Fig. 14.3. The circuit in Fig. 14.3a produces an increasing output voltage level for increasing light intensity levels and is suitable for ambient light measurement systems used in camera exposure metres and brightness control circuits. The circuit in Fig. 14.3b is a DC relay system that can be utilized in a range of applications including smoke detectors and night lights. Light above a certain intensity causes the resistance of the LDR to fall sufficiently such that

© Springer Nature Switzerland AG 2021
S. J. G. Gift, B. Maundy, *Electronic Circuit Design and Application*,
https://doi.org/10.1007/978-3-030-46989-4_14

the voltage across the base-emitter junction of the transistor falls below 0.7 volts thereby turning off the transistor. If the light falls below this intensity, the resistance of the LDR increases thereby turning on the transistor and activating

Fig. 14.1 Symbol of light-dependent resistor

the relay that can be used to switch on a desired system.

Example 14.1

Design a system using a photocell and an op-amp as a comparator to switch on a light when ambient light falls below a certain level.

Solution One approach using a photocell is shown in Fig. 14.4. The potentiometer $VR_1 = 20$ k in conjunction with the light-dependent resistor sets a potential at the non-inverting input of an op-amp. The op-amp is powered from the single-ended +15 V supply. A reference potential $V_{REF} = 7.5$ V is set by $R_1 = 10$ k and $R_2 = 10$ k at the inverting input. When the light is sufficiently bright, the resistance of the LDR is

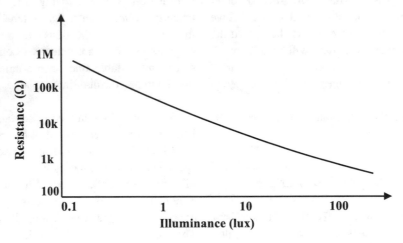

Fig. 14.2 Resistance vs illuminance for photocell

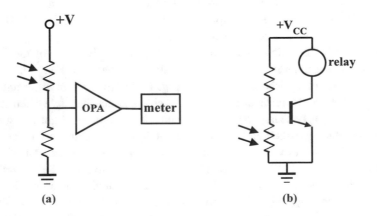

Fig. 14.3 Two applications of the photoconductive cell

Fig. 14.4 Light turn-on application

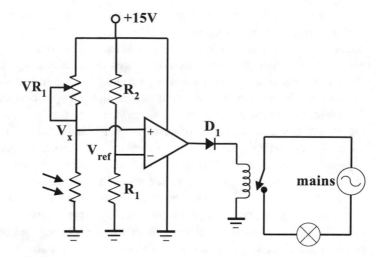

low, and the voltage V_X is less than the reference potential V_{REF}.

The output of the op-amp saturates negatively, and the diode D_1 is reverse biased, thereby ensuring that the relay remains inactive. When the light intensity falls, the resistance of the LDR increases such that $V_X > V_{REF}$. The output of the op-amp then saturates positively switching on the diode and activating the relay, thereby turning on the light. The light level at which the system switches may be set by adjusting the potentiometer.

Example 14.2

Design a system using a photocell and a transistor to switch on a LED when ambient light falls below a certain level. Indicate the changes needed for the system to indicate the presence rather than the absence of light.

Solution The basic system is shown in Fig. 14.5. and follows the approach of Fig. 14.3b. The LDR is placed between the base terminal and ground, and a potentiometer $VR_1 = 50$ k is connected between the base and the supply voltage. This is used to set the reduced light level at which the system is activated and its value is not critical since from Fig. 14.2, the resistance change of the LDR in response to varying light intensity is between about 1 k and 1 M. Resistor $R_1 = 1$ k limits the current through the led when the transistor turns on to less than 9 V/1 k = 9 mA. Of course other

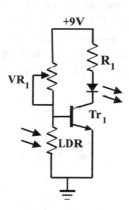

Fig. 14.5 Circuit for Example 14.2

supply voltage levels can be used. For the system to indicate the presence instead of the absence of light, the positions of the LDR and potentiometer must be interchanged such that the system will turn on the LED when sufficient light falls on the LDR instead of when there is reduced light.

14.2 Photodiode

The cadmium sulphide light-dependent resistor discussed above is sensitive to light but responds slowly to changes in light intensity. It is therefore suitable for slow-acting light-intensity-sensing applications only. For use as optical sensors in medium- to high-speed applications, the photodiode and the phototransistor are more suitable.

A photodiode is a diode that generates a current or a voltage when exposed to light. Its symbol is shown in Fig. 14.6. It operates as a reverse-biased pn junction and therefore experiences a reverse current arising from thermally generated carriers. As the reverse-bias voltage is increased, this current increases to its saturation value when the current flow equals the rate of thermal generation of the carriers. When such a junction is exposed to increased illumination, additional electron-hole pairs are generated which result in an increased reverse saturation current. This current is approximately proportional to the intensity of the light. The current vs voltage characteristic of the photodiode in the dark state is essentially that of a conventional diode as shown in curve 1 of Fig. 14.7. Upon illumination however the characteristic shifts downward to curve 2 and then to curve 3 as light intensity increases. These curves do not pass through the origin since

reverse saturation current arising from exposure to incident light continues to flow for zero reverse bias. In curves 2 and 3 for short-circuited diode terminals, a photo current I_{SC} that is proportional to the light intensity flows through the diode, while for open-circuited terminals, a voltage V_{OC} appears across the diode terminals. I_{SC} and V_{OC} are indicated on Fig. 14.7.

As is evident from the curves, the reverse current increases as the incident luminous flux increases. For a constant level of illumination, the reverse current increases only slightly with increasing reverse bias indicating that the reverse current is largely independent of the reverse-bias voltage. It can also be seen from the curve that reverse current flows even as the reverse bias is reduced to zero and then to a small forward bias up to about 0.3 V. Forward bias above this value causes the reverse current to fall, and it goes to zero when the forward bias is approximately 0.5 V. "Dark current" is the reverse saturation current of the device in the absence of incident light.

The short-circuit current has a very linear relationship with incident light intensity. A typical reverse (short-circuit) current vs illumination characteristic for this device is shown in Fig. 14.8. Here luminous flux is the total amount of light emitted or received by a surface, and its unit of measurement is the lumen (lm) where

Fig. 14.6 Symbol of photodiode

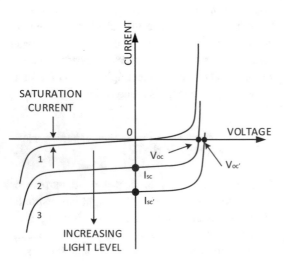

Fig. 14.7 Current vs voltage characteristics for a photodiode

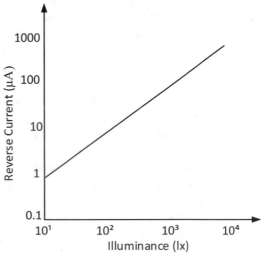

Fig. 14.8 Typical reverse current vs illuminance for photodiode

Fig. 14.9 Spectral
response for photodiode

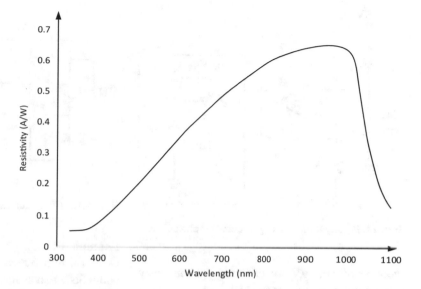

1 lm = 1.46 mW. The luminous flux incident on a surface is measured in lux (lx) or in mW/m^2 where 1 lx = 1 mW/m^2. The specifications of a photo diode include the following:

- Responsivity in *A/W* is the ratio of generated photo current to incident light power.
- Sensitivity is the reverse current in amperes resulting from unit incident luminous flux. It is generally quoted for a reverse bias that is less than the maximum allowable value. A typical sensitivity specification is 100 nA/lx at reverse bias of 12 V with maximum reverse bias of 15 V.
- The maximum reverse-bias voltage and current, e.g., 22 V and 18 mA.
- Maximum forward current, typically 8 mA.

The spectral response relating sensitivity or responsivity to wavelength of incident light is shown in Fig. 14.9.

Photodiodes provide excellent linearity with respect to incident light, low noise and wide spectral response. Additionally, the device is mechanically rugged, compact and lightweight and has a long life. They are used in consumer electronic devices such as smoke detectors and the receivers in remote controllers for electronic and other home appliances.

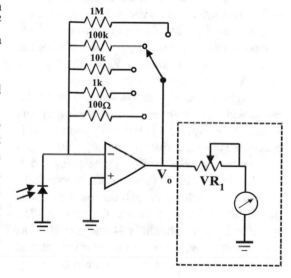

Fig. 14.10 Light metre using photodiode

Photodiodes can be operated in either the zero bias or reverse-bias state. A typical application in the zero bias state is shown in Fig. 14.10. The circuit is a light metre using an op-amp in the transimpedance mode. As light falls on the photodiode, current is generated. This current flows into the inverting input of the op-amp thereby producing a voltage at the output of the op-amp that is proportional to the light intensity. The various values of resistors allow a nominal output

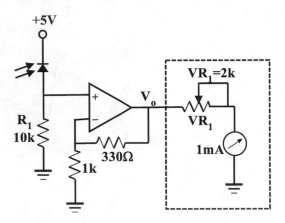

Fig. 14.11 Light metre using reverse-bias photodiode

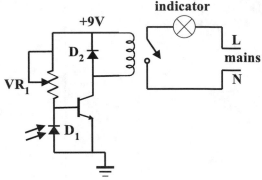

Fig. 14.12 Circuit for Example 14.3

voltage of 100 mV for different light intensity. Thus the ranges correspond to 1 lx, 10 lx, 100 lx, 1000 lx and 10, 000 lx for the highest range. The higher resistance values correspond to lower lux ranges which enable detection of reduced light intensity. In actual use, $VR_1 = 2$ k is adjusted to give full-scale deflection for $V_o = 100$ mV on a 100 μA microammeter.

A light metre can be constructed using the photodiode in the reverse mode. Such a system is shown in Fig. 14.11. Here the photodiode is connected in series with a resistor $R_1 = 10$ k and a reverse bias applied to the system using a 5 V supply. When light falls on the photodiode, reverse current flows through the resistor thereby developing a voltage across the resistor. The specifications for the BPW34 indicate that the output voltage V_o across the resistor is related to the lux of the light falling on the photodiode by

$$\text{lux} = 1333V_o \qquad (14.1)$$

Thus, a light intensity of 1000 lux corresponds to an output voltage of $V_o = (1000/1333)V$. If this voltage is scaled up by a factor 1.333, then $V_o = \frac{1000}{1333} \times 1.333 = 1$ V. Therefore, a 1 V output now corresponds to 1000 lux. This amplification factor is provided by the non-inverting amplifier which has a gain of 1.33. The output drives a 1 mA milliameter through a potentiometer $VR_1 = 2$ k which is adjusted such that 1 V out results in full-scale deflection of the milliameter.

Example 14.3

Design a light-activated relay system using the photodiode that switches on a light when ambient light level is low and switches off the light when ambient light rises above a certain level.

Solution One solution is shown in Fig. 14.12 where the photodiode is used in the reverse-bias mode. When light falls on diode, a reverse current flows through the diode causing a voltage drop across the resistor $VR_1 = 10$ k. This lowers the base voltage of the transistor causing it to switch off, deactivating the relay. When the light level falls, current flow through the diode drops and the voltage across the base-emitter junction of the transistor increases thereby turning on the transistor and activating the relay. This switches on the light. The potentiometer $VR_1 = 10$ k adjusts the sensitivity of the system to the light level at which switching occurs.

Example 14.4

Design the system in Example 14.3 using an op-amp driving a LED.

Solution One solution is shown in Fig. 14.13. Here the photodiode D_1 is again operated in the reverse-bias mode. The diode and resistor $R_1 = 1$ M sets a voltage V_X at the inverting input of the op-amp, while the wiper of potentiometer $VR_1 = 100$ k sets a variable reference voltage V_{REF} at the non-inverting terminal. When light falls on diode, current flow through

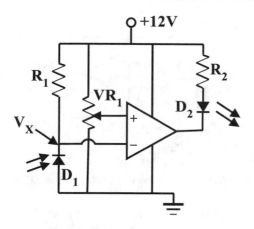

Fig. 14.13 Circuit for Example 14.4

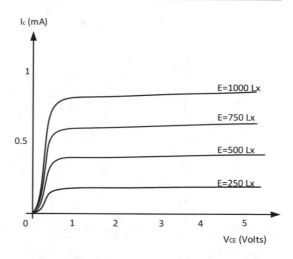

Fig. 14.15 Phototransistor output characteristics

Fig. 14.14 Phototransistor

the diode increases, and the voltage V_X falls below V_{REF}. This causes the op-amp output voltage to go high, turning off the LED. When the light level falls, the diode current falls and V_X exceeds V_{REF}. The op-amp therefore goes low, turning on the LED D_2. The potentiometer $VR_1 = 100$ k adjusts the light level at which switching occurs. In order to limit the current in the LED to about 10 mA, set $R_2 = 1$ k.

14.3   Phototransistor

The phototransistor, the symbol for which is shown in Fig. 14.14, is a transistor that is sensitive to light. It is essentially a transistor that includes a transparent window that allows light to fall on the collector-base junction. The collector-base junction of the device is photosensitive, and therefore exposure to light generates electron-hole pairs resulting in a small base current.

The phototransistor may have an external base lead that is usually left unconnected. Even though

the base is open, the light-induced base current results in a collector leakage current $I_{CEO} = \beta I_{CBO}$. Thus, light exposure causes a proportional flow of collector leakage current. The response to light intensity changes is much faster than the light-dependent resistor. Typical output characteristics are shown in Fig. 14.15. Here the varying parameter producing the different curves is light flux density. These characteristics are very similar to those of the normal bipolar junction transistor with light flux density instead of base current being the changing parameter. A small collector current flows even in the absence of light and is referred to as a dark current. Its value is of the order of 100 nA and is temperature dependent. Phototransistors respond only to light of specific frequencies. The spectral response factor is normalized at the frequency of maximum response, and the device sensitivity is also stated at this frequency. It should be noted that while the current through the photodiode is linear over about eight decades of light intensity, the collector current of the phototransistor is linear over only about four decades of illumination. It follows therefore that the photodiode is preferred in linear applications while the phototransistor does especially well in switching applications.

The phototransistor can be applied in a wide variety of circuits including those associated with the normal transistor. Two basic applications are shown in Fig. 14.16. In the first (Fig. 14.16(a)),

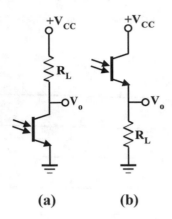

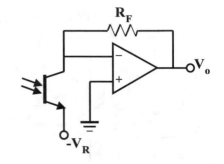

Fig. 14.18 Phototransistor transimpedance amplifier

(a)   (b)

Fig. 14.16 Two applications of a phototransistor

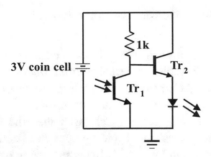

Fig. 14.17 Simple phototransistor application

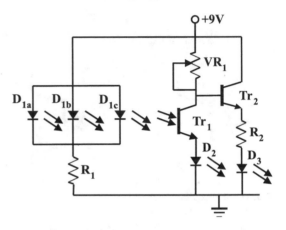

Fig. 14.19 Circuit for Example 14.5

the phototransistor is connected as a common emitter amplifier with a resistor connecting the transistor collector to the power supply. The circuit produces an output at the collector that transitions continuously from a high state to a low state when light in the near-infrared frequency falls on the phototransistor. A Schmitt trigger such as the 74LS14 can be added at the output to eliminate multiple switching for changing light levels that are close to the detection threshold. In the second circuit shown in Fig. 14.16(b), the phototransistor is connected in the emitter follower configuration with a resistor connecting the emitter to ground.

The circuit in Fig. 14.17 is another simple application. Here light falling on the phototransistor causes it to conduct. The current flow through the resistor results in a low voltage at the collector of the phototransistor, and hence Tr2 is turned off. As darkness falls and there is reduced light on the phototransistor, the reduced current causes the collector voltage to rise thereby turning on Tr2 and lighting the LED.

The phototransistor like the photodiode can also be used in a transimpedance amplifier. The basic circuit is shown in Fig. 14.18. It is very similar to the photodiode application except that the phototransistor requires a negative bias voltage $-V_R$ as shown. The circuit output is given by $V_O = I_C R_F$ and is a linear function of the light intensity falling on the phototransistor.

Similar to the phototransistor, there is the photofet which is a light-sensitive JFET. An example of this is the LS627 from linear systems. This device, like the phototransistor, can also be used in a variety of applications involving light activation.

Example 14.5

Use the photodiode and the phototransistor to design a system that provides an entry alarm at a doorway.

Solution The system shown in Fig. 14.19 comprises three infrared leds D_1 positioned on one side of a doorway and an opto-transistor receiver Tr_1 positioned on the other side. Resistor $R_1 = 220 \ \Omega$ sets the current through these leds. Both are positioned such that radiation from the leds falls on the phototransistor thereby activating it. The current flow turns on the green LED D_2 in the emitter circuit of the phototransistor. This pulls the base of Tr2 low, and this transistor turns off the red LED D_3 in its emitter circuit. The potentiometer $VR_1 = 2 \ k$ is adjusted to ensure that Tr2 is off under normal conditions. If the infrared radiation is interrupted by entry through the doorway, then the opto-transistor turns off causing the green LED to go off. The base of Tr2 goes high causing Tr2 to turn on and switching on the red LED. Resistor $R_2 = 1 \ k$ limits the current in the red LED. A buzzer may be included between the emitter and ground of Tr2 to provide an audible alarm.

There is also the photodarlington which is a Darlington pair that is photosensitive. In such a device, the first transistor is photosensitive, and its emitter is connected to the base of the second transistor. The resulting gain is therefore higher than the normal phototransistor, but the response of the photodarlington is slower.

14.4 Opto-isolator

The opto-isolator or opto-coupler is an integrated circuit that incorporates a light-emitting diode along with a light sensitive receiver such as a photoresistor, a photodiode, a phototransistor or a phototriac in the same package. They are used to isolate one part of a circuit from another. This may be to separate low-level signals from signals that may be potentially hazardous, or it may simply be to isolate signal that have different grounds. When current flows through the LED, light emitted by the LED activates the photosensitive device and results in a changed current flow through the device.

14.4.1 Photoresistor Opto-isolator

The opto-isolator comprises an LED and a light-dependent resistor at the output as shown in Fig. 14.20. This device provides low noise and a continuously variable resistance that varies from 15 Ω to 50 MΩ in response to a current input. Its characteristics are typically an isolation voltage of 2500 Vrms, an input current of 10 mA and an output voltage of 30 V. Applications of this device include audio limiting and compression, automatic gain control, circuit isolation and logic interfacing, SCR and triac drivers and noiseless switching.

One example shown in Fig. 14.21 is a remote gain control system. Here the light-dependent resistor is used as the feedback resistor in an inverting amplifier. This enables the remote control of the gain of the circuit using the associated light-emitting diode. The light-dependent resistor is activated by light from the LED, and then the reduced resistance of the light-dependent resistor causes the magnitude of the output voltage of the amplifier to be reduced. Another application example is the switching circuit of Fig. 14.22. In this circuit the light-dependent resistor is included in a potential divider arrangement with another resistor R that is connected to a 5-volt supply. When a logic signal is applied to the LED, light from the LED falls on the LDR, and then the resistance of the LDR falls and hence so does the voltage at the junction of these two resistors.

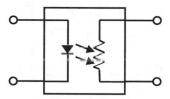

Fig. 14.20 Opto-isolator using light-dependent resistor

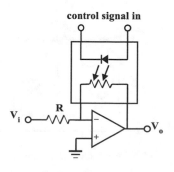

Fig. 14.21 Remote gain control

This voltage change causes the logic gate to change state.

Exercise 14.1

Explore the use of the photoresistor opto-coupler in the design of (i) a voltage-controlled Sallen-Key low-pass filter and (ii) a voltage-controlled Wien bridge oscillator. In both cases, two photoresistor opto-couplers would have to be used to simultaneously vary two resistors in these systems.

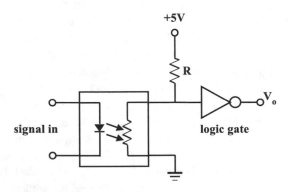

Fig. 14.22 Logic interface

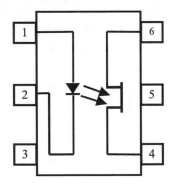

Fig. 14.23 Photofet opto-isolator

14.4.2   Photofet Opto-isolator

Another opto-isolator is one with a photofet at the output as shown in circuit Fig. 14.23. An example of this is the H11F series from Fairchild Semiconductor. Each device contains a gallium-aluminium arsenide infrared-emitting diode coupled to a bi-directional photosensitive FET that responds as a light-dependent resistor. The arrangement functions as an isolated FET with a resistance variation from $100~\Omega$ to $300~M\Omega$. Its characteristics include a maximum input diode forward current of 60 mA, isolation resistance of greater than 100 Gohms, continuous detector current of ±100 mA and a breakdown voltage of 30 V (Fig. 14.23).

The circuit element can function as a variable resistor as it has an extremely linear relationship between diode current and circuit resistance. It can also be used as an electronic switch. The bi-directional photofet and the optical isolation provided by the photodiode make this circuit element a very versatile solid-state relay. Two applications are shown in Fig. 14.24. They are both variable attenuators that allow complete isolation of the control signal.

14.4.3   Photodiode Opto-isolator

The photodiode opto-isolator shown in Fig. 14.25 utilizes an LED as a light source in the same package as a photodiode as light sensor. With the photodiode is in the reverse-bias mode, current flow through the LED causes light to be emitted from the LED, and this increases the reverse current flow through the photodiode.

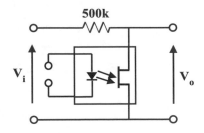
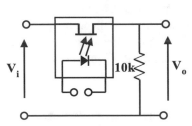
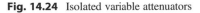

Fig. 14.24 Isolated variable attenuators

This is the photoconductive mode. This current flow through a resistor is easily detected as a voltage. This opto-coupler responds very quickly to changes in the LED current and is useful where opto-coupling with high speed is required.

Example 14.6

Outline a system in which the photodiode opto-coupler is used in a linear opto-isolator system.

Solution One IC that is designed for linear signal isolation is the IL300 linear opto-coupler from Vishay. It comprises an infrared (IR) LED D_3 irradiating two photodiodes D_1 and D_2 in one package. Photodiode D_1 is used to generate a feedback control signal that is used to establish the IR LED drive current necessary to produce a linear signal output from photodiode D_2. A simple implementation is shown in Fig. 14.26. Here op-amp A is connected in a non-inverting configuration and its output drives the IR LED D_3. Resistor R_3 assists in limiting the value of this current. The cathode of photodiode D_1 is connected to a reverse-bias supply V_{cc1}, and the anode is connected to feedback resistor R_1. Thus,

an input voltage signal V_i applied at the non-inverting input of op-amp A drives the IR LED such that a reverse current flows through diode D_1 that develops a voltage across R_1 which, because of the feedback is equal to V_i. Diode D_2 is similarly connected to a separate supply $V_{cc2} = V_{cc1}$ with the output connected to resistor $R_2 = R_1$ which is itself connected to a voltage follower op-amp B. The currents in diodes D_1 and D_2 are equal, and therefore the voltage developed across R_2 is equal to V_i resulting in the voltage $V_o = V_i$.

14.4.4 Phototransistor Opto-isolator

The opto-isolator using a photo transistor is shown in Fig. 14.27. The light from the LED activates the phototransistor and causes it to conduct. An example of such a device is the 4N25 from Vishay Semiconductors. Opto-isolators using a photodarlington are also available. The

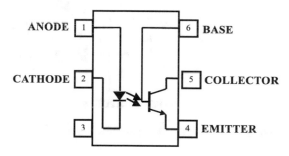

Fig. 14.27 Phototransistor opto-isolator

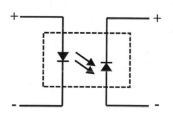

Fig. 14.25 Photodiode opto-isolator

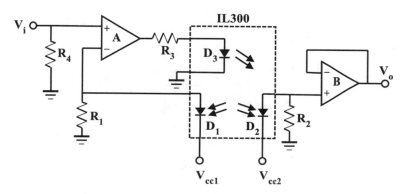

Fig. 14.26 Circuit for Example 14.6

4N32 is an example. Such circuits are ideal for interfacing one family of logic circuits to another or for isolating an analog or digital system from the physical world. The current to the LED must as usual be limited by an external resistor in series with the LED. The LED can be driven from an AC source providing a diode is connected in series with the LED. Such a diode will protect the LED from reverse DC voltages. The current through the phototransistor can be converted to a voltage by connecting a resistor to either the collector or the emitter as shown in Fig. 14.16.

A typical application is shown in Fig. 14.28. Here, the digital output of a microcontroller is coupled to the base of a transistor that drives the

opto-coupler LED. When the transistor is switched on by the microcontroller, the LED radiates and the opto-transistor is activated. This causes the output of the opto-transistor to go low, and this signal level is coupled to a separate circuit that can be completely electrically isolated from the microcontroller.

Example 14.7
Use the phototransistor opto-coupler to interface two analog audio systems.

Solution One approach to isolating two audio systems is shown in Fig. 14.29. Here a quiescent current is established through the LED of the

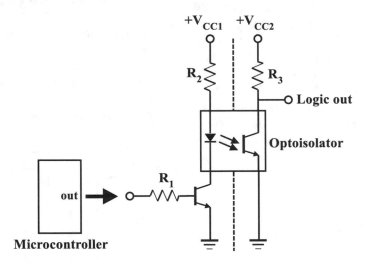

Fig. 14.28 Phototransistor opto-isolator application

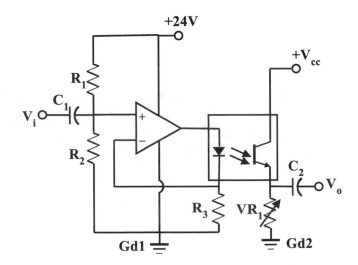

Fig. 14.29 Circuit for Example 14.7

opto-transistor, and this current is modulated by a signal from the associated op-amp. The op-amp is operated from a single-ended 24 V supply with $R_1 = R_2 = 100$ k setting the voltage at the non-inverting terminal at 12 V. The op-amp itself is connected in the voltage-follower mode with the LED of the opto-coupler inside the feedback loop. As a result, the voltage across resistor $R_3 = 5.6$ k follows the input voltage at the non-inverting terminal. The quiescent current through the LED is given by 12 V/ 5.6 k $= 2.1$ mA. This produces a current I_C in the opto-transistor depending on the transfer characteristics of the particular device used but may be about 1 mA. The opto-transistor is here connected in the emitter follower configuration and is powered by a separate supply V_{cc}. Potentiometer $VR_1 = 10$ k is adjusted to set the emitter of the opto-transistor to about half the separate power supply voltage in order to allow for maximum symmetrical swing. Coupling capacitors C_1 and C_2 are necessary because of the DC at input and output.

14.4.5 Solid-State Relay

The electronic switch or solid-state relay is a generic bilateral device supplied by several manufacturers that switches an electrical signal. It is activated by a control input and is useful in signal gating, modulator, demodulator and CMOS logic implementation. One device is the HSR312 solid-state relay from Fairchild Semiconductor shown in Fig. 14.30. A current through the diode connected between the terminals marked anode and cathode causes conduction between terminals 6 and 4. The device can be used as a direct replacement for mechanical relays. Solid-state relays are also available in quad and higher packages as well as in a multiplexer formulation. A simple application involving the switching of AC is shown in Fig. 14.31. Closing switch SW1 drives current through the LED. This closes the solid-state switch and therefore connects the load to the AC supply. Opening the switch results in the opening of the solid-state

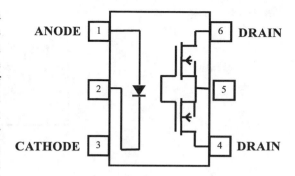

Fig. 14.30 Solid-state relay

switch and hence the switching off of the supply. The HSR312 provides 4000 Vrms isolation and can operate at 250 V. It can handle up to 190 mA and has an on resistance of 10 Ω series (when series connected).

14.5 Silicon-Controlled Rectifier

The *silicon-controlled rectifier (SCR) or* thyristor is a three-terminal semiconductor device used in the control of large flows of electrical power. It consists of four layers of semiconductor material alternatively n-type and p-type as shown in Fig. 14.32a. The outer p-type layer constitutes the anode while the outer n-type layer constitutes the cathode. The inner p-type layer forms a third terminal called the gate which is used to turn on the device. The symbol is shown in Fig. 14.32b. In order to understand the operation of the thyristor, it is instructive to view the device as a pnp transistor and an npn transistor connected as shown in Fig. 14.33. Application of a voltage across the device in the forward or reverse direction results in little or no current flow since both transistors are turned off until the voltage exceeds the breakdown or breakover voltage which is of the order of hundreds of volts.

In the reverse direction corresponding to the application of a negative voltage to the anode and a positive voltage to the cathode, the base-emitter junctions of Tr1 and Tr3 are reverse-biased as a result of which both transistors are off. The thyristor is in the reverse blocking region, and only a small leakage current flows. Apart from an

Fig. 14.31 Application of solid-state relay

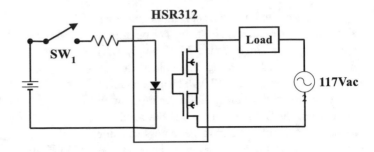

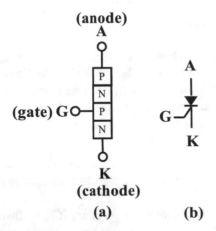

Fig. 14.32 Structure of silicon-controlled rectifier

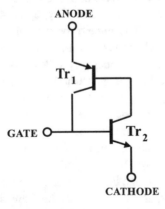

Fig. 14.33 SCR represented as two connected transistors

increased leakage current, the device remains nonconducting with an increased reverse voltage until a threshold voltage called the reverse breakover or breakdown value is reached. At or above this voltage, avalanche breakdown occurs in the semiconductor material and significant current flow occurs which, if not limited, could result in the destruction of the thyristor.

In the forward direction corresponding to the application of a positive voltage on the anode and a negative voltage on the cathode, the base-emitter junctions of both transistors become forward-biased, but the collector-base junctions are reverse-biased and the thyristor is off. Thus, only small leakage current flows and the device is in the forward blocking region. There is a gradual increase of this leakage current with increase of the applied forward voltage. However, the device remains effectively off until the forward voltage reaches a threshold value referred to as the forward breakover voltage V_S. At or above this value, conduction increases rapidly with a consequent fall in the terminal voltage. The thyristor is now in the forward conduction region with a maximum current of hundreds of amperes. The resulting characteristic is shown in Fig. 14.34.

It is possible to reduce the effective forward breakover voltage by injecting a triggering current at the gate terminal G corresponding to the base terminal of Tr2. The effect of this current is to forward-bias the base-emitter junction of Tr2 thereby producing an amplified current in the collector of Tr2. This current flows out of the base of Tr1 thereby resulting in a further amplified current flowing out of the collector of Tr1. This then enters the gate or base of Tr1 resulting in the regenerative switching on of both transistors and hence conduction between the anode and the cathode terminals. This action results in the curves of Fig. 14.35.

From these curves it can be seen that an increased gate current results in a reduced breakover voltage at which the thyristor switches on. The gate current pulse must be of sufficient duration such that the thyristor current exceeds a minimum value called the holding current I_H.

Fig. 14.34 SCR characteristic-SCR current vs voltage across SCR

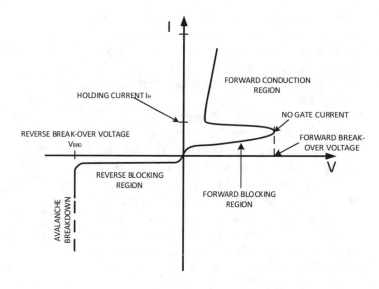

Fig. 14.35 SCR characteristic showing effect of gate current

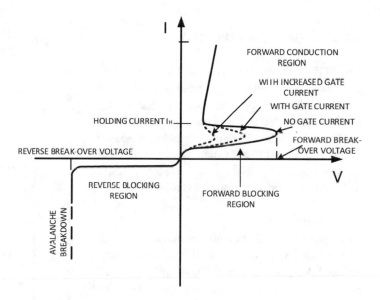

This minimum value is that which enables the device to remain conducting even after the removal of the gate current. Cessation of the gate current before the holding current level is attained will cause the thyristor to turn off again. Once the thyristor is turned on, the current must exceed the holding current in order to sustain the thyristor in the conducting state. In such circumstances the device remains in the conducting state independently of the gate which no longer exercises any control. The thyristor current may vary but must not fall below the holding current value. In the on state, the forward voltage across the thyristor is typically around 1 volt.

It is possible to turn on a thyristor by applying a forward voltage whose rate of change dV/dt exceeds some critical value specified by the manufacturer. In this turn-on mode, the voltage across the device does not have to exceed the breakover voltage for switch on to occur. At this critical value of dV/dt, the rapid change of voltage across the SCR results in a flow of current through the internal junction capacitances. In particular a flow of current to the gate terminal triggers the device. A snubber circuit consisting of a resistor of about 100 Ω in series with a capacitor of about 0.1 μF placed across the anode and cathode of the SCR can be used to protect against rapid increases in terminal voltage.

In order to turn off a thyristor, the current through the device must be reduced to a value below the holding current value. This can be accomplished by opening the thyristor circuit or short circuiting the thyristor. If the thyristor is operated from an AC mains supply, then the voltage across the device periodically goes through zero and hence turns off the device. If the operating voltage is DC, then special turn-off arrangements become necessary. An example of an SCR is the 2N5064.

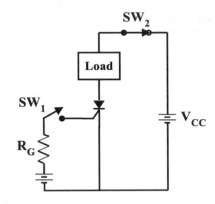

Fig. 14.36 Static switching of SCR

14.5.1 Gate Turn-On Methods

The SCR or thyristor can be switched on in several ways. The method to be used in a particular situation depends on the requirements of the application as well as the triggering specifications of the SCR. It should be noted that excessive gate voltage or current drive may destroy the SCR while inadequate gate-triggering levels may result in less than optimal performance of the system.

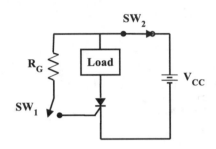

Fig. 14.37 Static switching of SCR omitting additional DC supply

Static Switching
SCR static switching circuits use either a constant or varying DC signal for SCR turn-on for both DC and AC power control. One simple approach is the use of manual switching and a DC bias supply to turn on the device as shown in Fig. 14.36. The closure of normally off switch SW1 delivers a voltage to the gate of the SCR via the resistor R_G thereby turning it on. Normally on switch SW2 is used to interrupt the current to the SCR in order to turn it off. The DC bias voltage may be removed and resistor R_G returned to the supply voltage as shown in Fig. 14.37. Several switches may be added in parallel with SW1 so that the SCR can be turned on from many locations. Switch SW1 can be moved to a gate-ground position where it must be normally closed as shown in Fig. 14.38. When opened, a triggering signal is applied through resistor R_G to the

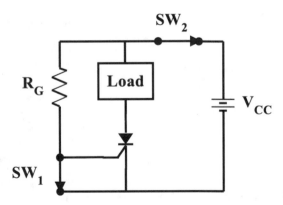

Fig. 14.38 Static switching of SCR using normally closed switch

gate of the SCR. Several switches can then be placed in series so that triggering can again take place at several locations.

Example 14.8
A thyristor has $I_G = 500$ μA, $V_{GC} = 0.7$ V, $V_{AC} = 0.2$ V and $I_H = 10$ mA. Using $V_{TRIG} = 3$ V and $V_{CC} = 24$ V, determine the resistor R_G in order to trigger the thyristor into conduction and

the resulting anode current for a load $R_L = 100\ \Omega$ in the circuit of Fig. 14.37.

Solution From Kirchoff's voltage law, $V_{TRIG} = I_G R_G + V_{GC}$ from which $R_G = (V_{TRIG} - V_{GC})/I_G = (3 - 0.7)/500\ \mu A = 4.6$ k. The anode current I_A is found from $V_{CC} = I_A R_L + V_{AC}$. Hence $I_A = (V_{CC} - V_{AC})/R_L = (24 - 0.2)/100 = 238$ mA.

A simple static switch for AC power control to a load is shown in Fig. 14.39. It is a basic half-wave circuit that is triggered by a signal through resistor R_1 to the gate of the SCR when SW1 is closed. The SCR conducts for one half-cycle and turns off for the other half as the waveform goes through zero. The load waveform is therefore a half-wave rectified AC as shown in Fig. 14.40. The reverse voltage rating of the SCR must exceed the peak reverse voltage to which it will be subjected. Diode D_1 prevents the gate cathode junction of the SCR from being subjected to reverse voltages. Again, the normally open switch SW1 may be replaced by a normally closed switch between the gate and the cathode.

Phase Control Switching

The basic system can be modified to introduce phase control as shown in Fig. 14.41. Here potentiometer VR_1 in conjunction with resistors R_1 and R_2 allows the adjustment of the triggering voltage to the SCR to switch the SCR at a specified point in the signal cycle. This is shown in Fig. 14.42

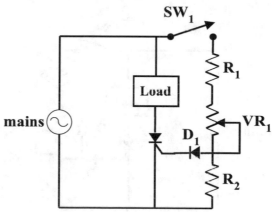

Fig. 14.41 Phase control switching of SCR

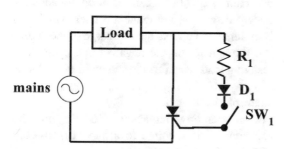

Fig. 14.39 Static AC power switch

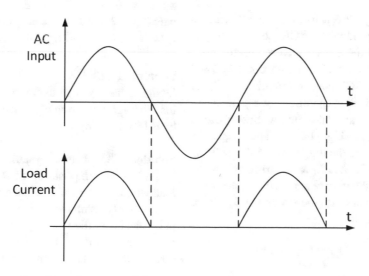

Fig. 14.40 Load waveform for static power switch

Fig. 14.42 Waveform
resulting from phase control

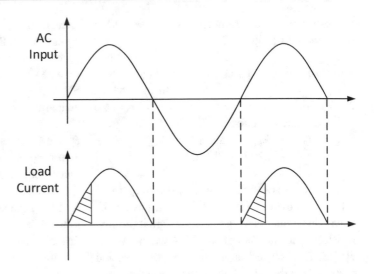

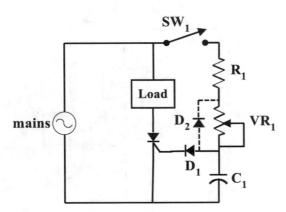

Fig. 14.43 Extended phase control switching of SCR

where phase adjustment from $0°$ to $90°$ can be effected corresponding to SCR conduction for $180°$ to $90°$ and the shaded area represents non-conduction. For phase adjustment from $0°$ to $180°$ corresponding to SCR conduction for $180°–0°$, the circuit must be further modified by replacing R_2 by a capacitor C_1 and introducing another diode D_2 as shown in Fig. 14.43. The capacitor introduces delay in the triggering of the SCR as C_1 charges via VR_1 and R_1 to the triggering voltage level. Since the voltage across the capacitor in a series RC network lags the driving AC voltage by an angle θ given by

$$\theta = \tan^{-1}\left(\frac{R}{X_C}\right) \qquad (14.2)$$

where θ is the lagging phase angle of the capacitor voltage in degrees, X_C is the capacitor reactance and R is the resistance in series with the capacitor, Eq. (14.2) can be used to approximately determine component values. Diode D_2 is optional and allows the discharge of the capacitor through R_1 on the negative half cycle. The resulting load waveform is shown in Fig. 14.44.

Example 14.9
For an operating frequency of 60 Hz, find the capacitor value required to achieve a phase lag of $50°$ with a resistor of 12 k.

Solution From Eq. (14.2), $\tan 50° = 12 \times 10^3 \times 2 \times \pi \times 60 \times C$. Therefore $C = \frac{\tan 50°}{12 \times 10^3 \times 2 \times \pi \times 60} = 0.26\,\mu\text{F}$.

Example 14.10
Design a static switch to control power to an AC load through the action of an SCR being powered by a 15 vAC supply.

Solution The basic system is shown in Fig. 14.39. It comprises an SCR in series with a load with the arrangement being connected to the mains supply. With the switch open, no activating pulse can be applied to the gate, and therefore the SCR remains off. If the switch is closed on the positive half-cycle, a current pulse is supplied to

Fig. 14.44 Waveform resulting from extended phase control

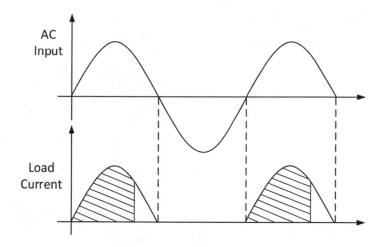

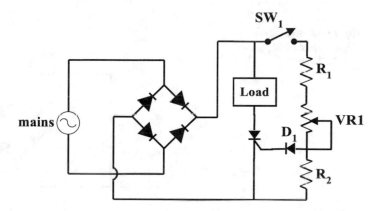

Fig. 14.45 Phase control of AC power to a DC load

the gate through the resistor causing it to turn on and thereby connect the load to the mains supply. The device will turn off during the negative half-cycle of the supply but will turn on again on the positive half-cycle. The result is that the supply to the load is a half-wave supply. The diode ensures unidirectional gate current flow. The value of the resistor R_1 is calculated using the peak supply voltage say 4 volts at which turn occurs and the gate current value required to trigger the thyristor. For the 2N3669, the trigger current has a maximum value of 40 mA giving $R = \frac{4}{40 \times 10^{-3}} = 100\,\Omega$. This value of gate resistor allows for the unknown load resistance which will reduce the gate current below the 40 mA value used in the calculation.

Full-Wave Control

The circuit in Fig. 14.45 is designed to control AC power to a DC load using an SCR. The load and the SCR are connected after the diode bridge.

Therefore, only unidirectional flow occurs through the load and SCR. As the rectified AC voltage rises across the resistor-potentiometer potential divider, the potentiometer is adjusted so that triggering of the SCR occurs at an appropriate point in the AC cycle thereby realizing phase control. The resulting load waveform is shown in Fig. 14.46. The potentiometer enables adjustment of the firing angle from 0° to 90° for each half cycle. For phase adjustment from 0° to 180° corresponding to SCR conduction for 180–0°, the circuit must be modified by replacing R_2 by a capacitor C_1 as shown in Fig. 14.47. The capacitor introduces delay in the triggering of the SCR as C_1 charges via VR_1 and R_1 to the triggering voltage level. The resulting load waveform is shown in Fig. 14.48.

A simple application of this system is a temperature controller as shown in Fig. 14.49. The

Fig. 14.46 Waveform
resulting from phase control
of AC power to DC load

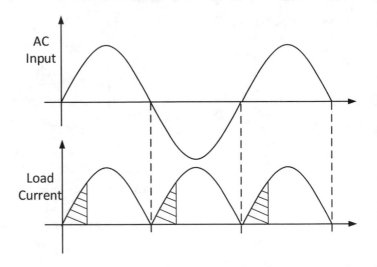

Fig. 14.47 Extended
phase control of AC power
to a DC load

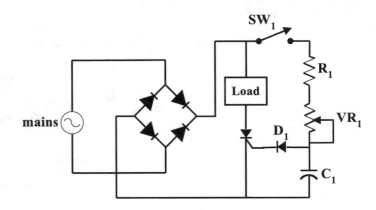

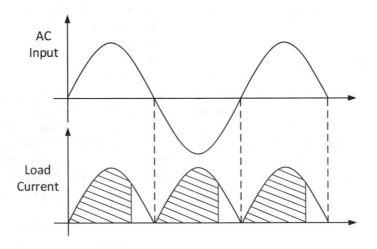

Fig. 14.48 Waveform resulting from extended phase control of AC power to DC load

diode bridge arrangement produces unidirectional current flow through the thyristor, while the resistor, capacitor and thermostat together regulate the gate drive to the device. At low temperatures when the thermostat is open, for each half-cycle of the supply voltage, the capacitor is charged through the resistor to a potential that triggers the SCR, thereby allowing current flow to the heater. As the ambient temperature rises, the thermostat closes at a specific temperature and short circuits the gate-cathode terminals as a result of which the SCR cannot be triggered. No power is then delivered to the heater and the temperature falls.

The full wave circuit in Fig. 14.50 is designed to control AC power to an AC load using an SCR. The load is connected before the diode bridge, while the SCR is connected after the bridge. Therefore, only unidirectional flow occurs through the SCR. As the full-wave rectified AC voltage rises across the resistor-potentiometer

potential divider, the potentiometer is adjusted so that triggering of the SCR occurs at a specified point in the AC cycle thereby effecting phase control. The effect on the load waveform is shown in Fig. 14.51. Using the potentiometer, the firing angle can be varied from $0°$ to $90°$ for each half cycle. Again, for phase adjustment from $0°$ to $180°$ corresponding to SCR conduction for $180–0°$, the circuit must be modified by replacing R_2 by a capacitor C_1.

14.5.2 Gate Turn-Off Switch

The symbol for the gate turn-off switch is shown in Fig. 14.52. This device is effectively an SCR that can be turned off (and on) at the gate-cathode junction. This feature is very useful in many power-switching applications. This device can both be turned on and off at the gate which results in a simplification of the associated circuitry.

14.5.3 Light-Activated SCR

The *light-activated SCR* is an SCR that conducts when exposed to light. The symbol for this device is shown in Fig. 14.53. These devices usually have the standard gate where they can also be triggered by an electrical pulse. Once turned on, the LSCR will continue to conduct even though the light source is removed. Maximum sensitivity to light is achieved by having the gate open. However, the sensitivity can be adjusted by connecting a resistor from the gate to the cathode.

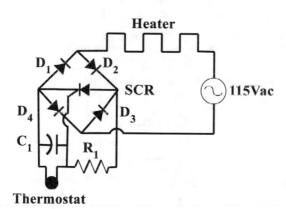

Fig. 14.49 Temperature controller

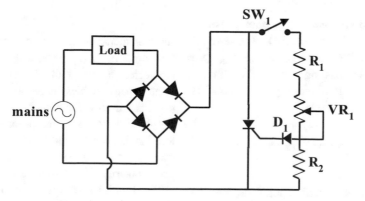

Fig. 14.50 Phase control of AC power to AC load

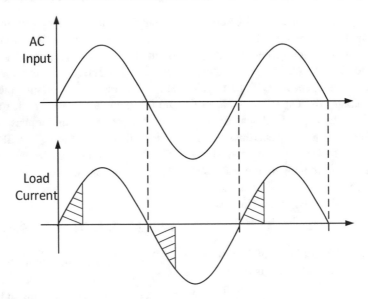

Fig. 14.51 Waveform resulting from phase control of AC power to AC load

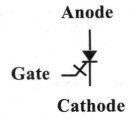

Fig. 14.52 Gate turn-off switch

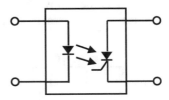

Fig. 14.54 Photo-SCR opto-coupler

Fig. 14.53 Light-activated SCR

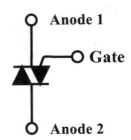

Fig. 14.55 Triac

The *photo-SCR opto-coupler* is a photosensitive SCR in a package with an LED as shown in Fig. 14.54. The SCR can therefore be triggered into conduction by passing current through the LED. It can be used to trigger high current SCRs supplying high power loads.

14.6   Triac

While a thyristor is a unidirectional device, the triac allows current flow in either direction. It is a

semiconductor device that can be viewed as two thyristors in parallel with the anode of one connected to the cathode of the other. The arrangement enables large bi-directional current flow. It is used in the switching of AC power and for phase control of different loads. Only one gate terminal is utilized, and this is used to turn on the triac in either direction with a gate pulse of appropriate polarity. The symbol for the triac is shown in Fig. 14.55, and construction of a typical triac is shown in Fig. 14.56. The terminals T1 and T2 are connected to both p-type and n-type material as shown. This enables conventional current flow in either direction, i.e., from T1(+) to T2(−) via p1, n3, p2 and n2 and from T2(+) to T1(−) via p2, n3, p1 and n1. The gate terminal G is also connected to both semiconductor types namely p1 and n4,

and this enables gate-T1 voltage turn-on to be of either polarity. Like the thyristor, the current through the triac must exceed the latching current before the cessation of the gate current if the device is to stay on. When turned on, the device current must not fall below the holding current. The static characteristic of the triac is shown in Fig. 14.57. Again, similar to the thyristor, the forward breakover voltage falls as the gate trigger current is increased. The device operates in either the first quadrant when terminal T1 is positive relative to terminal T2 and in the third quadrant when T2 is positive relative to T1.

The triac like the SCR will be turned on if the rate of change of the terminal voltage exceeds a specified value. A snubber circuit can again be used to prevent turn-on by sharp voltage changes.

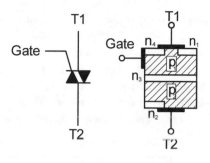

Fig. 14.56 Triac construction

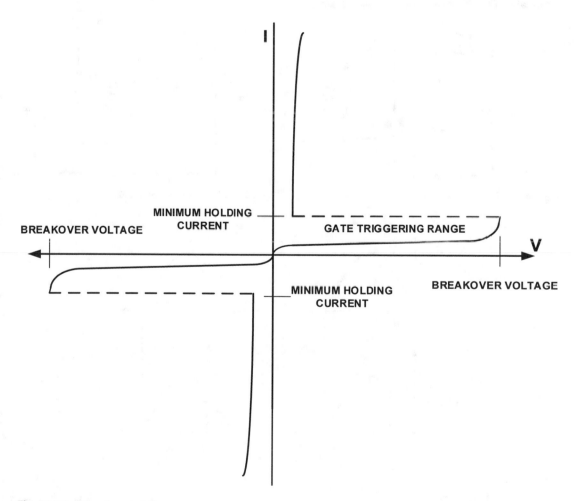

Fig. 14.57 Triac characteristics

14.6.1 Triggering Methods

Several triggering methods are presented in this section.

Static AC Switch

A popular application of the triac is as a static AC switch for AC loads as shown in Fig. 14.58. Closing the switch S1 causes the triac to be triggered by the flow of current through the gate. Resistor R_1 provides the gate current required to trigger the triac, and its exact value determines where in the signal half-cycle turn-on actually occurs. Let the phase angle at turn-on be θ and the corresponding peak supply voltage be V_T. Then for $\theta = 2°$, since $\sin\theta = V_T/115\sqrt{2}$, it follows that $V_T = 115\sqrt{2}\sin 2° = 5.7$ volts. Let the gate current be I_{GT} and the associated gate-to-terminal $T1$ voltage be V_{GT}. Therefore

$V_T = I_{GT}(R_L + R_1) + V_{GT}$ where R_L is the resistance of the load. Using $V_{GT} = 1$ V, $V_T = 5.7$ V, $I_{GT} = 20$ mA and $R_L = 115$ V/5 A $= 23$ Ω, we get $R_1 = \frac{5.7-1}{0.02} - 23 = 212$ Ω. Once the triac is on, the full supply voltage is available across the load.

Phase Control Switching

The basic principle in triac phase control circuits is to vary the firing angle of the triac relative to the cycle of the supply and as a result change the effective voltage across the circuit load. The simple resistive gate triggering circuit in Fig. 14.59 allows the variation of the firing angle from almost zero up to a maximum of 90° as shown in Fig. 14.60. This is accomplished by simply

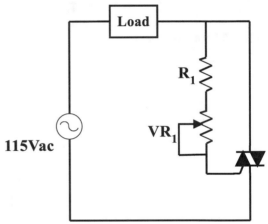

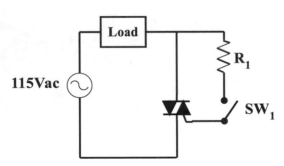

Fig. 14.58 Static AC Switch

Fig. 14.59 Triac phase control circuit

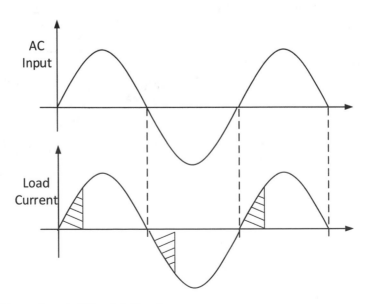

Fig. 14.60 Gate firing angle up to 90° each half-cycle

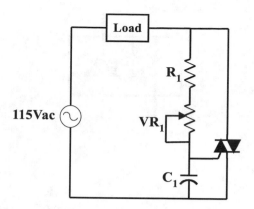

Fig. 14.61 Triac RC phase control circuit

Fig. 14.62 Gate firing
angle up to 180° each
half-cycle

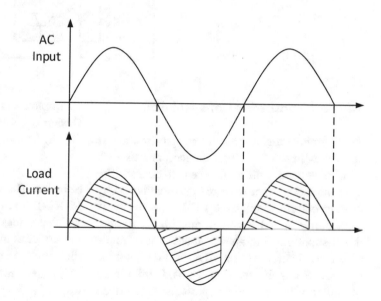

varying the gate resistor value and hence the time in the half-cycle when the triac switches on. By adding a capacitor as shown in Fig. 14.61, the resulting RC phase-shifting network enables the increase of the firing angle to almost 180° as shown in Fig. 14.62.

The capacitor introduces delay in the triggering of the triac as C_1 charges via R_1 and VR_1 to the triggering voltage level. Since the voltage across a capacitor in an RC network lags the input AC voltage by an angle θ given by

$$\theta = \tan^{-1}\left(\frac{R}{X_C}\right) \qquad (14.2)$$

where θ is the phase angle of the capacitor voltage in degrees, X_C is the capacitor reactance and R is the sum of R_1 and VR_1, Eq. (14.2) can be used to approximately determine component values.

When driving inductive loads, a snubber circuit consisting of a resistor (100 Ω) in series with a capacitor (0.1 μF) placed across the terminals of the triac may be used to protect against the effects of rapid increases in terminal voltage.

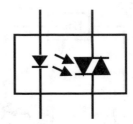

Fig. 14.63 Phototriac opto-isolator

Fig. 14.64 Remote
control of AC load

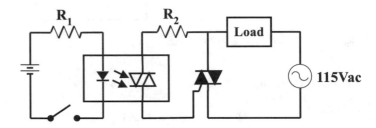

14.6.2   Phototriac Opto-isolator

The symbol for a phototriac opto-isolator is shown in Fig. 14.63. This device consists of a LED packaged with a photosensitive triac. The triac can therefore be triggered into conduction by driving current through the LED. This device is particularly suited for triggering higher current triacs supplying high-power loads. An example of an application of the MOC 3011 phototriac opto-isolator is shown in Fig. 14.64. In this circuit the input of the phototriac opto-coupler is driven by the digital output of a microcontroller or personal computer, represented here by a switch in series with a battery. Resistor R_1 is chosen to limit the current through the light-emitting diode, while resistor R_2 limits the current to the gate of the triac. The output of the MOC 3011 is used to provide gate drive to the high-power triac 2N6071 which is connected to the load and the mains supply. Thus closing the switch to the input of the MOC 3011 triggers the internal triac which in turn triggers the external 2N6071 triac into conduction. The result is that the load is connected to the mains supply. This arrangement isolates the low-power computer circuit from the high-power load, thereby minimizing the risk of shock.

Example 14.11
Design a system to remotely switch on a large (10 A) lighting load using the phototriac and a triac.

Solution The system is based on Fig. 14.64. The MOC3011 is connected through a resistor R_2, which provides the gate current required to trigger the triac. Let the phase angle at turn-on be θ and the corresponding peak supply voltage be V_T. Then for $\theta = 2^\circ$, since $\sin \theta = V_T / 115\sqrt{2}$, it follows that $V_T = 115\sqrt{2} \sin 2^\circ = 5.7$ volts. Let the gate current be I_{GT} and the associated gate-to-terminal $T1$ voltage be V_{GT}. Therefore $V_T = I_{GT}(R_L + R_2) + V_{GT}$ where R_L is the resistance of the load. Using $V_{GT} = 1$ V, $V_T = 5.7$ V, $I_{GT} = 20$ mA and $R_L = 115$ V/10 A $= 11.5$ Ω, we get $R_2 = \frac{5.7-1}{0.02} - 11.5 = 224 \Omega$. Current to activate the diode in the phototriac is supplied via resistor R_1 and long low-voltage wiring with a 5 V source and a switch S1. If the required diode current is 15 mA, then resistor $R_1 = 5$ V/ 15 mA $= 333$ Ω. Such wiring need not be in a conduit since it is at low voltage and therefore significant cost savings can be realized in installing a lighting system in commercial or residential buildings. Other loads such as motors,

Fig. 14.65 Shockley diode symbol

Fig. 14.66 Shockley diode characteristic

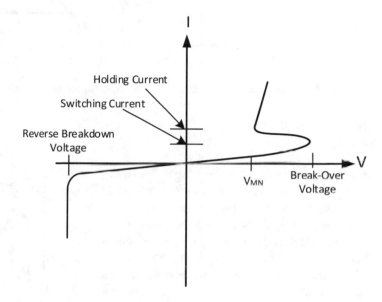

fans or pumps can be switched, but a snubber circuit may be necessary.

14.7 Shockley Diode

The Shockley diode in Fig. 14.65 is essentially an SCR without the gate terminal. Its characteristic therefore corresponds to that of the SCR with zero gate current as shown in Fig. 14.66. From this characteristic it can be seen that for a voltage applied across the anode and the cathode, little or no current flows. In the case of a reverse voltage, a value exceeding the reverse breakdown voltage of the device (typically 10–20 volts) causes the device to turn on. It is not intended that the device operates in this mode. When the forward voltage is increased beyond the breakover voltage V_{BR}, current flow increases rapidly and the device turns on when this current exceeds the latching value. The switching current at this point is I_S. The terminal voltage drops significantly to a voltage V_{MN}, and conduction continues providing the current exceeds the

holding current I_H. Once the device begins to conduct, it will continue until the anode current is reduced to a value less than the holding current. This device is generally designed for unidirectional operation and is often used in SCR triggering circuits.

One simple application is in a relaxation oscillator as shown in Fig. 14.67a. Closure of the switch results in the charging of the capacitor and increase of the voltage across the capacitor. When the voltage reaches the breakover voltage of the Shockley diode, the diode conducts thereby discharging the capacitor and the capacitor voltage falls. Once the diode current drops below the holding current of the diode, the diode turns off and the capacitor begins to charge again. The capacitor waveform is shown in Fig. 14.67b. The rising waveform can be made linear by using a constant current source to charge the capacitor.

Example 14.12
Design a relaxation oscillator to produce a linear waveform with a slope of 10 mV/µs. Use a

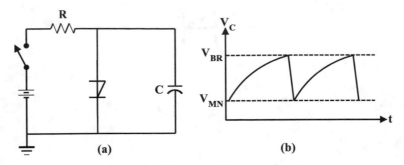

Fig. 14.67 Shockley diode. (**a**) Relaxation oscillator and (**b**) output waveform

Shockley diode having $V_{BR} = 12$ V, $V_{MN} = 1$ V and $I_H = 2$ mA.

Solution One circuit for achieving this is shown in Fig. 14.68. Zener diode D_1 has $V_Z = 2.7$ V and with $R_2 = 2.2$ k, a current of $(15 - 2.7)/2.2$ k $= 5.6$ mA flows through the Zener diode. Transistor Tr1 is connected as a constant current source delivering a constant current of $(2.7 - 0.7)/R_1$. For $R_1 = 2$ k the constant current is 1 mA. This constant current charges capacitor C_1 for which $\frac{I}{C_1} = \frac{dV}{dt}$. Now 10 mV/μs $= 10^4$ V/s, and therefore $C_1 = I/\frac{dV}{dt} = 10^{-3}$ A$/10^4$ Vs^{-1} $= 0.1$ μF. A rail-to-rail op-amp such as the LMC6482 is used in the voltage follower mode to buffer the capacitor voltage and deliver an output voltage. The peak output voltage is $V_{BR} = 12$ V with a minimum value less than $V_{MN} = 1$ V.

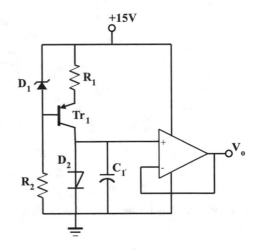

Fig. 14.68 Circuit for Example 14.12

Fig. 14.69 Symbol for diac

14.8 Diac

The diac in Fig. 14.69 is essentially a triac without the gate terminal. Its behaviour under an applied voltage therefore corresponds to the triac characteristic for zero gate current. Thus, as the terminal voltage of the diac is increased, there is initially little or no current flow until the voltage attains the breakover value. At this point the diac conducts heavily and the terminal voltage falls to a lower value. Its characteristic is shown in Fig. 14.70. The diac is often used for reliable triggering of triacs. The two phase control triac circuits in Figs. 14.59 and 14.61 are rarely used in practice. This is because of variations of gate-trigger signal levels between triacs of the same type, quadrant of operation and device case temperature. This variation prevents the reliable calibration of the phase control resistor for specified firing angles.

This basic problem is eliminated by using a diac as part of the triggering circuit as shown in

Fig. 14.70 Diac
characteristic

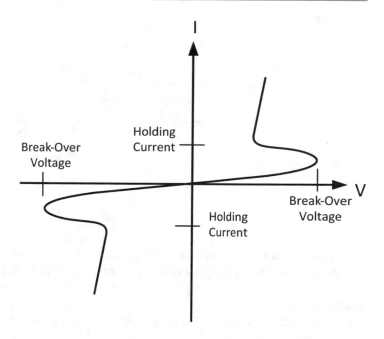

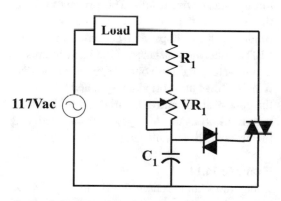

Fig. 14.71 Triac phase control using diac

Fig. 14.71. Here the voltage across the capacitor must exceed the breakover voltage V_{BO} of the diac. When this occurs, the diac turns on, and the voltage across it falls to a lower voltage V_D. This presents a fast-rising trigger signal of amplitude $(V_{BO} - V_D)$ to the gate of the triac which turns on the triac in a reliable and predictable manner. This circuit provides excellent speed control for universal motors used in many shop tools and kitchen appliances, fans, sewing machines and food blenders. It can also serve as a very effective light dimmer controller. An example of a diac is the 1N5758 with a 16–24 V breakover voltage.

With the potentiometer, the charging resistance R can be varied from a minimum $R = R_1$ to a maximum $R = VR_1 + R_1$. At low values of charging resistance, the phase angle is small, and the capacitor voltage will be almost in phase with the supply voltage. At higher values of charging resistance, the phase lag increases. However, the amplitude of the voltage V_C across the capacitor decreases relative to the amplitude of the supply voltage V_S. From circuit theory, the relationship is given by

$$V_C = \frac{X_C}{\sqrt{R^2 + X_C^2}} V_S \qquad (14.3)$$

While this was not an issue before because of the lower triggering voltages involved, the introduction of the diac means that the capacitor voltage needs to rise to higher values such that the diac breakover voltage V_{BO} can be overcome. Therefore, for triac conduction, the charging resistance R must be such that

$$V_C = \frac{X_C}{\sqrt{R^2 + X_C^2}} V_S \geq V_{BO} \qquad (14.4)$$

Fig. 14.72 Telephone beacon

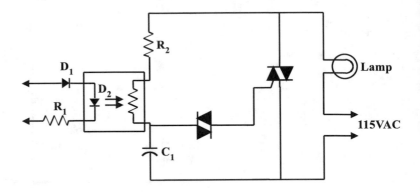

This means that this resistance cannot exceed a maximum value lest the triac fail to be triggered.

Example 14.13

Using the system in Fig. 14.71, design an AC motor speed controller.

Solution Let capacitor $C_1 = 0.1\ \mu F$ and use $R_1 = 3.3$ k with potentiometer $VR_1 = 250$ k. Then the minimum value of R is $R_{mn} = 3.3$ k, while the maximum value is $R_{mx} = 250$ k + 3.3 k $= 253.3$ k. Noting that $X_C = 1/2 \times \pi \times 60 \times 0.1 \times 10^{-6} = 26.5 \times 10^3\ \Omega$, the minimum phase lag is $\theta = \tan^{-1}(R_{mn}/X_C) = \tan^{-1}(3.3 \times 10^3/26.5 \times 10^3) = 7°$, and the maximum phase lag is $\theta = \tan^{-1}(R_{mn}/X_C) = \tan^{-1}(253.3 \times 10^3/26.5 \times 10^3) = 84°$. The minimum value of V_C occurs when $R = R_{mx} = 253.3$ k and is given by $V_{Cmn} = \frac{X_C}{\sqrt{R^2 + X_C^2}}V_S = \frac{26.5}{10^3\sqrt{(253.3)^2 + (26.5)^2}} = 0.1V_S$. The peak value of the supply voltage is $V_S(pk) = 115\sqrt{2} = 162.6$ V. This gives $V_{Cmn} = 0.1V_S = 16.3$ V. Therefore the diac breakover voltage V_{BO} must be less than 16.3 V if triac triggering is to occur for all values of VR_1. A higher breakover voltage can of course be used, but the triac will remain off for high settings of the potentiometer resistance.

Another application is a telephone beacon shown in Fig. 14.72. Here the appearance of ringing voltage on the telephone line causes a current to flow through the diode of the opto device. Diode D_1 ensures unidirectional current flow through D_2. This current flow causes the resistance of the LDR to fall thereby resulting in a voltage across the capacitor C_1. When the breakdown voltage of the diac is exceeded, a voltage is applied to the gate of the triac, and conduction through this device results. This process turns on the triac, and the lamp is turned on. When the ringing voltage stops and current through the LED ceases, the resistance of the LDR returns to a high value, and the voltage applied to the diac falls. The diac turns off as a result, and this causes the voltage at the gate of the triac to fall. The triac then turns off as the AC passes through zero and thereby switches off the lamp.

Example 14.14

Using the configuration shown in Fig. 14.72, design the system to activate a 100 W lamp when there is ringing on the telephone system.

Solution Ringing voltage is typically about 105 V giving a peak voltage of about 150 V. Therefore diode D_1 must have a PIV of at least 200 V, and the 1N4003 is suitable. Using $R_1 = 10$ k allows about 10 mA to flow through the opto-coupler LED. This causes the resistance of the photoresistor to fall allowing C_1 to be charged through R_2. Let the diac breakover voltage be 20 V. Using $R_2 = 10$ k and $C_1 = 0.1\ \mu F$ ensures that the breakover voltage of the diac is reached early in the half-cycle thereby switching on the triac.

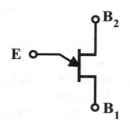

Fig. 14.73 Unijunction transistor

14.9   Unijunction Transistor

The unijunction transistor whose symbol is shown in Fig. 14.73 may be considered a special member of the thyristor family. It is made of a silicon lightly doped n-type silicon bar with contacts at either end as the base terminals. Like the JFET, some p-type material is positioned in an area between the base 1 and base 2 terminals as shown in Fig. 14.74a and the associated terminal is referred to as the emitter. The equivalent circuit of this device is shown in Fig. 14.74b comprising a variable resistor R_{B1} and a fixed resistor R_{B2} along with a silicon diode that represents the junction between the p-type region and n-type. The sum of these two resistors is known as the interbase resistance R_{BB} given by $R_{BB} = R_{B1} + R_{B2}$.

A potential V_{BB} across the two base terminals B_2, B_1 with B_2 at the higher potential as shown in Fig. 14.75 (under conditions of zero emitter current) will produce only a small current flow as a result of the light doping of the channel. The effective resistance then is of the order of units of kilo-ohms (4–9 k). An important UJT parameter η referred to as the Intrinsic Standoff Ratio is defined as the ratio of the resistor R_{B1} to the interbase resistance R_{BB} when the emitter current is zero, i.e., $\eta = R_{B1}/R_{BB}$. Its value varies from about 0.5–0.8.

Now if the emitter is at zero potential, the voltage drop along the channel will result in a voltage V_{RB1} across R_{B1} given by

$$V_{RB1} = \frac{R_{B1}}{R_{BB}} V_{BB} = \eta V_{BB} \qquad (14.5)$$

which is at the cathode of the diode. This voltage V_{RB1} reverse biases the pn junction D_1 as occurs in an n-channel JFET, and there is no emitter current flow. As the emitter voltage V_E rises, only a tiny emitter current of a few microamps flows. When it eventually exceeds V_{RB1} by an

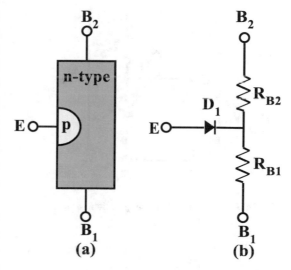

Fig. 14.74 UJT. (**a**) Physical composition and (**b**) equivalent circuit

amount equal to the diode threshold voltage 0.7 V, then the diode becomes forward biased resulting in an emitter current flow into the base 1 region comprising majority carrier holes into the light-doped n-channel. These positive charge carriers are attracted to the terminal B1, this causing the effective resistance of R_{B1} to fall. This results in an increase in current which causes a further reduction in resistance in a regenerative action. This action constitutes a negative resistance since a reduction in voltage is associated with an increase in current. The device goes into saturation as the resistance R_{B1} falls to its lowest value of about 50 Ω referred to as saturation resistance. This low resistance results in a reduced voltage drop with a consequent fall in the emitter voltage to a value less than 1 V. This is shown in Fig. 14.76.

The UJT is now triggered (sometimes referred to as the ON state), and the maximum current flows through the channel. When the emitter voltage falls to a sufficiently low value, the diode turns off and the effective resistance of R_{B1} rises

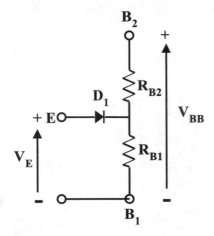

Fig. 14.75 UJT operating configuration

to its former value. While the triggering action of the UJT involves the channel region close to B1 that is represented by resistor R_{B1}, the channel region close to B2 that is represented by resistor R_{B2} is largely unaffected by this action and the associated emitter current flow. The emitter voltage that results in the forward-biasing of the pn junction and the flow of an emitter current is usually designated V_P and is given by

$$V_P = \eta V_{BB} + 0.7 \qquad (14.6)$$

From Fig. 14.76 at the peak point, the peak emitter voltage V_P is associated with a peak emitter current I_P, while at the valley point, the minimum emitter voltage V_V is associated with a minimum emitter current I_V.

Example 14.15
A UJT with $\eta = 0.75$ is powered by a 10-volt supply. If the emitter is at zero potential, determine the reverse bias across the pn junction.

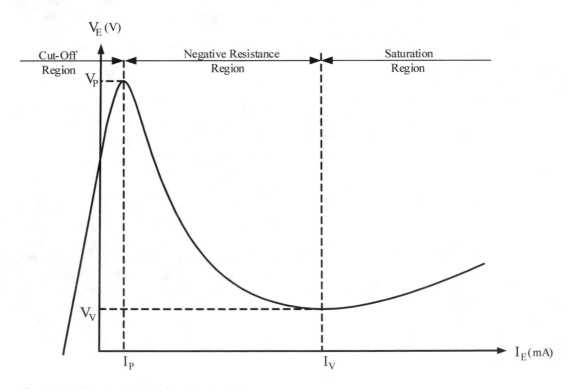

Fig. 14.76 Characteristic of unijunction transistor

Solution From Eq. (14.5), $V_{RB1} = \eta V_{BB} = 0.75 \times 10 = 7.5$ V. Hence the reverse bias across the pn junction is 7.5 V.

Unijunction transistors are available in power ratings up to about 500 mW with voltage ratings up to about 60 V and peak currents ranging between 50 mA and 2 A. A very common application of this device is in a relaxation oscillator. The basic system using the widely available 2N2646 UJT is shown in Fig. 14.77.

When the switch is closed, the capacitor charges exponentially through resistor R_1 with a time constant $R_1 C_1$. As a result, the voltage at the emitter rises. When $V_E = V_P$, the unijunction is

triggered and the capacitor discharges through the emitter through R_{B1} and R_2 with time constant $(R_{B1} + R_2)C_1$ that is usually much smaller than $R_1 C_1$. Therefore, the charge time is usually much larger than the discharge time. When the capacitor voltage falls to $V_E = V_V$, the diode turns off, the UJT returns to its original state, capacitor charging again commences, and the cycle is repeated. The waveform at the emitter and at B1 are shown in Fig. 14.78. The former results from the capacitor charge and is an approximate sawtooth wave. The latter signal results from the current flow through resistor R_2 during the discharge process and is a short duration pulse.

The peak value of the voltage at the emitter is V_P, while the peak value of the voltage at B1 is given by

$$V_{B1}(pk) = \frac{R_2}{R_2 + R_3 + R_{BB}(mn)} V_{BB} \quad (14.7)$$

where $R_{BB}(mn)$ is the minimum value of the ON state resistance between B1 and B2 given approximately by $R_{BB}(mn) \simeq \eta R_{BB}$. The time t_1 to charge the capacitor C using V_{BB} through resistor R_1 from an initial voltage of V_V to a final voltage of V_P is given by

$$t_1 = R_1 C \ln\left(\frac{V_{BB} - V_V}{V_{BB} - V_P}\right)$$

$$= R_1 C \ln\left(\frac{V_{BB} - V_{BB}/10}{V_{BB} - \eta V_{BB}}\right)$$

$$= R_1 C \ln\left(\frac{0.9}{1-\eta}\right) \quad (14.8)$$

where $V_V \simeq V_{BB}/10$ is used. The period is given by $T = t_1 + t_2$ where t_2 is the discharge time. Since $t_1 >> t_2$, the period is approximately equal to t_1, and hence

$$T = R_1 C \ln\left(\frac{0.9}{1-\eta}\right) \quad (14.9)$$

giving

$$f = 1/R_1 C \ln\left(\frac{0.9}{1-\eta}\right) \quad (14.10)$$

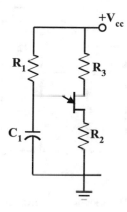

Fig. 14.77 Relaxation oscillator using UJT

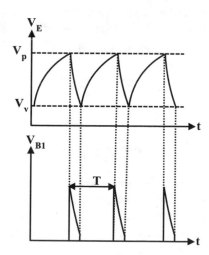

Fig. 14.78 Output waveforms for UJT relaxation oscillator

Example 14.16
Using the 2N2926 unijunction transistor, design a relaxation oscillator with a frequency of oscillation of 1 kHz and operating from a 9-volt supply.

Solution The 2N2926 has $\eta = 0.56 - 0.75$, $I_P(mx) = 5$ μA and $I_V(mx) = 4$ mA. Now resistor R_1 must allow the emitter current to rise to I_P when the emitter voltage equals V_P and limit the current to less than I_V when the emitter voltage falls to V_V. From this, we can find the maximum value of R_1 as $R_{1mx} = (V_{BB} - V_P)/I_P$ and the minimum value of R_1 as $R_{1mn} = (V_{BB} - V_V)/I_V$. Using the average value of η which is $\eta_{avg} = (0.56 + 0.75)/2 = 0.66$, the peak voltage V_P is given by $V_P = \eta V_{BB} + 0.7 = 0.66 \times 9 + 0.7 = 6.6$ V. Assuming $V_V \simeq V_{BB}/10 = 0.9$ V, then $R_{1mx} = (9 - 6.6)/5 \times 10^{-6} = 480$ k and $R_{1mn} = (9 - 0.9)/4$ mA $= 2$ k. An intermediate value $R_1 = 10$ k can be chosen. From (14.10), $f = 1/R_1 C \ln\left(\frac{0.9}{1-\eta}\right) = 1/R_1 C \ln\left(\frac{0.9}{1-0.66}\right) \simeq 1/R_1 C$. This gives $C = 1/R_1 f = 1/10 \times 10^3 \times 10^3 = 0.1$ μF. A small resistor $R_2 = 100$ Ω enables a pulse voltage out while not affecting V_P. Resistor R_3 limits the dissipation in the unijunction transistor and can be set to a value $R_3 = 1$ k.

14.10 Programmable Unijunction Transistor

The programmable unijunction transistor shown in Fig. 14.79 has three terminals designated anode, cathode and gate. It has a 4-layer semiconductor structure similar to the thyristor but with the gate connected to the (n-type) layer close to the (p-type) anode and not to the (p-type) layer close to the (n-type) cathode. Its operational characteristics are similar to those of the UJT except that the intrinsic stand-off ratio η can be externally programmed using a potential divider connected to the gate. This enables the control of the peak value V_P of the anode voltage. Examples of PUTs are the 2N6027 and 2N6028 from ON Semiconductor (Fig. 14.79).

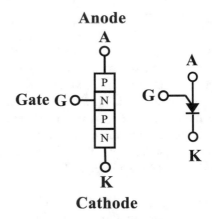

Fig. 14.79 Programmable unijunction transistor

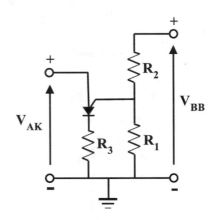

Fig. 14.80 PUT operating configuration

The system operation can be described using Fig. 14.80 where a variable voltage V_{AK} is applied to the anode of the PUT. Here, the potential divider $R_1 - R_2$ holds the gate at a fixed voltage

$$V_{R1} = \frac{R_1}{R_1 + R_2} V_{BB} = \eta V_{BB} \qquad (14.11)$$

where

$$\eta = \frac{R_1}{R_1 + R_2} \qquad (14.12)$$

and the cathode is connected through a resistor R_3 to ground. As the supply voltage is increased from zero, only a small reverse leakage current flows through the PUT as the anode-gate junction is reverse-biased and therefore is in a

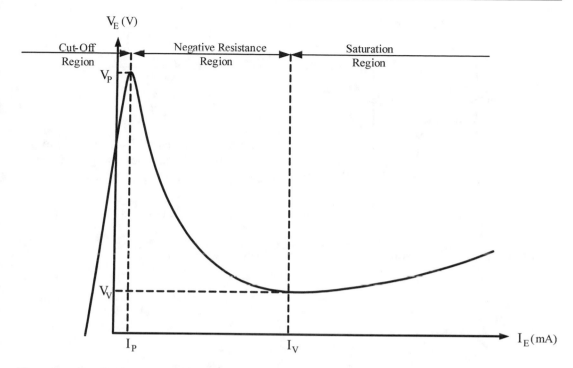

Fig. 14.81 Characteristic of PUT

nonconducting state. When the anode voltage V_A exceeds the gate voltage by $0.7V$ such that its value equals V_P where

$$V_P = \eta V_{BB} + 0.7 \qquad (14.13)$$

the anode-gate junction becomes forward-biased and an anode current I_A begins to flow. This value of anode voltage is called the peak voltage V_P as shown on the anode characteristic in Fig. 14.81. The PUT now rapidly turns on and the anode voltage falls as the current increases down to the valley point. This corresponds to negative resistance over this portion of the characteristic. Similar to the UJT, at the peak point the peak emitter voltage V_P is associated with a peak emitter current I_P, while at the valley point, the minimum emitter voltage V_V is associated with a minimum emitter current I_V. External resistors R_1 and R_2 in the PUT correspond to the internal resistors R_{B1} and R_{B2} in the UJT. By adjusting these resistor values, the PUT can be programmed to have different values of V_P and I_P. The output is usually taken at the cathode with the current I_A developing a voltage drop across resistor R_3. Thus,

when the PUT is off, the output voltage V_o is zero. When the device is triggered on, the rapid rise in current causes a sudden rise in V_o to a value that can be controlled.

A relaxation oscillator can also be implemented using a PUT such as the 2N6027. The basic system is shown in Fig. 14.82. When power is applied, the capacitor charges exponentially through resistor R_4 with a time constant R_4C. As a result, the voltage at the anode rises. When $V_A = V_P$, the PUT is triggered and the capacitor discharges through the PUT with time constant R_3C that is usually much smaller than R_4C. As a result, the time to charge is usually much larger than the time to discharge. When the capacitor voltage falls to $V_A = V_V$, the pn junction becomes reverse-biased and the PUT turns off. It then returns to its original state at which time capacitor charging again commences and the cycle is repeated. The waveforms at the anode and cathode are shown in Fig. 14.83. The former results from the charging and discharging of the capacitor and is an approximate sawtooth wave. The latter signal results from the current flow

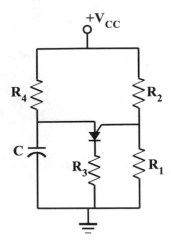

Fig. 14.82 PUT relaxation oscillator

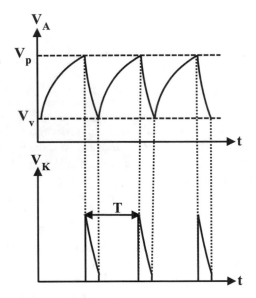

Fig. 14.83 Anode (V_A) and cathode (V_K) waveforms for PUT relaxation oscillator

through resistor R_3 during the discharge process and is a short duration pulse.

The peak value of the voltage at the anode is V_P, while the peak value of the voltage at the cathode depends on the discharge current through the PUT. The time t_1 to charge the capacitor C using V_{BB} through resistor R_4 from an initial voltage of $V_V \simeq 0$ to a final voltage of V_P is given by

$$t_1 = R_4 C \ln \left(\frac{V_{BB} - V_V}{V_{BB} - V_P} \right)$$

$$= R_4 C \ln \left(\frac{V_{BB}}{V_{BB} - \eta V_{BB}} \right)$$

$$= R_4 C \ln \left(\frac{1}{1 - \eta} \right) \qquad (14.14)$$

The period is given by $T = t_1 + t_2$ where t_2 is the discharge time. Since $t_1 > > t_2$, then the period is approximately equal to t_1, and hence

$$T = R_4 C \ln \left(\frac{1}{1 - \eta} \right) \qquad (14.15)$$

giving

$$f = 1/R_4 C \ln \left(\frac{1}{1 - \eta} \right) \qquad (14.16)$$

Example 14.17
Using the 2N6027 PUT, design a relaxation oscillator with a frequency of oscillation of 1 kHz and operating from a 9-volt supply.

Solution The 2N6027 gives parameters for $R_G = R_1 // R_2 = 10$ k. Therefore, we use $R_1 = 27$ k and $R_2 = 16$ k which give $\eta = 27$ k/ $(27$ k $+ 16$ k$) = 0.63$. Typical current values are $I_P = 4$ μA and $I_V = 150$ μA. Now resistor R_4 must allow the anode current to rise to I_P when the anode voltage equals V_P and limit the current to less than I_V when the anode voltage falls to V_V. From this, we can find the maximum value of R_4 as $R_{4mx} = (V_{BB} - V_P)/I_P$ and the minimum value of R_4 as $R_{4mn} = (V_{BB} - V_V)/I_V$. The peak voltage V_P is given by $V_P = \eta V_{BB} + 0.7 = 0.63 \times 9 + 0.7 = 6.37$ V. Now at $I_F = 50$ mA, $V_F = 0.8$ V, and therefore we will use $V_V \simeq 0$. Then $R_{4mx} = (9 - 6.37)/4 \times 10^{-6} = 658$ k and $R_{4mn} = (9 - 0.0)/0.15$ mA $= 60$ k. Choosing $R_4 > 60$ k ensures that the PUT can reset from the valley point following capacitor discharge while choosing $R_4 < 658$ k ensures that the capacitor can charge up to the peak point. An intermediate value $R_4 = 100$ k can be chosen. From (14.16), $f =$

Fig. 14.84 (a) UJT
symbol; (b) UJT equivalent
circuit; (c) PUT circuit
as UJT

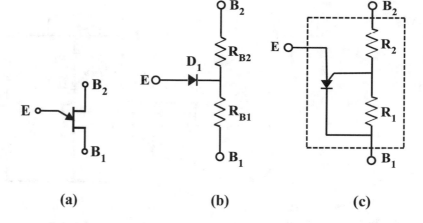

(a) (b) (c)

$1/R_4 C \ln\left(\frac{1}{1-\eta}\right) = 1/R_4 C \ln\left(\frac{1}{1-0.63}\right) \simeq 1/R_4 C.$

This gives $C = 1/R_4 f = 1/100 \times 10^3 \times 10^3 = 0.01$ µF. A small resistor $R_3 = 100\ \Omega$ enables a pulse voltage out while not affecting V_P.

Finally, the close relationship between the UJT and the PUT is demonstrated in Fig. 14.84 where the UJT symbol and the UJT equivalent circuit are compared with a PUT drawn as a UJT.

14.11 Applications

In this section, applications of the various devices are presented.

Battery Charger
This SCR application is a battery charger as shown in Fig. 14.85. In this circuit, a low-voltage transformer supplies power through a thyristor which results in a half-wave rectified voltage being delivered to a battery through $R_5 = 1\ \Omega$. This resistor serves as a current limiting resistor. The resistor R_1 and the diode D_3 provide unidirectional gate current drive that turns on the SCR on the positive half-cycle of the supply voltage. For the 2N3668 SCR, the trigger current has a maximum value of 40 mA giving $R_1 = \frac{15\sqrt{2}-6}{40\times10^{-3}} = 380\,\Omega$ where a discharge battery voltage of 6 volts is used in the

calculation. The battery is charged during this part of the cycle. The transistor Tr1 is used to stop the charging process when the battery is fully charged. It is switched on when the voltage at its base exceeds the turn-on value. The battery voltage at which this occurs is set by the potential divider comprising $R_3 = 8.2$ k, potentiometer $VR_1 = 1$ k and $R_4 = 1$ k in conjunction with the diode D_2. The turn-on voltage is the sum of the transistor base-emitter voltage and the diode voltage giving 1.4 V. The maximum battery voltage V_B at which the transistor turns on occurs when the potentiometer wiper is at the lower position and is given by $\frac{1.4}{1k} = \frac{V_B}{(1+1+8.2)k}$. This gives $V_B = 14.3$ V. Adjusting the potentiometer will give lower battery voltages. Diode D_1 allows one-directional current flow through the transistor. Of course, as the battery voltage increases, the available gate current will fall and eventually will be insufficient to turn on the SCR. Resistor $R_2 = 100\ \Omega$ and diode D_4 allow a small trickle charge to the battery after the SCR turns off.

Ideas for Exploration Introduce a diode bridge such that battery charging occurs on both half-cycles and not just the positive half-cycle.

Temperature Controller Using a Triac
This application is a temperature controller using a 555 timer and a triac as show in Fig. 14.86. The 555 is connected in the monostable mode with the output driving the gate of a triac which controls

Fig. 14.85 Battery charger

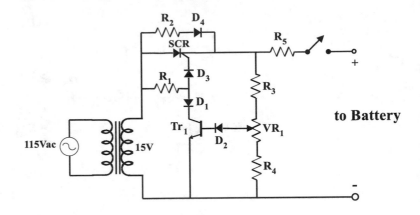

Fig. 14.86 Temperature controller using triac

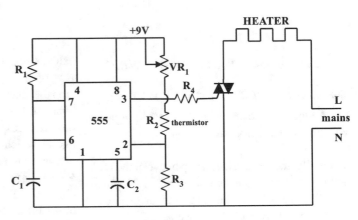

AC power to a heater. Potentiometer VR_1 along with the NTC (negative temperature coefficient) thermistor R_2 and resistor R_3 sets a voltage at the trigger input pin 2. In the steady state with a temperature above a set point such that the resistance of the thermistor results in a potential at pin 2 which is above the threshold voltage of $V_{cc}/3$, the timing capacitor C_1 is clamped at ground potential by the internal transistor connected to pin 7, and the output at pin 3 is low thereby ensuring that the triac is off. If the temperature falls below the specified value, the resistance of the thermistor increases resulting in the voltage at the trigger input falling below $V_{cc}/3$. This starts the timing cycle and the output goes high, turning on the triac and power to the heater. If as a result the temperature rises above the set point and the thermistor resistance falls causing the voltage at pin 2 to rise above $V_{cc}/3$ before the end of the timing cycle, then the output goes low at the end of the cycle thereby switching off power to the

heater as the supply voltage to the triac goes through zero. If, however, at the end of the cycle the temperature has not exceeded the set point, then the voltage at pin 2 will be less than $V_{cc}/3$ and the 555 keeps the triac on until the set point temperature is exceeded after which turn-off occurs. At the specified temperature, the thermistor must have a resistance R_2 such that $VR_1 + R_2 = 2R_3$. This corresponds to a voltage at pin 2 of $V_{cc}/3$. Using a thermistor with a base resistance of 5 k (at 25°C) along with $VR_1 = 2k$ and $R_3 = 3.3$ k allows variation of the set point by adjustment of the potentiometer. Decreasing VR_1 results in a higher potential at pin 2 and therefore a lower set point temperature. Increasing VR_1 towards its maximum value reduces the voltage at pin 2 and hence increases the set point temperature. The cycle time T is set by R_1 and C_1 based on $T = 1.1R_1C_1$. For $R_1 = 1$ M and $C_1 = 10$ μF, we get $T = 1.1 \times 10^6 \times 10 \times 10^{-6} = 11$ s. Using a 2N6071 triac which has a gate trigger current requirement of 5 mA and a gate trigger voltage of 1.4 V and noting that the peak output from the 555 is about 7 V when powered by $V_{cc} - 9$ V, then resistor $R_4 = (7 - 1.4)/5$ mA $\simeq 1$ k.

Ideas for Exploration (i) The 9 V supply for the circuit was intended to be a 9 V battery. Design a suitable 9 V regulated supply using a step-down transformer to power this system, the ground for which would be the neutral line of the mains supply; (ii) use an opto-coupler to completely isolate the 555 circuit from the triac.

Universal Motor Speed Controller

This application enables the speed of a universal motor such as used in power drills to be varied. The circuit is shown in Fig. 14.87. On positive half cycles of the mains voltage, a positive voltage is applied across the anode and cathode of the SCR. The potential divider formed by $R_1 = 6.8$ k and $VR_1 = 2$ k along with diode D_1 provides a positive gate voltage to turn the SCR on. This occurs when the gate voltage is approximately 0.7 V relative to the cathode and about 10 mA of current flows into the gate. Potentiometer VR_1 controls the voltage applied to the gate and therefore the time during the positive half cycle when the SCR turns on. The on time is variable from almost $180°$ when the wiper of the potentiometer is close to the upper end in the diagram to $90°$ when the wiper is close to the anode of diode D_1 in which case the line voltage has to reach its maximum in order to turn on the SCR. This mechanism enables control of the power delivered to the motor. Diode D_2 provides protection for the SCR against high reverse voltages. The switch SW_1 is used to bypass the SCR so that the full supply voltage is applied to the motor for full-speed operation.

Ideas for Exploration Examine circuit performance using different SCRs.

Sawtooth Generator

An application of the PUT in a sawtooth generator is shown in Fig. 14.88. The circuit is essentially an integrator with the PUT connected across

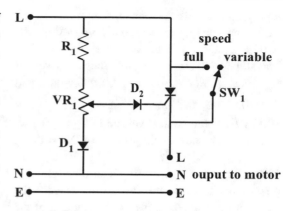

Fig. 14.87 Universal motor speed controller

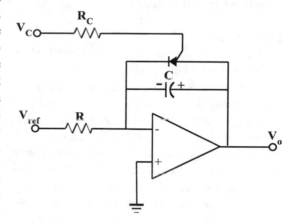

Fig. 14.88 PUT-controlled sawtooth generator

the capacitor. Initially with an uncharged capacitor C the PUT is off. With a negative voltage V_{ref} at the input of the circuit, current flows from the output of the operational amplifier onto the plates of the capacitor charging it with a polarity as shown. Since one end of the capacitor is at the inverting input of the op-amp (virtual earth), the rising voltage on the capacitor appears at the output of the op-amp. This voltage increases until the capacitor voltage is 0.7 volts above the control voltage V_C at the gate of the PUT at which time the PUT turns ON. As a result, the capacitor is discharged through the PUT and therefore the output voltage falls. The PUT switches OFF when the current through the device falls below the holding current after which the cycle is repeated. The voltage across the terminals of the PUT when this occurs is about 1 volt. Thus, the output voltage varies between 1 V and V_C + 0.7. The slope of the rising output voltage is given by $\frac{dV_o}{dt} = \frac{V_{ref}}{RC}$. Using $V_{ref} = -1$ V, $R = 100$ k and $C = 0.01$ μF, the slope is $\frac{dV_o}{dt} = \frac{1}{10^5 \times 0.01 \times 10^{-6}} = 1000$ V/s. If the control voltage is $V_C = 10.3$ V, then the peak output voltage is $10.3 + 0.7 = 11$ V.

Ideas for Exploration (i) Determine the frequency of the sawtooth waveform; (ii) since the slope of the waveform is a direct function of the reference voltage, introduce a variable reference voltage in order to convert the system into a voltage-controlled oscillator.

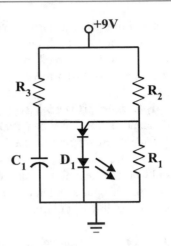

Fig. 14.89 PUT flasher circuit

Flasher Circuit

Another application of the PUT is the flasher circuit in Fig. 14.89. Resistors, $R_1 = 18$ k and $R_2 = 10$ k, set a voltage at the gate of the PUT such that the device is off when capacitor C_1 is discharged. As C_1 charges up through resistor R_3, the voltage at B_2 exceeds the gate voltage by 0.7 volts. This triggers the PUT thereby discharging C_1 through the LED which turns on. When the current through the PUT falls below the holding current, the device switches off and the cycle is repeated. For this circuit $\eta = \frac{R_1}{R_1+R_2} = \frac{18}{18+10} = 0.64$. Using this, the frequency is given by $f = 1/R_3C_1 \ln\left(\frac{1}{1-\eta}\right) = 1/R_3C_1 \ln\left(\frac{1}{1-0.64}\right) \simeq 1/R_3C_1$

. For a flashing frequency of 1 Hz, using $C_1 = 10$ μF, then $R_3 = 1/fC_1 = 1/10 \times 10^{-6} = 100$ k.

Ideas for Exploration Replace the LED by a buzzer such that the system can be used as a metronome.

Research Project 1

The circuit shown in Fig. 14.90 is a linear opto-coupled DC amplifier using two op-amps and two opto-couplers. The output of the op-amp A drives both leds of the two opto-couplers which are in series with potentiometer $VR_1 = 10$ k. Opto-transistor 1 is connected as an emitter follower

with its output driving the inverting input of op-amp A, while the opto-transistor 2 is connected as an emitter follower with is output driving the non-inverting input of op-amp B. The feedback via opto-transistor 1 ensures that the input signal V_i to op-amp A is delivered to resistor $R_1 = 4.7$ k and since opto-transistor 2 receives the same signal as opto-transistor 1, both leds being activated by the same current, then the signal V_i will appear across $R_2 = R_1 = 4.7$ k and delivered at the output of op-amp B. The potentiometer is used to adjust the effective loop gain around op-amp A. The input signal can vary between zero and about 5 V but cannot go negative as the leds can only accommodate unidirectional currents. Both supplies can be +12 V.

Ideas for Exploration (i) Implement this system using the LTV826 comprising two opto-transistors in a package; (ii) vary potentiometer VR_1 and see how that affects the performance of the system.

Research Project 2

The project to be undertaken is the design of a stairstep generator that can be used in the development of a transistor curve tracer for displaying the output characteristics of a transistor on an oscilloscope. A stairstep waveform is one that increases in discrete steps with time giving an appearance of physical stairsteps. The basic

Fig. 14.90 Linear opto-coupled DC amplifier

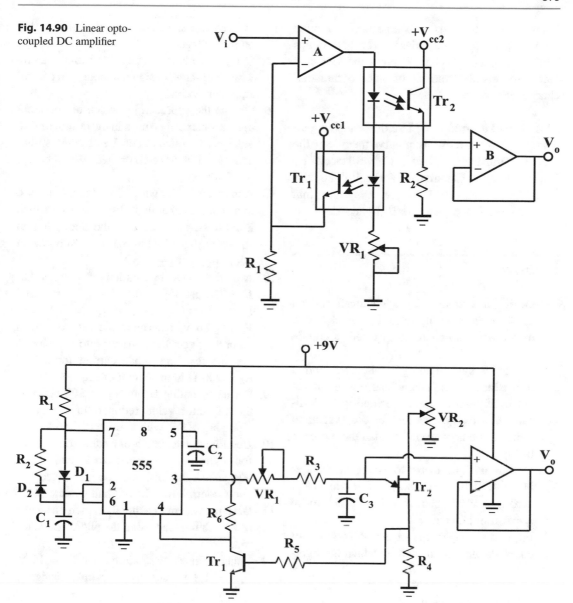

Fig. 14.91 Stairstep generator

system is shown in Fig. 14.91. Here a 555 timer is connected as an astable multivibrator. Capacitor C_1 is charged via resistor R_1 and diode D_1 and discharges via resistor R_2 and diode D_2. By making R_2 appropriately larger than R_1, the output of the 555 is a set of short duration pulses. These pulses charge capacitor C_3 through VR_1 and R_3. These pulses cause the voltage across C_3 to increase in steps thereby producing a stairstep waveform. This waveform is buffered by the op-amp which is a rail-rail device so that it can be operated from a single-ended supply while transmitting low voltages to its output. When the stairstep voltage across C_3 reaches a threshold value, the unijunction Tr_2 fires and discharges this capacitor. This discharge produces a voltage pulse across resistor R_4 which turns on transistor Tr1 and resets the 555. The cycle then starts again with pulses from the 555 delivering charge to capacitor C_3 and building a stairstep of voltages.

Potentiometer VR_1 sets the charge current into the capacitor and therefore varies the size of the steps while VR_2 sets the voltage across the UJT and therefore sets the number of steps before discharge occurs.

Ideas for Exploration (i) Use this stairstep generator to develop a transistor curve tracer. See the article "Transistor and FET Curve Tracer" by Daniel Metzger, Electronics World, Vol. 86, No. 2, p52, August 1971, where a system that employs a stairstep generator is discussed.

Problems

1. Design a system using a photocell and an op-amp as a comparator to switch on a motor when ambient light exceeds a certain level.
2. Design a system using a photocell and a transistor to switch on an alarm when ambient light rises above a specified level. Indicate the changes needed for the system to indicate the absence rather than the presence of light.
3. Design a light metre to cover the ranges 10–10,000 lx using the photo diode in the zero bias mode shown in Fig. 14.92 and a 1 mA milliameter.
4. Explain the operation of the phototransistor and indicate one method by which the light-induced current can be converted to a useable output voltage.
5. Using the photoresistor opto-isolator along with an op-amp, design a remote gain control amplifier system.
6. Outline the approach by which the photofet opto-coupler can be used in the design of (i) a voltage-controlled Sallen-Key low-pass filter and (ii) a voltage-controlled Wien bridge oscillator.
7. A microcontroller with a 0–5 V output is used to activate another digital system from which it must be isolated. Use the configuration shown in Fig. 14.93 to realize such a system with $V_{CC1} = V_{CC2} = 5$ V.
8. A thyristor used in a static power switch has $I_G = 750$ μA, $V_{GC} = 0.7$ V, $V_{AC} = 0.4$ V and $I_H = 15$ mA. Using $V_{TRIG} = 4$ V and $V_{CC} = 36$ V, determine the resistor R_G in order to trigger the thyristor into conduction and the resulting anode current for a load $R_L = 700$ Ω in the circuit of Fig. 14.37.
9. For an operating frequency of 300 Hz, find the capacitor value required to achieve a phase lag of 75° with a resistor of 56 k.
10. Describe the operation of a static triac switch for controlling AC power to a load. Show how phase control can be introduced to allow control over the full half cycle.
11. Design a system to remotely switch on a large (8 A) heating load using the phototriac and a triac.
12. Using a Shockley diode having $V_S = 15$ V, $V_H = 1.3$ V and $I_H = 5$ mA, design a

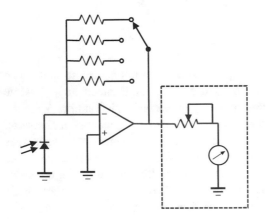

Fig. 14.92 Circuit for question 3

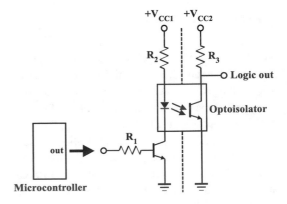

Fig. 14.93 Circuit for question 7

Fig. 14.94 Circuit for question 16

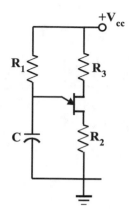

relaxation oscillator to produce a linear waveform with a slope of 100 mV/μs.

13. Explain how a diac can be used to improve the turn-on performance of a triac-controlled power switch.

14. A UJT with $\eta = 0.65$ is powered by a 12 V supply. If the emitter is at zero potential, determine the reverse bias across the pn junction.

15. Using the 2N2926 unijunction transistor, design a relaxation oscillator with a frequency of oscillation of 12 kHz and operating from a 15 V supply. Provide a buffered voltage output from the system. For this UJT, $\eta = 0.56 \rightarrow 0.75$, $I_P(mx) = 5$ μA and $I_V(mx) = 4$ mA.

16. Determine the oscillating frequency of the relaxation oscillator shown in Fig. 14.94 where the UJT has $\eta = 0.63$.

17. Explain how the PUT differs from the UJT.

18. Using the 2N6027 PUT, design a relaxation oscillator with a frequency of oscillation of 5 kHz and operating from a 24 V supply. For the 2N6027, $R_G = R_1//R_2 = 10$ k, and typical current values are $I_P = 4$ μA and $I_V - 150$ μA.

Bibliography

J. Graeme, *Photodiode Amplifiers: Op Amp Solutions* (McGraw-Hill, New York, 1996)

Index

© Springer Nature Switzerland AG 2021
S. J. G. Gift, B. Maundy, *Electronic Circuit Design and Application*,
https://doi.org/10.1007/978-3-030-46989-4

Printed in the United States
by Baker & Taylor Publisher Services